18th International Symposium on Particles Strings and Cosmology

(PASCOS 2012)

AA002937

Journal of Physics: Conference Series Volume 485

Merida, Mexico
3-8 June 2012

ISBN: 978-1-63266-706-9
ISSN: 1742-6588

Printed from e-media with permission by:

Curran Associates, Inc.
57 Morehouse Lane
Red Hook, NY 12571

Some format issues inherent in the e-media version may also appear in this print version.

Copyright© by the Institute of Physics
All rights reserved.

Printed by Curran Associates, Inc. (2014)

For permission requests, please contact the Institute of Physics at the address below.

Institute of Physics
Dirac House, Temple Back
Bristol BS1 6BE UK

Phone: 44 1 17 929 7481
Fax: 44 1 17 920 0979

techtracking@iop.org

Additional copies of this publication are available from:

Curran Associates, Inc.
57 Morehouse Lane
Red Hook, NY 12571 USA
Phone: 845-758-0400
Fax: 845-758-2634
Email: curran@proceedings.com
Web: www.proceedings.com

TABLE OF CONTENTS

PLENARIES

PARALLELS

Journal of Physics: Conference Series **485** (2014) 011001 doi:10.1088/1742-6596/485/1/011001

Preface

The XVII International Conference on Strings, Particles and Cosmology, PASCOS 2012, was held in the City of Mérida, Mexico, from June 3-8, 2012. The conference series is aimed at exploring the interface and interplay between particle physics, string theory and cosmology. With the advent of new data, the emphasis of the XVIII edition of PASCOS was on phenomenology and the interpretation of recent observational and experimental results.

The conference followed the format of previous conferences in this series, with plenary reviews and contributed presentations in parallel sessions. The lectures covered a wide range of subjects which included: Dark matter and dark energy, flavor physics and CP violation, neutrino physics, supersymmetry, Higgs physics, baryogenesis and EDMs, supergravity, high energy cosmic rays, string and F-theory GUTs, and string phenomenology.

This is the first time that PASCOS was held in Latin America. The aim to do it in Mexico was to engage the Latin American community and thus to bring the conference to a wider and different audience, a goal which was thoroughly achieved.

The venue was held at the Hotel Fiesta Americana in the beautiful city of Mérida. The social events included a reception with typical local food at the Katun restaurant, conference dinner at the historical Quinta Montes Molina, and an excursion to the archeological site of Dzibilchaltún including a swim at the famous cenote.

PASCOS 2012 was possible thanks to the generous support of the following sponsors: CONACyT (Consejo Nacional de Ciencia y Tecnología), UNAM (Universidad Nacional Autónoma de México: Consejo Técnico de la Investigación Científica, Instituto de Ciencias Nucleares, Instituto de Física), Cinvestav, (Centro de Estudios Avanzados del IPN: U. Zacatanco, U. Mérida and Secretaría General), ICyTDF (Instituto Científico y Tecnológico del D.F.), PIFI (Programa Integral de Fortalecimiento Institucional, Universidad de Guanajuato, Campus León), SMF (Sociedad Mexicana de Física), ICTP (International Centre for Theoretical Physics), BUAP (Benemérita Universidad Autónoma de Puebla), the Government of the State of Yucatán, the University of Hamburg, and Telmex.

We also want to acknowledge the invaluable help of the staff of the Mexican Physical Society, in particular Lic. Santos Zúñiga Sánchez and Ms. Claudia Velasco Marín, and of the conference secretaries, Ms. Lizette Ramírez Bermúdez (UNAM) and Ms. Mariana del Castillo Sánchez (Cinvestav), for their support before, during and after the organization of PASCOS 2012.

Last but not least, we would like to thank all the PASCOS 2012 participants for their attendance and for contributing to make the conference an engaging and stimulating event.

The organizers,

Myriam Mondragón, Adnan Bashir, David Delepine, Francisco Larios, Oscar Loaiza, Axel de la Macorra, Lukas Nellen, Sarira Sahu, Humberto Salazar and Liliana Velasco-Sevilla.

Content from this work may be used under the terms of the Creative Commons Attribution 3.0 licence. Any further distribution of this work must maintain attribution to the author(s) and the title of the work, journal citation and DOI.
Published under licence by IOP Publishing Ltd

Peer review statement

All papers published in this volume of *Journal of Physics: Conference Series* have been peer reviewed through processes administered by the proceedings Editors. Reviews were conducted by expert referees to the professional and scientific standards expected of a proceedings journal published by IOP Publishing.

Content from this work may be used under the terms of the Creative Commons Attribution 3.0 licence. Any further distribution of this work must maintain attribution to the author(s) and the title of the work, journal citation and DOI.
Published under licence by IOP Publishing Ltd

Journal of Physics: Conference Series **485** (2014) 012001 doi:10.1088/1742-6596/485/1/012001

Gamma-ray bursts: Recent results and connections to very high energy cosmic rays and neutrinos

Péter Mészáros[1], **Katsuaki Asano**[2] **and Péter Veres**[1]

[1] Center for Particle and Gravitational Astrophysics, Dept. of Astronomy & Astrophysics and Dept. of Physics, Pennsylvania State University, University Park, PA 16802, USA

[2] Interactive Research Center of Science, Tokyo Institute of Technology, 2-12-1 Ookayama, Meguro-ku, Tokyo 152-8550, Japan

E-mail: pmeszaros@astro.psu.edu, asano@phys.titech.ac.jp, puv2@psu.edu

Abstract. Gamma-ray bursts are the most concentrated explosions in the Universe. They have been detected electromagnetically at energies up to tens of GeV, and it is suspected that they could be active at least up to TeV energies. It is also speculated that they could emit cosmic rays and neutrinos at energies reaching up to the $10^{18} - 10^{20}$ eV range. Here we review the recent developments in the photon phenomenology in the light of *Swift* and *Fermi* satellite observations, as well as recent IceCube upper limits on their neutrino luminosity. We discuss some of the theoretical models developed to explain these observations and their possible contribution to a very high energy cosmic ray and neutrino background.

1. Introduction

Gamma-ray bursts (GRB) are detected at an average rate of a one per day, lasting in gamma-rays from fractions of a second to tens of minutes. We know that these objects are distributed almost isotropically throughout the Universe, out to the largest cosmological distances yet sampled. Yet, while they are on, they far outshine all other sources of gamma-rays in the sky, including the Sun. They are, in fact, the most concentrated and brightest electromagnetic explosions in the Universe, the GRB prompt electromagnetic energy output during tens of seconds being comparable to that of the Sun over $\sim$ few $\times\, 10^{10}$ years, or to that of the entire Milky Way over a few years. Their initial $\gamma-$ ray emission is followed by an X-ray and an optical afterglow lasting for weeks, which during the first day can outshine the brightest quasars and active galactic nuclei in the Universe.

It is thought that GRB result as a consequence of a cataclysmic "end game" in the life of very evolved stars, where about a solar rest mass worth of gravitational energy is released in a matter of seconds or less within a a small region of the order of tens of kilometers. This is likely to be the result of the collapse of the core of a massive star, or in some cases from the merger of two compact stellar remnants, either of these scenarios ultimately leading to a stellar mass black hole. Only a small fraction of this energy needs to be converted into electromagnetic radiation to satisfy the observations. It is thought that this conversion occurs through the

Content from this work may be used under the terms of the Creative Commons Attribution 3.0 licence. Any further distribution of this work must maintain attribution to the author(s) and the title of the work, journal citation and DOI.
Published under licence by IOP Publishing Ltd

dissipation of the kinetic energy of a collimated relativistic jet outflow, a "fireball", whose bulk Lorentz factors are in the range of $\Gamma \sim 10^2 - 10^3$, expanding from the central engine which is powered by the gravitational accretion of surrounding matter into the collapsed core.

The generic scenario for the production of the observed non-thermal photons typically invokes synchrotron radiation and/or inverse Compton (IC) scattering by relativistic electrons which have been accelerated to a power-law distribution in the shocks expected in the optically thin regions of the outflow. These may be internal shocks, resulting in prompt γ-ray emission, and also external shocks at the termination of the relativistic outflow, which can explain many of the properties of the afterglows. Other mechanisms considered for the prompt emission are, e.g., magnetic dissipation or reconnection in the outflow, jitter radiation in shocks, or dissipative effects in the photosphere where the outflow transitions to optical thinness.

In the past few years, the *LAT* instrument on the *Fermi* spacecraft has shown that a substantial fraction of GRBs have photon spectra which extend at least to tens of GeV [1, 2]. These could be due either to leptonic mechanisms (such as the electron synchrotron or inverse Compton mentioned above), or they might be due to hadronic cascades. Various uncertainties hamper the analysis and modeling of these systems, including our the lack of knowledge about two important parameters of the outflow. These are the baryon load of the outflow, and the magnetic ratio σ between magnetic stresses and kinetic energy, which affect not only the bulk dynamics but also the mechanisms responsible for accelerating electrons and protons in the shocks or the dissipation region. It is the accelerated protons which could lead, in principle, to GRBs being luminous in cosmic rays and neutrinos. Under optimistic scenarios, they could be even more luminous in these particle channels than in the commonly observed MeV electromagnetic channels.

2. Electromagnetic phenomenology: MeV and sub-MeV

The prompt photon emission of GRBs, as documented already in the 1990s by the *Compton* (CGRO) satellite, shows often rapidly variable γ-ray light-curves, leading to a classification into "long" GRBs (LGRBs) whose γ-ray light curve lasts $2\text{ s} \lesssim t_\gamma \lesssim 10^3$ s, and "short" GRBs (SGRBs) for which $t_\gamma \lesssim 2$ s, although the latter can be longer at softer energies. The first X-ray afterglows, lasting weeks or more, were discovered by the Italian-Dutch *Beppo-SAX* satellite in 1997, being acquired typically 8 hours after the initial γ-ray trigger. The NASA *Swift* satellite, launched in 2004, was designed to study afterglows within a minute after the trigger.

The *Swift* satellite has three instruments, covering the soft gamma-ray, the X-ray and the ultraviolet/optical ranges. The gamma-ray detector locates bursts to ~ 2 arcminute accuracy and the position is used to repoint the onboard X-ray and UV/O instruments, as well as being rapidly relayed to Earth so ground telescopes can follow the afterglows. Measurements of the redshift distance and studies of host galaxies are generally done with large ground-based telescopes which receive the alerts from the spacecraft. The average *Swift* burst detection rate is ~ 90 per year, of which approximately $\sim 90\%$ have detected X-ray afterglows, mostly among the longer GRBs. The shorter bursts, however, are harder to detect in X-rays, since often they fade rapidly below the X-ray sensitivity limit. With help from ground-based optical observations, about $\sim 60\%$ of *Swift* GRBs yield also an optical afterglow detection. One of the interesting findings of *Swift* was that the afterglow lightcurve has a complex structure. Often a fast decay is seen in the first 1000 s, followed by a shallow decay and then a re-steepening.

The long GRBs are found in the brightest regions of galaxies where intense star formation occurs. Those which occur near enough are generally found in association with a (simultaneous) supernova of Type Ib or Ic (in more distant LGRB, the supernova is expected to be too faint to be detected). These facts support the view that LGRB are caused by the central core of a massive star collapsing to a compact object such as a black hole, or possibly a magnetar. LGRBs are extremely bright in both their gamma-ray prompt emission and their multiwavelength afterglow. This makes them unique tools for studying the high-redshift universe. For instance GRB 090423, at a redshift $z = 8.2$ is the source with the largest spectroscopically determined redshift. Such high redshift bursts provide information about the universe at a time when it was only a few percent of its current age, providing information about the process of re-ionization of the intergalactic medium and the chemical evolution of the universe. LGRBs also contribute to determining the star formation history history of the Universe, since they are the endpoints of the lives of massive stars and their rate is approximately proportional to the star formation rate.

Previous to *Swift*'s launch the origin of short GRBs was very poorly constrained, since no afterglows had been detected to localize them. This changed in 2005 when *Swift* and *HETE-2* detected afterglows leading to a precise localization for several SGRBs, the total number by now being significant. Unlike long bursts, the evidence indicates that SGRBs originate in host galaxies with both low and high star formation rates, i.e. old and young stellar populations. These host properties are substantially different than those of long bursts, indicating a different origin, and, furthermore, nearby SGRBs show no evidence for simultaneous supernovae, both properties being very different than for long bursts. These results support the interpretation that SGRBs arise from an old population of stars and are probably due to mergers of compact binaries, such as double neutron stars or neutron star - black hole binaries. *Swift* observations have also revealed, in about 25% of SGRBs, long (~ 100 s) lightcurve tails with softer spectra than the first short prompt emission episode, which has a harder spectrum. The localization by *Swift* of short GRBs, if they are indeed compact binary mergers, could also help narrow the search window for gravitational waves from such binaries, which would lead to a great scientific payoff for gravitational physics, as well as for studies of progenitor stellar types and neutron star equations of state.

3. Electromagnetic phenomenology: GeV and above

The *Fermi* spacecraft, launched in late 2008, started detecting GRBs with two instruments, the Large Area Telescope (*LAT*, 20 MeV to > 300 GeV) and the Gamma-ray Burst Monitor (*GBM*, 8 keV to 40 MeV), which jointly cover more than seven decades in energy. The low energy *GBM* triggers on bursts at a rate of about 250 yr^{-1}, of which $\sim 80\%$ are LGRBs and $\sim 20\%$ are SGRBs, while the high energy instrument, *LAT*, detects bursts at a rate of ~ 10 yr^{-1}. Of the latter, the *LAT* detects about twice as many at energies ≥ 100 MeV than it does at energies ≥ 1 GeV. An interesting and unexpected behavior is that in many cases the GeV emission starts with a noticeable delay after the MeV emission. E.g. in GRB 080916C, the GeV emission appears only in a second pulse, delayed by ~ 4 s relative to the first pulse (visible only in MeV); [3], see Fig. 1. Such a delay is present also in short bursts, such as GRB 090510 [4, 5], where it is a fraction of a second. Such a soft-to-hard spectral evolution is clearly seen in the brightest *LAT* bursts, and also to various degrees in most other weaker *LAT* bursts.

Perhaps the most exciting, or exotic, consequence of the observed delays between the *LAT* GeV emission and the *GBM* MeV emission is that it can be used to set robust constraints on effective field theory formulations of quantum gravity. In particular it rules out a first order dependence on $(E_\gamma/E_{\rm Planck})$ of any Lorentz invariance violating (LIV) terms, using

Journal of Physics: Conference Series **485** (2014) 012001 doi:10.1088/1742-6596/485/1/012001

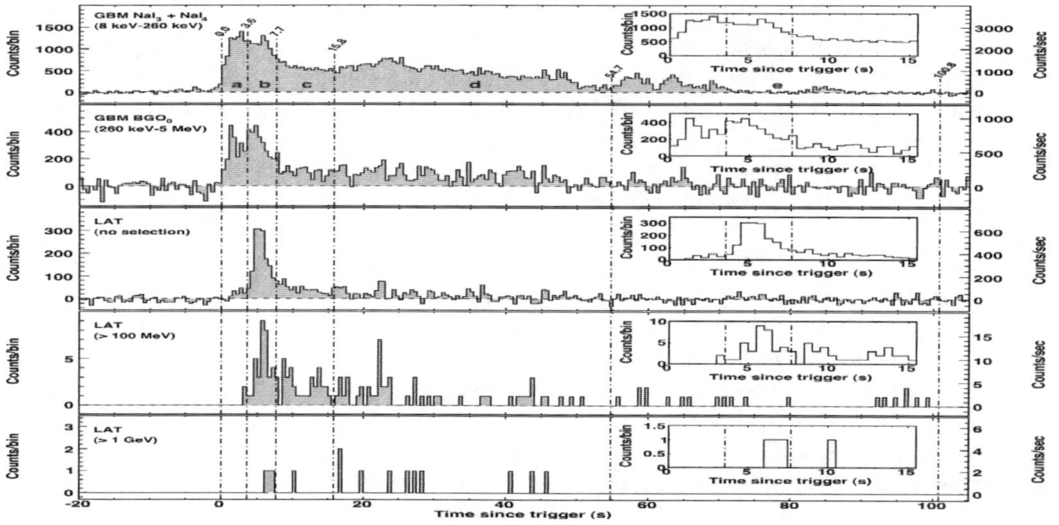

Figure 1. Light curves of GRB080916C with the *GBM* (top two curves) and *LAT* (bottom three curves)[3].

.

GRB 090510 data [4]. This result is robust, the limits getting even more stringent if there additional astrophysics causes for the delay, which indeed are expected.

In some burst, such as GRB 080916C, the broad-band gamma-ray spectra appear, to within statistical accuracy, as simple "Band" type (broken power law) functions in *all* time bins (similar to the spectrum in time bin [a] of Fig. 2). The absence of statistically significant evidence for a distinct second high energy spectral component in this and some other *LAT* bursts was initially puzzling, since naively such an extra component is expected from inverse Compton up-scattering or from hadronic cascades. However, subsequent observations, e.g. of GRB090510 [4, 5] and GRB 090902B [6], have shown a second hard spectral component, extending above 10 GeV, in addition to the common Band spectral component dominant in the 8 keV-10 MeV band. In some cases, such as GRB 090926A (Fig. 2), this second hard component also shows a downturn around a few GeV.

An exciting discovery by the *Fermi LAT* was the detection of GeV emission from two short bursts (GRB 081024B [8] and GRB 090510 [5]), whose general behavior (including a GeV delay) is qualitatively similar to that of long bursts. The ratio of short to long GRBs is $\sim 10-20\%$, and while the statistics on short GRBs are still small, it appears that the ratio of the *LAT* energy fluence to the *GBM* fluence is $> 100\%$ for the short bursts as compared to $\sim 5-60\%$ for the long bursts. Thus, although fewer in number, future large ground-based Cherenkov such as *CTA* [9] and *HAWC* [10] may be able to detect short bursts.

A remarkable feature of both long and short GRBs is that the ≥ 100 MeV emission generally lasts much longer than the $\lesssim 1$ MeV emission. The flux of the long-lived *LAT* emission decays as a power law with time, which is more reminiscent of the smooth temporal decay of the afterglow X-ray and optical fluxes, rather than the variable temporal structure in the prompt keV–MeV flux.

Interestingly, the *LAT* detects only $\lesssim 10\%$ of the bursts triggered by the *GBM* which are

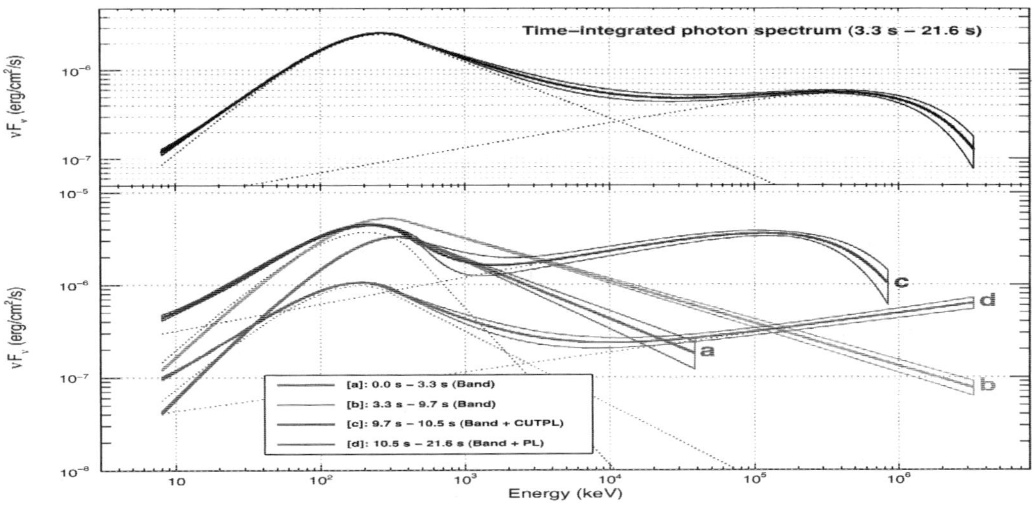

Figure 2. Spectra of GRB090926A from *Fermi* at four different time intervals, a= [0.0-3.3s], b= [3.3-9.7s], c= [9.7-10.5s], d= [10.5-21.6s] [7].

in the common *GBM-LAT* field of view. This may be related to the fact that the *LAT*-detected GRBs, both long and short, are generally among the highest fluence bursts, as well as being among the intrinsically most energetic GRBs. For instance, GRB 080916C was at $z = 4.35$ and had an isotropic-equivalent energy of $E_{\gamma,\rm iso} \approx 8.8 \times 10^{54}$ ergs in γ rays, the largest ever measured from any burst [3]. The long *LAT* bursts GRB 090902B [6] at $z = 1.82$ had $E_{\gamma,\rm iso} \approx 3.6 \times 10^{54}$ ergs, while GRB 090926A [7] at $z = 2.10$ had $E_{\gamma,\rm iso} \approx 2.24 \times 10^{54}$ ergs. Even the short burst GRB 090510 at $z = 0.903$ produced, within the first 2 s, an $E_{\gamma,\rm iso} \approx 1.1 \times 10^{53}$ ergs [5].

4. Leptonic GRB Models

The standard scenario of the mechanics of a burst is that the rotating debris falling into the central black hole leads, via a shear or turbulent dynamo mechanism, to extremely strong magnetic fields, which couple the debris to the rotating black hole and, like super-strong rubber bands, extract the rotational energy of the black hole and pump it into a jet, which becomes highly relativistic and collimated into a 5-10 degree angular extent (a similar jet is possible if the central engine is a highly magnetized neutron star).

The energy of the jet is initially mainly in the kinetic energy of its motion, and as it moves away from the black hole, the initially large particle density in it decreases until at a "photospheric" radius the photon mean free path becomes larger than the jet dimension and the photons trapped in the jet can escape freely. However, if the jet energy is still mainly bulk kinetic energy at the photosphere, this escaping radiation would be weak, unless a substantial fraction of the bulk kinetic energy has been dissipated into random energy of charged particles and radiated. This can be achieved if the kinetic energy is dissipated beyond the photosphere in shocks, either internal shocks within the jet itself [11], or external shocks [12], as the jet is decelerated by external matter. Charged electrons bouncing across these shocks can then

Journal of Physics: Conference Series **485** (2014) 012001 doi:10.1088/1742-6596/485/1/012001

be accelerated via the Fermi mechanism to a relativistic power law energy distribution, and can produce a non-thermal photon spectrum via synchrotron or inverse Compton radiation, which approximates the observed Band type broken power law spectra. This is the "standard shock leptonic" model. The simple internal shock interpretation of the prompt MeV emission, however, has typically a low radiative efficiency, while the observed spectra sometimes disagree with a straightforward synchrotron interpretation, which has motivated searches for alternative interpretations of the origin of the prompt MeV emission (see below).

The external shock is generally expected to be accompanied by a reverse shock, which can produce a prompt optical emission [13] at the time the jet deceleration begins. This has been detected in a number of bursts, with robotic ground-based telescopes such as ROTSE and others [14]. As the jet keeps decelerating by virtue of sweeping up increasing amounts of external matter, the bulk Lorentz factor of the shock decreases as a power law of the distance, and the resulting synchrotron radiation becomes a long lasting, fading X-ray, optical and radio afterglow [13], whose predicted detection allowed the first measurements of host galaxies and redshift distances [14]. The external shock interpretation of the late afterglow is quite robust overall. However, there is debate about some of the more detailed features seen in the first few hours of the afterglow by *Swift*, such as X-ray steep decays followed by a flat plateau and occasional large flares, and various proposed extensions of the basic external shock picture continue to be tested.

The above internal and external standard leptonic shock model is generally used also for interpreting the *Fermi* data on individual *Fermi LAT* bursts, e.g. [15, 16], etc. Broader formulations of the shock leptonic scenario attempting to cover *LAT* bursts in general were discussed by [17], where the GeV emission arises from a fast cooling forward shock, and by [18, 19], where the forwards shock is assumed to be slow cooling (referring to the ratio of radiative cooling time to jet dynamic time). In such models, it is argued that the external shock GeV emission would naturally start with a delay relative to the prompter MeV emission, assuming the latter arise from internal shocks (see also [20]).

However, taking into account the constraints provided by the Swift MeV and X-ray observations, it is clear that at least during the prompt emission, there must be an interplay between the shorter lasting mechanism providing the MeV radiation and the mechanism, or emission region, responsible for the bulk of the longer lasting GeV radiation [22, 21]. The interplay between the two regions involves a number of subtleties, and taking into account the spatial structure by means of multi-zone models, the inverse Compton scattering by an outer shock of the MeV radiation from an assumed inner source of Band-like MeV photons can give raise to the right delays [23]. These studies avoid specifying a model of the prompt emission origin. If one assumes that this is due to an internal shock, that is open to critique because of its radiative inefficiency and sometimes inconsistent predicted spectra. This problem can be resolved if the prompt MeV Band spectrum is due to an efficient dissipative photosphere with an internal shock upscattering the MeV photons at a lower efficiency, giving the delayed GeV spectrum [22]. Alternatively, for a magnetically dominated outflow, where internal shocks are not expected, an efficient dissipative photospheric Band spectrum can be up-scattered by the external shock and produce the observed delayed GeV spectrum [21] (see Fig. 3).

5. Hadronic GRB Models

If GRB jets are baryon loaded, the charged baryons are likely to be co-accelerated in shocks, reconnection zones, etc., and hadronic processes would lead to both secondary high energy photons and neutrinos. Monte Carlo codes have been developed to model hadronic effects in

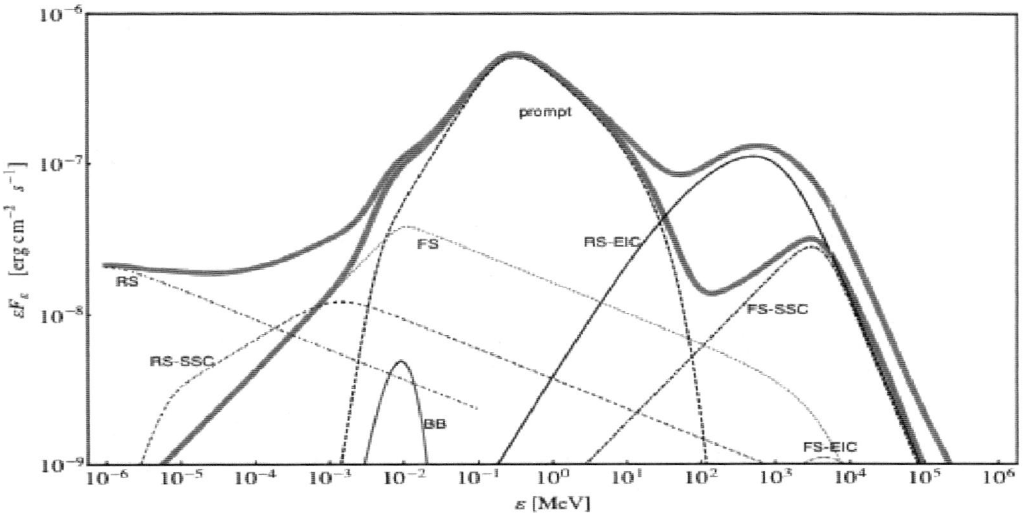

Figure 3. A magnetically dominated leptonic model, with parameters typical of *Fermi* LAT GRBs. The MeV Band spectrum is due to photospheric emission, there are no internal shocks, and the external reverse and forward shock upscatter the MeV spectrum into the GeV range [21].

relativistic flows, including p,γ cascades, Bethe-Heitler interactions, etc. E.g., one such code [24, 25] was used to calculate the photon spectra in GRBs from secondary leptons resulting from hadronic interactions following the acceleration of protons in the same shocks that accelerate primary electrons. The code uses an escape probability formulation to compute the emerging spectra in a steady state, and provides a detailed quantification of the signatures of hadronic interactions, which can be compared to those arising from purely leptonic acceleration. Spectral fits of the *Fermi LAT* observations of the short GRB 090510 were modeled by [25] as electron synchrotron for the MeV component and photohadronic cascade radiation for the GeV distinct power law component. More generally, calculations used an advanced version of the above code show that hadronic models can describe GRB spectra where a second, harder photon spectral component arrives later (Fig. 4 because of the time delay needed for hadrons to be accelerated to high energies and for cascades to develop. A prediction of such calculations is that the ν_μ spectrum is harder than that in the models assumed in the recent IceCube papers [26, 27], and satisfies the constraints posed by those papers. The neutrino light curve expected from the charged pion decays also shows a delay relative to the MeV photon light curve (Fig. 5).

Hadronic interactions can also have interesting implications for GRB prompt optical flashes, observed in some bursts. As discussed in [29], besides the usual Band MeV spectrum produced by leptonic mechanisms, the acceleration of hadrons leads to secondaries whose radiation results not only in a GeV component but also to prompt synchrotron radiation in the optical band. This could, in principle, explain the observed "naked eye" 5th magnitude optical flash of GRB 080319B discussed in [30].

In addition to photo-hadronic interactions, also hadronic binary collisions may be important, both for an efficient bulk kinetic energy dissipation and for shaping the photon spectrum. The

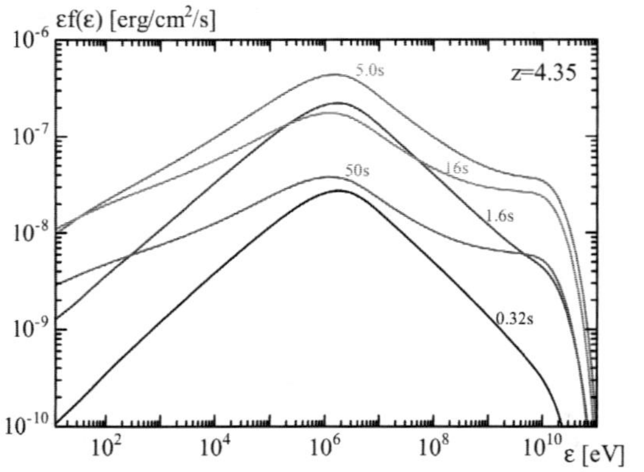

Figure 4. Hadronic GRB model Monte Carlo simulations: time evolution of the observable photon spectral radiation from hadronic cascades for typical *Fermi LAT* parameters, electron synchrotron producing a Band MeV spectrum and hadronic cascade secondaries producing the GeV spectrum as well as a low energy component [28].

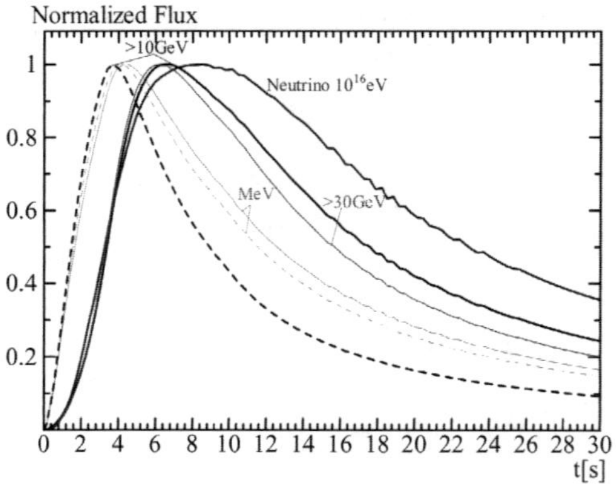

Figure 5. Monte Carlo photon and neutrino light curves of the same hadronic model (full lines) and a similar leptonic model (dashed lines) for a bright *Fermi-LAT* burst at $z = 4.35$, showing the expected delay between the MeV and GeV photon lightcurves and that of the neutrinos [28].

baryons in a jet will be mainly protons (p) and neutrons (n), especially if heavy elements are photo-dissociated. The protons are coupled to the radiation during the acceleration phase but the neutrons are carried along only thanks to nuclear (p, n) elastic collisions, whose characteristic timescale at some point becomes longer than the expansion time. At this point the p and n relative drift velocity v approaches c, leading to the collisions becoming inelastic, $p + n \rightarrow \pi^+, \pi^0$, in turn leading to positrons, gamma-rays and neutrinos [31]. Such inelastic (p, n) collisions can also arise in jets where the bulk Lorentz factor is transversely inhomogeneous [32], e.g. going from large to small as the angle increases, as expected intuitively from a jet experiencing friction against the surrounding stellar envelope. In such cases, the neutrons from the slower, outer jet regions can diffuse into the faster inner regions, leading to inelastic (p, n) and (n, n) collisions resulting again in pions. An interesting consequence of either radial or tangential (n, p) drifts is that the decoupling generally occurs below the scattering photosphere, and the resulting positrons and gamma-rays deposit a significant fraction of the relative kinetic energy into the flow, reheating it [33]. Internal dissipation below the photosphere has been advocated, e.g. [34] to explain the MeV peaks as quasi-thermal photospheric peaks [35, 36], while having a large radiative efficiency. Such internal dissipation is naturally provided by (p, n) decoupling, and numerical simulations [33] indicate that a Band spectrum and a high efficiency is indeed obtained, which remains the case even when the flow is magnetized up to $\varepsilon_B = 2$ [37], while keeping the dynamics dominated by the baryons. These numerical results were obtained for nominal cases based on a specific radial (n, p) velocity difference, although the phenomenon is generic.

6. Cosmic rays from GRB

A GRB origin of extragalactic ultra-high energy cosmic rays (UHECR) at energies $10^{18.5}-10^{20.5}$ eV is in principle possible, if the lower energy cosmic rays are provided by other, e.g. galactic sources, such as supernova remnants [38, 39]. This is because the The maximum energy for a particle of shock accelerated particle of charge Z is $E \leq \beta ZeBR$, which is $\sim 10^{20}$ eV for a proton in a typical GRB shock. However, only the highest energy range can be supplied by GRB, mainly because the spectrum is expected to be $\propto E^{-2}$. This, combined with the low energy galactic component would yield approximately the observed $E^{-2.7}$ in the sub-GZK range. A steeper production spectrum in GRB, such as $E^{-2.7}$ or even $E^{-2.3}$, in amounts enough to explain the UHECR observations in the $10^{15} - 10^{18}$ eV range would be too energy demanding for a stellar collapse or merger GRB model, under usual intergalactic propagation conditions (c.f. [40]).

In the most common version of the GRB scenario the UHECR are considered to be protons accelerated in GRB internal shocks [38, 41], while another version attributes them to external shocks [39, 42]. An important caveat of the internal shock UHECR scenario is that it assumes that the GRB prompt gamma-ray emission is due to such internal shocks. Although this is the leading work-horse scenario, there is no strong proof so far for this (as there is for external shocks). In fact, there are problems with the radiative efficiency and the electrons synchrotron spectrum of internal shocks (see previous sections), which put internal shocks into doubt, at least in their simple form usually assumed for UHECR GRB models. (Such problems may be solved in alternative internal shocks or slow magnetic dissipation models [43]).

A relevant development is the evidence accumulating from the Pierre Auger Observatory on the depth of shower penetration, as well as on the fluctuations of this quantity and the muon content of the showers, which suggests that in the range $10^{18} - 10^{20}$ eV the UHECR chemical composition acquires an increasing contribution from heavy nuclei [44, 44]. In a baryonic GRB jet, where magnetic fields are dynamically sub-dominant, the pressure is provided by radiation, and this is expected to photo-dissociate any heavy elements down to p, n and He [45, 46]. However, if magnetic fields are dynamically dominant, they would provide the dominant pressure, the internal radiation field being lower, and in such GRB jets nuclei can survive [47, 48]. If GRB are the sources of UHECR, this is another argument suggesting that they are magnetically dominated.

7. High energy neutrinos from GRB

If protons are accelerated in GRB shocks, these would interact with the observed photons mainly near the ~MeV peak energy in the GRB spectrum, chiefly through the Δ^+ resonance, $p\gamma \to \Delta^+$. The threshold condition to produce a Δ^+ is $E_p E_\gamma = 0.2\Gamma_i^2$ GeV^2 in the observer frame, which corresponds to a proton energy of $E_p = 1.8\times 10^7 E_{\gamma,\mathrm{MeV}}^{-1}\Gamma_{300}^2$ GeV. The short-lived Δ^+ decays either to $p\pi^0$ or to $n\pi^+ \to n\mu^+\nu_\mu \to ne^+\nu_e\bar{\nu}_\mu\nu_\mu$ with roughly equal probability. It is the latter process that produces high energy neutrinos in the GRB fireball, contemporaneous with the γ-rays [49]. The secondary π^+ receive $\sim 20\%$ of the proton energy in such an $p\gamma$ interaction and each secondary lepton roughly shares 1/4 of the pion energy. Hence each flavor (ν_e, $\bar{\nu}_\mu$ and ν_μ) of neutrino is emitted with $\sim 5\%$ of the proton energy. Using the standard internal shock model, as in [49], the neutrino spectrum has a spectral break at an energy $E_{\nu,b} \sim 10^{15}$ eV, where the neutrino production efficiency is high. This break is related via the Δ-resonance condition and the bulk Lorentz factor to the photon spectral break energy $E_{\gamma,b} \sim 1$ MeV of the Band spectrum. For a generic photon spectrum with slopes $dN(E_\gamma)/dE_\gamma \propto E_\gamma^{-1,-2}$ below/above $E_{\gamma,b}$, and protons with a spectrum $N(E_p) \propto E_p^{-2}$, the neutrino spectrum coincidentally has slopes $dN(E_\nu)/dE_\nu \propto E_\nu^{-1,-2}$ below/above $E_{\nu,b} \sim$ PeV; for other photon and proton spectral slopes the neutrino slopes are different, but qualitatively similar. The fluxes of all three neutrino flavors (ν_e, ν_μ and ν_τ) are expected to be equal after oscillation in vacuum over astrophysical distances. The diffuse muon neutrino flux from GRB internal shocks in this scenario, using an average GRB photon luminosity, spectrum and bulk Lorentz factor as well as standard GRB occurrence statistics, is expected to be comparable or somewhat below the so-called Waxman-Bahcall (WB) diffuse neutrino flux for optically (neutrino) thin sources implied by the observed cosmic ray flux [50].

This GRB ultra-high energy neutrinos (UHENU) scenario, based on internal shocks, was used by [51] to predict a diffuse neutrino flux which scales with the MeV photon fluxes of GRBs detected by a spacecraft. This can be done having observed the photon spectra, and using these for predicting a neutrino flux (same duration as the photon burst) by assuming neutrino production via the Δ-resonance, a proton spectrum $\propto E_p^{-2}$ and a relativistic proton energy scaling with the relativistic electron energy by a factor $f_p = 1/f_e = (\mathcal{E}_p/\mathcal{E}_e) \sim 10$, where the electron energy is assumed, due to the fast cooling, to be equal to the observed MeV photon energy. The neutrino spectral shape has a break which scales with observed photon spectral break, and thus using the actual MeV fluxes and spectra of GRBs measured by a spacecraft they predict a cumulative neutrino flux over the period of observations, say a year.

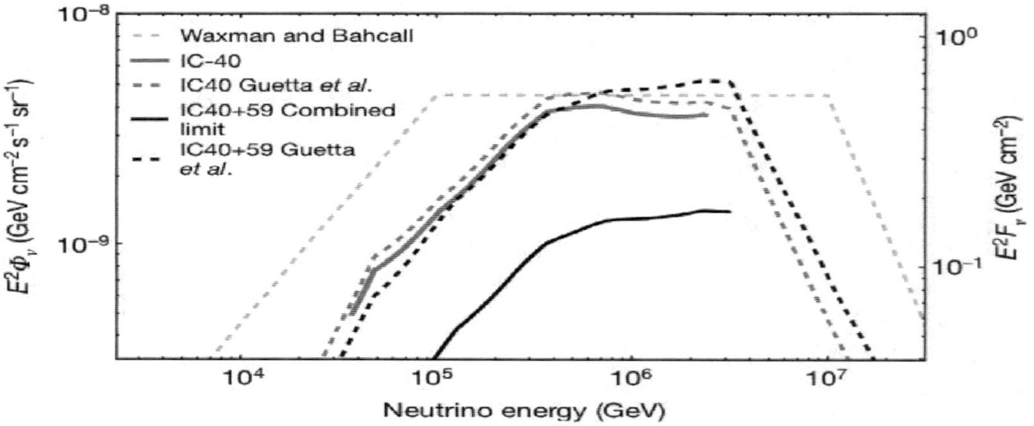

Figure 6. IceCube upper limits [27] for the 40-string array and the 40+59 string array on 190 GRBs localized by *Swift*, compared to an internal shock model [51] scaled to the photon luminosity (right y-axis). The data is a factor 3.7 below this model. The left y-axis is the diffuse flux calculated for 677 bursts/year, the WB limit is in light dashed lines.

Recently, the IceCube group has been analyzing the TeV-PeV UHENU data accumulated by the 40-string array and subsequently the 59-string array [26, 27], and comparing it to the GRB neutrino model flux expectations based on [51]. In the combined 40 and 59 string data (IC-40+59) they used 190 electromagnetically detected bursts (Fig. 6), with their specific photon spectral slopes and breaks to predict, modulo the factor $f_p = 10$, the predicted neutrino spectra, adding them all up. The predicted neutrino cumulative spectrum is shown in Fig. 6 for the 40 string array as the gray dashed (model) and gray solid (data) lines; and for the 40+59 string array as the dark dashed (model) and dark solid (data) lines; the original generic

Waxman-Bahcall spectrum is shown by the very light gray dashed lines. It is seen that the IC-40+59 observed data fall a factor 0.27 below the predicted model neutrino flux. This means that this internal shock model over-predicts the data by a factor 3.7, assuming a proton to electron energy ratio of 10, interactions via the Δ-resonance and buck Lorentz factors in the usual range $\Gamma \sim 300 - 600$. Using similar assumptions but not using the spectral information, if these GRB supply the GZK cosmic ray flux, the implied neutrino flux is $2 - 3\sigma$ above the level allowed by the data. The conclusion is that either $f_p = (\mathcal{E}_p/\mathcal{E}_e)$ is significantly below 10, or the production efficiency of neutrinos is lower than was assumed - or the specific model as used can be largely ruled out. This is a major landmark, being the first time that a specific extragalactic astrophysical source model is being tested through neutrino observations, at this $\sim 95\%$ level of significance. IceCube is doing exciting astrophysics, and this is a major step towards testing a candidate astrophysical source of GZK cosmic rays.

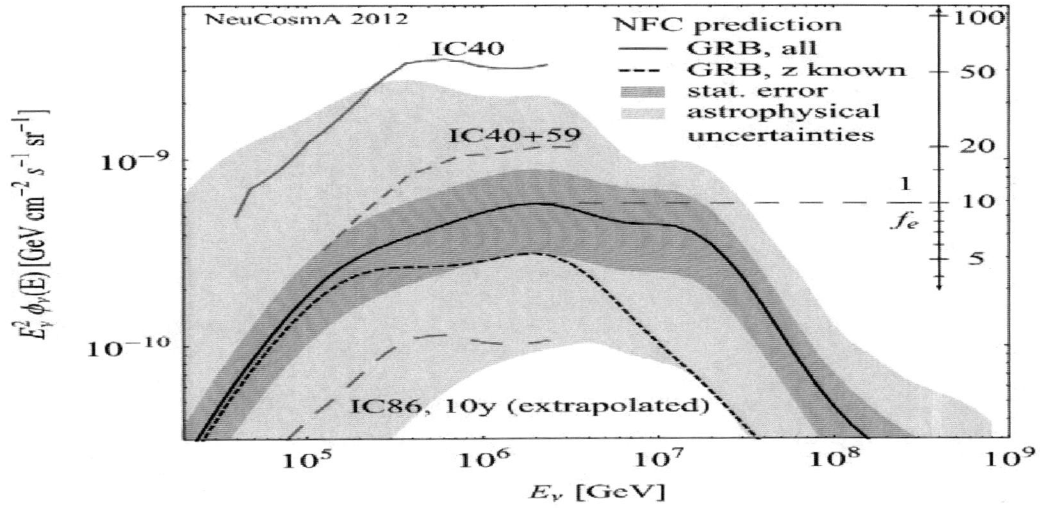

Figure 7. Predictions of the same internal shock model neutrino flux without several of the previous approximations (solid and dashed dark lines), compared to IC-40 and IC-40+59 data. The statistical error is given by the darker shaded region, the astrophysical uncertainties is given by the lighter shaded area [52].

There are two additional points in this connection. One is that in using the standard internal shock model to predict the expected neutrino flux [26, 27] made a number of simplifications. As discussed in [53], the neutrino production efficiency at all energies was assumed to be the same as that evaluated at the break energy, but integrating the efficiency over the actual photon spectrum below and above the break yields a total neutrino flux a factor five lower than that used in the comparison with the data. Also, as discussed in [52, 54], including production not just via Δ-resonance but also via multi-pion and Kaon production, the spectrum is harder and this results in lower predicted fluxes at the lower energies sampled. Allowing for the statistical uncertainty, the $\pm 1\sigma$ limits of the predictions are calculated to be still below the current IC-40+59 data, with $f_p = 1/f_e = 10$ (see Fig. 7). Considering the range of variability of the various astrophysical parameters, it appears that at least several years of observations with the full array may be needed to reach conclusive results about the standard internal shock model. The other point is that internal shocks, in their standard form used above, have been known for a while to have some problems as far as efficiency and spectrum of the prompt γ-ray emission (see discussion in Sections 4 and 5). For this reason, alternative models of MeV photon production have been considered, and the neutrino fluxes and spectra from such models are still in the process of evaluation (e.g. [28]).

Acknowledgments
We acknowledge partial support from NASA NNX08AL40G, NSF PHY0757155 and Grant-in-Aid for Scientific Research No.22740117 from the Ministry of Education, Culture, Sports, Science and Technology (MEXT) of Japan.

References

[1] Ackermann, M and the Fermi-LAT collab. 2012 *Astrophys.J.* **754** 121–+ (*Preprint* `1201.3948`)
[2] Pe'er A 2011 *ArXiv e-prints* (*Preprint* `1111.3378`)
[3] Abdo A and the Fermi Collaboration 2009 *Science* **323** 1688
[4] Abdo A A, and the Fermi LAT Collaboration 2009 *Nature* **462** 331–334
[5] Ackermann M and the Fermi collab 2010 *Astrophys.J.* **716** 1178–1190
[6] Abdo A A and the Fermi collaboration 2009 *Astrophys.J.Lett.* **706** L138–L144 (*Preprint* `0909.2470`)
[7] Ackermann M and the Fermi collaboration 2011 *Astrophys.J.* **729** 114–+ (*Preprint* `1101.2082`)
[8] Abdo A A and the Fermi collab 2010 *Astrophys.J.* **712** 558–564
[9] CTA 2012 homepage URL `http://www.cta-observatory.org/`
[10] HAWC 2012 homepage URL `http://hawc.umd.edu/`
[11] Rees M J and Mészáros P 1994 *Astrophys.J.Lett.* **430** L93–L96 (*Preprint* `arXiv:astro-ph/9404038`)
[12] Rees M J and Meszaros P 1992 *M.N.R.A.S* **258** 41P–43P
[13] Mészáros P and Rees M J 1997 *Astrophys.J.* **476** 232–+ (*Preprint* `arXiv:astro-ph/9606043`)
[14] Gehrels N, Ramirez-Ruiz E and Fox D B 2009 *Annu.Rev.Astron.Astrophys.* **47** 567–617 (*Preprint* `0909.1531`)
[15] De Pasquale M, and the Fermi/GBM collaborations 2010 *Astrophys.J.Lett.* **709** L146–L151 (*Preprint* `0910.1629`)
[16] Corsi A, Guetta D and Piro L 2010 *Astrophys.J.* **720** 1008–1015 (*Preprint* `0911.4453`)
[17] Ghisellini G, Ghirlanda G, Nava L and Celotti A 2010 *M.N.R.A.S.* **403** 926–937 (*Preprint* `0910.2459`)
[18] Kumar P and Barniol Duran R 2009 *M.N.R.A.S.* **400** L75–L79 (*Preprint* `0905.2417`)
[19] Kumar P and Barniol Duran R 2010 *M.N.R.A.S.* **409** 226–236 (*Preprint* `0910.5726`)
[20] Mészáros P and Rees M J 1994 *M.N.R.A.S.* **269** L41+ (*Preprint* `arXiv:astro-ph/9404056`)
[21] Veres P and Mészáros P 2012 *Astrophys.J.* **755** 12–23 (*Preprint* `1202.2821`)
[22] Toma K, Wu X F and Mészáros P 2011 *M.N.R.A.S.* **415** 1663–1680 (*Preprint* `1002.2634`)
[23] Asano K and Mészáros P 2011 *Astrophys.J.* **739** 103 (*Preprint* `1107.4825`)
[24] Asano K, Inoue S and Mészáros P 2009 *Astrophys.J.* **699** 953–957 (*Preprint* `0807.0951`)
[25] Asano K, Guiriec S and Mészáros P 2009 *Astrophys.J.Lett.* **705** L191–L194 (*Preprint* `0909.0306`)
[26] Abbasi R, Abdou Y, Abu-Zayyad T, Adams J, Aguilar J A, Ahlers M, Altmann D, Andeen K, Auffenberg J, Bai X and et al 2011 *Phys.Rev.D* **84** 082001 (*Preprint* `1104.5187`)
[27] Abbasi R, Abdou Y, Abu-Zayyad T, Ackermann M, Adams J, Aguilar J A, Ahlers M, Altmann D, Andeen K, Auffenberg J and et al 2012 *Nature* **484** 351–354 (*Preprint* `1204.4219`)
[28] Asano K and Mészáros P 2012 *Ap.J., in press* (*Preprint* `1206.0347`)
[29] Asano K, Inoue S and Mészáros P 2010 *Astrophys.J.Lett.* **725** L121–L125 (*Preprint* `1009.5178`)
[30] Racusin J L, and the Switf collab. 2008 *Nature* **455** 183–188 (*Preprint* `0805.1557`)
[31] Bahcall J N and Mészáros P 2000 *Physical Review Letters* **85** 1362–1365 (*Preprint* `arXiv:hep-ph/0004019`)
[32] Mészáros P and Rees M J 2000 *Astrophys.J.Lett.* **541** L5–L8 (*Preprint* `arXiv:astro-ph/0007102`)
[33] Beloborodov A M 2010 *M.N.R.A.S.* **407** 1033–1047 (*Preprint* `0907.0732`)
[34] Rees M J and Mészáros P 2005 *Astrophys.J.* **628** 847–852 (*Preprint* `arXiv:astro-ph/0412702`)
[35] Ryde F, et al. 2010 *Astrophys.J.Lett.* **709** L172–L177 (*Preprint* `0911.2025`)
[36] Pe'er A and Ryde F 2011 *Astrophys.J.* **732** 49

[37] Vurm I, Beloborodov A M and Poutanen J 2011 *Astrophys.J.* **738** 77–+ (*Preprint* 1104.0394)
[38] Waxman E 1995 *Physical Review Letters* **75** 386–389 (*Preprint* arXiv:astro-ph/9505082)
[39] Vietri M 1995 *Astrophys. J.* **453** 883–889 (*Preprint* astro-ph/9506081)
[40] Wick S D, Dermer C D and Atoyan A 2004 *Astropart. Phys.* **21** 125–148 (*Preprint* astro-ph/0310667)
[41] Waxman E 2004 *Astrophys.J.* **606** 988–993 (*Preprint* arXiv:astro-ph/0210638)
[42] Dermer C D 2002 *Astrophys.J.* **574** 65–87 (*Preprint* arXiv:astro-ph/0005440)
[43] Murase K, Asano K, Terasawa T and Mészáros P 2012 *Astrophys.J.* **746** 164 (*Preprint* 1107.5575)
[44] Abraham J and the Pierre Auger Collaboration 2010 *Physical Review Letters* **104** 091101–+ (*Preprint* 1002.0699)
[45] Lemoine M 2002 *Astron.Astrophys.* **390** L31–L34 (*Preprint* arXiv:astro-ph/0205093)
[46] Beloborodov A M 2003 *Astrophys.J.* **588** 931–944 (*Preprint* arXiv:astro-ph/0210522)
[47] Metzger B D, Giannios D and Horiuchi S 2011 *M.N.R.A.S.* **415** 2495–2504 (*Preprint* 1101.4019)
[48] Horiuchi S, Murase K, Ioka K and Mészáros P 2012 *Astrophys.J.* **753** 69 (*Preprint* 1203.0296)
[49] Waxman E and Bahcall J 1997 *Phys. Rev. Lett.* **78** 2292–2295
[50] Waxman E and Bahcall J N 2000 *Astrophys.J.* **541** 707–711 (*Preprint* arXiv:hep-ph/9909286)
[51] Guetta D, Hooper D, Alvarez-Muñiz J, Halzen F and Reuveni E 2004 *Astroparticle Physics* **20** 429–455 (*Preprint* arXiv:astro-ph/0302524)
[52] Hümmer S, Baerwald P and Winter W 2012 *Physical Review Letters* **108** 231101 (*Preprint* 1112.1076)
[53] Li Z 2012 *Phys.Rev.D* **85** 027301 (*Preprint* 1112.2240)
[54] He H N, Liu R Y, Wang X Y, Nagataki S, Murase K and Dai Z G 2012 *Astrophys.J.* **752** 29 (*Preprint* 1204.0857)

Gravitino Dark Matter confronts LHC

Laura Covi

Institut für theoretische Physik, Friedrich-Hund-Platz 1, D-37077 Göttingen, Germany

E-mail: laura.covi@theorie.physik.uni-goettingen.de

Abstract. We review the scenario of gravitino LSP and Dark Matter and its cosmological constraints both in the case of R-parity conservation and R-parity violation. We discuss the possible signals and compare with the recent LHC results.

1. Introduction

Dark Matter is surely one of the stronger indications for physics beyond the Standard Model and its identification within the particle zoo is still an open question for both cosmology and particle physics. Indeed from astrophysical and cosmological observations, we have a very clear idea of the Dark Matter density (the most recent WMAP determination is $\Omega_{CDM}h^2 = 0.1126 \pm 0.0036$ [1]) and what it is not: not baryonic, not made of Massive Compact Halo Objects or Hot Dark Matter (and therefore the only neutral massive SM particles, the neutrinos are unfortunately excluded...), not collisional, possibly only interacting gravitationally, but maybe also participating in the weak interaction [2].

Since most Dark Matter candidates "looks" the same from the cosmological point of view (all sufficiently non-relativistic particles are just matter for a cosmologist !), one way to select a Dark Matter particles is theoretical prejudice based on the well-trodden path of an enhanced symmetry. In that respect the leading candidate for enlarging the Standard Model symmetries is supersymmetry, the maximal extension of the Poincaré symmetry. Supersymmetry, especially in its low-energy version, has also other important theoretical advantages since it stabilized the electroweak scale, it can give rise even to radiative symmetry breaking and allows for the possibility of Grand Unification. Still, even within "minimal" Supersymmetry many different Dark Matter candidates are present, i.e. all the superpartners of neutral SM particles like the sneutrino, the neutralino and also the gravitino. The gravitino is often not considered in the supersymmetric phenomenology since gravity is usually neglected: indeed the gravitino is the superpartner of the graviton and appears inevitably in any supersymmetrization of gravity (for an introduction to supersymmetry and supergravity see [3]). It has both Lorentz and spinor indices, corresponding to a spin $3/2$ particle, and is in some sense the gauge field of local supersymmetry and it becomes massive as soon as supersymmetry is broken by absorbing the Goldstino field. Observing the gravitino would therefore be for supergravity the equivalent of the observation of the $W^{\pm}$ boson for the electroweak symmetry. Due to such particular nature, the interactions of the gravitino are completely fixed by supersymmetry and since it belong to the gravity sector all couplings are suppressed by the Planck scale.

Gravitinos are often a nuisance in cosmology, since if they are not the LSP and therefore unstable towards a decay into a lighter superparticle, their lifetime is naturally long on

Content from this work may be used under the terms of the Creative Commons Attribution 3.0 licence. Any further distribution of this work must maintain attribution to the author(s) and the title of the work, journal citation and DOI.
Published under licence by IOP Publishing Ltd

cosmological scales, due to the Planck scale suppression. If the gravitinos are not sufficiently heavy, i.e. well beyond 30 TeV, they decay during or after Big Bang Nucleosynthesis (BBN) giving rise to the famous "gravitino problem" [4; 5]. In that case, in order to reduce the damage, the gravitino density has to be small enough at the cost of an upper bound on the reheat temperature of the Universe (see e.g. [6]). On the other hand, if gravitinos are the LSP, they can be very good Dark Matter candidates, either as stable particles or as metastable particles with lifetime much longer than the age of the Universe. Then gravitino Dark Matter can be in some cases compatible with thermal leptogenesis (for a review see [7]), a mechanism for baryogenesis that in the simplest realization requires reheat temperatures above 10^9 GeV [7; 8]. We will discuss in the following different scenarios for gravitino Dark Matter with and without R-parity breaking and their phenomenology at the LHC.

2. Gravitinos as Dark Matter

Gravitinos were the first supersymmetric Dark Matter candidates ever proposed: they were discussed in a paper by Pagels & Primack [9] as relativistic thermal relics whose energy density is given by [1]

$$\Omega_{3/2}h^2 \sim 0.1 \left(\frac{m_{3/2}}{0.1\text{keV}}\right)\left(\frac{g_*}{106.75}\right)^{-1} \tag{1}$$

where $m_{3/2}$ is the gravitino mass and g_* is the effective number of degrees of freedom thermalized at the time of gravitino decoupling. We see that in this case to have the right energy density gravitinos have to be light and therefore they are Warm/Hot Dark Matter and excluded by structure formation.

To be a Cold Dark Matter (CDM) candidate, the gravitinos must be heavier and then their number density smaller than the equilibrium density. Scattering processes in the primordial plasma can produce a sufficiently large gravitino population for gravitino CDM. Such scatterings are mediated by non-renormalizable interactions, in particular dimension 5 operators involving the gauge interactions.

Then the gravitino energy density can be computed by integrating a Boltzmann equation without back-reaction and it turns out to be proportional to the reheat temperature due to the non-renormalization nature of the interaction. Different groups [10–12] have computed the production rate for gravitinos including one-loop HTML resummation and thermal corrections. Assuming that the Goldstino component dominates the production, i.e. when the gravitino is sufficiently lighter than the other superpartners, the energy density of gravitinos is given by

$$\Omega_{3/2}h^2 \sim 0.3 \left(\frac{m_{3/2}}{1\text{GeV}}\right)^{-1}\left(\frac{T_{RH}}{10^{10}\text{GeV}}\right)\sum_i c_i \left(\frac{M_i}{100\text{GeV}}\right)^2 \tag{2}$$

where c_i are coefficients of order 1 and M_i denote the three gaugino masses at EW temperature (RGE effects to T_{RH} can be included in the coefficients c_i).

We see then that the Dark Matter density is set by the model parameters (i.e. the sparticle masses) and the reheat temperature of the universe T_{RH}. It is therefore possible to arrange for the right density by setting such temperature, e.g. from a particular inflationary model and reheating dynamics. A large reheat temperature around 10^{10} GeV is possible only for gravitino masses above approximately 1-10 GeV and in general an upper bound on T_{RH} is obtained to avoid overclosure.

[1] We give energy densities as a function of the critical energy density ρ_c, i.e. as $\Omega_i h^2 = \rho_i h^2/\rho_c$. Ωh^2 corresponds to a density in units of 1.05×10^{-5}GeV cm^{-3}.

Note that another population of gravitinos is generated by the decay of the NLSP out of equilibrium in the SuperWIMP mechanism [13; 14], as

$$\Omega_{3/2}h^2 = \frac{m_{3/2}}{m_{WIMP}}\Omega_{WIMP}h^2 \tag{3}$$

where we are assuming the NLSP to be a WIMP particle [15]. But if such decay happens during or after BBN, the energy density of the NLSP is very strongly constrained by BBN [16]. Then it has often to be smaller than the present DM density and therefore this gravitino population is effectively negligible.

3. Stable gravitino CDM

In the case of R-parity conservation, the gravitino LSP is stable, and the NLSP decays into gravitinos and Standard Model particles with a very long lifetime. For the case of a Bino NLSP one has simply

$$\tau_{\tilde{B}} = 5.7 \times 10^6 \text{ s} \left(\frac{m_{\tilde{B}}}{100\text{GeV}}\right)^{-5} \left(\frac{m_{3/2}}{10\text{GeV}}\right)^2 . \tag{4}$$

This means that the decay takes place after BBN has already started, apart if the Bino is sufficiently heavy: $\tau_{\tilde{B}} < 0.1$ s requires

$$m_{\tilde{B}} \geq 3.6 \text{ TeV} \left(\frac{m_{3/2}}{10\text{GeV}}\right)^{2/5} \tag{5}$$

well beyond the reach of the LHC if the gravitino is not much lighter than 10 GeV. For longer lifetimes, one has to take into account the different BBN constraints [16], depending on the branching ratios of the neutralino and the energy released in electromagnetic or hadronic showers. We will discuss here only the case of a general neutralino NLSP and comment on the case of a colored NLSP like the stop or the gluino.

The neutralino NLSP number density at freeze-out in the CMSSM is quite large and therefore most parameter space is excluded, as it has been found by different groups [17; 18]. We recently considered the bounds for a general neutralino NLSP in [19], computing the neutralino branching ratio into hadrons and comparing with the bounds for neutral relics in [20]. We found that even if it possible to suppress the hadronic branching ratio tuning the neutralino composition, this does not allow to relax the bounds very much and so only very special cases are still allowed. For example for a neutralino with substantial Higgsino component the energy density can be sufficiently reduced by resonant annihilation or in the case of a very light Wino, the hadronic branching ratio is small enough due to phase-space suppression.

Another mechanism that helps bypassing the bounds and allows also to keep the reheat temperature as small as possible, is to consider the case of quasi-degenerate gauginos [21]. Then in fact the efficient coannihilation between neutralino and gluino reduces the neutralino density by more than three orders of magnitude. Including as well the Sommerfeld enhancement gains approximately four orders of magnitude suppression at the cost of a mass degeneracy of the order of few percent. These scenarios with very compressed spectrum are a challenge for the LHC, since very soft particles are produced in the chain decays.

Since the lower the NLSP number density the weaker the constraints, it may seem a good idea to select a colored NLSP instead of the neutralino. Unfortunately the BBN bounds on colored relics are much stronger than those for charged or neutral particles due to bound state effects that modify strongly the rates of the nuclear processes during BBN. Such constraints have been computed by [22] and they approximately exclude NLSP lifetimes longer than ~ 100 s. Even increasing the colored state annihilation cross-section including the Sommerfeld effect does not relax those bounds substantially since the curves are very steep in the number density.

The only other way to relax the bound is to consider a sufficiently heavy NLSP so that the decay happens early enough. For the stop NLSP, considering as dominant decay channel the one to top and gravitino, we obtain

$$m_{\tilde{t}} \geq 1 \text{ TeV} \left(\frac{m_{3/2}}{10\text{GeV}} \right)^{2/5} . \tag{6}$$

A slightly weaker bound can be obtained for the gluino NLSP, which annihilates a bit more efficiently.

4. Unstable gravitino CDM

Another simple way to satisfy the BBN constraints is to break R-parity and allow the NLSP to decay only to SM particles (long) before BBN [23]. Note that the gravitino is a good Dark Matter candidate also for broken R-parity, as long as the breaking is sufficiently small. In fact the gravitino decay rate is suppressed both by the Planck mass and by the small R-parity breaking coupling and so can be naturally many orders of magnitude longer than the age of the Universe.

The requirement for the NLSP to decay early enough and for the lepton- or baryon-violating processes connected to R-parity violation not to be in equilibrium at the same time as the sphaleron processes [24; 25], singles out a particular window in the range of couplings:

$$10^{-12} \leq \lambda \leq 10^{-7} \tag{7}$$

where λ is a generic R-parity breaking coupling. Of course once R-parity is broken, the danger of too fast proton decay reappears in the MSSM, but it can be avoided if e.g. only the lepton flavour violating couplings are switched on.

The general R-parity breaking renormalizable MSSM superpotential is given by [26]

$$W_{R\!\!\!/_p} = \mu_i H_u L_i + \lambda_{ijk} L_i L_j E_k + \lambda'_{ijk} L_i Q_j D_k + \lambda''_{ijk} U_i D_j D_k . \tag{8}$$

One particular class of models is that of *bilinear* R-parity violation, where the dominant R-parity breaking coupling is the first one in the expression above. In that case the main effect is to mix the neutrinos with the neutralinos and the charged leptons with the chargino and this opens up the decay of the gravitino into neutral gauge (γ, Z) or Higgs boson and neutrino or $W^{\pm}$ and charged lepton [23; 27]. If instead the trilinear R-parity breaking couplings dominate, the gravitino decays at three level in the three-bodies, while the 2-body decay is realized only at one loop [28].

If the R-parity breaking couplings are large enough such a decay could be observable in different channels of indirect DM detection observations, [29–35]. At the moment the FERMI data set a lower bound on the DM lifetime in photons of the order of 5×10^{28} s, already excluding part of the R-parity breaking parameter space [36; 37]. In the case of non-universal gaugino masses, the gravitino signal in the gamma-line can be suppressed [38], while allowing also R-parity breaking couplings sufficiently large to explain the neutrino masses.

5. Gravitinos and the LHC

The supersymmetric models with gravitino Dark Matter can give at the LHC quite different signals than the neutralino DM case: in fact, the different types of signatures depend strongly on the nature of the NLSP particle and its lifetime. In many of the cosmology inspired scenarios we discussed, either because the gravitino is sufficienly heavy or because the R-parity breaking coupling is small, the NLSP may even appear stable at colliders. Then, if the NLSP is charged, the very striking signature of a charged track passing through the whole detector would indicate immediately that such particle is not the Dark Matter and that a very weakly interacting particle like the gravitino could play such role. The LHC collaborations have already performed searches for exotic metastable particles, setting strong constraints on the masses of the stop and gluino NLSP, reaching a lower limits of the order of about 800 and 1200 GeV respectively [39]. Weaker constraints apply on the other hand to the stau as well as the other particles of the electroweak spectrum.

On the other hand, also displaced vertices are possible, with the NLSP decaying inside the detector. In that case, of example in the R-parity violating models, the LHC can probe the case of neutralino NLSP also beyond the parameter space already excluded by FERMI [40].

If the gaugino spectrum is nearly-degenerate, as in many other compressed spectrum scenarios, the most promising channel is the monojet signature with a large missing E_T. In the case we discussed, the jet may be coming either from Initial State Radiation or from the associate production of gluino and squark. The monojet channel has been recently studied by the LHC collaborations giving bounds for the case of graviton production in extra-dimensional scenarios and for Dark Matter production at colliders [41; 42]. Those results have been also reinterpreted for the case of a degenerate gaugino spectrum and they give important constraints, excluding gluino masses up to 450-500 GeV [43]. Still these bounds are much weaker than the gluino mass bound for non-degenerate spectra, which exceeds now 1 TeV.

6. Conclusion

The gravitino is a very good Dark Matter particle, both in the case of conserved and broken R-parity. In the first case, BBN imposes some cosmological restrictions on the scenario, depending on the nature of the NLSP and its lifetime and number density. For neutralino NLSP, these constraints possibly point to particular supersymmetric parameter ranges, where the neutralino density is suppressed. One example is the case of quasi-degenerate gauginos. These mass spectra are more difficult to disentangle at colliders than the normal case of large mass differences, but are nevertheless starting to be explored at the LHC, until now without success. On the other hand, for heavy NLSP the BBN constraints are relaxed and the gravitino can be a viable DM candidate also if the superparticles spectrum is beyond the reach of the LHC. The price to pay in this case is a much lower reheat temperature than that required by thermal leptogenesis.

In the case of R-parity violation, the gravitino DM can decay and therefore indirect Dark Matter searches can offer a complementary channel of discovery. So far no signal of gravitino Dark Matter has been found and the present FERMI data already exclude part of the parameter space. Further constraints or even a discovery are still possible at the LHC, for example in channels with displaced vertices.. The R-parity violating models have in fact only recently started to be studied systematically at the LHC.

The search for gravitino Dark Matter at LHC and in indirect DM detection is only at the beginning and, even if no evidence has been found so far, let us hope for a new signal soon !

Acknowledgments

The author acknowledges partial financial support by the German-Israeli Foundation for scientific research and development (GIF) and by the EU FP7 ITN Invisibles (Marie Curie Actions, PITN-GA-2011-289442).

The author would like to thank the organizers for the invitation to a very exciting conference and all her collaborators, W. Buchmüller, F. Dradi, M. Grefe, K. Hamaguchi, J. Hasenkamp, A. Ibarra, M. Olechowski, S. Pokorski, J. Roberts, D. Tran, K. Turzynski, J. D. Wells and T. Yanagida, for the enjoyable work together.

References

[1] Komatsu E *et al.* (WMAP Collaboration) 2011 *Astrophys.J.Suppl.* **192** 18 (*Preprint* 1001.4538)

[2] Bertone G, Hooper D and Silk J 2005 *Phys.Rept.* **405** 279–390 (*Preprint* hep-ph/0404175)

[3] Wess J and Bagger J 1992

[4] Khlopov M Y and Linde A D 1984 *Phys.Lett.* **B138** 265–268

[5] Ellis J R, Nanopoulos D V and Sarkar S 1985 *Nucl.Phys.* **B259** 175

[6] Kawasaki M, Kohri K, Moroi T and Yotsuyanagi A 2008 *Phys.Rev.* **D78** 065011 (*Preprint* 0804.3745)

[7] Davidson S, Nardi E and Nir Y 2008 *Phys.Rept.* **466** 105–177 (*Preprint* 0802.2962)

[8] Buchmuller W, Di Bari P and Plumacher M 2004 *New J.Phys.* **6** 105 (*Preprint* hep-ph/0406014)

[9] Pagels H and Primack J R 1982 *Phys.Rev.Lett.* **48** 223

[10] Bolz M, Brandenburg A and Buchmuller W 2001 *Nucl.Phys.* **B606** 518–544 (*Preprint* hep-ph/0012052)

[11] Pradler J and Steffen F D 2007 *Phys.Rev.* **D75** 023509 (*Preprint* hep-ph/0608344)

[12] Rychkov V S and Strumia A 2007 *Phys.Rev.* **D75** 075011 (*Preprint* hep-ph/0701104)

[13] Feng J L, Rajaraman A and Takayama F 2003 *Phys.Rev.Lett.* **91** 011302 (*Preprint* hep-ph/0302215)

[14] Feng J L, Rajaraman A and Takayama F 2003 *Phys.Rev.* **D68** 063504 (*Preprint* hep-ph/0306024)

[15] Griest K and Kamionkowski M 2000 *Phys.Rept.* **333** 167–182

[16] Jedamzik K and Pospelov M 2009 *New J.Phys.* **11** 105028 (*Preprint* 0906.2087)

[17] Ellis J R, Olive K A, Santoso Y and Spanos V C 2004 *Phys.Lett.* **B588** 7–16 (*Preprint* hep-ph/0312262)

[18] Cerdeno D G, Choi K Y, Jedamzik K, Roszkowski L and Ruiz de Austri R 2006 *JCAP* **0606** 005 (*Preprint* hep-ph/0509275)

[19] Covi L, Hasenkamp J, Pokorski S and Roberts J 2009 *JHEP* **0911** 003 (*Preprint* 0908.3399)

[20] Jedamzik K 2006 *Phys.Rev.* **D74** 103509 (*Preprint* hep-ph/0604251)

[21] Covi L, Olechowski M, Pokorski S, Turzynski K and Wells J D 2011 *JHEP* **1101** 033 (*Preprint* 1009.3801)

[22] Kusakabe M, Kajino T, Yoshida T and Mathews G J 2009 *Phys.Rev.* **D80** 103501 (*Preprint* 0906.3516)

[23] Buchmuller W, Covi L, Hamaguchi K, Ibarra A and Yanagida T 2007 *JHEP* **0703** 037 (*Preprint* hep-ph/0702184)

[24] Dreiner H K and Ross G G 1993 *Nucl.Phys.* **B410** 188–216 (*Preprint* hep-ph/9207221)

[25] Endo M, Hamaguchi K and Iwamoto S 2010 *JCAP* **1002** 032 (*Preprint* 0912.0585)

[26] Barbier R, Berat C, Besancon M, Chemtob M, Deandrea A *et al.* 2005 *Phys.Rept.* **420** 1–202 (*Preprint* hep-ph/0406039)

[27] Takayama F and Yamaguchi M 2000 *Phys.Lett.* **B485** 388–392 (*Preprint* hep-ph/0005214)

[28] Lola S, Osland P and Raklev A 2007 *Phys.Lett.* **B656** 83–90 (*Preprint* 0707.2510)

[29] Bertone G, Buchmuller W, Covi L and Ibarra A 2007 *JCAP* **0711** 003 (*Preprint* 0709.2299)

[30] Ishiwata K, Matsumoto S and Moroi T 2008 *Phys.Rev.* **D78** 063505 (*Preprint* 0805.1133)

[31] Covi L, Grefe M, Ibarra A and Tran D 2009 *JCAP* **0901** 029 (*Preprint* 0809.5030)

[32] Buchmuller W, Ibarra A, Shindou T, Takayama F and Tran D 2009 *JCAP* **0909** 021 (*Preprint* 0906.1187)

[33] Choi K Y, Lopez-Fogliani D E, Munoz C and de Austri R R 2010 *JCAP* **1003** 028 (*Preprint* 0906.3681)

[34] Bomark N E, Lola S, Osland P and Raklev A 2010 *Phys.Lett.* **B686** 152–161 (*Preprint* 0911.3376)

[35] Bobrovskyi S, Buchmuller W, Hajer J and Schmidt J 2010 *JHEP* **1010** 061 (*Preprint* 1007.5007)

[36] Abdo A, Ackermann M, Ajello M, Atwood W, Baldini L *et al.* 2010 *Phys.Rev.Lett.* **104** 091302 (*Preprint* 1001.4836)

[37] Vertongen G and Weniger C 2011 *JCAP* **1105** 027 (*Preprint* 1101.2610)

[38] Restrepo D, Taoso M, Valle J and Zapata O 2012 *Phys.Rev.* **D85** 023523 (*Preprint* 1109.0512)

[39] Chatrchyan S *et al.* (CMS Collaboration) 2012 *Phys.Lett.* **B713** 408–433 (*Preprint* 1205.0272)

[40] Bobrovskyi S, Buchmuller W, Hajer J and Schmidt J 2011 *JHEP* **1109** 119 (*Preprint* 1107.0926)

[41] Aad G *et al.* (ATLAS Collaboration) 2012 (*Preprint* 1209.4625)

[42] Chatrchyan S *et al.* (CMS Collaboration) 2012 (*Preprint* 1206.5663)

[43] Dreiner H K and Tattersall M K J 2012 (*Preprint* 1207.1613)

Journal of Physics: Conference Series **485** (2014) 012003 doi:10.1088/1742-6596/485/1/012003

SUSY dark matter in nonuniversal gaugino mass models

D P Roy
Homi Bhabha Centre for Science Education, Tata Institute of Fundamental Research, Mumbai 400088, India
E-mail: dproy@tifr.res.in

Abstract. We discuss the SUSY dark matter phenomenology in some simple and predictive models of nonuniversal gaugino masses at the GUT scale. Assuming the gaugino masses to transform as a sum of singlet and a nonsinglet representation of the GUT group SU(5), one can evade the LEP constraints to access the bulk annihilation region of the bino dark matter relic density. Besides, with this assumption one can also have a mixed gaugino-higgsino dark matter, giving the right relic density over large parts of the parameter space. We consider the model predictions for LHC and dark matter experiments in both the cases. Finally we consider the AMSB model prediction of wino dark matter giving the right relic density for TeV scale wino mass. Assuming this wino dark matter mass to be at the first Sommerfeld resonance of $\sim$ 4 TeV one can simultaneously reproduce the right relic density as well as the hard positron spectrum observed by the PAMELA experiment.

1. Introduction

The most phenomenologically attractive feature of SUSY and in particular the minimum supersymmetric standard model (MSSM) is that it provides a natural candidate for the dark matter (DM) in terms of the lightest neutralino $\tilde{\chi}_1^0$ (abbreviated as χ). In the constrained version of this model (CMSSM), corresponding to universal gaugino and scalar masses at the GUT scale, the dark matter is dominantly bino. And since the bino carries no gauge charge, its main annihilation mechanism is via sfermion exchange. This is called the bulk annihilation process; and the region of parameter space giving the right DM relic density via this process is called the bulk annihilation region. Unfortunately, the LEP limit on Higgs boson mass imposes rather stringent limits on bino and sfermion masses in the CMSSM, which rule out the bulk annihilation region of its parameter space [1]. Therefore the model can give the right DM relic density only over a few narrow strips of parameter space, corresponding to fairly high degrees of fine-tuning.

We consider here the DM phenomenology of a set of simple and predictive models of nonuniversal gaugino masses, which are assumed to transform like the sum of the singlet and a nonsinglet representation of the GUT group SU(5). With this assumption, one can evade the LEP constraint to access the bulk annihilation region of a bino DM. Besides one can also have a mixed bino-higgsino DM giving the right relic density over a large part of the parameter space. In section 2 we shall introduce the basic model assumptions and these two types of DM models resulting from it. Then in sections 3 and 4 we shall discuss the implications of these two types

Content from this work may be used under the terms of the Creative Commons Attribution 3.0 licence. Any further distribution of this work must maintain attribution to the author(s) and the title of the work, journal citation and DOI.
Published under licence by IOP Publishing Ltd

of DM models for LHC and DM experiments. Finally in section 5 we shall discuss the AMSB model prediction of a TeV scale wino DM along with its implications for DM experiments.

2. Nonuniversality of Gaugino Masses in SU(5) GUT

The gauge kinetic function responsible for GUT scale gaugino masses arises from the vacuum expectation value of the F term of the chiral superfield Φ responsible for SUSY breaking,

$$\frac{\langle F_\Phi \rangle_{ij}}{M_{Planck}} \lambda_i \lambda_j \ , \tag{1}$$

where $\lambda_{1,2,3}$ are the U(1), SU(2) and SU(3) gaugino fields – bino, wino and gluino. Since the gauginos belong to the adjoint representation of the GUT group SU(5), the Φ and F_Φ can belong to any of the irreducible representations appearing in their symmetric product,

$$(24 \times 24)_{symm} = 1 + 24 + 75 + 200 \ . \tag{2}$$

Thus the GUT scale gaugino masses for a given representation of the SUSY breaking superfield Φ are given in terms of a single mass parameter [2],

$$M^G_{1,2,3} = C^n_{1,2,3} m^n_{1/2} \ , \tag{3}$$

where

$$C^1_{1,2,3} = (1,1,1); \quad C^{24}_{1,2,3} = (-1,-3,2); \quad C^{75}_{1,2,3} = (-5,3,1); \quad C^{200}_{1,2,3} = (10,2,1) \ . \tag{4}$$

The CMSSM assumes Φ to be a singlet, implying universal gaugino masses at the GUT scale. On the other hand, any of the nonsinglet representations of Φ would imply nonuniversal gaugino masses via (3) and (4). These nonuniversal gaugino mass models are known to be consistent with the universality of gauge couplings at the GUT scale, with $\alpha_G \sim 1/25$; and their phenomenology has been widely studied [3]. The superparticle masses at the EW scale are related to these GUT scale gaugino masses and the universal scalar mass parameter m_0 via the RGE. In particular the gaugino masses evolve like the corresponding gauge couplings at the one-loop level of the RGE, i.e.

$$\begin{aligned} M_1 &= (\alpha_1/\alpha_G) \simeq (25/60) C^n_1 m^n_{1/2} \ , \\ M_2 &= (\alpha_2/\alpha_G) \simeq (25/30) C^n_2 m^n_{1/2} \ , \\ M_3 &= (\alpha_3/\alpha_G) \simeq (25/9) C^n_3 m^n_{1/2} \ . \end{aligned} \tag{5}$$

The corresponding higgsino mass μ is obtained from the EW symmetry breaking condition along with the one-loop RGE for the Higgs scalar mass. i.e.

$$\mu^2 + M_Z^2/2 \simeq -m^2_{H_u} \simeq -0.1 m_0^2 + 2.1 M_3^{G^2} - 0.22 M_2^{G^2} - 0.19 M_3^G M_2^G \tag{6}$$

neglecting the contribution from the GUT scale trilinear coupling term A_0 [4]. The numerical coefficients on the right correspond to a representative value of $\tan\beta = 10$; but they show only mild variations over the moderate $\tan\beta$ region.

Although our results are based on exact numerical solutions to the two-loop RGE [5], the approximate formulae (5) and (6) are very useful in understanding the essential features of the models. Combining these formulae with (4) shows that for the 1 and 24-plet representations $\mu > M_1$ implying bino DM, while for the 75 and 200-plet representations $\mu < M_1$ implying a

Journal of Physics: Conference Series **485** (2014) 012003 doi:10.1088/1742-6596/485/1/012003

higgsino DM. As mentioned earlier, the bino DM generically gives an over-abundance of relic density except for the bulk annihilation region. On the other hand, the higgsino DM can efficiently annihilate via their isospin gauge coupling, leading to an under-abundance of relic density for sub-TeV DM mass. (A higgsino DM mass of about 1 TeV gives the right relic density, but will be out of reach of LHC.) One gets phenomenologically promising DM models by assuming a combination of two SUSY breaking superfields belonging to singlet and one of the above nonsinglet representations of SU(5). It was shown in [6] that one can evade the above mentioned LEP constraint to access the bulk annihilation region of right DM relic density for the 1+24, 1+75 and 1+200 representations. Moreover it was shown in [7] that one can also get mixed bino-higgsino DM for the 1+75 and 1+200 representations, leading to right relic density over large parts of the parameter space. We shall discuss the implications of these two types of models for LHC and DM detection experiments in the next two sections.

3. Bino DM in the Bulk Annihilation Region in 1+75 and 1+200 Models

For the bulk annihilation region in the 1+75 and 1+200 models the singlet contributions to gaugino masses dominate over the nonsinglet ones, while it is the other way around in the 1+24 model [6]. We shall concentrate here on the first two models. For a general nonuniversal gaugino mass model the bulk annihilation region of bino DM spans over the parameter range

$$M_1^G = 150 - 250 \text{ GeV} \quad \text{and} \quad m_0 = 50 - 80 \text{ GeV} , \tag{7}$$

practically independent of the other SUSY parameters [8]. This is because the main annihilation process for the bino DM pair proceeds via right handed slepton exchange,

$$\chi\chi \xrightarrow{\tilde{l}_R} \bar{l}l ; \tag{8}$$

and the bino mass is determined by M_1^G via the RGE (5), while the right handed slepton mass is determined via its RGE by M_1^G and m_0. For our models we have one more gaugino mass parameter at our disposal. We choose this to be M_3^G, because it makes the dominant contribution to the squark/gluino masses via their RGE as well as the higgsino mass via (6).

Table 1 shows the SUSY mass spectrum in the 1+75 and 1+200 models for the bino DM mass near the upper end of the bulk annihilation region, where we have chosen $M_3^G = 800$ GeV [9]. The resulting values of squark/gluino masses in the range of 1500-1700 GeV are outside the reach of the 7-8 TeV LHC, but well within that of 14 TeV LHC data. Note that the M_2^G values for the two models are determined via (3) and (4). Therefore the resulting values of wino and left slepton masses are different for the two models. However, the small value of m_0 ensures that the left sleptons are lighter than the wino for each model. Consequently the models predict the SUSY cascade decay at LHC to proceed via these slepton states resulting in a distinctive signal with a hard electon/muon or tau-jet accompanying the missing-E_T.

The direct DM detection experiments are based on their elastic scattering on a heavy nucleus like Germanium or Xenon, which is dominated by the spin-independent χp scattering mediated by the Higgs boson exchange. Since the Higgs coupling to the lightest neutralino χ is proportional to the product of its higgsino and gaugino components, the direct detection cross-section is expected to be small for a bino dominated χ state. Numerical result using DarkSUSY [10] gives a cross-section well below the current experimental limits [11–13], but within the reach of a future 1 ton Xenon experiment [14].

The indirect DM detection experiments are based on observing the decay products of their pair annihilation at the present time. Since the DM particles are highly non-relativistic ($v \sim 10 - 3$), only the S-wave annihilation cross-sections are of any significance at the present time. One can then show from symmetry considerations that for Majorana particles like the

Table 1. SUSY mass spectrum in the 1+75 and 1+200 models for the bino DM mass near the upper end of the bulk annihilation region and a representative value of $\tan\beta = 10$ [9].

$M_1^G = 250$ GeV, $M_3^G = 800$ GeV, $m_0 = 67$ GeV

Particle	Mass (GeV)	
	(1+75) model	(1+200) model
$\tilde{\chi}_1^0$ (bino)	100	99.6
$\tilde{\chi}_2^0$ (wino)	772	586
$\tilde{\chi}_3^0$ (higgsino)	933	970
$\tilde{\chi}_4^0$ (higgsino)	955	979
$\tilde{\chi}_1^+$ (wino)	772	586
$\tilde{\chi}_2^+$ (higgsino)	955	979
M_1	102	102
M_2	778	579
M_3	1718	1723
μ	928	965
$\tilde{g}$	1766	1766
$\tilde{\tau}_1$	100	104
$\tilde{\tau}_2$	627	474
$\tilde{e}_R, \tilde{\mu}_R$	119	119
$\tilde{e}_L, \tilde{\mu}_L$	628	474
$\tilde{t}_1$	1221	1251
$\tilde{t}_2$	1541	1507
$\tilde{b}_1$	1512	1480
$\tilde{b}_2$	1529	1528
$\tilde{q}_{1,2,R}$	~ 1528	~ 1533
$\tilde{q}_{1,2,L}$	~ 1640	~ 1593

neutralino χ the cross-section for the annihilation process (8) is helicity suppressed by a factor of $(m_l/M_W)^2$. In contrast the cross-section for the radiative annihilation process

$$\chi\chi \xrightarrow{\tilde{l}_R} \bar{l}l\gamma \tag{9}$$

is only suppressed by a factor α, so that it constitutes the dominant annihilation process at the present time [15]. We have computed the resulting hard positron signal for the SUSY spectrum of table 1 using the DarkSUSY [10] and Galprop [16] codes. As expected the result depends only on the input parameters M1G and m0 irrespective of the other model parameters. Hence it provides a robust prediction for the bino DM models in the bulk annihilation region [9]. The figure 1 compares the model prediction with the hard positron spectrum observed by the PAMELA experiment [17]. There is good agreement between the shape of the predicted spectrum with this data, with only the last data point overshooting the prediction by two standard deviations. Note that the annihilation process (9) produces no antiprotons and hence predicts no antiproton excess over the cosmic ray background in agreement with the PAMELA data. However, the model has no explanation for the huge boost factor of 7000 required to match the size of the PAMELA positron signal. Therefore one needs to attribute this factor to astrophysical sources like a local population of intermediate mass black holes leading to spikes in the DM density distribution [18], or a nearby DM clump [19].

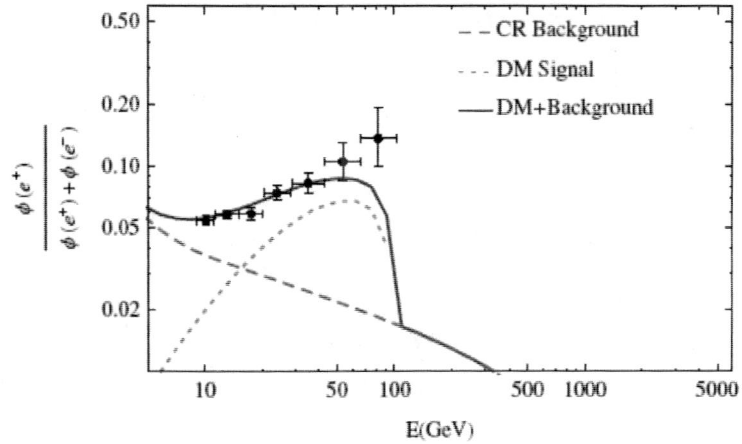

Figure 1. The predicted positron flux of the bino DM in the bulk annihilation region is compared with the PAMELA data assuming a boost factor of ∼ 7000 [9].

4. Mixed Neutralino DM in the 1+75 and 1+200 Models

In these models the relative values of the singlet and nonsinglet gaugino mass parameters of (3) can be described in terms of a dimensionless parameter α [7], i.e.

$$M^1_{1/2} = (1-\alpha_{75})m_{1/2},\ m^{75}_{1/2} = \alpha_{75}m_{1/2} \quad \text{and} \quad m^1_{1/2} = (1-\alpha_{200})m_{1/2},\ m^{200}_{1/2} = \alpha_{200}m_{1/2}\,. \tag{10}$$

Then one sees from (3-6) that for the 1+75 model,

$$\alpha_{75} = 0.5 \Rightarrow |M_1| \simeq |\mu| \simeq m_{1/2} < |M_2| \ \text{ for } \ m_0 \geq m_{1/2}\,. \tag{11}$$

This corresponds to a mixed neutralino DM with bino and higgsino components of roughly similar size. It represents a simple realization of the so called well-tempered neutralino scenario [20]. It is analogous to the mixed bino-higgsino DM in the focus point region of the CMSSM [21], except that it holds over a wide region of the parameter space here. Figure 2 shows the $m_0 - m_{1/2}$ parameter space of the 1+75 model for an optimal choice of $\alpha_{75} = 0.475$ with contours of M_1 and μ. It also shows the contours of the bino component of DM and the DM relic density. The regions satisfying the cold DM relic density measured by WMAP [22]

$$\Omega h^2 = 0.110 \pm 0.019(3\sigma) \tag{12}$$

are indicated by the bands of red dots. The upper band corresponds to the DM being an almost equally mixed bino-higgsino state, where the gauge coupling of the higgsino component leads to the pair annihilation of $\chi\chi$ into WW, ZZ and fermion-antifermion pairs (via Z) as in the focus point region of the CMSSM. On the other hand, the lower band represents the upper edge of the resonant annihilation region of $\chi\chi$ into a heavy quark (lepton) pairs via the pseudo scalar Higgs boson A. The lower edge appears as a narrow strip at the lower right corner, while the region in between has under abundance of DM relic density due to efficient resonant annihilation. Thus the lower branch enclosed by these two edges is similar to the so called funnel region of the

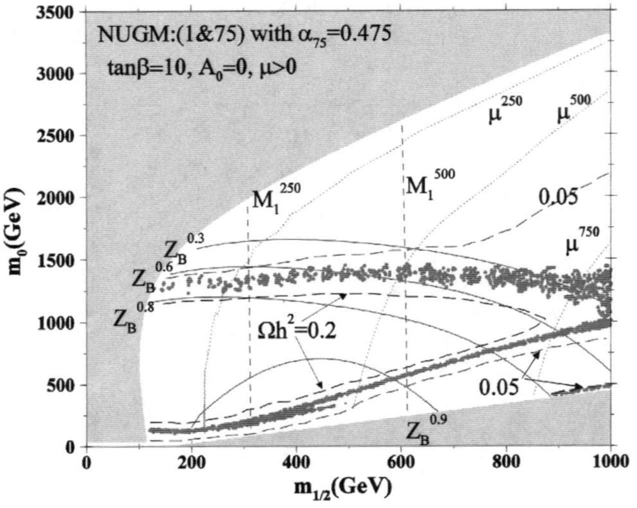

Figure 2. The composition and relic density of the mixed netralino DM of the 1+75 model [7].

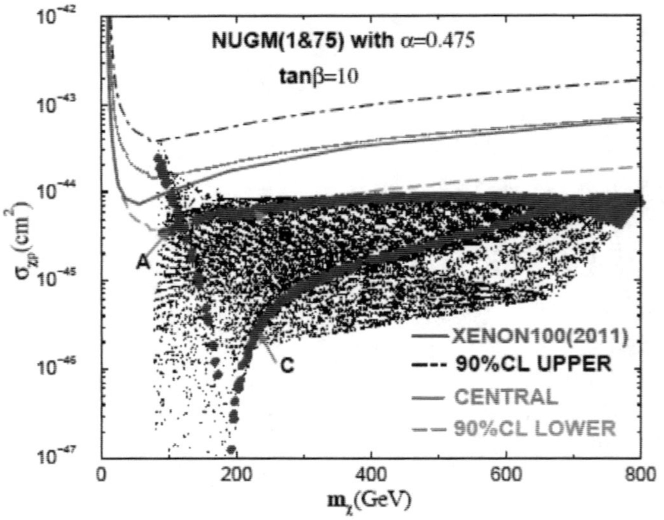

Figure 3. The direct detection cross-section prediction of the 1+75 model. The upper and lower band of blue dots represent the WMAP relic density satisfying points corresponding to the respective red bands of figure 2 [23].

the CMSSM. But since the higgsino component of ∼10% here is still significantly larger than in CMSSM, the funnel region occurs even for a moderate $\tan\beta = 10$ and spans a wide region of parameter space.

The model predicts a large direct DM detection cross-section, since the dominant contribution from Higgs exchange is proportional to the product of the higgsino and gaugino components of DM. Fig 3 compares the model prediction with the experimental band corresponding to the central and 90% CL limits of the two CDMS candidate events [11]. The 90% CL upper limit of the XENON100 (2011) data [12] is also shown for comparison. The upper band of WMAP relic density satisfying points are already being probed by these experiments. In fact the very recent XENON100 (2012) data seems to give an upper limit very close to this upper band [13]. The SUSY mass spectra corresponding to the three representative points A,B,C of Fig. 3 are listed in Table 2 . One clearly sees an inverted hierarchy in the squark masses, where the stop is significantly lighter than the other squarks. Thus one expects the gluino decay to proceed mainly via the stop, resulting in four b jets from the decay of the gluino pair. Therefore one can use ≥ 3 b-tags to search for the SUSY signal of this model with the available 7 TeV LHC data up to a DM mass of 250 GeV, corresponding to a gluino mass of 800 GeV [23].

Model	$\tilde{g}$	$\tilde{q}_L$	$\tilde{q}_R$	$\tilde{t}_1$	$\tilde{b}_1$	$\tilde{e}_l$	$\tilde{\tau}_1$	χ_1^0	χ_2^0	χ_1^+	χ_2^+
A	433	1280	1274	759	1054	1263	1246	104	122	123	271
B	793	1480	1440	902	1246	1375	1327	227	256	257	501
C	722	750	660	483	649	437	237	231	301	302	490

Table 2. The SUSY mass spectra corresponding to the three representative points of fig 3 [23].

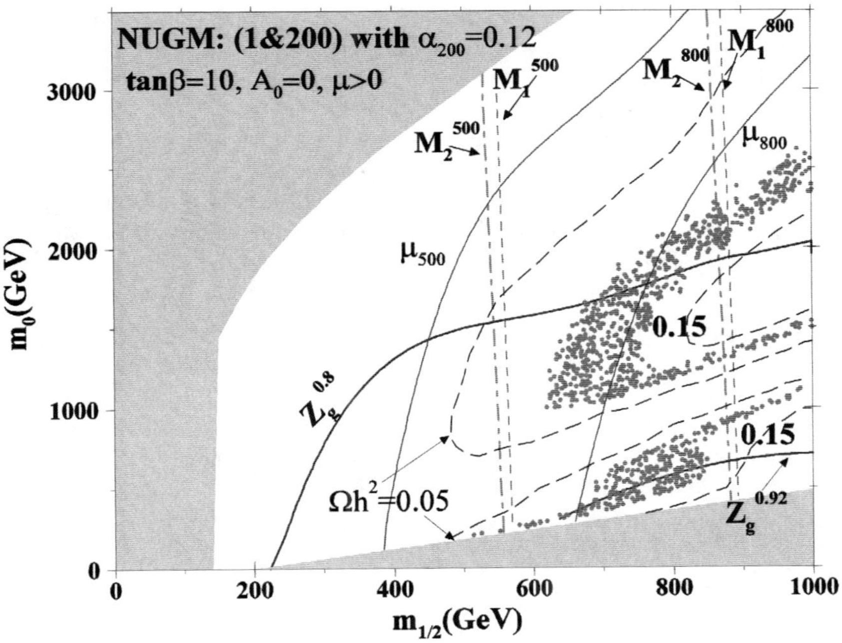

Figure 4. The composition and relic density of DM in the 1+200 model [7].

Figure 4 shows the $m_0 - m_{1/2}$ parameter space of the 1+200 model for an optimal choice of $\alpha_{200} = 0.12$ with contours of M_1, M_2 and μ. The near equality of these three masses shows that DM is a mixed neutralino state of comparable bino, wino and higgsino components. It also shows the DM relic density contours, with the WMAP relic density satisfying regions indicated by the bands of red dots. Like figure 2, the uppermost band corresponds to the annihilation of $\chi\chi$ into WW, ZZ and fermion-pairs (via Z), while the two lower bands together represent their resonant annihilation into heavy fermion pairs via A.

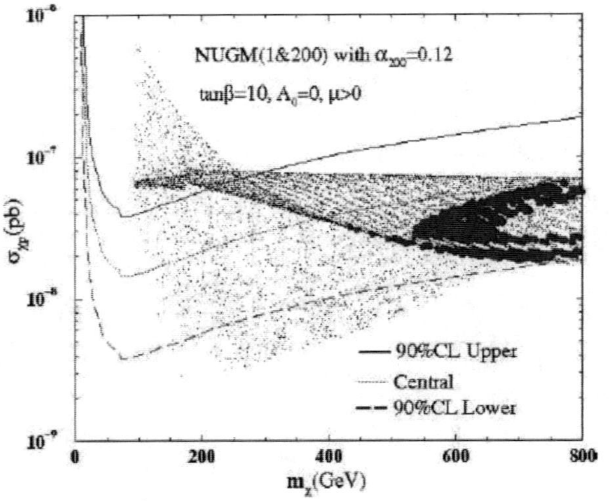

Figure 5. The predicted direct detection cross-section of the 1+200 model [24]. The blue dots represent the WMAP relic density satisfying points corresponding to the red dots of figure 4.

Fig 5 compares the 1+200 model prediction with the experimental band representing the central value and the 90% CL upper and lower limits corresponding to the two candidate events of the CDMS data [11]. The model predicts very large direct detection cross-section at the level suggested by these two candidate events. Unfortunately, the recent XENON100 (2012) data [13] gives a stringent 90%CL upper limit, roughly coinciding with the shown lower limit from the two candidate events, which is in conflict with this model prediction.

It may be added here that both these models of mixed neutralino DM predict large indirect detection signal in terms of high energy neutrinos coming from their pair annihilation inside the sun, which can be detected by the IceCube experiment [7].

5. Wino DM in the AMSB Model

Finally let us consider the most popular model of nonuniversal gaugino masses, i.e. the anomaly mediated SUSY breaking model [25]. Here the GUT scale gaugino masses arise from SUSY breaking in the hidden sector via the super-Weyl anomaly contributions. Being loop effects these contributions would be small compared to the tree-level contributions from eqs. (1-4). The AMSB model tacitly assumes that the SUSY breaking superfield in the hidden sector does not belong to one of the representations appearing in the symmetric product of the two adjoint representation of the GUT group in (2); so that the tree-level contributions to gaugino masses are forbidden by symmetry considerations. The GUT scale gaugino masses in the AMSB model are given in terms of the gravitino mass m3/2 via the Callan-Symanzik β functions of their respective gauge couplings, i.e.

$$M_1^G = \frac{33}{5}\frac{\alpha_1}{4\pi}m_{3/2}, \quad M_2^G = \frac{\alpha_2}{4\pi}m_{3/2}, \quad M_3^G = -3\frac{\alpha_3}{4\pi}m_{3/2} \,. \tag{13}$$

With the universality of the gauge couplings at the GUT scale one clearly sees the wino to be the lightest gaugino. Evolving the gaugino masses down via the RGE one sees that the wino remains the lightest gaugino at the EW scale as well. Thus one has a wino DM over most of the parameter space of the AMSB model [26]. Like the higgsino, the wino can efficiently pair-annihilate via its isospin gauge coupling to W boson, leading to an under-abundance of DM relic density for sub-TeV wino mass. In fact wino has twice as large an isospin as higgsino, so that one gets the right DM relic density for a wino mass twice as large as the higgsino case, i.e. ~ 2 TeV. This is too large to be probed by LHC experiments. Besides the direct DM detection rate is severely suppressed by the large mass as well as the absence of a significant higgsino component in the DM.

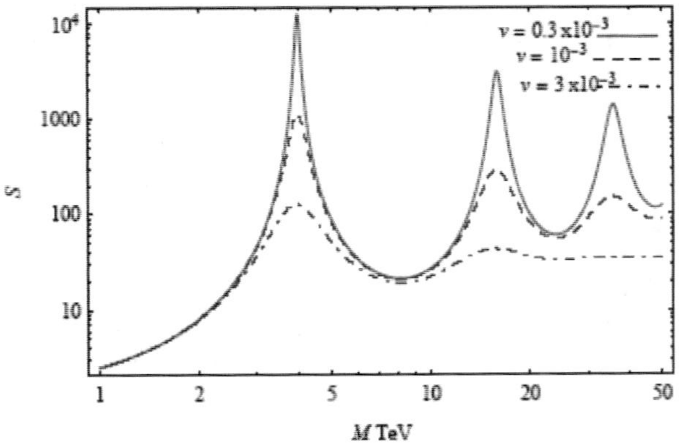

Figure 6. Sommerfeld enhancement factor from W boson exchange as a function of the DM mass at different relative velocities [29].

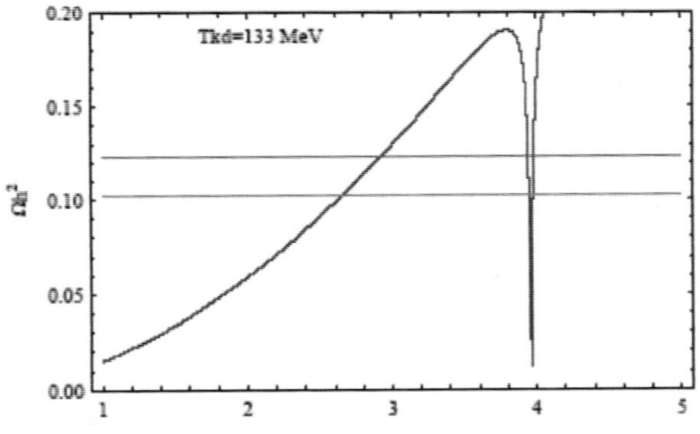

Figure 7. The suppression of the DM relic density at the Sommerfeld resonance. The horizontal band represents the WMAP relic density (12).

By far the best chance of detecting a TeV scale wino DM lies in the indirect detection experiments looking for the products of their pair annihilation,

$$\chi\chi \xrightarrow{\tilde{\chi}_1^+} W^+W^- \ . \tag{14}$$

In particular the leptonic decay of W bosons can give a hard positron spectrum of the type observed by PAMELA experiment. Moreover, a possible source of the large boost factor required to match the size of the PAMELA data is provided by the Sommerfeld effect [27, 28]. To understand this effect one notes that the annihilation process (14), mediated by a TeV scale charged wino exchange, has a negligibly small range of interaction. So the annihilation cross-section is proportional to the modulus square of the $\chi\chi$ wave function at negligibly small relative distance, $|\psi(0)|^2$. On the other hand, the incoming $\chi\chi$ pair has attractive interaction via a relatively long range W boson exchange which distorts the wave function, resulting in the enhancement of $|\psi(0)|^2$. One can solve the nonrelativistic scattering problem nonperturbatively by solving the Schrodinger equation. One finds that the effect is modest except for the resonance points, where one can have enhancement factors of several orders of magnitude, as shown in figure 6 [29].

The Sommerfeld enhancement of the DM pair annihilation cross-section results in a significant suppression of the DM relic density at the resonance point [30], as shown in figure 7 for our wino DM [29]. Thus one can simultaneously have a large enhancement factor of the annihilation cross-section, S $\sim 10^4$, and also reconcile the DM relic density with the WMAP value (12) by choosing the wino DM mass to be very close to the first resonance value of $\sim$ 4 TeV. The resulting spectrum of the positron flux ratio is compared with PAMELA experiment in figure 8. It may be added here that the model prediction for the anti-proton spectrum is also in reasonable agreement with the PAMELA data [29].

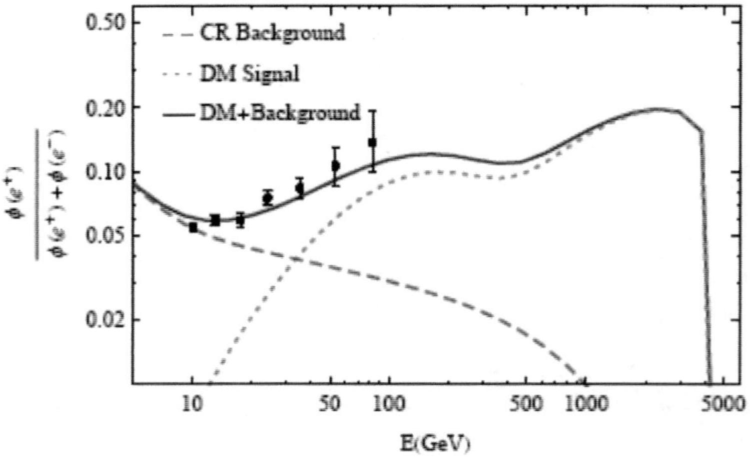

Figure 8. The predicted positron flux ratio of a 3.98 TeV wino DM is compared with the PAMELA data using a Sommerfeld enhancement factor of $\sim 10^4$ [29].

Acknowledgments
I thank the organizers of the PASCOS 2012 and in particular Myriam Mondragon and Liliana Velasco-Sevilla for their wonderful hospitality. The works reported here were supported in part by a senior scientist fellowship of the Indian National Science Academy.

References

[1] Nakamura K et al. [Particle Data Group] 2010 *J. Phys. G* **37** 075021
[2] Ellis J R, Enqvist K, Nanopoulos D V and Tamvakis K 1985 *Phys. Lett.* **B155** 381; Drees M 1985 *Phys. Lett.* **B158** 409
[3] Anderson G, Baer H, Chen C H and Tata X 2000 *Phys. Rev.* **D61** 095005; Huitu K, Kawamura Y, Kobayashi T and Puolamaki K 2000 *Phys. Rev.* **D61** 035001; Chatto padhyay U and Nath P 2002 *Phys. Rev.* **D65** 075009; Chattopadhyay U and Roy D P 2003 *Phys. Rev.* bf D68 033010
[4] Carena M S, Olechowski M, Pokorski S and Wagner C E M 1994 *Nucl. Phys.* **B426** 269; Komine S and Yamaguchi M 2001 *Phys. Rev.* **D63** 035005
[5] Djouadi A, Kneur J L and Moultaka G 2007 *Comput. Phys. Commun.* **176** 426
[6] King S F, Roberts J P and Roy D P 2007 *JHEP* **0710** 106
[7] Chattopadhyay U, Das D and Roy D P 2009 Phys. Rev. D 79 095013
[8] King S F and Roberts J P 2006 *JHEP* **0609** 036
[9] Mohanty S, Rao S and Roy D P (*Preprint* 1208.0894 [hep-ph])
[10] Gondolo P, Edsjo J, Ullio P, Bergstrom L, Schelke M and Baltz E A 2004 *JCAP* **0407** 008

[11] Ahmed Z et al. [CDMS-II Collaboration] 2010 *Science* **327** 1619; Ahmed Z et al. [CDMS-II Collaboration] 2011 *Phys. Rev. Lett.* **106** 011301
[12] Aprile E et al. [XENON100 Collaboration] 2011 Phys. Rev. Lett. 107 131302
[13] Aprile E et al. [XENON100 Collaboration] 2012 (*Preprint* 1207.5988 [astro-ph.CO])
[14] http://dmtools.brown.edu/
[15] Bergstrom L, Bringmann T and Edsjo J 2008 *Phys. Rev.* **D78** 103520
[16] Moskalenko I V and Strong A W 1998 *Astrophys. J.* **493** 694; Strong A W, Moskalenko I V and Ptuskin V S 2007 *Ann. Rev. Nucl. Part. Sci.* **57** 285
[17] Adriani O et al. [PAMELA Collaboration] 2009 *Nature* **458** 607
[18] Brun P, Bertone G, Lavalle J, Salati P and Taillet R 2007 *Phys. Rev.* **D76** 083506
[19] Hooper D, Taylor J E and Silk, J 2004 *Phys. Rev.* **D69** 103509
[20] Arkani-Hamed N, Delgada A and Giudice G F 2006 *Nucl. Phys.* **B741** 108
[21] Chan K L, Chattopadhyay U and Nath P 1998 *Phys. Rev.* **D58** 096004; Feng J L, Matchev K T and Moroi T 2000 *Phys. Rev.* **D61** 075005
[22] Komatsu E et al. [WMAP Collaboration] 2009 *Astrophys. J. Suppl. Ser.* **180** 330
[23] Guchait M, Sengupta D and Roy D P 2012 *Phys. Rev.* **D85** 035024
[24] Roy D P 2010 *Phys. Rev.* **D81** 057701
[25] Giudice G F, Luty M A, Murayama H and Rattazzi R 1998 *JHEP* **9812** 027; Gherghetta T, Giudice G F and Wells J D 1999 *Nucl. Phys.* **B559** 27
[26] Chattopadhyay U, Das D, Konar P and Roy D. P. 2007 *Phys. Rev.* **D75** 073014
[27] Hisano J, Matsumoto S and Nojiri M M 2004 *Phys. Rev. Lett.* **92** 031303; Profumo S 2005 *Phys. Rev.* **D72** 103521; J Hisano, Matsumoto S, Saito O and Senami M 2006 *Phys. Rev.* **D73** 055004
[28] Lattanzi M and Silk J 2009 *Phys. Rev.* **D79** 083523
[29] Mohanty S, Rao S and Roy D P 2012 *Int. J. Mod Phys.* **A27** 1250025
[30] Feng J J, Kaplinghat M and Yu H B 2010 *Phys. Rev.* **D82** 083525

Journal of Physics: Conference Series **485** (2014) 012004 doi:10.1088/1742-6596/485/1/012004

Effective SUSY modified after LHC run

Jihn E Kim

Department of Physics, Seoul National University, Seoul 151-747, Korea, and
GIST College, Gwangju Institute of Science and Technology, Gwangju 500-712, Korea

E-mail: jihnekim@gist.ac.kr

Abstract. Some supersymmetric models after the recent LHC reports are discussed. Especially, the light Higgs boson around 125 GeV is attempted to be accommodated. An extra Z' from GUTs, and the scalar and pseudoscalar Higgs bosons in supersymmetric extension of the standard model are also discussed.

1. Introduction

I will discuss a beyond the standard model (BSM) view based on my recent works.

The most recent interest in high energy physics is the LHC report on the Higgs boson discovery at the mass range of 125 GeV [1]. This small Higgs boson mass compared to the Planck mass needs a huge hierarchy of mass scales, inviting solutions of the hierarchy problem. The attempt to explain the Higgs boson mass within one GeV limit has been possible in supersymmetry (SUSY) extension of the standard model (SM). But the LHC data is not consistent with the constrained minimal supersymmetric standard model (CMSSM) prediction in the region $m_{\rm gluino} m_{\rm squark} < 1\,{\rm TeV}^2$. A small Higgs boson mass ($m_h \simeq 0.125$ TeV) needs a large stop mass or/and a large A-term in the CMSSM.

Of course, there have been several important issues to be considered in the BSM: a new CP, axions, SUSY, string, etc. For example, if there exists a new force around the electroweak scale, an extra Z' is expected at the TeV scale. As was done for the Z boson [2], the indirect limits for the extra force have been studied for a long time. For example, some string inspired Z' mass limit has been given as 1.36 TeV [3]. Recently, the LHC data seems to give limits on the Z' mass limit in the SUSY SM(SSM) as 2.2 TeV [4].

For the extra Z', the first interest resides in whether it can arise from a GUT or not. For GUT groups, we consider a few interesting simple and semi-simple groups such as SU(5), SO(10), SU(6), trinification $\rm SU(3)^3$. These are subgroups of $\rm E_6$. Therefore, the Z' issue of those interesting GUT groups is contained in the study of $\rm E_6$. Then, it is sufficient to study the diagonal generators of $\rm E_6$, because any Cartan subalgebra generator of SO(10), SU(6), and trinification $\rm SU(3)^3$ can be expressed as a linear combination of the $\rm E_6$ diagonal generators. With this philosophy, we consider any rank 6 subgroup of $\rm E_6$, and SU(6)$\times$SU(2) is particularly useful for this purpose because one can see all the quantum numbers of representations as displayed

Content from this work may be used under the terms of the Creative Commons Attribution 3.0 licence. Any further distribution of this work must maintain attribution to the author(s) and the title of the work, journal citation and DOI.
Published under licence by IOP Publishing Ltd

below:

$$
(\mathbf{15},\mathbf{1}) \equiv \begin{pmatrix} 0 & u^c & -u^c & u & d & D \\ -u^c & 0 & u^c & u & d & D \\ u^c & -u^c & 0 & u & d & D \\ -u & -u & -u & 0 & e^c & H_u^+ \\ -d & -d & -d & -e^c & 0 & H_u^0 \\ -D & -D & -D & -H_u^+ & -H_u^0 & 0 \end{pmatrix},
$$

$$
(\overline{\mathbf{6}},\mathbf{2}^{\uparrow}) \equiv \begin{pmatrix} d^c \\ d^c \\ d^c \\ -\nu_e \\ e \\ N \end{pmatrix}, \quad (\overline{\mathbf{6}},\mathbf{2}^{\downarrow}) = \begin{pmatrix} D^c \\ D^c \\ D^c \\ -H_d^0 \\ H_d^- \\ N' \end{pmatrix}. \tag{1}
$$

The vertical (SU(6)) and the horizontal (SU(2)) directions of the **15** and $\overline{\mathbf{6}}$ introduce six diagonal generators of E_6.

Firstly, we note that baryon number B cannot be a U(1) generator of E_6 as discussed by many people and as shown recently in [5]. If it is not baryon number, this $U(1)'$ may be constructed such that Z' does not couple to leptons, which has been known as a leptophobic $U(1)'$. Considering Z' from E_6 subgroups, the ρ parameter of the neutral current study from the LEPII data does not allow the Z' mass below 10 TeV [5]. But if we do not restrict to a subgroup of E_6, a TeV scale Z' is perfectly allowed from string, which is because the SSM in this case is basically obtained from $E_8 \times E_8'$, *i.e.* beyond Eq. (1) of E_6. Still the extra Z' with a large Z' mass from E_6 can contribute to low energy physics. In SUSY models, they can work as the messengers of SUSY breaking sector to the SSM physics at TeV scale.

2. Effective SUSY

Effective SUSY has been proposed in the middle of 1990s. In this regard, we note that the third generation and Higgs doublets are the necessary ingredients of the gauge hierarchy solution through SUSY [6]. It is also useful to fit the $(g-2)_\mu$ data [7].

To remove too large FCNC effects among the light two family members by raising the 1st and 2nd family squark masses a little bit, one needs a symmetry among the first two family squarks such as a U(2) symmetry. But, if the 1st and 2nd family squark masses are raised above 10^{5-6} GeV, one does not need such a symmetry among the light family members.

We heard here on the LHC results interpreted in the constrained MSSM and hence a simplified SUSY model [4]. The latest CMS results from Razor analysis at the 4.4 fb^{-1} level exclude squark and gluino masses up to 1.35 TeV, and the latest ATLAS results from $m_{\rm eff}$ on 0-lepton at 4.7 fb^{-1} is mapped on with the massless LSP assumption on $m_{\rm gluino} < 940$ GeV and $m_{\rm squark} < 1380$ GeV. For 0-lepton + high multiplicity jets (≥ 6 to ≥ 9) gave $m_{\rm gluino} < 880$ GeV. So, squark masses seems to be heavier than TeV in a simplified MSSM. But, if this simplified assumption is not taken, these limits do not apply. One attractive such case is effective SUSY (effSUSY).

Z' mediation: The effSUSY was a phenomenological condition for the squark and gluino mass spectra [6]. For a long time a theoretical model has not been proposed. One needs a model distinguishing families. For example, in the gravity mediation scenario, one needs to assign specifically the first two family and the third family members in topologically different manifolds. This needs a complete model in the ultra-violet completion, e.g. in string compactification in the bulk and different fixed points. So, in the gravity mediation, one needs an explicit string model.

In the gauge mediated SUSY breaking (GMSB) scenario, one can pursue the effSUSY at the field theory level since the gravity mediation is assumed to be sub-dominant. However, at present there has not appeared a serious GMSB model. If an SSM is obtained from string compactification, the group nature is well worked out from the $E_8 \times E_8'$ heterotic string. Suppose that the hidden sector is SU(4)′ or SU(5)′ which are favored most. For SU(4)′, an anomaly-free chiral representation needs a gigantic structure. For SU(5)′, however, it is easy to obtain an anomaly-free chiral representation, for example $\mathbf{10}' + \overline{\mathbf{5}}'$. If we include $\mathbf{10}' + \overline{\mathbf{5}}'$, there should be 45 visible sector fermions and at least 15 (from $\mathbf{10}' + \overline{\mathbf{5}}'$) hidden sector fermions. These require at least 60 chiral fermions from string compactification. In addition, we need the messenger sector fermions carrying the SU(5)′ charge. Namely, the messenger sector in a general GMSB must carry color, weak and $U(1)_Y$ charges in addition to the hidden sector gauge charge. For $\mathbf{5}' + \overline{\mathbf{5}}'$ messenger, the minimum number of the messengers are therefore 50 (ten times color 3 and weak 2). Thus, at least we need 110 chiral fields (with the DSB sector $\mathbf{10}' + \overline{\mathbf{5}}'$ and the messenger sector $\mathbf{5}' + \overline{\mathbf{5}}'$) from the string compactification. Experience tells us that roughly 150 fermions appear in the spectrum from string compactification. So, we notice that the GMSB scenario needs more chiral fields to realize a complete mediation scheme, which tightens the model building possibilities from string.

One interesting GMSB scenario is the Z'-mediation where SUSY breaking at the hidden sector is manifested in the visible sector (SM) by the U(1)′ charged fields. If the messenger sector has only $\mathbf{5}' + \overline{\mathbf{5}}'$, we need only 10 instead of 50 of the preceding paragraph. The Z'-mediation has been known for some time [8], but the recent suggestion by Langacker *et al.* is relevant for this talk [9]. Furthermore, we find that it can be obtained from a string compactification [10].

Langacker *et al.* discusses mediation with one Z' and Mohapatra and Nandi [8] discuss mediation through Z' and Z. In the latter case, it is proper to call it the mixed mediation (MM). In our study, we find that the MM needs a fine-tuning of parameters [11]. So, we discuss Z' mediation the type considered by Langacker *et al.* [9]. However, ours is different from that of Langacker *et al.* in that we consider a family-dependent Z' mediation while Langacker *et al.* considered the family-universal Z' mediation [9].

Effective SUSY:

The U(1)′ charge is carried by the mediators and the first two family members, but NOT by the 3rd family members and the Higgs doublets. Because the U(1)′ charge is the differentiating one between the first two and the third family members, it is a CALCULABLE MODEL with only a SUSY gauge theory knowledge. In particular, we need a light Higgs doublets pair for the hierarchy solution via SUSY. The SU(2)×U(1) breaking is achieved radiatively as pointed out in [12]. The spectrum of H_u, H_d and the 3rd family members being light is called effSUSY or some other words [6]. [Note added after the talk: After the discovery of the Higgs boson [1], the opposite view 'Inverted effective SUSY' where the 3rd family members are considered to be the heaviest family members has been proposed [13].]

In this regard, we note that the experimental limits on Higgs boson searches from the ATLAS and CMS data give the light Higgs boson mass around 125 GeV if it is light [14]. Here, the combined constraint from ATLAS and CMS is given by Jinnouchi [4].

The idea of Z' mediation toward effSUSY is pictorially depicted in Fig. 1. Here, the superpartner of Z', *i.e.* Zprimino $\tilde{Z}'$ is the one mediating SUSY breaking. The SUSY breaking source is shown as dynamical SUSY braking(DSB), probably a hidden-color sector. The DSB matter does not carry the weak hypercharge Y. The messenger sector carries the DSB hidden-color and also the Z' charge but does not carry the weak hypercharge Y. The first two families carry the Z' charge but the third family does not carry the Z' charge. Finally, the Higgs fields do not carry the Z' charge. To obtain this kind, we assign different quantum numbers to the third family members from those of the first two families.

The mediation mechanism is called the Z' meditation but ours is different from that of Ref.

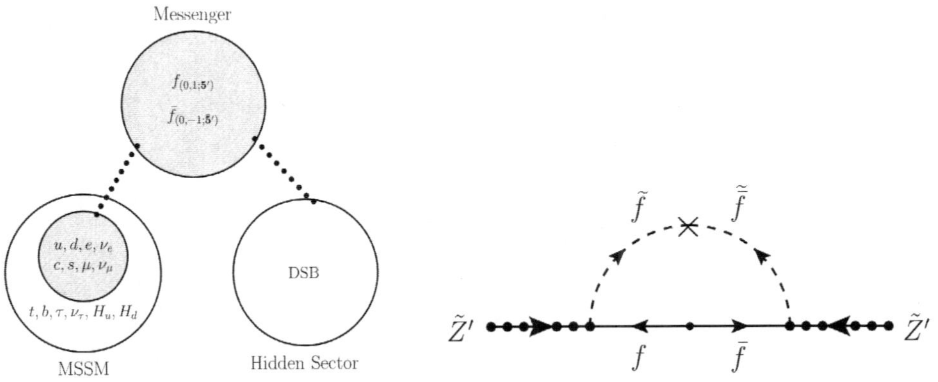

Figure 1. An effSUSY through the Z' mediation and the mass diagram of Zprimino. The Z' is the bulletted dots and the Zprimino is the bulleted line. The SUSY breaking insertion from DSB is ×. This soft mass shown as the Feynman diagram is added to the SUSY mass.

[9] in that here we need family dependent Z' charges while that of Ref. [9] introduces the family universal Z' charges.

The largest SUSY splitting mass is the one arising in the Zprimino mass as shown as the Feynman diagram in Fig. 1. The SUSY splitting mass is

$$\Delta M_{\tilde{Z}'} \simeq \frac{g_{Y'^2}}{16\pi^2}\frac{F'_{\rm mess}}{M'_{\rm mess}} \tag{2}$$

The next largest splitting appears for the first two family sfermions which appear at one-loop level beyond $\Delta M_{Z'}$, as shown in the left panel of Fig. 2,

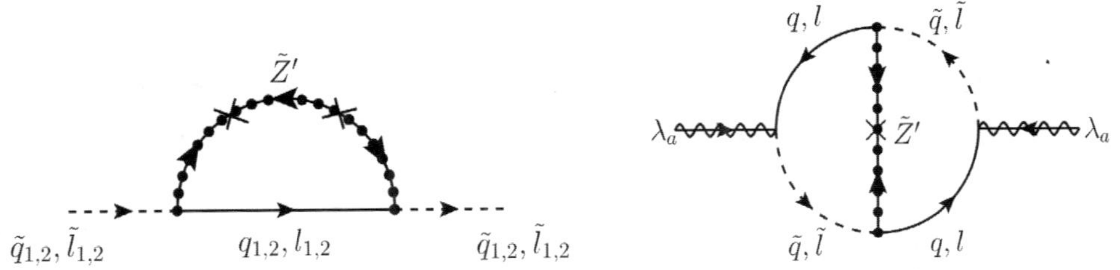

Figure 2. The first two family sfermion($\tilde{q}_{1,2}, \tilde{l}_{1,2}$) mass diagrams(left panel) and the mass diagram of the SM gauginos(right panel). The SUSY breaking from Zprimino sector is shown as ×. The $\tilde{Z}'$ line is a bulleted line.

$$\Delta M_{\tilde{q}_{1,2},\tilde{\ell}_{1,2}}(\mu) \simeq \frac{g_{Y'}Y'_{\tilde{q}_{1,2},\tilde{\ell}_{1,2}}}{4\pi} M_{\tilde{Z}'}\left(\ln\frac{M_{\tilde{Z}'}}{\mu}\right)^{3/2}. \tag{3}$$

This achieves our objective of raising the first two family squark and slepton masses beyond the gaugino and the third family sfermion masses.

Since the messengers are not charged under the SM gauge group, the MSSM gaugino masses are induced only through RG running from the loops shown here (two-loop beyond the Zprimino

light families	Y	Y'	3rd family and $H_{d,u}$	Y	Y'
$q_{1,2}$	$\frac{1}{6}$	$\frac{1}{3}$	(t,b)	$\frac{1}{6}$	0
$u^c_{1,2}$	$\frac{-2}{3}$	$\frac{-1}{3}$	t^c	$\frac{-2}{3}$	0
$d^c_{1,2}$	$\frac{1}{3}$	$\frac{-1}{3}$	b^c	$\frac{1}{3}$	0
$l_{1,2}$	$\frac{-1}{2}$	-1	(ν_τ,τ)	$\frac{-1}{2}$	0
$e^c_{1,2}$	1	1	τ^c	1	0
$N^c_{1,2}$	0	1	N^c_3	0	0
			H_d	$\frac{-1}{2}$	0
			H_u	$\frac{1}{2}$	0

Table 1. The Y' charges of the SM fermions, Higgs doublets and heavy neutrinos.

mass), as shown in the right panel of Fig. 2,

$$\Delta M_{\lambda_a}(\mu) \simeq \frac{g_{Y'}^2 g_a^2}{(16\pi^2)^2} M_{\tilde{Z}'} \left(\ln \frac{M_{\tilde{Z}'}}{\mu}\right). \tag{4}$$

The smallest is the third family members and the Higgs doublet pair. They are one more loop beyond the gaugino masses (three loop beyond the Zprimino mass) as shown in Fig. 3.

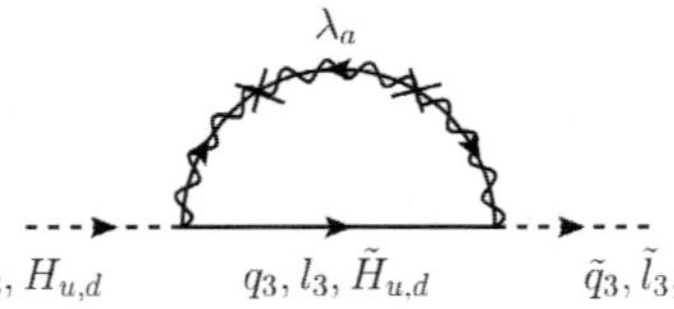

Figure 3. The mass diagrams for the third family sfermion(q_3, l_3) and Higgs bosons. The SUSY breaking from the SM gauginos are shown as $\times$.

The above hierarchical mass splitting is the effective SUSY scenario, through Z' mediation [9, 11]. In the literature, it has been pointed out that if the first two family sfermions are too heavy, the stop tends to be tachyonic at the two loop level [15]. At the moment, therefore, it seems that only a little hierarchy between the first two family sfermion members and those of the third family is the working one. But we speculate that some so-far unknown physics may remove this tachyonic problem as the tachyonic problem of the anomaly mediation was saved in the mirage mediation scenario.[1]

For an explicit calculation at the field theory level, we take $Z' = B - L$ as shown in Table 1. The mass splitting is shown in Fig. 4. However, a realistic string models may give a bit different spectra from that of Fig. 4.

In this kind of Z' mediation, the LSP is most probably the gravitino. So, CDM may be cold axions and/or axinos. Anyway, axion seems needed also for a solution of the strong CP problem, and we proceed to comment on the axion solution of the strong CP problem.

[1] The effective SUSY with the combined effects of gravity and Z' mediations resolves this problem [13].

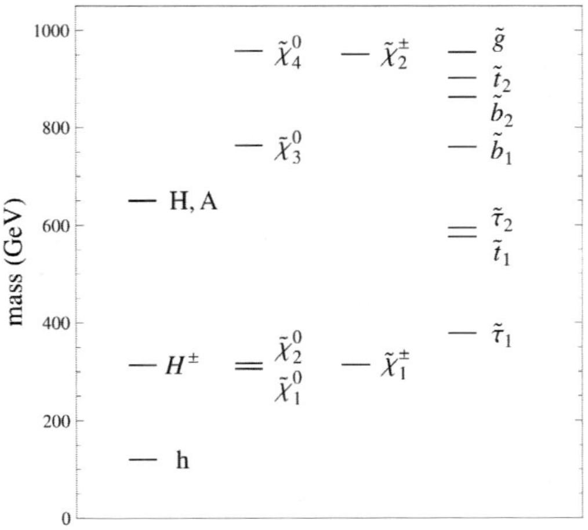

Figure 4. The sparticle and Higgs boson mass spectra in the Z' mediation. We have taken $M_{\rm mess} = 10^{14}$ GeV, $M_{Z'} = 10^8$ GeV and $M_{\tilde{Z}'}(M_{\rm mess}) = 1.8 \times 10^6$ GeV, for which the squark masses of the first two families are above 10^6 GeV [11].

3. Strong CP and axions

A brief history of QCD:

Quantum chromodynamics (QCD) started with the prediction of Ω^- [16]. In terms of the current understanding, it is the completely symmetric states of three s quarks and the spin $= \frac{3}{2}$ states is $s^\uparrow s^\uparrow s^\uparrow$. If s quark is a fermion, Ω^- should be completely antisymmetric under the exchange of two s quarks [17]. Greenberg tried a paraquark model toward this spin-statistics problem [18]. The real solution of the spin statistics problem was suggested by Han and Nambu by introducing another SU(3) degrees but with integer charged quarks [19]. Soon after the Han-Nambu paper, the fractionally charged quarks were suggested by Hori [20] which has been named later as QCD by Fritzsch and Gell-Mann [21]. The discovery of the asymptotic freedom crowned the QCD as the theory of strong interaction theory [22].

This has led to the consideration of eight gluons interacting strongly with color triplet quarks. The current quark masses were first estimated in [23] and after the advent of QCD it has been calculated again in [24]. But, QCD with MeV order quark masses gives a light η' [25], which has been known as the U(1) problem. It is known that an effective interaction of gluons $G^a_{\mu\nu}$ at the QCD θ-vacuum is present,

$$\frac{\overline{\theta} g_c^2}{64\pi^2} \epsilon^{\mu\nu\rho\sigma} G^a_{\mu\nu} G^a_{\rho\sigma} \tag{5}$$

which is a total derivative, but it cannot be neglected due to the surface contributions at Euclidian space infinity (due to instanton solutions). Weinberg's U(1) problem is solved by this gluon anomaly term [26]. This suggest that the vacuum angle term is working. Otherwise, the U(1) problem is not solved.

This is the brief QCD history.

The strong CP problem and axions:

The anomaly term (5) breaks P and T and hence breaks the CP symmetry. It has led to the strong CP problem and three types of natural solutions [27]. Among these the axion solution which results from the spontaneously broken Peccei-Quinn(PQ) symmetry [28] is the beautifully realized one. The currently accepted axion solutions are the so-called *invisible axions* [29, 30] which live long enough to have survived until now. The axion agenda has been focussed on many different points during the last 35 years: massless up quark in 1975, the PQWW axion in 1977, the invisible axion in 1979, cosmological axions in 1983, and the axion detection experiments in 1988.

The leading contribution of the θ term is the η' mass solving the U(1) problem. But it has led to the strong CP problem which is solved by making θ as a dynamical variable. If there is an additional almost massless dynamical field with the same anomaly term, *i.e.* if θ is made dynamical as axion a, the next order contribution proportional to some light quark mass gives mass to θ. This is the axion mass. Diagonalizing the π^0, η' and a mass, we obtain [27],

$$m_{\pi^0}^2 \simeq \frac{\tilde{v}^4}{f_{\pi^0}^2},\ m_{\eta'}^2 \simeq \frac{\Lambda_{\rm inst}^4+\tilde{v}^4}{f_{\eta'}^2},\ m_a^2 \simeq \frac{Z}{(1+Z)^2}\frac{f_{\pi^0}^2 m_{\pi^0}^2}{F^2}(1+\Delta), \tag{6}$$

$$\text{with } \Delta = \frac{m_-}{m_+}\frac{\Lambda_{\rm inst}^4(m_+v^3+\mu\Lambda_{\rm inst}^3)}{f_{\pi^0}^4 m_{\pi^0}^4}$$

where $m_+ = m_u + m_d$, $m_- = m_d - m_u$, $Z = m_u/m_d$, and $\tilde{v}^4 \equiv m_+v^3 + 2\mu\Lambda_{\rm inst}^3$.

The dark matter particle in the universe is the most looked-for one(s) in cosmology and at the LHC, and also at low temperature labs: at the mass scales of 100 GeV (for WIMP) and 10–1,000 μeV (for axion). The axion potential is almost flat for a large axion decay constant F. In this axion potential, the minimum point $\theta = 0$ is the CP conserving point. In the evolving universe, at some temperature, say T_1, a starts to roll down the hill to the CP conserving point $\theta = 0$. This analysis depends on the axion decay constant and the initial VEV (the so-called misalignment angle) of a at T_1. A recent analysis of these effects has been given in [31].

Fortunately, the invisible axion is at the level of possible detection through $\gamma-\gamma-a$ and/or $\bar{e}-e-a$ interaction. By discovering the axion, QCD is completed satisfactorily. If the axion is ruled out at the axion window, either its mass is too small to be detected, or the fine-tuning problem of θ remains unsolved theoretically.

A new form of the CKM matrix:

I have discussed some issues related to SUSY and axion. Related to axion, I discussed a brief history of QCD, the need for a solution of the strong CP problem, and the invisible axion. Since the discussions here are related to CP, a good weak CP violation parametrization may be useful. One recent CKM parametrization allows one to look at the Jarlskog triangle from the CKM matrix itself [32]. Studying in this new parametrization, the maximum CP violation in the quark sector is apparent as shown in Fig. 5. In the recent study [32], we showed that by making the whole determinant of the CKM matrix V real, the imaginary part of any one term of the determinant of V (e. g. $|{\rm Im}\, V_{31}V_{22}V_{13}|$) is the Jarlskog determinant J. This method of calculating J is much easier than calculating [33]

$$J = \frac{-{\rm Det.}\, C}{2F(m_{t,c,u})F(m_{b,s,d})}$$

where $F(m_{t,c,u}) = (m_t-m_c)(m_t-m_u)(m_c-m_u)$ and $F(m_{b,s,d}) = (m_b-m_s)(m_b-m_d)(m_s-m_d)$, and C is $iC = [M_u, M_d]$, where $M_{u,d} = L^{(u,d)\dagger}M^{0(u,d)}L^{(u,d)}$ with the diagonal $M^{0(u,d)}$. It can be also written as $iC = [M_uM_u^\dagger, M_dM_d^\dagger]$. With this method, it is easy to convince oneself that the weak CP violation data shows the maximality as shown in Fig. 5.

4. Higgs boson at 125 GeV ?

It seems that some new particle is present around 125 GeV. Most probably, it looks like a Higgs boson. Is it the SM Higgs boson or the lightest or the next lightest scalar Higgs boson of the

Journal of Physics: Conference Series **485** (2014) 012004 doi:10.1088/1742-6596/485/1/012004

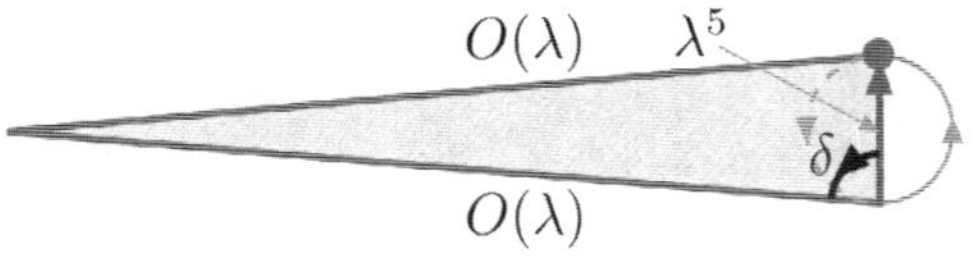

Figure 5. The Jarlskog triangle. This triangle is for two long sides of $O(\lambda)$. Rotating the $O(\lambda^5)$ side (the red arrow), the area becomes maximum as the CP phase δ becomes approximately $\frac{\pi}{2}$ [32].

NMSSM, or something else?

In SUSY models, the probability for it to be the MSSM Higgs boson is low [14], and a singlet X is introduced to fit the LHC preliminary data. These SUSY models with one more singlet is next MSSM (NMSSM). Supergravity models with one more singlet X is quite plausible, in particular in string compactifications toward SUSY standard models [34]. Here, we point out that the pseudoscalar in X can be at the electroweak scale (even if it is not the 125 GeV particle) and its two photon decay mode is discussed here.

We require that the strong CP problem is solved in this NMSSM. So, the PQ symmetry in the NMSSM is needed, which we call N_{PQ}MSSM. The MSSM cannot achieve this because $\mu H_u H_d$ in the MSSM breaks the PQ symmetry. Then, the question is, "What is the spectrum of the N_{PQ}MSSM?" In Refs. [35] and [36], N_{PQ}MSSM has been discussed. The axino mass in the N_{PQ}MSSM has been discussed in [37] where the heavy axino is most likely in contrast to the light axino study of [38]. The light pseudoscalar in the N_{PQ}MSSM has been emphasized in [35].

$$W = -H_u H_d X + mX\overline{X} - \eta\overline{X}\, S_1^2 - \xi H_u H_d X' + m' X'\,\overline{X} + Z_1(S_1 S_2 - F_1^2) + Z_2(S_1 S_2 - F_2^2) \quad (7)$$

where the Z_1 and Z_2 terms are introduced to parametrize SUSY breaking and the PQ symmetry breaking [39]. Z_1 and Z_2 of (7) is of order TeV but we assume them sequestered from the other light fields since they are assumed to mimick the DSB scheme. Beyond Z_1 and Z_2, there can be electroweak scale singlets. If we introduce just one, say X_{ew}, the electroweak scale superpotential from the N_{PQ}MSSM is

$$W_{\text{ew}} = -\mu H_u H_d - f_h H_u H_d X_{\text{ew}} + \cdots \quad (8)$$

For example, X_{ew}, X_{ew}^2 and X_{ew}^3 terms cannot be present [35, 37]. How X_{ew} survives down to the electroweak scale depends on the details of the scheme. In fact, finding a reasonable scheme is pain in the neck. At the electroweak scale, we use the μ term generated at the high energy scale [40], and also the $H_u H_d X_{\text{ew}}$ renormalizable term. The important thing is that X_{ew} survives down to the TeV scale, which we do not question here. Treating μ as a perturbation, there exists a symmetry:

$$X_{\text{ew}} \to e^{-2i\delta} X_{\text{ew}},\ H_u \to e^{+i\delta} H_u,\ H_d \to e^{+i\delta} H_d. \quad (9)$$

Then, the chargino (more generally the Higgsino pair) has the anomaly current, and there results the pseudoscalar $a_{X_{\text{ew}}}$ coupling to the anomaly term of the electromagnetic field, $a_{X_{\text{ew}}} F^{\text{em}}_{\mu\nu} \tilde{F}^{\mu\nu}_{\text{em}}$. The symmetry (9) originating from the PQ symmetry gives a light pseudoscalar if $\mu \simeq 0$. The photon (or weak gauge boson) fusion gives the two photon mode which can be detected in the future.

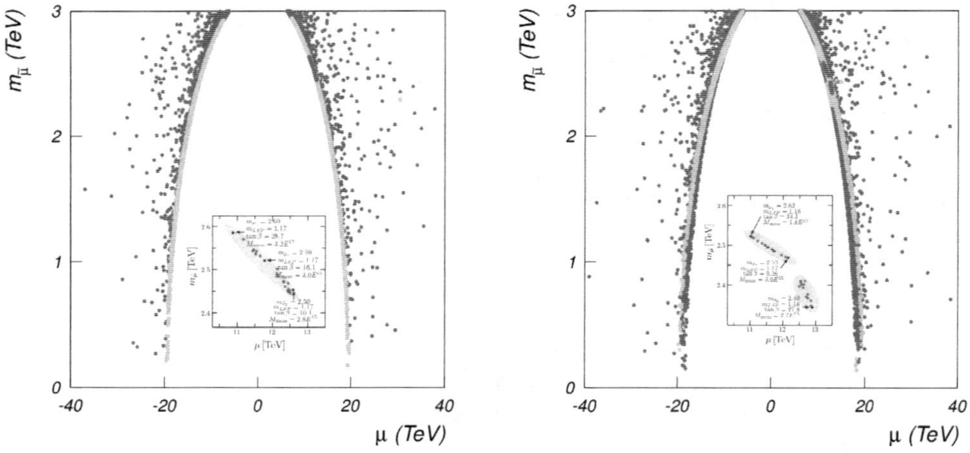

(a) (b)

Figure 6. The scatter plot in the $m_{\tilde{\mu}} - \mu$ space for $\Lambda_h = 3 \times 10^{13}\,\text{GeV}$ out of 10^5 trial points: (a) for $\lambda_f = \frac{1}{6}$, $\lambda_h = 0$, and (b) for $\lambda_f = -1$, $\lambda_h = 0$ with the notation of Ref. [13]. The scanned parameters are μ and $M_{\text{mess}} = (0.1 \sim 10) \times 10^{15}$ GeV. The top quark mass corresponds to $m_t = (173.5 \pm 1.4)$ GeV. The gray dots are the trial points. The green dots are those satisfying the LHC constraints ($m_{\tilde{q}_{1,2}} > 1.5$ TeV and the LHC gluino mass bound) and $m_h = (125 \sim 127)$ GeV. The pink dots are filtered by $(g-2)_\mu$. The red dots are those satisfying all the constraints including $(g-2)_\mu$. In the enlarged insets, some selected red points are shown again with more information in blue dots .

5. Conclusion

In view of the recent discovery of the Higgs boson, we paid attention to some effective supersymmetric models, *i.e.* the light Higgs boson around 125 GeV is attempted to be accommodated. A note on an extra Z' from GUTs, and the scalar and pseudoscalar Higgs bosons in supersymmetric extension of the standard model are also discussed. We also discussed a few topics related to QCD such as the strong CP problem and the recent parametrization of the weak CP violation where the physical quantity related to the Jarlskog triangle is easy to identify.

[Note added after the talk: In this talk I presented my recent works related to the effSUSY. But after the Higgs boson discovery, it became urgent to interpret the 3.6 σ data on $(g-2)_\mu$ of the Brookhaven National Laboratory(BNL) [41]. The 125 Higgs boson with the BNL $(g-2)_\mu$ is attempted to be explained in the MSSM framework with the inverted effSUSY(IeffSUSY) [13]. The related parameters in the IeffSUSY are shown in Fig. 6. This figure is based on the Z' and gravity mediations with the third family sfermions are the heaviest, which is possible with the Z' quantum numbers of Table 2.]

Acknowledgements

This work is supported in part by the National Research Foundation (NRF) grant funded by the Korean Government (MEST) (No. 2005-0093841).

1st f.	Y	Z'	2nd f.	Y	Z'	3rd f.	Y	Z'
(u,d)	$\frac{1}{6}$	$\frac{1}{3}\lambda_f$	(c,s)	$\frac{1}{6}$	$\frac{1}{3}\lambda_f$	(t,b)	$\frac{1}{6}$	$\frac{1}{3}$
u^c	$\frac{-2}{3}$	$\frac{-1}{3}\lambda_f$	c^c	$\frac{-2}{3}$	$\frac{-1}{3}\lambda_f$	t^c	$\frac{-2}{3}$	$\frac{-1}{3}$
d^c	$\frac{1}{3}$	$\frac{-1}{3}\lambda_f$	s^c	$\frac{1}{3}$	$\frac{-1}{3}\lambda_f$	b^c	$\frac{1}{3}$	$\frac{-1}{3}$
(ν_e,e)	$\frac{-1}{2}$	$-\lambda_f$	(ν_μ,μ)	$\frac{-1}{2}$	0	(ν_τ,τ)	$\frac{-1}{2}$	$-(1+\lambda_f)$
e^c	1	λ_f	μ^c	1	0	τ^c	1	$1+\lambda_f$
N_1^c	0	λ_f	N_2^c	0	0	N_3^c	0	$1+\lambda_f$

$H_{d,u}$	Y	Z'
H_d	$\frac{-1}{2}$	$\frac{-1}{2}\lambda_h$
H_u	$\frac{1}{2}$	$\frac{1}{2}\lambda_h$

Table 2. The Z' charges of the SM fermions, Higgs doublets and heavy neutrinos. The choice $\lambda_f = \lambda_h = 0$ is the simplest case. For the BNL $(g-2)_\mu$, the muonic leptons carry vanishing Z' charges while the others carry nonzero Z' charges with nonzero λ_f.

References

[1] The CMS Collaboration 2012 Observation of a new boson at a mass of 125 GeV with the CMS experiment at the LHC [arXiv:1207.7235];
The ATLAS Collaboration 2012 Observation of a New Particle in the Search for the Standard Model Higgs Boson with the ATLAS Detector at the LHC [arXiv:1207.7214].

[2] Glashow S L 1961 Partial Symmetries of Weak Interactions *Nucl. Phys.* **22** 579. For the neutral current confirming this, see a review
Kim J E, Langacker P, Levine M, and Williams H H 1981 A Theoretical and Experimental Review of the Weak Neutral Current: A Determination of Its Structure and Limits on Deviations from the Minimal $SU(2)_L \otimes U(1)$ Electroweak Theory *Rev. Mod. Phys.* **51** 211.

[3] Erler J *et al* 2009 [arXiv:0906.2435[hep-ph]].

[4] Jinnouchi O (ATLAS Collaboration) 2012 Searches for BSM and Higgs boson at LHC *Talk at GUT 2012, Kyoto, Mar. 15, 2012.*

[5] Kim J E and Shin S 2012 Z' from $SU(6)\times SU(2)_h$ GUT, Wjj anomaly and Higgs boson mass bound *Phys. Rev. D***85** 015012 [arXiv: 1104.5500[hep-ph]].

[6] Dimopoulos S and Giudice G G 1995 Naturalness constraints in supersymmetric theories with nonuniversal soft terms *Phys. Lett. B***357** 573 [hep-ph/9507282];
Pomarol A and Tommasini D 1995 Horizontal symmetries for the supersymmetric flavor problem *Nucl. Phys. B***466** 3 [hep-ph/9507462];
Cohen A, Kaplan D B and Nelson A E 1996 The More minimal supersymmetric standard model *Phys. Lett. B***388** 588 [hep-ph/9607394];
Giudice G G, Nardecchia M and Romanino A 2009 Hierarchical Soft Terms and Flavor Physics *Nucl. Phys. B***813** 156 [arXiv:0812.3610 [hep-ph]].

[7] Kim J E, Kyae B and Lee H M 2001 Effective supersymmetric theory and $(g-2)_\mu$ with R parity violation *Phys. Lett. B***520** 298 [arXiv:hep-ph/0103054].

[8] Mohapatra R N and Nandi S 1997 A New messenger sector for gauge mediated supersymmetry breaking *Phys. Rebv. Lett* **79** 181;
Kikuchi T and Kubo T 2008 Radiative B-L symmetry breaking and the Z' mediated SUSY breaking *Phys. Lett. B***666** 262 [arXiv:0804.3933 [hep-ph]].

[9] Langacker P, Paz G, Wang L T, and Yavin I 2008 Z'-mediated Supersymmetry Breaking *Phys. Rev. Lett.* **100** 041802 [arXiv:0710.1632 [hep-ph]].

[10] Kim J E 2007 GMSB at a stable vacuum and MSSM without exotics from heterotic string *Phys. Lett. B***656**207 [arXiv:0707.3292[hep-ph]].

[11] Jeong K, Kim J E and Seo M S 2011 Gauge mediation to effective SUSY through U(1)s with a dynamical SUSY breaking, and string compactification *Phys. Rev D***84** 075008 [arXiv:1107.5613[hep-ph]].

[12] Ibanez L E and Ross G G 1982 $SU(2)_L \times U(1)$ symmetry breaking as a radiative effect of supersymmetry breaking in GUTs *Phys. Lett. B***110** 215;
Inoue K, Kakuto A, Komatsu H, and Takeshita S 1982 Aspects of grand unified models with softly broken supersymmetry *Prog. Theor. Phys.* **68** 927, [Erratum-ibid. **70** (1983) 330];
Ellis J R, Hagelin S, Nanopoulos D V, and Tamvakis K 1983 Weak symmetry breaking by radiative corrections in broken supergravity *Phys. Lett. B***125** 275.

[13] Kim J E 2012 Inverted effective SUSY with combined Z' and gravity mediation, and muon anomalous magnetic moment [arXiv:1208.5484.v2[hep-ph]].

[14] Erler J 2012 Weighing in on the Higgs [arXiv:1201.0695 [hep-ph]].

[15] Arkani-Hamed N and Murayama H 1997Can the supersymmetric flavor problem decouple? *Phys. Rev. D***56** 6733 [arXiv: hep-ph/9703259];
Tamarit C 2012 Decoupling heavy sparticles in Effective SUSY scenarios: Unification, Higgs masses and tachyon bounds [arXiv:1204.2645 [hep-ph]].

[16] Okubo S 1962 Note on unitary symmetry in strong interactions *Prog. Theor. Phys.* **27** 949.

[17] Gürsey F and Radicati L A 1964 Spin and unitary spin independence of strong interactions *Phys. Rev. Lett.* **13** 173.

[18] Greenberg O W 1964 Spin and unitary spin independence in a paraquark model of baryons and mesons *Phys. Rev. Lett.* **13** 598.

[19] Han M Y and Nambu Y 1965 Three triplet model with double SU(3) symmetry *Phys. Rev.* **139** B1006.

[20] Hori S 1966 The fundamental triplet and its new degree of freedom *Prog. Theor. Phys.* **36** 131.

[21] Fritzsh H, Gell-Mann M, and Lewtwyler H 1973 Advantages of the color octet gluon picture *Phys. Lett. B***47** 365.

[22] Gross D and Wilczek F 1973 Ultraviolet behavior of nonabelian gauge theories *Phys. Rev. Lett.* **30** 1343;
Politzer H D 1973 Reliable perturbative results for strong interactions? *Phys. Rev. Lett.* **30** 1346.

[23] Gell-Mann M, Oakes R J and Renner B 1968 Behavior of current divergences under SU(3) x SU(3) *Phys. Rev.* **175** 2195.

[24] Langacker P and Pagels H 1973 Chiral perturbation theory *Phys. Rev. D***8** 4595.

[25] Weinberg S 1975 The U(1) problem *Phys. Rev. D***11** 3853.

[26] 't Hooft G 1986 How instantons solve the U(1) problem *Phys. Rep.* **142** 357.

[27] Kim J E and Carosi G 2010 Axions and the strong CP problem *Rev. Mod. Phys.* **82** 557 [arXiv: 0807.3125[hep-ph]].

[28] Peccei R D and Quinn H R 1977 CP conservation in the presence of instantons *Phys. Rev. Lett.* **38** 1440.

[29] Kim J E 1979 Weak interaction singlet and strong CP invariance *Phys. Rev. Lett.* **43** 103;
Shifman M A , Vainshtein V I, and Zakharov V I 1980 Can confinement ensure natural CP invariance of strong interactions? *Nucl. Phys. B***166** 493.

[30] Dine M, Fischler W and Srednicki M 1981 A simple solution to the strong CP problem with a harmless axion 1981 *Phys. Lett. B***104** 199;
Zhytnitskii A P 1980 Possible suppression of axion-hadron interactions *Sov. J. Nucl. Phys.* **31** 260.

[31] Bae K J, Huh J H and Kim J E 2008 Update of axion CDM energy density *J. Cosmol. Atropart. Phys.* **0809** 005 [arXiv: 0806.0497].

[32] Kim J E and Seo M S 2011 Parametrization of the CKM matrix *Phys. Rev. D***84** 037303 [arXiv:1105.3304[hep-ph]], and 2011 A simple expression of the Jarlskog determinant [arXiv:1201.3005 [hep-ph]].

[33] Jarlskog C 1985 Commutator of the Quark Mass Matrices in the Standard Electroweak Model and a Measure of Maximal CP Violation *Phys. Rev. Lett.* **55** 1039.

[34] See, for example, Kim J E, Kim J H and Kyae B 2007 Superstring standard model from Z_{12-I} orbifold compactification with and without exotics, and effective *R*-parity *J. High Energy Phys.* **0706** 034 [hep-ph/0702278].

[35] Kim J E, Nilles H P and Seo M S 2012 Singlet superfield extension of the minimal supersymmetric standard model with Peccei-Quinn symmetry and a light pseudoscalar Higgs boson at the LHC *Mod. Phys. Lett. A***27** 1250166 [arXiv: 1201.6547[hep-ph]].

[36] Jeong K S, Shoji Y, and Yamaguchi M 2011 Peccei-Quinn invariant extension of the NMSSM [arXiv: 1112.1014].

[37] Kim J E and Seo M S 2012 Mixing of axino and goldstino, and axino mass *Nucl. Phys. B***864** 296 [arXiv:1204.5495 [hep-ph]].

[38] Chun E J, Kim H B and Kim J E 1994 Dark matters in axino–gravitino cosmology *Phys. Rev. Lett* **72** 1956 [arXiv:hep-ph/9305208];
Kim H B and Kim J E 1995 Dark Matter and Structure Formation with Late Decaying Particles *Nucl. Phys. B***433** 421 [arXiv:hep-ph/9405385].

[39] Kim J E 1984 A common scale for the invisible axion, local SUSY GUTs and saxino decay *Phys. Lett. B***136**

378.
[40] Kim J E and Nilles H P 1984 The μ-problem and the strong CP problem *Phys. Lett.* B**138** 150.
[41] Bennett G W *et al* 2002 Measurement of the positive muon anomalous magnetic moment to 0.7 ppm *Phys. Rev. Lett.* **89** 101804 and **89** 129903(E).

Journal of Physics: Conference Series **485** (2014) 012005 doi:10.1088/1742-6596/485/1/012005

Neutrino masses: evidences and implications

J. W. F. Valle

AHEP Group, Instituto de Física Corpuscular – C.S.I.C./Universitat de València Edificio de Institutos de Paterna, Apartado 22085, E–46071 València, Spain

E-mail: valle@ific.uv.es

Abstract.
I give an overview of the evidences for neutrino masses and mixing, the associated neutrino mass generation schemes, as well as the resulting implications in particle physics experiments and cosmology.

1. Introduction

Ever since the first historic hints for neutrino oscillations the evidences for neutrino masses and mixing have mounted, and have by now become overwhelming. The bulk of the current evidence fits nicely into a three-neutrino paradigm which we will tacitly adopt in this talk. The basic tool to describe oscillations is the lepton mixing matrix, characterized by three angles, one phase which affects oscillations, and two other phases intrinsic to the Majorana nature of neutrinos [1]. The latter are important in the description of lepton-number-violating processes such as neutrinoless double beta decay, better described in terms of the original [1, 2] symmetric parametrization over its equivalent presentation adopted by the PDG [1].

Neutrino oscillation data come from a variety of solar (Homestake, SAGE, GALLEX/GNO, Super-K, SNO and Borexino), atmospheric (mainly Super-K), reactor (in particular the recent Double-Chooz, Daya-Bay and RENO experiments) and accelerator experiments (mainly MINOS and the recent T2K) [4]. To describe them one assumes the simplest unitary approximation for the lepton mixing matrix and uses state-of-the-art solar and atmospheric flux calculations. Here we summarize the results of the global analysis of neutrino oscillation parameters (for details see the review [5] and references therein).

The results obtained for $\sin^2\theta_{13}$ and δ are summarized in Fig. 1. In the upper panels we show the $\Delta\chi^2$ profile as a function of $\sin^2\theta_{13}$ for normal (left panel) and inverted (right panel) neutrino mass hierarchies. The solid blue/dark line corresponds to the result obtained from the combination of all the data samples while the others correspond to the individual reactor data samples, as indicated. One sees from the constraints on $\sin^2\theta_{13}$ coming from each of the new reactor experiments separately [2] that, as expected, the global constraint on θ_{13} is dominated by the recent reactor measurements. One finds overwhelming 10.2σ exclusion of $\theta_{13} = 0$ for both neutrino mass hierarchies. In contrast, one sees that there is no "preferred region" for the CP phase δ at $\Delta\chi^2 = 1$ (note that, given the approximations adopted in the atmospheric

[1] In seesaw-type schemes the lepton mixing has a rectangular form, which leads to lepton-flavour violation involving charged leptons [3] and non-standard neutrino interactions (talk by Miranda).
[2] Here we have fixed all the other oscillation parameters to their best fit values.

Content from this work may be used under the terms of the Creative Commons Attribution 3.0 licence. Any further distribution of this work must maintain attribution to the author(s) and the title of the work, journal citation and DOI.
Published under licence by IOP Publishing Ltd

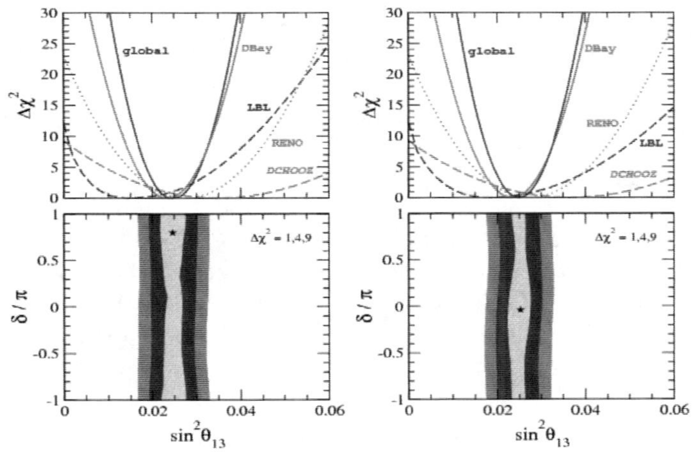

Figure 1. Upper panels: Determination of $\sin^2\theta_{13}$ from the analysis of the total event rate in Daya Bay (solid magenta/light line), RENO (dotted line) and Double Chooz (dashed line) as well as from the global neutrino data analysis (solid blue/dark line). Lower panels: contours of $\Delta\chi^2 = 1, 4, 9$ in the $\sin^2\theta_{13} - \delta$ plane from the global oscillation fit. Left (right) panels correspond to normal (inverted) neutrino mass hierarchy.

neutrino analysis in Ref. [6], the sensitivity to the parameter δ in our global fit comes only from long-baseline neutrino data). In the lower panels of Fig. 1 we show the contours of $\Delta\chi^2 = 1, 4, 9$ in the $\sin^2\theta_{13} - \delta$ plane from the global fit to the neutrino oscillation data. We minimize over all undisplayed oscillation parameters.

Besides θ_{13} and δ, the global neutrino data analysis yields best fit values and allowed ranges for all the other neutrino oscillation parameters, as summarized in Fig. 2

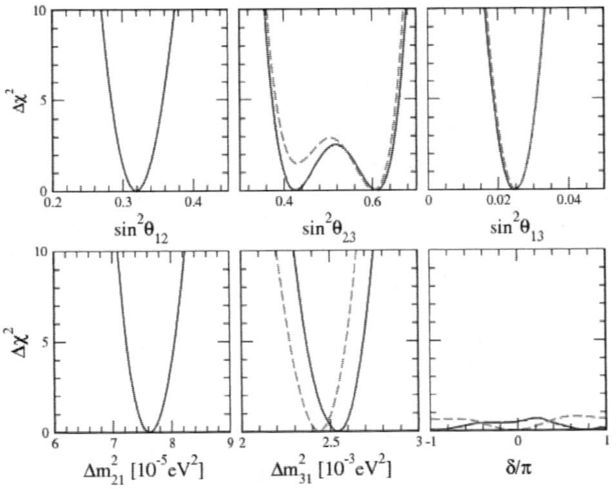

Figure 2. Global determination of the neutrino oscillation parameters $\sin^2\theta_{12}$, $\sin^2\theta_{23}$, $\sin^2\theta_{13}$, Δm^2_{31}, Δm^2_{12} and δ, from [7]. The solid (dashed) lines in the central and right panels correspond to normal (inverted) mass hierarchy. The $\Delta\chi^2$ profile for the CP phase δ is nearly flat.

Note that at 3σ we find approximately a 17% accuracy in the determination of Δm^2_{31}, an improvement that follows from the new long-baseline neutrino oscillation data. Also the best fit values for the atmospheric mass splitting parameter Δm^2_{31} have been shifted to somewhat larger values mainly due to the new MINOS disappearance data.

Turning to the atmospheric mixing angle we note a slight rejection for maximal θ_{23} values. In particular, our global fit shows a preference for the second octant. This preference is very weak for the normal mass hierarchy case (see the symmetric solid $\Delta\chi^2$ profile in the middle-top panel of Fig. 2). However, for inverse hierarchy, the profile is asymmetric (see details in [7]). Note that maximal mixing, i.e. $\theta_{23} = \pi/4$ is disfavored at $\sim 90\%$ C.L. for both mass hierarchies. As discussed in Ref [7] the preference for non-maximal values of the atmospheric mixing angle comes directly from the new MINOS data, while the choice of a particular octant comes from the interplay of long-baseline, reactor and atmospheric neutrino data.

2. Origin of neutrino mass

Now we turn to the big question: What is the origin of neutrino mass? The answer remains as elusive as ever. The simplest way to induce neutrino masses makes use of Weinberg's dimension-five operator, illustrated in Fig. 3 (left panel).

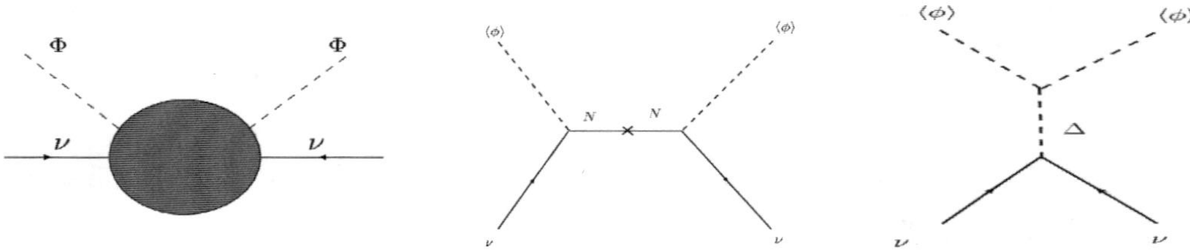

Figure 3. Weinberg operator and type-1 and type-2 seesaw schemes.

2.1. Seesaw mechanism

As illustrated by the other panels in Fig. 3, the dimension-five operator may be induced by the exchange of "messenger" states, either fermions or Higgs scalars. Effective neutrino masses arise when the electroweak symmetry breaks through a non-zero vacuum expectation value (vev) of the Higgs doublet. Their smallness relative to the charged standard model fermion masses would arise from the messengers being super-heavy. This is the idea behind the simplest seesaw mechanism [1] [8, 9, 10, 11, 12, 13, 14]. In type-1 seesaw the messenger states are SU(3) $\otimes$ SU(2) $\otimes$ U(1) singlet "right"-handed neutrinos, while in type-2 a heavy triplet scalar boson is the messenger particle, as indicated in the right panel of Fig. 3. The associated large scale characterizing the violation of lepton number symmetry will depend on the precise model realization considered.

As a result of the arbitrary number of gauge singlets that may be added in the seesaw [1] one can realize the type-1 version also at low-scale, the smallness of the lepton-number-violating parameter being natural in 't Hooft sense. Moreover such small L-violating parameter may arise dynamically through radiative corrections [15]. The inverse [16] and linear seesaw schemes [17, 18, 19] provide examples of these constructions. In the presence of supersymmetry such low-scale seesaw schemes allow for a sneutrino-like state to be the lightest supersymmetric particle and play the role of cold dark matter, instead of the standard neutralino of minimal supergravity [20, 21] (talk by De Romeri).

If the breaking of lepton number takes place spontaneously there are additional states associated to neutrino mass generation. For instance, if lepton number is ungauged, there is a remnant physical Goldstone boson, generically called Majoron. In contrast to the case of high-scale seesaw [12, 22], in low-scale versions [23] the Majoron may show up as Higgs boson decay products [24].

On the other hand in such extended schemes with gauged lepton number [3] [17, 18, 19] there is (at least) an additional neutral gauge boson coupled to neutrinos. For example one can show that in $SO(10)$ the B-L breaking scale need not be the one that determines the lightness of neutrinos [19]. As a result it can lie low enough to be within reach of current and future experiments. The associated Z' may be searched for either directly through Drell-Yan production at high-energy hadron colliders [26], such as the LHC, and also through low-energy electroweak precision measurements [27] (talk by Garces).

2.2. *Supersymmetry as origin of neutrino mass: probing neutrinos at the LHC*

There are also neutrino mass generation schemes based on weak-scale physics. An important example is "low-energy" supersymmetry without a conserved R-parity [28]. If the breaking of R-parity occurs spontaneously [4] [29, 30, 31, 32] the associated scale is of the same order as the supersymmetry breaking scale. Below that scale one is left with a very interesting scheme called bilinear R-parity violation (BRpV) [33].

In this case one finds that the neutralino decays, typically inside collider detectors, such as those at the Large Hadron Collider (LHC) (see Fig. 4), decreasing the missing energy expected in the unbroken R-parity supersymmetry.

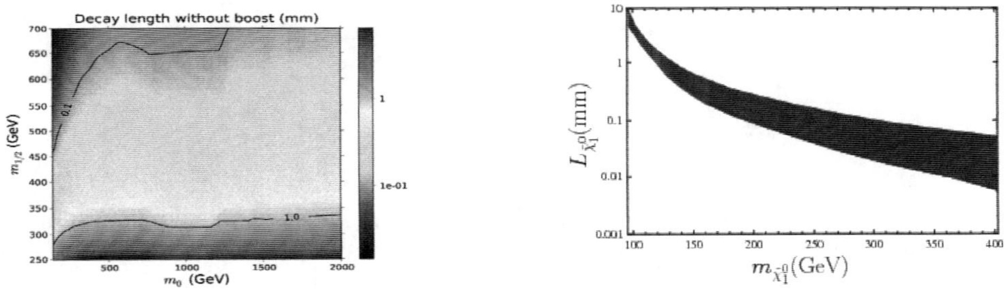

Figure 4. Neutralino decays typically produce displaced vertices inside the detector in the mSUGRA BRpV model. For illustration we show the decay length with A_0=100 GeV, sign(μ)=+1, $\tan\beta$=10. Left: in the $m_{1/2}, M_0$ plane. Right: as a function of the LSP mass, for $200\leq M_0 \leq 1000$ GeV and $240\leq m_{1/2} \leq 1000$ GeV.

The neutrino mass spectrum is expected to be normal hierarchy with the atmospheric neutrino mass scale generated by tree seesaw-like diagrams such as the one in the left plot in Fig. 5 involving the exchange of supersymmetric fermions, while the solar scale is induced radiatively (right). This provides a *rationale* for the smallness of the solar squared mass splitting with respect to the atmospheric. The model provides a successful phenomenological scheme for neutrino masses and mixings [34, 35].

The generic CP conserving form of this model is characterized by six free parameters, which correspond to the three mixing angles and three neutrino masses. As a result it does not predict the neutrino mixing angles [5]. However, the same parameters that determine neutrino masses and mixing also induce LSP decays, leading to tight correlations between both, whose details depend on the nature of the LSP [37]. An important consequence of this is that in this model

[3] For an early paper see, e.g. [25].
[4] In contrast to the seesaw Majoron which has $|L| = 2$, in this case the Majoron carries $|L| = 1$.
[5] This changes if one adds flavor symmetries. For a predictive "flavored" BRpV scheme see Ref. [36].

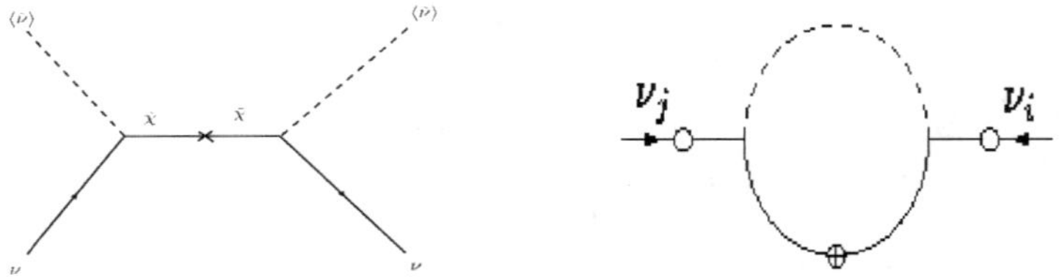

Figure 5. Neutrino mass generation diagrams from spontaneous R-parity violation.

neutrino properties can be directly probed in high energy hadron colliders [38, 39, 40, 41, 42, 43], illustrating the complementarity of accelerator and non-accelerator approaches in elementary particle physics. As an example, in the minimal supergravity BRpV model the semi-leptonic neutralino decay branching ratios correlate with the neutrino mixing angles describing oscillations, as illustrated in Fig. 6.

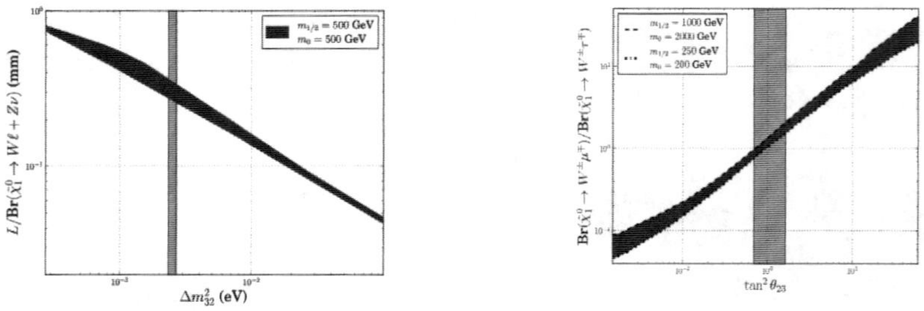

Figure 6. Probing neutrino properties at the LHC: correlation between LSP decays and neutrino oscillation parameters, from Ref. [43]. The left panel shows the LSP decay length parameter versus the atmospheric squared mass scale. The right panel gives the relation between the indicated ratio of LSP branching ratios and the atmospheric mixing angle. The vertical bands are allowed at 2σ by global oscillation fits [7].

Finally we note that in this case the neutralino can not be dark matter, as it decays with a short lifetime. However one can show that the gravitino can play the role of dark matter [44] while keeping the above phenomenological tests.

3. Flavour symmetries and neutrino mixing angles

Non-Abelian continuous and discrete flavour symmetries have been extensively used to account for the observed pattern of neutrino mixing [45, 46, 47]. The tri-bimaximal (TBM) mixing *ansatz* [48], with an effective bimaximal mixing of ν_μ and ν_τ at the atmospheric scale and trimaximal mixing of ν_e with ν_μ and ν_τ at the solar scale, was proposed in 2002 by Harrison, Perkins and Scott as a good approximation for the observed neutrino mixing angles,

$$U = \begin{pmatrix} 2/\sqrt{6} & 1/\sqrt{3} & 0 \\ -1/\sqrt{6} & 1/\sqrt{3} & 1/\sqrt{2} \\ -1/\sqrt{6} & 1/\sqrt{3} & -1/\sqrt{2} \end{pmatrix}. \tag{1}$$

Although there are other viable possibilities, the TBM ansatz has often been take as reference for model building. After the recent reactor and accelerator neutrino results presented at Neutrino 2012, the global analysis of neutrino oscillation data now indicates a robust measurement of a relatively "large" value of θ_{13} and non-maximal θ_{23} [7], which challenges the validity of the

TBM *ansatz* as a good first approximation to the neutrino mixing pattern. However in concrete models there can be large corrections to the TBM pattern, so that we can still take it as a useful reference *ansatz*.

Assigning the three known generations of leptons to non trivial irreducible representations (*irreps*) of a non-Abelian discrete flavour symmetry group one can make predictions for masses and mixings in the lepton sector. In general the number of free parameters in models based on Abelian flavour symmetries is typically larger than the corresponding number of free parameters needed to describe non-Abelian flavour symmetry theories. Moreover there are non-Abelian discrete groups with triplet irreps, exactly as the number of the standard model generations.

The smallest group that contains triplet irreps is A_4, the group of the even permutations of four objects, isomorphic to the group of the symmetries of the tetrahedron T [6]. A realistic A_4 model was proposed in Ref. [49] predicting maximal atmospheric mixing and vanishing θ_{13} to first approximation. Although expected to be sizeable, the solar mixing angle was not predicted and neutrino masses were (quasi)degenerate.

However one can obtain the full tri-bimaximal pattern in Eq. (1) from a non-Abelian flavor symmetry. Consider A_4 as an example. A_4 contains two Abelian subgroups, namely Z_2 and Z_3. When broken into Z_3 in the charged sector, and into Z_2 in the neutrino sector, A_4 leads to a lepton mixing matrix of tri-bimaximal form, Eq. (1).

In order to yield the tri-bimaximal pattern in Eq. (1), the neutrino mass matrix must be

$$M_\nu = \begin{pmatrix} y & x & x \\ x & y+z & x-z \\ x & x-z & x+z \end{pmatrix}, \tag{2}$$

where x, y, z are free parameters. The above matrix has two properties:

- it is $\mu - \tau$ invariant giving maximal atmospheric and zero reactor angles;
- it satisfies the relation $(M_\nu)_{11} + (M_\nu)_{12} = (M_\nu)_{22} + (M_\nu)_{23}$ giving trimaximal solar angle.

Note that Eq. (2) is diagonalized by the TBM mixing matrix in Eq. (1) independently of the mass eigenvalues. The derivation of TBM mixing from a flavour symmetry was first given in Refs. [50, 51, 52]. Note that A_4 is totally broken, therefore deviations to TBM are expected. In general one can not align A_4 in the Z_3 and Z_2 directions in the charged and neutral lepton sectors respectively, this is known as the *alignment* problem. This may be circumvented by using extra dimensions and/or supersymmetry [50, 51] or by assuming a suitably chosen soft breaking sector. Alternatively one may use a large discrete group, such as $Z_2^3 \times U_L(1)^3 \rtimes S_3$ [53].

4. Implications of a "large" reactor angle

At the Neutrino 2012 conference in Kyoto, reactor and accelerator experiments have presented important results. In particular

$$\sin^2 2\theta_{13} = 0.092 \pm 0.016(\text{stat}) \pm 0.005(\text{syst}) \quad \text{at } 5.2\sigma \quad (\text{DayaBay [54]}) \tag{3}$$
$$\sin^2 2\theta_{13} = 0.113 \pm 0.013(\text{stat}) \pm 0.019(\text{syst}) \quad \text{at } 4.9\sigma \quad (\text{RENO [55]}) \tag{4}$$

indicating that $\sin\theta_{13} \sim \lambda_C$ where $\lambda_C \approx 0.2$ is the Cabibbo angle, and casting doubts on the validity of the TBM *ansatz*. As we saw, in order to have the TBM mixing pattern we need to break separately our flavour group (for instance in A_4) into Z_3 in the charged sector and into Z_2 in the neutrino sector. Therefore the flavour group is completely broken. Since the flavour symmetry leading to the TBM *ansatz* is in general broken we expect deviations from TBM

[6] For a classification of the irreps of different non Abelian discrete groups see for instance Ref. [45].

which, in particular, could generate a nonzero reactor angle. As a result TBM can still be taken as a good first order approximation.

However many models with TBM at leading order are ruled out because they give corrections of the same order δ_θ to the three angles θ_{13}, θ_{12} and θ_{23}. If $\delta_\theta \sim \lambda_C$ we have

$$\sin^2 2\theta_{13} = 0.087 \quad \text{for } \delta_\theta = 0.15, \tag{5}$$
$$\sin^2 2\theta_{13} = 0.152 \quad \text{for } \delta_\theta = 0.20 \tag{6}$$

close to the best fits. However the deviations of the solar mixing from its trimaximal values $\sin^2\theta_{12}^{TBM} \equiv 1/3$ will be too large if we take $\delta_\theta = 0.15 \sim \lambda_C$, namely

$$\sin^2(\theta_{12}^{TBM} + \delta_\theta) = 0.48 \quad (0.38\,@3\sigma), \tag{7}$$
$$\sin^2(\theta_{12}^{TBM} - \delta_\theta) = 0.20 \quad (0.27\,@3\sigma). \tag{8}$$

This in principle rules out many TBM schemes. Indeed, most extensions of TBM models which allow for a large reactor angle also predict a deviations of the atmospheric and solar mixing angle from their TBM values [56, 57]. Therefore a burning question is to evaluate the extent to which solar and atmospheric mixing angles *deviate* from their TBM values.

However, not all TBM models proposed in the past are excluded, for example in the model of Ref. [58], based on A_4, large reactor angle $\theta_{13} \sim \lambda_C$ has been obtained with deviation of θ_{12}^{TBM} of order of λ_C^2 in agreement with data. There are other examples in the literature of models where such deviations are small, despite the relatively large reactor angle value, see for instance [59, 60, 61, 62, 63]. Needless to say that this poses no problem for the model in [49], as it does not predict the solar angle.

Finally we note that alternative *ansatze* have been suggested to circumvent this problem. An interesting possibility is that the leading order neutrino mass matrix is not diagonalized by the TBM *ansatz*, but rather by the *bi-maximal* one (where both solar and atmospheric mixing angles are maximal from the start) [64] or simply *bi-large* [65], or by the *golden ratio* [63, 66]. Clearly the "large" reactor angle will not only open a new world of CP violation in the lepton sector, but may also shed light into the flavour problem, one of the most challenging puzzles in particle physics.

5. Neutrinoless double beta decay and neutrino properties

Given that neutrinos have mass one expects $0\nu\beta\beta$ to receive a contribution from tree-level neutrino exchange if they are, as expected, Majorana fermions. The amplitude is proportional to an effective mass parameter combination $\langle |m_{ee}| \rangle$ illustrated in Fig. 7. In contrast to neutrino oscillations, this is sensitive also to the absolute scale of neutrino masses, which is independently tested also in searches for tritium beta decay [70] and cosmology [71, 72]. For example, the model in [49] implies a lower bound on the absolute neutrino mass $m_\nu \gtrsim 0.4$ eV and therefore will be tested fairly soon. In addition, this amplitude can bring complementary information on the underlying structure of flavour in the lepton sector, as we now discuss.

Many of the models based on non–Abelian discrete flavour symmetries are characterized by a specific (complex) relation between neutrino mass eigenvalues, leading to mass sum rules, e. g.

$$A) \quad \chi\, m_2^\nu + \xi\, m_3^\nu = m_1^\nu, \tag{9}$$
$$B) \quad \frac{\chi}{m_2^\nu} + \frac{\xi}{m_3^\nu} = \frac{1}{m_1^\nu}, \tag{10}$$
$$C) \quad \chi\,\sqrt{m_2^\nu} + \xi\,\sqrt{m_3^\nu} = \sqrt{m_1^\nu} \tag{11}$$
$$D) \quad \frac{\chi}{\sqrt{m_2^\nu}} + \frac{\xi}{\sqrt{m_3^\nu}} = \frac{1}{\sqrt{m_1^\nu}}\,. \tag{12}$$

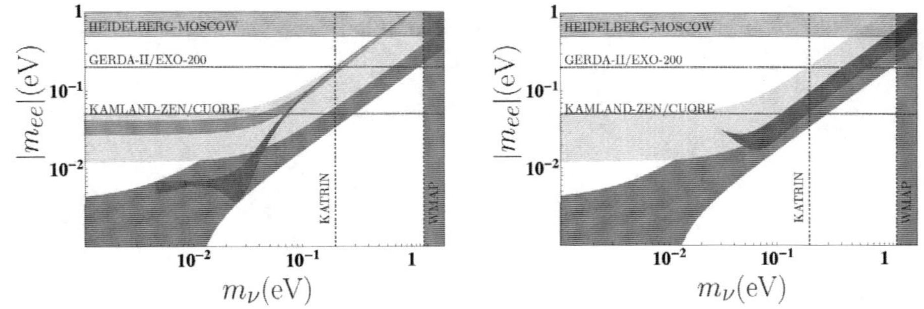

Figure 7. The broad upper and lower branches correspond the *flavor-generic* inverse (yellow) and normal (gray) hierarchy neutrino spectra, respectively. Flavor model predictions are indicated by the green and red (darker-shaded) regions. Only these sub-bands are allowed by the corresponding mass sum rule. The left panel is taken from Ref. [67] while the right plot is from Ref. [68]. We also give the current limit and future sensitivities on $\langle|m_{ee}|\rangle$ [69] and m_ν [70].

Here m_i^ν denote neutrino mass eigenvalues, up to a Majorana phase factor, while χ and ξ are free parameters characterizing the model, taken as positive without loss of generality.

The effective neutrino mass parameter $|m_{ee}|$ determining the $0\nu\beta\beta$ decay amplitude in a flavour-generic scheme is given in Fig. 7, as a function of the lightest neutrino mass. Allowing the neutrino oscillation parameters to vary in their allowed ranges one obtains two regions in the $(|m_{ee}|, m^\nu_{\rm light})$ plane corresponding to normal and inverse hierarchy spectra, as shown by the broad bands in Fig. 7. There is a lower bound for the parameter $|m_{ee}|$ only in the case of inverse mass hierarchy.

Let us turn to the case where relations like *(A),(B),(C)* and *(D)* hold, such as in the tri-bimaximal mixing pattern. In Fig 7 we show the prediction for $|m_{ee}|$ as function of m_{light} for the indicated models [7]. Note that when deriving a lower bound on the effective $0\nu\beta\beta$ decay amplitude parameter $|m_{ee}|$, we include explicitly the effects of non-vanishing θ_{13} as indicated by recent experiments [73, 74] as well as global neutrino oscillation fits [7] by taking the 3 σ oscillation parameter ranges [8]. Note that within specific flavor models a lower bound on $|m_{ee}|$ can be set even for the case of normal neutrino mass hierarchy, since a destructive interference among the light neutrinos [75, 76, 77] is prevented in this case by the flavor symmetry. Of course, all inverse hierarchy schemes corresponding to various choices of (χ, ξ) within sum-rules A-D have a lower bound for the parameter $|m_{ee}|$. However, the numerical value obtained depends on the flavor scheme.

Before closing let me stress that the importance of $0\nu\beta\beta$ lies in that it so far provides the only feasible way to test the nature of neutrinos. Indeed, the *black box theorem* [78, 79] states that the observation of the $0\nu\beta\beta$ would point towards the Majorana nature of neutrinos irrespective of the mechanism generating their mass and irrespective of the mechanism inducing the decay, as illustrated in Fig. 8 [9].

6. Lepton flavour violating phenomena

The phenomenon of lepton flavour violation required to account for neutrino oscillation data may show up in processes involving charged leptons. A remarkable fact is that their associated

[7] A compilation is summarized in Tab. I of Ref. [67]. See talks by Dorame and Rojas.

[8] Corrections leading to $\theta_{13} \neq 0$ may arise from higher dimensional operators or the charged lepton sector.

[9] In principle CP and electromagnetic properties of neutrinos are also sensitive to their nature [76, 80, 81, 82].

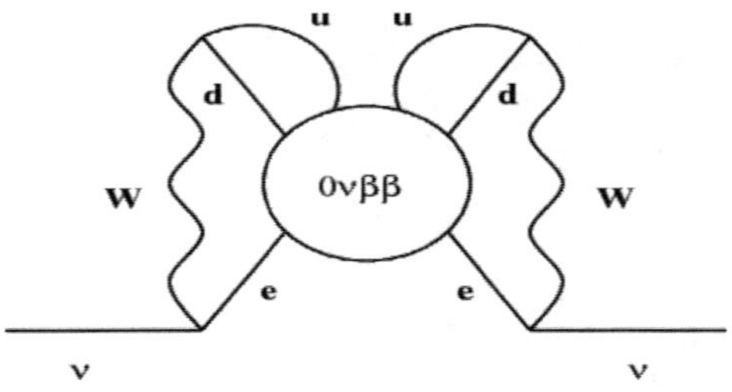

Figure 8. The observation of $0\nu\beta\beta$ implies the Majorana nature of a neutrino [78].

strength need not be suppressed by the smallness of neutrino masses with similar results for leptonic CP violation [83, 84, 85, 86]. For example, in the presence of supersymmetry, high-scale seesaw schemes may induce sizeable lepton flavour violation decays such as $\mu^- \to e^-\gamma$ and flavour violating tau decays (Fig. 9) as well as nuclear $\mu^- - e^-$ conversion (Fig. 10) as a result of the exchange of supersymmetric leptons. The existence of such loop effects, illustrated in Fig. 9, has been known for a while [87, 88]. The resulting rates will be accessible to the upcoming

Figure 9. Supersymmetric Feynman diagrams for $l_i^- \to l_j^- \gamma$.

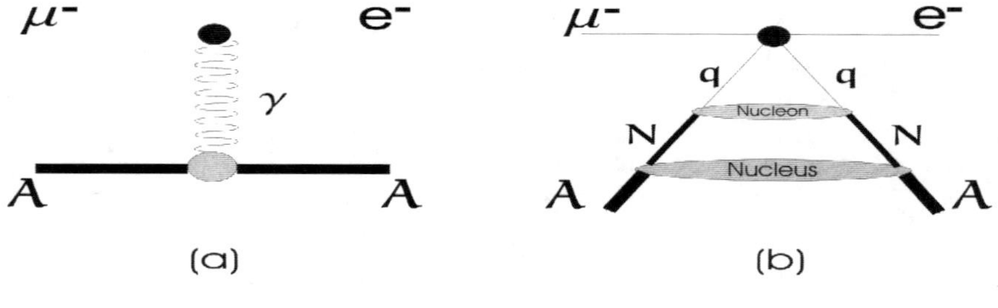

Figure 10. Long-distance (a) and short-distance (b) contributions to nuclear $\mu^- - e^-$ conversion.

generation of experiments [89, 90], the pattern of lepton flavour violation processes depending on the model details.

6.1. Heavy neutrino exchange in low-scale seesaw

Low-scale seesaw schemes generate neutrino masses from fermion messengers ("right"-handed neutrinos) with mass at the TeV scale. As interesting examples we have the inverse and the linear seesaw schemes [16, 17, 18, 19]. These are potentially accessible to the LHC, especially in the presence of a new gauge boson "portal" associated, for example, to left-right symmetry [91, 92].

Within low-scale seesaw mechanisms lepton flavour violation and/or CP violating effects arise at the one–loop level. In these models the 9×9 Majorana neutrino mass matrix is diagonalized by a unitary matrix $U_{\alpha\beta}$, α, $\beta = 1...9$, leading to the 3 light eigenstates ν_i with $i = 1, 2, 3$ and 6 heavy ones N_j with $j = 4, .., 9$. The leptonic mixing matrix is given as a rectangular matrix [1],

$$\mathcal{L}_{CC} = \frac{g}{\sqrt{2}} K_{i\alpha} \overline{L}_i \gamma_\mu (1 + \gamma_5) N_\alpha W^\mu, \tag{13}$$

where $i = 1, 2, 3$ label the left-handed charged leptons and α the neutrals. The well-known one–loop contribution to this branching ratio is given by

$$Br(l_i \to l_j \gamma) = \frac{\alpha^3 s_W^2}{256\pi^2} \frac{m_{l_i}^5}{M_W^4} \frac{1}{\Gamma_{l_i}} |G_{ij}|^2 , \tag{14}$$

where G_{ij} is the appropriate loop function [93, 94]. We note that the resulting branching ratio can be sizeable, since it need not be suppressed by the smallness of neutrino masses [83, 84, 85, 86, 95] Similar results hold other lepton flavour violation processes, including nuclear muon-electron conversion [96] whose expected rates correlate with those of $\mu \to e\gamma$, see Fig. 11. In contrast to the general case considered in [96, 97], in the presence of a flavor symmetry [95] it is easier to display the dependence of $B(\mu \to e\gamma)$ upon the parameters μ, v_L characterizing the low-scale violation of lepton number, and the heavy neutrino mass M, fixed as $M = 100,\ 200,\ 1000$ GeV (continuous, dashed and dot-dashed lines) in Fig. 11. Note also that, in a given flavor model, there may be predictions for ratios of lepton flavour violation branching ratios such as

$$\frac{Br(\tau \to e\gamma)}{Br(\mu \to e\gamma)} = \left(\frac{m_\tau}{m_\mu}\right)^5 \frac{\Gamma_\mu}{\Gamma_\tau} \approx 0.18. \tag{15}$$

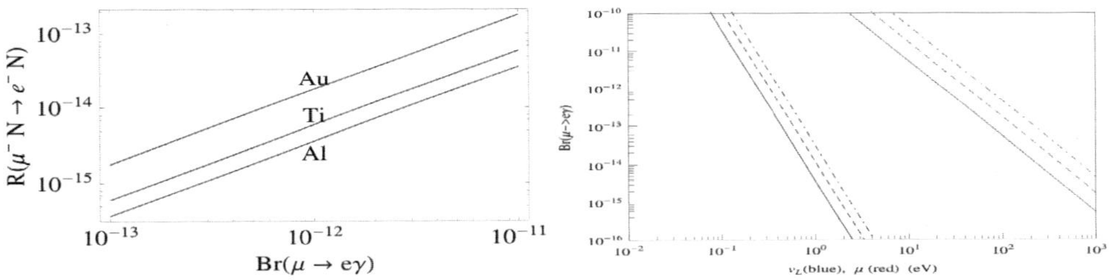

Figure 11. Left: $Br(\mu \to e\gamma)$ correlates with muon-electron conversion in nuclei. From Ref. [96]. Right: $Br(\mu \to e\gamma)$ versus the lepton number violation scale characterizing the inverse (red), and the linear seesaw (blue) of Ref. [95].

6.2. Lepton flavour violation in supersymmetric low-scale seesaw

We now consider lepton flavour violation rates in the framework of a generic supersymmetric inverse seesaw model [97]. The left panel in Fig. 12 displays the dependence of the branching ratios for $\mu^- - e^-$ conversion in Au, Ti and Al (left) and $\mu^- \to e^-\gamma$ (right) with the small neutrino mixing angle θ_{13}, for different values of θ_{12}. The inverse seesaw parameters are given by: $M = 1$ TeV and $\mu = 30$ eV. The light neutrino parameters used are from [5], except for θ_{13} which is varied as shown in the plots. The vertical lines indicate the $\sin^2\theta_{13}$ values indicated by recent experiments. [7]. These rates may be testable in the upcoming lepton flavour violation experiments [89, 90] [10]. A novel feature present in low-scale (M in the TeV range) supersymmetric seesaw models is the interplay between heavy neutrino and supersymmetric

[10] The situation for very large M ("double seesaw" limit) has been recently considered in Ref. [98] and they show how the Z-penguin diagram can enhance the $l_i \to 3l_j$ decays, in contrast to naive estimates.

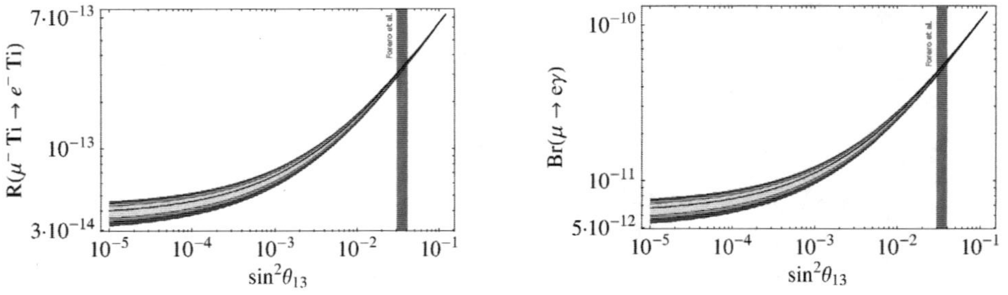

Figure 12. LFV branching ratios in the supersymmetric inverse seesaw model (see text).

contributions to $Br(\mu \to e\gamma)$ and other lepton flavour violation processes such as nuclear $\mu^- - e^-$ conversion, depicted in Fig. 10, which fall within the sensitivity of future experiments [89]. Note that large lepton flavour violation rates are possible even in the massless neutrino limit, hence the allowed lepton flavour and CP violation rates are unsuppressed by the smallness of neutrino masses [83, 84, 85, 86]. Finally, for low enough M the corresponding heavy neutrinos can be searched directly at particle accelerators [99, 100]. In the presence of new gauge bosons, such as right-handed gauge bosons these heavy neutrinos can be searched at the LHC [91, 92].

6.3. Supersymmetric high-scale A_4-based seesaw

The exchange of supersymmetric leptons induces flavour violating processes, as illustrated in Fig. 10 and discussed in [101, 102, 103, 104, 105]. Instead of a generic "flavour-blind" supersymmetric high-scale seesaw scheme we now consider, as an example, the A_4-based model introduced in Ref. [49]. The allowed parameter space is determined by scanning the multi-dimensional parameter space, keeping all supersymmetric masses real and in the range 100 GeV to 1000 GeV in Ref. [106].

The spectra fall into two different groups. The normal hierarchy having two low mass sleptons ($\sim$ 150 GeV) and one heavy (above $\sim$ 500 GeV), and the inverted hierarchy case having two heavy sleptons and one light. In both cases at least one slepton mass lies below about 200 GeV, detectable at the LHC. Most points fall into the case of normal hierarchy, corresponding to a normal hierarchy for the neutrinos as well, with one small and two large mixing angles.

One finds a prediction for the charged lepton decays $\ell_i \to \ell_j\gamma$ seen in Fig.13, with a lower bound of 10^{-9} for BR($\tau \to \mu\gamma$), which lies within reach of B-factories. Similarly, BR($\mu \to e\gamma$) is found to be larger than about 10^{-15} which may be observed in the future [90].

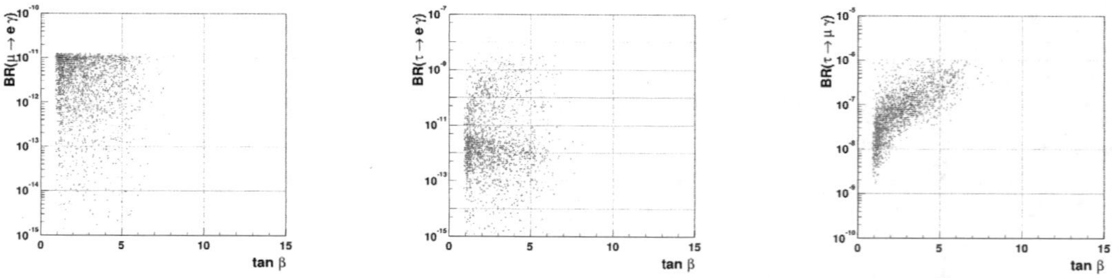

Figure 13. Expected branching ratios for the processes $\ell_i \to \ell_j\gamma$ as a function of $\tan(\beta)$.

7. Neutrino mass and dark matter

Dark matter and neutrino masses are two sectors that clearly require physics beyond the Standard Model. These may, in fact, be closely related [107, 108]. I now consider some "neutrino-motivated" dark matter candidates and discuss their direct/indirect detection prospects. While they are all cold insofar as the properties of the Cosmic microwave background (CMB) are concerned, some, like the Majoron, behave as warm dark matter regarding structure formation.

7.1. Sneutrino-like dark matter in low-scale seesaw

If supersymmetrized, the simplest SU(3) ⊗ SU(2) ⊗ U(1) inverse seesaw model of neutrino masses [16, 23] may lead to a sneutrino-like state as the lightest super-particle, instead of the lightest neutralino [20], as illustrated in the left plot in Fig. 14. This happens even in the constrained Minimal Supersymmetric Standard Model (CMSSM) picture inspired by minimal supergravity [20]. As discussed in Ref. [15] the low-scale breaking of lepton number may arise dynamically.

Indeed, in the shaded (red) and light (yelow) areas in Fig. 14 (plotted for a given choice for $\tan\beta = 35$, $A_0 = 0$ and $\mu > 0$) one finds that, for a suitable choice of parameters associated to the inverse seesaw scheme a sneutrino-like state is the LSP [20]. Note also that the light (yellow) region includes all the region where the $\tilde{\tau}$ is the LSP in the standard mSUGRA case. On the other hand the dark (blue) region is excluded by experimental and theoretical constraints. This region is now substantially larger due to the recent LHC results of ATLAS and CMS [109], as illustrated by the extended dark region in Fig. 14 (left).

As shown in the mid panel in Fig. 14, the model can reproduce the correct relic sneutrino-like dark matter abundance Ωh^2, when the supersymmetric parameters lie within the ranges $100\,\mathrm{GeV} < m_0 < 3\,\mathrm{TeV}$, $100\,\mathrm{GeV} < m_{1/2} < 3\,\mathrm{TeV}$, $A_0 = 0$, $3 < \tan\beta < 50$ (see Ref. [20] for details). Finally, the right panel in Fig. 14 illustrates how one can obtain an accessible direct detection rates in nuclear recoil experiments.

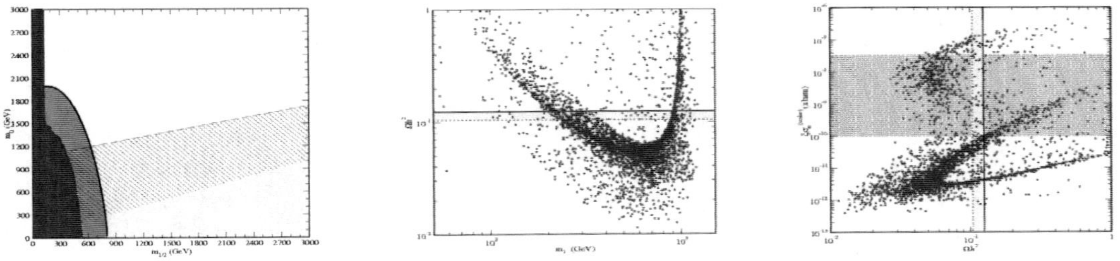

Figure 14. Left: regions of mSUGRA parameters where the "sneutrino" is the LSP, see [20] for details. Middle: Blue points give the LSP relic abundance Ωh^2 versus the LSP mass. Right: "sneutrino"–nucleon scattering cross section vs. its relic abundance, for the same scan of parameters. The horizontal (light blue) band denotes the current sensitivity of direct detection experiments. The yellow bands delimit the WMAP cold dark matter 3σ range [110].

7.2. Dark matter stabilized by flavour symmetry

The observed pattern of mixing angles radically differs from that which characterizes the Cabibbo-Kobayashi-Maskawa angles [111, 112]. Attempts to understand this in terms of basic flavour symmetries have brought in the idea that perhaps the symmetry explaining flavour may also account for the required dark matter stability. A non-Abelian flavour symmetry model where dark matter is stabilized through a discrete unbroken Z_2 subgroup has been recently proposed in Ref. [113]. The model is based on A_4 and extends the Higgs sector of the Standard Model with three scalar doublets. After electroweak symmetry breaking two of the scalars of the model acquire vacuum expectation values (vevs), leaving a residual parity symmetry. The lightest neutral odd scalar must be stable and is a viable dark matter candidate. The phenomenology of this model has been studied in [114]. The dark matter particle is thermally produced via a Higgs portal (Fig. 15, left) and obeys all constraints from current laboratory experiments and astrophysical observations. The right panel in Fig. 15 gives the spin-independent dark matter scattering cross section off-protons as a function of the dark matter mass. Direct detection prospects are promising for the near future as many collaborations are closing in on the low mass region of the WIMP parameter space. The orange regions show the DAMA/LIBRA annual modulation regions including (neglecting) the channeling effect (dashed and solid, respectively) [115]. The green region corresponds to the COGENT data [116]. Dashed and dotted lines correspond to the upper bound from CDMS (respectively from [117] and [118]). XENON100 bounds [119] are indicated by a solid black line. Coming to indirect detection [114] one finds that the sensitivity of current Fermi-LAT observations are starting to probe the model for low-intermediate dark matter masses (see e.g. [120, 121]). Predictions for the neutrino sector are discussed in [114]. This scheme is just an example of a class which may possibly realize a deeper connection between neutrino and dark matter physics (see talk by Peinado).

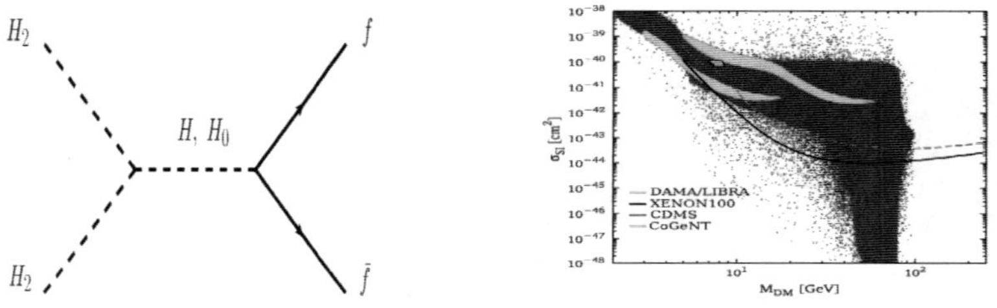

Figure 15. Left: Higgs portal. Right: Spin-independent dark matter scattering cross section off-protons versus the dark matter mass. Current experimental sensitivities indicated.

7.3. Majoron as decaying dark matter

A wide class of neutrino mass models is based upon the spontaneous breaking of ungauged lepton number [22, 12]. The associated Goldstone boson - the Majoron - will pick up a mass as a result of quantum gravity effects [122]. The massive Majoron will necessarily decay to a pair of neutrinos [123]. The lifetime and mass are constrained by cosmic microwave background observations as indicated in Fig. 16 [124]. The scenario fits nicely in models where neutrinos get mass *a la seesaw*, and may lead to other possible cosmological implications associated to structure formation. Indeed, although the Majoron in this mass range behaves as cold dark matter as far as the CMB is concerned, its small thermal velocities might play a role in structure formation.

In a number of schemes the decaying Majoron has also a sub-leading decay mode into two photons see, e.g. Refs. [108] and [125, 126]. There are upper limits on the decay rate arising from NGC3227 (red), the Milky Way halo observed with a prototype cryogenic spectrometer (salmon), XMM observations of the Milky Way (sand), Chandra observations of the Bullet Cluster and M31 (orange), HEAO-1 observations of the diffuse X-ray background (aquamarine), INTEGRAL SPI line search in the Milky Way halo (blue). These exclude the various shaded regions, which have been compiled and extended in Ref. [125] where further details and references can be found. Improved sensitivities are expected by the proposed XENIA mission [127].

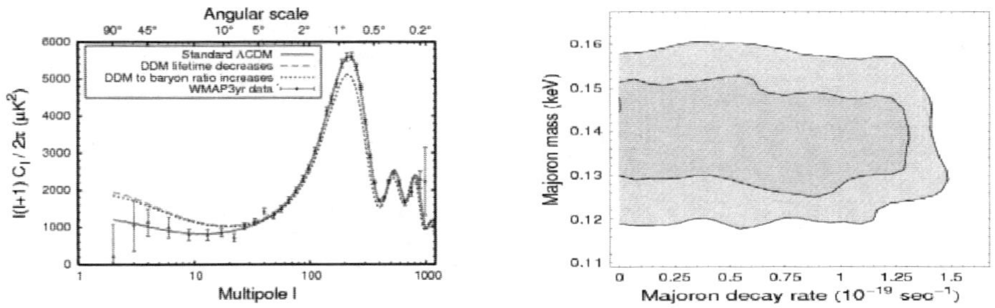

Figure 16. Left panel: (Color online) Distortion of CMB anisotropy spectrum arising from decaying dark matter Majoron, see Ref. [124] for details. Right panel: (Color online) Contours of the 68% (green/dark) and 95% (yellow/light) CL allowed mass-decay-rate regions.

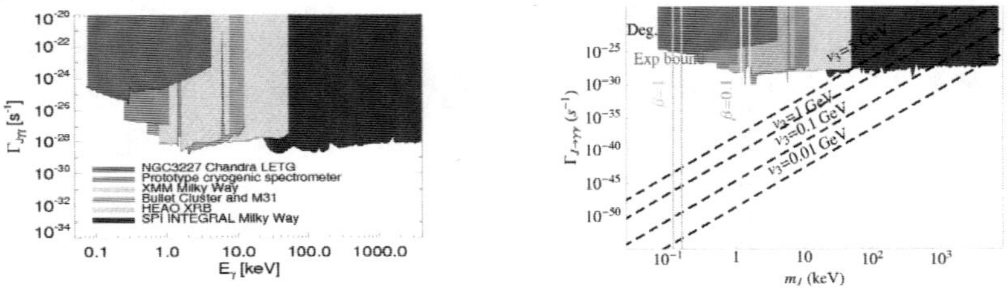

Figure 17. Majoron decay rate to photons versus Majoron mass, for different values of the triplet vev, v_3. We assume the invisible decay bound to be saturated and a (quasi)degenerate neutrino mass spectrum. Shaded regions are excluded by observations [125] and vertical lines characterize the cosmological Majoron production mechanism [124].

7.4. *Decaying gravitino as dark matter in BRPV*

Weak-scale supersymmetry with broken R-parity [28] provides an attractive way to induce neutrino masses and mixing in an intrinsically supersymmetric way, as indicated in Fig. 5. In the presence of SU(3) ⊗ SU(2) ⊗ U(1) singlet "right-handed" neutrino superfields one can indeed induce the spontaneous breaking of R-parity in agreement with LEP measurements of the invisible Z width [29, 30]. This leads to bilinear R-parity violation (BRpV) as an effective model [33], the simplest extension of the minimal supersymmetric standard model (MSSM) with lepton number violation that can account for the present neutrino oscillation data [34, 35].

One of its main features is that the lightest neutralino can not be dark matter, as it decays typically inside collider detectors, as illustrated in Fig. 4. A natural possibility is to assume that the LSP is the gravitino [44]. In this picture the same lepton number violating superpotential which generate neutrino masses and mixing, also induce R-parity breaking gravitino decays, into a photon plus a neutrino. The latter is doubly suppressed, (i) by the smallness of the R-parity violating couplings, and (ii) by its gravitational origin. The allowed gravitino mass-lifetime region consistent with neutrino oscillation data and astrophysical bounds from negative searches for gamma-ray lines from dark matter decay is indicated in Fig. 18 (left). The yellow region is excluded by gamma-ray line searches at Fermi and EGRET. The lower and upper black lines correspond to $m_{1/2}$=240 and 3000 GeV respectively. The right plot shows the parameters in terms of reheat temperature and gravitino mass [44]. In the orange area a consistent gravitino relic abundance would require gluino masses already excluded by present collider searches, while the blue area corresponds to very large gluino masses, M_3 >6000 GeV. The yellow region is excluded by astrophysical gamma-ray line searches.

Note that the BRpV gravitino dark matter model can be probed at future collider experiments, like the LHC or the Next Linear Collider, since the decay pattern of the next-to-lightest supersymmetric particle provides a direct connection with the lepton mixing angles determined by neutrino experiments [38, 39, 40, 41, 42], as illustrated in Fig. 4.

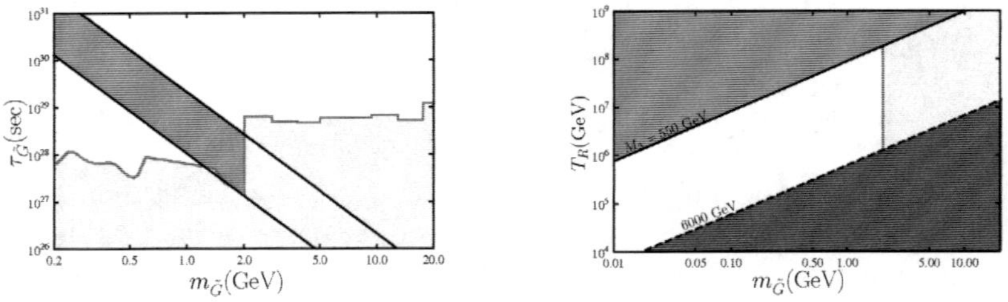

Figure 18. Left: Allowed gravitino mass-lifetime region (grey color) consistent with neutrino oscillation data and astrophysical bounds on gamma-ray lines from dark matter decay. Right: The allowed parameters (white band) in terms of reheat temperature and gravitino mass [44].

Acknowledgments
I thank the organizers for organizing a very pleasant and fruitful conference at Merida, and Lucho Dorame for reading the manuscript. This work was supported by the Spanish MINECO under grant FPA2011-22975 and MULTIDARK CSD2009-00064 (Consolider-Ingenio 2010 Programme), by Prometeo/2009/091 (Generalitat Valenciana), by the EU ITN UNILHC PITN-GA-2009-237920.

References
[1] Schechter J and Valle J W F 1980 *Phys. Rev.* **D22** 2227
[2] Rodejohann W and Valle J W F 2011 *Phys.Rev.* **D84** 073011 (*Preprint* 1108.3484)
[3] Forero D *et al.* 2011 *JHEP* **1109** 142
[4] Nakamura K *et al.* 2010 *Journal of Physics G: Nuclear and Particle Physics* **37** 075021
[5] Maltoni M, Schwetz T, Tortola M A and Valle J W F 2004 *New J. Phys.* **6** 122 hep-ph/0405172
[6] Wendell R *et al.* (Super-Kamiokande Collaboration) 2010 *Phys.Rev.* **D81** 092004 (*Preprint* 1002.3471)
[7] Forero D, Tortola M and Valle J W F 2012 (*Preprint* arXiv:1205.4018) to appear in Phys. Rev. D. This updates New J.Phys. 13 (2011) 109401 and New J.Phys. 13 (2011) 063004 by including the most recent neutrino oscillation data presented at Neutrino 2012. For other analyses see Fogli et al Phys.Rev. D86 (2012) 01301 and M. C. Gonzalez-Garcia et al, arXiv:1209.3023.
[8] Minkowski P 1977 *Phys. Lett.* **B67** 421
[9] Gell-Mann M, Ramond P and Slansky R 1979 Print-80-0576 (CERN)
[10] Yanagida T KEK lectures, 1979 Ed. O. Sawada and A. Sugamoto (KEK, 1979)
[11] Mohapatra R N and Senjanovic G 1981 *Phys. Rev.* **D23** 165
[12] Schechter J and Valle J W F 1982 *Phys. Rev.* **D25** 774
[13] Lazarides G, Shafi Q and Wetterich C 1981 *Nucl. Phys.* **B181** 287
[14] Valle J W F 2006 *J. Phys. Conf. Ser.* **53** 473–505 these review lectures were given at Corfu, 2005 and contain extensive references to the early papers on the seesaw mechanism (*Preprint* hep-ph/0608101)
[15] Bazzocchi F, Cerdeno D, Munoz C and Valle J W F 2010 *Phys.Rev.* **D81** 051701
[16] Mohapatra R N and Valle J W F 1986 *Phys. Rev.* **D34** 1642
[17] Akhmedov E *et al.* 1996 *Phys. Lett.* **B368** 270–280
[18] Akhmedov E *et al.* 1996 *Phys. Rev.* **D53** 2752–2780
[19] Malinsky M, Romao J C and Valle J W F 2005 *Phys. Rev. Lett.* **95** 161801
[20] Arina C, Bazzocchi F, Fornengo N, Romao J and Valle J W F 2008 *Phys. Rev. Lett.* **101** 161802
[21] De Romeri V and Hirsch M 2012 (*Preprint* 1209.3891)
[22] Chikashige Y, Mohapatra R N and Peccei R D 1981 *Phys. Lett.* **B98** 265

[23] Gonzalez-Garcia M C and Valle J W F 1989 *Phys. Lett.* **B216** 360
[24] Joshipura A S and Valle J W F 1993 *Nucl. Phys.* **B397** 105–122
[25] Valle J W F 1987 *Phys. Lett.* **B196** 157
[26] Chatrchyan S *et al.* (CMS Collaboration) 2012 *Phys.Lett.* **B714** 158–179 (*Preprint* `ATLAS results in http://atlas.web.cern.ch/Atlas/GROUPS/PHYSICS/CONFNOTES/ATLAS-CONF-2012-129/`)
[27] Garces E, Miranda O, Tortola M and Valle J W F 2012 *Phys.Rev.* **D85** 073006 (*Preprint* `1112.3633`)
[28] Ross G G and Valle J W F 1985 *Phys. Lett.* **B151** 375
[29] Masiero A and Valle J W F 1990 *Phys. Lett.* **B251** 273–278
[30] Romao J C, Santos C A and Valle J W F 1992 *Phys. Lett.* **B288** 311–320
[31] Romao J C, Ioannisian A and Valle J W F 1997 *Phys. Rev.* **D55** 427–430
[32] Bhattacharyya G and Pal P B 2010 *Phys.Rev.* **D82** 055013 (*Preprint* `1006.0631`)
[33] Hirsch M and Valle J W F 2004 *New J. Phys.* **6** 76
[34] Hirsch M *et al.* 2000 *Phys. Rev.* **D62** 113008 err-ibid. **D65**:119901,2002 (*Preprint* `hep-ph/0004115`)
[35] Diaz M A *et al.* 2003 *Phys. Rev.* **D68** 013009
[36] Bazzocchi F *et al.* 2012 (*Preprint* `1202.1529`)
[37] Hirsch M *et al.* 2002 *Phys. Rev.* **D66** 095006 (*Preprint* `hep-ph/0207334`)
[38] Datta A, Mukhopadhyaya B and Vissani F 2000 *Phys. Lett.* **B492** 324–330 (*Preprint* `hep-ph/9910296`)
[39] Porod W *et al.* 2001 *Phys. Rev.* **D63** 115004
[40] de Campos F *et al.* 2005 *Phys. Rev.* **D71** 075001 (*Preprint* `hep-ph/0501153`)
[41] de Campos F *et al.* 2008 *JHEP* **05** 048
[42] De Campos F *et al.* 2010 *Phys. Rev.* **D82** 075002
[43] de Campos F, Eboli O, Magro M, Porod W, Restrepo D *et al.* 2012 (*Preprint* `1206.3605`)
[44] Restrepo D, Taoso M, Valle J and Zapata O 2012 *Phys.Rev.* **D85** 023523 (*Preprint* `1109.0512`)
[45] Ishimori H *et al.* 2010 *Prog. Theor. Phys. Suppl.* **183** 1–163 (*Preprint* `1003.3552`)
[46] Morisi S and Valle J 2012 (*Preprint* `1206.6678`)
[47] Hirsch M *et al.* 2012 (*Preprint* `1201.5525`)
[48] Harrison P F, Perkins D H and Scott W G 2002 *Phys. Lett.* **B530** 167
[49] Babu K S, Ma E and Valle J W F 2003 *Phys. Lett.* **B552** 207–213 (*Preprint* `hep-ph/0206292`)
[50] Altarelli G and Feruglio F 2005 *Nucl. Phys.* **B720** 64–88 (*Preprint* `hep-ph/0504165`)
[51] Altarelli G and Feruglio F 2006 *Nucl. Phys.* **B741** 215–235 (*Preprint* `hep-ph/0512103`)
[52] Babu K and He X G 2005 (*Preprint* `hep-ph/0507217`)
[53] Grimus W and Lavoura L 2009 *JHEP* **0904** 013 (*Preprint* `0811.4766`)
[54] An F *et al.* (DAYA-BAY Collaboration) 2012 *Phys.Rev.Lett.* **108** 171803 (*Preprint* `1203.1669`)
[55] Ahn J *et al.* (RENO collaboration) 2012 (*Preprint* `1204.0626`)
[56] Antusch S, King S F, Luhn C and Spinrath M 2012 *Nucl.Phys.* **B856** 328–341 (*Preprint* `1108.4278`)
[57] Ishimori H and Ma E 2012 (*Preprint* `1205.0075`)
[58] Lin Y 2010 *Nucl.Phys.* **B824** 95–110 (*Preprint* `0905.3534`)
[59] Morisi S, Patel K M and Peinado E 2011 *Phys.Rev.* **D84** 053002 (*Preprint* `1107.0696`)
[60] Araki T 2011 *Phys.Rev.* **D84** 037301 (*Preprint* `1106.5211`)
[61] Toorop R d A, Feruglio F and Hagedorn C 2011 *Phys.Lett.* **B703** 447–451 (*Preprint* `1107.3486`)
[62] Bazzocchi F 2011 (*Preprint* `arXiv:1108.2497`)
[63] King S 2012 (*Preprint* `arXiv:1205.0506`)
[64] Altarelli G, Feruglio F and Merlo L 2009 *JHEP* **05** 020 (*Preprint* `0903.1940`)
[65] Boucenna S *et al.* 2012 *Phys.Rev.* **D86** 051301 (*Preprint* `1206.2555`)
[66] Ding G J, Everett L L and Stuart A J 2011 (*Preprint* `1110.1688`)
[67] Dorame L *et al.* 2012 *Nucl.Phys.* **B861** 259–270 (*Preprint* `1111.5614`)
[68] Dorame L *et al.* 2012 *Phys.Rev.* **D86** 056001 (*Preprint* `1203.0155`)
[69] Rodejohann W 2011 *Int.J.Mod.Phys.* **E20** 1833–1930 (*Preprint* `1106.1334`)
[70] Drexlin G (KATRIN collaboration) 2005 *Nucl. Phys. Proc. Suppl.* **145** 263–267
[71] Lesgourgues J and Pastor S 2006 *Phys. Rep.* **429** 307–379 (*Preprint* `astro-ph/0603494`)
[72] Hannestad S 2006 *Ann. Rev. Nucl. Part. Sci.* **56** 137–161 (*Preprint* `hep-ph/0602058`)
[73] Abe K *et al.* (T2K Collaboration) 2011 *Phys.Rev.Lett.* **107** 041801 (*Preprint* `1106.2822`)
[74] Abe Y *et al.* (DOUBLE-CHOOZ Collaboration) 2012 *Phys.Rev.Lett.* **108** 131801 (*Preprint* `1112.6353`)
[75] Wolfenstein L 1981 *Phys. Lett.* **B107** 77
[76] Schechter J and Valle J W F 1981 *Phys. Rev.* **D24** 1883 err. D25, 283 (1982)
[77] Valle J W F 1983 *Phys.Rev.* **D27** 1672–1674
[78] Schechter J and Valle J W F 1982 *Phys. Rev.* **D25** 2951
[79] Duerr M, Lindner M and Merle A 2011 *JHEP* **1106** 091 (*Preprint* `1105.0901`)
[80] Wolfenstein L 1981 *Nucl. Phys.* **B186** 147

[81] Pal P B and Wolfenstein L 1982 *Phys. Rev.* **D25** 766
[82] de Gouvea A, Kayser B and Mohapatra R N 2003 *Phys.Rev.* **D67** 053004 (*Preprint* hep-ph/0211394)
[83] Bernabeu J *et al.* 1987 *Phys. Lett.* **B187** 303
[84] Branco G C, Rebelo M N and Valle J W F 1989 *Phys. Lett.* **B225** 385
[85] Rius N and Valle J W F 1990 *Phys. Lett.* **B246** 249–255
[86] Gonzalez-Garcia M C and Valle J W F 1992 *Mod. Phys. Lett.* **A7** 477–488
[87] Hall L J, Kostelecky V A and Raby S 1986 *Nucl. Phys.* **B267** 415
[88] Borzumati F and Masiero A 1986 *Phys. Rev. Lett.* **57** 961
[89] Kurup A (COMET Collaboration) 2011 *Nucl.Phys.Proc.Suppl.* **218** 38–43
[90] Kutschke R K 2011 (*Preprint* 1112.0242)
[91] Das S *et al.* 2012 *Phys.Rev.* **D86** 055006 (*Preprint* 1206.0256)
[92] Aguilar-Saavedra J *et al.* 2012 *Phys.Rev.* **D85** 091301 (*Preprint* 1203.5998)
[93] Ilakovac A and Pilaftsis A 1995 *Nucl. Phys.* **B437** 491 (*Preprint* hep-ph/9403398)
[94] He B, Cheng T and Li L F 2003 *Phys.Lett.* **B553** 277–283 (*Preprint* hep-ph/0209175)
[95] Hirsch M, Morisi S and Valle J W F 2009 *Phys. Lett.* **B679** 454 (*Preprint* 0905.3056)
[96] Deppisch F, Kosmas T S and Valle J W F 2006 *Nucl. Phys.* **B752** 80–92 (*Preprint* hep-ph/0512360)
[97] Deppisch F and Valle J W F 2005 *Phys. Rev.* **D72** 036001 (*Preprint* hep-ph/0406040)
[98] Hirsch M, Staub F and Vicente A 2012 *Phys.Rev.* **D85** 113013 (*Preprint* 1202.1825)
[99] Dittmar M *et al.* 1990 *Nucl. Phys.* **B332** 1
[100] Abreu P *et al.* (DELPHI collaboration) 1997 *Z. Phys.* **C74** 57–71
[101] Antusch S, Arganda E, Herrero M J and Teixeira A M 2006 *JHEP* **11** 090 (*Preprint* hep-ph/0607263)
[102] Calibbi L, Faccia A, Masiero A and Vempati S 2006 *Phys.Rev.* **D74** 116002 (*Preprint* hep-ph/0605139)
[103] Joaquim F R and Rossi A 2006 *Phys. Rev. Lett.* **97** 181801 (*Preprint* hep-ph/0604083)
[104] Hirsch M *et al.* 2008 *Phys. Rev.* **D78** 013006 (*Preprint* 0804.4072)
[105] Esteves J N *et al.* 2009 *JHEP* **05** 003
[106] Hirsch M *et al.* 2004 *Phys. Rev.* **D69** 093006 (*Preprint* hep-ph/0312265)
[107] Gelmini G, Schramm D N and Valle J W F 1984 *Phys. Lett.* **B146** 311
[108] Berezinsky V and Valle J W F 1993 *Phys. Lett.* **B318** 360–366 (*Preprint* hep-ph/9309214)
[109] Aad G *et al.* (Atlas Collaboration) 2011 *Phys.Lett.* **B701** 186–203 (*Preprint* 1102.5290)
[110] Komatsu E *et al.* (WMAP) 2009 *Astrophys. J. Suppl.* **180** 330–376 (*Preprint* 0803.0547)
[111] Schwetz T, Tortola M and Valle J W F 2011 *New J. Phys.* **13** 063004
[112] Schwetz T, Tortola M and Valle J 2011 *New J.Phys.* **13** 109401 (*Preprint* 1108.1376)
[113] Hirsch M, Morisi S, Peinado E and Valle J 2010 *Phys.Rev.* **D82** 116003 (*Preprint* 1007.0871)
[114] Boucenna M *et al.* 2011 *JHEP* **1105** 037 (*Preprint* 1101.2874)
[115] Fornengo N, Scopel S and Bottino A 2011 *Phys.Rev.* **D83** 015001 (*Preprint* 1011.4743)
[116] Aalseth C, Barbeau P, Colaresi J, Collar J, Diaz Leon J *et al.* 2011 *Phys.Rev.Lett.* **107** 141301
[117] Ahmed Z *et al.* (CDMS-II Collaboration) 2010 *Science* **327** 1619–1621 (*Preprint* 0912.3592)
[118] Ahmed Z *et al.* (CDMS-II Collaboration) 2011 *Phys.Rev.Lett.* **106** 131302 (*Preprint* 1011.2482)
[119] Aprile E *et al.* (XENON100 Collaboration) 2011 *Phys.Rev.Lett.* **107** 131302 (*Preprint* 1104.2549)
[120] Ackermann M *et al.* (Fermi-LAT collaboration) 2011 *Phys.Rev.Lett.* **107** 241302 (*Preprint* 1108.3546)
[121] Geringer-Sameth A and Koushiappas S M 2011 *Phys.Rev.Lett.* **107** 241303 (*Preprint* 1108.2914)
[122] Coleman S R 1988 *Nucl. Phys.* **B310** 643
[123] Schechter J and Valle J W F 1982 *Phys. Rev.* **D25** 774
[124] Lattanzi M and Valle J W F 2007 *Phys. Rev. Lett.* **99** 121301
[125] Bazzocchi F *et al.* 2008 *JCAP* **0808** 013 (*Preprint* 0805.2372)
[126] Esteves J, Joaquim F, Joshipura A, Romao J, Tortola M *et al.* 2010 *Phys.Rev.* **D82** 073008
[127] den Herder J W *et al.* 2009 (*Preprint* 0906.1788)

Some aspects of Physics beyond the Standard Model at LHC

J Alberto Casas
Instituto de Fisica Teorica, IFT-UAM/CSIC
E-mail: alberto.casas@uam.es

Abstract. The LHC is constraining BSM physics at an impressive efficience, but no sign of new physics has been found yet. SUSY (and other BSM scenarios) are starting to be in trouble, though there is still room for new physics able to solve the Hierarchy Problem. This situation poses new challenges to optimize the LHC discovery potential using smarter strategies for analysis. In this sense, direct and indirect searches of new physics can play complementary roles.

1. Introduction

The two main purposes of the LHC are to probe the Higgs mechanism and to look for physics beyond the Standard Model (BSM). In both goals the LHC performance has been impressively good, although of course the outputs are so far very different.

Concerning the Higgs mechanism, in only one year the LHC has been able, first, to exclude most of the mass range of the Higgs boson; second, find possible signals for a Higgs boson in the 120 GeV - 130 GeV region (this was the situation at the time of this PASCOS conference in June-2012); and finally give $> 5\ \sigma$ evidence of a Higgs-like particle at $125 - 126$ GeV [1, 2].

On the other hand, the job of the LHC probing paradigmatic BSM scenarios has been equally impressive. However nature has not been as friendly to us as in the Higgs case, and all we have at the moment are strong exclusion limits for supersymmetric models and other scenarios of new physics (and maybe weak hints coming from the Higgs-boson properties). In spite of this, the LHC results have already profound implications in BSM physics.

Under the title "Physics beyond the Standard Model" there are many different scenarios. On the one hand, we have those related to a solution of the Hierarchy Problem (stability of the electroweak scale). The paradigmatic one is Supersymmetry (SUSY), especially the Minimal Supersymetric Standard Model (MSSM), which is the minimalistic supersymmetric extension of the SM without assuming new degrees of freedom, except those obliged by SUSY. But there are many MSSM scenarios depending on the assumptions about the unknown parameters (typically the soft terms). The most popular and extensively studied one is the Constrained MSSM (i.e. CMSSM, sometimes called mSUGRA) [3], in which one starts with universal soft terms at the gauge unification scale, $M_X \sim 10^{16}$ GeV. This scenario is in part motivated by the strong constraints in flavour-changing neutral-currents and the empirical and theoretical hints of a unification scale close to M_P. Moreover it takes place in s ome well-motivated theoretical scenarios, like gravity-mediated ones. But there are other MSSMs, like NUHM, gauge-mediated MSSM, string-inspired MSSMs, etc. All of them represent particular (and quite constrained)

Content from this work may be used under the terms of the Creative Commons Attribution 3.0 licence. Any further distribution of this work must maintain attribution to the author(s) and the title of the work, journal citation and DOI.
Published under licence by IOP Publishing Ltd

possibilities of the full MSSM. Besides, we have supersymmetric models beyond the MSSM, like the NMSSM (characterized by the presence of an extra singlet) and the BMSSM (characterized by SUSY breaking at low-scale). Apart from SUSY, there are other frameworks of new physics, like the different versions of Extra-Dimensions scenarios: ADD, RS, ...; Composite and Little Higgs models; etc.

On the other hand, we have additional scenarios or chances for new physics, motivated by other phenomenological and theoretical facts or possibilities: dark matter candidates, flavour violation, extra massive-gauge-bosons, 4th generation, etc.

Roughly speaking there are two main strategies to constrain new physics: direct searches, i.e. production and detection of BSM particles; and indirect searches, i.e. search of fingerprints of the new physics in the effective (SM-like) theory. In this note we will concentrate on direct searches, but we will address indirect ones in sect. 5. We will focus mostly (but not only) on SUSY scenarios, although many strategies and results reviewed here are essentially valid for other frameworks of new physics.

SUSY has been the paradigmatic scenario of physics BSM, and for good theoretical and phenomenological reasons. SUSY is a beautiful symmetry, strongly suggested by string theories. It provides an elegant solution to the Hierarchy Problem since the dangerous quadratic divergences contributing to the Higgs mass (and thus to the electroweak scale) cancel. In addition, SUSY presents nice features which was not designed for. Namely, the supersymmetric running of the gauge couplings makes them unify at a high scale, M_X, with great precision. Also, the electroweak (EW) breaking occurs in a quite natural way since the square-mass of one of the two Higgs doublets is driven towards negative values along the renormalization group (RG) running from the high scale. This is called radiative EW breaking. Let us also mention that SUSY models offer a natural WIMP candidate for dark matter (usually the lightest neutralino). SUSY is indeed beautiful... but maybe false! Now it is the time for SUSY to be probed by the experiment (LHC).

2. The search of SUSY

2.1. Typical SUSY signals

The highest cross sections of SUSY production are normally gluino and/or squark pair-production. By squark we mean here a squark state of the first or second generation since the production of stops and sbottoms is very suppressed. In a typical SUSY process the gluinos ($\tilde{g}$) and squarks ($\tilde{q}$) decay along cascades with diverse topology. In models with some kind of R-parity (desirable though not mandatory to avoid proton decay and other baryon/lepton number violating processes), each cascade always produces one lightest supersymmetric particle (LSP), typically a neutralino χ_1^0, among the final states. In addition, one or more jets, with or without leptons, are created in each cascade. Therefore the most direct search of SUSY is to look for events with

- jets with high p_T
- $E_T^{\rm miss}$
- 0-N leptons

Normally, multijet events with large $E_T^{\rm miss}$ and 0-leptons are the most efficient channel, but channels with leptons may play an important (sometimes capital) role, especially for particularly elusive types of SUSY spectrum (see subsect. 3.2 below).

It is not straightforward to translate LHC data about these types of events into concrete limits on SUSY (MSSM) parameters. Let us recall that the MSSM has ~ 100 independent parameters, mainly soft terms related to the unknown mechanism of SUSY breaking and its transmission to the observable sector: $\{m_{ij}^2, M_a, A_{ij}, B, \mu\}$. Here m, M and A are scalar masses, gaugino masses and trilinear scalar couplings; i, j and a are family and gauge group indices respectively; B is the bilinear scalar coupling and μ is the usual Higgs mass term in the superpotential. Requiring no flavour or CP violation in the first and second generations reduces the number of parameters to ~ 20 (plus those already present in the SM), still a huge number. It is certainly cumbersome to translate the LHC data into constraints on such complex parameter-space. A usual strategy is to present the LHC data as constraints in a simplified version of the MSSM, typically the CMSSM. Then the previous parameters are reduced to $\{m, M, A, B, \mu\}$, i.e. the universal scalar mass, gaugino mass and trilinear scalar coupling; plus the B and μ parameters. All quantities are to be understood at the high scale M_X. Using the EW breaking conditions, coming from the minimization of the Higgs potential, one can eliminate μ (except for the sign) and trade B by $\tan\beta = \langle H_u\rangle/\langle H_d\rangle$ (the ratio of the expectation values of the two Higgs doublets). So the usual set of CMSSM parameters is

$$\{m, M, A, B, \tan\beta, \text{sign}\mu\}$$

Due to the remarkable growth of the gluino and squark masses along the RG running, the typical CMSSM spectrum is

$$\begin{aligned} M_{\tilde{g}} &\sim m_{\tilde{q}} > m_{\tilde{l}} \\ M_{\tilde{g}} &> M_{\chi^\pm} \gtrsim M_{\chi_1^0} \\ \chi_1^0 &\equiv \text{LSP} \end{aligned}$$

where $\chi^\pm$ and χ_1^0 denote the lightest chargino and neutralino states (the squark can be much heavier than the gluino if $m \gg M$). The values of A and $\tan\beta$ are not very important for the multijet signal, as they play almost no role in the gluino and squark production. Because of that, the impact of the LHC results on the CMSSM are usually presented as exclusion limits in the $m - M$ plane (or in the $m_{\tilde{q}} - M_{\tilde{g}}$ plane) for A and $\tan\beta$ fixed (often $A = 0$, $\tan\beta = 10$). Fig. 1 shows the ATLAS limits with an integrated luminosity of ~ 5.8 fb^{-1} at $\sqrt{s} = 8$ TeV [4]. The results for CMS are similar.

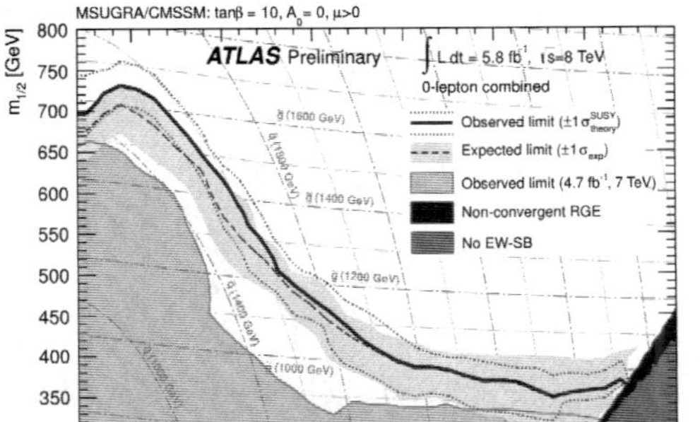

Figure 1. ATLAS constraints on the $m - M$ plane of the CMSSM.

2.2. *The MSSM in trouble*

Roughly speaking, the above-mentioned ATLAS and CMS results imply that for $M_{\tilde{g}} \sim m_{\tilde{q}}$, then $M_{\tilde{g}}$, $m_{\tilde{q}} \gtrsim 1100$ GeV (recall that $M_{\tilde{g}} \sim 2.5M$) . And this means that the CMSSM is *in trouble.* The reason is that with such large masses the EW breaking is rather fine-tuned. We cannot just "forget" about the fine-tuning problem, since the main reason to consider low-scale SUSY was to avoid the Hierarchy Problem, i.e. the big fine-tuning problem of the EW breaking in the SM.

Let us briefly review how the fine-tuning appears. The minimization of the tree-level Higgs potential leads to expectation values for the Higgses, which must be of the right size to reproduce the $Z-$ and $W-$boson masses. More precisely, working with the Higgs potential at tree-level, the $Z-$mass can be expressed as

$$M_Z^2 = \frac{m_{H_d}^2 - m_{H_u}^2 \tan^2\beta}{\tan^2\beta - 1} - 2\mu^2 \tag{1}$$

where $m_{H_u}^2, m_{H_d}^2$ are the mass parameters of the two Higgses. Clearly, the correct value of M_Z requires an unnatural cancellation of the various terms unless $m_{H_u}, m_{H_d} \lesssim \mathcal{O}(1\text{TeV})$. One cannot just take $m_{H_u}^2, m_{H_d}^2$ small while $M_{\tilde{g}}$, $m_{\tilde{q}}$, etc. are large, since the Higgs mass parameters receive large radiative contributions from the latter along the RG running from M_X to M_{EW} [3]. In other words, large $M_{\tilde{g}}$, $m_{\tilde{q}}$,... imply large m_{H_u}, m_{H_d} (in absolute value). Thus the EW breaking is fine-tuned unless the typical soft masses are $\mathcal{O}(1\text{TeV})$ or less.

It has been often observed that, since the RG running of $m_{H_u}^2$ (the most relevant term in eq.(1) for non-small $\tan\beta$) is mostly driven by stop masses, which in turn are mostly driven by the gluino mass, then if $M_{\tilde{g}}$, $m_{\tilde{t}}$ are small (less than 1 TeV) and the first and second generations are heavy, the EW breaking can still be natural while SUSY remains undetected at LHC. However, this scenario, sometimes called "natural-SUSY" [5], is also in trouble. First, the direct LHC constraints, though still not very constraining for the stops (more about this later), are already severe for the gluino, implying $M_{\tilde{g}} \gtrsim 1$ TeV. Second, the observation of the Higgs mass around $125 - 126$ GeV makes natural-SUSY very hard to implement and, more generally, makes the fine-tuning problem more severe, as we discuss next.

2.3. The impact of the Higgs discovery

In the effective theory coming from the MSSM, the quartic coupling of the Higgs field has (at tree-level) a purely gauge origin, namely $\lambda_{\text{tree}} = \frac{1}{4}(g^2 + g'^2)\cos^2 2\beta$, where g, g' are the $SU(2) \times U(1)$ gauge couplings. This relation means, in particular, that at tree-level the mass of the Higgs in the MSSM is bounded by the mass of the Z-boson (91.1 GeV). As it is well-known, radiative corrections increase m_h, which can then get compatible with the experimental observation, at the expense of requiring a relatively heavy spectrum ($\gtrsim 1$ TeV) of superpartners. An approximate analytic formula for m_h [6, 7, 8, 9] reads

$$m_h^2 \simeq M_Z^2 \cos^2 2\beta + \frac{3}{4\pi^2}\frac{m_t^4}{v^2}\left[\log\frac{M_{\text{SUSY}}^2}{m_t^2} + \frac{X_t^2}{M_{\text{SUSY}}^2}\left(1 - \frac{X_t^2}{12M_{\text{SUSY}}^2}\right)\right] + \cdots \tag{2}$$

where m_t is the top running mass, M_{SUSY} represents a certain average of the stop masses and $X_t = A_t - \mu\cot\beta$ [7, 8, 9]. The first term of (2) is the tree-level Higgs-mass and the second are the dominant radiative and threshold contributions. Note that the radiative corrections grow logarithmically with the stop masses while the threshold correction has a maximum for $X_t = \pm\sqrt{6}M_{\text{SUSY}}$.

The important point is that, in order to achieve $m_h = 125 - 126$ GeV, one typically needs (besides non-small $\tan\beta$) stop masses $\gtrsim 3$ TeV, unless X_t happens to be close to the above-mentioned maximizing value. Then the stop masses could be around 1 TeV, but not less than that. This is also well illustrated in the scatter-plot shown in Fig.2 (from ref.[10]). Recalling now the discussion of the previous subsection, the obvious consequence is that the EW breaking in *any* MSSM is fine-tuned unless X_t is close to its maximazing value. The latter possibility represents itself another kind of fine-tuning unless one has some theoretical reason to expect

X_t in the favorable range. Incidentally, note that for moderately large $\tan\beta$ the value of X_t is essentially given by A_t, so one would essentially need a value of A_t in the favorable range. In any case, this amounts to go beyond the MSSM, e.g. invoking a particular SUSY breaking and mediati on mechanism that could lead to this result for A_t; see ref.[11] for a recent paper arguing that this can actually occur in a class of superstring scenarios.

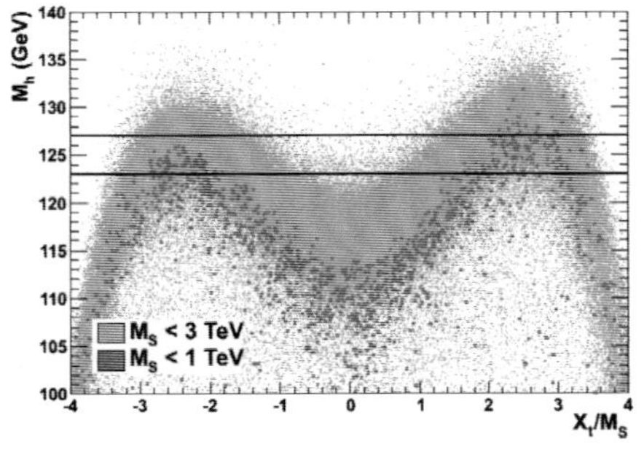

Figure 2. Scatter plt in the $X_t - m_h$ plane of the CMSSM.

We can go a bit further and evaluate the amount of fine-tuning required for a correct EW breaking when the stop masses are large. A quick and useful estimate of the fine-tuning can be found in ref.[12], namely $c \simeq 2\frac{\mu^2}{m_h^2}$. Then, for $m_{\tilde{t}} \sim 3$ TeV one gets $c \sim 100$, i.e. SUSY is fine-tuned at $\sim 1\%$ (other alternative estimators of fine-tuning give a similar value). Does this mean that the CMSSM, or even the general MSSM, is dead?

Before addressing this question, let us mention another interesting impact of the Higgs discovery on SUSY theories. The previous discussion reflects the fact that, as it is clear from eq.(2), a measurement of m_h implies a lower bound on $M_{\rm SUSY}$. But it is also true that a measurement of m_h (or, more generally, an upper bound on m_h) implies an *upper* bound on $M_{\rm SUSY}$. Actually, eq.(2) is only valid for $M_{\rm SUSY} \lesssim 1$ TeV. When it is larger (as it happens for the task of evaluating the upper bound) it is necessary to run numerically the effective quartic coupling, λ, from the threshold scale, $M_{\rm SUSY}$, till the EW scale (a nearly optimal choice is $Q_{\rm EW} = m_t$ [7]). Then $m_h \simeq 2\lambda(Q_{\rm EW})v^2$. Besides, in order to to get the Higgs pole-mass one has to add (pretty small) radiative corrections. The previously-sketched calculation of m_h using the 2-loop SM RG equation of λ was performed in refs.[13], [14].

Fig.3 shows m_h as a function of $M_{\rm SUSY}$ for three representative values of $\tan\beta$, namely $\tan\beta = 1, 3, 10$. Due to the parametric dependence on $\cos^2 2\beta$ shown in eq.(2), the results remain almost unchanged for larger values of $\tan\beta$. From the plots it becomes clear that one cannot reproduce an arbitrary large value of m_h by just increasing $M_{\rm SUSY}$. Actually, there is an absolute upper bound of ~ 145 GeV, which becomes more stringent as $\tan\beta$ decreases. It should be kept in mind that $M_{\rm SUSY}$ essentially stands for "stop masses". Indeed, in the usual MSSM scenarios the masses of all supersymmetric particles are of the same order, say within a factor of 10 or less. (A notable exception are split-SUSY models, to be discussed shortly.)

The width of the bands comes from the various sources of uncertainty, including experimental errors and the uncertainty about the value of X_t, which has been left varying within its range, $0 \leq X_t \leq 6$. Clearly a Higgs around 125-126 GeV implies that SUSY (if true) should take place at a scale $\lesssim 10^{10}$ GeV. Of course, this limit is out of the scope of any realistic experiment. On the other hand, one can take the attitude of *only* considering low $M_{\rm SUSY} \lesssim \mathcal{O}(\rm TeV)$ as reasonable, in order to avoid the above-mentioned fine-tuning to get the correct electroweak scale. Still, these results are important from a more philosophical point of view. It has been suggested that in a landscape scenario such fine-tuning can be largely compensated by the overabundance of vacua with SUSY broken at a high scale, in which the anthropic principle would operate, see e.g. ref.

[15]. But, as we have just seen, this kind of framework does not wo rk in the simplest scenario where the typical scale of SUSY breaking is around M_P.

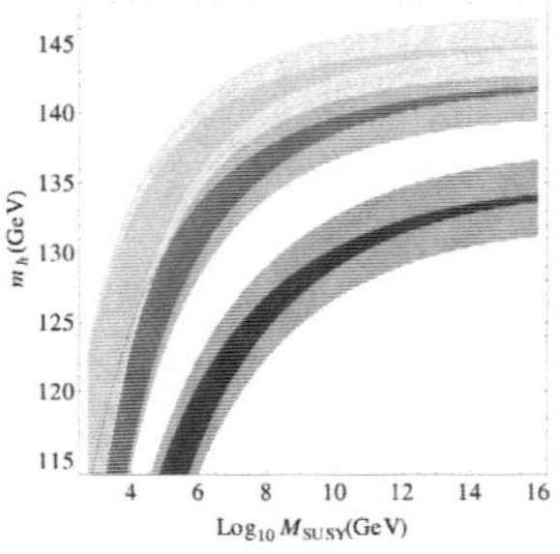

Figure 3. m_h as a function of $\log_{10} M_{\rm SUSY}$ in the MSSM.

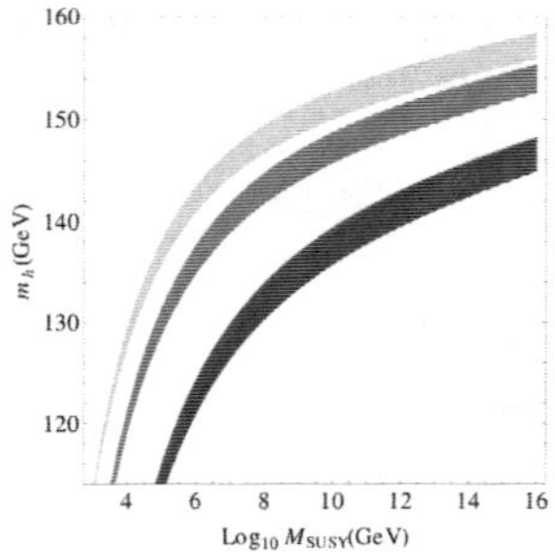

Figure 4. The same as Fig.3 for split-SUSY

The split-SUSY framework [16] is in fact a popular variant of the above-mentioned landscape scenario. In split-SUSY the masses of scalar superpartners are very high but the gauginos and higgsinos are relatively light (allowing for gauge unification and dark matter candidates). Concerning the calculation of m_h, the main difference with respect to the MSSM case is that in split-SUSY the gluinos remain active and contribute significantly to the RG equations between the upper and the lower SUSY-thresholds. The results are shown in Fig.4, which is analogous to the previous Fig.3 for the ordinary MSSM. The limits are even more stringent now. Namely, the scale of the heavy SUSY states should be at most at 10^7 GeV. I.e. the other side of SUSY cannot remotely be at $\sim M_P$.

3. Prospects for SUSY

3.1. Global fits and Bayesian forecasts

One can be more precise about the situation and prospects of the CMSSM (or any other MSSM or BSM model) by performing global fits. I.e., one can use all the available experimental information (dominated by LHC data) to show favoured/disfavoured regions in the CMSSM (or the BSM model under consideration) parameter-space. There are two basic approaches to do this task: the frequentist one and the Bayesian one.

In the frequentist approach one scans the parameter-space of the CMSSM evaluating the likelihood (based on the usual χ^2). This leads to zones of estimated probability (inside contours of constant χ^2) around the best fit points in the parameter space. A recent analysis of this kind (which takes the Higgs signal into account) has been performed in ref.[17]. The results are illustrated in Fig.5, which shows 68% and 95% c.l. regions in the $m - M$ plane. Clearly the favoured region occurs at quite high values of the soft terms, $m \sim M \sim 1500$ GeV, essentially beyond the actual LHC reach (recall e.g. that the gluino mass is $\sim 2.5M$); though there survive regions at lower scale. On the other hand, in the Bayesian approach, for a model defined by some parameters, θ_i and given some experimental data, the goal is to evaluate the probability density (or "posterior") in the parameter space, $p(\theta_i|\text{data})$. This is given by the Bayesian relation

$$p(\theta_i|\text{data}) \;=\; \frac{p(\text{data}|\theta_i)\; p(\theta_i)}{p(\text{data})}$$

where $p(\text{data}|\theta_i)$ is the likelihood, i.e. the probability density of measuring the given data for the chosen point in the parameter space (this is exactly the quantity used in frequentist approaches);

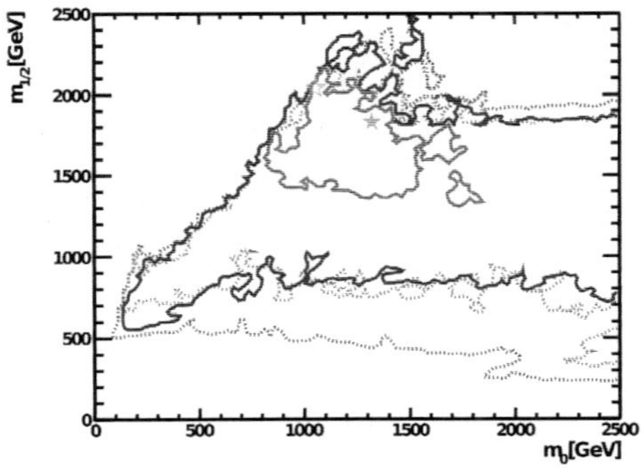

Figure 5. Likelihood distribution in the $m-M$ plane of the CMSSM.

$p(\theta_i)$ is the prior, i.e. the "theoretical" probability density that we assign a priori to the point in the parameter space; and finally, $p(\text{data})$ is a normalization factor which plays no role unless one wishes to compare different classes of models.

Some Bayesian analyses of the CMSSM previous to the Higgs discovery can be found in refs.[18, 19]. The impact of the Higgs discovery has been studied in refs.[20, 21]. Here we present the results of the improved analyses of refs.[19] and [21] [1].

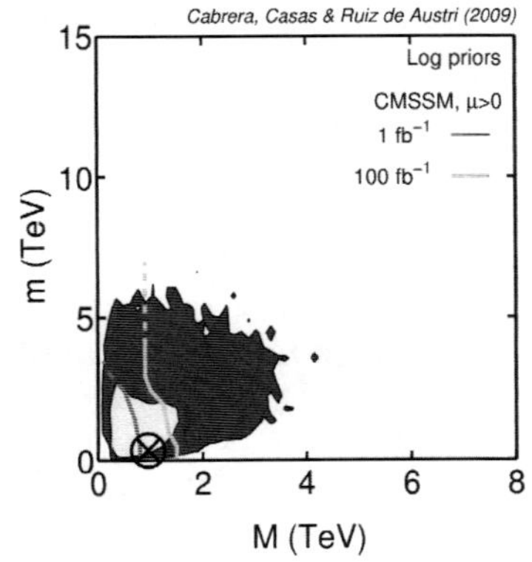

Figure 6. Probability density before Higgs discovery.

Fig.6 and Fig.7 show the posterior probability in the $M-m$ plane (the other SUSY and SM parameters have been marginalized) before and after the inclusion of the Higgs discovery respectively. As expected the bulk of the probability has been pushed to higher scales, making problematic (though by no means impossible) the discovery of the CMSSM in the LHC. If one further includes dark matter constraints, then the co-annihilation region is essentially excluded (by the Higgs mass) and one is essentially left with the focus point region (part of it is also excluded by XENON data). This pushes further the probability to high-energy scales, as shown in Fig.8. The pesimistic conclusion is that not only the CMSSM is fine-tuned at $\sim 1\%$, but that, even if the CMSSM is true, its chances to be discovered at the LHC have decreased dramatically. Of course, there are some caveats, e.g. supersymmetric dark matter could be inestable due to a tiny amount of R-parity violation (without modi fying the rest of the analysis). In that case the

[1] These analyses incorporates automatically (without any ad hoc consideration) a penalization of the fine-tuned regions. This is a logical result in a careful Bayesian study since these regions have little statistical weight. A welcome consequence of this is that the results are quite independent of the ranges chosen for the parameters and even of the choice of prior (e.g. flat or logarithmic).

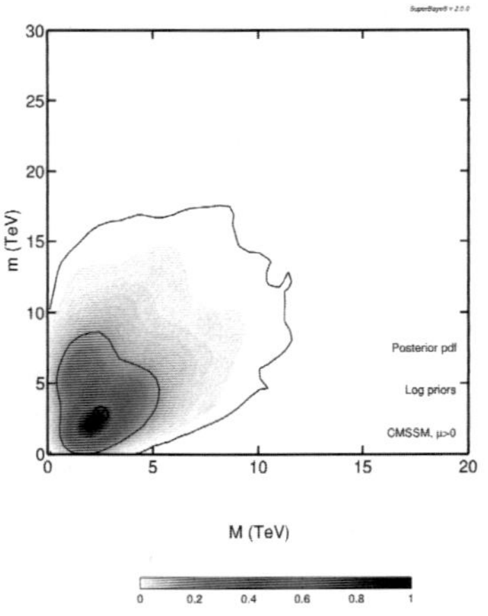

Figure 7. The same as Fig.6 after Higgs discovery.

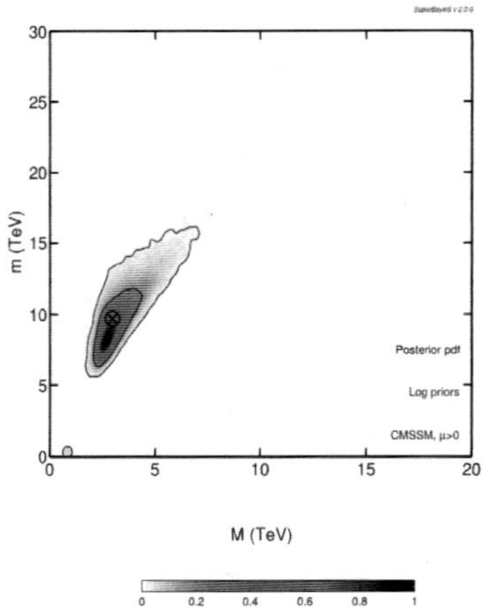

Figure 8. The same as Fig.7 with dark matter constraints.

actual dark matter would have a different origin and dark matter constraints would be irrelevant for SUSY, so Fig.8 would not apply.

3.2. And now what? (some present directions)
Let us now re-take the question we made a few pages ago. Do these results mean that SUSY is dead? We can add some complementary questions: To which extent the problems of the CMSSM remain in general MSSMs? Are there natural way-outs to this situation (maybe beyond the MSSM)?

To address these questions let us recall first the original motivations for the CMSSM. These were 1) Minimal Flavour and CP violation, 2) Simplicity, 3) The fact that it arises in some theoretically motivated scenarios (like minimal SUGRA or Dilaton-dominated SUSY breaking). From these motivations only the first one is robust, but the experimental results do not require completely universal soft terms. E.g. the third generation of squarks and sleptons could have very different masses. The degeneracy of gaugino masses at M_X is not experimentally justified either. Therefore, going beyond the CMSSM is very plausible. Does this solve the problems of the CMSSM?

This question can be explored by promoting the CMSSM to more general MSSMs compatible with the flavour and CP constraints. In this spirit, a pretty general and popular model is the so-called "phenomenological MSSM" (or pMSSM). It is directly defined at low-energy as an MSSM with no additional CP phases (apart from the CKM ones), and flavour-diagonal sfermion mass-matrices and trilinear couplings. Furthermore, the first and second generation are degenerate (with negligible $A-$terms) and the lightest neutralino is the LSP. This includes the possibility of a lighter third generation, as invoked in "natural SUSY" models. It also allows certain types of supersymmetric spectrum that can evade the detection at LHC. In particular if the LSP is heavy or the masses of the supersymmetric states are not far from each other ("compressed spectrum" [22]), then the supersymmetric events have small p_Ts and are triggered out in the experimental analyses. Let us briefly comment o n all these possibiities.

First of all, as discussed in a previous section, the third generation cannot be too light, since it is needed to be rather heavy in order to reproduce (thanks to the radiative corrections) the experimental mass of the Higgs. This requirement is alleviated if A_t is close to its maximizing value, but still the required stop masses imply that the EW breaking is quite fine-tuned. The situation is better if one goes *beyond* the MSSM. Then the tree-level Higgs mass can be larger, so one does not need such important radiative contributions. In consequence the third generation is allowed to be lighter, thus improving the naturalness of the EW breaking. One possibility in this direction is the next-to minimal MSSM (i.e. NMSSM), which has an extra singlet in the spectrum coupled to H_u, H_d in the superpotential, which provides an extra source for the effective quartic coupling of the (scalar componentes of the) Higgses. Another possibility, keeping the degrees of freedom of the MSSM, is to assume that the scale of SUSY breaking is quite close to the size of the soft terms. This is called low-scale SUSY breaking or BMSSM ("B" stands for "beyond"). Let us recall here that the size of the soft terms typically goes like $M_{\rm SUSY} \sim F/\Lambda$, where F is the SUSY breaking scale, which corresponds to the dominant VEV among the auxiliary fields in the SUSY breaking sector (it can be an $F-$term or a $D-$term) and Λ is the messenger scale, associated to the interactions that transmit the breaking to the observable sector. If Λ is close to $M_{\rm SUSY}$, say $\lesssim \mathcal{O}(10)$ TeV, then beside the usual soft terms there appear extra dimension-4 operators (e.g. quartic couplings beteween scalars) in the Lagrangian, suppressed by powers of F/Λ^2. Even if they are not vey sizeable, they contribute to the tree-level Higgs mass, thus allowing smaller radiative correc tions to m_h and hence smaller stop masses. In addition, since the RG running is much shorter, the "contamination" of the Higgs-mass parameters by the other soft terms is less important. This scenario, however, requires the existence of new physics (beyond SUSY!) at relatively low scales, thus reducing the beauty of SUSY as a solution of the Hierarchy Problem.

Regarding the chance of a compressed SUSY spectrum or a heavy LSP, it is undoubtedly possible to arrange the MSSM spectrum in these ways in order to fool the LHC. But, in the absence of a solid theoretical reason for them, it sounds a rather artificial possibility.

In any case, all this represents new challenges for the data analysis:

- Test more general MSSMs (like the pMSSM)
- Test a light third generation of SUSY particles
- Test a compressed SUSY spectrum or a heavy LSP
- Detect heavy SUSY

Regarding the search of a light third generation, the main problem is of course that the production of stops or sbottoms is very suppressed with respect to the first two generations, and there is nothing one can do about that. The present strategies are to look for direct stop and sbottom pair production or, if kinematically allowed, by their production through gluino decays. Roughly speaking, the present experimental limits imply that stop and sbottom masses can be anything $\gtrsim$ 450 GeV, so there is still a lot of room for a relatively light third generation. These limits will improve much with the future up-gradings in energy of the LHC. They can also improve with the developing of (even) smarter strategies for the data analysis, e.g. exploring new and clever kinematical variables and the like.

Concerning the probe of a compressed SUSY spectrum or a heavy LSP, the study of events with $E_T^{\rm miss}$ + jets + multileptons may play a crucial role in these scenarios since it is much easier to detect and characterise low-p_T leptons than jets. With respect to heavy SUSY, the only way of improving (beside of course the progress in the luminosity and the energy of the LHC) is to design more efficient strategies of discovery. This includes to look for alternative channels, like chargino-neutralino production, that could be relevant (or even dominant) if the squarks and gluinos are too heavy. Let us recall here that the chargino and neutralino states arise from dimension-two and dimension-four mass matrices, which contain other parameters apart from the gaugino masses (particularly the $\mu-$parameter). This makes natural that the lightest states could be much lighter than the typical scalar and gluino masses. As mentioned above, progress can also occur by improving the kinematical variables used in the analyses and also by performing more complete confrontations of theory vs. experiment (e.g. using the whole richness of the theoretical and experimental histograms).

Concerning the test of more general MSSMs (like the pMSSM), an strategy that is gaining relevance is the use of so-called "simplified models". A simplified model is defined by an effective Lagrangian describing the interactions of a small number of new particles. They can also be described by a small number of masses and cross-secctions. The latter parameters are directly related to collider-physics observables. The idea is to mimic the collider signatures of a particular physical scenario (e.g. the pMSSM) with a dominant simplified model (or a reduced set of them) in each region of the parameter space. This makes more efficient the exploration of such complex models. E.g. squark or gluino decays $\tilde{q} \to q\chi_1^0$, $\tilde{g} \to q\bar{q}\chi_1^0$ are dominant if the other relevant super-particles are heavier. In a simplified model their masses can be just sent to infinity. Of course, additional complexity can be built-in. The strategy is also efficie nt for non-supersymmetric models of new physics.

4. The Hierarchy Problem under scrutiny

In the previous sections we have focused on SUSY scenarios, but of course there are alternatives to SUSY that may also solve or alleviate the Hierarchy Problem, such as Extra Dimensions or Composite and Little Higgs models. We have no room to review how the LHC has already constrained these scenarios. Roughly speaking, the situation is similar to that of SUSY, i.e. the scale of new physics has been pushed upwards, making them rather fine-tuned. Of course, using more or less ingenuity and cookery this pushing can be more or less dramatic, but it is a rather general fact. The reason is that the Hierarchy Problem requires, on very general grounds, the existence of new physics (with sizeable couplings to the SM states) at energies within the reach of the LHC. The argument is simple and well-known:

In the SM (treated as an effective theory valid below $\Lambda_{\rm SM}$) the mass parameter m^2 in the Higgs potential

$$V = \frac{1}{2}m^2h^2 + \frac{1}{2}\lambda h^4 \tag{3}$$

receives important quadratically-divergent contributions. At one-loop,

$$\delta_{\rm q}m^2 = \frac{3}{64\pi^2}\Lambda_{\rm SM}^2(3g^2 + g'^2 + 8\lambda - 8\lambda_t^2)\ , \tag{4}$$

where g, g', λ and λ_t are the $SU(2) \times U(1)_Y$ gauge couplings, the quartic Higgs coupling and the top Yukawa coupling, respectively. The requirement of no fine-tuning between the above contribution and the tree-level value of m^2 sets an upper bound on $\Lambda_{\rm SM}$. Using $m_h = 125$ GeV,

$$\left|\frac{\delta_{\rm q}m^2}{m^2}\right| \leq 10\ \Rightarrow\ \Lambda_{\rm SM} \lesssim 2\ {\rm TeV} \tag{5}$$

where we have implicitely used $v^2 = -m^2/\lambda$, $m_h^2 = 2\lambda v^2$ (with $v^{\rm exp} = 246$ GeV). Consequently, we should expect new physics at $\lesssim 2$ TeV able to cancel the quadratic contributions to the Higgs mass. Since the latter are dominated by the top, one would expect some new states that are somehow related to the top, namely interacting with the strong force and with large Yukawa couplings. This is exactly what happens for the stops in the SUSY scenario, but in general it will occur for other BSM candidates able to solve this Hierarchy Problem.

One of the lessons from the analysis of the CMSSM is that those new states could be quite light (~ 500 GeV) and still remain so far undetected at LHC. In the case of the CMSSM this possibility does not work very well because one needs heavy stops in order to reproduce the experimental Higgs mass. But this problem may be avoided in other scenarios of BSM physics (and in other SUSY scenarios, as discussed above). So it is still too soon to claim that the LHC is putting the Hierarchy Problem argument in trouble. Nevertheless, if with higher luminosity and after subsequent upgradings in energy, we continue with no observations of new physics, the Hierarchy Problem should be re-visited.

5. Indirect searches

Let us now briefly consider the other way to search for new physics, namely to look for fingerprints in the effective theory. In the past this approach was used to put very important bounds on BSM physics from (LEP) EW precision tests. The procedure is conceptually simple. One starts with a Lagrangian

$$\mathcal{L} \;=\; \mathcal{L}_{\rm SM} \;+\; \text{Higher} - \text{Dim Operators}$$

The second term of the r.h.s. carries the fingerprints of the BSM physics. In the LEP case, all data were consistent with the SM Lagrangian, so we were only capable to put exclusion-limits on the undetected new physics. But, had we been lucky, we could have discovered new physics at LEP. We can repeat now this scheme for the Higgs physics, which is starting to be probed at the LHC. In other words, we can use the information about Higgs-couplings from LHC data on Higgs production and decay, to constrain (or detect) BSM operators involving the Higgs. The approach is less spectacular than performing direct searches of new physics, but it may be more powerful (depending on the nature of the new physics) and it is more model-independent.

So one has to confront the observed rate of Higgs decays in different channels: $\gamma\gamma$, ZZ, WW, $b\bar{b}$, $\tau\bar{\tau}$,..., with the SM prediction. Actually, at the moment there are discrepancies with the SM predictions, in particular there too many $\gamma\gamma$ events (and also too few $b\bar{b}$ events); but of course the error-bars are still too large and the statistics is too low to claim any BSM signal. However, if these deviations persist as the statistics increases, they could be an herald of new physics.

Assuming that in the effective theory there is just one Higgs state (as in the SM) whose couplings keep minimal flavour violation, not inducing flavour-changing-neutral-currents, one can write the effective Lagrangian for the Higgs, which is quite complex, see ref.[24]. A usual simplifying assumption is to take all the fermionic couplings proportional to the Yukawa coupling matrices with a universal factor, c and neglect higher orders. Similarly the couplings to the gauge bosons are as in the SM times a factor a. In this way the BSM physics is simply parametrized by the $\{c, a\}$ parameters and the choice $c = a = 1$ corresponds to the SM.

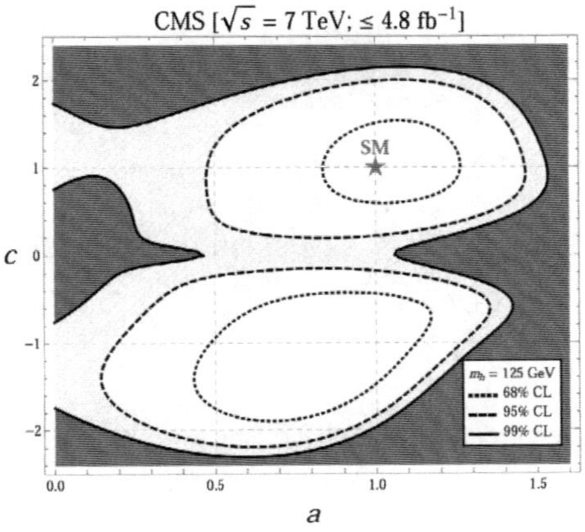

Figure 9. Favoured regions in the $a - c$ plane, see text.

Fig.9 (from ref.[24]) shows the favoured regions in the $a - c$ plane. There is an island around the SM prediction and a second, *more favoured* island around $c = -1$. The reason is that, due to the interference of diagrams with fermion and gauge-bosons loops, the branching-ratio to $\gamma\gamma$

goes like $BR(h \to \gamma\gamma) \sim |1.8c - 8.3a|^2$. Consequently an excess in $\gamma\gamma-$events can be reproduced with a negative c. Keeping $|c| \sim 1$, the rest of the predictions are essentially as in the SM. The increase of statistics in the next years will tell us if this possible hint is reinforced or disappears.

6. Conclusions

The LHC is constraining BSM physics at an impressive efficience, but no sign of new physics has been found yet. SUSY (and other BSM scenarios) are starting to be in trouble, though there is still room for new physics able to solve the Hierarchy Problem. This situation poses new challenges to optimize the LHC discovery potential using smarter strategies for analysis (in addition to the increasing LHC luminosity and energy). In this sense, direct and indirect searches of new physics can play complementary roles.

Acknowledgements

I am are extremely grateful to my collaborators María E. Cabrera, Roberto Ruiz de Austri and Antonio Delgado. I am also extremely grateful to the organizers of PASCOS 2012 for the warm and effective atmosphere, in especial to Myriam Mondragón. This work has been partially supported by the MICINN, Spain, under contract FPA2010-17747; Consolider-Ingenio PAU CSD2007-00060, CPAN CSD2007-00042. We thank as well the Comunidad de Madrid through Proyecto HEPHACOS S2009/ESP-1473 and the European Commission under contract PITN-GA-2009-237920.

References

[1] G. Aad *et al.* [ATLAS Collaboration], Phys. Lett. B **716** (2012) 1 [arXiv:1207.7214 [hep-ex]].
[2] S. Chatrchyan *et al.* [CMS Collaboration], Phys. Lett. B **716** (2012) 30 [arXiv:1207.7235 [hep-ex]].
[3] See e.g. S. P. Martin, In *Kane, G.L. (ed.): Perspectives on supersymmetry II* 1-153 [hep-ph/9709356].
[4] ATLAS note, ATLAS-CONF-2012-103 (13 Aug 2012), http://cdsweb.cern.ch/record/1472710
[5] See e.g. C. Brust, A. Katz, S. Lawrence and R. Sundrum, JHEP **1203** (2012) 103 [arXiv:1110.6670 [hep-ph]].
[6] J. R. Ellis, G. Ridolfi and F. Zwirner, *Phys. Lett.* **B257** (1991) 83–91; H. E. Haber and R. Hempfling, *Phys. Rev. Lett.* **66** (1991) 1815–1818;
[7] J. A. Casas, J. R. Espinosa, M. Quiros and A. Riotto, *Nucl. Phys.* **B436** (1995) 3–29 [arXiv:hep-ph/9407389].
[8] M. S. Carena, J. R. Espinosa, M. Quiros and C. E. M. Wagner, Phys. Lett. B **355** (1995) 209 [arXiv:hep-ph/9504316].
[9] H. E. Haber, R. Hempfling and A. H. Hoang, Z. Phys. C **75** (1997) 539 [arXiv:hep-ph/9609331].
[10] A. Arbey, M. Battaglia and F. Mahmoudi, Eur. Phys. J. C **72** (2012) 1847 [arXiv:1110.3726 [hep-ph]].
[11] L. Aparicio, D. G. Cerdeno and L. E. Ibanez, JHEP **1204** (2012) 126 [arXiv:1202.0822 [hep-ph]].
[12] J. A. Casas, J. R. Espinosa and I. Hidalgo, JHEP **0401** (2004) 008 [hep-ph/0310137].
[13] M. E. Cabrera, J. A. Casas and A. Delgado, Phys. Rev. Lett. **108** (2012) 021802 [arXiv:1108.3867 [hep-ph]].
[14] G. F. Giudice and A. Strumia, Nucl. Phys. B **858** (2012) 63 [arXiv:1108.6077 [hep-ph]].
[15] L. Susskind, [arXiv:hep-th/0405189].
[16] N. Arkani-Hamed and S. Dimopoulos, JHEP **0506**, 073 (2005) [arXiv:hep-th/0405159]; G. F. Giudice and A. Romanino, Nucl. Phys. B **699** (2004) 65 [Erratum-ibid. B **706** (2005) 65] [arXiv:hep-ph/0406088].
[17] O. Buchmueller, R. Cavanaugh, A. De Roeck, M. J. Dolan, J. R. Ellis, H. Flacher, S. Heinemeyer and G. Isidori *et al.*, Eur. Phys. J. C **72** (2012) 2020 [arXiv:1112.3564 [hep-ph]].
[18] B. C. Allanach and C. G. Lester, Phys. Rev. D **73** (2006) 015013 [arXiv:hep-ph/0507283]; R. R. de Austri, R. Trotta and L. Roszkowski, JHEP **0605** (2006) 002 [arXiv:hep-ph/0602028]; M. E. Cabrera, J. A. Casas and R. Ruiz de Austri, JHEP **0903**, 075 (2009) [arXiv:0812.0536 [hep-ph]].
[19] M. E. Cabrera, J. A. Casas and R. Ruiz d Austri, JHEP **1005** (2010) 043 [arXiv:0911.4686 [hep-ph]].
[20] S. Akula, P. Nath and G. Peim, Phys. Lett. B **717** (2012) 188 [arXiv:1207.1839 [hep-ph]];
[21] M. E. Cabrera, J. A. Casas and R. Ruiz d Austri, to appear.
[22] S. P. Martin, Phys. Rev. D **75** (2007) 115005 [hep-ph/0703097 [HEP-PH]].
[23] G. Aad *et al.* [ATLAS Collaboration], arXiv:1208.2590 [hep-ex]].
[24] D. Carmi, A. Falkowski, E. Kuflik and T. Volansky, JHEP **1207** (2012) 136 [arXiv:1202.3144 [hep-ph]]; A. Azatov, R. Contino and J. Galloway, JHEP **1204** (2012) 127 [arXiv:1202.3415 [hep-ph]]; J. R. Espinosa, C. Grojean, M. Muhlleitner and M. Trott, JHEP **1205** (2012) 097 [arXiv:1202.3697 [hep-ph]].

PASCOS 2012 – 18th International Symposium on Particles Strings and Cosmology IOP Publishing
Journal of Physics: Conference Series **485** (2014) 012007 doi:10.1088/1742-6596/485/1/012007

Fundamental physics from the sky: Cosmic rays, gamma rays and the hunt for dark matter

Stefano Profumo

Santa Cruz Institute for Particle Physics, Santa Cruz, CA 95064, USA, and
Department of Physics, University of California, 1156 High St., Santa Cruz, CA 95064, USA

E-mail: profumo@ucsc.edu

Abstract. Can we learn about New Physics with astronomical and astro-particle data? Understanding how this is possible is key to unraveling one of the most pressing mysteries at the interface of cosmology and particle physics: the fundamental nature of dark matter. I will discuss some of the recent puzzling findings in cosmic-ray electron-positron data and in gamma-ray observations that might be related to dark matter. I will argue that recent cosmic-ray data, most notably from the Pamela and Fermi satellites, indicate that previously unaccounted-for powerful sources in the Galaxy inject high-energy electrons and positrons. Interestingly, this new source class might be related to new fundamental particle physics, and specifically to pair-annihilation or decay of galactic dark matter. This exciting scenario is directly constrained by Fermi gamma-ray observations, which also inform us on astrophysical source counterparts that could also be responsible for the high-energy electron-positron excess. Observations of gamma-ray emission from the central regions of the Galaxy as well as claims on a gamma-ray line at around 130 GeV also recently triggered a wide-spread interest: I will address the question of whether we are really observing signals from dark matter annihilation, how to test this hypothesis, and which astrophysical mechanisms constitute the relevant background.

1. Introduction

The standard picture for galaxies such as our own Milky Way is one where the visible matter is embedded in a halo of *dark matter* of as of yet unknown particle nature. Many theoretically well-motivated particle theories beyond the Standard Model predict the existence of particles that could be the dark matter. In many of these theories, such particles can pair-annihilate, producing as a result Standard Model particles that can be detected. These detectable final states include antimatter particles, neutrinos, gamma rays, electrons and positrons ($e^{\pm}$), and secondary products of $e^{\pm}$ energy losses, such as radiation at X-ray frequencies from Inverse Compton and at radio frequencies from synchrotron.

Collectively, the search for a signature from dark matter looking at its annihilation products is known as *indirect detection*. The key question for a particle physicist is: can we do fundamental physics with indirect dark matter detection? Specifically, I will address the sub-question of whether this is possible employing cosmic ray $e^{\pm}$ and gamma rays.

It is relevant to note that historically big leaps in our understanding of the underlying structure of particle physics models have been made possible by cosmic ray studies. These leaps include the discovery of antimatter, of a second generation of matter fermions, of pions and the first hints of strong interactions, and more recently of neutrino masses and mixing.

Content from this work may be used under the terms of the Creative Commons Attribution 3.0 licence. Any further distribution of this work must maintain attribution to the author(s) and the title of the work, journal citation and DOI.
Published under licence by IOP Publishing Ltd

Indirect dark matter detection is a multi-wavelength and multi-messenger endeavor. Key to doing fundamental physics is to control astrophysical backgrounds to signatures possibly stemming from dark matter. Not surprisingly, the most exciting places to look for a dark matter signature correspond to some of the places with the worse astrophysical background. In this contribution, I will address the high-energy positron excess, dark matter annihilation in the Galactic Center, and the 130 GeV line.

2. The high-energy positron excess

The Pamela telescope has observed a significant excess of positrons in the 10-100 GeV range, well above the standard Galactic cosmic-ray expectations [1]. A notable recent result is the confirmation by the Fermi Large Area Telescope (LAT) of the excess [2], a result which extends the rise in the positron fraction (the fraction of positrons to electrons plus positrons as a function of energy) up to 200 GeV. The notable experimental technique employed by the Fermi-LAT team hinges upon the use of the Earth shadow, as a function of the cosmic ray $e^{\pm}$ arrival direction and energy: For particular directions, electrons or positrons are completely forbidden. The result of Ref. [2] is notable from at least three standpoints:

(i) It confirms the Pamela result, that was potentially plagued by positron mis-identification (for every 50 GeV cosmic-ray positron, one has about 10 electrons and 10^4 protons, the latter being easily mis-identified with positrons)
(ii) It extends the Pamela result to higher energy, with no evidence for a turnover or suppression in the source of high-energy positrons
(iii) It rules out certain models for the positron fraction excess, for example supernova remnant inhomogeneities in the vicinity of the Sun

The question of the nature of the additional source of big-energy $e^{\pm}$ remains to be addressed: numerous studies have advocated dark matter annihilation, despite serious issues, including:

(i) The need for very large annihilation cross sections, $10^2 - 10^3$ times larger than the natural scale needed for a standard thermal freeze-out scenario to produced dark matter with the correct relic density
(ii) The need for rather large dark matter masses (even larger after the Fermi result)
(iii) The need to suppress excess antiprotons and therefore to select very special annihilation final states, such as $\mu^+\mu^-$

Despite these shortcomings, several possible scenarios have been proposed. Any such scenario is constrained by the necessary occurrence of final state radiation and of inverse Compton from the $e^{\pm}$ needed to explain the positron fraction. One of the best handles to constrain dark matter explanations of the high-energy positron excess is observing nearby galaxy clusters at gamma-ray frequencies [3]. The resulting limits depend on assumptions on substructure, but provide rather stringent constraints on the range of masses and cross sections that could explain the $e^{\pm}$ excess. Additional constraints come from observations of the cosmic microwave background and of the diffuse isotropic extragalactic gamma-ray background.

An alternate scenario, proposed among other studies in Ref. [4], postulates that the excess $e^{\pm}$ originate from existing nearby radio pulsars, in certain ranges of age and of distance. The Fermi-LAT discovery of multiple radio-quiet pulsars has further strengthen this possibility [5], that remains compatible with the new Fermi determination of the positron fraction [2].

Can we discriminate between dark matter and pulsars as the origin of the $e^{\pm}$ excess? A smoking gun for the pulsar interpretation is the detection of an anisotropy in the arrival direction of the cosmic ray $e^{\pm}$, especially if such anisotropy pointed towards a known pulsar [6]. For dark matter, the general expectation is a high-latitude inverse Compton emission from annihilation of dark matter in the halo, with a peculiar spectrum. None of these smoking guns has been observed so far, although limits on the anisotropy of the $e^{\pm}$ arrival directions are close to the level expected if the excess were due to one single nearby pulsar [6]. At present, both interpretations are compatible with data, and the origin of the positron excess remains still unsettled (albeit with increasingly tight constraints on the dark matter explanation).

Looking at the nearby future, AMS, successfully launched and deployed on the international Space Station, is undergoing calibration and alignment. The expectation is from AMS to deliver preliminary data on the positron fraction in early 2013, with the next priorities antiproton data and high-Z antimatter searches [7].

3. The Galactic Center

Multiple studies have identified a possible signal from dark matter annihilation at the center of the Galaxy [8, 9, 10]. The results of these studies illustrate how crucial a proper modeling of the diffuse emission and of the emission from the very center of the Galaxy is to reliable pinpoint an "anomalous" signal from dark matter. An *under-fitting* of the astrophysical background leads to the possibly misleading identification of excesses to be attributed to dark matter. The opposite issue is the *over-fitting* of the diffuse gamma-ray model, that would lead to fine-tuning the astrophysical background to incorporate at all costs any possible excess.

A key element in searching for dark matter in the Galactic center is understanding the gamma-ray source associated with Sag A*. This source, if fueled by protons, can be responsible for a significant diffuse emission from the protons accelerated by the central supermassive black hole. A careful treatment of the target density of gas in the region shows that a key diagnostic to unveil whether Sag A* accelerates protons is the gamma-ray source morphology [11]. Only future Cherenkov Telescope Arrays will have the needed angular resolution to pinpoint a hadronic origin for Sag A*, leading to possibly understanding some of the diffuse gamma-ray emission in the region [12].

If the dark matter interpretation of the diffuse excess in the Galactic center is correct, a likely mass range is about 10 GeV. This mass range resonates with a series of recent direct and indirect dark matter detection results (notably, recently, at radio frequencies [13, 14]). Albeit there is no conclusive evidence for new physics for any of these signals, this is an interesting possibility [15] worth investigating on a case-by-case basis and definitely worth of model-building efforts (see e.g. [16]).

4. The 130 GeV line

Evidence for a monochromatic gamma-ray line at 130 GeV has emerged from the analysis, in Ref. [17] of optimized signal-to-noise regions in the gamma-ray sky, at about a 3σ level once look-elsewhere effect is taken into account [17]. In Ref. [18], myself and Linden pointed out that the relevant region of interest overlaps with the so-called Fermi bubbles [19], which are thus an important source of background to the line. It was also noticed that a broken power-law spectrum (as what observed in the Fermi bubbles [19]) could be mis-identified, or could create confusion, with a gamma-ray line.

It is important to note that the outline of the relevant regions for the Fermi bubbles only partly overlaps with where most of the 130 GeV photons originate from. However, the Fermi bubble regions were specifically identified outside the Galactic plane, while the observed X-ray counterpart indicates that most of the gamma-ray emission associated with the mechanism fueling the bubbles should live exactly in the Galactic center region where the 130 GeV line is most prominent.

It is currently unclear whether the 130 GeV line can be associated with an instrumental effect - perhaps issues with the energy-reconstruction algorithm employed by the current version of the Fermi Tools. Likely, the forthcoming "Pass 8" Fermi Tools software suite will address relevant aspects associated with instrumental systematics. Future observations with ground based Cherenkov Telescopes could also be decisive in deciding of the validity of the 130 GeV line observation.

5. Conclusions

The search for dark matter is in an extraordinarily exciting phase:

- Direct dark matter experiments are tapping into the exciting low-mass region, and carving into interesting regions of particle dark matter parameter space
- the Large Hadron Collider at CERN is producing results that are sharpening the predictions of certain dark matter models (e.g. Universal Extra Dimesions) and could soon provide close of the correct beyond-the Standard Model framework
- results from AMS-02 will help understanding cosmic ray propagation much better, and provide more information on cosmic ray antimatter, especially positrons, antiprotons and antideuterons
- post-Fermi-LAT high-energy telescopes will cover the high-mass end of the particle dark matter parameter space (CTA), with the low-mass end possibly under scrutiny with the new hard X-ray facilities [20]

The indirect search for dark matter resonates with a famous quote by painter René Magritte[1]: "Everything we see hides another thin, we always want to see what is hidden by what we see".

Acknowledgments

SP is partly supported by an Outstanding Junior Investigator Award from the US Department of Energy and by Contract DE-FG02-04ER41268, and by NSF Grant PHY-0757911.

References

[1] Adriani O *et al.* (PAMELA Collaboration) 2009 *Nature* **458** 607 (*Preprint* 0810.4995 [astro-ph])
[2] Ackermann M *et al.* (Fermi LAT Collaboration) 2012 *Phys. Rev. Lett.* **108** 011103 (*Preprint* 1109.0521 [astro-ph.HE])
[3] Ackermann M, Ajello M, Allafort A, Baldini L, Ballet J, Barbiellini G, Bastieri D and Bechtol K *et al.*, 2010 *JCAP* **1005** 025 (*Preprint* 1002.2239 [astro-ph.CO]); Dugger L, Jeltema T E and Profumo S 2010 *JCAP* **1012** 015 (*Preprint* 1009.5988 [astro-ph.HE])
[4] Profumo S 2011 *Central Eur. J. Phys.* **10** 1 (*Preprint* 0812.4457 [astro-ph])
[5] Gendelev L, Profumo S and Dormody M 2010 *JCAP* **1002** 016 (*Preprint* 1001.4540 [astro-ph.HE])
[6] Ackermann M *et al.* (Fermi-LAT Collaboration) 2010 *Phys. Rev.* **D82** 092003 (*Preprint* 1008.5119 [astro-ph.HE])
[7] Gebauer I, private communication
[8] Goodenough L and Hooper D 2009 "Possible Evidence For Dark Matter Annihilation In The Inner Milky Way From The Fermi Gamma Ray Space Telescope" (*Preprint* 0910.2998 [hep-ph])
[9] Hooper D and Goodenough L 2011 *Phys. Lett.* **B697** 412 (*Preprint* 1010.2752 [hep-ph])
[10] Hooper D and Linden T 2011 *Phys. Rev.* **D83** 083517 (*Preprint* 1011.4520 [astro-ph.HE])
[11] Linden T, Lovegrove E and Profumo S 2012 *Astrophys. J.* **753** 41 (*Preprint* 1203.3539 [astro-ph.HE])
[12] Linden T and Profumo S "Exploring the Nature of the Galactic Center Gamma-Ray Source with the Cherenkov Telescope Array" (*Preprint* 1206.4308 [astro-ph.HE])
[13] Linden T, Hooper D and Yusef-Zadeh F 2011 *Astrophys. J.* **741** 95 (*Preprint* 1106.5493 [astro-ph.HE])
[14] Hooper D, Belikov A V, Jeltema T E, Linden T, Profumo S and Slatyer T R 2012 "The Isotropic Radio Background and Annihilating Dark Matter" (*Preprint* 1203.3547 [astro-ph.CO])
[15] Hooper D 2012 "The Empirical Case For 10 GeV Dark Matter" (*Preprint* 1201.1303 [astro-ph.CO])
[16] Boucenna M S and Profumo S 2011 *Phys. Rev.* **D84** 055011 (*Preprint* 1106.3368 [hep-ph])
[17] Weniger C 2012 *JCAP* **1208** 007 (*Preprint* 1204.2797 [hep-ph])
[18] Profumo S and Linden T 2012 *JCAP* **1207** 011 (*Preprint* 1204.6047 [astro-ph.HE])

[1] I thank Pasquale Serpico for bringing this quote to my attention

[19] Su M, Slatyer T R and Finkbeiner D P 2010 *Astrophys. J.* **724** 1044 (*Preprint* 1005.5480 [astro-ph.HE])
[20] Jeltema T E and Profumo S 2011 "Dark Matter Detection with Hard X-ray Telescopes" (*Preprint* 1108.1407 [astro-ph.HE])

Journal of Physics: Conference Series **485** (2014) 012008 doi:10.1088/1742-6596/485/1/012008

Current and future constraints on neutrino physics from cosmology

Steen Hannestad[1], Jan Hamann[1] and Yvonne Y Y Wong[2]

[1]Department of Physics and Astronomy
University of Aarhus, DK-8000 Aarhus C, Denmark
[2]Institut für Theoretische Teilchenphysik und Kosmologie
RWTH Aachen, D-52056 Aachen, Germany

E-mail: sth@phys.au.dk, hamann@phys.au.dk,ywong@physik.rwth-aachen.de

Abstract. In recent years precision cosmology has become an increasingly powerful probe of particle physics. Perhaps the prime example of this is the very stringent cosmological upper bound on the neutrino mass. However, other aspects of neutrino physics, such as their decoupling history and possible non-standard interactions, can also be probed using observations of cosmic structure. Here, we review the current status of cosmological bounds on neutrino properties and discuss the potential of future observations, for example by the recently approved EUCLID mission, to precisely measure neutrino properties.

1. Introduction and bounds on standard model neutrinos

It has long been known that neutrinos are important for cosmological structure formation. Because neutrinos have finite mass and no electromagnetic interactions they are a source of dark matter, albeit with properties very different from cold dark matter. The problem with neutrinos as the sole source of dark matter is that their very small masses and high thermal velocities prevent them from clustering on small scales, in stark contrast to the known properties of the dominant dark matter component.

On the other hand, the fact that neutrinos behave so differently from other sources of matter in the universe also means that precision cosmology can be used to set very tight constraints on the properties of neutrinos, such as their mass and cosmological number density (see e.g. [1, 2] for recent reviews).

With the advent of very large scale cosmological observations, constraints on some neutrino parameters have now reached a precision where they outperform any current or planned terrestrial experiments. For example the combination of cosmic microwave background observations by WMAP [3] with large scale structure surveys such as the Sloan Digital Sky Survey (SDSS) [4] have been able to constrain the mass of standard model neutrinos to below the 1 eV scale, compared to the current upper bound on the sum of neutrino masses from beta decay measurements by the Mainz experiment [5] of around 7 eV.

However, a drawback of using cosmology to probe neutrino masses has consistently been that cosmological neutrino constraints are susceptible to the exact cosmological model used. Depending on the data sets used this may still allow some room for example for the existence of a partly (or perhaps even fully) thermalised sterile neutrino species [6, 7, 8, 9]. Another potential

Content from this work may be used under the terms of the Creative Commons Attribution 3.0 licence. Any further distribution of this work must maintain attribution to the author(s) and the title of the work, journal citation and DOI.
Published under licence by IOP Publishing Ltd

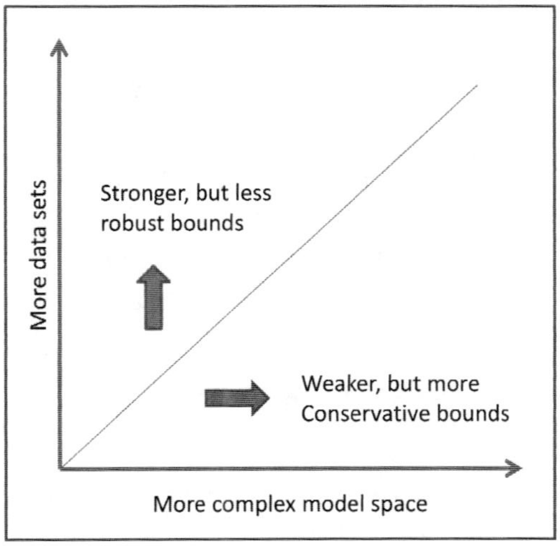

Figure 1. The dependence of cosmological neutrino mass bounds on the choice of model framework and the number of data sets used.

drawback has been that a useful constraint initially could only be derived by combining a variety of different data sets. If just one of these data sets has a systematic bias the result will formally be either a very stringent constraint on the neutrino mass or, conversely, a spurious detection of the neutrino mass. This general trend is illustrated in Figure 1.

One of the main developments during the past decade has been that a useful constraint on neutrino properties can now be gained from either just a single set of data or at most the combination of two data sets with well-understood systematics. An example of this is that an upper bound of typically $\sum m_\nu \sim 0.3 - 0.5$ eV [10, 11] can now be derived from just CMB data combined with almost linear large scale structure data from the SDSS survey and a measurement of the Hubble parameter [12] (see also e.g. [13, 14, 15, 16, 17, 18, 19, 20] for additional recent constraints on the neutrino mass from large scale structure probes). Previously, bounds of the same formal strength were typically derived with the addition of Lyman-α and other auxillary data (see e.g. [21]). However, this relied on an understanding of baryonic physics in the semi- or non-linear regime of structure formation. More recent Lyman-α analyses have been much more conservative and hence not been able to add much extra information on the properties of light neutrinos [22]. We note that although this is true for light neutrinos, the Lyman-α data still provides an extremely strong constraint on possible keV mass sterile neutrinos which would constitute warm dark matter. [23, 24, 25].

2. Sterile neutrinos

While constraints on pure standard model neutrinos are by now very stringent, there is still the possibility that non-standard neutrino physics could be present. For example, one or more additional sterile neutrinos might be present. Even though such neutrinos have no standard model interactions they can thermalize through mixing with the active flavours (see e.g. [26, 27] for recent treatments). Full thermalisation occurs for large mass differences and large mixing angles and the usual bound on light neutrino masses strongly constrains the mass of such neutrinos. However, for sufficiently large masses the contribution to the energy density at present can be large even though thermalisation is incomplete - the sterile neutrino warm dark matter limit. This particular case is not subject to the usual cosmological constraints, simply because the free streaming scale is much smaller than the scale measured by CMB and large scale structure surveys. For warm dark matter, more useful constraints can be derived from the Lyman-alpha forest and from the dynamics of small scale non-linear structures such as dwarf galaxies.

From a neutrino oscillation perspective there is no indication of very high mass neutrinos. The mixing angle required for sterile neutrino warm dark matter is much smaller than what is required for detection through oscillation experiments. However, there is an indication from a number of different experiments that a fourth light neutrino species might be needed. Since it must necessarily interact much more weakly than standard model neutrinos it is by construction a sterile neutrinos. The evidence for such neutrinos comes through mixing with the active flavours and in fact the best fitting mixing angle is relatively large, leading to complete or almost complete thermalisation in the early Universe. The question is therefore whether sterile neutrinos are compatible with current cosmological bounds. Contrary to naive expectations eV mass sterile neutrinos might actually be allowed, depending somewhat on the specifics of the analysis [6, 7, 8, 9]. In Figure 2 we show an example of a recent analysis of mass constraints on light sterile neutrinos, taken from Ref. [6].

3. Dark radiation

Another question which has recently received significant attention is the possible presence of extra relativistic degrees of freedom in the Universe, not necessarily related to light neutrinos. While some types of sterile neutrinos would behave essentially as dark radiation there are many other model predicting the existence of extra radiation. Examples are new, massless or very light particles such as Majorons, or the presence of early dark energy.

The reason for the current interest is that the relativistic energy measured by WMAP and other CMB experiments point towards a value of higher than predicted in the standard model. The CMB anisotropy spectrum is sensitive to additional relativistic energy density mainly because of two effects: The early integrated Sachs-Wolfe effect leads to an increase in power around the first acoustic peak because matter-radiation equality is delayed, and the small scale peaks are damped because of diffusion [28].

Usually, relativistic, weakly interacting energy density is quantified in terms of the effective number of neutrino species, $N_{\rm eff} \equiv \rho/\rho_{\nu,0}$ where $\rho_{\nu,0} = (4/11)^{1/3}\rho_\gamma$. Because of finite temperature corrections to the photon propagator as well as a small amount of entropy transfer from electrons and positrons to neutrinos the standard model prediction is $N_{\rm eff} = 3.046$, instead of exactly 3 (see e.g. [29, 30, 31, 32]). The most recent analysis of WMAP plus auxillary data gives a result of $N_{\rm eff} = 4.29^{+1.05}_{-0.96}$ at 95% C.L., i.e. an indication of beyond standard model physics at slightly more than 2σ [33] (see also [3, 11, 34, 35, 8]).

4. Future constraints

The rapid improvement of cosmological neutrino constraints is set to continue in coming years, with new CMB observations by Planck covering both large and small angular scales currently covered by several different experiments (thus with much more well controlled systematics). Even more astounding will be the development in terms of large scale structure surveys. The recently approved ESA mission EUCLID [36] will measure a substantial fraction of the entire

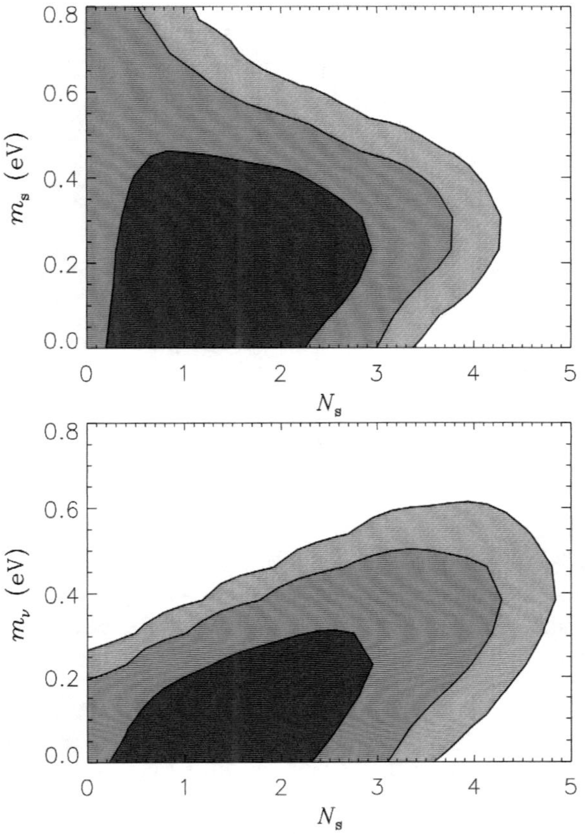

Figure 2. 2D marginalized 68%, 95% and 99% credible regions for the neutrino mass and thermally excited number of degrees of freedom N_s. *Top:* The $3 + N_s$ scheme, in which ordinary neutrinos have $m_\nu = 0$, while sterile states have a common mass scale m_s. *Bottom:* The $N_s + 3$ scheme, where the sterile states are taken to be massless $m_s = 0$, and 3.046 species of ordinary neutrinos have a common mass m_ν. Figure taken from [6].

Hubble volume with a deep photometric survey. In practise, the magnitude and wavelength limitations of the instrument means that the median redshift of galaxies measured by EUCLID will be around 0.8-1, with a substantial tail up to redshifts higher than 2. Because it is a space-based experiment it is not limited by seeing in the atmosphere and therefore the shape of galaxies can be measured exceedingly accurately, making it a perfect instrument for weak lensing. Together with the photometric survey, EUCLID will feature a spectroscopic survey of around 10^7 galaxies which will be important both for calibration of the much larger photometric survey and for precision measurements of the baryon acoustic oscillation feature.

However, in terms of constraining the neutrino mass the most important parameters in the power spectrum are the shape and amplitude, not the acoustic feature. The photometric survey is therefore the most important EUCLID data deliverable. As it turns out the combination of Planck data with either the photometric shear data or the galaxy power spectrum are able to constrain the neutrino mass to approximately 0.02-0.03 eV at 1σ [37]. While this is enough to guarantee that the effect of neutrino mass will be seen in the data (and of course also enough that it is crucial to include massive neutrinos in the analysis), it is not enough to be sure that a genuine detection of the neutrino mass will be made.

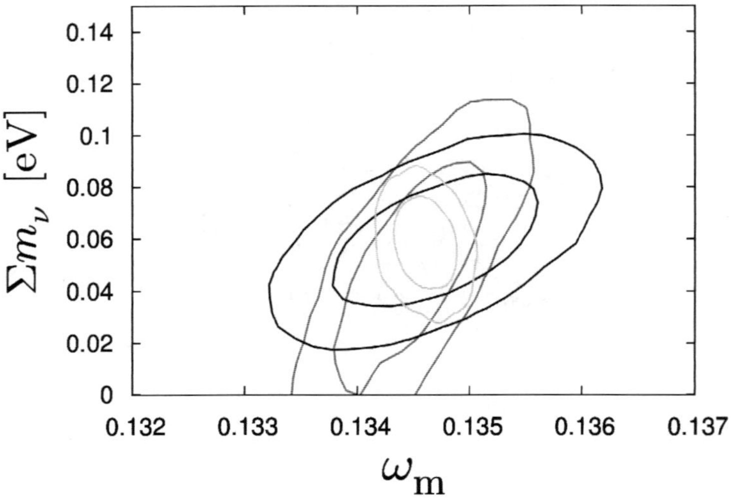

Figure 3. 68% and 95% constraints on the matter density, $\omega_m = \Omega_m h^2$, and the sum of neutrino masses, $\sum m_\nu$, for the combination of Planck CMB data with EUCLID shear data (red curves) and EUCLID galaxy data (black curves), as well as the combination of all three (green curves). The data used corresponds to the "csgx" case in [37].

However, it turns out that the combination of EUCLID shear *and* galaxy measurements breaks the degeneracy between $\sum m_\nu$, ω_m, and h, leading to a significantly tighter constraint on all three parameters [37]. In fact, if galaxy bias can be reliably estimated from the spectroscopic survey, the combined EUCLID data, together with CMB data leads to a much tighter constraint than naively expected. In fact, the combination of CMB, shear, and galaxy data might allow for a 5σ detection of non-zero neutrino masses, even in the normal hierarchy. Figure 3 (taken from [37]) shows the projected constraint on the sum of neutrino masses for the combination of Planck data with EUCLID galaxy and shear data.

Preliminary studies using the dataset called "csgx" in [37] show that going to more complex model spaces does not significantly affect this conclusion. Adding the equation of state parameter for dark energy, w, and the effective number of relativistic species, $N_{\rm eff}$, as free parameters beyond ΛCDM plus $\sum m_\nu$ only marginally relaxes the constraint which can be achieved from 0.011 eV to 0.012 eV at the 1σ level. Table 1 also shows the corresponding constraints on w and $N_{\rm eff}$. Interestingly, as can be seen in Figure 4, the degeneracy between $\sum m_\nu$ and w is no longer very strong when the data reaches this level of precision. A similar thing happens for the correlation between $\sum m_\nu$ and $N_{\rm eff}$. In Figure 5 we show correlations between $\sum m_\nu$, ω_m, and $N_{\rm eff}$ for Planck CMB data, EUCLID shear plus galaxy data, and the two data sets combined. As can be seen in the top panel, the correlation between $\sum m_\nu$ and $N_{\rm eff}$ is almost entirely broken by the combination of the two data sets.

Even though the neutrino mass is the main parameter of interest because it very directly affects the formation of structure, the effective number of neutrino species is an equally important parameter from a neutrino physics point of view. Not only is it a sensitive probe of physics beyond the standard model, the standard model prediction itself is interesting. The predicted value of 3.046 relies on a very precise understanding of standard model physics to an energy scale of more than 1 MeV in the early universe. If data becomes precise enough to distinguish 3.046 from 3 it de facto means that we will, through observations of structure formation in the late time universe, be able to see effects of finite temperature field theory in the early Universe.

Table 1. Estimated 1σ sensitivity of Planck+EUCLID to $\sum m_\nu$, w, and $N_{\rm eff}$. The data used correspond to the csgx case in [37].

Model	$\sum m_\nu$	w	$N_{\rm eff}$
Standard ΛCDM + $\sum m_\nu$	0.011	–	–
Standard ΛCDM + $\sum m_\nu$ + w + $N_{\rm eff}$	0.012	0.02	0.055

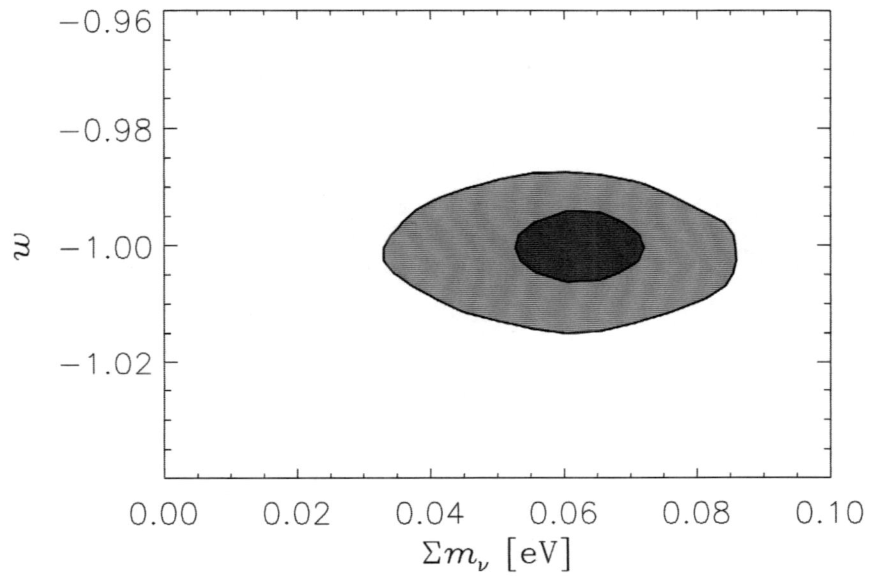

Figure 4. 68% and 95% constraints on the sum of neutrino masses, $\sum m_\nu$, and the dark energy equation of state, w, for the combination of Planck CMB data with EUCLID shear and galaxy data. The data used corresponds to the "csgx" case in [37].

The estimated sensitivity of EUCLID plus Planck data is almost, but not quite at the required level. At 1σ the combined data might reach $\sigma(N_{\rm eff}) \sim 0.05$ which means that the standard model prediction can only be tested at approximately 1σ.

5. Conclusions

We have reviewed both the current status of neutrino constraints from large scale structure formation, as well as possible future bounds from experiments such as EUCLID.

Even at present epoch late time cosmology provides extremely useful constraints on light neutrinos. The bound on the masses of light, standard model neutrinos is significantly stronger than from direct experiments. While an exact value of the current bound is hard to quantify it is somewhere in the range of 0.3-0.5 eV, depending on the exact data and model space used in the analysis. Any value close to the current bound from beta decay experiments is certainly ruled out at very high significance. However, in the presence of sterile neutrinos the bound does become significantly weaker, and the existence of an additional eV-mass sterile neutrino, as is hinted at by current short baseline oscillation data, may not be excluded.

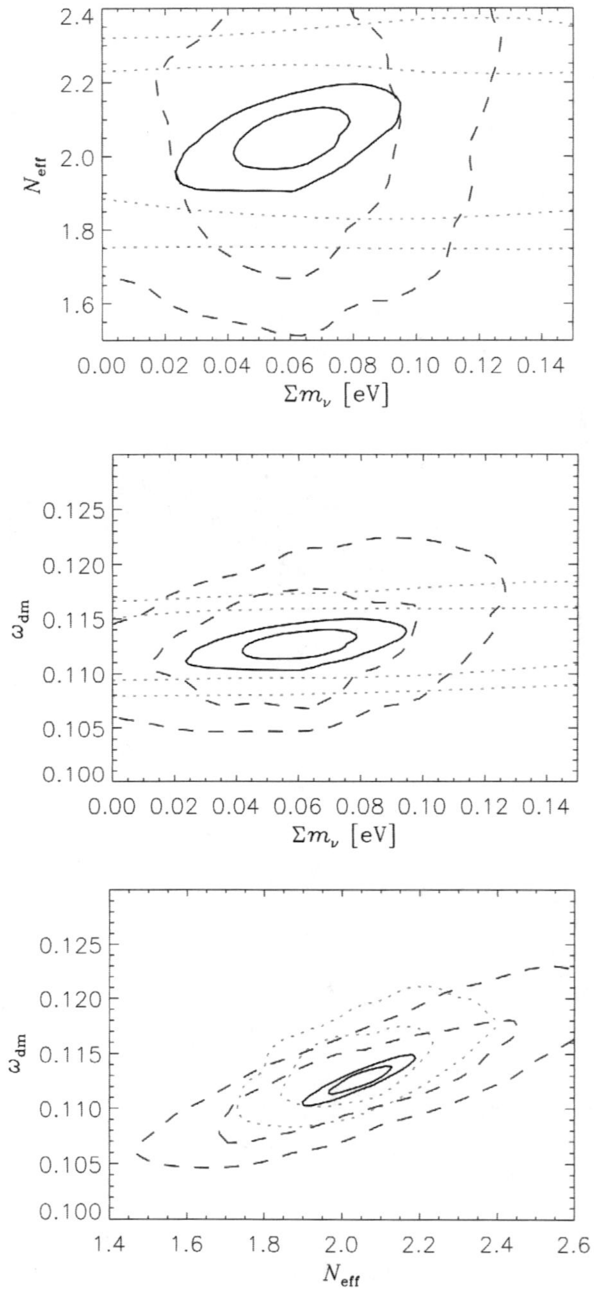

Figure 5. 68% and 95% constraints on the sum of neutrino masses, $\sum m_\nu$, the effective number of relativistic species, $N_{\rm eff}$, and the dark matter density, $\omega_{\rm dm}$, for Planck CMB data (dotted lines), EUCLID shear and galaxy data (dashed lines), and the two combined (full lines). The data used corresponds to the "csgx" case in [37].

In fact, as was also discussed above, there might be an indication from mainly CMB data of the existence of extra radiation beyond what is predicted by the standard model. Light neutrinos would contribute to this energy density because they would still be relativistic at the epoch of recombination where the CMB is formed. Thus, sterile neutrinos might yet be discovered by precision measurements in cosmology. The upcoming data release from the Planck satellite should give a strong indication of whether sterile neutrinos are needed or not.

Finally, we also discussed the ability of the recently approved EUCLID satellite to measure neutrino masses. Interestingly, with Planck and EUCLID data combined it might be possible to not only put an upper limit on the neutrino mass, but make a real measurement at the 5σ level even if the sum of neutrino masses is as low as possible (i.e. in the normal hierarchy with a massless ν_1).

One remaining problem, however, is that it may be difficult to ascertain, based on cosmological data alone, whether the effect seen is due to standard model neutrinos. What the measurement will measure with great precision is the presence of the expected relativistic energy density and rest mass in *some* species that is weakly interacting. That it is weakly interacting follows from the requirement that the energy density is in a component with anisotropic stress. Without an anisotropic stress term the light particle species will behave as a perfect fluid, i.e. a fluid with no viscosity. This in turn leads to a strong enhancement of structure on all scales within the sound horizon on the fluid because free-streaming is absent (see e.g. [38, 39, 40, 41, 42, 43, 44]). Already with the current WMAP data such a scenario is very strongly disfavoured, and with new data from Planck an even stronger bound can be imposed on the interaction strength of the light species.

Since cosmological data based on structure formation can never be sensitive to flavour, it is not possible to "directly" measure the cosmic neutrino background. Therefore it will of course be highly desirable with a corresponding laboratory measurement of neutrino mass even when a non-zero neutrino mass has been detected at more than 5σ significance.

Acknowledgments
We acknowledge computing resources from the Danish Center for Scientific Computing (DCSC) as well as funding via grant R45-A4316 from the Lundbeck Foundation.

References

[1] Hannestad S 2010 *Prog.Part.Nucl.Phys.* **65** 185–208 (*Preprint* 1007.0658)
[2] Wong Y Y 2011 *Ann.Rev.Nucl.Part.Sci.* **61** 69–98 (*Preprint* 1111.1436)
[3] Komatsu E *et al.* (WMAP) 2011 *Astrophys.J.Suppl.* **192** 18 (*Preprint* 1001.4538)
[4] Aihara H *et al.* (SDSS) 2011 *Astrophys.J.Suppl.* **193** 29 (*Preprint* 1101.1559)
[5] Kraus C, Bornschein B, Bornschein L, Bonn J, Flatt B *et al.* 2005 *Eur.Phys.J.* **C40** 447–468 (*Preprint* hep-ex/0412056)
[6] Hamann J, Hannestad S, Raffelt G G, Tamborra I and Wong Y Y 2010 *Phys.Rev.Lett.* **105** 181301 (*Preprint* 1006.5276)
[7] Giusarma E, Corsi M, Archidiacono M, de Putter R, Melchiorri A *et al.* 2011 *Phys.Rev.* **D83** 115023 (*Preprint* 1102.4774)
[8] Hamann J, Hannestad S, Raffelt G G and Wong Y Y 2011 *JCAP* **1109** 034 (*Preprint* 1108.4136)
[9] Joudaki S, Abazajian K N and Kaplinghat M 2012 (*Preprint* 1208.4354)
[10] de Putter R, Mena O, Giusarma E, Ho S, Cuesta A *et al.* 2012 (*Preprint* 1201.1909)
[11] Hamann J, Hannestad S, Lesgourgues J, Rampf C and Wong Y Y 2010 *JCAP* **1007** 022 (*Preprint* 1003.3999)
[12] Riess A G, Macri L, Casertano S, Sosey M, Lampeitl H *et al.* 2009 *Astrophys.J.* **699** 539–563 (*Preprint* 0905.0695)
[13] Sekiguchi T, Ichikawa K, Takahashi T and Greenhill L 2010 *JCAP* **1003** 015 (*Preprint* 0911.0976)
[14] Reid B A, Percival W J, Eisenstein D J, Verde L, Spergel D N *et al.* 2010 *Mon.Not.Roy.Astron.Soc.* **404** 60–85 (*Preprint* 0907.1659)
[15] Saito S, Takada M and Taruya A 2011 *Phys.Rev.* **D83** 043529 (*Preprint* 1006.4845)

[16] Riemer-Sorensen S, Blake C, Parkinson D, Davis T M, Brough S *et al.* 2012 *Phys.Rev.* **D85** 081101 (*Preprint* 1112.4940)
[17] Xia J Q, Granett B R, Viel M, Bird S, Guzzo L *et al.* 2012 *JCAP* **1206** 010 (*Preprint* 1203.5105)
[18] Joudaki S 2012 (*Preprint* 1202.0005)
[19] Moresco M, Verde L, Pozzetti L, Jimenez R and Cimatti A 2012 *JCAP* **1207** 053 (*Preprint* 1201.6658)
[20] Reid B A, Verde L, Jimenez R and Mena O 2010 *JCAP* **1001** 003 (*Preprint* 0910.0008)
[21] Seljak U, Slosar A and McDonald P 2006 *JCAP* **0610** 014 (*Preprint* astro-ph/0604335)
[22] Viel M, Haehnelt M G and Springel V 2010 *JCAP* **1006** 015 (*Preprint* 1003.2422)
[23] Viel M, Lesgourgues J, Haehnelt M G, Matarrese S and Riotto A 2005 *Phys.Rev.* **D71** 063534 (*Preprint* astro-ph/0501562)
[24] Abazajian K 2006 *Phys.Rev.* **D73** 063513 (*Preprint* astro-ph/0512631)
[25] Boyarsky A, Lesgourgues J, Ruchayskiy O and Viel M 2009 *JCAP* **0905** 012 (*Preprint* 0812.0010)
[26] Hannestad S, Tamborra I and Tram T 2012 *JCAP* **1207** 025 (*Preprint* 1204.5861)
[27] Mirizzi A, Saviano N, Miele G and Serpico P D 2012 (*Preprint* 1206.1046)
[28] Hou Z, Keisler R, Knox L, Millea M and Reichardt C 2011 (*Preprint* 1104.2333)
[29] Mangano G, Miele G, Pastor S, Pinto T, Pisanti O *et al.* 2005 *Nucl.Phys.* **B729** 221–234 (*Preprint* hep-ph/0506164)
[30] Dodelson S and Turner M S 1992 *Phys.Rev.* **D46** 3372–3387
[31] Hannestad S and Madsen J 1995 *Phys.Rev.* **D52** 1764–1769 (*Preprint* astro-ph/9506015)
[32] Hannestad S 2002 *Phys.Rev.* **D65** 083006 (*Preprint* astro-ph/0111423)
[33] Archidiacono M, Giusarma E, Melchiorri A and Mena O 2012 *Phys.Rev.* **D86** 043509 (*Preprint* 1206.0109)
[34] Hamann J 2012 *JCAP* **1203** 021 (*Preprint* 1110.4271)
[35] Nollett K M and Holder G P 2011 (*Preprint* 1112.2683)
[36] Laureijs R, Amiaux J, Arduini S, Augueres J L, Brinchmann J *et al.* 2011 (*Preprint* 1110.3193)
[37] Hamann J, Hannestad S and Wong Y Y 2012 (*Preprint* 1209.1043)
[38] Hannestad S 2005 *JCAP* **0502** 011 (*Preprint* astro-ph/0411475)
[39] Serra P, Zalamea F, Cooray A, Mangano G and Melchiorri A 2010 *Phys.Rev.* **D81** 043507 (*Preprint* 0911.4411)
[40] Friedland A, Zurek K M and Bashinsky S 2007 (*Preprint* 0704.3271)
[41] Sawyer R F 2006 *Phys.Rev.* **D74** 043527 (*Preprint* astro-ph/0601525)
[42] Bell N F, Pierpaoli E and Sigurdson K 2006 *Phys.Rev.* **D73** 063523 (*Preprint* astro-ph/0511410)
[43] Trotta R and Melchiorri A 2005 *Phys.Rev.Lett.* **95** 011305 (*Preprint* astro-ph/0412066)
[44] Bashinsky S and Seljak U 2004 *Phys.Rev.* **D69** 083002 (*Preprint* astro-ph/0310198)

PASCOS 2012 – 18th International Symposium on Particles Strings and Cosmology IOP Publishing
Journal of Physics: Conference Series **485** (2014) 012009 doi:10.1088/1742-6596/485/1/012009

Baryon and lepton numbers: Life in the desert[1]

Pavel Fileviez Perez[a,b]

[a]Particle and Astro-Particle Physics Division
Max Planck Institute for Nuclear Physics (MPIK)
Saupfercheckweg 1, 69117 Heidelberg, Germany
[b]Center for Cosmology and Particle Physics (CCPP)
New York University, 4 Washington Place, NY 10003, USA

E-mail: fileviez@mpi-hd.mpg.de

Abstract. The simplest theories where we can understand the origin of the baryon and lepton number violating interactions are discussed. We discuss the desert hypothesis in particle physics and the different scenarios where there is no need to assume it. It is shown that the minimal supersymmetric B-L theory predicts lepton number violation at the Large Hadron Collider if supersymmetry is realized at the low scale. We present the BLMSSM where both symmetries, B and L, can be spontaneously broken at the TeV scale.

1. Introduction

It is well-known that in physics beyond the Standard Model (SM) very often one has to think about the violation of two SM symmetries, Baryon (B) and Lepton (L) numbers, and postulate the existence of a large desert between the low and high scales. It was pointed out by S. Weinberg [1] long time ago that one can write down operators such as, $\mathcal{O}_5 = c_\nu LLHH/\Lambda$, and $\mathcal{O}_6 = c_{\rm BL} {\rm QQQL}/\Lambda^2$, where the first one breaks total lepton number and the second operator violates both symmetries. Typically, one can compute the coefficient in front of these operators in a grand unified theory defined at the high scale. The scale Λ in $\mathcal{O}_6$ has to be very large, i.e. $\Lambda > 10^{14-16}$ GeV, in order to satisfy the bounds on the proton decay lifetime. See Fig.1 for a cartoon illustration of the desert hypothesis in particle physics.

If low energy supersymmetry is realized in nature the desert picture is different since at the TeV or multi-TeV scale we can have for example the Minimal Supersymmetric Standard Model (MSSM). In this case one can even predict the unification of gauge couplings at the unified scale, $M_{GUT} \approx 10^{16}$ GeV, if the desert hypothesis is true. See Fig.2 for a possible illustration of the desert picture in the case when we have Supersymmetry. Unfortunately, in this case one has to face the fact that there are new interactions allowed by the SM gauge symmetry which violate B or L at the renormalizable level. These interactions are the well-known R-parity breaking interactions: $\hat{L}\hat{H}_u$, $\hat{Q}\hat{L}\hat{d}^c$, $\hat{L}\hat{L}\hat{e}^c$, and $\hat{u}^c\hat{d}^c\hat{d}^c$.

The main goal of this review is to discuss the possible origin of the B and L violating interactions in the Minimal Supersymmetric Standard Model and show the possibility to define a simple theory where one can have the spontaneous breaking of B and L at the low scale. Therefore, in this case there is no need to postulate of the existence of a large desert. One has to say that the most of the people in the particle physics community got used to the idea of

[1] The slides can be found in http://www.mpi-hd.mpg.de/personalhomes/fileviez/PASCOS-2012-Slides.pdf

Content from this work may be used under the terms of the Creative Commons Attribution 3.0 licence. Any further distribution of this work must maintain attribution to the author(s) and the title of the work, journal citation and DOI.
Published under licence by IOP Publishing Ltd

Figure 1. Desert Hypothesis in Particle Physics.

having a desert. In my opinion, it is hard to believe that there is no new physics between the low and high scales, and the fact that the people got used to it does not mean that this idea is true.

What do we know about experimental evidences for B and L violation? Unfortunately, after many experimental efforts still there is no direct evidences in a low energy process for the violation of baryon number. As it is well-known, many experimental collaborations have been looking for proton decay signals (a $\Delta B = 1$ and $\Delta L =$ odd process) and today the lower bounds on the proton decay lifetimes are very impressive. See Fig.3 for a summary of the proton decay lifetime bounds from different collaborations and Ref.[3] for a review on proton decay. An interesting process where B must be broken in two units is $n - \bar{n}$ oscillation. In this case the bounds on the lifetime for free neutron oscillations are very weak, $\tau_{n-\bar{n}} > 10^8$ s [4]. There is a second $\Delta B = 2$ process which can be used to set relevant constraints on new physics scenario, the di-nucleon decay $p\, p \to K^+ K^+$. Recently, a new bound has been found in Ref. [5] and the current lower bound is $\tau_{pp \to K^+K^+} > 1.7 \times 10^{32}$ years. What about lepton number violation?. We know from neutrino oscillation experiments that the lepton number defined for each SM family is broken in nature, but still the total lepton number could be conserved. There are many experimental searches for neutrinoless double beta decay. In this case the total lepton number must be broken in two units and if it is discovered we can know about the Majorana nature of the neutrinos. For a review on neutrinoless double beta decay see Ref. [6]. Then, what do we know from cosmology? It is well-known that in order to explain the baryon asymmetry in the universe we need to have baryon number violation. Therefore, even if there is no a direct connection between the low energy processes mentioned above and the baryogenesis mechanism, we expect baryon number violating processes in nature.

2. B and L in the Superworld

We have mentioned before that in the context of Supersymmetry we have interactions which break B and L at the renormalizable level and dimension five operators which give rise to proton decay:

$$\mathcal{W}_{BL} = \epsilon \hat{L} \hat{H}_u + \lambda \hat{L} \hat{L} \hat{e}^c + \lambda^{'} \hat{Q} \hat{L} \hat{d}^c + \lambda^{''} \hat{u}^c \hat{d}^c \hat{d}^c \tag{1}$$

Journal of Physics: Conference Series **485** (2014) 012009 doi:10.1088/1742-6596/485/1/012009

$$+ \frac{c_\nu}{\Lambda}\hat{L}\hat{L}\hat{H}_u\hat{H}_u + \frac{c_L}{\Lambda}\hat{Q}\hat{Q}\hat{Q}\hat{L} + \frac{c_R}{\Lambda}\hat{u}^c\hat{d}^c\hat{u}^c\hat{e}^c. \quad (2)$$

The main difference between the terms in the first and second lines in this equation is that only the first interactions break matter parity. Matter parity is defined as $M = (-1)^{3(B-L)}$, being -1 for all matter chiral superfields and $+1$ for all other superfields. Let us discuss some possible scenarios where these interactions could play a role. The possibility to generate Majorana neutrino masses through the bilinear interaction, $\hat{L}\hat{H}_u$, has been studied by many groups. See for example Ref.[7] for a detailed analysis. A second scenario which give us very fast proton decay corresponds to the case when we combine the interactions $\hat{Q}\hat{L}\hat{d}^c$ and $\hat{u}^c\hat{d}^c\hat{d}^c$. In this case one has the so-called dimension four contributions to nucleon decay. If the relevant coefficients are of order one and the sfermion masses are at the TeV scale the lifetime of the proton is too short, $\tau_p \sim 10^{-20}$ years. Therefore, one has two options: a) we assume small couplings, $\lambda'\lambda'' < 10^{-26}$, or b) we try to understand the origin of these interactions to see if they are suppressed. It is important to mention that even if renormalizable interactions are absent still we get severe constraints on the dimension five operators mediating proton decay. See Ref.[3] for a detailed discussion.

Now, let us focus on the first goal and try to understand the origin of the B and L violating interactions at the renormalizable level. We have mentioned that these interactions break M-parity and of course B-L. Therefore, if we define a theory where B-L is conserved at the scale Λ_{B-L} these interactions are not allowed before symmetry breaking and once we break the symmetry we can understand the size of these interactions. This idea has been studied by many different groups [8, 9, 10, 11, 12, 13, 14, 15, 16, 17, 18, 19]. We investigated this issue for the first in Ref. [16] where we defined the simplest supersymmetric left-right model. In this case the only way to break the gauge symmetry $SU(2)_R \bigotimes U(1)_{B-L}$ to $U(1)_Y$ is to give a vacuum expectation value to the right-handed sneutrinos. Therefore, one can say that the minimal supersymmetric left-right model predicts that R-parity should be spontaneously broken and one expects lepton number violating signals at the Large Hadron Collider. In this short review I will discuss mainly the simplest supersymmetric theory which predicts spontaneous breaking of R-parity and lepton number violation at the TeV scale.

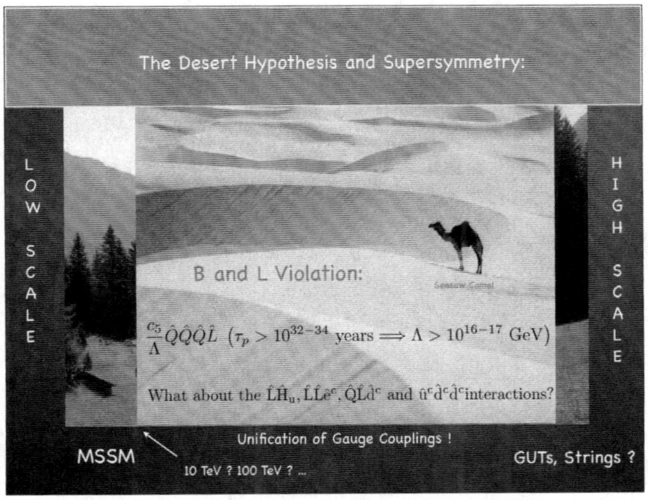

Figure 2. Supersymmetry and the Desert Hypothesis.

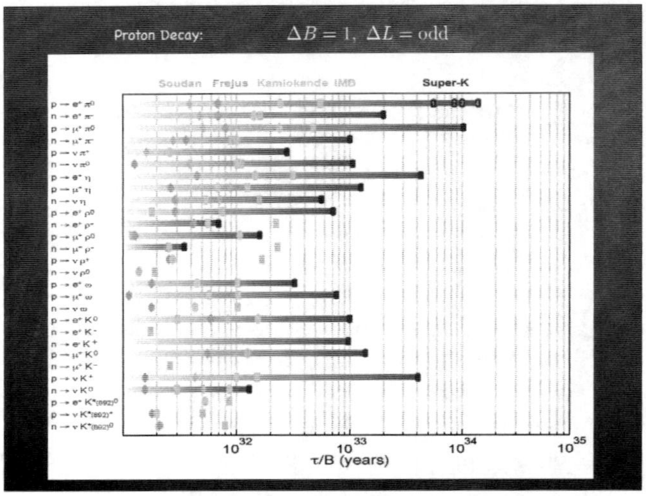

Figure 3. Proton decay bounds [2].

3. The Minimal Gauged $U(1)_{B-L}$ Model with Spontaneous R-parity Violation

We have mentioned before that if we want to understand the possible origin of the lepton and baryon number violating interactions in the MSSM one has to define a theory where B-L is part of the gauge symmetry. Now, the simplest theory was proposed in Ref.[17] and here we discuss the main features and predictions:

- This theory is based on the gauge group $SU(3)_C \bigotimes SU(2)_L \bigotimes U(1)_Y \bigotimes U(1)_{B-L}$.
- The main prediction is that R-parity must be broken.
- The only needed extra matter fields with B-L quantum numbers are the right-handed superfields.
- The B-L and R-parity breaking scales are defined by the SUSY breaking scale.
- One predicts lepton number violation at the LHC.
- The theory predicts two light sterile neutrinos below the eV scale.

The superpotential of this theory is very simple, one has the MSSM superpotential and an extra Yukawa interaction for the neutrinos,

$$\mathcal{W}_{B-L} = \mathcal{W}_{RpC} \;+\; Y_\nu \, \hat{L}\hat{H}_u\hat{\nu}^c, \tag{3}$$

where

$$\mathcal{W}_{RpC} = Y_u\hat{Q}\hat{H}_u\hat{u}^c - Y_d\hat{Q}\hat{H}_d\hat{d}^c - Y_e\hat{L}\hat{H}_d\hat{e}^c. \tag{4}$$

The new soft terms relevant for our discussion are given by

$$V_{soft} \supset M^2_{\tilde{\nu}^c}|\tilde{\nu}^c|^2 + \left(A_\nu \tilde{L} H_u \tilde{\nu}^c \;+\; \text{h.c.}\right). \tag{5}$$

Now, in order to break the extra gauge symmetry, B-L, there is only one solution: *The right-handed sneutrino must get a vacuum expectation value !*

It is very easy to show how the right-handed sneutrino gets a VEV. We compute the scalar potential and from the minimization conditions one finds

$$v_R^2 = -8 \frac{M_{\tilde{\nu}^c}^2}{g_{BL}^2}, \tag{6}$$

where $M_{\tilde{\nu}^c}^2$ has to be negative. It has been shown in Ref. [18] that it is very easy to have a realistic spectrum for all sfermions in the theory. As one can expect the sfermion masses have a new contribution from the B-L D-term and typically the left-handed sleptons should be lighter if the soft mass for all the particles are similar at the TeV scale.

The neutrino spectrum in this theory is quite peculiar, it is easy to prove that only one of the right-handed neutrinos get a large mass around TeV scale and the rest are light. Therefore, one can have a 3 + 2 system with masses below the eV scale. See Refs. [10, 20, 21] for the detailed analysis. One can think about the possible new contributions coming from proton decay in this theory. Using the lepton number violating interactions above and the terms coming from the higher-dimensional operator $\hat{u}^c \hat{d}^c \hat{d}^c \hat{\nu}^c / \Lambda_B$ one finds new contributions to proton decay [18] but assuming a large cutoff, around the GUT scale, one can satisfy the experimental bounds on the proton decay lifetime. Notice that here one needs to assume again a large desert from the TeV scale to the GUT scale.

This theory makes very interesting predictions for the LHC because the B-L symmetry is broken at the TeV scale and one has lepton number violating signals. Therefore, in order to test this theory one needs to look for the B-L gauge boson, identify the signals from the supersymmetric particles but in this case there is no missing energy signals associated to a stable LSP because R-parity is broken. We have investigated the most striking signals from this theory in Ref. [18] where we have pointed out the properties of the channels with four leptons, three of them with the same electric charge, and four jets. See Fig.4 for the topology of these signals and Ref. [18] for the detailed analysis of these signals at the LHC. Before we finish this

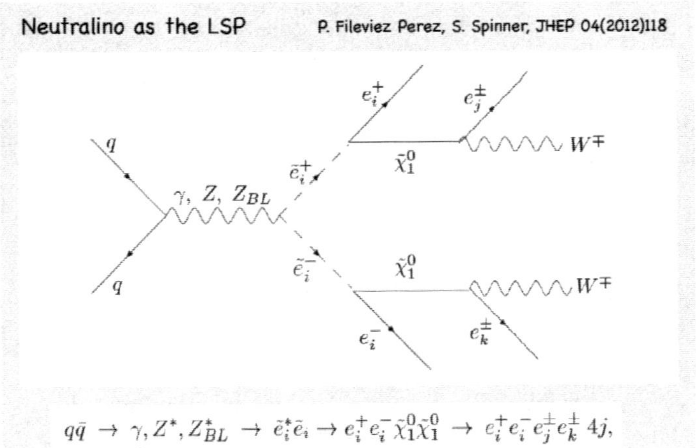

Figure 4. Lepton Number Violating Signals [18].

section let us mention the main idea behind the simplest theory for R-parity conservation. In order to write down a simple theory where local B-L is broken to matter parity, we need to go beyond the minimal model discussed above and add extra Higgses. Since the right-handed neutrinos are present in the theory we always have the vacuum which corresponds to their vacuum expectation value. Therefore, we need to assume that only the extra Higgses with an even number of B-L break the gauge symmetry. See also Ref. [22, 23] for the application of the Stueckelberg mechanism in this context. Unfortunately, in the context of grand unified theories we cannot find a simple scenario for R-parity conservation. It is well-known that in SUSY $SU(5)$ we need to impose matter parity by hand as in the MSSM, and in the context of $SO(10)$ we need huge representations to achieve R-parity conservation. Therefore, one can say that there is no a simple scenario based on GUTs for R-parity conservation. Here, we would like to emphasize again that the minimal supersymmetric theory based on B-L makes a hard prediction: R-parity must be spontaneously broken and one expects striking lepton number violating signals at the LHC if supersymmetry is realized at the low scale.

4. Life in the Desert: Breaking B and L at the TeV scale

We have discussed the possibility to understand the origin of the lepton and baryon number violating interactions present in the MSSM at the renormalizable level. Unfortunately, in this case one needs to assume a large cutoff in order to satisfy the bounds on the proton decay lifetime. Here we will discuss a different theory where B and L can be broken at the TeV scale and there is no need to assume a desert.

In order to prove that B and L can be broken at the low scale we can define a theory where we have two extra local symmetries. Then, the theory is based on $G_{SM} \bigotimes U(1)_B \bigotimes U(1)_L$[24] where G_{SM} is the gauge group of the Standard Model. We refer to this model as "BLMSSM". The main features of this theory are:

- The baryon and lepton numbers are local symmetries broken at the SUSY scale.
- In order to cancel all anomalies we add a vector-like generation which is composed of $\hat{Q}_4$, $\hat{u}_4^c$, $\hat{d}_4^c$, $\hat{L}_4$, $\hat{e}_4^c$, $\hat{\nu}_4^c$, and $\hat{Q}_5^c$, $\hat{u}_5$, $\hat{d}_5$, $\hat{L}_5^c$, $\hat{e}_5$, $\hat{\nu}_5$. Here the baryon and lepton numbers of the new chiral superfields are different since one has to satisfy the conditions $B_{Q_4} + B_{Q_4^c} = -1$ and $\ell_{L_4} + \ell_{L_4^c} = -3$.
- In order to break the local baryonic symmetry and generate vector-like masses for the extra quarks we need to add new Higgses. In this case we need to have for example the term $\hat{Q}_4 \hat{Q}_4^c \hat{S}_B$. Therefore, this term defines the baryonic quantum number of $\hat{S}_B \sim 1$, and we have $\hat{\bar{S}}_B \sim -1$. As we pointed out in Ref. [24] one cannot avoid the operator $\hat{u}^c \hat{d}^c \hat{d}^c \hat{S}_B / \Lambda_B$, and after symmetry breaking one has baryon number violating interactions which can be suppressed by the cutoff of the theory.
- In the leptonic sector the situation is more involved since one can have different scenarios: a) In the first case if one sticks to the seesaw mechanism the new Higgses should have an even leptonic number, $\hat{S}_L, \hat{\bar{S}}_L \sim \pm 2$. The new leptons have only chiral masses and there is no proton decay after symmetry breaking because all the interactions violate the total lepton number in an even number. This case was investigated in details in Ref. [24] and Ref. [25].
  b) One can generate vector-like masses for the new leptons, $\hat{L}_5^c \hat{L}_4 \hat{S}_L$, if the Higgses $\hat{S}_L, \hat{\bar{S}}_L \sim \pm 3$. In this case one gets operators mediating proton decay since $|\Delta B| = 1$ and $|\Delta L| = 3$. One of the operators that mediate proton decay in this case has dimension eleven, $\hat{Q}\hat{Q}\hat{Q}\hat{L}(\hat{L}\hat{H}_u)^2 \hat{S}_B \hat{\bar{S}}_L / \Lambda^7$. Therefore, the cutoff scale in this case does not need to be very large.
- There is no extra flavour violation at tree level because the new fermions do not mix with the SM fermions.
- There is no Landau pole at the low scale.

It is important to mention that in the context of the BLMSSM one could modify the LHC bounds on the supersymmetric particles because one has baryon number violation at the low scale. We have investigated in detail the impact of the new fermions on the Higgs mass [26] and decays in Ref. [25].

5. Summary

We have discussed in a general way the need to postulate a desert between the low and high scales in order to satisfy the bounds on the proton decay lifetime. In the first part of this review we have shown that the minimal supersymmetric B-L theory predicts that R-parity should be spontaneously broken and one expects lepton number violation at the LHC. The simplest scenarios for the conservation of R-parity were briefly discussed. It has been mentioned that there is no simple grand unified theory defined in four dimensions where we understand the conservation of R-parity. A simple theory where there is no need for a desert has been discussed. In this case B and L can be broken at the low scale and we do not get any operator mediating proton decay. It is important to emphasize that the two scenarios discussed here can be realized at the TeV scale and one can have very interesting supersymmetric signals at the Large Hadron Collider.

Acknowledgments

I would like to thank the organizers of the 18th International Symposium on Particles, Strings and Cosmology (PASCOS2012) in Merida for the invitation and warm hospitality. It is a great pleasure to thank my collaborators J. M. Arnold, V. Barger, B. Fornal, P. Nath, S. Spinner, and M. B. Wise for many discussions and enjoyable collaborations. This work was partially supported by the James Arthur Fellowship, CCPP, New York University.

References

[1] S. Weinberg, "Baryon and Lepton Nonconserving Processes," Phys. Rev. Lett. **43** (1979) 1566.
[2] M. Miura, Talk given at the 2011 International Workshop on Baryon and Lepton Number Violation (BLV2011), UTK, USA, 2011.
[3] P. Nath and P. Fileviez Perez, "Proton stability in grand unified theories, in strings and in branes," Phys. Rept. **441** (2007) 191.
[4] Y. Kamyshkov, Talk given at the Spontaneous Workshop VI, Cargese, May 11, 2012.
[5] Michael D. Litos, "A search for dinucleon decay into kaons using the SK water cherenkov detector", Ph.D. Thesis, Boston University, 2010.
[6] J. Beringer *et al.* [Particle Data Group Collaboration], "Review of Particle Physics (RPP)," Phys. Rev. D **86** (2012) 010001.
[7] M. Hirsch, M. A. Diaz, W. Porod, J. C. Romao and J. W. F. Valle, "Neutrino masses and mixings from supersymmetry with bilinear R parity violation: A Theory for solar and atmospheric neutrino oscillations," Phys. Rev. D **62** (2000) 113008 [Erratum-ibid. D **65** (2002) 119901] [hep-ph/0004115].
[8] C. S. Aulakh and R. N. Mohapatra, "Neutrino as the Supersymmetric Partner of the Majoron," Phys. Lett. B **119** (1982) 136.
[9] M. J. Hayashi and A. Murayama, "Radiative Breaking of $SU(2)_R \times U(1)_{B-L}$ gauge symmetry induced by broken $N = 1$ Supergravity in a Left-Right symmetric model," Phys. Lett. B **153** (1985) 251.
[10] R. N. Mohapatra, "Mechanism For Understanding Small Neutrino Mass In Superstring Theories," Phys. Rev. Lett. **56** (1986) 561.
[11] L. M. Krauss and F. Wilczek, "Discrete Gauge Symmetry in Continuum Theories," Phys. Rev. Lett. **62** (1989) 1221.
[12] A. Font, L. E. Ibanez and F. Quevedo, "Does Proton Stability Imply the Existence of an Extra Z0?," Phys. Lett. B **228** (1989) 79.
[13] A. Masiero and J. W. F. Valle, "A Model For Spontaneous R Parity Breaking," Phys. Lett. B **251** (1990) 273.
[14] S. P. Martin, "Some simple criteria for gauged R-parity," Phys. Rev. D **46** (1992) 2769 [hep-ph/9207218].
[15] C. S. Aulakh, A. Melfo, A. Rasin and G. Senjanovic, "Seesaw and supersymmetry or exact R-parity," Phys. Lett. B **459** (1999) 557 [hep-ph/9902409].

[16] P. Fileviez Perez and S. Spinner, "Spontaneous R-Parity Breaking and Left-Right Symmetry," Phys. Lett. B **673** (2009) 251 [arXiv:0811.3424 [hep-ph]].
[17] V. Barger, P. Fileviez Perez and S. Spinner, "Minimal gauged U(1)(B-L) model with spontaneous R-parity violation," Phys. Rev. Lett. **102** (2009) 181802 [arXiv:0812.3661 [hep-ph]].
[18] P. Fileviez Perez and S. Spinner, "The Minimal Theory for R-parity Violation at the LHC," JHEP **1204** (2012) 118 [arXiv:1201.5923 [hep-ph]].
[19] P. Fileviez Perez and S. Spinner, "The Fate of R-Parity," Phys. Rev. D **83** (2011) 035004 [arXiv:1005.4930 [hep-ph]].
[20] D. K. Ghosh, G. Senjanovic and Y. Zhang, "Naturally Light Sterile Neutrinos from Theory of R-parity," Phys. Lett. B **698** (2011) 420 [arXiv:1010.3968 [hep-ph]].
[21] V. Barger, P. Fileviez Perez and S. Spinner, "Three Layers of Neutrinos," Phys. Lett. B **696** (2011) 509 [arXiv:1010.4023 [hep-ph]].
[22] D. Feldman, P. Fileviez Perez and P. Nath, "R-parity Conservation via the Stueckelberg Mechanism: LHC and Dark Matter Signals," JHEP **1201** (2012) 038 [arXiv:1109.2901 [hep-ph]].
[23] P. Nath, "SUGRA Grand Unification, LHC and Dark Matter," arXiv:1207.5501 [hep-ph].
[24] P. Fileviez Perez and M. B. Wise, "Breaking Local Baryon and Lepton Number at the TeV Scale," JHEP **1108** (2011) 068 [arXiv:1106.0343 [hep-ph]].
[25] J. M. Arnold, P. Fileviez Perez, B. Fornal and S. Spinner, "On Higgs Decays, Baryon Number Violation, and SUSY at the LHC," Phys. Rev. D **85** (2012) 115024 [arXiv:1204.4458 [hep-ph]].
[26] P. Fileviez Perez, "SUSY Spectrum and the Higgs Mass in the BLMSSM," Phys. Lett. B **711** (2012) 353 [arXiv:1201.1501 [hep-ph]].

Tests of the Electroweak Standard Model

Jens Erler

Departamento de Física Teórica, Instituto de Física, Universidad Nacional Autónoma de México, México D.F. 04510, México

E-mail: erler@fisica.unam.mx

Abstract. Electroweak precision tests of the Standard Model of the fundamental interactions are reviewed ranging from the lowest to the highest energy experiments. Results from global fits are presented with particular emphasis on the extraction of fundamental parameters such as the Fermi constant, the strong coupling constant, the electroweak mixing angle, and the mass of the Higgs boson. Constraints on physics beyond the Standard Model are also discussed.

1. Introduction

The Standard Model (SM) of the electroweak (EW) interactions has been developed mostly in the 1960s, where the gauge group $SU(2)_L \times U(1)_Y$ was suggested [1], the Higgs mechanism for spontaneously broken gauge theories developed, and the model for leptons constructed explicitly [2]. Subsequently, key predictions of the SM were observed in the 1970s, including neutral currents and parity non-conservation in atoms and in deep-inelastic electron scattering (DIS) [3]. The basic structure of the SM was established in the 1980s after mutually consistent values of the weak mixing angle, $\sin^2\theta_W$, were determined from many different processes. The 1990s saw the highly successful Z-factories, LEP and SLC, and the confirmation of the SM at the loop level. It thus became clear that any new physics beyond the SM could at most be a perturbation. The previous decade added precision measurements in the neutrino and quark sectors (including a 0.5% measurement of the top quark mass [4]), as well as ultra-high precision determinations of the W-boson mass, M_W (to 2×10^{-4}) [5], the anomalous magnetic moment of the muon [6], and the Fermi constant, G_F [7]. These results suggest that the new physics must be separated by at least a little hierarchy from the EW scale unless one considers the possibility that a conspiracy is at work. The current decade will elucidate the EW symmetry breaking sector at the LHC and witness a new generation of experiments at the intensity frontier with sensitivities to the multi-TeV scale and beyond.

The next section reviews some recent developments in slightly more detail. Interpretations of these and other results for the mass of the Higgs boson, M_H, and new physics, are discussed, respectively, in the two sections thereafter.

2. Recent Developments

2.1. Properties of charged leptons

The MuLan Collaboration at the PSI in Switzerland [7] has measured the μ-lifetime to parts-per-million precision, $\tau_\mu = 2.1969803(2.2) \times 10^{-6}$ s, which translates into a determination of

$$G_F = 1.1663787(6) \times 10^{-5} \text{ GeV}^{-2}. \quad (1)$$

Content from this work may be used under the terms of the Creative Commons Attribution 3.0 licence. Any further distribution of this work must maintain attribution to the author(s) and the title of the work, journal citation and DOI.
Published under licence by IOP Publishing Ltd

The Higgs vacuum expectation value is given by $\langle 0|H|0\rangle = (\sqrt{2}G_F)^{-1/2} = 246.22$ GeV. This measurement is so precise that even the error in the definition of the atomic mass unit (u) can shift G_F (MuLan quotes $G_F = 1.1663788(7) \times 10^{-5}$ GeV^{-2}). Moreover, it is so precise that the effect of the finite M_W in the W-propagator is no longer negligible. One may either choose to correct for it, *i.e.*, absorb it in Δq defined through $\tau_\mu^{-1} \propto G_F^2 m_\mu^5(1+\Delta q)$, or else not to do so [8], *i.e.*, absorb it in Δr [9] defined in terms of the accurately known fine structure constant, α, and Z-boson mass, M_Z,

$$\sqrt{2}G_F M_W^2 \left(1 - \frac{M_W^2}{M_Z^2}\right) \equiv \frac{\pi\alpha}{1-\Delta r}\ . \tag{2}$$

The latter convention is motivated by an effective Fermi theory point of view, and used by MuLan and since this year also by the PDG [10].

What τ_μ is to G_F is the τ-lifetime to the strong coupling constant, α_s. At least one low-energy α_s-value is needed to promote the Z-width and related Z-pole observables from a quantitative measurement in QCD to an EW SM test (or to constrain physics beyond the SM). Perturbative QCD has recently been extended to 4-loop order [12], but there is a controversy whether the perturbative series should be truncated, *i.e.*, fixed order perturbation theory (FOPT) should be used [13], or whether higher order terms from the running strong coupling in the complex plane should be re-summed in what is called contour-improved perturbation theory (CIPT) [14]. Unfortunately, FOPT and CIPT appear to converge to different values. There are also non-perturbative contributions parametrized by condensate terms which can be constrained by experimentally determined spectral functions. There are two different approaches [15, 16] which at present give very similar numerical results. Using FOPT and the condensates from Ref. [16],

$$\alpha_s[\tau] = 0.1193 \pm 0.0021, \qquad\qquad \alpha_s[Z] = 0.1197 \pm 0.0028, \tag{3}$$

is found, where the latter determination from the Z-pole is the only extraction of α_s with a very small theory uncertainty. The two values can be seen to agree perfectly.

The anomalous magnetic moment of the muon was measured to extreme precision,

$$a_\mu \equiv \frac{g_\mu - 2}{2} = (1165920.80 \pm 0.63) \times 10^{-9}, \tag{4}$$

by the BNL–E821 Collaboration [6]. The prediction, $a_\mu = (1165918.41 \pm 0.48) \times 10^{-9}$, from the SM includes e^+e^- as well as τ-decay data in the dispersion integral needed to constrain the two- and three-loop vacuum polarization contributions and differs by 3.0 σ. The data based on τ-decays requires an isospin rotation and a corresponding correction to account for isospin violating effects and suggests a smaller (2.4 σ) discrepancy, while the e^+e^--based data sets (from annihilation and radiative returns) by themselves would imply a 3.6 σ conflict. Indeed, there is a 2.3 σ discrepancy between the experimental branching ratio, $B(\tau^- \to \nu\pi^0\pi^-)$, and its SM prediction using the e^+e^- data [17]. In view of this, it is tempting to ignore the τ-decay data and blame the difference to the e^+e^- data on unaccounted for isospin violating effects. However, there is also a 1.9 σ experimental conflict between KLOE and BaBar (both using the radiative return method [18]) the latter not being inconsistent with the τ-data.

The above results include an additional uncertainty from hadronic 3-loop light-by-light scattering diagrams which contribute, $\Delta a_\mu(\gamma\times\gamma) = (1.05 \pm 0.26) \times 10^{-9}$ [19]. This is consistent with the 95% CL upper bound, $\Delta a_\mu(\gamma\times\gamma) < 1.59 \times 10^{-9}$, found in Ref. [20]. One may point out that if the three dominant errors from experiment (6.3×10^{-10}), hadronic vacuum polarization (3.6×10^{-10}) and light-by-light scattering (2.6×10^{-10}) can be pushed below 3×10^{-10}, then a 5 σ discovery would be established (if the central value persists). The nominal $\Delta a_\mu(\gamma\times\gamma)$ error is already there, but it is also the hardest to defend.

As for the question whether the deviation in a_μ may arise from physics beyond the SM (especially supersymmetry [21]), my personal take is that I am less concerned about these hadronic issues than the absence of convincing new physics hints at the Tevatron or the LHC.

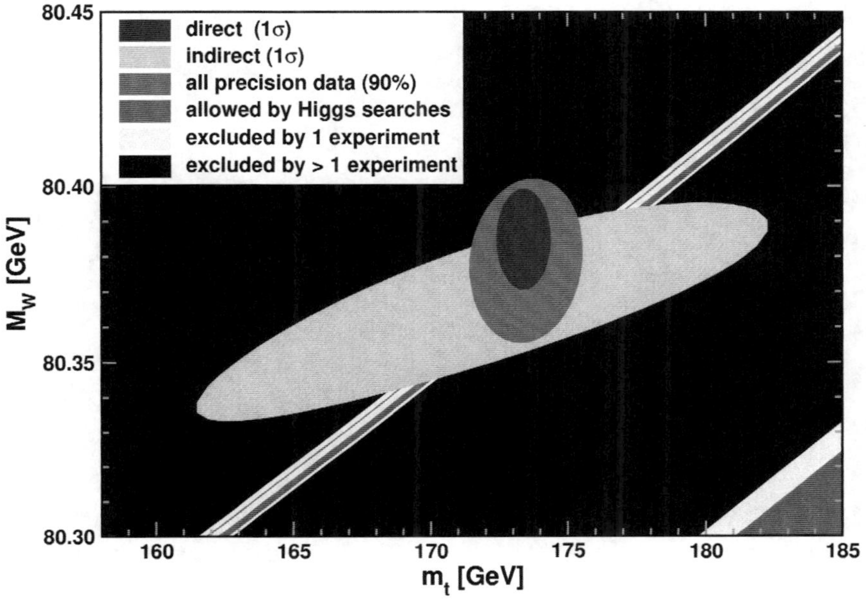

Figure 1. 1 σ region in M_W *vs.* m_t for the direct and indirect EW precision data, and the 90% CL region ($\Delta\chi^2 = 4.605$) allowed by both data sets. The SM predictions are also indicated, where the dark (green) bands correspond to M_H which are currently allowed at the 95% CL. The bright (yellow) bands are excluded by one experiment and the remaining (black) regions are ruled out by more than one experiment.

2.2. $\sin^2\theta_W^{\text{on-shell}}$

The most precise derived and purely EW precision observable is no longer the Z-pole combination of $\sin^2\theta_W$, but rather $M_W = 80.387 \pm 0.016$ GeV from the CDF and DØ Collaborations at the Tevatron [5] which is dominated by a ± 19 MeV determination by CDF using only 2.2 fb^{-1} of their data. Together with the LEP 2 combination [22], $M_W = 80.376 \pm 0.033$ GeV, one obtains for the on-shell definition of $\sin^2\theta_W$,

$$\sin^2\theta_W^{\text{on-shell}} \equiv 1 - \frac{M_W^2}{M_Z^2} = 0.22290 \pm 0.00028, \tag{5}$$

from which and can extract $M_H = 96^{+29}_{-25}$ GeV. The prospects for the full 10 fb^{-1} dataset are a ± 13 MeV M_W determination from CDF alone, even when no reduction of the parton distribution function (PDF) (± 10 MeV) and QED (± 4 MeV) uncertainties is assumed. In the most optimistic scenario, CDF could shrink the error to ± 10 MeV, which is to be compared with the ± 6 MeV accuracy expected from a threshold scan at a future International Linear Collider. The current direct and indirect determinations of M_W and the top quark mass, m_t, are compared in Figure 1.

Values of $\sin^2\theta_W^{\text{on-shell}}$ are also often quoted from R_ν, $R_{\bar\nu}$ or R^-, which are combinations of

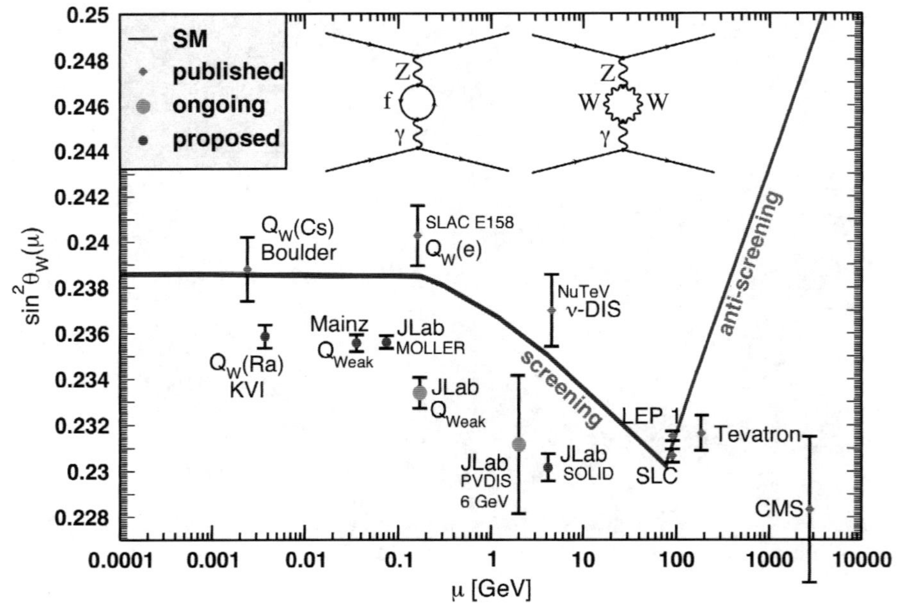

Figure 2. Current and future measurements of the running weak mixing angle. The uncertainty in the prediction is small except possibly in the hadronic transition region roughly between 0.1 and 2 GeV [37]. The relevant Q^2 of the Tevatron and CMS values make them effectively additional Z-pole measurements, but for clarity they have been shifted horizontally to the right.

ν-DIS and $\bar{\nu}$-DIS neutral-current (NC) and charged-curent (CC) cross sections,

$$R_\nu \equiv \frac{\sigma_{\nu N}^{\rm NC}}{\sigma_{\nu N}^{\rm CC}}, \qquad R_{\bar{\nu}} \equiv \frac{\sigma_{\bar{\nu} N}^{\rm NC}}{\sigma_{\bar{\nu} N}^{\rm CC}}, \qquad R^- \equiv \frac{\sigma_{\nu N}^{\rm NC} - \sigma_{\bar{\nu} N}^{\rm NC}}{\sigma_{\nu N}^{\rm CC} - \sigma_{\bar{\nu} N}^{\rm CC}} \approx \frac{1}{2} - \sin^2\theta_W. \tag{6}$$

It should be cautioned, however, that $\sin^2\theta_W^{\rm on-shell}$ is affected very differently by variations of M_H and by physics beyond the SM than the quantities defined in Eq. (6). The most precise determination has been obtained by the NuTeV Collaboration [23] at Fermilab,

$$\sin^2\theta_W^{\rm on-shell} = 0.2277 \pm 0.0016, \tag{7}$$

which is 3.0 σ above the SM prediction, $\sin^2\theta_W^{\rm on-shell} = 0.22296 \pm 0.00028$. The deviation sits in the left-handed effective quark coupling, g_L^2, which is 2.7 σ off. Various SM effects have been suggested, such as an asymmetric strange sea, isospin violation (both, from QED splitting effects [24] and through the PDFs [25]), or nuclear effects (*e.g.*, the so-called isovector EMC effect [26]), as well as QED [27], QCD [28] and EW [29] radiative corrections. We stress that the precise impact of these effects need to be evaluated carefully by the collaboration with a new and self-consistent set of PDFs, including new radiative corrections, while simultaneously allowing isospin breaking and asymmetric strange seas. This effort is currently on its way.

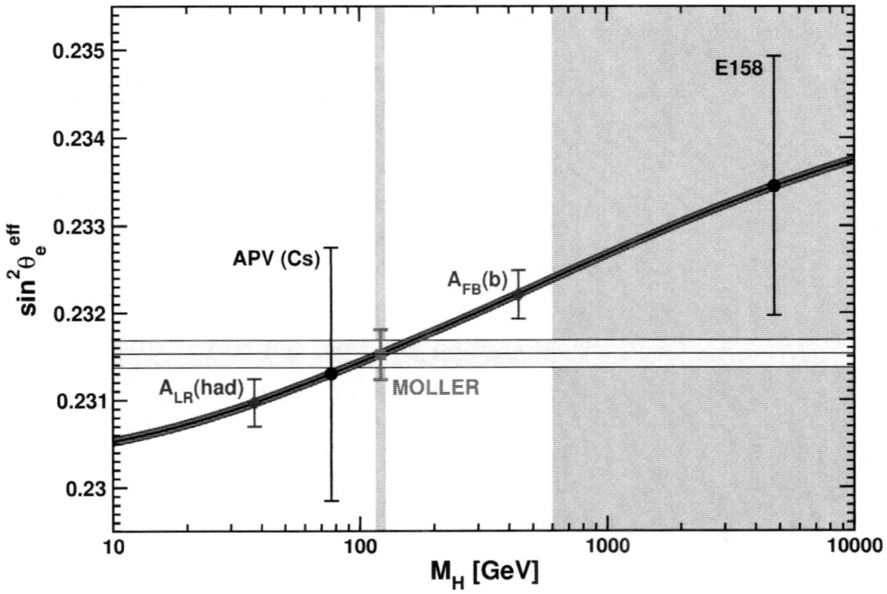

Figure 3. Implication of $\sin^2\theta_W$ measurements for M_H. Shown are the most precise determinations from LEP 1 and the SLC, and the extractions from APV and from current (E158) and future (MOLLER) polarized Møller scattering. Also indicated are the non-excluded intervals from direct Higgs searches.

2.3. Parity-violating electron scattering

High precision measurements in the EW sector are also possible at the intensity frontier, when QED and QCD effects are filtered out by using parity-violating observables.

The JLab Qweak detector [30] at the 6 GeV CEBAF was dedicated to a measurement of the weak charge of the proton, $Q_W^p \propto 1-4\sin^2\theta_W$, to 4% precision in elastic polarized e^-p scattering at $Q^2 = 0.026$ GeV^2. Data taking is complete and the analysis is in progress. Q_W^p is similar to the weak charges of heavy nuclei measured in atomic parity violation (APV) but at a different kinematics. This circumstance results in a re-enhancement of the $\gamma - Z$ box contribution [31, 32] introducing an extra theory uncertainty.

The $\gamma - Z$ box is less of an issue at lower Q^2 which is one of the reasons why a similar experiment is also planned at a future facility (MESA) in Mainz at $Q^2 = 0.0022$ GeV^2. The projected uncertainties for Q_W^p and the extracted $\sin^2\theta_W$ are 2.1% and ± 0.00037, respectively.

MOLLER [33] is an ultra-high precision measurement of $\sin^2\theta_W$ in polarized Møller scattering at the 12 GeV upgraded CEBAF [34]. It aims at a factor of 5 improvement over a similar experiment at SLAC by the E158 Collaboration [35], and would be one of the worlds most precise determinations of $\sin^2\theta_W$ and the most accurate at low energies.

PVDIS was a deep-inelastic polarized e^- scattering experiment using the 6 GeV CEBAF and is currently in the analysis phase [36]. Together with SOLID (at 12 GeV) an array of kinematics

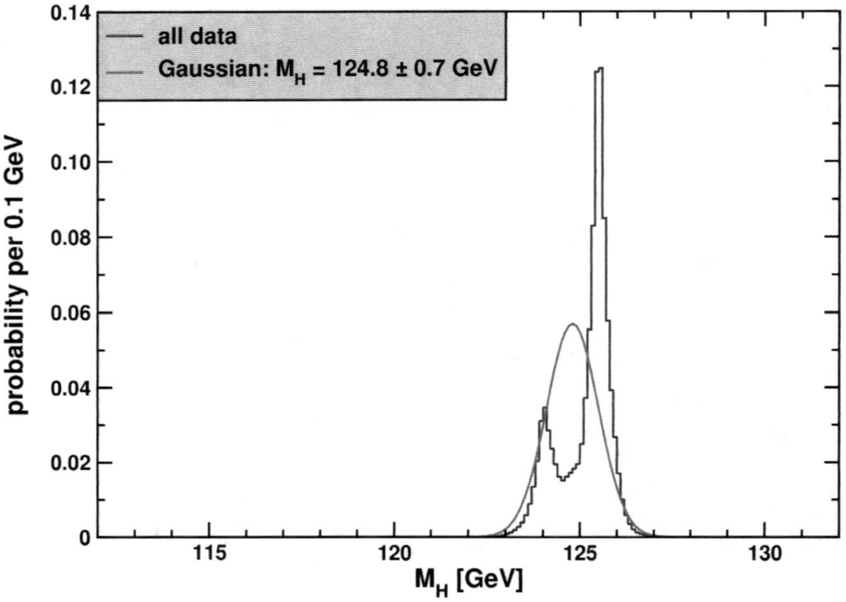

Figure 4. The histogram shows the normalized probability distribution of M_H. The bell shaped curve is a reference Gaussian density defined to contain the same probability as the histogram over the region of bins which are higher than the tail bins. The significance of this region corresponds to 3.4 σ.

points will be measured to test strong, EW, and new physics. Figure 2 summarizes these and other current and future (projected) determinations of $\sin^2\theta_W$ as a function of energy scale μ.

3. SM Interpretation: M_H

The various $\sin^2\theta_W$ measurements discussed above can be used to constrain M_H and compare it with the results obtained at the LHC. It is important to recall that the most precise determinations at LEP 1 (from the forward-backward cross-section asymmetry of Z-bosons decaying into $b\bar{b}$ pairs, $A_{FB}(b)$) and at the SLC (from the polarization asymmetry for hadronic final states, $A_{LR}(\text{had})$), both of which being mostly sensitive to the initial state (electron) coupling, are discrepant by three standard deviations. Their average, on the other hand, corresponds to values of M_H that are in perfect agreement with the Higgs boson candidates seen by the ATLAS [38] and CMS [39] Collaborations at the LHC. This is illustrated in Figure 3 together with the low-energy determinations from E158 [35] and APV which is dominated by the experiment in Cs [40] and makes use of the atomic theory calculation[1] of Ref. [42].

Estimating the significance of the LHC data [38, 39] by themselves requires a "look elsewhere

[1] After the conference had adjourned there appeared an update of the atomic structure calculation [41] finding significant corrections to formally subleading terms. Taking this into account moves the extracted Cs weak charge 1.5 σ below the SM prediction, which then favors lower values of M_H.

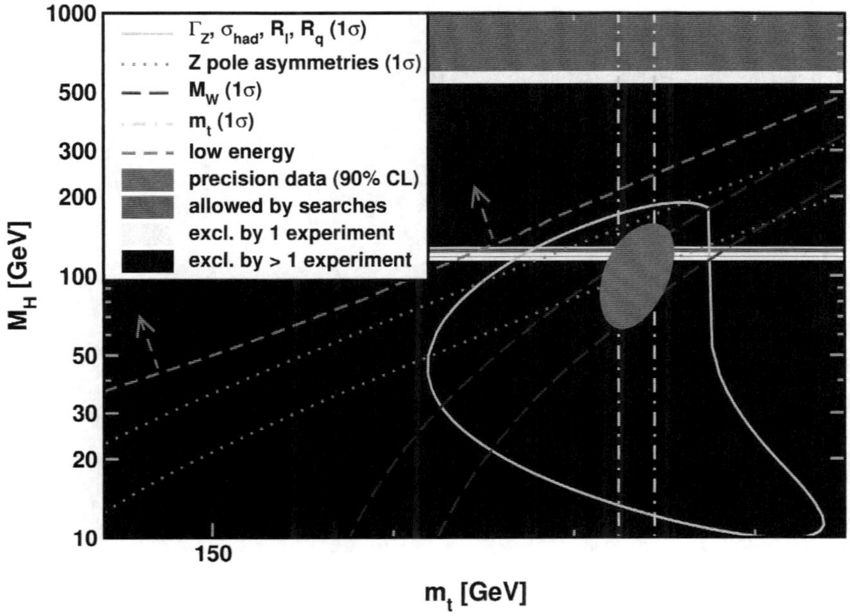

Figure 5. 1 σ contours of M_H as a function of the top quark pole mass for various inputs, and the 90% CL region allowed by all data. The color codes for the bands are as in Figure 1.

effect correction" which is, however, poorly defined. It can be avoided when they are combined with the Higgs search results from LEP 2 [43] and the Tevatron [44] as well as with the EW precision data [45], the latter providing a normalizable probability distribution[2] shown in Figure 4. This requires the validity of the SM which used to be a very strong assumption in the past. But with the absence of clear new physics signals at the energy frontier this can now be seen as a reasonable approximation.

As for the global EW fit without the collider events, I currently find,

$$M_H = 102^{+24}_{-20} \text{ GeV}. \tag{8}$$

To extract values of M_H from the EW precision data, as for example in Eq. (8), it is important to know the top quark mass, m_t, to very high precision. The various measurements from the Tevatron [49] and the LHC [50] (strongly dominated by the CMS μ+jets channel) combine to,

$$m_t = 173.21 \pm 0.51_{\text{uncorr}} \pm 0.75_{\text{corr}} \pm 0.5_{\text{theo}} \text{ GeV} = 173.2 \pm 1.0 \text{ GeV}, \tag{9}$$

where I assumed that the Tevatron systematic error is common to both colliders, and where I have added a theory uncertainty from the relation [51] between the top quark pole mass and the

[2] After the conference had adjourned there appeared updates of the LHC [46, 47] and Tevatron [48] Higgs search results including the announcement of the observation of a new boson consistent with the SM Higgs. For an update of Figure 4 reflecting the new data, see Ref. [45].

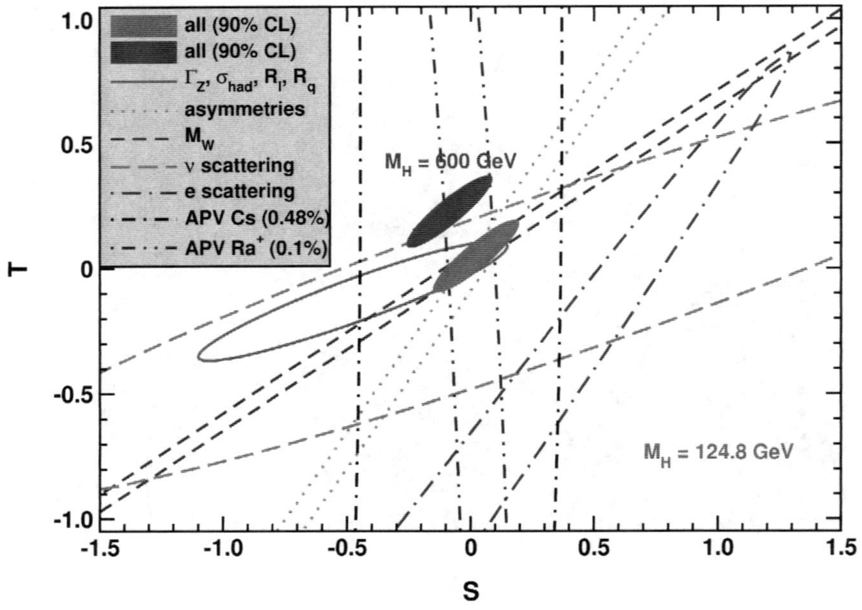

Figure 6. 1 σ contours in S and T from various inputs assuming $M_H = 124.8$ GeV except for the upper (violet) one for all data which is for $M_H = 600$ GeV. The contour labeled APV Ra^+ refers to a *future* experiment on a single trapped Ra ion which is in preparation at the KVI in Groningen [62]. The atomic structure of Ra^+ is alkali-like so that the atomic theory parallels that of Cs, but due to its greater neutron excess, Ra constrains a linear combination of S and T which is different from Cs and quite orthogonal to the M_W and $\sin^2\theta_W$ contours.

$\overline{\text{MS}}$-mass definitions (the size of the three-loop term). It is moreover assumed that the kinematic mass extracted from the collider events corresponds within this uncertainty to the pole mass (see also Ref. [52]). Further improvements will eventually require to extract definitions of m_t which are easier to interpret theoretically. For example, by combining the effective theories for soft collinear (SCET) and heavy quarks (HQET) one can form the jet mass of Ref. [53]. Alternatively, one may extract the $\overline{\text{MS}}$-mass, $\bar{m}_t(\bar{m}_t) = 160.0 \pm 3.3$ GeV, which is the one actually entering the global EW analysis, directly from the $t\bar{t}$ cross-section [54] (this corresponds to a pole mass of $m_t = 169.6 \pm 3.5$ GeV and would yield $M_H = 81^{+32}_{-24}$ GeV instead of Eq. (8)). Constraints on M_H as a function of m_t are shown in Figure 5 for various data sets.

4. New Physics Interpretations

4.1. Oblique new physics

The EW precision tests also set strong constrains on models of new physics. *E.g.*, an extra generation or anti-generation of fermions is severely constrained [55], and if the Higgs hints are real, these are ruled out[3] at the 99.6% CL [56]. This leaves us with basically three scenarios, all

[3] After the conference had adjourned there appeared a new analysis [57] raising the exclusion to the 5.3 σ CL.

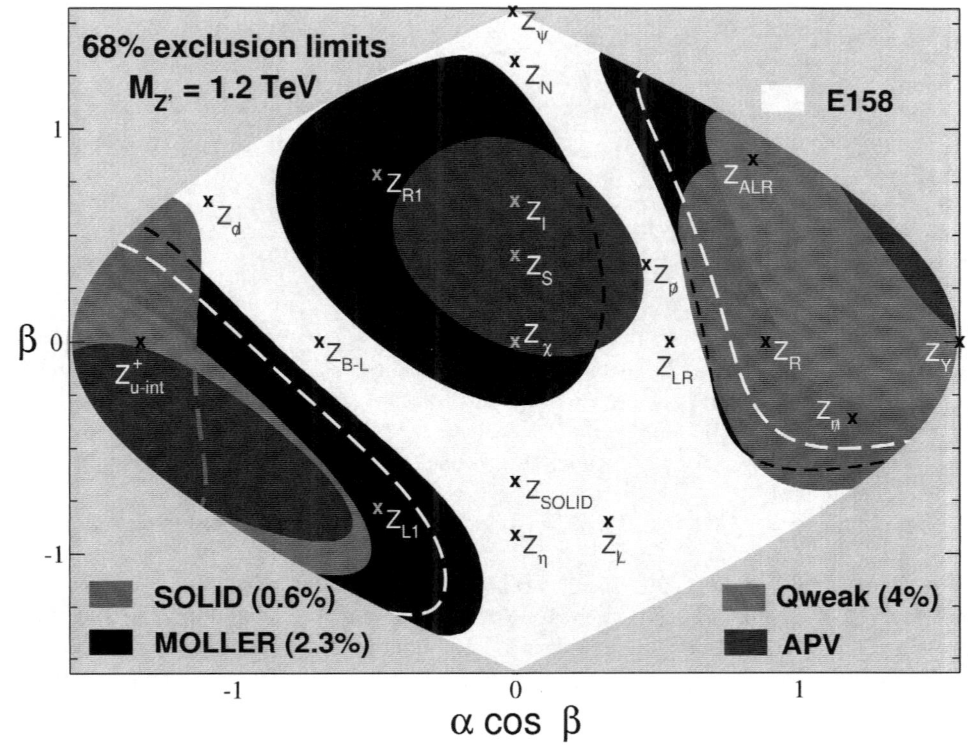

Figure 7. 68% exclusion constraints on the E_6 parameters α and β for $M_{Z'} = 1.2$ TeV from various actual and hypothetical low energy measurements (for future measurements it is assumed that the central values will coincide with the SM).

of which in need of some tuning and faith (the mass spectra are generally quite similar): (i) One ignores the collider bumps (or assigns them to something else) and assumes $M_H \lesssim 120$ GeV (see e.g., Ref. [58]); (ii) one assumes instead $M_H \gtrsim 450$ GeV [59]; (iii) or one accepts $M_H \approx 125$ GeV and introduces new physics beyond a fourth generation, such as an extra Higgs doublet [60].

More generally, whenever the new physics is rather heavy and mostly affects the gauge boson self-energies, one can parametrize it in terms of the oblique parameters S and T [61] (a third parameter, U, is usually small). The constraints on S and T from various data sets are shown in Figure 6 (where $U = 0$ is assumed). The three-parameter fit result is shown in Table 1.

4.2. Extra Z bosons

Among the best motivated kinds of physics beyond the SM are additional neutral Z' bosons [63] which are *not* well described by oblique parameters. They easily appear in top-down scenarios like Grand Unified Theories or superstring constructions. In fact, it often requires extra assumptions if one wants to *avoid* an additional $U(1)'$ gauge symmetry or decouple the associated Z' from observation. This is even more true in bottom-up approaches where $U(1)'$ symmetries are a standard tool to alleviate problems in models of dynamical symmetry breaking, supersymmetry, large or warped extra dimensions, little Higgs, *etc.* And as all these models are linked to electroweak symmetry breaking, the Z' mass, $M_{Z'}$, should be in the TeV region, providing a rationale why they might be accessible at current or near future experiments.

Table 1. Result (including correlations) of the global fit to S, T, and U for $M_H = 125$ GeV.

parameter	fit result	correlations		
S	0.00 ± 0.10	1.00	0.89	−0.55
T	0.02 ± 0.11	0.89	1.00	−0.80
U	0.04 ± 0.09	−0.55	−0.80	1.00

Z' discovery would most likely occur as an s-channel resonance at a collider, but interference with the photon or the standard Z provides leverage also at lower energies. Once discovered at a collider, angular distributions may give an indication of its spin to discriminate it against states of spin 0 (*e.g.*, the sneutrino) and spin 2 (like the Kaluza-Klein graviton in extra dimension models). The diagnostics of its charges would be of utmost importance as they can hint at the underlying principles.

An interesting class of models is related to E_6, a plausible gauge group for unified model building. All representations of E_6 are free of anomalies so that its $U(1)'$ subgroups correspond to Z' candidates. Z' bosons with the same charges for the SM fermions as in E_6 also arise within a bottom-up approach [64] when anomaly cancellation is demanded in supersymmetric extensions of the SM together with a set of fairly general requirements. The breaking chain, $E_6 \to SO(10) \times U(1)_\psi \to SU(5) \times U(1)_\chi \times U(1)_\psi$, defines a 2-parameter class of models,

$$Z' = \cos\alpha \cos\beta Z_\chi + \sin\alpha \cos\beta Z_Y + \sin\beta Z_\psi, \tag{10}$$

where Y denotes hypercharge, and $\alpha \neq 0$ corresponds to the presence of a kinetic mixing term $\propto F_Y^{\mu\nu} F'_{\mu\nu}$. Figure 7 shows how the combined data from e^- scattering alone may cover the entire parameter space for a Z' boson from E_6 with a reference mass of 1.2 TeV. Moreover, the low energy constraints are complementary to other precision constraints on Z' bosons [65] as well as to collider searches [66].

4.3. Charged current observables

The observables discussed so far are mostly related to the weak NC, but tests of CC universality can also provide information on new physics. Deviations from first row unitarity of the CKM matrix are subject to the constraint, $\Delta_{\rm CKM} \equiv |V_{ud}|^2 + |V_{us}|^2 + |V_{ub}|^2 - 1 = -0.0001 \pm 0.0006$ [67] while those from lepton universality in $\pi^+ \to \ell^+ \nu_\ell(\gamma)$ decays ($\ell = e, \mu$) are constrained by [68],

$$\Delta_{e/\mu} \equiv \frac{\Delta R_{e/\mu}}{R_{e/\mu}} \equiv \frac{R^{\rm exp}_{e/\mu} - R^{\rm SM}_{e/\mu}}{R^{\rm SM}_{e/\mu}} = -0.0034 \pm 0.0030 \pm 0.0001, \tag{11}$$

and one finds for the minimal supersymmetric standard model (MSSM) the results in Figure 8. $\Delta_{\rm CKM}$ is enhanced when there is a large difference between the masses of the first generation squarks and the second generation sleptons. Similarly, $\Delta_{e/\mu}$ is enhanced when the first and second generation slepton masses are significantly split.

5. Conclusions

Precision tests have reached per-mille and sub per-mille accuracy in derived quantities. The data are in very good agreement with the SM with the only tantalizing deviation sitting in a_μ. When combined with the absence of any observation challenging the SM at the LHC, this provides tight constraints on new physics and it becomes increasingly likely that its energy scale is separated from the SM by at least a little hierarchy.

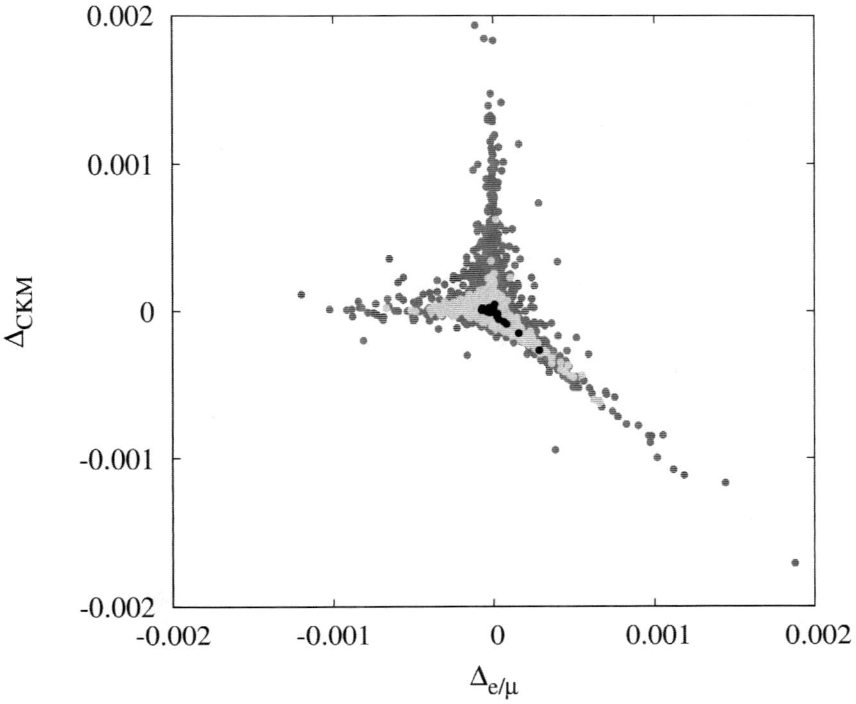

Figure 8. Scatter plot [69] of MSSM points satisfying the $\Delta_{\rm CKM}$ and $\Delta_{e/\mu}$ constraints. Points satisfying in addition the EW precision data including (excluding) LHC bounds are shown in black (green).

Acknowledgments
It is a pleasure to thank the organizers of PASCOS 2012 for the invitation to a very enjoyable symposium. I would also like to thank Sky Bauman, Leo Bellantoni, Jonathan Heckman, Paul Langacker, Shoaib Munir, Michael Ramsey-Musolf and Eduardo Rojas for collaboration on some of the covered topics. This work was supported by the CONACyT projects 82291–F and 15 1234.

References

[1] Glashow S L 1961 *Nucl. Phys.* **22** 579
[2] Weinberg S 1967 *Phys. Rev. Lett.* **19** 1264
[3] Prescott C Y *et al.* 1978 *Phys. Lett.* B **77** 347
[4] Aaltonen T *et al.* (CDF and DØ Collaborations) 2012 *Preprint* arXiv:1207.1069 [hep-ex]
[5] Tevatron EW Working Group (CDF and DØ Collaborations) 2012 *Preprint* arXiv:1204.0042 [hep-ex]
[6] Bennett G W *et al.* (Muon g-2 Collaboration) 2006 *Phys. Rev.* D **73** 072003
[7] Webber D M *et al.* (MuLan Collaboration) 2011 *Phys. Rev. Lett.* **106** 041803
[8] van Ritbergen T and Stuart R G 2000 *Nucl. Phys.* B **564** 343
[9] Sirlin A 1980 *Phys. Rev.* D **22** 971
[10] Erler J and Langacker P 2012 Electroweak Model and Constraints on New Physics *Reference* [11]
[11] Beringer J *et al.* (Particle Data Group) 2012 *Phys. Rev.* D **86** 010001
[12] Baikov P A, Chetyrkin K G and Kühn J H 2008 *Phys. Rev. Lett.* **101** 012002
[13] Beneke M and Jamin M 2008 *J. High Energy Phys.* JHEP09(2008)044
[14] Le Diberder F and Pich A 1992 *Phys. Lett.* B **286** 147
[15] Davier M *et al.* 2008 *Eur. Phys. J.* C **56** 305

[16] Boito D *et al.* 2012 *Phys. Rev.* D **85** 093015
[17] Davier M, Höcker A, Malaescu B and Zhang Z 2011 *Eur. Phys. J.* C **71** 1515
[18] Arbuzov A B, Kuraev E A, Merenkov N P and Trentadue L 1998 *J. High Energy Phys.* JHEP12(1998)009
[19] Prades J, de Rafael E and Vainshtein A 2009 *Preprint* arXiv:0901.0306 [hep-ph]
[20] Erler J and Toledo G 2006 *Phys. Rev. Lett.* **97** 161801
[21] Ellis J R, Hagelin J S and Nanopoulos D V 1982 *Phys. Lett.* B **116** 283
[22] Alcaraz J *et al.* (ALEPH, DELPHI, L3, OPAL and LEP EW Working Group) *Preprint* hep-ex/0612034
[23] Zeller G P *et al.* (NuTeV Collaboration) 2002 *Phys. Rev. Lett.* **88** 091802
[24] Glück M, Jimenez-Delgado P and Reya E 2005 *Phys. Rev. Lett.* **95** 022002
[25] Sather E 1992 *Phys. Lett.* B **274** 433
Rodionov E N, Thomas A W and Londergan J T 1994 *Mod. Phys. Lett.* A **9** 1799
Martin A D, Roberts R G, Stirling W J and Thorne R S 2004 *Eur. Phys. J.* C **35** 325
[26] Cloët I C, Bentz W and Thomas A W 2009 *Phys. Rev. Lett.* **102** 252301
[27] Arbuzov A B, Bardin D Y and Kalinovskaya L V 2005 *J. High Energy Phys.* JHEP06(2005)078
Park K, Baur U and Wackeroth D 2009 *Preprint* arXiv:0910.5013 [hep-ph]
Diener K P O, Dittmaier S and Hollik W 2004 *Phys. Rev.* D **69** 073005
[28] Dobrescu B A and Ellis R K 2004 *Phys. Rev.* D **69** 114014
[29] Diener K P O, Dittmaier S and Hollik W 2005 *Phys. Rev.* D **72** 093002
[30] Armstrong D S *et al.* (Qweak Collaboration) 2007 *Preprint* arXiv:1202.1255 [physics.ins-det]
[31] Gorshteyn M and Horowitz C J 2009 *Phys. Rev. Lett.* **102**, 091806
[32] Blunden P G, Melnitchouk W and Thomas A W 2011 *Phys. Rev. Lett.* **107**, 081801
[33] Mammei J (for the MOLLER Collaboration) 2012 *Preprint* arXiv:1208.1260 [hep-ex]
[34] Dudek J *et al.* 2012 *Preprint* arXiv:1208.1244 [hep-ex]
[35] Anthony P L *et al.* (SLAC–E158 Collaboration) 2005 *Phys. Rev. Lett.* **95** 081601
[36] Subedi R R *et al.* 2011 *AIP Conf. Proc.* **1374**, 602
[37] Erler J and Ramsey-Musolf M J 2005 *Phys. Rev.* D **72** 073003
[38] Aad G *et al.* (ATLAS Collaboration) 2012 *Phys. Rev.* D **86** 032003
[39] Chatrchyan S *et al.* (CMS Collaboration) 2012 *Phys. Lett.* B **710** 26
[40] Wood C S *et al.* 1997 *Science* **275** 1759
[41] Dzuba V A, Berengut J C, Flambaum V V and Roberts B 2012 *Preprint* arXiv:1207.5864 [hep-ph]
[42] Porsev S G, Beloy K and Derevianko A 2010 *Phys. Rev.* D **82** 036008
[43] ALEPH, DELPHI, L3, OPAL and LEP Working Group for Higgs Boson Searches 2003 *Phys. Lett.* B **565** 61
[44] CDF, DØ and Tevatron New Phenomena and Higgs Working Group 2012 *Preprint* arXiv:1203.3774 [hep-ex]
[45] Erler J 2012 *Preprint* arXiv:1201.0695 [hep-ph]
[46] Aad G *et al.* (ATLAS Collaboration) 2012 *Phys. Lett.* B **716** 1
[47] Chatrchyan S *et al.* (CMS Collaboration) 2012 *Phys. Lett.* B **716** 30
[48] CDF, DØ and Tevatron New Physics and Higgs Working Group 2012 *Preprint* arXiv:1207.0449 [hep-ex]
[49] CDF, DØ and Tevatron Electroweak Working Group 2011 *Preprint* arXiv:1107.5255 [hep-ex]
[50] ATLAS and CMS Collaborations 2012
`https://twiki.cern.ch/twiki/bin/view/CMSPublic/PhysicsResultsTOP12001`
[51] Chetyrkin K G and Steinhauser M 2000 *Nucl. Phys.* B **573** 617
[52] Skands P Z and Wicke D 2007 *Eur. Phys. J.* C **52** 133
[53] Fleming S, Hoang A H, Mantry S and Stewart I W 2008 *Phys. Rev.* D **77** 074010
[54] Langenfeld U, Moch S and Uwer P 2009 *Phys. Rev.* D **80** 054009
[55] Erler J and Langacker P 2010 *Phys. Rev. Lett.* **105** 031801
[56] Kuflik E, Nir Y and Volansky T 2012 *Preprint* arXiv:1204.1975 [hep-ph]
[57] Eberhardt O *et al.* 2012 *Preprint* arXiv:1209.1101 [hep-ph]
[58] Dighe A, Ghosh D, Godbole R M and Prasath A 2012 *Phys. Rev.* D **85** 114035
[59] Buchkremer M, Gérard J M and Maltoni F 2012 *J. High Energy Phys.* JHEP06(2012)135
[60] Bellantoni L, Erler J, Heckman J J and Ramirez-Homs E 2012 *Preprint* arXiv:1205.5580 [hep-ph]
[61] Peskin M E and Takeuchi T 1992 *Phys. Rev.* D **46** 381
[62] Wansbeek L W *et al.* 2008 *Phys. Rev.* A **78** 050501(R)
[63] Langacker P 2009 *Rev. Mod. Phys.* **81** 1199
[64] Erler J 2000 *Nucl. Phys.* B **586** 73
[65] Erler J, Langacker P, Munir S and Rojas E 2009 *J. High Energy Phys.* JHEP08(2009)017
[66] Erler J, Langacker P, Munir S and Rojas E 2011 *J. High Energy Phys.* JHEP11(2011)076
[67] Towner I S and Hardy J C 2010 *Rept. Prog. Phys.* **73** 046301
[68] Cirigliano V and Rosell I 2007 *J. High Energy Phys.* JHEP10(2007)005
[69] Bauman S, Erler J and Ramsey-Musolf M J 2012 *Preprint* arXiv:1204.0035 [hep-ph]

Journal of Physics: Conference Series **485** (2014) 012011 doi:10.1088/1742-6596/485/1/012011

Search for Higgs and Supersymmetry

Dirk Zerwas
LAL, CNRS/IN2P3, Univ. Paris-Sud, Orsay, FR
E-mail: zerwas@lal.in2p3.fr

Abstract. The search for the Higgs boson and the search for supersymmetry at the LHC collider are summarized. Particular attention is payed to the LHC potential to study the properties of the Higgs boson as well as the potential of future upgrades of the LHC and the ILC to determine supersymmetric parameters.

1. Introduction

The Standard Model of particle physics [1–3] describes successfully matter and its interactions to unprecedented precision. The only missing particle is also the only fundamental scalar: the Higgs boson [4–8] whose mass is not predicted by theory but constrained to the intermediate mass range.

Even with a discovery of the Higgs boson, other questions remain open to which a supersymmetric extension of the Standard Model could provide an answer. Supersymmetry can ensure the unification of the gauge couplings at a high scale. This scale is close to the Planck scale, potentially opening the route to the unification with gravity. Supersymmetry also stabilizes the mass of the Higgs boson at the electroweak scale. Last but not least, many supersymmetric models provide a candidate for dark matter.

The LHC [9] has started operations with the experiments ATLAS [10] and CMS [11] recording roughly 5 fb^{-1} of data in 2010 and 2011 at a center-of-mass energy of 7 TeV. In 2012 the center-of-mass energy was increased to 8 TeV. The prospects for the discovery of the Higgs boson and/or new phenomena have been summarized in many studies [12, 13].

In the following, first the search for the Higgs will be summarized. Then the determination of the couplings of the Higgs boson with the current data as well as the projection to the future is discussed. Searches for Higgs bosons in models extending the Standard Model are then discussed. Finally the status of the search for supersymmetry will be presented and as outlook the bottom-up reconstruction of supersymmetric parameters will be discussed. All experimental methods are illustrated at a center-of-mass energy of 7 TeV. The results are summarized including 8 TeV.

2. Standard Model Higgs boson (low mass)

The search for the Standard Model Higgs is generally separated into a low mass region and a high mass region. In the high mass region the partial widths for decays to pairs of W and Z bosons dominate the total width. A low mass Higgs boson will decay predominantly to b-quark pairs. The precision measurements of the parameters of the Standard Model at LEP and the TeVatron are compatible with a low mass Higgs boson [15]. For the sake of concreteness, in the following only the results for a low mass of 120 – 130 GeV will be discussed.

Content from this work may be used under the terms of the Creative Commons Attribution 3.0 licence. Any further distribution of this work must maintain attribution to the author(s) and the title of the work, journal citation and DOI.
Published under licence by IOP Publishing Ltd

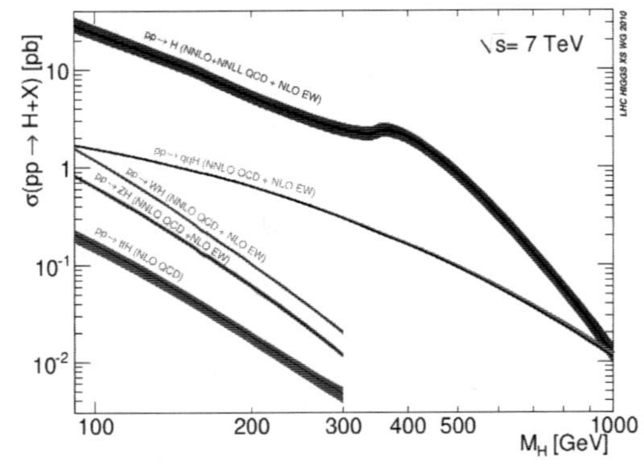

Figure 1. The Higgs production cross section is shown as function of the mass for a center-of-mass energy of 7 TeV [14].

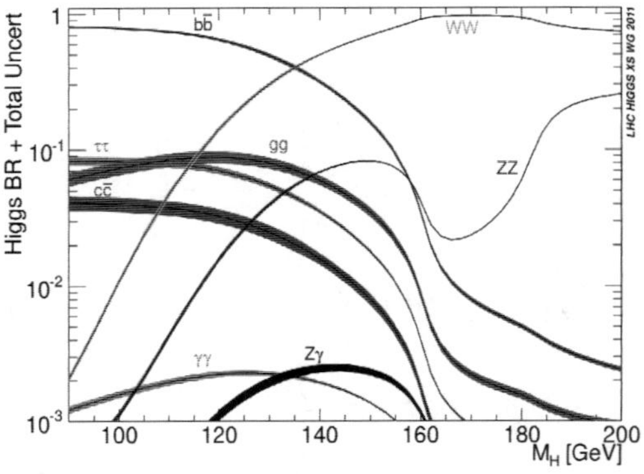

Figure 2. The branching ratios of the Standard Model Higgs boson are shown as function of mass [14].

The Standard Model Higgs boson is produced at the LHC predominantly in the gluon fusion process as shown in Figure 1 from [14, 16–24]. The weak boson fusion (VBF) is about an order of magnitude smaller. Higgs-strahlung, the associated production of the Higgs boson together with a weak boson is possible, but suffers from a very large background. The abundant production of top quark pairs opens up the channel ttH. This channel however is more promising at higher center-of-mass energies.

The branching ratios of the Standard Model Higgs boson are shown in Figure 2 from [14, 25]. At low mass the largest branching ratio is to pairs of b-quarks. The branching ratio to τ-leptons is about an order of magnitude smaller. Decays to pairs of W and Z bosons also have a non-negligible contribution. At a mass of 125 GeV, rare decays such as the decay to pairs of photons, are typically of the order of 10^{-3}.

At low mass the QCD background is so overwhelmingly large that the search for the hadronic decays of the Higgs boson is very difficult, unless it is accompanied by a vector boson or a pair of top quarks. Therefore the most promising channels for this mass region are the leptonic decay modes, either via the decay of the vector bosons or through τ pairs. The golden channels are the decay to a pair of leptonically decaying Z bosons and the rare decay to two photons as the mass of the Higgs boson can be reconstructed with good precision. The general strategy is to devise an analysis dedicated to an inclusive final state and then further enhance the separation between signal and background by separating the selected final state according to the production mode, e.g., gluon fusion with 0-jets, VBF with two jets etc, where further dedicated cuts are applied.

The different production and decay modes which will be discussed in the following are the fruit of a long dedicated effort of theorists and experimentalists. All of the discovery channels shown have been identified and scrutinized for many years, see e.g. [26–31].

The Higgs boson decay to pairs of τ-leptons is difficult because the neutrinos of the τ decays make the mass information is less precise than in other channels. The presence of a jet recoiling against the Higgs boson such as in the VBF production at LO or in gluon fusion at NLO, improve the information for mass reconstruction. CMS has three event categories and reconstructs the $\tau\tau$ mass with a likelihood technique to improve the mass resolution [32]. However the 0/1-jet channel remains difficult. The situation improves, e.g. shown by ATLAS [33], for the VBF category where two jets are reconstructed. The signal can be separated from dominant background of the Z boson decays, but unfortunately no clear excess is observed.

The Higgs-strahlung process, where the Higgs boson decays to a pair of b quarks and the

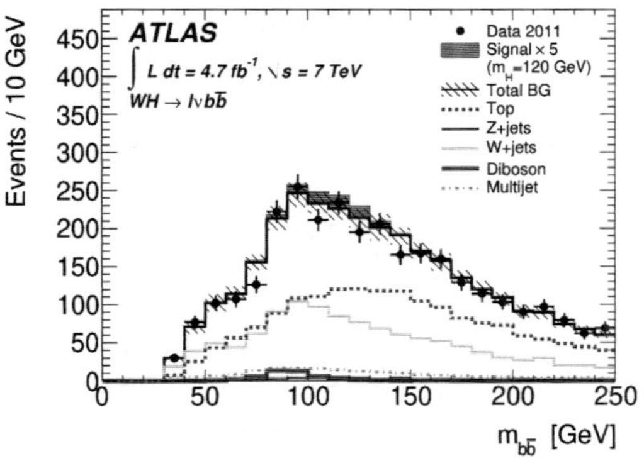

Figure 3. The distribution of the reconstructed Higgs boson mass in b-jets is shown for the production in association with a leptonically decaying W boson for ATLAS [35].

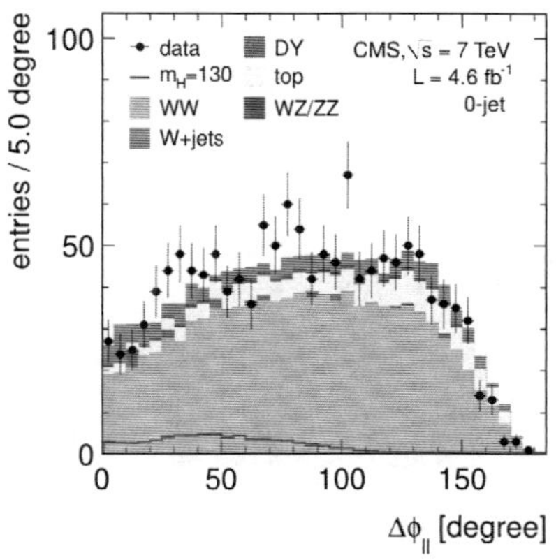

Figure 4. The distribution of the azimuthal angular separation between the leptons is shown for the CMS search in the WW final state [39].

weak boson does not decay hadronically, provides access to the main decay channel of a light Higgs boson. In contrast to the $\tau\tau$ channel the mass information can be used directly via the reconstructed b-tagged jets to reconstruct the Higgs boson mass. The channel is nevertheless difficult as the background is very large and the mass resolution for hadronically decaying resonances is less good than for final states with leptons. ATLAS and CMS have searched for this channel in the data [34, 35]. Even after the application of sophisticated methods such as boosted decision trees [34] and requiring b-tagged jets, the background remains dominant and in some of the categories, as shown in Figure 3, it has almost the same form as the signal. This makes the determination of the background from the data challenging. The future will tell whether subjet analyses [36–38] will allow to improve significantly the search for decays to b quarks.

Turning now to the three most promising channels at low mass involving leptons, or photons, the first channel to be considered is the decay to W boson pairs followed by leptonic decays. In the following leptons will be restricted to mean electrons and muons. Here the mass information is very weak as two neutrinos are present in the final state.

As in the other channels the events are divided into different categories as function of the jet multiplicity as well as the lepton flavor by ATLAS and CMS [39, 40]. The background dominating this search is the pair production of W bosons. To differentiate between signal and background one variable is of particular importance, the azimuthal separation of the reconstructed leptons. As shown by CMS in Figure 4 the background is flattish while the signal, due to the scalar nature of the Higgs boson, the angle is small as a result of helicity conservation. Instead of reconstructing the visible mass, inevitably spoiled by the neutrinos, the transverse mass has a peak structure. The peak is not at the nominal mass, but only roughly 5 GeV below as the two leptons are reconstructed close together. For this channel the background is determined from control regions. It is important to have an excellent control of the transverse missing energy.

The decay of the Higgs boson to a pair of Z bosons, one of the golden channels, where one of the Z bosons is off-shell, is intrinsically clean when concentrating on the leptonic decays of the Z bosons. Here only the electroweak production of Z boson pairs (on and off-shell, the latter including a photon) contribute to the background. The main difficulty is to have good lepton identification with a good efficiency and low misidentification down to low transverse momenta

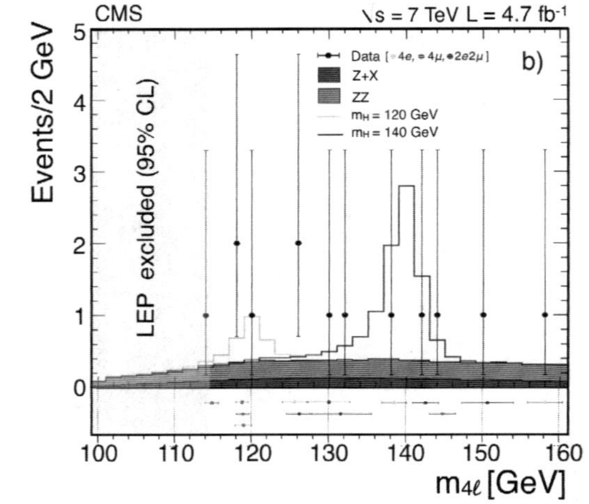

Figure 5. The invariant mass of the ZZ final state decaying to leptons is shown for CMS [41].

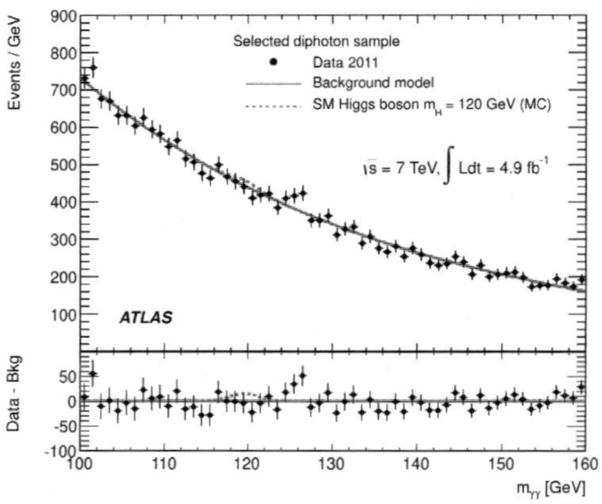

Figure 6. The final state $\gamma\gamma$ is shown for [44] together with the line corresponding to the background model.

of order 5 GeV. While the branching ratio of the Higgs boson to Z pairs is large, the leptonic branching ratio of the Z boson is small. At high mass the background is dominated by the leptonic decays of the Z boson pair production, whereas at low mass the relative contribution from the production of a Z boson plus jets is stronger. The ATLAS and CMS results in this channel [41, 42], shown as an example in Figure 5 for CMS, show at low mass an intriguing accumulation of data around 125 GeV. The Standard Model background is well described.

The final channel is a rare decay channel, the decay to a pair photons. The small branching ratio is offset with the excellent mass resolution for this channel which is better than 2 GeV. The calorimeter and tracker information can then be used to reduce the background due to the misidentification of jet as a photon below the level of the background from the Standard Model continuum production of two photons. The resolution of the invariant mass reconstruction is affected by the knowledge of the position of the primary vertex. ATLAS and CMS employ different techniques [43, 44] in this respect. While CMS relies strongly on the tracker, ATLAS uses the directional information in the calorimeter in order to improve the resolution. The material in front of the calorimeters in ATLAS and CMS is not constant as function of the polar angle. This leads to a variation of the resolution of the photons as function of the polar angle. The photons can convert before reaching the detector. It is therefore natural to split the event sample according to the expected resolution and into converted/non-converted photons. This categories differ naturally between the two experiments. The combined spectrum is shown in Figure 6 for ATLAS. As expected for the low branching ratio of this final state, the background is very large with respect to the expected signal. However the use of the data outside the probed Higgs mass allows to determine the background in the Higgs mass window from the data in a theory independent way. Nevertheless the form of the background distribution has also been checked with with the Monte Carlo. Around 125 GeV an intriguing excess of events is observed in the data of the two experiments above the background expectation as expected from the properties of a Standard Model Higgs boson.

The search for the Higgs boson at the TeVatron is complementary to the search at the LHC. Here the dominant channels are the production of the Higgs boson in association with an electroweak boson, followed by the decay of the Higgs boson to a pair of b-quarks. This

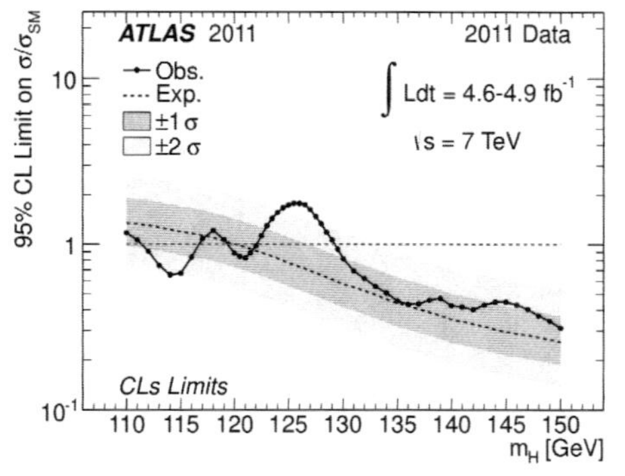

Figure 7. The 95% C.L. limit on the excluded Higgs boson cross section normalized to the Standard Model Higgs production cross section from ATLAS is shown as function of the Higgs boson mass [46].

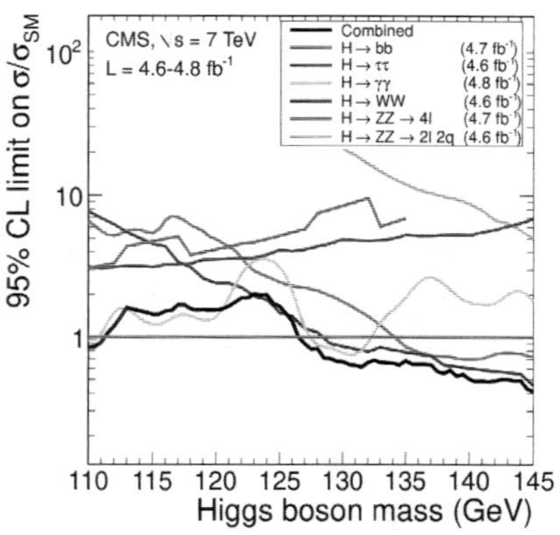

Figure 8. The 95% C.L. limit on the excluded Higgs boson cross section normalized to the Standard Model Higgs production cross section from CMS is shown as function of the Higgs boson mass [47].

search is promising as the Standard Model backgrounds for these channels are much lower than at the LHC due to the lower center-of-mass energy. The full combination of the collaborations CDF and D0 for all channels has been performed [45]. In the low mass region, as from roughly 115 GeV to 140 GeV, an excess with respect to the Standard Model background expectation is observed. The maximal (local) excess is 2.5 standard deviations, confirming the results at the LHC.

The combination of the Higgs searches is shown in Fig. 7 for ATLAS collaboration [46] and in Fig. 8 for the CMS collaboration [47] for the full 7 TeV dataset. Even without a combination of the two experiments around 125 GeV the Higgs boson cannot be excluded. Stated differently, around 125 GeV an excess is observed in both experiments which is compatible with the TeVatron excess and compatible overall with the properties expected for a Standard Model Higgs boson. The excess is mainly driven by the ZZ and $\gamma\gamma$ final states. For ATLAS the largest local significance at 126 GeV is almost three standard deviations, taking into account the "look elsewhere effect" (LEE) in a very large window from 110 to 600 GeV, the excess is about one standard deviations. For CMS the largest local significance at 124 GeV is slightly higher than three standard deviations which is reduced to about one and a half standard deviations after inclusion of the LEE effect for the same large mass window as ATLAS.

After the conclusion of the conference on July 4, 2012, ATLAS and CMS announced a local significance each of 5 standard deviations each, compatible with the properties of a Standard Model Higgs boson of mass around 126 GeV [48, 49], including data recorded at 8 TeV. The highest deviations from the Standard Model background expectation in the search for the Higgs boson are within 2 GeV from two independent experiments, each internally compatible with at least two different search channels, and compatible with the TeVatron results.

3. Beyond the Standard Model Higgs boson

In a supersymmetric theory the Higgs sector is extended with respect to the Standard Model as two complex doublets are necessary to give mass to up and down type particles [51]. The additional degrees of freedom lead to a richer particle spectrum. There are at least two additional

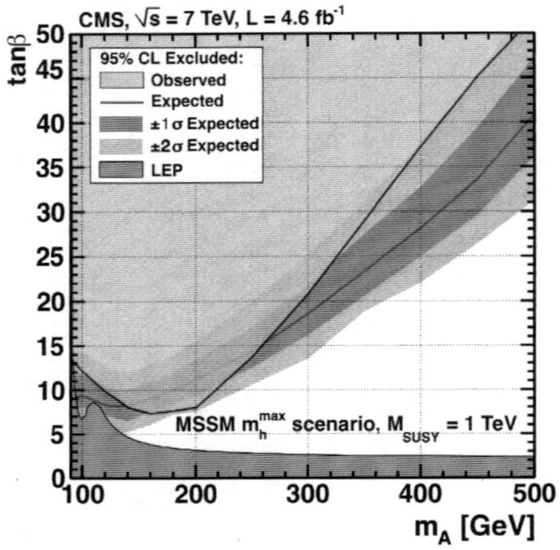

Figure 9. The region in the $\tan\beta$-m$_A$ plan excluded in the search for neutral MSSM Higgs bosons by CMS is shown [32].

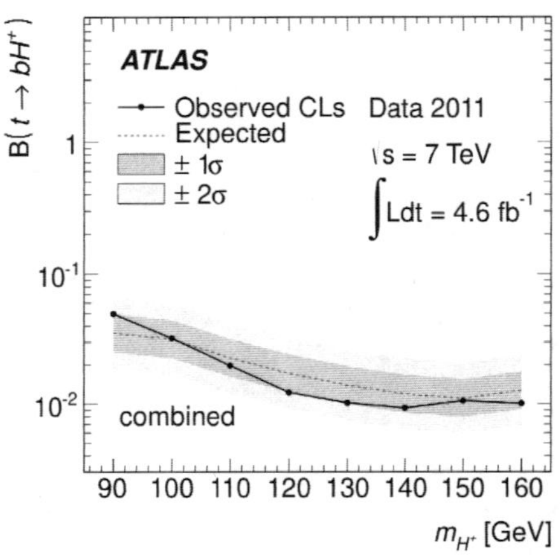

Figure 10. The limit on the top quark branching ratio to a charged Higgs boson is shown as function of the charged Higgs boson mass for ATLAS [50].

neutral Higgs bosons H and A as well as a pair of charged Higgs bosons. The ratio of the vacuum expectation values $\tan\beta$ and the mass of the A boson parametrize the Higgs sector to a good approximation. As $\tan\beta$ increases the coupling of the Higgs bosons H and A to the τ lepton and the b quark increases. Thus the search for the Higgs boson decaying to a pair of τ leptons as performed by CMS and ATLAS [32, 52] is a probe of the extended Higgs sector. In absence of a signal, a large fraction of the $\tan\beta$ versus m_A plane is excluded by this search when the mass of the A is restricted to less than 500 GeV. The limit gets weaker as function of the mass of the A as the cross section decreases as shown in Figure 9.

The charged Higgs boson, decaying to a τ lepton and its neutrino, with a mass less than the top quark could manifest itself as a modification of the branching ratio of the top quark. Additionally such events would have a larger transverse mass than the decay of the W boson (from the top quark) to τ and neutrino. The absence of a signal for charged Higgs boson at the LHC [50, 53] can be used to determine limits as low as 1% on the barnching ratio of the top quark as shown in Figure 10.

4. Higgs boson couplings and implications

It is a formidable task to determine the properties of the Higgs boson such as the spin and its couplings experimentally, e.g. [54, 55]. The Higgs boson will be measured in several channels and in each of these measurements several couplings intervene, e.g., for the two photon decay channel on the production side the effective coupling to gluons, on the decay side the effective coupling to photons as well as the total width, dominated by the coupling to b-quarks intervene. The effective coupling to gluons has a large contribution from the tree-level coupling to the top quark while the effective coupling to photons also receives contributions from the tree-level coupling to the W boson.

To measure the couplings a parameter Δ_x is defined as deviation from the Standard Model value:

$$g_{XXH} = g_x = (1 + \Delta_x) g_x^{\rm SM} \tag{1}$$

where $g_x^{\rm SM}$ is the Standard Model coupling of particle X to the Higgs boson. As the square of the couplings enters in the observables, $\Delta_x = 0$ as well as $\Delta_x = -2$ will lead to the same prediction. For effective couplings the definition is slightly more evolved with:

$$g_{XXH} = g_x = (1 + \Delta_x^{SM} + \Delta_x) g_x^{\rm SM} \tag{2}$$

where Δ_x^{SM} is the sum of the contributions of the non-standard tree-level couplings defined in Eq. 1. For example, if Δ_t is non-zero, the impact of this modification is transported to g_g in the term Δ_g^{SM} and only a genuine new contribution, e.g., a stop, is measured in Δ_g.

The total width of the Higgs boson enters in all branching ratios in the common denominator. Not all couplings, e.g., the one to charm quarks, can be determined at the LHC. Thus for $\Delta_c > 0$ the total width would be increased, leading to a reduction of the detected events. Determining the non-c-quark couplings such as Δ_t, Δ_b, Δ_τ, Δ_W and Δ_Z from the observed events would result in a determination of anomalous couplings shifted systematically to smaller values. Therefore it is essential to specify the model and the underlying assumptions when determining the couplings.

In Refs. [56, 57] the charm quark coupling is linked to the top quark coupling via: $g_c = m_c/m_t \cdot g_t^{SM}(1 + \Delta_t)$. No invisible decays of the Higgs boson are allowed and the total width is restricted to be less than 2 GeV, of the order of mass resolution in the golden channels:

$$\Gamma_{\rm tot} = \sum_{\rm obs} \Gamma_x(g_a) + \text{2nd generation} < 2\text{GeV} \tag{3}$$

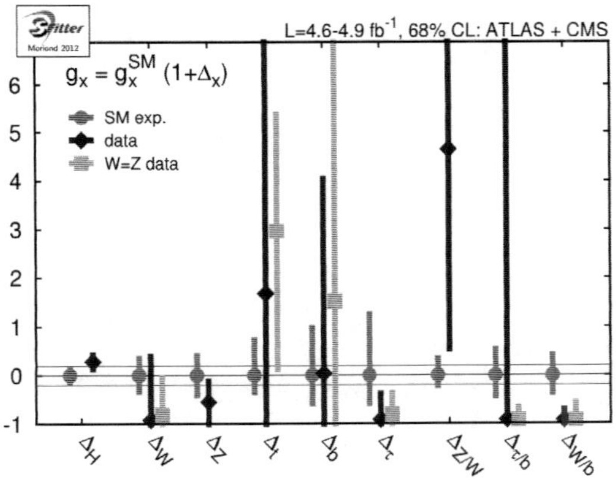

Figure 11. The Higgs coupling precision for the data as well as for the Standard Model expectation at 7 TeV is shown for a Higgs boson mass of 125 GeV in different scenarios [57].

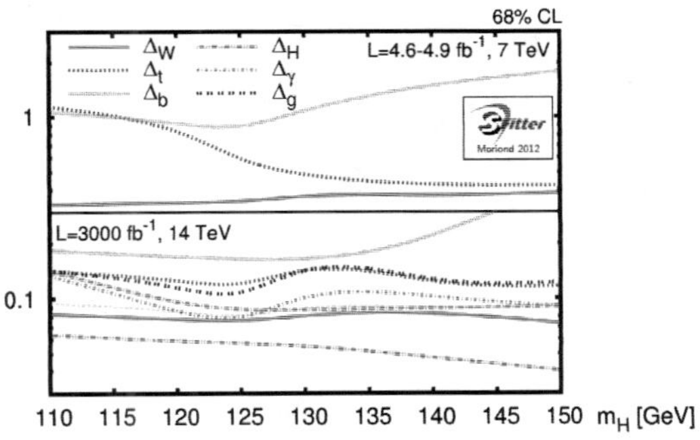

Figure 12. The precision of the Higgs coupling measurements at the LHC is shown for the data at 7 TeV as well as for a projection for 3000 fb^{-1} at 14 TeV [57].

In Fig. 11 the result of the determination of the couplings from the 7 TeV data is shown [57]. In red the Standard Model expectation is shown. In dark blue the result for the data are shown. For the light blue points the coupling of the Higgs boson to the weak bosons is forced to be fixed. The central values of the couplings are compatible with the Standard Model expectation of $\Delta_x = 0$. The negative value for Δ_W can be understood from the data as the channels for the Higgs decay to W bosons is more background like. The overall picture does not change when calculating ratios of couplings. In coupling ratios the correlated systematic errors and theoretical errors cancel, however with the current low statistics the uncorrelated statistical error dominates the coupling determination. A global deviation of all couplings is parametrized with Δ_H [58–60]. As in this case all measurements are used to determine a single parameter the best possible precision is achieved.

In Fig. 12 two luminosity scenarios are studied for the determination of the Higgs boson couplings as function of the Higgs boson mass. In the first scenario the data at 7 TeV (up to 4.9 fb^{-1}) are shown without effective couplings. Here the precision is typically 50%. In the lower half of the figure the expectation for 3000 fb^{-1} at 14 TeV is shown including effective couplings. The coupling precision is improved to roughly order 10%. For the Higgs portal the best precision can be achieved: 5%. This is close to the ultimate precision given by the theoretical error on the cross section.

In addition to the couplings the mass of the Higgs boson can be used to constrain new phenomena beyond the Standard Model. In particular in the supersymmetric models the Higgs boson mass, after including radiative corrections, depends on the mass of the top quark as well as the masses and the mixing of the spin-0 partners of the top quark, the stops. Using the measured mass of the Higgs boson in Ref. [61], the parameter space of the supersymmetric models can be restricted. Some supersymmetry breaking models, such as minimal gauge mediation (mGMSB), are disfavored. In the minimal supersymmetric extension of the Standard Model (MSSM) part of the regions compatible with a Higgs boson mass of 126 GeV are characterized by large mixing which could mean that a light stop could be just around the corner at the LHC.

5. Supersymmetric particles

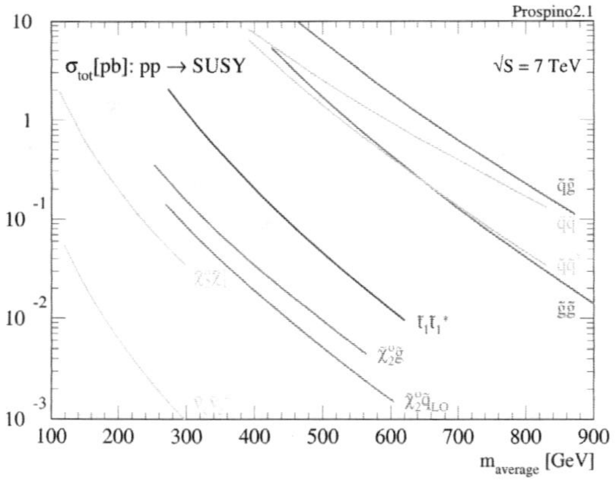

Figure 13. The production cross sections for supersymmetric particles at 7 TeV are shown as calculated by Prospino [62–66].

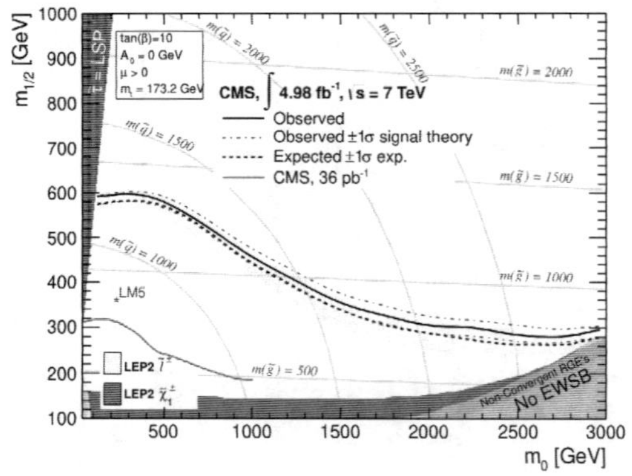

Figure 14. The m_0-$m_{1/2}$ plane with the exclusion in mSUGRA as obtained by CMS is shown [71].

In supersymmetric theories [67, 68] in addition to the extension of the Higgs sector, the matter particles as well as the carrier of the forces each have a partner differing by a half-unit of spin. In particular gluinos (fermions) are the partners of the gluons and the sleptons (scalars) are the partners of the leptons. In the gauge and Higgs sector, the neutralinos are mixtures of the partners of the neutral Higgs and gauge bosons, the charginos are the partners of the charged Higgs and gauge bosons. The lightest neutralino is the lightest supersymmetric particles, and as a stable weakly interacting particle also a candidate for dark matter. R-parity [69] is defined to be +1 for Standard Model particles and −1 for supersymmetric particles. This new quantum

Journal of Physics: Conference Series **485** (2014) 012011 doi:10.1088/1742-6596/485/1/012011

number is assumed to be conserved in the following, so that the sparticles are produced in pairs, decaying to the LSP in possibly long cascades.

The largest cross section for the production of supersymmetric particles [62–66] are those for gluinos and squarks as shown in Figure 13. As the integrated luminosity recorded at the LHC increases, the sensitivity to non-colored particles also increases. An increase of the center-mass-energy is even more powerful.

The search for squarks and gluinos as performed by ATLAS [70] and CMS [71] has not yielded a signal, even though performed in many different channels. The main analysis is the search for multiple jets with large transverse momenta and large missing transverse momentum due to the LSPs escaping the detectors undetected. As the LHC has opened a new large mass regime to be explored, assumptions have to be made about the particle content when presenting the results in terms of limits. An example of a theory with many different final states is mSUGRA. The limits shown in the plane $m_{1/2}$ versus m_0, the common fermion and scalar SUSY breaking parameters defined at the unification (=high scale) by CMS in Figure 14 depend only weakly on $\tan\beta$. As an example, for equal gluino and squark masses the limit is at 1.4 TeV by the ATLAS experiment.

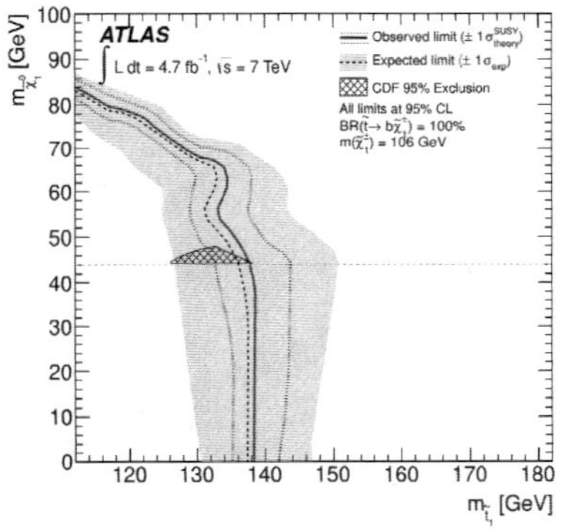

Figure 15. The limit on the stop quark mass as function of the LSP mass is shown for ATLAS [73].

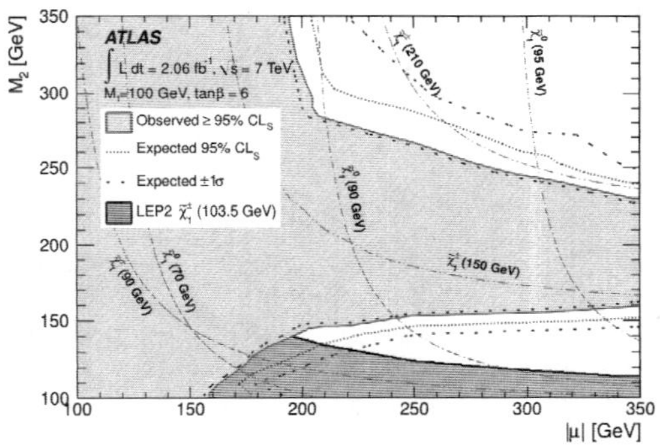

Figure 16. The excluded region in M_2 versus μ is shown for the ATLAS tri-lepton search [75].

The third generation plays a special role in supersymmetry. Here mixing effects can be become large due to the large masses of the top quark and the b quark. The mixing can lead to sbottom and stops which are much lighter than the (degenerate) squarks of the first and second generation. While a lower mass increases the cross section, this effect cannot compensate the loss due to the fact that one now probes only a single squark. On the other hand, the use of b-tagging, as the stops and sbottoms will lead to a stronger production of b-jets than first and second generation squarks, will reduce the background.

The search for sbottoms at the LHC has not lead to a signal so far [72, 74]. The search by CMS for the production of gluinos decaying to sbottoms can be interpreted as an excluded region of the gluino mass as function of the LSP mass. Gluino masses up to order 900 GeV can be excluded in this model-dependent scenario.

For the stops ATLAS has performed a search for the direct production of stops [73]. Here the stops decay to a bottom and a chargino. The search yields limits roughly of the order of 137 GeV as shown in Figure 15, depending on the LSP mass. As the mass difference between stop and LSP decreases the limits become weaker.

While the LHC as a proton-proton collider has a natural sensitivity to colored sparticles, it is interesting to note that the search for the production and decay of neutralinos and charginos has started. Typically the most promising channel for such a search are the leptonic decays, i.e., final states with several leptons and missing transverse momentum [75, 77]. As shown in Figure 16 the absence of a signal is translated into a new exclusion domain in the plane of M_2, the SUSY breaking parameter associated to the $SU(2)$ group and μ, the Higgs-higgsino mass parameter. Both parameters are defined at the electroweak scale. While it is impressive to see that these searches in the electroweak sector extend significantly the limits from LEP, a caveat must be added: in order to obtain these limits, sleptons are presumed to have a mass between the produced chargino and neutralino and the LSP. Such a hypothesis leads to a strong enhancement of the leptonic branching ratios.

The search for supersymmetry at the LHC goes well beyond mSUGRA or models parametrize supersymmetry breaking via gravity mediation. One example is the search for gauge mediated supersymmetry performed by ATLAS [76]. This analysis is particularly interesting as the final state contains jets, missing transverse momentum and τ leptons, which are more difficult to identify the electrons or muons. While unfortunately no signal has been found, the excluded region is extended significantly beyond the region excluded by the searches at LEP for the first time.

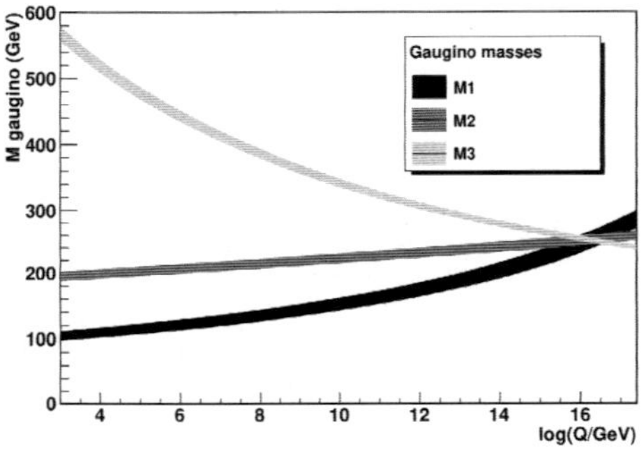

Figure 17. The gaugino SUSY breaking parameter unification is it could be measured at the LHC [78] is shown as function of the scale.

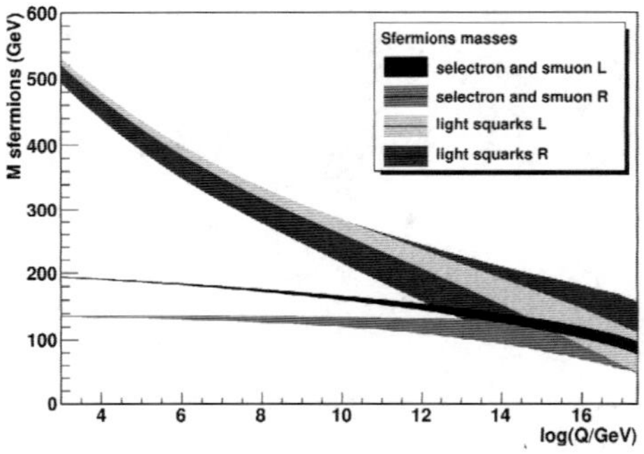

Figure 18. The evolution of the scalar SUSY breaking parameters determined at LHC and ILC is shown as function of the scale [78].

While no evidence for supersymmetry has turned up so far at the LHC, it is still interesting to see what information on the fundamental parameters could be obtained from measurements at the LHC. This question was studied in particular in [78–81] among others. In a very optimistic scenario, the gluinos, squark of the first two generations the sleptons, the sbottoms as well as three of the four neutralinos could be measured at the LHC. While this is sufficient information for a highly constrained high scale model such as mSUGRA, the picture changes significantly when going to the more interesting question of reconstructing the parameters of the general MSSM at the low scale. The incomplete information in the neutralino/chargino sector leads to a 12-fold ambiguity when determining M_1, M_2, μ. However, evolving the MSSM parameters of the gaugino sector, i.e., M_1, M_2 and M_3 via the renormalization group equations to the high scale, leads to unification (in this particular scenario) in only one of the cases. The precision with which the unification can be probed at the LHC is shown in Figure 17. As pointed out in [82] at this conference the Higgs mass from ATLAS and CMS is compatible with light gauginos in SUGRA models. To get a complete picture, an ILC will be necessary to complete the spectrum and resolve the ambiguities. In this case even the unification of the scalar SUSY breaking parameters can be measured from the data as shown in Figure 18.

6. Conclusions

ATLAS and CMS have observed a particle compatible with the properties of a Standard Model Higgs boson. It will now be essential to determine its spin and its couplings. Ultimately the couplings can be measured with a precision of about 10% at the LHC.

No significant deviation from the Standard Model has been observed in the search for supersymmetry. This means that easy supersymmetry was easily ruled out. More difficult searches are now being started for supersymmetric processes with lower cross sections. The increase of the center-of-mass energy of the LHC to 14TeV$-\varepsilon$ after a longer shutdown will open up a new mass range to be explored.

Acknowledgments

It is a pleasure to thank the organizers of PASCOS 2012, in particular Myriam Mondragon, for the wonderful conference in Merida. I would like to thank P.M. Zerwas, T. Plehn and Michael Rauch for discussions and the careful reading of the manuscript.

References

[1] Glashow S 1961 *Nucl.Phys.* **22** 579–588
[2] Salam A 1968 *Conf.Proc.* **C680519** 367–377
[3] Weinberg S 1967 *Phys.Rev.Lett.* **19** 1264–1266
[4] Englert F and Brout R 1964 *Phys.Rev.Lett.* **13** 321–323
[5] Higgs P W 1964 *Phys.Rev.Lett.* **13** 508–509
[6] Higgs P W 1964 *Phys.Lett.* **12** 132–133
[7] Higgs P W 1966 *Phys.Rev.* **145** 1156–1163
[8] Guralnik G, Hagen C and Kibble T 1964 *Phys.Rev.Lett.* **13** 585–587
[9] Evans Lyndon e and Bryant Philip e 2008 *JINST* **3** S08001
[10] Aad G *et al.* (ATLAS Collaboration) 2008 *JINST* **3** S08003
[11] Chatrchyan S *et al.* (CMS Collaboration) 2008 *JINST* **3** S08004
[12] Morrissey D E, Plehn T and Tait T M 2012 *Phys.Rept.* **515** 1–113 (*Preprint* `0912.3259`)
[13] Nath P, Nelson B D, Davoudiasl H, Dutta B, Feldman D *et al.* 2010 *Nucl.Phys.Proc.Suppl.* **200-202** 185–417 (*Preprint* `1001.2693`)
[14] LHC Higgs Cross Section Working Group, Dittmaier S, Mariotti C, Passarino G and Tanaka (Eds) R CERN, Geneva, 2011 *CERN-2011-002* (*Preprint* `1101.0593`)
[15] ALEPH, CDF, D0, DELPHI, L3, OPAL, SLD Collaborations, the LEP Electroweak Working Group, the Tevatron Electroweak Working Group, and the SLD electroweak and heavy flavour groups 2010
[16] Georgi H, Glashow S, Machacek M and Nanopoulos D V 1978 *Phys.Rev.Lett.* **40** 692
[17] Djouadi A, Spira M and Zerwas P 1991 *Phys.Lett.* **B264** 440–446
[18] Dawson S 1991 *Nucl.Phys.* **B359** 283–300
[19] Spira M, Djouadi A, Graudenz D and Zerwas P 1995 *Nucl.Phys.* **B453** 17–82 (*Preprint* `hep-ph/9504378`)
[20] Harlander R V and Kilgore W B 2002 *Phys.Rev.Lett.* **88** 201801 (*Preprint* `hep-ph/0201206`)
[21] Ravindran V, Smith J and van Neerven W L 2003 *Nucl.Phys.* **B665** 325–366 (*Preprint* `hep-ph/0302135`)
[22] Ahrens V, Becher T, Neubert M and Yang L L 2009 *Eur.Phys.J.* **C62** 333–353 (*Preprint* `0809.4283`)
[23] Ciccolini M, Denner A and Dittmaier S 2008 *Phys.Rev.* **D77** 013002 (*Preprint* `0710.4749`)
[24] Arnold K, Bellm J, Bozzi G, Brieg M, Campanario F *et al.* 2011 (*Preprint* `1107.4038`)
[25] Djouadi A, Kalinowski J and Spira M 1998 *Comput.Phys.Commun.* **108** 56–74 (*Preprint* `hep-ph/9704448`)
[26] Richter-Was E and Sapinski M 1999 *Acta Phys.Polon.* **B30** 1001–1040
[27] Dittmar M and Dreiner H K 1997 *Phys.Rev.* **D55** 167–172 (*Preprint* `hep-ph/9608317`)
[28] Ellis R K, Hinchliffe I, Soldate M and van der Bij J 1988 *Nucl.Phys.* **B297** 221
[29] Abdullin S, Dubinin M, Ilyin V, Kovalenko D, Savrin V *et al.* 1998 *Phys.Lett.* **B431** 410–419 (*Preprint* `hep-ph/9805341`)
[30] Kauer N, Plehn T, Rainwater D L and Zeppenfeld D 2001 *Phys.Lett.* **B503** 113–120 (*Preprint* `hep-ph/0012351`)
[31] Plehn T, Rainwater D L and Zeppenfeld D 1999 *Phys.Lett.* **B454** 297–303 (*Preprint* `hep-ph/9902434`)

[32] CMS Collaboration 2012 *Phys.Lett.* **B713** 68–90 (*Preprint* 1202.4083)
[33] ATLAS Collaboration 2012 (*Preprint* 1206.5971)
[34] CMS Collaboration 2012 *Phys.Lett.* **B710** 284–306 (*Preprint* 1202.4195)
[35] ATLAS Collaboration 2012 (*Preprint* 1207.0210)
[36] Butterworth J M, Davison A R, Rubin M and Salam G P 2008 *Phys.Rev.Lett.* **100** 242001 (*Preprint* 0802.2470)
[37] Plehn T, Salam G P and Spannowsky M 2010 *Phys.Rev.Lett.* **104** 111801 (*Preprint* 0910.5472)
[38] Altheimer A, Arora S, Asquith L, Brooijmans G, Butterworth J *et al.* 2012 *J.Phys.* **G39** 063001 (*Preprint* 1201.0008)
[39] CMS Collaboration 2012 *Phys.Lett.* **B710** 91–113 (*Preprint* 1202.1489)
[40] ATLAS Collaboration 2012 *Phys.Lett.* **B716** 62–81 (*Preprint* 1206.0756)
[41] CMS Collaboration 2012 *Phys.Rev.Lett.* **108** 111804 (*Preprint* 1202.1997)
[42] ATLAS Collaboration 2012 *Phys.Lett.* **B710** 383–402 (*Preprint* 1202.1415)
[43] CMS Collaboration 2012 *Phys.Lett.* **B710** 403–425 (*Preprint* 1202.1487)
[44] ATLAS Collaboration 2012 *Phys.Rev.Lett.* **108** 111803 (*Preprint* 1202.1414)
[45] CDF and D0 Collaborations and the Tevatron New Physics and Higgs Working Group (Tevatron New Physics Higgs Working Group, CDF Collaboration, D0 Collaboration) 2012 (*Preprint* 1207.0449)
[46] ATLAS Collaboration 2012 *Phys.Rev.* **D86** 032003 (*Preprint* 1207.0319)
[47] CMS Collaboration 2012 *Phys.Lett.* **B710** 26–48 (*Preprint* 1202.1488)
[48] ATLAS Collaboration 2012 *Phys.Lett.* **B716** 1–29 (*Preprint* 1207.7214)
[49] CMS Collaboration 2012 *Phys.Lett.* **B716** 30–61 (*Preprint* 1207.7235)
[50] ATLAS Collaboration 2012 *JHEP* **1206** 039 (*Preprint* 1204.2760)
[51] Martin S P 1997 (*Preprint* hep-ph/9709356)
[52] ATLAS Collaboration 2011 *Phys.Lett.* **B705** 174–192 (*Preprint* 1107.5003)
[53] CMS Collaboration 2012 *JHEP* **1207** 143 (*Preprint* 1205.5736)
[54] Choi S, Muhlleitner M and Zerwas P 2012 (*Preprint* 1209.5268)
[55] Plehn T and Rauch M 2012 (*Preprint* 1207.6108)
[56] Lafaye R, Plehn T, Rauch M, Zerwas D and Duhrssen M 2009 *JHEP* **0908** 009 (*Preprint* 0904.3866)
[57] Klute M, Lafaye R, Plehn T, Rauch M and Zerwas D 2012 *Phys.Rev.Lett.* **109** 101801 (*Preprint* 1205.2699)
[58] Bock S, Lafaye R, Plehn T, Rauch M, Zerwas D *et al.* 2010 *Phys.Lett.* **B694** 44–53 (*Preprint* 1007.2645)
[59] Englert C, Plehn T, Zerwas D and Zerwas P M 2011 *Phys.Lett.* **B703** 298–305 (*Preprint* 1106.3097)
[60] Englert C, Plehn T, Rauch M, Zerwas D and Zerwas P M 2012 *Phys.Lett.* **B707** 512–516 (*Preprint* 1112.3007)
[61] Arbey A, Battaglia M, Djouadi A, Mahmoudi F and Quevillon J 2012 *Phys.Lett.* **B708** 162–169 (*Preprint* 1112.3028)
[62] Beenakker W, Hopker R, Spira M and Zerwas P 1997 *Nucl.Phys.* **B492** 51–103 (*Preprint* hep-ph/9610490)
[63] Beenakker W, Kramer M, Plehn T, Spira M and Zerwas P 1998 *Nucl.Phys.* **B515** 3–14 (*Preprint* hep-ph/9710451)
[64] Beenakker W, Klasen M, Kramer M, Plehn T, Spira M *et al.* 1999 *Phys.Rev.Lett.* **83** 3780–3783 (*Preprint* hep-ph/9906298)
[65] Spira M 2002 217–226 (*Preprint* hep-ph/0211145)
[66] Plehn T 2005 *Czech.J.Phys.* **55** B213–B220 (*Preprint* hep-ph/0410063)
[67] Golfand Y and Likhtman E 1971 *JETP Lett.* **13** 323–326
[68] Wess J and Zumino B 1974 *Nucl.Phys.* **B70** 39–50
[69] Farrar G R and Fayet P 1978 *Phys.Lett.* **B76** 575–579
[70] ATLAS Collaboration 2012 (*Preprint* 1208.0949)
[71] CMS Collaboration 2012 (*Preprint* 1207.1898)
[72] CMS Collaboration 2012 (*Preprint* 1208.4859)
[73] ATLAS Collaboration 2012 (*Preprint* 1208.4305)
[74] ATLAS Collaboration (ATLAS Collaboration) 2012 *Phys.Rev.Lett.* **108** 181802 (*Preprint* 1112.3832)
[75] ATLAS Collaboration 2012 *Phys.Rev.Lett.* **108** 261804 (*Preprint* 1204.5638)
[76] ATLAS Collaboration 2012 *Phys.Lett.* **B714** 180–196 (*Preprint* 1203.6580)
[77] CMS Collaboration 2012 *JHEP* **1206** 169 (*Preprint* 1204.5341)
[78] Adam C, Kneur J L, Lafaye R, Plehn T, Rauch M *et al.* 2011 *Eur.Phys.J.* **C71** 1520 (*Preprint* 1007.2190)
[79] Allanach B, Blair G, Kraml S, Martyn H, Polesello G *et al.* 2004 (*Preprint* hep-ph/0403133)
[80] Blair G, Porod W and Zerwas P 2003 *Eur.Phys.J.* **C27** 263–281 (*Preprint* hep-ph/0210058)
[81] Lafaye R, Plehn T, Rauch M and Zerwas D 2008 *Eur.Phys.J.* **C54** 617–644 (*Preprint* 0709.3985)
[82] Nath P 2012 (*Preprint* 1207.5501)

Journal of Physics: Conference Series **485** (2014) 012012 doi:10.1088/1742-6596/485/1/012012

SUGRA Grand Unification, LHC and dark matter

Pran Nath

Department of Physics, Northeastern University, Boston, MA 02115, USA

E-mail: nath@neu.edu

Abstract. A brief review is given of recent developments related to the Higgs signal and its implications for supersymmetry in the supergravity grand unification framework. The Higgs data indicates that the allowed parameter space largely lies on Focal Curves and Focal Surfaces of the Hyperbolic Branch of radiative breaking of the electroweak symmetry where TeV size scalars naturally arise. The high mass of the Higgs leads to a more precise prediction for the allowed range of the spin independent neutralino-proton cross-section which is encouraging for the detection of dark matter in future experiments with greater sensitivity. Also discussed is the status of grand unification and a natural solution to breaking the GUT group at one scale and resolving the doublet-triplet problem. It is shown that the cosmic coincidence can be compatible within a supersymmetric framework in a muticomponent dark matter picture with one component charged under $B - L$ while the other component is the conventional supersymmetric dark matter candidate, the neutralino.

1. Introduction

In December 2011 the ATLAS [1] and CMS [2] Collaborations reported the result of their updated search for the Higgs boson [3] which showed indications of a signal around the 3σ level. These results lead to considerable theoretical activity to extract the implications of the signal [4, 5, 6]. More recently the analyses of [1, 2] have received further support from the Tevatron analyses [7] and from the combined 7 TeV and 8 TeV CMS and ATLAS data [8]. Thus the CMS Collaboration indicates a signal for a boson with mass of 125.3 ± 0.6 GeV at 4.9σ and the ATLAS Collaboration indicates a signal for a boson with a mass ~ 126.5 GeV at 5.0σ in the combined 7 TeV and 8 TeV analysis [8] which confirm the previous indication of the boson signal by ATLAS [1] and CMS [2]. The properties of the boson still need to be fully established including its spin. Assuming the particle discovered is a spin 0 CP even boson, it is pertinent to ask if it is indeed the Higgs particle which enters in spontaneous breaking of the electroweak symmetry and gives mass to gauge bosons and to quarks and leptons[3]. Further, if indeed the discovered particle is the Higgs boson, it is of interest to determine its properties and any possible deviations from the standard model predictions which would be indicators of new physics beyond the standard model. Definitive answers to these questions will have to wait for much more data. However, partial answers already exist in the data from the Tevatron and from the CMS/ATLAS experiments. Thus there is evidence already that the observed particle emulates the most obvious property of the Higgs boson, i.e., that its coupling to heavier fermions is larger than its coupling to lighter fermions. Thus, if the Higgs boson coupled with equal strength to fermions one would have seen an equal abundance of $\mu^+\mu^-$ as of $b\bar{b}$. However, the Tevatron data suggests otherwise. While there is evidence of $b\bar{b}$ decay of the Higgs there is

Content from this work may be used under the terms of the Creative Commons Attribution 3.0 licence. Any further distribution of this work must maintain attribution to the author(s) and the title of the work, journal citation and DOI.
Published under licence by IOP Publishing Ltd

no evidence of Higgs decaying into $\mu^+\mu^-$ [7].

.

2. Higgs and SUSY

An experimental determination of the Higgs couplings are essential in testing the standard model and to find any deviations from the standard model prediction which would be an indication of new physics. The relevant couplings to test are the Higgs boson couplings to femion-anti-fermion pairs and to dibosons, i.e., $hf\bar{f}$, hWW, hZZ, $h\gamma\gamma$, $hZ\gamma$. Thus at the LHC with $\sqrt{s} = 14$ TeV, and $\mathcal{L} = 300$ fb^{-1} up to 10% accuracy for couplings of h^0 with dibosons WW, ZZ and up to $\sim 20\%$ for couplings to b, τ appear possible [9]. Additionally, of course, one needs to test the overall consistency of the total production cross-section with the SM prediction and here currently CMS has $\sigma/\sigma_{SM} = 0.80 \pm 0.22$ and ATLAS has $\sigma/\sigma_{SM} = 1.2 \pm 0.3$ [8]. Important clues to new physics can emerge by looking at deviations of the Higgs boson couplings from the standard model predictions. The possibility that CMS and ATLAS data may have an excess in the diphoton channel has already led to significant interest in the diphoton decay of the Higgs. In the SM the Higgs boson can decay into photons via W^+W^- and $t\bar{t}$ loops. The W-loop gives the larger contribution which is partially cancelled by the top-loop. To enhance the decay one could add extra charged particles in the loops and various works have appeared along these lines (see, e.g., [10]) More data is needed to confirm the diphoton excess as the result could be due to QCD uncertainties [11].

We discuss now the implications of the 125 GeV Higgs for supersymmetry (see also [12]). There are cogent arguments for the 125 GeV Higgs being the first evidence for supersymmetry at the LHC. The argument assumes the following: (i) perturbative physics up to the grand unification scale; (ii) no large hierarchy problem. Item (i) disfavors composite Higgs models which require non-perturbative physics below the grand unification scale. Thus under constraint (i) the standard model is a possible candidate since the SM is a renormalizable theory and is valid up to the grand unification scale. However, the SM gives large loop corrections to the Higgs mass, so that at one loop one has $m_h^2 = m_0^2 + O(\Lambda^2)$. Since $\Lambda \sim O(10^{16})$ GeV a large fine tuning is involved, i.e., one part in 10^{14}. Thus the SM while satisfying (i) does not satisfy (ii). On the other hand SUGRA models have all the merits of the SM, but are free of the large hierarchy problem and give perturbative physics up to the GUT scale. Here Λ is essentially replaced by the stop masses.

The second strong hint that the newly discovered particle is a SUSY Higgs is its mass. In the standard model the upper limit on the Higgs mass could be much larger. Using the condition that the weak interactions not become strong one derives roughly the upper bound [13] $M_H < (8\pi\sqrt{2}/3G_F)^{1/2} \simeq 1\ TeV$. In SUSY, the Higgs quartic couplings are governed by gauge interactions and there is an upper limit which is much smaller, i.e., $M_H < 156$ GeV [14] and in well motivated models such as mSUGRA, GMSB, AMSB it is even smaller. Specifically in mSUGRA [15] $M_H \leq 130$ GeV [4, 16]. It is important to realize that SUSY could have been excluded or come under a shadow if the Higgs mass was found to be, say 200 GeV. Thus it is very meaningful that experimentally the Higgs mass ended up below the upper limit predicted in SUSY and more specifically below the upper limit predicted in SUGRA models. In SUGRA the dominant one loop contribution to the Higgs mass arises from the top/stop sector and is given by $\Delta m_h^2 \simeq 3m_t^4/(2\pi^2v^2)\ln(M_S^2/m_t^2) + 3m_t^4/(2\pi^2v^2)(X_t^2/M_S^2 - X_t^4/(12M_S^4))$, where $v = 246$ GeV, M_S is an average stop mass, and X_t is given by $X_t \equiv A_t - \mu\cot\beta$ (for a review see [17]). The loop correction is maximized when $X_t \sim \sqrt{6}M_S$. Thus the 'large' Higgs mass, i.e., 125 GeV points to the SUSY scale being rather high. Indeed most of the allowed parameter space consistent with all the experimental constraints appears to lie on the Hyperbolic Branch [18, 19, 20, 21], and specifically on Focal Curves and on Focal Surfaces [21] (see Sec.4) which is another indication of the overall SUSY scale being high. The 125 GeV Higgs boson also strongly constrains the sparticle landscape [22]) (for a review see [23]). The analysis of the sparticle spectra under the constraints of the high Higgs boson mass allows for light sparticle masses, i.e., the neutralino, the chargino, the gluino, the stau and the stop. The light gluino possibility has been extensively discussed in the literature (see. e.g., [24, 25]). Thus a low lying gluino along with the neutralino, the chargino, and the stop appear prime candidates for discovery at the LHC, although a more quantitative analysis is needed correlated with the energy of the future runs at the LHC. The discovery potential for SUSY in mSUGRA at LHC at $\sqrt{s} = 14$ TeV and at high luminosity with 300 (3000) fb^{-1} of integrated luminosity is discussed in [26].

3. GUTS and SUGRA GUTS

Gauge symmetry based on $SO(10)$ provides a framework for unifying the $SU(3)_C \times SU(2)_L \times U(1)_Y$ gauge groups and for unifying quarks and leptons in a single 16–plet spinor representation. Additionally, the 16–plet also contains a right–handed singlet state, which is needed to give mass to the neutrino via the seesaw mechanism. Supersymmetric $SO(10)$ models [27] have the added attraction that they predict correctly the unification of gauge couplings, and solve the hierarchy problem by virtue of supersymmetry. However, SUSY $SO(10)$ models, as usually constructed, have two drawbacks, both related to the symmetry breaking sector. First, two different mass scales are involved in breaking of the GUT symmetry, one to reduce the rank and the other to reduce the symmetry all the way to $SU(3)_C \times SU(2)_L \times U(1)_Y$. Thus typically three types of Higgs fields are needed (for a review see [28]) e.g., $16 \oplus \overline{16}$ or $126 \oplus \overline{126}$ for rank reduction, and a 45, 54 or 210 for breaking the symmetry down to the standard model symmetry, and a 10-plet for electroweak symmetry breaking. Multiple step breaking requires additional assumptions relating VEVs of different breakings to explain gauge coupling unification of the electroweak and the strong interactions. A single step breaking does not require such an assumption and can be achieved with $144 \oplus \overline{144}$ [29, 30] of Higgs fields. The reason for this is easily understood by looking at the decomposition of 144 and $\overline{144}$ multiplets under $SU(5) \times U(1)$. Thus one has $\overline{144} = \bar{5}(3) + 5(7) + 10(-1) + 15(7) + 24(-5) + 40(-1) + \overline{45}(3)$. Now the 24-plet carries a $U(1)$ quantum number and thus a VEV formation of it will reduce the rank of the group as well as break $SU(5)$. Additionally, one can obtain a pair of light Higgs doublets needed for electroweak symmetry breaking from the same irreducible $144 \oplus \overline{144}$ Higgs multiplet. Thus $SO(10)$ reduces to $SU(3)_C \times U(1)_{em}$ with just one pair of $< 144 \oplus \overline{144} >$ multiplets.

Second, GUT theories typically have the doublet-triplet problem, i.e., one must do an extreme fine–tuning at the level of one part in 10^{14} to get the Higgs doublets of MSSM light, while color-triplets remain superheavy. Some possible solutions to the doublet-triplet problem include (i) missing VEV mechanism where $SO(10)$ breaks in the B-L direction, (ii) flipped $SU(5) \times U(1)$, (iii) missing partner mechanism, (iv) orbifold GUTs. The flipped $SU(5) \times U(1)$, missing partner mechanism and the orbifold GUTs are rather compelling in that some doublets are forced to be massless. We will focus on the missing partner mechanism [31] and discuss how it works in $SU(5)$ and then discuss how one can extend to $SO(10)$. In SU(5) to obtain Higgs doublets which are naturally light one uses an array of light and heavy Higgs multiplets so that the heavy multplets consist of $50, \overline{50}, 75$ multiplets and the light multiplets consist of $5, \bar{5}$, i.e., one has mass terms for $75, 50, \overline{50}$ and no mass terms for $5 + \bar{5}$. The 75-plet breaks the GUT symmetry to $SU(3) \times SU(2) \times U(1)$. The key element is that $50 + \overline{50}$ have no doublet pairs (D) and only triplet/anti-triplet pairs (T). Thus $50 + \overline{50}$ have $0D + 1T$ while $5 + \bar{5}$ have $1D + 1T$. The Higgs triplets/anti-triplets of $50 + \overline{50}$ mix with the Higgs triplets/anti-triplets of $5 + \bar{5}$ to become heavy. This is accomplished via the superpotential $W(\text{higgs}) = W_0(75) + M 50 \otimes \overline{50} + \lambda_1 50 \otimes 75 \otimes \bar{5} + \lambda_2 \overline{50} \otimes 75 \otimes 5$. The doublets in $5 \oplus \bar{5}$ have nothing to pair up with and remain light. Next we extend the missing partner mechanism to

$SO(10)$. Here one case has previously been worked out, i.e., the case with the heavy sector consisting of $126 \oplus \overline{126}$ and the light sector consisting of $10 \oplus 120$ [32]. In recent works several other cases have been found [33]. These models are anchored in $126 + \overline{126}$ or $560 \oplus \overline{560}$ which play the role of $50 \oplus \overline{50}$. Analysis of these larger representations require special tools which utilize oscillator techniques [34, 35].

However, to connect the high scale physics to low energies we need to break supersymmetry. As is well known in global supersymmetry, it is difficult to achieve the breaking of supersymmetry in a phenomenologically acceptable fashion. To do so one must consider local supersymmetry which involves gravity [36, 37]. Thus one needs to construct supergravity grand unified models [15], coupled with radiative breaking of the electroweak symmetry (for a review see [38]) using renormalization group evolution of gauge, Yukawa and of soft terms [39]. A broad class of models fall under this rubric. These include mSUGRA (CMSSM), and SUGRA models with non-universalities in the Higgs sector and in the gaugino sector. mSUGRA has the parameter space $m_0, m_{1/2}, A_0, \tan\beta, \text{sign}(\mu)$ where m_0 is the universal scalar mass, $m_{1/2}$ is the universal gaugino mass, A_0 is the universal trilinear coupling, $\tan\beta$ is the ratio $< H_2 > / < H_1 >$ where, $< H_2 >$ gives mass to the up quarks and $< H_1 >$ gives mass to the down quarks and leptons and μ is the Higgs mixing parameter. The mSUGRA model assumes a flavor independent Kahler potential. However, the nature of physics at the Planck scale is not fully understood. Thus it is useful to consider more general Kahler potentials, and indeed in string models and in D brane models one often encounters more general Kahler potentials (see, e.g., [40]). Such choices give rise to non-universal supergravity models which we may label as NUSUGRA models. These include models with non-universal gaugino masses [41] (see also [42]) where the gaugino sector is characterized by the three parameters $\tilde{m}_1, \tilde{m}_2, \tilde{m}_3$ which characterize the gaugino masses for the gauge groups $U(1)$, $SU(2)_L$ and $SU(3)_C$. Similarly, one could have non-uinversality in the Higgs boson sector [43] with different masses m_{H_1}, m_{H_2} for the Higgses H_1 and H_2, as well as non-universalities in the scalar sector.

4. Natural TeV Size Scalars

Radiative breaking of the electroweak symmetry naturally allows TeV size scalars without the necessity of excessive fine tuning. To illustrate this phenomenon we consider the generic form of the radiative breaking of the electroweak symmetry (REWSB) which can be written in the form $\mu^2 + \frac{1}{2}M_Z^2 = m_0^2 C_1 + A_0^2 C_2 + m_{\frac{1}{2}}^2 C_3 + m_{\frac{1}{2}} A_0 C_4 + \Delta\mu_{loop}^2$. With a redefinition of parameters one can write this equation in the form $\mu^2 + \frac{1}{2}M_Z^2 = m_0^2 C_1 + \bar{A}_0^2 C_2 + m_{1/2}^2 \bar{C}_3 + \Delta\mu_{loop}^2$ where $\bar{A}_0 \equiv A_0 + (C_4/2C_2)m_{1/2}$ and $\bar{C}_3 = C_3 - C_4^2/(4C_2)$. To simplify matters one can go the renormatlization group (RG) scale Q where the loop contribution is minimized (and ideally vanishes). In this case we can use the tree part of the REWSB reliably. Typically $C_1, C_2, \bar{C}_3$ are all positive which implies that the soft parameters sit on the surface of an ellipsoid given by the REWSB equation. Thus for fixed μ none of the soft parameters can get too large. One may call this the Ellipsoidal Branch [15]. Suppose now that for certain choices of the inputs such as the top Yukawa, $\tan\beta$, the gauge couplings etc., that the co-efficient C_1 turns negative. In this case large value of m_0 can occur due to the possibility of cancellation of the C_1 term with the rest of the terms in REWSB. This is the Hyperbolic Branch [18, 19, 20, 21] which contains the Focal Point (HB/FP), Focal Curves (HB/FC), and Focal Surfaces (HB/FS). The Focal Point corresponds to the case when C_1 vanishes [19], Focal Curves are when two parameters get large, i.e., m_0, A_0 (HB/FC1) or $m_0, m_{1/2}$ (HB/FC2), and Focal Surfaces (HB/FS) occur when all the three soft parameters $m_0, m_{1/2}, A_0$ get large while μ remains fixed and small. In all these cases one finds that large scalar masses exist. Using the LHC constraints the HB/FP region is seen to be depleted [21] while most of the remaining parameter space appears to lie on Focal Curves and Focal Surfaces [21, 44].

5. Dark Matter

In supergravity models the neutralino is the lightest R parity odd particle over most of the parameter space of the model and with R parity it is a possible candidate for dark matter. The annihilation of dark matter proceeds via the Z and the Higgs poles (h^0, H^0, A^0) in the s-channel and via the sfermion exchange in the t-channel. Thus the relic density is sensitive to the Higgs masses as well as to the masses of the sfermions. Similarly, the spin independent neutralino-proton cross-section arises from the scattering of the neutralino from quarks in a nucleus and the scattering proceeds via squark poles in the s-channel and via the Z, h^0, H^0, A^0 poles in the t-channel. Consequently again the spin independent neutralino-proton cross-section is sensitive to the scale of squark masses and to the Higgs masses. Thus one might ask as to the impact of the lightest Higgs mass at 125 GeV on dark matter. As discussed above, the direct implication of a 125 GeV Higgs mass is to raise the scale of supersymmetry. The larger scale leads to heavier sfermion masses which implies that the relic density will be governed mostly by the s-channel Z and Higgs poles and by coannihilation. Analogously the spin independent neutralino-proton cross-section will be largely governed by the t-channel Z and Higgs pole exchanges. These constraints lead to further limits on the allowed parameter space of the model consistent with WMAP [45]. In this context we mention that often the upper limit WMAP constraint is imposed so as to allow for the possibility of multicomponent dark matter [46]. An analysis of the Higgs constraint on the spin independent neutralino-proton cross-section was given in [4, 47] (see also [44]) and it is found that most of the allowed parameter space of mSUGRA model lies between the current sensitivity of the XENON100 experiment [48] and the expected sensitivity in the XENON-1T [49] and SuperCDMS [50]. Thus the prospects for the discovery of dark matter look bright if the proposed sensitivities in the new generation of dark matter experiments can be reached.

6. $g_\mu - 2$ constraint

The anomalous magnetic moment of the muon $g_\mu - 2$ is a sensitive test of new physics. In supersymetric models $g_\mu - 2$ can receive important contributions from supersymmetric loops. In the standard model electroweak contributions arise from the exchange of the W and Z boson. Analogously in supersymmtric models one has contributions arising from the exchange of charginos and sneutrinos, and from the exchange of neutralinos and smuons [51]. If the sparticle spectrum is light, one finds that these contributions can be comparable to or even exceed the standard model contribution. The experimental determination of the correction to $g_\mu - 2$ or to a_μ defined by $a_\mu = (g_\mu - 2)/2$, i.e., $\delta a_\mu = a_\mu^{exp} - a_\mu^{SM}$ is sensitively dependent on the hadronic corrections in the standard model. Thus there are two main estimations of the hadronic error: one using the e^+e^- annihilation and the other using τ decay. The analysis using e^+e^- estimation of hadronic corrections gives [52] $\delta a_\mu = (28.7 \pm 8.0) \times 10^{-10}$ which is a 3.6σ deviation from the standard model, while the analysis using τ decay gives [52] $\delta a_\mu = (19.5 \pm 8.3) \times 10^{-10}$ (2.4σ) which is 2.4σ deviation from the standard model result. As discussed above the high Higgs mass leads to a high scale for the sfermions and with universal boundary conditions on soft parameters they lead to a rather small correction to a_μ. Thus there is a tension between the $g_\mu - 2$ experimental result and the high Higgs mass. Assuming the $g_\mu - 2$ result does not undergo further oscillations, this tension could be relieved if one assumes that part of the Higgs mass arises from sources other than from MSSM. Thus extra matter could give additional contribution to the Higgs mass [53] effectively lowering the component of the Higgs mass arising from MSSM which would allow an a_μ correction consistent with experiment. Similarly, the presence of an extra gauge group under which the Higgs is charged would lead to a contribution from the D term to the Higgs mass again effectively lowering the MSSM Higgs mass. Yet another possibility is that $g_\mu - 2$ contribution arises from new physics other than SUSY. Thus, for example, a Z' which couples to the muon will produce a contribution to a_μ. For, instance a Z' arising from a gauged $L_\mu - L_\tau$ can have a significantly lower mass than the current experimental limits on a generic Z' and can generate a correction to $g_\mu - 2$ of the right size [46] (see the discussion in Sec.7).

7. Cosmic coincidence

An interesting cosmic coincidences is the fact that the ratio of dark matter to baryonic matter is roughly 5, or more precisely $\Omega_{DM}/\Omega_B = 4.99 \pm 0.20$ [45]. The fact that they are about the same size points to a possible common origin for them. The so called asymmetric dark matter (AsyDM) idea proposes that dark matter originated from baryonic matter in the early universe by the transfer of a net $B-L$ charge from the visible sector to the dark sector [54] (For a review see [55]). The implementation of this idea has two basic ingredients: first is to find a mechanism for the transfer of $B-L$ from the visible sector to the dark sector, and second to find a mechanism to deplete the symmetric component of dark matter produced by thermal processes. The standard procedure for the transfer of $B-L$ from the visible sector to the dark sector is to consider an interactions of the type [56] $M_a^{-n} O_{DM} O_{asy}^{SM}$ which is operative at temperatures T_{int} so that $T_{int} > (M_a^{2n} M_{Pl}^{-1})^{\frac{1}{2n-1}}$, $M_{Pl} = 2.4 \times 10^{18}$ GeV where O_{asy}^{SM} is constituted of only standard model particles and O_{DM} is constituted of dark matter fields and they have opposite $B-L$ charges so that overall the interaction is $B-L$ charge neutral. Concerning the depletion of the symmetric component of dark matter, one needs a mechanism for an efficient annihilation of the thermally produced dark matter.

The early universe can be viewed as a weakly interacting plasma in which each particle carries a chemical potential μ_i. In such a plasma the particle-anti-particle asymmetries are given by $n_i - \bar{n}_i \simeq (g_i \beta T^3/6)(\mu_i(\text{fermi}), 2\mu_i(\text{bose}))$ where g_i is the degrees of freedom, and $\beta = 1/T$. The chemical potentials are constrained by (i)sphaleron interactions, (ii) conservation of charge and hypercharge, (iii) Yukawa and gauge interactions. When the transfer interaction is in equilibrium (see, e.g., [57]), one can solve for the ratio Ω_{DM}/Ω_B so that $\Omega_{DM}/\Omega_B = (X/B)(m_{DM}/m_B) \simeq 5$, where X is the dark matter number density and B is the baryon number density. One can generate a variety of models such as models where the visible sector is taken to be the standard model, the two Higgs doublet model, or the MSSM [58]. For the MSSM many variations are possible depending on the scale of sparticle masses. For each model there are various interactions that allow a transfer of the $B-L$ asymmetry from the standard model sector to the dark matter sector. The relic density in AsyDM will have two components: a thermal and a non-thermal component. Thus one may write the total as a sum $\Omega_{\rm DM} = \Omega_{\rm DM}^{\rm asy} + \Omega_{\rm DM}^{\rm sym}$. For AsyDM to work we need $\Omega_{\rm DM}^{\rm sym} << \Omega_{\rm DM}^{\rm asy}$. Thus we need an efficient mechanism for the annihilation of thermally produced dark matter. We accomplish this via the exchange of a gauge field using the Stueckelberg formalism where the gauge field couples to $L_\mu - L_\tau$ [58]. In the unitary gauge the massive vector boson field will be called Z' and its interaction with fermions is given by $L_{\rm int} = Q^\psi g_C \bar{\psi} \gamma^\mu \psi Z'_\mu + Q^f g_C \bar{f} \gamma^\mu f Z'_\mu$, $f = \mu, \tau$. where f runs over μ and τ families and $Q_C^\mu = -Q_C^\tau$. The LEP constraints on the $M_{Z'}$ mass are not valid since Z' does not couple with the first generation leptons. This result also holds at the loop level to a good approximation. The strongest constraint comes from $g_\mu - 2$. Thus the correction to $g_\mu - 2$ arising from the exchange of Z' boson associated with the $L_\mu - L_\tau$ symmetry is given by $\Delta(g_\mu - 2) = (\frac{1}{2} g_C Q_C^\mu)^2 m_\mu^2/(6\pi^2 M_{Z'}^2)$. Imposing the constraints $\Delta a_\mu = \Delta(g_\mu - 2)/2 \leq 3 \times 10^{-9}$ one finds the restriction $M_{Z'}/(g_C Q_C^\mu) \geq 90$ GeV. The above constraint allows for a low lying Z' which couples only to muons and taus and allows for a rapid annihilation of symmetric dark matter via the Z' pole.

To obtain relic densities at current temperatures for ψ and $\bar{\psi}$ one must solve the Boltzman equations in the presence of asymmetries. The Boltzmann equations obeyed by f_ψ and $f_{\bar{\psi}}$ take the form $df_\psi/dx = \alpha\langle\sigma v\rangle(f_\psi f_{\bar{\psi}} - f_\psi^{\rm eq} f_{\bar{\psi}}^{\rm eq})$, $df_{\bar{\psi}}/dx = \alpha\langle\sigma v\rangle(f_\psi f_{\bar{\psi}} - f_\psi^{\rm eq} f_{\bar{\psi}}^{\rm eq})$, where $x = k_B T/m_\psi$ and $f_\psi = n_\psi/hT^3$, $f_{\bar{\psi}} = n_{\bar{\psi}}/hT^3$ where h is the entropy degrees of freedom. One finds that $\gamma = f_\psi - f_{\bar{\psi}}$, is a constant independent of temperature. The relic densities for ψ and $\bar{\psi}$ are then given by $\Omega_\psi h_0^2/(\Omega_\psi h_0^2)_{\xi=0} \simeq \xi J(x_f)/(1 - exp(-\xi J(x_f)) \to 1$ as $\xi \to 0$, where $\xi = \gamma C$ and where C is a numerical constant and $J(x_f) \equiv \int_{x_0}^{x_f} \langle\sigma v\rangle\, dx$. Similarly, $\Omega_{\bar{\psi}} h_0^2/(\Omega_\psi h_0^2) \simeq exp(-\xi J(x_f)) \to 1$ as $\xi \to 0$. We need to show that $\Omega_{\bar{\psi}} h_0^2 << (\Omega h_0^2)_{\rm WMAP}$ and that $\Omega_\psi h_0^2$ is the major component of WMAP. This has been exhibited explicitly in [58]. We discuss now collider implications of the model. In a muon collider there would be final states with muons and taus and their neutrinos but no e^+e^- final states providing a smoking gun signature for the model [58]. The analysis is done including one loop corrections arising from the first and the second generation leptons in the loop. One can carry out a direct extension of AsyDM to the supersymmetric case. The basic interaction responsible for the asymmetry has the form $W_{\rm asy} = M_a^{-n} O_{DM} O_{\rm asy}^{\rm mssm}$. In general there are many possibilities for the operators $O_{\rm asy}^{\rm mssm}$ such as $LH_2, LLE^C, QLD^C, U^C D^C D^C$ or any product thereof. Obviously $O_{\rm DM}$ will carry the opposite quantum numbers to those of $O_{\rm asy}^{\rm mssm}$. In this case there are two dark matter particles, i.e., ψ and the neutralino $\tilde{\chi}^0$ and here the total relic density is $\Omega_{\rm DM} = \Omega_{\rm DM}^{\rm asy} + \Omega_{\rm DM}^{\rm sym} + \Omega_{\tilde{\chi}^0}$, where $\Omega_{\tilde{\chi}^0}$ is the relic density from the neutralino. One must show that the neutralino contribution is subdominant, i.e., it is no more than 10% of the WMAP value. An interesting question is if a subdominant neutralino is detectable. This appears to be the case as demonstrated in [58].

8. R parity and Proton Stability

R parity is needed to get rid of baryon and lepton number violating dimension 4 operators which can generate unacceptably fast proton decay. Within MSSM R parity is ad hoc. Further, R parity as a global symmetry is not desirable since it can be broken by wormhole effects. This problem can be evaded if a model possesses a gauge symmetry so that R parity arises as a discrete remnant of this gauge symmetry (see, e.g., [59]). Since $R = (-1)^{2S+3(B-L)}$ the obvious extended symmetry is $SU(3)_C \times SU(2)_L \times U(1)_Y \times U(1)_{B-L}$. In this case the $U(1)_{B-L}$ gauge symmetry will forbid R parity violating interactions such as $QLd^c, LLe^c, u^c d^c d^c, L\bar{H}$. Now an unbroken $U(1)_{B-L}$ gauge symmetry is undesirable since there would be associated with it a massless gauge boson which can generate an unacceptable long range force and thus one must generate a mass for the $B - L$ gauge boson. If the $B - L$ gauge symmetry is broken spontaneously, R parity is no longer guaranteed. Specifically R parity is protected if $3(B-L)$ is an even integer but is not protected if $3(B - L)$ is an odd integer. To illustrate this point in a specific example one may consider an extension of MSSM with a $U(1)_{B-L}$ symmetry which includes three right handed neutrinos fields ν^c for anomaly cancellation. The extended superpotential in this case is $\mathcal{W} = \mathcal{W}_{\rm MSSM} + h_\nu\, LH_u\nu^c + h_{\nu^c}\nu^c\nu^c\Phi + \mu_\Phi\Phi\bar{\Phi}$ where the new fields are assigned the $B - L$ quantum numbers as $(\nu^c, \Phi, \bar{\Phi}) : (-1, -2, 2)$. Since the $B - L$ of the Φ field is even, a VEV growth for it will not violate R parity. However, the VEV growth for ν^c for which $B - L$ is odd, will violate R parity. It turns out that the latter possibility can be realized in radiative breaking. This is so because the beta functions due to the coupling of the Φ and ν^c can turn the mass of ν^c tachyonic which leads to a VEV growth for ν^c and a violation of R parity [60, 61].

A violation of R parity by the ν^c VEV growth can be avoided if one uses the Stueckelberg mechanism [62, 63] (see also [64]) for the mass growth of the $B - L$ gauge boson rather than spontaneous breaking. Specifically let us assume a Stueckelberg mechanism for the $B - L$ gauge boson mass growth with the additional assumptions $\langle\tilde{q}\rangle = 0$, $\langle\tilde{e}_L\rangle = 0 =$

$\langle\tilde{e}^c\rangle$ which are just the conditions for charge conservation. Since $\tilde{\nu}_L$ and $\tilde{e}_L$ belong to the same $SU(2)_L$ multiplet, the vanishing of $\langle\tilde{e}_L\rangle$ also implies the vanishing of $<\tilde{\nu}_L>$, i.e., $<\tilde{\nu}_L>=0$ since the RG evolution of $M_{\tilde{e}_L}$ and of $M_{\tilde{\nu}_L}$ are very similar. It then follows after integrating over the remaining Stueckelberg fields that the scalar potential takes the form $V_{\nu^c}=M^2_{\tilde{\nu}^c}\tilde{\nu}^{c\dagger}\tilde{\nu}^c+(g^2_{BL}M^2_\rho)/[2(M^2_{BL}+M^2_\rho)](\tilde{\nu}^{c\dagger}\tilde{\nu}^c)^2$. In this case there are no beta functions to turn $M^2_{\tilde{\nu}^c}$ negative in the renormalization group evolution. As a consequently the potential cannot support spontaneous breaking to generate a VEV of $\tilde{\nu}^c$ and $\langle\tilde{\nu}^c\rangle=0$. Thus with the Stueckelberg mechanism the $B-L$ gauge boson gains a mass but R parity remains unbroken.

We discuss now briefly proton decay. While R parity can eliminate baryon and lepton number violating dimension 4 operators, baryon and lepton number violating dimension five and dimension six operators do exist in unified models. The baryon and lepton number violating dimension five operators arise from color higgsino exchange while the baryon and lepton number violating dimension six operators arise from the exchange of lepto-quarks. The baryon and lepton number violating dimension six operators lead to the dominant proton decay mode $p\to e^+\pi^0$ and estimates give a lifetime which could be in the 10^{35} yrs range (For a recent review see [65]). The current experimental limit from SuperKamiokande for this mode is $\tau(p\to e^+\pi^0)>1.4\times10^{34}$yrs. It is expected that at Hyper-Kamiokande [66] one will reach the limit $\tau(p\to e^+\pi^0)>1\times10^{35}$yrs, and thus there is a chance for the observation of the $e^+\pi^0$ mode. Proton decay from baryon and lepton number violating dimension five operators is more model dependent. Thus the baryon and lepton number violating dimension five operators must be dressed by chargino, gluino, and neutralino exchanges to produce baryon and lepton number violating dimension six operators responsible for proton decay. Since the sparticle spectrum enters in the dressing loops, the proton lifetime is dependent on the nature of the sparticle spectrum as well on CP phases (for a review see [67]) and thus proton decay provides a test of models of strings and branes via proton decay branching ratios (see, e.g., [68],[28]). Specifically a light sparticle spectrum will lead to a shorter proton decay while a heavier spectrum will lead to a longer proton decay lifetime. The current experimental limit on p decay from SuperKamiokande gives $\tau(p\to\bar{\mu}K^+)>4\times10^{33}$yrs which puts severe constraints on supersymmetric models. These constraints can be relieved either by GUT models which invoke a cancellation mechanism for the baryon and lepton number violating dimension five operators or a heavy sparticle spectrum. The data from the large hadron collider which indicates that the scale of SUSY breaking for the scalar sector may be large helps stabilize the proton against too fast a decay from dimension five operators. It is expected that Hyper-Kamiokande [66] will reach a sensitivity of 2×10^{34}yrs for the $\tau(p\to\bar{\mu}K^+)$ mode. In any case proton decay from dimension five operators could become visible even with modest increase in sensitivity for the detection of this mode in the future.

9. Conclusion

We have given a brief summary of the current status of supergravity grand unification. A very significant constraint on the model arises from the recent discovery of the Higgs boson by the ATLAS and CMS detectors and the determination that the mass of the Higgs is high, i.e., around 125 GeV. It has been known for some time that in mSUGRA the mass of the Higgs must lie below 130 GeV, and thus it is interesting that the mass of the Higgs as determined by experiment lies below the predicted limit. However, the high mass of the Higgs implies typically that the average mass of the stops must be significantly large and further that the trilinear coupling A_0 typically must be large. These results imply that the part of the parameter space of mSUGRA which gives rise to the large Higgs mass would lie on the Hyperbolic Branch of radiative breaking of the electroweak symmetry and more specifically on Focal Curves or Focal Surfaces. A theoretical analysis using the Higgs mass as a constraint also implies that there is a significant region of the parameter space consistent with all experimental constraints where the gaugino masses are relatively light. Additionally in some restricted regions of the parameter space on may also have relatively light stops and staus. Thus several light sparticles appear to be prime candidates for discovery at the LHC. These include the neutralino, the chargino, the gluino, and the stop. Another implication of the heavy Higgs mass pertains to neutralino dark matter. The Higgs mass constraint significantly restricts the allowed range of the spin independent neutralino-proton cross section. Interestingly most of the allowed range lies between the current sensitivity of the XENON100 and the expected sensitivity of XENON-1T and SuperCDMS experiments. This result is very encouraging for the discovery of supersymmetric dark matter. The issue of the $g_\mu - 2$ constraint was also discussed. The excess of $g_\mu - 2$ over the standard model result is dependent on the hadronic corrections. The e^+e^- annihilation data and the τ decay data give somewhat different results for this quantity leading to either 3.6σ or 2.4σ discrepancy. If the deviation from the standard model stays there would be tension between the $g_\mu - 2$ result and the high Higgs mass in models with universal boundary conditions on the soft parameters. This tension can be relieved in several ways, such as by having part of the Higgs mass arise from extra matter at the loop level, or from D terms if there is an extra gauge group under which the Higgs is charged. Alternately flavored supergravity models with non-universalities in the flavor sector could account for the discrepancy.

Also discussed was the status of grand unification. Currently $SO(10)$ is the favored group for the unification of the electroweak and the strong interactions. $SO(10)$ has the advantage over $SU(5)$ in that it unifies a full generation of quarks and leptons in a single 16-plet representation. However, the current models based on $SO(10)$ suffer from a couple of drawbacks. The first one concerns the breaking of the GUT group. Here in conventional models more than one scale enters in the breaking of the group. Thus one scale enters in reducing the rank and the other to break the symmetry all the way down to the standard model gauge group. This drawback can be overcome if one considers $144 + \overline{144}$ of Higgs whose VEV formation can break the group down to the standard model gauge group in one step. A second draw back which is a generic problem in GUT models concerns the doublet-triple splitting problem. A new class of $SO(10)$ models using $560 + \overline{560}$ of Higgs fields not only break the gauge symmetry at one scale they also solve the doublet-triplet problem by the missing partner mechanism. Further, work using these models is thus desirable. The baryon and lepton number violating dimension five and dimension six operatos which give rise to proton decay and the prospects for the discovery of proton decay in future experiments were discussed. Finally we discussed the so called cosmic coincidence which pertains to the fact that the dark matter and the baryonic matter are in the ratio $\sim 5:1$. The realization of this possibility in the context of superymmetry was discussed in a multicomponent dark matter framework consisting of a Dirac fermion (carrying $B-L$ charge) and a Majorana fermion (the neutralino).

Acknowledgments

This research is supported in part by the U.S. National Science Foundation (NSF) grants PHY-0757959 and PHY-0969739 and through XSEDE under grant number TG-PHY110015, and NERSC grant DE-AC02-05CH11231.

References

[1] F. Gianotti [on behalf of ATLAS], "Update on the Standard Model Higgs searches in ATLAS", joint CMS/ATLAS seminar, December 13, 2011; [ATLAS Collaboration] ATLAS-CONF-2011-163. [ATLAS Collaboration], arXiv:1202.1408 [hep-ex].
[2] G. Tonelli [on behalf of CMS], "Update on the Standard Model Higgs searches in CMS", joint CMS/ATLAS seminar, December 13, 2011; [CMS Collaboration] CMS PAS HIG-11-032.
[3] F. Englert and R. Brout, Phys. Rev. Lett. **13**, 321 (1964); P. W. Higgs, Phys. Lett. **12**, 132 (1964). Phys.

Rev. Lett. **13**, 508 (1964); G. S. Guralnik, C. R. Hagen and T. W. B. Kibble, Phys. Rev. Lett. **13**, 585 (1964).

[4] S. Akula, B. Altunkaynak, D. Feldman, P. Nath and G. Peim, Phys. Rev. D **85**, 075001 (2012).

[5] H. Baer *et al* Phys. Rev. D **85**, 075010 (2012); A. Arbey *et al* Phys. Lett. B **708**, 162 (2012); J. L. Feng *et al* Phys. Rev. D **85**, 075007 (2012); M. Carena *et al* JHEP **1203**, 014 (2012); S. Heinemeyer *et al* Phys. Lett. B **710**, 201 (2012); P. Draper *et al* Phys. Rev. D **85**, 095007 (2012); J. Ellis and K. A. Olive, Eur. Phys. J. C **72**, 2005 (2012); O. Buchmueller *et al* arXiv:1112.3564 [hep-ph]; U. Ellwanger, JHEP **1203**, 044 (2012).

[6] M. Kadastik *et al* JHEP **1205**, 061 (2012); J. Cao *et al* Phys. Lett. B **710**, 665 (2012); B. Batell *et al* arXiv:1112.5180 [hep-ph]; L. J. Hall *et al* JHEP **1204**, 131 (2012); J. F. Gunion *et al* Phys. Lett. B **710**, 454 (2012); P. Fileviez Perez, Phys. Lett. B **711**, 353 (2012); S. F. King *et al* Nucl. Phys. B **860**, 207 (2012); J. Cao *et al* JHEP **1203**, 086 (2012); C. -F. Chang *et al* JHEP **1206**, 128 (2012); L. Aparicio *et al* JHEP **1204**, 126 (2012); H. Baer *et al* JHEP **1205**, 091 (2012); D. Ghosh *et al* arXiv:1202.4937 [hep-ph]. N. Desai *et al* arXiv:1202.5190 [hep-ph]; D. Carmi *et al* arXiv:1202.3144 [hep-ph]; L. Maiani *et al* arXiv:1202.5998 [hep-ph]; T. Cheng *et al* arXiv:1202.6088 [hep-ph]; J. Cao *et al* JHEP **1206**, 145 (2012); F. Jegerlehner, arXiv:1203.0806 [hep-ph]; N. D. Christensen *et al* arXiv:1203.3207 [hep-ph]; A. Choudhury and A. Datta, JHEP **1206**, 006 (2012); B. S. Acharya *et al* Int. J. Mod. Phys. A **27**, 1230012 (2012); M. A. Ajaib *et al* arXiv:1204.2856 [hep-ph]; J. Ellis and T. You, arXiv:1204.0464 [hep-ph]; C. Balazs *et al* arXiv:1205.1568 [hep-ph]; M. Badziak, E. Dudas, M. Olechowski and S. Pokorski, JHEP **1207**, 155 (2012) [arXiv:1205.1675 [hep-ph]]; J. L. Feng and D. Sanford, arXiv:1205.2372 [hep-ph]; N. Okada, arXiv:1205.5826 [hep-ph]; M. Carena *et al* arXiv:1205.5842 [hep-ph]; E. Dudas *et al* arXiv:1205.5988 [hep-ph]; A. Fowlie *et al* arXiv:1206.0264 [hep-ph]; M. Hirsch *et al* arXiv:1206.3516 [hep-ph]; R. M. Chatterjee *et al* arXiv:1206.5770 [hep-ph]; M. W. Cahill-Rowley *et al* arXiv:1206.5800 [hep-ph]; A. Albaid and K. S. Babu, arXiv:1207.1014 [hep-ph]; C. Boehm, J. Da Silva, A. Mazumdar and E. Pukartas, arXiv:1205.2815 [hep-ph].

[7] [The TeVatron], arXiv:1207.0449 [hep-ex].

[8] J. Incandela [on behalf of CMS], "Status of the CMS SM Higgs Search", and F. Gianotti [on behalf of ATLAS], "Status of the SM Higgs Search in ATLAS", joint CMS/ATLAS seminars at CERN, July 4, 2012; CMS Collaboration, CMS-PAS-HIG-12-020; ATLAS Collaboration, ATLAS-CONF-2012-093.

[9] M. E. Peskin, arXiv:1207.2516 [hep-ph].

[10] M. Carena, I. Low and C. E. M. Wagner, arXiv:1206.1082 [hep-ph].

[11] J. Baglio, A. Djouadi and R. M. Godbole, arXiv:1207.1451 [hep-ph].

[12] D. Zerwas, Talk at PASCOS2012, Merida Mexico, June 3-8, 2012.

[13] B. W. Lee, C. Quigg and H. B. Thacker, Phys. Rev. Lett. **38**, 883 (1977). See also, D. A. Dicus and V. S. Mathur, Phys. Rev. D7, 3111(1973); M. Veltman, Acta. Phys. Polon. B8, 475 (1977).

[14] G. L. Kane, C. F. Kolda and J. D. Wells, Phys. Rev. Lett. **70**, 2686 (1993) [hep-ph/9210242].

[15] A. H. Chamseddine, R. L. Arnowitt, P. Nath, Phys. Rev. Lett. **49**, 970 (1982); Nucl. Phys. **B227**, 121 (1983); L. J. Hall, J. D. Lykken, S. Weinberg, Phys. Rev. **D27**, 2359-2378 (1983); R. L. Arnowitt and P. Nath, Phys. Rev. Lett. **69**, 725 (1992). For a review see, P. Nath, [hep-ph/0307123].

[16] A. Arbey, M. Battaglia, A. Djouadi and F. Mahmoudi, arXiv:1207.1348 [hep-ph].

[17] A. Djouadi, Phys. Rept. **459**, 1 (2008); M. S. Carena and H. E. Haber, Prog. Part. Nucl. Phys. **50**, 63 (2003)

[18] K. L. Chan *et al* Phys. Rev. D **58** (1998) 096004; U. Chattopadhyay, A. Corsetti and P. Nath, Phys. Rev. D **68**, 035005 (2003). H. Baer, C. Balazs, A. Belyaev, T. Krupovnickas and X. Tata, JHEP **0306**, 054 (2003).

[19] J. L. Feng, K. T. Matchev and T. Moroi, Phys. Rev. Lett. **84**, 2322 (2000).

[20] D. Feldman, G. Kane, E. Kuflik and R. Lu, arXiv:1105.3765 [hep-ph].

[21] S. Akula, M. Liu, P. Nath and G. Peim, Phys. Lett. B **709**, 192 (2012) [arXiv:1111.4589 [hep-ph]].

[22] D. Feldman, Z. Liu and P. Nath, Phys. Rev. Lett. **99**, 251802 (2007); Phys. Lett. B **662**, 190 (2008); JHEP **0804**, 054 (2008); Phys. Rev. D **78**, 083523 (2008); Phys. Rev. D **80**, 015007 (2009); Phys. Rev. D **81**, 117701 (2010).

[23] P. Nath, B. D. Nelson, H. Davoudiasl, B. Dutta, D. Feldman, Z. Liu, T. Han and P. Langacker *et al.*, Nucl. Phys. Proc. Suppl. **200-202**, 185 (2010) [arXiv:1001.2693 [hep-ph]].

[24] D. Feldman, K. Freese, P. Nath, B. D. Nelson and G. Peim, Phys. Rev. D **84**, 015007 (2011); N. Chen, D. Feldman, Z. Liu, P. Nath and G. Peim, Phys. Rev. D **83**, 035005 (2011); D. Feldman, Z. Liu and P. Nath, Phys. Rev. D **80**, 015007 (2009).

[25] M. A. Ajaib, T. Li and Q. Shafi, Phys. Lett. B **705**, 87 (2011) [arXiv:1107.2573 [hep-ph]].

[26] H. Baer, V. Barger, A. Lessa and X. Tata, arXiv:1207.4846 [hep-ph].

[27] H. Georgi, in Particles and Fields (edited by C.E. Carlson), A.I.P., 1975; H. Fritzch and P. Minkowski, Ann.

Phys. **93**, 193 (1975).
[28] P. Nath and P. Fileviez Perez, Phys. Rept. **441**, 191 (2007).
[29] K. S. Babu, I. Gogoladze, P. Nath and R. M. Syed, Phys. Rev. D **72**, 095011 (2005) [arXiv:hep-ph/0506312]; Phys. Rev. D **74**, 075004 (2006) [arXiv:hep-ph/0607244].
[30] P. Nath and R. M. Syed, JHEP **0602**, 022 (2006) [arXiv: hep-ph/0511172]; Phys. Rev. D **81**, 037701 (2010) [arXiv:0909.2380 [hep-ph]].
[31] A. Masiero, D. V. Nanopoulos, K. Tamvakis and T. Yanagida, Phys. Lett. B **115**, 380 (1982); B. Grinstein, Nucl. Phys. B **206**, 387 (1982).
[32] K. S. Babu, I. Gogoladze and Z. Tavartkiladze, Phys. Lett. B **650**, 49 (2007) [arXiv:hep-ph/0612315].
[33] K. S. Babu, I. Gogoladze, P. Nath and R. M. Syed, Phys. Rev. D **85**, 075002 (2012).
[34] R.N. Mohapatra and B. Sakita, Phys. Rev. **D21**, 1062 (1980).
[35] P. Nath and R. M. Syed, Phys. Lett. B **506**, 68 (2001); Nucl. Phys. B **618**, 138 (2001); Nucl. Phys. B **676**, 64 (2004); R. M. Syed, arXiv: hep-ph/0411054; arXiv: hep-ph/0508153.
[36] P. Nath and R. L. Arnowitt, Phys. Lett. B **56**, 177 (1975); R. L. Arnowitt, P. Nath and B. Zumino, Phys. Lett. B **56**, 81 (1975).
[37] D. Z. Freedman, P. van Nieuwenhuizen and S. Ferrara, Phys. Rev. D **13**, 3214 (1976).
[38] L. E. Ibanez and G. G. Ross, Comptes Rendus Physique **8**, 1013 (2007).
[39] S. P. Martin and M. T. Vaughn, Phys. Rev. D **50**, 2282 (1994)
[40] B. Kors and P. Nath, Nucl. Phys. B **681**, 77 (2004); D. Lust, S. Reffert and S. Stieberger, Nucl. Phys. B **706**, 3 (2005).
[41] U. Chattopadhyay and P. Nath, Phys. Rev. D **65**, 075009 (2002).
[42] D. P.Roy, Talk at PASCOS2012, Merida Mexico, June 3-8, 2012
[43] P. Nath, R. L. Arnowitt, Phys. Rev. **D56**, 2820-2832 (1997); J. R. Ellis, K. A. Olive, Y. Santoso, Phys. Lett. **B539**, 107-118 (2002).
[44] S. Akula, D. Feldman, Z. Liu, P. Nath and G. Peim, Mod. Phys. Lett. A **26**, 1521 (2011) [arXiv:1103.5061 [hep-ph]].
[45] E. Komatsu *et al.* [WMAP Collaboration], Astrophys. J. Suppl. **192**, 18 (2011).
[46] D. Feldman, Z. Liu, P. Nath and G. Peim, Phys. Rev. D **81**, 095017 (2010).
[47] S. Akula, P. Nath and G. Peim, arXiv:1207.1839 [hep-ph].
[48] E. Aprile *et al.* [XENON100 Collaboration], arXiv:1104.2549 [astro-ph.CO]; Phys. Rev. Lett. **105**, 131302 (2010); arXiv:1103.0303 [hep-ex].
[49] E. Aprile, The XENON Dark Matter Search, WONDER Workshop, LNGS, March 22, 2010.
[50] B. Cabrera, "SuperCDMS Development Project", 2005.
[51] T. C. Yuan, R. L. Arnowitt, A. H. Chamseddine and P. Nath, Z. Phys. C **26**, 407 (1984); D. A. Kosower *et al* Phys. Lett. B **133**, 305 (1983).
[52] A. Hoecker, Nucl. Phys. Proc. Suppl. **218**, 189 (2011); K. Hagiwara *et al* J. Phys. G G **38**, 085003 (2011).
[53] K. S. Babu, I. Gogoladze, M. U. Rehman and Q. Shafi, Phys. Rev. D **78**, 055017 (2008) [arXiv:0807.3055 [hep-ph]].
[54] S. Nussinov, Phys. Lett. B **165**, 55 (1985); K. Griest and D. Seckel. Nucl. Phys. B **283**, 681 (1987); R.S. Chivukula and T.P. Walker, Nucl. Phys. B **329**, 445 (1990); S. Dodelson, B. R. Greene and L. M. Widrow, Nucl. Phys. B **372**, 467 (1992); S. M. Barr, Phys. Rev. D **44**, 3062 (1991); D. B. Kaplan, Phys. Rev. Lett. **68**, 741 (1992).
[55] H. Davoudiasl and R. N. Mohapatra, arXiv:1203.1247 [hep-ph].
[56] D. E. Kaplan, M. A. Luty and K. M. Zurek, Phys. Rev. D **79**, 115016 (2009).
[57] J. A. Harvey and M. S. Turner Phys. Rev. D **42**, 3344 (1990).
[58] W. -Z. Feng, P. Nath and G. Peim, Phys. Rev. D **85**, 115016 (2012).
[59] S. P. Martin, Phys. Rev. D **46**, 2769 (1992).
[60] V. Barger, P. Fileviez Perez and S. Spinner, Phys. Rev. Lett. **102**, 181802 (2009).
[61] P. Fileviez Perez, Talk at PASCOS2012, Merida Mexico, June 3-8, 2012.
[62] B. Körs and P. Nath, Phys. Lett. B **586** (2004) 366; JHEP **0412** (2004) 005; JHEP **0507**, 069 (2005).
[63] D. Feldman, Z. Liu and P. Nath, Phys. Rev. Lett. **97**, 021801 (2006); D. Feldman, B. Kors and P. Nath, Phys. Rev. D **75**, 023503 (2007); D. Feldman, Z. Liu and P. Nath, Phys. Rev. D **75**, 115001 (2007); D. Feldman, P. Fileviez Perez and P. Nath, JHEP **1201**, 038 (2012).
[64] A. Ringwald, Talk at PASCOS2012, Merida Mexico, June 3-8, 2012.
[65] J. L. Hewett, H. Weerts, R. Brock, J. N. Butler, B. C. K. Casey, J. Collar, A. de Govea and R. Essig *et al.*, arXiv:1205.2671 [hep-ex].
[66] K. Abe, T. Abe, H. Aihara, Y. Fukuda, Y. Hayato, K. Huang, A. K. Ichikawa and M. Ikeda *et al.*, "Letter of Intent: The Hyper-Kamiokande Experiment — Detector Design and Physics Potential —," arXiv:1109.3262 [hep-ex].

Journal of Physics: Conference Series **485** (2014) 012012 doi:10.1088/1742-6596/485/1/012012

[67] T. Ibrahim and P. Nath, Rev. Mod. Phys. **80**, 577 (2008).
[68] R. L. Arnowitt and P. Nath, Phys. Rev. D **49**, 1479 (1994).

Journal of Physics: Conference Series **485** (2014) 012013 doi:10.1088/1742-6596/485/1/012013

Searching for axions and ALPs from string theory

Andreas Ringwald
Deutsches Elektronen-Synchrotron DESY, Notkestr. 85, D-22607 Hamburg, Germany
E-mail: andreas.ringwald@desy.de

Abstract. We review searches for closed string axions and axion-like particles (ALPs) in IIB string flux compactifications. For natural values of the background fluxes and TeV scale gravitino mass, the moduli stabilisation mechanism of the LARGE Volume Scenario predicts the existence of a QCD axion candidate with intermediate scale decay constant, $f_a \sim 10^{9\div 12}$ GeV, associated with the small cycles wrapped by the branes hosting the visible sector, plus a nearly massless and nearly decoupled ALP associated with the LARGE cycle. In setups where the visible sector branes are wrapping more than the minimum number of two intersecting cycles, there are more ALPs which have approximately the same decay constant and coupling to the photon as the QCD axion candidate, but which are exponentially lighter. There are exciting phenomenological opportunities to search for these axions and ALPs in the near future. For $f_a \sim 10^{11\div 12}$ GeV, the QCD axion can be the dominant part of dark matter and be detected in haloscopes exploiting microwave cavities. For $f_a \sim 10^{9\div 10}$ GeV, the additional ALPs could explain astrophysical anomalies and be searched for in the upcoming generation of helioscopes and light-shining-through-a-wall experiments.

1. Introduction

The QCD axion arises [1, 2] in the course of the arguably most plausible solution of the strong CP puzzle [3], that is the non-observation of a θ-angle term in QCD. In this context, the axion field a is introduced as a dynamical θ-angle term, enjoying a shift symmetry, $a \to a + \text{constant}$, broken only by anomalous CP-violating couplings to gauge fields. Correspondingly, its most general low-energy effective Lagrangian below the weak scale has the form [4],

$$\begin{aligned} \mathcal{L} &= \tfrac{1}{2}\,\partial_\mu a\,\partial^\mu a - \tfrac{g_3^2}{32\pi^2}\left(\bar\theta + \tfrac{a}{f_a}\right) F^b_{3,\mu\nu}\tilde F_3^{b,\mu\nu} - \tfrac{e^2}{32\pi^2} C_{a\gamma}\tfrac{a}{f_a}\,F^{\rm em}_{\mu\nu}\tilde F^{\mu\nu}_{\rm em} \\ &\quad + \textstyle\sum_\Psi \left[\tfrac{1}{2}(\tilde X_{\psi_R} + \tilde X_{\psi_L})\overline\Psi\gamma^\mu\gamma_5\Psi + \tfrac{1}{2}(\tilde X_{\psi_R} - \tilde X_{\psi_L})\overline\Psi\gamma^\mu\Psi\right]\tfrac{\partial_\mu a}{f_a}\,, \end{aligned} \tag{1}$$

where $F^b_{3,\mu\nu}$ is the bth component of the gluon field strength tensor, $\tilde F_3^{b,\mu\nu} = \frac{1}{2}\epsilon^{\mu\nu\rho\sigma}F^b_{3,\rho\sigma}$ its dual, g_3 is the strong coupling, $F^{\rm em}_{\mu\nu}$ is the electromagnetic field strength, Ψ denotes standard model matter fields, and f_a is the axion decay constant. The θ-term in the QCD Lagrangian can then be eliminated by absorbing it into the axion field, $a = \bar a - \bar\theta f_a$. Moreover, the topological charge density $\propto \langle F^b_{3,\mu\nu}\tilde F_3^{b,\mu\nu}\rangle \neq 0$, induced by topological fluctuations of the gluon fields such as QCD instantons, provides a nontrivial potential for the axion field $\bar a$ which is minimized at zero expectation value, $\langle \bar a\rangle = 0$, thus wiping out strong CP violation, and giving the fluctuating field around this minimum, the QCD axion, a mass

$$m_a = \frac{m_\pi f_\pi}{f_a}\frac{\sqrt{m_u m_d}}{m_u + m_d} \simeq 0.6\,\text{meV} \times \left(\frac{10^{10}\,\text{GeV}}{f_a}\right), \tag{2}$$

Content from this work may be used under the terms of the Creative Commons Attribution 3.0 licence. Any further distribution of this work must maintain attribution to the author(s) and the title of the work, journal citation and DOI.
Published under licence by IOP Publishing Ltd

in terms of the light (u, d) quark masses, the pion mass m_π and the pion decay constant f_π [1], For large axion decay constant f_a, we see that the axion is a very weakly interacting (cf. Eq. (1)) slim particle [5, 6, 7, 8]. In particular, its coupling to photons [9, 10, 11] and electrons,

$$\mathcal{L} \supset -\frac{\alpha}{2\pi f_a}\left(C_{a\gamma} - \frac{2}{3}\frac{m_u + 4m_d}{m_u + m_d}\right)\frac{a}{4}\,F^{\rm em}_{\mu\nu}\tilde{F}^{\mu\nu}_{\rm em} + \frac{C_{ae}}{2f_a}\,\bar{e}\gamma^\mu\gamma_5 e\partial_\mu a\,, \tag{3}$$

are very much suppressed, e.g.

$$g_{a\gamma} \equiv \frac{\alpha}{2\pi f_a}\underbrace{\left(C_{a\gamma} - \frac{2}{3}\frac{m_u + 4m_d}{m_u + m_d}\right)}_{c_{a\gamma}} \sim 10^{-13}\ {\rm GeV}^{-1}\left(\frac{10^{10}\,{\rm GeV}}{f_a}\right)c_{a\gamma}. \tag{4}$$

Laboratory experiments as well as astrophysics, in particular the non-observation of solar axions by the helioscope CAST and the non-observation of drastic energy losses in horizontal branch stars or white dwarfs, constrain the axion decay constant, divided by the appropriate dimensionless coupling constants, to a scale much above the weak scale (see also Fig. 1),

$$\frac{f_a}{C_{a\gamma}} > 10^7\ {\rm GeV} \Leftrightarrow g_{a\gamma} < 10^{-10}\ {\rm GeV}^{-1}, \quad \frac{f_a}{C_{ae}} > 10^9\ {\rm GeV}. \tag{5}$$

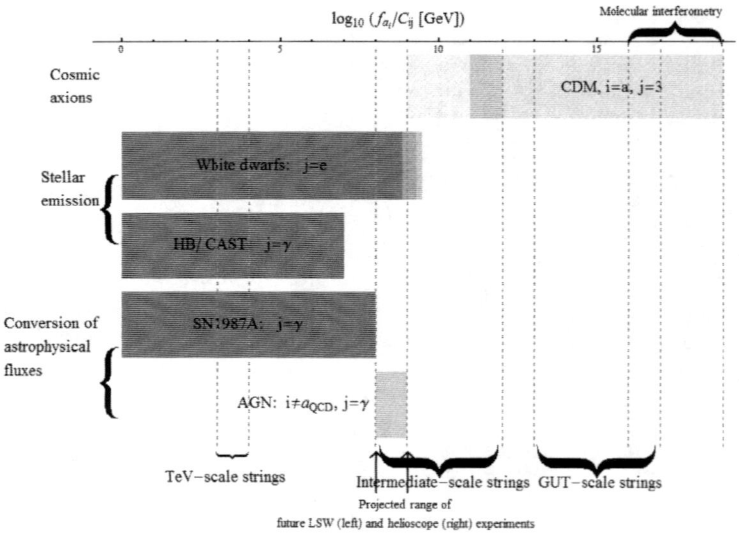

Figure 1. A summary of constraints on and hints for the couplings f_{a_i}/C_{ij} of axion-like particles a_i to standard model particles j [22]. Red regions are excluded, and the orange region would be excluded by red giants but is compatible with the hints from white dwarfs. The green regions from top to bottom correspond respectively to the classic 'axion dark matter window', hints of an axion from white dwarf cooling and transparency of the Universe to very high energy gamma rays. The blue region would be excluded by dark matter overproduction in the absence of a dilution mechanism or tuning of the initial misalignment angle.

Intriguingly, for an even higher decay constant, $f_a \gtrsim 10^{11}$ GeV, the QCD axion can contribute significantly to cold dark matter (CDM), being non-thermally produced in the early universe via initial misalignment of the axion field, resulting in coherent field oscillations corresponding to a condensate of non-relativistic axions [12, 13, 14]. In fact, assuming that the reheating temperature after inflation is below f_a and that there is no dilution by, e.g., late decays of particles beyond the standard model, the expected cosmic mass fraction in QCD axion CDM is

$$\Omega_a h^2 \approx 0.71 \times \left(\frac{f_a}{10^{12}\ \mathrm{GeV}}\right)^{7/6} \left(\frac{\Theta_a}{\pi}\right)^2 , \tag{6}$$

where Θ_a is the initial misalignment angle.

Therefore, the QCD axion is necessarily associated with a very high energy scale and so it is natural to search for it in ultra-violet completions of the standard model such as string theory. Indeed, it has long been known that the low-energy effective field theory of string compactifications predicts natural candidates for the QCD axion [15, 16, 17, 18, 19], often even an 'axiverse' [20], containing many additional light axion-like particles (ALPs) whose masses are logarithmically hierarchical. But only very recently explicit moduli-stabilised string theoretic examples with a viable QCD axion candidate and possibly additional light ALPs have been constructed (cf. Sec. 2), with decay constants ranging from the GUT scale, $f_a \sim 10^{16}$ GeV [21] down to the intermediate scale, $f_a \sim 10^{9\div 12}$ GeV [22] – the latter offering exciting opportunities to detect effects of axions and ALPs in astrophysics and in the upcoming generation of axion experiments (cf. Sec. 3).

2. Axions and ALPs in IIB string flux compactifications

String theory requires the existence of six extra space dimensions, which appear to be unobservable because they are supposed to be compact and of very small size. String phenomenology is then the attempt to make contact between the perturbative ten dimensional (10D) effective field theories (EFTs) describing the massless degrees of freedom of string theory at very high energies – say, the heterotic or type II (with D-branes) EFT – and the low energy physics in our 4D real world. In fact, different 4D low energy EFTs emerge depending on the EFT to start with in 10D. One of the fundamental tasks of string phenomenology is to find a compactification whose low energy EFT reproduces (a suitable extension of) the standard model. Along this way it is also mandatory to understand moduli stabilisation: expectation values of the moduli fields, which parameterise the shape and size of the extra dimensions, determine many parameters of the low energy EFT, such as gauge and Yukawa couplings. Often, one needs to incorporate quantum corrections in order to fix many of these expectation values and to give their associated particle excitations a non-zero mass.

Moduli stabilisation is best understood in type IIB string flux compactifications, on which we concentrate in the following. They provide explicit and well motivated realisations of brane world scenarios: the standard model is supposed to live on a stack of space-time filling branes wrapping cycles in the compact dimensions, while gravity propagates in the bulk, leading to a string scale $M_s \sim M_P/\sqrt{\mathcal{V}}$ possibly much smaller than the Planck scale M_P, at the expense of a large compactification volume $\mathcal{V} \gg 1$ (in units of the string length). Importantly, axion-like fields emerge in those compactifications inevitably as Kaluza-Klein zero modes of ten dimensional form fields, as we will review in the following subsection.

2.1. *Tree-level candidates for axions and ALPs*

We will consider here type IIB flux compactifications on Calabi-Yau orientifolds X in the presence of space-time filling $D7$ branes and $O7$ planes. Below the Kaluza-Klein scale $M_{\rm KK}$, they lead to a low-energy $\mathcal{N}=1$ (supergravity) EFT in 4D. Their 4D closed string moduli, comprised by the axio-dilaton $S = e^{-\phi} + iC_0$, the complex structure moduli U_α, $\alpha = 1, ..., h^{2,1}_-(X)$, and the

Kähler moduli,

$$T_i = \tau_i + i\,c_i, \quad \tau_i = \mathrm{Vol}(D_i), \quad c_i = \int_{D_i} C_4, \quad i = 1, ..., h^{1,1}, \tag{7}$$

are obtained via the Kaluza-Klein reduction of the massless bosonic fields of the original 10D theory – the latter including, in the Ramond-Ramond sector, the forms C_0 and C_4, and, in the Neveu Schwarz-Neveu Schwarz sector, the dilaton ϕ. Importantly, the number of Kähler moduli T_i is determined by the topology of X, namely the number of inequivalent four-cycles D_i of X (we have specialised for convenience to orientifold projections such that $h_-^{1,1} = 0 \Rightarrow h_+^{1,1} = h^{1,1}$). Their imaginary parts c_i have all the properties of axion-like fields, as we will see next.

As already mentioned, (a suitable extension of) the standard model is realised in such compactifications through (stacks of) space-time filling $D7$-branes wrapping some of the four-cycles D_i. At low energies, the dynamics of a $D7$-brane reduces to a $U(1)$ gauge theory that lives on its eight-dimensional world-volume. Moreover, $D7$-branes can be magnetised by turning on internal magnetic fluxes $\mathcal{F}$. Taking all these effects into account, the 4D low energy effective Lagrangian of the axion-like fields c_i, including the brane-localised $U(1)$ gauge bosons A_i, is obtained from the Kaluza-Klein reduction of the $D7$-brane action as follows [23, 24]:

$$\begin{aligned}\mathcal{L} \supset & -\left(dc_\alpha + \frac{M_P}{\pi}A_i q_{i\alpha}\right)\frac{\mathcal{K}_{\alpha\beta}}{8}\wedge\star\left(dc_\beta + \frac{M_P}{\pi}A_j q_{j\beta}\right) + \frac{1}{4\pi M_P}r^{i\alpha}c_\alpha\,\mathrm{tr}(F\wedge F)\\ & + \frac{M_P^2}{2(2\pi)^2}A_i A_j q_{i\alpha}\mathcal{K}_{\alpha\beta}q_{j\beta} - \frac{r^{i\alpha}\tau_\alpha}{4\pi M_P}\mathrm{tr}(F_i\wedge\star F_i),\end{aligned} \tag{8}$$

where $M_P = (8\pi G_N)^{-1/2} \simeq 2.4\cdot 10^{18}$ GeV is the reduced Planck mass, while the various other quantities in Eq. (8) are defined as follows:

- The Kähler metric $K_{\alpha\beta}$, describing the kinetic mixing of the axion-like fields, is obtained via $\mathcal{K}_{\alpha\beta} \equiv \frac{\partial^2 K}{\partial\tau_\alpha\partial\tau_\beta}$ from the Kähler potential, the latter taking, at tree-level, the following form:

$$K_{\mathrm{tree}} = -2\ln\mathcal{V} - \ln\left(S+\bar{S}\right) - \ln\left(-i\int_X \Omega\wedge\bar{\Omega}\right). \tag{9}$$

  It depends implicitly on the complex structure moduli via the holomorphic (3,0)-form Ω and on the Kähler moduli via the Calabi-Yau volume $\mathcal{V}$, measured by an Einstein frame metric $g^E_{\mu\nu} = e^{-\phi/2}\,g^s_{\mu\nu}$, and expressed in units of the string length $\ell_s = 2\pi\sqrt{\alpha'}$, in terms of its tension α',

$$\mathcal{V} = \frac{1}{6}\int_X J\wedge J\wedge J = \frac{1}{6}k_{\alpha\beta\gamma}t^\alpha t^\beta t^\gamma. \tag{10}$$

  Here, the Kähler form J has been expanded, in a basis $\{\hat{D}_\alpha\}_{\alpha=1}^{h^{1,1}}$ of $H^{1,1}(X,\mathbb{Z})$ of two-forms which are Poincaré dual to D_α, as $J = t^\alpha\hat{D}_\alpha$, and we denoted the triple intersection numbers of X by $k_{\alpha\beta\gamma}$. The volume, the Kähler form and ultimately the Kähler metric can then be obtained as a function of the τ_α by inverting the following relations:

$$\tau_\alpha = \frac{1}{2}\int_X \hat{D}_\alpha\wedge J\wedge J = \frac{\partial\mathcal{V}}{\partial t^\alpha} = \frac{1}{2}k_{\alpha\beta\gamma}\,t^\beta\,t^\gamma. \tag{11}$$

  The string scale $M_s = 1/\ell_s$ is obtained from the dimensional reduction of the IIB supergravity action,

$$M_s = M_P/\sqrt{4\pi\mathcal{V}}. \tag{12}$$

- The couplings $r^{i\alpha}$ appearing in the gauge kinetic term and in the axionic couplings to gauge bosons, the respective last terms in both lines of Eq. (8), are the expansion coefficients of the two form $\hat{D}_i = r^{i\alpha}\omega_\alpha$,

$$r^{i\alpha} \equiv \ell_s^{-4} \int_{D_i} \tilde{\omega}^\alpha = \ell_s^{-4} \int \hat{D}_i \wedge \tilde{\omega}^\alpha, \quad \alpha = 1, ..., h^{1,1}, \tag{13}$$

where the basic forms satisfy $\ell_s^{-4} \int \omega_\beta \wedge \tilde{\omega}^\alpha = \delta^\alpha_\eta$. The gauge coupling can be inferred from the gauge kinetic term as

$$\frac{1}{g_i^2} = \frac{r^{i\alpha}\tau_\alpha}{2\pi M_P} \times \begin{cases} 1 & U(1) \\ 1/2 & SU(N) \end{cases} . \tag{14}$$

- The couplings $q_{i\alpha}$ appearing in the (Stückelberg) mass terms for the $U(1)$ gauge fields A_i in Eq. (8) are given by

$$q_{i\alpha} \equiv \ell_s^{-2} \int_{D_i} \omega_\alpha \wedge \frac{\mathcal{F}}{2\pi} = \ell_s^{-4} \int_{D_i} \omega_\alpha \wedge \ell_s^2 \mathcal{F}, \tag{15}$$

where $\mathcal{F}$ is the gauge flux on D_i. Axions c_α experiencing such a coupling disappear from the low energy EFT because they are eaten by the corresponding $U(1)$ gauge boson [25, 26].

Thus, it appears from Eq. (8), that IIB string flux compactifications have potentially many, $h^{1,1} - d$, axion candidates c_i, where d is the number of $U(1)$ bosons A_i which get a Stückelberg mass by eating the associated axions. However, before reaching this conclusion one has to consider perturbative and non-perturbative corrections to this tree-level result. In fact, such corrections are necessarily to be taken into account in the course of the stabilisation of the associated Kähler moduli τ_i. Importantly, the mechanisms to fix the τ_i may also generate large masses for the corresponding axions c_i [27, 28], as we will see next.

2.2. Axions and ALPs in moduli stabilised IIB string flux compactifications

In IIB string flux compactifications, the tree-level superpotential [29]

$$W_{\rm tree} = \int_X G_3 \wedge \Omega . \tag{16}$$

which is generated by turning on background fluxes of the form $G_3 = F_3 + iSH_3$, where $F_3 = dC_2$ and $H_3 = dB_2$, does not depend on the Kähler moduli, but on the dilaton C and the complex structure moduli U_α. This implies that the dilaton and the complex structure moduli can be fixed at tree-level by imposing vanishing F-term conditions [30]. By appropriate tuning of the internal fluxes, one can always fix the dilaton such that the string coupling, $g_s = 1/\mathrm{Re}(S)$, is in the perturbative regime. Further effects from fixing S and U are then parametrised by the flux-dependent constant $W_0 = \langle W_{\rm tree} \rangle$ and by an overall factor in the F-term scalar potential arising from the S and U dependent part in the tree-level Kähler potential (9).

The Kähler moduli τ_i, however, remain precisely massless at leading semiclassical order, because of the no-scale structure of $K_{\rm tree}$. Their stabilisation requires taking into acount perturbative (p) and nonperturbative (np) α' and g_s corrections to the tree-level result,

$$W = W_{\rm tree} + \delta W_{\rm np}, \qquad K = K_{\rm tree} + \delta K_{\rm p} + \delta K_{\rm np}, \tag{17}$$

eventually leading also to non-trivial potentials, i.e. masses, for the axions c_i associated with the scalars τ_i. These masses can arise, however, only via the non-perturbative corrections to the superpotential, $\delta W_{\rm np}$, and to the Kähler potential, $\delta K_{\rm np}$: only those can break the shift symmetry of the axions c_i.

The respective order of magnitude of the perturbative versus the nonperturbative corrections to the scalar potential is set by W_0. The main mechanisms proposed for τ moduli stabilisation and their consequences for the physics of their associated c axions can be characterised as follows:

- The first scenario of Kähler moduli stabilisation in IIB string compactifications neglected the corrections to the Kähler potential and considered only the non-perturbative corrections to the superpotential [31],

$$\delta W_{\rm np} = \sum_{i=1}^{h^{1,1}} A_i(S,U)\, e^{-a_i T_i}\,, \tag{18}$$

which arises by $ED3$ instantons (in which case $a_i = 2\pi$) or by stacks of $D7$ branes supporting a condensing gauge theory (for which $a_i = 6\pi/b_0$ with b_0 being the coefficient of the one-loop beta function), wrapping the four-cycles D_i. The threshold effects A_i can be considered as $\mathcal{O}(1)$ constants since they depend on the complex structure moduli which are flux-stabilised at tree-level. The four-cycle volumina are then fixed at

$$\tau_i \sim \frac{1}{a_i} \ln\left(\frac{W_0}{A_i}\right). \tag{19}$$

Thus, W_0 has to be fine-tuned to extremely small values, $W_0 \ll 1$, in order that the volume is large, $\mathcal{V} \sim \tau_i^{3/2} \gg 1$, the latter being a prerequisite of the underlying supergravity approximation. Therefore, in these scenarios it is extremely hard to lower the string scale (12) much below the Planck scale. Moreover, since the axionic shift symmetry $c_i \to c_i + \text{constant}$ of all axions is broken non-perturbatively by Eq. (18), all the axion candidates get a large mass of the order of the mass of the particle excitations of the associated Kähler moduli (the so-called "saxions"),

$$m_{c_i} \sim m_{\tau_i} \sim a_i W_0 M_P/\mathcal{V},$$

leaving no candidate for a QCD axion [16], let alone a light ALP.

- This can be avoided if the non-pertubative effects arise from wrapping an *ample* four-cycle,

$$D_{\rm am} = \sum_{i=1}^{h^{1,1}} \lambda_i D_i, \text{ with } \lambda_i > 0\ \forall\, i = 1, ..., h^{1,1}, \tag{20}$$

in terms of a basis $\{D_i\}$ of $H_4(X,\mathbb{Z})$, resulting in a contribution to the superpotential of the form

$$\delta W_{\rm np} = A\, e^{-\,a\,T_{\rm am}} = A\, e^{-\,a\sum_{i=1}^{h^{1,1}} \lambda_i T_i}. \tag{21}$$

In this case, by fine-tuning $W_0 \sim A\, e^{-\,a\,T_{\rm am}} \ll 1$, this single non-perturbative effect can generate a minimum for all the Kähler moduli τ_i within the regime of validity of the EFT [32]. However it can lift only one axion corresponding to the imaginary part of the ample divisor modulus: $c_{\rm am} = {\rm Im}(T_{\rm am})$. All the remaining $h^{1,1} - 1$ axions are massless at leading order and develop a potential only via tiny higher order instanton effects of the form [21]:

$$\delta W_{\rm np} = A\, e^{-\,a\,T_{\rm am}} + \sum_{i=1}^{h^{1,1}-1} A_i\, e^{-\,n_i a_i T_i}, \tag{22}$$

where T_i is a combination of moduli orthogonal to $T_{\rm am}$ $\forall\, i = 1, ..., h^{1,1} - 1$.

Thus, this moduli stabilisation scenario gives rise to an axiverse [20]: possibly many, $h^{1,1}-1$, axions which acquire a mass spectrum which is logarithmically hierarchical. This number may still be diminished in case that some of the axions are eaten by the Stückelberg mechanism to provide masses to brane-localised $U(1)$ gauge bosons.
Two main concerns regarding the microscopic realisation of this scenario are related to the difficulty to find an ample divisor which is rigid (and so definitely receives non-perturbative effects), and the possibility to choose gauge fluxes that avoid chiral intersections between the instanton and the visible sector. Moreover, it should be noted that in this case again very large volumina are not possible: the string scale (12) is expected not much below the Planck scale, of order the GUT scale.

- The latter difficulties are avoided in the LARGE volume scenario (LVS), which realizes the possibility of exponentially large volumina for generic values of $W_0 \sim \mathcal{O}(1)$ [33] and allows for the construction of explicit globally and locally consistent Calabi-Yau examples with magnetised D7-branes, realising MSSM or GUT like chiral extensions of the standard model [34, 35].
The LVS requires the existence of a single *del Pezzo* four-cycle $\tau_{\rm dP}$ which guarantees the emergence of a non-perturbative contribution to the superpotential

$$\delta W_{\rm np} = A\, e^{-\,a\, T_{\rm dP}} \tag{23}$$

via an $ED3$ instanton or gaugino condensation. This effect fixes $\tau_{\rm dP}$ at a small size and gives the corresponding saxion and axion a large mass [36],

$$m_{\tau_{dP}} \sim m_{c_{\rm dP}} \sim \frac{W_0\sqrt{\ln \mathcal{V}}}{\mathcal{V}}\, M_P\,. \tag{24}$$

All the other τ moduli are stabilised perturbatively by α' or g_s effects or by D-terms arising from magnetised branes. The exponential large volume emerges from an interplay between the non-perturbative contribution associated with del Pezzo cycle and the leading α' correction [37]:

$$\delta K_{\rm p} \simeq -\frac{\zeta}{g_s^{3/2}\mathcal{V}}\,, \quad \text{with } \zeta \propto (h^{1,2}-h^{1,1})\,, \tag{25}$$

which yields, for $h^{1,2} > h^{1,1} > 1$ (i.e. negative Euler number), a supersymmetry-breaking anti de Sitter (AdS) minimum at exponentially large volume [33]:

$$\mathcal{V} \sim W_0\, e^{a\,\tau_{\rm dP}}\,. \tag{26}$$

Hence only one axion, $c_{\rm dP} = {\rm Im}(T_{\rm dP})$, becomes heavy whereas all the other ones (except those eaten up by anomalous $U(1)$s, whose scalar partners are fixed by the above mentioned D-terms) remain light and develop a potential via subleading higher order instanton effects.
The simplest version of a standard LVS is build upon a Swiss-cheese Calabi-Yau three-fold with volume given by [38]:

$$\mathcal{V} = \alpha\left(\tau_b^{3/2} - \sum_{i=1}^{h^{1,1}-1} \gamma_i \tau_i^{3/2}\right). \tag{27}$$

It is dominated by the exponentially large volume τ_b of one of the four-cycles, D_b, while all the other, $h^{1,1}-1$, four-cycles are small.
Importantly, the realisation of an LVS requires $h^{1,1} \geq 4$ (cf. Fig. 2): a single del Pezzo divisor $D_{\rm dP}$ to support the non-perturbative effects, one big four-cycle $\tau_{\rm b}$ to parametrise

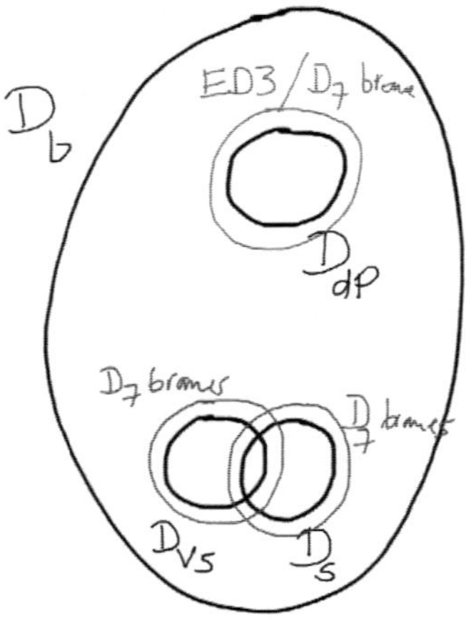

Figure 2. Schematic picture of the simplest Swiss-cheese setup of a compactification in the LARGE volume scenario: a big four-cycle (=divisor) D_b with exponentially large volume $\tau_b \sim \mathcal{V}^{2/3}$; a del Pezzo divisor $D_{\rm dP}$ supporting the leading non-perturbative effect ($ED3$ instanton or gaugino condensation of a pure Yang-Mills theory described by a stack of $D7$ branes); a rigid divisor $D_{\rm vs}$ supporting the stack of $D7$ branes describing the visible sector; a rigid divisor $D_{\rm s}$ intersecting with $D_{\rm vs}$ and supporting a stack of magnetised $D7$ branes for D-term stabilisation.

the large volume, $\mathcal{V} \sim \tau_{\rm b}^{3/2}$, one small rigid four-cycle $D_{\rm vs}$ to support the stack of space-time filling magnetised D7-branes corresponding to (a suitable extension of) the MSSM and intersecting with another small cycle $D_{\rm s}$ whose brane setup is there to provide D-terms which stabilise $\tau_{\rm vs}$ and thus the visible sector gauge coupling $g_{\rm vs}^{-2} \sim \tau_{\rm vs}$. Two of the $h^{1,1} \geq 4$ axion-like fields disappear from the low energy spectrum, however: $c_{\rm dP}$, as explained above, and $c_{\rm s}$, which is eaten by the Stückelberg mechanism. Thus, realisations of the LVS axiverse involve at least two light axions: one QCD axion candidate plus one ALP [22]. More generic models for $h^{1,1}$ very large will include an arbitrarily large number of ALPs.
The main scales in the model are:

$$\begin{aligned}
M_s &= \frac{M_P}{\sqrt{4\pi\mathcal{V}}} \sim 10^{10}\,{\rm GeV}, m_{\tau_{\rm s}} \sim \frac{M_P}{\mathcal{V}^{1/2}} \sim 10^{10}\,{\rm GeV}, M_{\rm KK} \sim \frac{M_P}{\mathcal{V}^{2/3}} \sim 10^{9}\,{\rm GeV}, \\
m_{\tau_{\rm dP}} &\sim \sqrt{g_s}\, W_0 \frac{M_P}{\mathcal{V}} \ln \mathcal{V} \sim 30\,{\rm TeV}, m_{\rm soft} \sim m_{3/2} \sim \sqrt{g_s}\, W_0 \frac{M_P}{\mathcal{V}} \sim 1\,{\rm TeV}, \qquad (28) \\
m_{\tau_{\rm vs}} &\sim \alpha_{\rm vs} m_{3/2} \sim 40\,{\rm GeV}, m_{\tau_b} \sim \frac{M_P}{\mathcal{V}^{3/2}} \sim 0.1\,{\rm MeV}.
\end{aligned}$$

The numerical values have been given for generic values of the underlying parameters, $g_s \sim 0.1, W_0 \sim 1$, and for a volume $\mathcal{V} \sim 10^{14}$, demonstrating that, for an intermediate string scale, the LVS naturally realises TeV-scale SUSY.

- Finally, the contributions from both δK_p and non-perturbative terms in the superpotential δW_{np} from gaugino condensation on stacks of 4-cycle wrapping $D7$ branes allow for a 2nd branch of supersymmetry breaking "Kähler uplifted" vacua distinct from the LVS branch. These vacua can be either AdS, Minkowski or dS by themselves without any need for further 'external' sources of uplifting. Their possible existence was first pointed out in [39], while they were shown to be viable in producing controlled large-volume vacua for all moduli in a supergravity analysis [40, 41]. Recently, [42] provided the first explicit global and consistent F-theory constructions of such Kähler uplifted dS vacua. On Swiss-cheese Calabi-Yau threefolds the Kähler uplifting branch generates a class of minima for the volume moduli, where $\tau_b \sim N_b$ with N_b the rank of the condensing gauge group of the $D7$ brane stack wrapping the large divisor D_b. The blow-up Kähler moduli $\tau_i \sim N_i$ are stabilised at smaller volume dictated by the rank of the corresponding condensing gauge groups. For ranks $N_b \sim 30 \div 100$ this leads to stabilisation of all Kähler moduli at an overall volume $\mathcal{V} \sim N_b^{3/2} \sim 10^{2\div 3}$, with volume moduli masses suppressed by a factor $1/\mathcal{V}$

compared the masses of the complex structure moduli and the axio-dilaton from fluxes. So far, stabilisation of the $h^{1,1}$ τ_i moduli utilises an identical number of non-perturbative contributions to the superpotential from gaugino condensation. Hence, all associated c_i axions are rendered massive with mass scales tied to their scalar partners. Kähler uplifting with an (partially) ample divisor utilising less than $h^{1,1}$ instanton effects for stabilisation and in turn realising an axiverse represents an open question.

2.3. Axion and ALP decay constants and their couplings to visible sector particles in the LVS

Let us know for the remainder of this review concentrate on the LVS, because, as we have seen, it predicts the existence of a QCD axion candidate plus at least one ALP. Their decay constants f_{a_i} and their couplings C_{ij} to gauge bosons (or matter fields) j can be read off from the matching of the prediction (8) with the generalisation of the generic low energy effective Lagrangian (1) to the case of many axion-like fields, by transforming in the former from the original basis $\{c_i\}$ to a basis of fields $\{a_i\}$ with canonically normalised kinetic terms [22]. For the simple Swiss-cheese Calabi-Yau setup discussed above and illustrated in Fig. 2, the original axion fields c_b and $c_{\rm vs}$ can be written in terms of the canonically normalised fields a_b and $a_{\rm vs}$ as:

$$\frac{c_b}{\tau_b} \simeq \mathcal{O}(1)\, a_b + \mathcal{O}\left(\tau_{\rm vs}^{3/4}\mathcal{V}^{-1/2}\right) a_{\rm vs}\,, \qquad \frac{c_{\rm vs}}{\tau_{\rm vs}} \simeq \mathcal{O}(1)\, a_b + \mathcal{O}\left(\tau_{\rm vs}^{-3/4}\mathcal{V}^{1/2}\right) a_{\rm vs}\,, \tag{29}$$

leading to

$$f_{a_b} \simeq \frac{M_{\rm KK}}{4\pi}\,, \qquad f_{a_{\rm vs}} \simeq \frac{M_s}{\sqrt{4\pi}\tau_{\rm vs}^{1/4}}\,, \tag{30}$$

$$C_{bb} \simeq \mathcal{O}(1)\,, \qquad C_{{\rm vs}b} \simeq \mathcal{O}\left(\mathcal{V}^{-1/3}\right), \qquad C_{b{\rm vs}} \simeq \mathcal{O}\left(\mathcal{V}^{-2/3}\right), \qquad C_{\rm vs\,vs} \simeq \mathcal{O}(1)\,. \tag{31}$$

The axion $a_{\rm vs}$ can be identified with the QCD axion. A range of values for the proper QCD axion decay constant, spanning the classic QCD axion window,

$$f_a \equiv \frac{f_{a_{\rm vs}}}{C_{\rm vs\,vs}} \sim M_s \sim \sqrt{\frac{M_P\, m_{3/2}}{g_s^{1/2} W_0}} \sim 10^{9\div12}\,{\rm GeV}\,, \tag{32}$$

are then possible depending upon the exact numerical coefficients, the value of the gravitino mass $m_{3/2} \gtrsim$ TeV, and the tuning of $g_s \lesssim 1$ and W_0.

The large cycle ALP a_b has a smaller decay constant $f_{a_b} \sim M_{\rm KK}$, but its coupling to the standard Model gauge bosons is completely negligible, $C_{b{\rm vs}} \simeq \mathcal{O}\left(\mathcal{V}^{-2/3}\right)$.

More ALPs, but with decay constant similar to the one of the QCD axion,

$$f_{\rm ALP_i} \sim f_{a_{\rm vs}} \sim M_s, \tag{33}$$

and coupling to gauge bosons other than the gluon similar to the one of the QCD axion,

$$C_{\rm ALP_i\,vs} \sim C_{\rm vs\,vs} \sim \mathcal{O}(1) \Rightarrow g_{i\gamma} \equiv \frac{\alpha}{2\pi f_{a_i}} C_{i\gamma} \sim 10^{-15} \div 10^{-11}\ {\rm GeV}^{-1}, \tag{34}$$

are obtained, if there are more small cycles $D_{\rm vs_i}$ intersecting the visible branes, but without introducing additional D-term conditions. Their masses are expected to be smaller than the mass of the QCD axion and to be distributed logarithmically hierarchical [22],

$$m_{\rm ALP_i} \sim e^{-n\pi\tau_{\rm ALP_i}} \times \begin{cases} M_P, & \text{for } \delta W_{\rm np} \text{ terms or QCD-like masses,} \\ m_{3/2}, & \text{for } \delta K_{\rm np} \text{ terms.} \end{cases} \tag{35}$$

These findings have been reproduced also in explicit constructions of LVS examples, exploiting concrete Calabi-Yau orientifolds and semi-realistic $D7$-brane and flux setups [22, 35]. That paper considers also LVS variants with asymmetric fibred Calabi-Yaus and sequestered SUSY breaking, allowing for more freedom in $m_{3/2}$ and $m_{\rm soft}$, at the expense of more fine tuning in W_0, and discusses briefly the cosmological evolution of the different LVS incarnations.

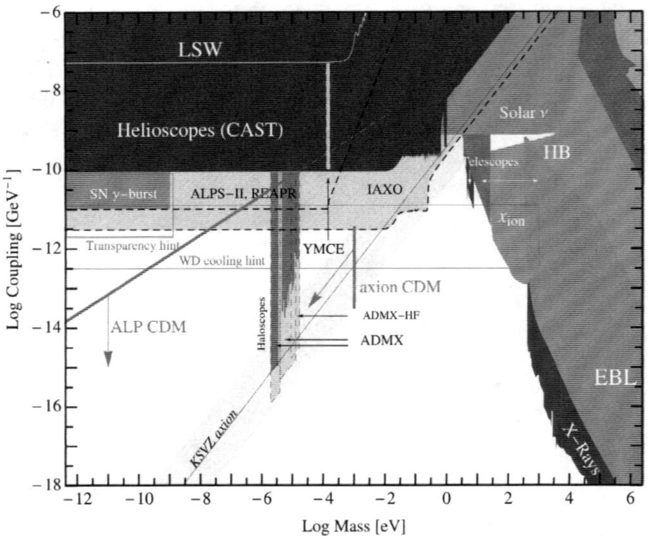

Figure 3. Axion and ALP coupling to photons, $g_{i\gamma} \equiv \alpha C_{i\gamma}/(2\pi f_{a_i})$, vs. its mass (adapted by Javier Redondo [49] from Refs. [50, 51]). The yellow band is the generic prediction for the QCD axion, exploiting Eqs. (2) and (4), which relate its mass with its coupling to photons.

3. Opportunities to probe the intermediate string scale LVS

3.1. Haloscope searches

We have seen, that the LVS predicts – for the least fine-tuning of fluxes, such that $g_s \sim 0.1$ and $W_0 \sim 1$, and a TeVish gravitino mass – an intermediate string scale and thus a QCD axion in the classic window, cf. Eq. (32). For decay constants in the upper part of this window, $f_a \gtrsim 10^{11\div12}$ GeV, the QCD axion is expected to contribute substantially to the cold dark matter in the universe, see Eq. (6). Therefore, the intermediate string scale LVS can be probed by haloscope searches for axion cold dark matter [43] such as ADMX [44, 45, 46, 47]. These experiments exploit the coupling (4) by searching for the signal of dark matter axion to photon conversions in a narrow bandwidth microwave cavity sitting in a strong magnetic field. As can be seen from the light green area in Fig. 3 labelled as "Haloscopes", a substantial range of the QCD axion dark matter parameter range will be probed by ADMX and other haloscopes [48] in the next decade.

3.2. Helioscope searches

A complementary search for the QCD axion in the lower part of the classic window, $f_a \gtrsim 10^{9\div10}$ GeV, can be conducted with the next generation of axion helioscopes [43], in which one tries to detect solar axions by their conversion into photons inside of a strong magnet pointed towards the sun. Indeed, the projected sensitivity of the proposed International Axion Observatory IAXO [52] covers nicely a part of QCD axion parameter space which will not be covered by the haloscope searches, as can be seen in Fig. 3.

A very welcome feature of helioscopes is that they do not lose sensitivity towards low masses: their projected sensitivity are best and stay constant at small masses, see Fig. 3. That means, with IAXO one may also probe the LVS axiverse, in particular the possible existence of more ALPs with approximate the same coupling to photons as the QCD axion.

This is very important in view of recent tantalising astrophysical hints, such as the anomalous transparency of the Universe for TeV photons [53] and the anomalous cooling of white dwarfs [54, 55], which could be explained by the existence of an ALP with decay constants, couplings and mass [56, 57, 58, 59, 60, 61, 62]:

$$\frac{f_{a_i}}{C_{i\,e}} \simeq (0.7 \div 2.6) \times 10^9 \text{ GeV}\,, \quad \frac{f_{a_i}}{C_{i\gamma}} \sim 10^8 \text{ GeV}, \quad m_{a_i} \lesssim 10^{-9} \div 10^{-10} \text{ eV}\,, \tag{36}$$

compatible with the prediction of an intermediate string scale LVS axiverse with $C_{i\gamma}/C_{i\,e} \sim 10$. The projected sensitivity of IAXO nicely overlaps with the ALP parameter region required to explain these hints, see Fig. 3.

3.3. Light-shining-through-walls searches

Intriguingly, this parameter region for ALPs can also be partially probed by purely laboratory based light-shining-through-walls experiments [63], where laser photons are send along a strong magnetic field, allowing for their conversion into ALPs, which may then reconvert in the strong magnetic field behind a blocking wall into photons, apparently shining through the wall and susceptible to detection. The projected sensitivities of the proposed experiments ALPS-II at DESY and REAPR at Fermilab partially cover the expectations (34) from an intermediate string scale LVS axiverse and from the hints (36) from astrophysics, see Figs. 1 and 3.

4. Conclusions

String phenomenology holds the promise of an axiverse – the QCD axion plus a (possibly large) number of further ultralight axion-like particles, possibly populating each decade of mass down to the Hubble scale, 10^{-33} eV. However, although a plenitude of axion-like fields is a generic prediction of string theory, there may be few or no light axions remaining once constraints such as tadpole cancellation and moduli stabilisation are taken into account.

Interestingly, the promise of an axiverse seems to be fulfilled in the LARGE Volume Scenario (LVS) of IIB string flux compactifications. In fact, the simplest globally consistent LVS constructions with magnetised $D7$-branes and chirality have at least two light axions: a QCD axion candidate, with a decay constant of order the string scale, which is intermediate, $f_a \sim M_s \sim M_P/\sqrt{\mathcal{V}} \sim (M_P m_{3/2}/W_0)^{1/2} \sim 10^{9\div 12}$ GeV, for a TeV scale gravitino mass $m_{3/2}$ and an expectation value of the flux induced tree-level expectation value W_0 of the superpotential of order one, plus a nearly decoupled superlight axion-like particle. In setups where the small size branes describing the visible sector are wrapping more than the minimally required two intersecting four-cycles, there are more ultralight axion-like particles which have approximately the same decay constant and coupling to the photon as the QCD axion candidate.

At both ends of the above range of the decay constant there are exciting phenomenological opportunities. For $f_a \sim 10^{11\div 12}$ GeV, the QCD axion can be the dominant part of dark matter and be detected in haloscopes exploiting microwave cavities. For $f_a \sim 10^{9\div 10}$ GeV, the additional ALPs could explain astrophysical anomalies and be searched for in the upcoming generation of helioscopes or light-shining-through-a-wall experiments.

Acknowledgments

I would like to thank Michele Cicoli and Mark Goodsell for the great collaboration on the topics in this review and Alexander Westphal for many enlightning discussions.

References

[1] Weinberg S 1978 *Phys. Rev. Lett.* **40** 223
[2] Wilczek F 1978 *Phys. Rev. Lett.* **40** 279
[3] Peccei R D and Quinn H R 1977 *Phys. Rev. Lett.* **38** 1440
[4] Georgi H, Kaplan D B and Randall L 1986 *Phys. Lett.* B **169** 73
[5] Kim J E 1979 *Phys. Rev. Lett.* **43** 103

[6] Dine M, Fischler W and Srednicki M 1981 *Phys. Lett.* B **104** 199
[7] Shifman M A , Vainshtein A I and Zakharov V I 1980 *Nucl. Phys.* B **166** 493
[8] Zhitnitsky A R 1980 *Sov. J. Nucl. Phys.* **31** 260 [1980 *Yad. Fiz.* **31** 497]
[9] Bardeen W A and Tye S-H H 1978 *Phys. Lett.* B **74** 229
[10] Kaplan D B 1985 *Nucl. Phys.* B **260** 215
[11] Srednicki M 1985 *Nucl. Phys.* B **260** 689
[12] Preskill J, Wise M B and Wilczek F 1983 *Phys. Lett.* B **120** 127
[13] Abbott L F and Sikivie P 1983 *Phys. Lett.* B **120** 133
[14] Dine M and Fischler W 1983 *Phys. Lett.* B **120** 137
[15] Witten E 1984 *Phys. Lett.* B **149** 351
[16] Conlon J P 2006 *JHEP* **0605** 078
[17] Svrcek P and Witten E 2006 *JHEP* **0606** 051
[18] Conlon J P 2006 *Phys. Rev. Lett.* **97** 261802
[19] Choi K-S, Kim I-W and Kim J E 2007 *JHEP* **0703** 116
[20] Arvanitaki A *et al*, 2010 *Phys. Rev.* D **81** 123530
[21] Acharya B S, Bobkov K and Kumar P 2010 *JHEP* **1011** 105
[22] Cicoli M, Goodsell M and Ringwald A 2012 *Preprint* arXiv:1206.0819 [hep-th]
[23] Grimm T W and Louis J 2004 *Nucl. Phys.* B **699** 387
[24] Jockers H and Louis J 2005 *Nucl. Phys.* B **705** 167
[25] Goodsell M, Jaeckel J, Redondo J and Ringwald A 2009 *JHEP* **0911** 027
[26] Cicoli M, Goodsell M, Jaeckel J and Ringwald A 2011 *JHEP* **1107** 114
[27] Banks T, Dine M and Gorbatov E 2004 *JHEP* **0408** 058
[28] Donoghue J F 2004 *Phys. Rev.* D **69** 106012 [2004 *Erratum-ibid.* D **69** 129901]
[29] Gukov S, Vafa C and Witten E 2000 *Nucl. Phys.* B **584** 69
[30] Giddings S B, Kachru S and Polchinski J 2002 *Phys. Rev.* D **66** 106006
[31] Kachru S, Kallosh R, Linde A and Trivedi S P 2003 *Phys. Rev.* D **68** 046005
[32] Bobkov K, Braun V, Kumar P and Raby S 2010 *JHEP* **1012** 056
[33] Balasubramanian V, Berglund P, Conlon J P and Quevedo F 2005 *JHEP* **0503** 007
[34] Cicoli M, Mayrhofer C and Valandro R 2012 *JHEP* **1202** 062
[35] Cicoli M 2012 these proceedings
[36] Conlon J P, Quevedo F and Suruliz K 2005 *JHEP* **0508** 007
[37] Becker K, Becker M, Haack M and Louis J 2002 *JHEP* **0206** 060
[38] Collinucci A *et al* 2009 *JHEP* **0907** 074
[39] Balasubramanian V and Berglund P 2004 *JHEP* **0411** 085
[40] Westphal A 2007 *JHEP* **0703** 102
[41] Rummel M and Westphal A 2012 *JHEP* **1201** 020
[42] Louis J, Rummel M, Valandro R and Westphal A *Preprint* arXiv:1208.3208 [hep-th]
[43] Sikivie P 2983 *Phys. Rev. Lett.* **51** 1415 [1984 *Erratum-ibid.* **52** 695].
[44] Asztalos S J *et al* 2001 *Phys. Rev.* D **64** 092003
[45] Asztalos S J *et al* [The ADMX Collaboration] 2010 *Phys. Rev. Lett.* **104** 041301
[46] Asztalos S J *et al* 2011 *Nucl. Instrum. Meth.* A **656** 39
[47] Heilman J *et al* [ADMX Collaboration] 2010 *AIP Conf. Proc.* **1274** 115
[48] Baker O K *et al* 2012 *Phys. Rev.* D **85** 035018
[49] Redondo J 2012 private communication
[50] Arias P, Cadamuro D, Goodsell M, Jaeckel J, Redondo J and Ringwald A 2012 *JCAP* **1206** 013
[51] Hewett J L *et al.* 2012 *Preprint* arXiv:1205.2671 [hep-ex]
[52] Irastorza I G *et al* 2011 *JCAP* **1106** 013
[53] Horns D and Meyer M 2012 *JCAP* **1202** 033
[54] Isern J, Garcia-Berro E, Torres S and Catalan S 2008 *Astrophys. J.* **682** L109
[55] Isern J, Catalan S, Garcia-Berro E and Torres S 2009 *J. Phys. Conf. Ser.* **172** 012005
[56] De Angelis A, Mansutti O, Roncadelli M 2007 *Phys. Rev.* **D76** 121301
[57] Simet M, Hooper D, Serpico P D 2008 *Phys. Rev.* **D77** 063001
[58] Sanchez-Conde M A *et al* 2009 *Phys. Rev.* **D79** 123511
[59] Dominguez A, Sanchez-Conde M A and Prada F 2011 *JCAP* **1111** 020
[60] Tavecchio F, Roncadelli M, Galanti G and Bonnoli G 2012 *Preprint* arXiv:1202.6529 [astro-ph.HE]
[61] Isern J *et al*, 2012 *Preprint* arXiv:1204.3565 [astro-ph.SR]
[62] Horns D *et al* 2012 *Preprint* arXiv:1207.0776 [astro-ph.HE]
[63] Redondo J and Ringwald A 2011 *Contemp. Phys.* **52** 211

PASCOS 2012 – 18th International Symposium on Particles Strings and Cosmology IOP Publishing
Journal of Physics: Conference Series **485** (2014) 012014 doi:10.1088/1742-6596/485/1/012014

Constraints on neutrino physics from cosmology

Alessandro Melchiorri[1], **Maria Archidiacono**[1], **Erminia Calabrese**[2] **and Eloisa Menegoni**[3]

[1]Physics Department and INFN, Università di Roma "La Sapienza", Ple Aldo Moro 2, 00185, Rome, Italy

[2]Sub-department of Astrophysics, University of Oxford, Keble Road, Oxford, OX1 3RH, UK

[3] I.C.R.A. and INFN, Università di Roma "La Sapienza", Ple Aldo Moro 2, 00185, Rome, Italy

E-mail: alessandro.melchiorri@roma1.infn.it,maria.archidiacono@roma1.infn.it, maria.archidiacono@roma1.infn.it,eloisa.menegoni@roma1.infn.it

Abstract. We discuss the latest constraints from cosmology on the neutrino number and mass. We show that the cosmological data we considered are clearly suggesting the presence for an extra relativistic component with an effective neutrino number $N_{\rm eff} = 4.08^{+0.71}_{-0.68}$ at 95% c.l.. Performing an analysis on neutrino sound speed $c_{\rm eff}$ and viscosity $c_{\rm vis}$ parameters, we found $c^2_{\rm eff} = 0.312 \pm 0.026$ and $c^2_{\rm vis} = 0.29^{+0.21}_{-0.16}$ at 95% c.l., consistent with the expectations of a relativistic free streaming component ($c^2_{\rm eff}$=$c^2_{\rm vis}$=1/3). Assuming the presence of 3 relativistic neutrinos we constrain the extra relativistic component as $N^S_\nu = 1.10^{+0.79}_{-0.72}$ and $c^2_{\rm eff} = 0.24^{+0.08}_{-0.13}$ at 95% c.l. while $c^2_{\rm vis}$ results as unconstrained. Assuming a massive neutrino component we obtain further indications for an extra neutrino with $N^S_\nu = 1.12^{+0.86}_{-0.74}$ at 95% c.l. .

1. Introduction

The recent observations from Cosmic Microwave Background (CMB hereafter) satellite, balloon-borne and ground based experiments ([1, 2, 3, 4]), galaxy redshift surveys [5] and luminosity distance measurements, have fully confirmed the theoretical predictions of the standard ΛCDM cosmological model. This not only allows stringent constraints on the parameters of the model but can be fruitfully used to constrain non standard physics at the fundamental level, such as classes of elementary particle models predicting a different radiation content in the Universe.

One of the major theoretical predictions of the standard scenario is the existence of a relativistic energy component (see e.g. [6]), beside CMB photons, with a current energy density given by :

$$\rho_{rad} = \Big[1 + \frac{7}{8}(\frac{4}{11})^{4/3} N_{\rm eff}\Big]\rho_\gamma \ , \qquad (1)$$

where ρ_γ is the energy density of the CMB photons background at temperature $T_\gamma = 2.728K$ and $N_{\rm eff}$ is in principle a free parameter, defined as the effective number of relativistic degrees of freedom. Assuming standard electroweak interactions, three active massless neutrinos and including the (small) effect of neutrino flavour oscillations the expected value is $N_{\rm eff} = 3.046$ with a deviation from $N_{\rm eff} = 3$ that takes into account effects from the non-instantaneous neutrino decoupling from the primordial photon-baryon plasma (see e.g. [7]).

Content from this work may be used under the terms of the Creative Commons Attribution 3.0 licence. Any further distribution of this work must maintain attribution to the author(s) and the title of the work, journal citation and DOI.
Published under licence by IOP Publishing Ltd

In recent years, thanks to the continuous experimental advancements, the value of $N_{\rm eff}$ has been increasingly constrained from cosmology ([8], [9], [10], [11], [12], [1], [13], [14], [15], [16], [19], [20]), ruling out $N_{\rm eff} = 0$ at high significance.

However, especially after the new ACT [2] and SPT [4] CMB results, the data seem to suggest values higher than the "standard" one, with $N_{\rm eff} \sim 4-5$ (see e.g. [13], [14], [19], [20], [21]) in tension with the expected standard value at about two standard deviations.

A larger value for $N_{\rm eff}$ respect to the one expected in the standard scenario could be due to non-standard neutrino decoupling (see e.g. [22]) or could hint for the existence of a fourth (or fifth) sterile neutrino. The hypothesis of extra neutrino flavour could be also compatible with the recent results from short-baseline neutrino oscillation data from LSND [23] and MiniBooNE [24] experiments (see [13, 14] and references therein). Moreover $N_{\rm eff} \sim 4$ could also be produced by axions (see e.g. [25]), gravity waves ([26]), decaying particles (see e.g. [27]), extra dimensions [28, 29] and dark energy (see e.g. [30] and references therein).

Since there is a large variety of models that could enhance $N_{\rm eff}$ it is clearly important to investigate the possible ways to discriminate among them. If Dark Radiation is made of relativistic particles as sterile neutrinos it should behave as neutrinos also from the point of view of perturbation theory, i.e. it should have an effective sound speed $c_{\rm eff}$ and a viscosity speed $c_{\rm vis}$ such that $c^2_{\rm eff} = c^2_{\rm vis} = 1/3$. Free streaming of relativistic neutrino will indeed produce anisotropies in the neutrino background yielding a value of $c^2_{\rm vis} = 1/3$ while a smaller value would indicate possible non standard interactions (see e.g. [32]). A value of $c_{\rm vis}$ different from zero, as expected in the standard scenario, has been detected in [33] and confirmed in subsequent papers [34]. More recently, the analysis of [21] confirmed the presence of anisotropies from current cosmological data but also suggested the presence of a lower value for the effective sound speed with $c^2_{\rm eff} = 1/3$ ruled out at more than two standard deviations.

In these proceedings we present the results on $N_{\rm eff}$ reported in [17] from an analysis of the most recent cosmological data. We organize our work as follows: in Sec. II we describe the data and the data analysis method adopted. We present our results in the first two subsections of Sec. III, depending on two adopted different parametrizations for the Dark Radiation. Moreover a model independent analysis is also discussed in the last subsection of Sec. III. Finally we conclude in Sec. IV.

2. Analysis Method

We perform a COSMOMC [35] analysis combining the following CMB datasets: WMAP7 [1], ACBAR [3], ACT [2], and SPT [4], and we analyze datasets using out to $l_{\rm max} = 3000$. We also include information on dark matter clustering from the galaxy power spectrum extracted from the SDSS-DR7 luminous red galaxy sample [5]. Finally, we impose a prior on the Hubble parameter based on the last Hubble Space Telescope observations [36].

The analysis method we adopt is based on the publicly available Monte Carlo Markov Chain package `cosmomc` [35] with a convergence diagnostic done through the Gelman and Rubin statistic. We sample the following six-dimensional standard set of cosmological parameters, adopting flat priors on them: the baryon and cold dark matter densities $\Omega_{\rm b}$ and $\Omega_{\rm c}$, the ratio of the sound horizon to the angular diameter distance at decoupling θ, the optical depth to reionization τ, the scalar spectral index n_S, and the overall normalization of the spectrum A_S at $k = 0.002$ Mpc^{-1}. We consider purely adiabatic initial conditions and we impose spatial flatness. We vary the effective number of relativistic degrees of freedom $N_{\rm eff}$, the effective sound speed $c^2_{\rm eff}$, and the viscosity parameter $c^2_{\rm vis}$. In some cases, we consider only variations in the *extra* dark radiation component $N^S_\nu = N_{\rm eff} - 3.046$, varying the perturbation parameters $c_{\rm vis}$ and $c_{\rm eff}$ only for this extra component and assuming $c^2_{\rm eff} = c^2_{\rm vis} = 1/3$ for the standard 3 neutrino component.

In our analysis we always fix the primordial Helium abundance to the observed value

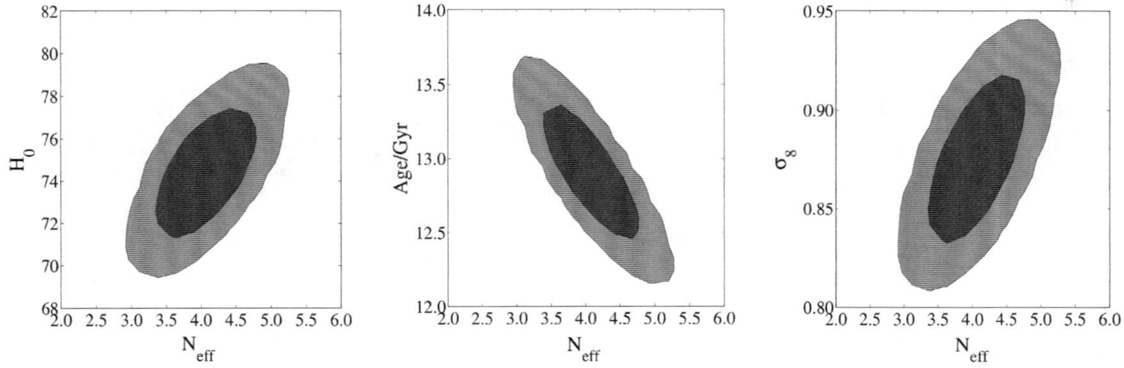

Figure 1. 68% and 95% c.l. constraints for the degeneracy between $N_{\rm eff}$ and the Hubble constant H_0, the age of the universe t_0, and the amplitude of mass fluctuations σ_8.

$Y_p = 0.24$. This procedure is different from the one adopted, for example, in [4], where the Y_p parameter is varied assuming Big Bang Nucleosynthesis for each values of $N_{\rm eff}$ and $\Omega_{\rm b}$ in the chain. Since the cosmological epoch and the energy scales probed by BBN are dramatically different from the ones probed by CMB and large scale structure we prefer to do not assume standard BBN in our analysis and to leave the primordial Helium abundance as fixed to a value consistent with current observations.

We account for foregrounds contributions including three extra amplitudes: the SZ amplitude A_{SZ}, the amplitude of clustered point sources A_C, and the amplitude of Poisson distributed point sources A_P. We marginalize the contribution from point sources only for the ACT and SPT data, based on the templates provided by [4]. We quote only one joint amplitude parameter for each component (clustered and Poisson distributed). Instead, the SZ amplitude is obtained fitting the WMAP data with the WMAP own template, while for SPT and ACT it is calculated using the [37] SZ template at 148 GHz. Again, this is different from the analysis performed in [4] where no SZ contribution was considered for the WMAP data.

3. Results

As stated in the previous section, we perform two different analyses. In the first analysis we vary the amplitude of the whole relativistic contribution changing $N_{\rm eff}$ and the corresponding perturbation parameters $c^2_{\rm vis}$ and $c^2_{\rm eff}$. In the second analysis we assume the existence of a standard neutrino background and vary only the extra component $N^S_\nu = N_{\rm eff} - 3.046$ considering only in this extra component the variations in $c^2_{\rm vis}$ and $c^2_{\rm eff}$.

3.1. Varying the number of relativistic degrees of freedom $N_{\rm eff}$.

In Table 4 we report the constraints on the cosmological parameters varying $N_{\rm eff}$ with and without variations in perturbation theory. We consider two cases: first we run our analysis fixing the perturbation parameters to the standard values, i.e. $c^2_{\rm eff} = c^2_{\rm vis} = 1/3$, then we let those parameters to vary freely.

As we can see from the results in the left column of Table 4, the WMAP7+ACT+SPT+DR7+H0 analysis is clearly suggesting the presence for Dark Radiation with $N_{\rm eff} = 4.08^{+0.71}_{-0.68}$ at 95% c.l.. When considering variations in the perturbation parameters (right column) the constraint is somewhat shifted towards smaller values with $N_{\rm eff} = 3.89^{+0.70}_{-0.70}$. The constraint on the sound speed, $c^2_{\rm eff} = 0.312 \pm 0.026$ is fully consistent with the expectations of a free streaming component. Anisotropies in the neutrino background are detected at high statistical significance with

$\Omega_b h^2$	0.02229 ± 0.00038	0.02206 ± 0.00081
$\Omega_c h^2$	0.1333 ± 0.0086	0.1313 ± 0.0094
τ	0.082 ± 0.012	0.083 ± 0.014
H_0	74.3 ± 2.2	74.2 ± 2.1
n_s	0.977 ± 0.011	0.972 ± 0.021
$log(10^{10} A_s)$	3.195 ± 0.035	3.196 ± 0.035
A_{SZ}	< 1.2	< 1.4
$A_C[\mu\mathrm{K}^2]$	< 14.3	< 14.6
$A_P[\mu\mathrm{K}^2]$	< 25.2	< 24.7
N_{eff}	$4.08^{+0.18+0.71}_{-0.18-0.68}$	$3.89^{+0.19+0.70}_{-0.19-0.70}$
c^2_{eff}	$1/3$	$0.312^{+0.008+0.026}_{-0.007-0.026}$
c^2_{vis}	$1/3$	$0.29^{+0.04+0.21}_{-0.06-0.16}$
χ^2_{min}	7594.2	7591.5

Table 1. MCMC estimation of the cosmological parameters assuming N_{eff} relativistic neutrinos. Upper bounds at 95% c.l. are reported for foregrounds parameters. We quote the one-dimensional marginalized 68% and 95% c.l. for the neutrino parameters.

$c^2_{\mathrm{vis}} = 0.29^{+0.21}_{-0.16}$ improving previous constraints presented in [33].

It is interesting to consider the possible degeneracies between N_{eff} and other "indirect" (i.e. not considered as primary parameters in MCMC runs) model parameters. In Figure 1 we therefore plot the 2D likelihood constraints on N_{eff} versus the Hubble constant H_0, the age of the universe t_0 and the amplitude of r.m.s. mass fluctuations on spheres of $8\mathrm{Mpc}h^{-1}$, σ_8.

As we can see from the three panels in the figure, there is a clear degeneracy between N_{eff} and those three parameters. Namely, an extra radiation component will bring the cosmological constraints (respect to the standard 3 neutrino case) to higher values of the Hubble constant and of σ_8 and to lower values of the age of the universe t_0. These degeneracies have been already discussed in the literature (see e.g. [38]) and could be useful to estimate the effect of additional datasets on our result. The 3% determination of the Hubble constant from the analysis of [19] plays a key role in our analysis in shifting the constraints towards larger values of N_{eff}. If future analyses will point towards lower values of the Hubble constant, this will make the standard 3 neutrino case more consistent with observations. If future observations will point towards values of the age of the universe significantly larger than 13 Gyrs, this will be against an extra dark radiation component, since it prefers $t_0 \sim 12.5$Gyrs. Clearly, adding cluster mass function data as presented in [39] and that points towards lower values of σ_8 renders the standard $N_{\mathrm{eff}} = 3.046$ case more consistent with observations. A future and precise determination of σ_8 from clusters or Lyman-α surveys could be crucial in ruling out dark radiation.

3.2. Varying only the excess in the relativistic component N^S_ν and assuming 3 standard neutrinos.

In Table 2 we report the constraints considering only an excess N^S_ν in the number of relativistic degrees of freedom over a standard 3 neutrinos background.

As we can see for the results in the table, the evidence for an extra background is solid with $N^S_\nu = 1.46^{+0.76}_{-0.74}$ at 95% c.l. when only variations in the c^2_{vis} component are considered, while the constraint is $N^S_\nu = 1.10^{+0.79}_{-0.72}$ when also variations in c^2_{eff} are considered. Again, the data provide a good determination for c^2_{eff} with $c^2_{\mathrm{eff}} = 0.24^{+0.08}_{-0.13}$ at 95% c.l., in marginal agreement at about 2σ with the standard $c^2_{\mathrm{eff}} = 1/3$ value. This lower value of c^2_{eff}, also found in [21], could hint for a dark radiation component with a varying equation of state, ruling out a a massless sterile

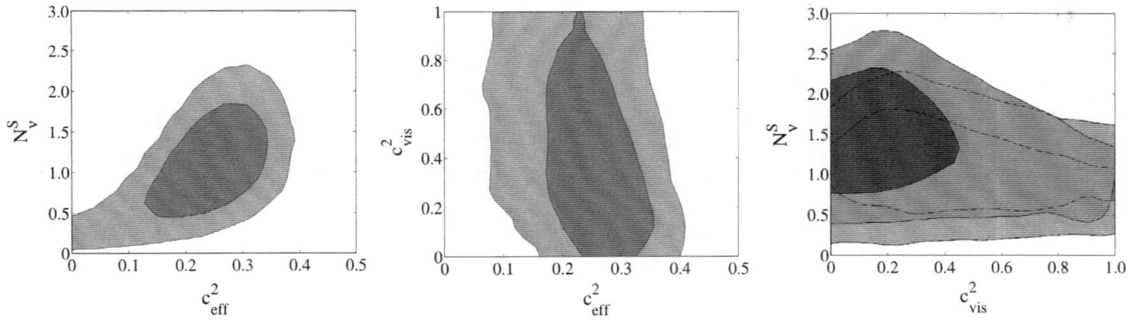

Figure 2. 68% and 95% c.l. constraints for the degeneracy between neutrinos parameters. Red contours refer to model (A) in Table 2, while blue contours show model (B).

Model :	**varying $c^2_{\rm eff}$, $c^2_{\rm vis}$ (A)**	**$c^2_{\rm eff} = 1/3$, varying $c^2_{\rm vis}$ (B)**
$\Omega_b h^2$	0.02177 ± 0.00066	0.02262 ± 0.00049
$\Omega_c h^2$	0.135 ± 0.010	0.143 ± 0.010
τ	0.086 ± 0.013	0.084 ± 0.013
H_0	72.8 ± 2.1	73.7 ± 2.2
n_s	0.989 ± 0.014	0.978 ± 0.014
$log(10^{10} A_s)$	3.178 ± 0.035	3.192 ± 0.035
A_{SZ}	< 1.6	< 1.4
$A_C[\mu\mathrm{K}^2]$	< 15.0	< 15.0
$A_P[\mu\mathrm{K}^2]$	< 24.8	< 24.8
N_ν^S	$1.10^{+0.19+0.79}_{-0.23-0.72}$	$1.46^{+0.21+0.76}_{-0.21-0.74}$
$c^2_{\rm eff}$	$0.24^{+0.03+0.08}_{-0.02-0.13}$	$1/3$
$c^2_{\rm vis}$	< 0.91	< 0.74
χ^2_{min}	7590.5	7592.0

Table 2. MCMC estimation of the cosmological parameters considering an extra component N_ν^S and assuming a standard background of 3 relativistic neutrinos. The perturbation parameters refer to the extra component. Both 68% and 95% confidence levels for the neutrino parameters are reported. Upper bounds are at 95% c.l. .

neutrino. It will be certainly interesting to investigate if this signal remains in future analyses. No significant constraint is obtained on $c^2_{\rm vis}$.

In Figure 2 we show the degeneracy between the parameters N_ν^S, $c^2_{\rm eff}$, and $c^2_{\rm vis}$ by plotting the 2D likelihood contours between them. As we can see a degeneracy is present between $c^2_{\rm eff}$ and N_ν^S: models with lower values of N_ν^S are more compatible with $c^2_{\rm eff} = 0$ since the effect of $c^2_{\rm eff}$ on the CMB spectrum is smaller. No apparent degeneracy is present between $c^2_{\rm vis}$ and the remaining parameters since $c^2_{\rm vis}$ is weakly constrained by current data.

Since oscillation experiments have clearly established that neutrino are massive, it is interesting to perform a similar analysis but letting the 3 neutrino standard background with $c^2_{\rm eff} = c^2_{\rm vis} = 1/3$ to be massive, and varying the parameter Σm_ν that consider the sum of masses of the 3 *active* neutrinos. The extra dark radiation component is assumed massless and we treat the perturbations in it as in the previous sections. In Table 3 we report the results of this analysis.

$\Omega_b h^2$	0.02174 ± 0.00063
$\Omega_c h^2$	0.135 ± 0.011
τ	0.087 ± 0.014
H_0	72.7 ± 2.1
n_s	0.989 ± 0.015
$log(10^{10}A_s)$	3.179 ± 0.036
A_{SZ}	< 1.6
$A_C[\mu\mathrm{K}^2]$	< 15.9
$A_P[\mu\mathrm{K}^2]$	< 26.1
$\sum m_\nu[\mathrm{eV}]$	< 0.79
N_ν^S	$1.12^{+0.21+0.86}_{-0.26-0.74}$
c^2_{eff}	$0.241^{+0.03+0.09}_{-0.02-0.12}$
c^2_{vis}	< 0.92
χ^2_{min}	7590.7

Table 3. MCMC estimation of the cosmological parameters considering $N_\nu = 3.04$ massive neutrinos. Values and 68% - 95% errors for the neutrino parameters are reported. Upper bounds are at 95% c.l. .

As we can see, when masses in the active neutrinos are considered, there is a slightly stronger evidence for the extra background with $N_\nu^S = 1.12^{+0.86}_{-0.74}$. This is can be explained by the degeneracy present between $\sum m_\nu$ and N_ν^S, well known in the literature (see e.g. [13]) and clearly shown in Figure 3 where we report the 2D marginalized contours in the plane $\sum m_\nu - N_\nu^S$.

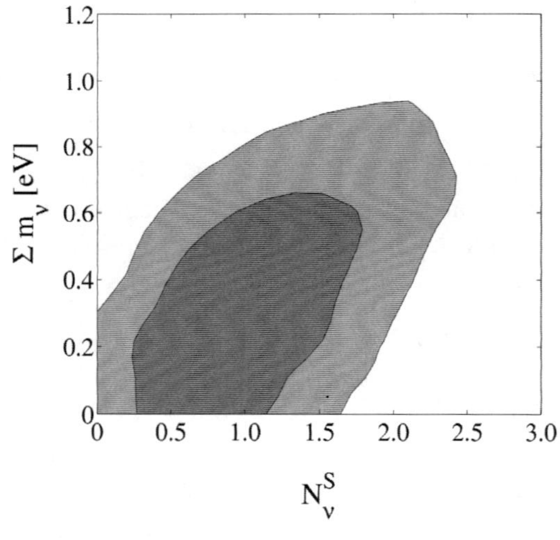

Figure 3. Degeneracy in the plane $\sum m_\nu - N_\nu^S$ at 68% and 95% c.l. .

4. A varying fine-structure constant ?

As we saw in the previous section the current data seems to prefer a larger value for N_{eff} at more than two standard deviation. However this could indicate the presence of a completely different process. As discussed in [18] a value for $N_{\mathrm{eff}} > 3$ could be compensated by a different value of the fine structure constant α at recombination. We have therefore perfomed a similar analysis but by also letting the value of the fine structure constant to be different from the current value of $\alpha_0 = 1/137.03599907$. To the 6 standard Λ-CDM parameters reported in the previous section we therefore add as additional free parameters the fine structure constant (first case), the number of relativistic degrees of freedom $N_{\rm eff}$ (second case) and the primordial Helium abundance Y_p (third case). In Table 4 we report the constraints on the cosmological parameters for these three scenarios.

Parameter	α/α_0	$\alpha/\alpha_0 + N_{\rm eff}$	$\alpha/\alpha_0 + N_{\rm eff} + Y_p$
$\Omega_b h^2$	0.0218 ± 0.0004	0.0224 ± 0.0005	0.0223 ± 0.0007
$\Omega_c h^2$	0.1144 ± 0.0034	0.1302 ± 0.0095	0.1303 ± 0.0094
τ	0.086 ± 0.014	0.088 ± 0.015	0.088 ± 0.016
H_0	68.9 ± 1.4	71.52 ± 2.0	71.8 ± 2.1
α/α_0	0.984 ± 0.005	0.990 ± 0.006	0.987 ± 0.014
n_s	0.976 ± 0.013	0.991 ± 0.015	0.992 ± 0.016
$log[10^{10} A_s]$	3.193 ± 0.037	3.169 ± 0.040	3.167 ± 0.042
A_{SZ}	< 2.00	< 2.00	< 2.00
A_C	< 16.0	< 15.8	< 14.8
A_P	< 24.7	< 24.9	< 22.4
Ω_Λ	0.7137 ± 0.0070	0.7020 ± 0.0094	0.704 ± 0.013
Age/Gyr	13.76 ± 0.24	13.18 ± 0.38	13.15 ± 0.37
Ω_m	0.2863 ± 0.0070	0.2980 ± 0.0094	0.296 ± 0.013
σ_8	0.836 ± 0.023	0.862 ± 0.028	0.859 ± 0.034
z_{re}	10.7 ± 1.2	11.0 ± 1.3	11.0 ± 1.3
$N_{\rm eff}$	–	$4.10^{+0.24}_{-0.29}$	$4.19^{+0.31}_{-0.35}$
Y_p	–	–	0.215 ± 0.096
χ^2_{min}	7600.2	7596.8	7596.5

Table 4. MCMC estimation of the cosmological parameters from the dataset described in the text. Results for the three analyses described in the text are reported. Upper bounds at 95% c.l. are reported for foregrounds parameters.

As we can see the dataset considered prefers a value of α/α_0 smaller than unity at more than two standard deviations when both the $N_{\rm eff}$ and Y_p are kept fixed at their standard values. This result, while interesting, is to be expected since is clearly driven by data preference for larger values of $N_{\rm eff}$. Allowing for variations in $N_{\rm eff}$ significantly shifts the best fit value for α/α_0, which is now consistent with the standard value. However, even in this case the best fit value for $N_{\rm eff}$ is still ~ 4, i.e. allowing for variation in the fine structure constant enlarges the error bars on $N_{\rm eff}$ of about $\sim 30\%$ but does not shift the best fit value towards the standard result. The largest effect on α comes however when also the helium abundance Y_p is let to vary. In this case, indeed, the errors on α are almost doubled.

5. Conclusions

In these proceedings we have reported the results obtained by a new search for Dark Radiation, parametrizing it with an effective number of relativistic degrees of freedom $N_{\rm eff}$. We have shown that the cosmological data we considered are clearly suggesting the presence for an extra dark radiation component with $N_{\rm eff} = 4.08^{+0.71}_{-0.68}$ at 95% c.l. . Performing an analysis on its effective sound speed $c_{\rm eff}$ and viscosity $c_{\rm vis}$ parameters, we found $c^2_{\rm eff} = 0.312 \pm 0.026$ and $c^2_{\rm vis} = 0.29^{+0.21}_{-0.16}$ at 95% c.l., consistent with the expectations of a relativistic free streaming component ($c^2_{\rm eff}=c^2_{\rm vis}=1/3$). Assuming the presence of 3 standard relativistic neutrinos we constrain the extra dark radiation component with $N_\nu^S = 1.10^{+0.79}_{-0.72}$ and $c_{\rm eff}^2 = 0.24^{+0.08}_{-0.13}$ at 95% c.l. while $c_{\rm vis}^2$ is practically unconstrained. Assuming a mass in the 3 neutrino component we obtain further indications for the dark radiation component with $N_\nu^S = 1.12^{+0.86}_{-0.74}$ at 95% c.l. . From these results we conclude that Dark Radiation currently represents one of the most relevant anomaly for the Λ-CDM scenario.

While a variation in the fine structure constant could explain a larger value for $N_{\rm eff}$, the data still prefer ths solution with $N_{\rm eff} > 3$ than a variation in α.

Dark Radiation will be severely constrained in the very near future by the Planck satellite data, where a precision on $N_{\rm eff}$ of about $\Delta N_{\rm eff} \sim 0.2$ is expected (see e.g. [41] and [42]) only from CMB data.

Acknowledgments

We thank Ryan Keisler for providing us with the likelihood code for the SPT data. We thank Luca Pagano for help. This work is supported by PRIN-INAF, "Astronomy probes fundamental physics". Support was given by the Italian Space Agency through the ASI contracts Euclid- IC (I/031/10/0).

References

[1] E. Komatsu *et al.*, arXiv:1001.4538 [astro-ph.CO].
[2] J. Dunkley *et al.*, arXiv:1009.0866 [astro-ph.CO].
[3] C. L. Reichardt *et al.*, Astrophys. J. **694** (2009) 1200 [arXiv:0801.1491 [astro-ph]].
[4] R. Keisler *et al.*, arXiv:1105.3182 [astro-ph.CO].
[5] B. A. Reid *et al.*, Mon. Not. Roy. Astron. Soc. **404** (2010) 60 [arXiv:0907.1659 [astro-ph.CO]].
[6] E. W. Kolb and M. S. Turner, Front. Phys. **69** (1990) 1.
[7] G. Mangano, G. Miele, S. Pastor, T. Pinto, O. Pisanti, P. D. Serpico, Nucl. Phys. **B729** (2005) 221-234. [hep-ph/0506164].
[8] R. Bowen, S. H. Hansen, A. Melchiorri, J. Silk and R. Trotta, Mon. Not. Roy. Astron. Soc. **334**, 760 (2002) [arXiv:astro-ph/0110636].
[9] U. Seljak, A. Slosar, P. McDonald, JCAP **0610** (2006) 014. [astro-ph/0604335].
[10] M. Cirelli and A. Strumia, JCAP **0612** (2006) 013 [arXiv:astro-ph/0607086].
[11] G. Mangano, A. Melchiorri, O. Mena, G. Miele, A. Slosar, JCAP **0703** (2007) 006. [astro-ph/0612150].
[12] K. Ichikawa, M. Kawasaki, F. Takahashi, JCAP **0705**, 007 (2007). [astro-ph/0611784].
[13] J. Hamann, S. Hannestad, G. G. Raffelt, I. Tamborra, Y. Y. Y. Wong, Phys. Rev. Lett. **105** (2010) 181301. [arXiv:1006.5276 [hep-ph]].
[14] E. Giusarma, M. Corsi, M. Archidiacono, R. de Putter, A. Melchiorri, O. Mena, S. Pandolfi, Phys. Rev. **D83** (2011) 115023. [arXiv:1102.4774 [astro-ph.CO]].
[15] L. M. Krauss, C. Lunardini, C. Smith, [arXiv:1009.4666 [hep-ph]].
[16] B. A. Reid, L. Verde, R. Jimenez, O. Mena, JCAP **1001**, 003 (2010). [arXiv:0910.0008 [astro-ph.CO]].
[17] M. Archidiacono, E. Calabrese and A. Melchiorri, Phys. Rev. D **84** (2011) 123008 [arXiv:1109.2767 [astro-ph.CO]].
[18] E. Menegoni, M. Archidiacono, E. Calabrese, S. Galli, C. J. A. P. Martins and A. Melchiorri, Phys. Rev. D **85** (2012) 107301 [arXiv:1202.1476 [astro-ph.CO]].
[19] A. G. Riess, L. Macri, S. Casertano, H. Lampeitl, H. C. Ferguson, A. V. Filippenko, S. W. Jha, W. Li *et al.*, Astrophys. J. **730**, 119 (2011). [arXiv:1103.2976 [astro-ph.CO]].
[20] Z. Hou, R. Keisler, L. Knox, M. Millea, C. Reichardt, [arXiv:1104.2333 [astro-ph.CO]].
[21] T. L. Smith, S. Das and O. Zahn, arXiv:1105.3246 [astro-ph.CO].
[22] G. Mangano, G. Miele, S. Pastor, T. Pinto, O. Pisanti, P. D. Serpico, Nucl. Phys. **B756** (2006) 100-116. [hep-ph/0607267].
[23] A. Aguilar *et al.* [LSND Collaboration], Phys. Rev. D **64** (2001) 112007 [arXiv:hep-ex/0104049].
[24] A. A. Aguilar-Arevalo *et al.* [The MiniBooNE Collaboration], Phys. Rev. Lett. **98** (2007) 231801 [arXiv:0704.1500 [hep-ex]]; A. A. Aguilar-Arevalo *et al.* [MiniBooNE Collaboration], Phys. Rev. Lett. **103** (2009) 111801 [arXiv:0904.1958 [hep-ex]].
[25] S. Hannestad, A. Mirizzi, G. G. Raffelt and Y. Y. Y. Wong, JCAP **1008** (2010) 001 [arXiv:1004.0695 [astro-ph.CO]]; A. Melchiorri, O. Mena and A. Slosar, Phys. Rev. D **76** (2007) 041303 [arXiv:0705.2695 [astro-ph]]. K. Nakayama, F. Takahashi and T. T. Yanagida, Phys. Lett. B **697** (2011) 275 [arXiv:1010.5693 [hep-ph]].

[26] T. L. Smith, E. Pierpaoli and M. Kamionkowski, Phys. Rev. Lett. **97** (2006) 021301 [arXiv:astro-ph/0603144].
[27] W. Fischler, J. Meyers, Phys. Rev. **D83** (2011) 063520. [arXiv:1011.3501 [astro-ph.CO]]; K. Ichikawa, M. Kawasaki, K. Nakayama, M. Senami, F. Takahashi, "Increasing effective number of neutrinos by decaying particles," JCAP **0705** (2007) 008. [hep-ph/0703034 [HEP-PH]]; K. Nakayama, F. Takahashi, T. T. Yanagida, "A theory of extra radiation in the Universe," Phys. Lett. **B697**, 275-279 (2011). [arXiv:1010.5693 [hep-ph]]
[28] P. Binetruy, C. Deffayet, U. Ellwanger and D. Langlois, Phys. Lett. B **477** (2000) 285 [arXiv:hep-th/9910219]; T. Shiromizu, K. i. Maeda and M. Sasaki, Phys. Rev. D **62** (2000) 024012 [arXiv:gr-qc/9910076]; V. V. Flambaum and E. V. Shuryak, Europhys. Lett. **74** (2006) 813 [arXiv:hep-th/0512038].
[29] A. Hebecker and J. March-Russell, Nucl. Phys. B **608** (2001) 375 [arXiv:hep-ph/0103214].
[30] E. Calabrese, D. Huterer, E. V. Linder, A. Melchiorri and L. Pagano, Phys. Rev. D **83**, 123504 (2011) [arXiv:1103.4132 [astro-ph.CO]]; J. S. Gagnon and J. Lesgourgues, JCAP **1109** (2011) 026 [arXiv:1107.1503 [astro-ph.CO]].
[31] W. Hu, Astrophys. J. **506** (1998) 485-494. [astro-ph/9801234]; W. Hu, D. J. Eisenstein, M. Tegmark, M. J. White, Phys. Rev. **D59**, 023512 (1999). [astro-ph/9806362].
[32] J. F. Beacom, N. F. Bell, S. Dodelson, Phys. Rev. Lett. **93** (2004) 121302. [astro-ph/0404585]; S. Hannestad, JCAP **0502** (2005) 011. [astro-ph/0411475]; A. Basboll, O. E. Bjaelde, S. Hannestad and G. G. Raffelt, Phys. Rev. D **79** (2009) 043512 [arXiv:0806.1735 [astro-ph]];
[33] R. Trotta, A. Melchiorri, Phys. Rev. Lett. **95** (2005) 011305. [astro-ph/0412066].
[34] A. Melchiorri, P. Serra, Phys. Rev. **D74**, 127301 (2006); F. De Bernardis, L. Pagano, P. Serra, A. Melchiorri, A. Cooray, JCAP **0806**, 013 (2008). [arXiv:0804.1925 [astro-ph]];
[35] A. Lewis and S. Bridle, Phys. Rev. D **66**, 103511 (2002) (Available from `http://cosmologist.info`.)
[36] A. G. Riess, L. Macri, S. Casertano, H. Lampeitl, H. C. Ferguson, A. V. Filippenko, S. W. Jha, W. Li *et al.*, Astrophys. J. **730** (2011) 119. [arXiv:1103.2976 [astro-ph.CO]].
[37] H. Trac, P. Bode and J. P. Ostriker, Astrophys. J. **727** (2011) 94 [arXiv:1006.2828 [astro-ph.CO]].
[38] F. de Bernardis, A. Melchiorri, L. Verde and R. Jimenez, JCAP **0803** (2008) 020 [arXiv:0707.4170 [astro-ph]].
[39] A. Vikhlinin *et al.*, Astrophys. J. **692** (2009) 1060 [arXiv:0812.2720 [astro-ph]].
[40] A. X. Gonzalez-Morales, R. Poltis, B. D. Sherwin and L. Verde, arXiv:1106.5052 [astro-ph.CO].
[41] S. Galli, M. Martinelli, A. Melchiorri, L. Pagano, B. D. Sherwin, D. N. Spergel, Phys. Rev. **D82** (2010) 123504. [arXiv:1005.3808 [astro-ph.CO]].
[42] M. Shimon, N. J. Miller, C. T. Kishimoto, C. J. Smith, G. M. Fuller, B. G. Keating, JCAP **1005** (2010) 037. [arXiv:1001.5088 [astro-ph.CO]].
[43] Y. Akrami, P. Scott, J. Edsjo, J. Conrad and L. Bergstrom, JHEP **1004**, 057 (2010) [arXiv:0910.3950 [hep-ph]]; F. Feroz, K. Cranmer, M. Hobson, R. Ruiz de Austri and R. Trotta, JHEP **1106** (2011) 042 [arXiv:1101.3296 [hep-ph]].
[44] S. Hannestad, JCAP **0305** (2003) 004 [astro-ph/0303076]; V. Barger, J. P. Kneller, H. -S. Lee, D. Marfatia and G. Steigman, Phys. Lett. B **566**, 8 (2003) [hep-ph/0305075]; P. Crotty, J. Lesgourgues and S. Pastor, Phys. Rev. D **67**, 123005 (2003) [astro-ph/0302337]; K. Ichikawa, M. Kawasaki and F. Takahashi, JCAP **0705**, 007 (2007) [astro-ph/0611784]; J. Hamann, arXiv:1110.4271 [astro-ph.CO].

Journal of Physics: Conference Series **485** (2014) 012015 doi:10.1088/1742-6596/485/1/012015

Yukawa unification: MSSM at large $\tan\beta$

Stuart Raby

The Ohio State University, 191 W. Woodruff Ave, Columbus, OH 43210, USA

E-mail: raby@mps.ohio-state.edu

Abstract. Talk given at PASCOS 2012, Merida, Mexico describing work in progress in collaboration with Archana Anandakrishnan, Christopher Plumberg and Akin Wingerter.

1. Introduction

In this talk we discuss SO(10) Yukawa unification and its ramifications for phenomenology. The initial constraints come from fitting the top, bottom and tau masses, requiring large $\tan\beta \sim 50$ and particular values for soft SUSY breaking parameters. We will show that this leads to a spectrum of squark and slepton masses with the first two families heavy while the third family is significantly lighter, i.e. an inverted scalar mass hierarchy. In addition, gauginos should be observable at the LHC. We perform a global χ^2 analysis, fitting the recently observed 'Higgs' with mass of order 125 GeV in addition to fermion masses and mixing angles and several flavor violating observables.

Fermion masses and mixing angles are manifestly hierarchical. The simplest way to describe this hierarchy is with Yukawa matrices which are also hierarchical. Moreover the most natural way to obtain the hierarchy is in terms of effective higher dimension operators of the form

$$W \supset 16_3\ 10\ 16_3 + 16_3\ 10\ \frac{45}{M}\ 16_2 + \cdots. \tag{1}$$

This version of SO(10) models has the nice features that it only requires small representations of SO(10), has many predictions and can, in principle, find an UV completion in string theory. There are a long list of papers by authors such as Albright, Anderson, Babu, Barr, Barbieri, Berezhiani, Blazek, Carena, Chang, Dermisek, Dimopoulos, Hall, Masiero, Murayama, Pati, Raby, Romanino, Rossi, Starkman, Wagner, Wilczek, Wiesenfeldt, and Willenbrock which have followed this line of model building.

The only renormalizable term in W is $\lambda\ 16_3\ 10\ 16_3$ which gives Yukawa coupling unification

$$\lambda = \lambda_t = \lambda_b = \lambda_\tau = \lambda_{\nu_\tau} \tag{2}$$

at M_{GUT}. Note, one CANNOT predict the top mass due to large SUSY threshold corrections to the bottom and tau masses, as shown in [1]. These corrections are of the form

$$\delta m_b/m_b \propto \frac{\alpha_3\ \mu\ M_{\tilde{g}}\ \tan\beta}{m_{\tilde{b}}^2} + \frac{\lambda_t^2\ \mu\ A_t\ \tan\beta}{m_{\tilde{t}}^2} + \text{log corrections}. \tag{3}$$

Content from this work may be used under the terms of the Creative Commons Attribution 3.0 licence. Any further distribution of this work must maintain attribution to the author(s) and the title of the work, journal citation and DOI.
Published under licence by IOP Publishing Ltd

So instead we use Yukawa unification to predict the soft SUSY breaking masses!! In order to fit the data, we need

$$\delta m_b/m_b \sim -2\%. \tag{4}$$

We take $\mu \; M_{\tilde{g}} > 0$, thus we need $\mu \; A_t < 0$. For a short list of references on this subject, see [2, 3].

Given the following GUT scale boundary conditions – universal squark and slepton masses, m_{16}, universal cubic scalar parameter, A_0, universal gaugino masses, $M_{1/2}$, and non-universal Higgs masses [NUHM] or 'just so' Higgs splitting, m_{H_u}, m_{H_d} or $m^2_{H_{u(d)}} = m^2_{10}[1 - (+)\Delta^2_{mH}]$, we find that fitting the top, bottom and tau mass forces us into the region of SUSY breaking parameter space with

$$A_0 \approx -2m_{16}, \;\; m_{10} \approx \sqrt{2}\, m_{16}, \;\; m_{16} > \text{ few TeV}, \;\; \mu, M_{1/2} \ll m_{16}; \tag{5}$$

and, finally,

$$\tan\beta \approx 50. \tag{6}$$

In addition, radiative electroweak symmetry breaking requires $\Delta^2_{mH} \approx 13\%$, with roughly half of this coming naturally from the renormalization group running of neutrino Yukawa couplings from M_G to $M_{N_\tau} \sim 10^{13}$ GeV [2].

It is very interesting that the above region in SUSY parameter space results in an inverted scalar mass hierarchy at the weak scale with the third family scalars significantly lighter than the first two families [4]. This has the nice property of suppressing flavor changing neutral current and CP violating processes. f

2. Heavy squarks and sleptons

Considering the theoretical and experimental results for the branching ratio $BR(B \to X_s\gamma)$, we argue that $m_{16} \geq 8$ TeV. The experimental value $BR(B \to X_s\gamma)_{\rm exp} = (3.55 \pm 0.26) \times 10^{-4}$, while the NNLO Standard Model theoretical value is $BR(B \to X_s\gamma)_{\rm th} = (3.15 \pm 0.23) \times 10^{-4}$. The amplitude for the process $B \to X_s\gamma$ is proportional to the Wilson coefficient, C_7. $C_7 = C_7^{SM} + C_7^{SUSY}$ and, in order to fit the data, we see that $C_7 \approx \pm C_7^{SM}$. Thus $C_7^{SUSY} \approx -2C_7^{SM}$ or $C_7^{SUSY} \approx 0$. The dominant SUSY contribution to the branching ratio comes from a stop - chargino loop with $C_7^{SUSY} \sim C_7^{\chi^+} \sim \frac{\mu \; A_t}{\tilde{m}^2} \; \tan\beta \times sign(C_7^{SM})$ (see Fig. 1). Hence, in our case $C_7 \approx -C_7^{SM}$.

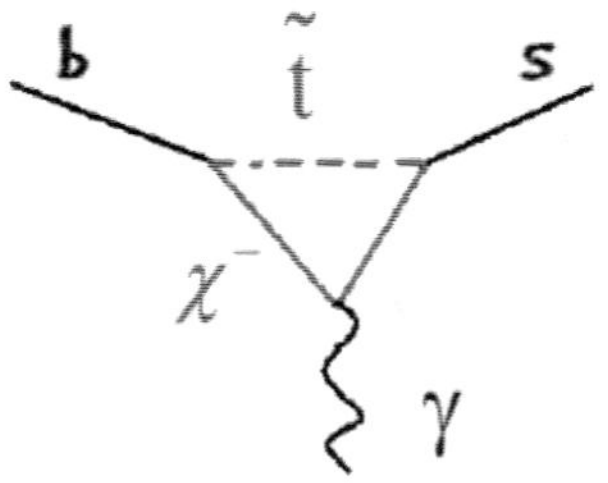

Figure 1. Dominant contribution to the process $b \to s \; \gamma$ in the MSSM.

Recent LHCb data on the $BR(B \to K^* \; \mu^+ \; \mu^-)$ now favors $C_7 \approx +C_7^{SM}$ [6]. See Fig. 2. This tension between the processes $b \to s\gamma$ and $b \to s \; \ell^+ \; \ell^-$ was already discussed by Albrecht et al. [5]. In order to be consistent with this data one requires $C_7^{\chi^+} \approx 0$ or $C_7 \approx C_7^{SM} + C_7^{SUSY} \approx +C_7^{SM}$ and therefore $m_{16} \geq 8$ TeV.

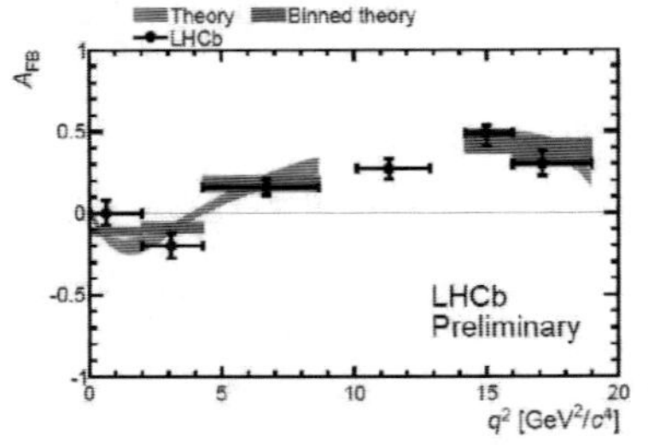

Figure 2. The forward-backward asymmetry for the process $B \to K^* \mu^+ \mu^-$ measured by LHCb.

3. Light Higgs mass

An approximate formula for the light Higgs mass is given by [7]

$$m_h^2 \approx M_Z^2 \cos^2 2\beta + \frac{3g^2 m_t^4}{8\pi^2 m_W^2}\left[\ln\left(\frac{M_{SUSY}^2}{m_t^2}\right) + \frac{X_t^2}{M_{SUSY}^2}\left(1 - \frac{X_t^2}{12M_{SUSY}^2}\right)\right] \quad (7)$$

where $X_t = A_t - \mu/\tan\beta$. The light Higgs mass is maximized as a function of X_t for $X_t/M_{SUSY} = \sqrt{6}$, referred to as maximal mixing. Hence we see that for large values of A_t and M_{SUSY} it is quite easy to obtain a light Higgs mass of order 125 GeV.

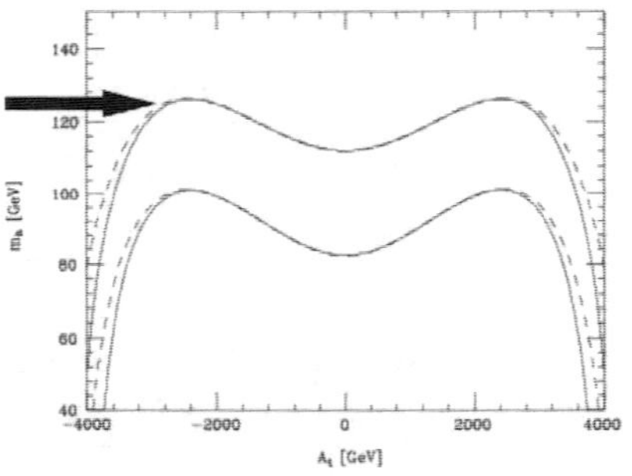

Figure 3. Light Higgs mass as a function of A_t for two values of $\tan\beta$. The upper curve corresponds to $\tan\beta \sim 50$. The arrow corresponds to maximal mixing.

4. $B_s \to \mu^+ \mu^-$

In this section we argue that the light Higgs boson must be Standard Model-like. To do this we show that the CP odd Higgs boson, A, must have mass greater than ~ 1 TeV and as a consequence this is also true for the CP even Higgs boson, H, and the charged Higgs bosons, $H^\pm$, as well. This is the well-known decoupling limit in which the light Higgs boson couples to matter just like the Standard Model Higgs.

Consider the branching ratio $BR(B_s \to \mu^+ \mu^-)$ which in the Standard Model is $\sim 3 \times 10^{-9}$. In the MSSM this receives a contribution proportional to $\sim \frac{\tan\beta^6}{m_A^4}$. Recent experimental results give [8]

$$CDF: \quad 1.8(+1.8, -0.9) \times 10^{-8}\,(95\%CL) \quad \text{with } 7fb^{-1} \quad (8)$$

$$LHCb\,\text{bound}: \quad < 4.5 \times 10^{-9}\,(95\%CL) \quad \text{with } 1fb^{-1}. \quad (9)$$

These two results seem to be inconsistent, so we will choose to try and satisfy the LHCb data which is also consistent with results from CMS. Since we have $\tan\beta \sim 50$, our only choice is to take the CP odd Higgs mass to be large with $m_A \geq 1$ TeV. Hence the light Higgs is SM-like.

5. 3 Family Model

The previous results depended solely on SO(10) Yukawa unification for the third family. We now consider a complete three family SO(10) model for fermion masses and mixing, including neutrinos [9, 5]. The model also includes a $D_3 \times [U(1) \times \mathbb{Z}_2 \times \mathbb{Z}_3]$ family symmetry which is necessary to obtain a predictive theory of fermion masses by reducing the number of arbitrary parameters in the Yukawa matrices. In the rest of this talk we will consider the new results due to the three family analysis. We shall consider the superpotential generating the effective fermion Yukawa couplings. We then perform a global χ^2 analysis, including precision electroweak data which now includes both neutral and charged fermion masses and mixing angles.

The superspace potential for the charged fermion sector of this model is given by:

$$\begin{aligned} W_{ch.fermions} = & \quad 16_3\ 10\ 16_3 + 16_a\ 10\ \chi_a \\ & + \bar{\chi}_a\ (M_\chi\ \chi_a + \ 45\ \tfrac{\phi_a}{\hat{M}}\ 16_3\ +\ 45\ \tfrac{\tilde{\phi}_a}{\hat{M}}\ 16_a + \mathbf{A}\ 16_a) \end{aligned} \tag{10}$$

where 45 is an $SO(10)$ adjoint field which is assumed to obtain a VEV in the B – L direction; and M is a linear combination of an $SO(10)$ singlet and adjoint. Its VEV $M_0(1 + \alpha X + \beta Y)$ gives mass to Froggatt-Nielsen states. Here X and Y are elements of the Lie algebra of $SO(10)$ with X in the direction of the $U(1)$ which commutes with $SU(5)$ and Y the standard weak hypercharge; and α , β are arbitrary constants which are fit to the data.

$$\phi_a, \quad \tilde{\phi}_a, \quad A, \tag{11}$$

are $SO(10)$ singlet 'flavon' fields, and

$$\bar{\chi}_a, \quad \chi_a \tag{12}$$

are a pair of Froggatt-Nielsen states transforming as a $\overline{16}$ and 16 under $SO(10)$. The 'flavon' fields are *assumed* to obtain VEVs of the form

$$\langle \phi_a \rangle = \begin{pmatrix} \phi_1 \\ \phi_2 \end{pmatrix}, \ \langle \tilde{\phi}_a \rangle = \begin{pmatrix} 0 \\ \tilde{\phi}_2 \end{pmatrix}. \tag{13}$$

After integrating out the Froggatt-Nielsen states one obtains the effective fermion mass operators in Fig. 4. We then obtain the Yukawa matrices for up and down quarks, charged leptons and

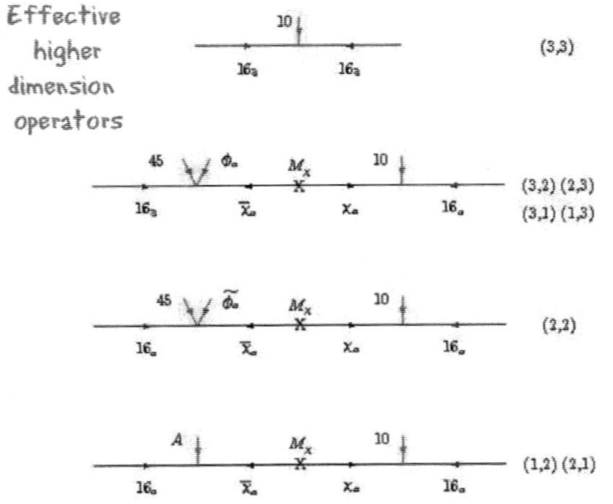

Figure 4. The effective fermion mass operators obtained after integrating out the Froggatt-Nielsen massive states.

neutrinos given in Fig. 5. These matrices contain 7 real parameters and 4 arbitrary phases.

$$Y_u = \begin{pmatrix} 0 & \epsilon'\,\rho & -\epsilon\,\xi \\ -\epsilon'\,\rho & \tilde{\epsilon}\,\rho & -\epsilon \\ \epsilon\,\xi & \epsilon & 1 \end{pmatrix} \lambda$$

$$Y_d = \begin{pmatrix} 0 & \epsilon' & -\epsilon\,\xi\,\sigma \\ -\epsilon' & \tilde{\epsilon} & -\epsilon\,\sigma \\ \epsilon\,\xi & \epsilon & 1 \end{pmatrix} \lambda$$

$$Y_e = \begin{pmatrix} 0 & -\epsilon' & 3\,\epsilon\,\xi \\ \epsilon' & 3\,\tilde{\epsilon} & 3\,\epsilon \\ -3\,\epsilon\,\xi\,\sigma & -3\,\epsilon\,\sigma & 1 \end{pmatrix} \lambda$$

$$Y_\nu = \begin{pmatrix} 0 & -\epsilon'\,\omega & \frac{3}{2}\,\epsilon\,\xi\,\omega \\ \epsilon'\,\omega & 3\,\tilde{\epsilon}\,\omega & \frac{3}{2}\,\epsilon\,\omega \\ -3\,\epsilon\,\xi\,\sigma & -3\,\epsilon\,\sigma & 1 \end{pmatrix} \lambda$$

Figure 5. The Yukawa matrices obtained from the effective fermion mass operators after taking into account the flavon VEVs.

Note, the superpotential (Eqn. 10) has many arbitrary parameters. However, at the end of the day the effective Yukawa matrices have many fewer parameters. This is good, because we then obtain a very predictive theory. Also, the quark mass matrices accommodate the Georgi-Jarlskog mechanism, such that $m_\mu/m_e \approx 9 m_s/m_d$.

We then add 3 real Majorana mass parameters for the neutrino see-saw mechanism. The anti-neutrinos get GUT scale masses by mixing with three $SO(10)$ singlets $\{N_a,\ a = 1, 2;\ N_3\}$ transforming as a D_3 doublet and singlet respectively. The full superpotential is given by $W = W_{ch.fermions} + W_{neutrino}$ with

$$\begin{aligned} W_{neutrino} = &\ \overline{16}\,(\lambda_2\ N_a\ 16_a\ +\ \lambda_3\ N_3\ 16_3) \\ &+\ \tfrac{1}{2}\,(S_a\ N_a\ N_a\ +\ S_3\ N_3\ N_3)\,. \end{aligned} \tag{14}$$

We assume $\overline{16}$ obtains a VEV, v_{16}, in the right-handed neutrino direction, and $\langle S_a \rangle = M_a$ for $a = 1, 2$ and $\langle S_3 \rangle = M_3$. The effective neutrino mass terms are given by

$$W = \nu\ m_\nu\ \bar{\nu} + \bar{\nu}\ V\ N + \frac{1}{2}\ N\ M_N\ N \tag{15}$$

with

$$V = v_{16} \begin{pmatrix} 0 & \lambda_2 & 0 \\ \lambda_2 & 0 & 0 \\ 0 & 0 & \lambda_3 \end{pmatrix}, \quad M_N = diag(M_1,\ M_2,\ M_3) \tag{16}$$

all assumed to be real. Finally, upon integrating out the heavy Majorana neutrinos we obtain the 3×3 Majorana mass matrix for the light neutrinos in the lepton flavor basis given by

$$\mathcal{M} = U_e^T\ m_\nu\ M_R^{-1}\ m_\nu^T\ U_e, \tag{17}$$

where the effective right-handed neutrino Majorana mass matrix is given by:

$$M_R = V\ M_N^{-1}\ V^T\ \equiv\ \mathrm{diag}(M_{R_1}, M_{R_2}, M_{R_3}), \tag{18}$$

with

$$M_{R_1} = (\lambda_2\ v_{16})^2/M_2, \quad M_{R_2} = (\lambda_2\ v_{16})^2/M_1, \quad M_{R_3} = (\lambda_3\ v_{16})^2/M_3. \tag{19}$$

6. Global χ^2 analysis

Just in the fermion mass sector we can see that the theory is very predictive. We have 15 charged fermion and 5 neutrino low energy observables given in terms of 11 arbitrary Yukawa parameters and 3 Majorana mass parameters. Hence there are 6 degrees of freedom in this

Journal of Physics: Conference Series **485** (2014) 012015 doi:10.1088/1742-6596/485/1/012015

Table 1. Parameters entering the global χ^2 analysis.

Sector	#	*Parameters*
gauge	3	$\alpha_G,\ M_G,\ \epsilon_3$
SUSY (GUTscale)	5	$m_{16},\ M_{1/2},\ A_0,\ m_{H_u},\ m_{H_d}$
textures	11	$\lambda,\ \epsilon,\ \epsilon',\ \rho,\ \sigma,\ \tilde{\epsilon},\ \xi$
neutrino	3	$M_{R_1},\ M_{R_2},\ M_{R_3}$
SUSY (EWscale)	2	$\mu,\ \tan\beta$

sector of the theory. However in order to include the complete MSSM sector we perform the global χ^2 analysis with 24 arbitrary parameters at the GUT scale given in Table 1. Note, this is to be compared to the 27 arbitrary parameters in the Standard Model or the 32 parameters in the CMSSM.

In 2007, Albrecht et al. [5] performed a global χ^2 analysis of this theory. Two of the tables from their paper are exhibited in Fig. 6. This analysis included 27 low energy observables and a reasonable fit to the data was only found for $m_{16} = 10$ TeV. Note, the Higgs mass was predicted to be 129 GeV. There is now a discovery by CMS and ATLAS of a boson with mass of order

Observable	Exp. value	Fit value	Pull (σ)
M_W	80.403	80.6	0.5
M_Z	91.1876	90.7	1.1
$G_F \times 10^5$	1.16637	1.16	0.3
$1/\alpha_{em}$	137.036	136.8	0.4
$\alpha_s(M_Z)$	0.1176	0.117	0.2
M_t	170.9	170.6	0.2
$m_b(m_b)$	4.2	4.22	0.3
$m_c(m_b)$	1.25	1.14	1.2
$m_s(2\,\mathrm{GeV})$	0.095	0.107	0.5
$m_d(2\,\mathrm{GeV})$	0.005	0.00741	1.2
$m_u(2\,\mathrm{GeV})$	0.00225	0.00461	**3.1**
M_τ	1.777	1.78	0.1
M_μ	0.10566	0.106	0.1
M_e	0.000511	0.000511	0.0
$\lvert V_{us}\rvert$	0.2258	0.225	0.6
$\lvert V_{ub}\rvert \times 10^3$	4.1	3.26	**2.1**
$\lvert V_{cb}\rvert$	0.0416	0.0416	0.1
$\sin 2\beta$	0.675	0.639	1.4
$\Delta m^2_{31} \times 10^{21}$	2.6	2.6	0.0
$\Delta m^2_{21} \times 10^{23}$	7.9	7.9	0.0
$\sin^2 2\theta_{12}$	0.852	0.852	0.0
$\sin^2 2\theta_{23}$	0.996	1.0	0.2
$\epsilon_K \times 10^3$	2.229	2.33	0.4
$\mathrm{BR}(B \to X_s\gamma) \times 10^4$	3.55	2.86	1.3
$\mathrm{BR}(B \to X_s\ell^+\ell^-) \times 10^6$	1.6	1.62	0.0
$\Delta M_s/\Delta M_d$	35.05	31.1	1.1
$\mathrm{BR}(B^+ \to \tau^+\nu) \times 10^4$	1.31	0.517	1.7
		total χ^2:	**27.4**

m_{16}	10000
μ	1200
$\mathrm{BR}(B_s \to \mu^+\mu^-) \times 10^8$	2.1
$\hat{s}_0$	0.14
$\mathrm{BR}(\mu \to e\gamma) \times 10^{13}$	0.0026
$\delta a_\mu^{\mathrm{SUSY}} \times 10^{10}$	+0.52
M_{h_0}	129
M_A	842
$m_{\tilde{t}_1}$	1903
$m_{\tilde{b}_1}$	2366
$m_{\tilde{\tau}_1}$	3933
$m_{\tilde{\chi}^0_1}$	60
$m_{\tilde{\chi}^+_1}$	120
$m_{\tilde{g}}$	506

Required $C_7 = +C_7^{SM}$

Figure 6. The results obtained by Albrecht et al. for $m_{16} = 10$ TeV.

125 GeV. This may well be the Higgs boson with Standard Model couplings, but further data is needed before such a conclusion can be confirmed. In the mean time we shall assume it is the lightest Higgs boson of the MSSM. In this case, we must conclude (based on our analysis) that it must look very much like the Standard Model Higgs boson, since as can be seen from Fig. 6, the CP odd Higgs mass, $m_A = 842$ GeV, so the theory is in the Higgs decoupling limit.

In this work we have decided to extend the analysis of Albrecht et al. to values of $m_{16} \geq 10$ TeV, including more low energy observables such as the light Higgs mass, the neutrino mixing angle θ_{13} and lower bounds on the gluino and squark masses coming from recent data. This is work in progress in collaboration with Archana Anandakrishnan, Christopher Plumberg and Akin Wingerter, along with significant help from Radovan Dermisek and Aditi Raval. We perform a three family global χ^2 analysis. We are using the code developed by Radovan Dermisek to renormalize the parameters in the theory from the GUT scale to the weak scale, perform electroweak symmetry breaking and calculate squark, slepton, gaugino masses, as well as quark and lepton masses and mixing angles. We also use the Higgs code of Pietro Slavich (suitably revised for our particular scalar spectrum) to calculate the light Higgs mass and SUSY_Flavor_v2.0 [10] to evaluate flavor violating B decays.

There are 24 arbitrary parameters defined mostly at the GUT scale and run down to the weak scale where the χ^2 function is evaluated. However the value of m_{16} has been kept fixed in our analysis, so that we can see the dependence of χ^2 on this input parameter. Thus with 23 arbitrary parameters we fit 37 observables, giving 14 degrees of freedom. The χ^2 function has been minimized using the CERN package, MINUIT. This work is on going so we just present the complete results for one value of $m_{16} = 20$ TeV (see Table 2). The fit is quite good with $\chi^2/d.o.f. = 1.8$. However, note that we have not taken into account correlations in the data, so we will just use χ^2 as a indicator of the rough goodness of the fit. In Table 3 we see that the value of χ^2 decreases as m_{16} increases, but our analysis shows that the $m_{16} \sim 20$ TeV minimizes χ^2, i.e. we have found that χ^2 slowly increases for $m_{16} > 20$ TeV.

7. Conclusion and plans for the future

We have presented an analysis of a theory satisfying Yukawa unification and large $\tan\beta$. The results are encouraging but as usual there are some good and some bad features.

Some of the good features are -

- gauge coupling unification is satisfied
- the Higgs mass is of order 125 GeV and Standard Model like
- there is an inverted scalar mass hierarchy, with the following consequence at the LHC -
- gluinos predominantly decay via $\tilde{g} \to t\,\bar{t}\,\tilde{\chi}^0$, $t\,\bar{b}\,\tilde{\chi}^-$, $b\,\bar{b}\,\tilde{\chi}^0$, etc.

And bad features -

- fine-tuned by at least 0.1 %
- and gluinos are light.

Work in progress -

- We would like to understand why χ^2 increases for values of $m_{16} > 20$ TeV or is there an upper bound on m_{16}?
- The gaugino masses are small. We want to evaluate the χ^2 dependence as a function of the gluino mass or, in other words, ascertain the sensitivity of χ^2 on varying the gaugino masses?
- Finally, we would like to consider alternative sets of boundary conditions at the GUT scale which may be consistent with Yukawa unification. In particular, we will consider the so-called "DR3" scheme and non-universal gaugino masses as may be obtained with (a) non-trivial gauge kinetic functions or (b) "mirage mediation."

In the "DR3" scheme [11] we have SO(10) boundary conditions at M_{GUT}, however the third family scalar masses are taken to be smaller than the first two, i.e. $m_{16}^{(3)} \leq m_{16}^{(1,2)}$. In addition, Higgs doublet mass splitting is accomplished with a $U(1)_X$ D-term, D_X, where $U(1)_X$ is the

Journal of Physics: Conference Series **485** (2014) 012015 doi:10.1088/1742-6596/485/1/012015

Table 2. Result of global χ^2 analysis for m_{16} fixed at 20 TeV.

Observable	Fit value	Exp. value	Pull	Sigma
M_Z	91.1876	91.1876	0.0000	0.0912
M_W	80.7955	80.4360	0.8938	0.4022
$1/\alpha_{em}$	136.5851	137.0360	0.6580	0.6852
$G_\mu \times 10^5$	1.1766	1.1664	0.8805	0.0117
α_3	0.1184	0.1184	0.0642	0.0007
M_t	173.2432	172.9000	0.3177	1.0800
$m_b(m_b)$	4.3155	4.1900	0.6973	0.1800
m_τ	1.7766	1.7770	0.0484	0.0089
$M_b - M_c$	3.4285	3.4300	0.0300	0.0500
$m_c(m_c)$	1.1723	1.2900	1.0697	0.1100
m_s	0.0969	0.1000	0.1039	0.0300
m_d/m_s	0.0693	0.0521	2.5690	0.0067
$1/Q^2$	0.0019	0.0019	0.3475	0.0001
M_μ	0.1056	0.1057	0.0386	0.0005
$M_e \times 10^4$	5.1125	5.1100	0.0978	0.0255
$\lvert V_{us}\rvert$	0.2250	0.2252	0.2407	0.0009
$\lvert V_{cb}\rvert$	0.0412	0.0406	0.4702	0.0013
$\lvert V_{ub}\rvert \times 10^3$	3.2245	3.8900	1.5124	0.4400
$\lvert V_{td}\rvert \times 10^3$	8.9128	8.4000	0.8547	0.6000
$\lvert V_{ts}\rvert$	0.0404	0.0429	0.9750	0.0026
sin2β	0.6370	0.6730	1.5642	0.0230
ϵ_K	0.0021	0.0022	0.3631	0.0002
$\Delta M_{Bs}/\Delta M_{Bd}$	31.7861	35.0000	0.4439	7.2400
$\Delta M_{Bd} \times 10^{13}$	4.2657	3.3370	1.3944	0.6660
$m_{21}^2 \times 10^5$	7.5243	7.5400	0.0340	0.4600
$m_{31}^2 \times 10^3$	2.4708	2.4600	0.0637	0.1700
$\sin^2\theta_{12}$	0.2984	0.3070	0.2445	0.0350
$\sin^2\theta_{23}$	0.4390	0.3980	0.5323	0.0770
$\sin^2\theta_{13}$	0.0132	0.0245	1.7062	0.0066
M_h	124.80	125.30	0.1664	3.0000
$BR(B \to X_s\gamma) \times 10^4$	3.4988	3.5500	0.0368	1.3900
$BR(B_s \to \mu^+\mu^-) \times 10^9$	4.2909	4.5000	0.0000	1.0000
$BR(B_d \to \mu^+\mu^-) \times 10^{10}$	1.3112	14.0000	0.0000	10.0000
$BR(B \to \tau\nu) \times 10^5$	6.8017	16.5000	1.0428	9.3000
$BR(B \to K^*\mu^+\mu^-)$(low) $\times 10^8$	5.0639	4.2000	0.3199	2.7000
$BR(B \to K^*\mu^+\mu^-)$(high) $\times 10^8$	7.5709	6.3000	0.4707	2.7000
$q_0^2(B \to K^*\mu^+\mu^-)$	4.6274	4.9000	0.2097	1.3000
Total χ^2			**24.2**	

Table 3. Results for the global χ^2 analysis with 3 different values of $m_{16} = 10, 15, 20$ TeV.

m_{16}	10 TeV	15 TeV	20 TeV
A_0	-20 TeV	-30 TeV	-41 TeV
μ	517	900	766
$M_{1/2}$	270	200	169
χ^2	32.12	25.87	24.51
M_A	2357	2743	5970
$m_{\tilde{t}_1}$	1699	2952	3738
$m_{\tilde{b}_1}$	2087	3481	3702
$m_{\tilde{\tau}_1}$	3919	5932	7409
$m_{\tilde{\chi}_1^0}$	145	133	136
$m_{\tilde{\chi}_1^+}$	277	267	279
$m_{\tilde{g}}$	925	850	850

is found in SO(10)/[SU(5) × U(1)$_X$]. Hence, $m_i^2 = Q_X^i\ D_X + (m_i^0)^2$ and the charges, Q_X^i, are given in Table 4.

Table 4. U(1)$_X$ charge and SO(10) invariant mass.

	Q	U	D	L	E	N	H_u	H_d
Q_X^i	1	1	-3	-3	1	5	-2	2
m_i^0			m_{16}				m_{10}	

One possible scheme for obtaining non-universal gaugino masses was contemplated in [12] using a non-trivial gauge kinetic function with a chiral 54 dimensional representation of SO(10) obtaining an F-term VEV, but NO scalar VEV. This gives gaugino masses in the ratio $M_3 : M_2 : M_1 = 1 : -3/2 : -1/2$. In addition they take $\mu < 0$ and U(1)$_X$ D-term splitting of squark, slepton and Higgs masses. In the end they obtain a totally different spectrum than with either "Just So" Higgs splitting or "DR3" splitting.

Finally, it has been shown that non-universal gaugino masses can be obtained with "mirage mediation" boundary conditions at M_{GUT} [13]. We would have gaugino masses given by

$$M_i(\mu)/g_i^2(\mu) = \left\{1 + \log(M_{Pl}/m_{3/2})\frac{\alpha_{GUT}\ b_i}{4\pi}\alpha\right\} M_0/g_{GUT}^2 \tag{20}$$

where α is an arbitrary parameter fixing the ratio of the contributions of gravity vs anomaly mediation. In addition we would take $\mu < 0$ and U(1)$_X$ D-term splitting for scalars.

Acknowledgements

I would like to thank the organizers of PASCOS 2012 for organizing such a wonderful conference. In addition, I am very thankful to Radovan Dermisek for letting us use his 3 family code which has been great and for the aid he gave in helping us implement the code along with his student, Aditi Raval. Finally I want to thank Pietro Slavich for sending us a copy of his code for calculating the light Higgs mass. This work has been partially supported by DOE grant - DOE/ER/01545-895.

References

[1] L.J. Hall, R. Rattazzi and U. Sarid, Phys. Rev. D bf 50, 7048 (1994); M. Carena, M. Olechowski, S. Pokorski and C. Wagner, Nucl. Phys. B **426**, 269 (1994); T. Blazek, S. Raby and S. Pokorski, Phys. Rev. D**52**, 4151 (1995).

[2] T. Blazek, R. Dermisek and S. Raby, Phys. Rev. Lett. **88**, 111804 (2002) [arXiv:hep-ph/0107097]; Phys. Rev. D **65**, 115004 (2002) [arXiv:hep-ph/0201081].

[3] H. Baer and J. Ferrandis, Phys. Rev. Lett. **87**, 211803 (2001); D. Auto, H. Baer, C. Balazs, A. Belyaev, J. Ferrandis and X. Tata, JHEP 0306:023 (2003); K. Tobe and J. D. Wells, Nucl. Phys. B **663**, 123 (2003); R. Dermisek, S. Raby, L. Roszkowski and R. Ruiz de Austri, JHEP 0304:037 (2003); ibid., JHEP 0509:029 (2005); H. Baer, S. Kraml, S. Sekmen and H. Summy, JHEP 0803:056 (2008); ibid., JHEP 0810:079 (2008).

[4] J. A. Bagger, J. L. Feng, N. Polonsky and R. J. Zhang, Phys. Lett. B **473**, 264 (2000).

[5] M. Albrecht, W. Altmannshofer, A.J. Buras, D. Guadagnoli and D.M. Straub, JHEP 0710:055 (2007).

[6] LHCb-Talk-2012-040 by Chris Parkinson, Imperial College (2012); LHCb arXiv:1208.3355 [hep-ex].

[7] M. Carena, M. Quiros and C. Wagner, Nucl. Phys. B **461**, 407 (1996).

[8] R. Aaij et al. [LHCb Collaboration], arXiv:1203.4493 [hep-ex]; T. Aaltonen et al. [CDF Collaboration], Phys. Rev. Lett. **107**, 239903 (2011) [Phys. Rev. Lett. **107**, 191801 (2011)] [arXiv:1107.2304].

[9] R. Dermisek and S. Raby, Phys. Lett. B bf 622, 327 (2005); R. Dermisek, M. Harada and S. Raby, Phys. Rev. D **74**, 035011 (2006).

[10] SUSY_Flavor_v2, arXiv:submit/0440384 [hep-ph].

[11] H. Baer, S. Kraml and S. Sekman, [arXiv:0908.0134].

[12] M. Badziak, M. Olechowski and S. Pokorski, arXiv:1107.2764 [hep-ph].

[13] O. Lebedev, H.P. Nilles and M. Ratz, Phys. Lett. B **636**, 126 (2006); K. Choi and K.S. Jeong, JHEP 0701, 103 (2007); K. Choi and H.P. Nilles, arXiv:hep-ph/0702146.

PASCOS 2012 – 18th International Symposium on Particles Strings and Cosmology IOP Publishing
Journal of Physics: Conference Series **485** (2014) 012016 doi:10.1088/1742-6596/485/1/012016

Thermalisation of light sterile neutrinos

Thomas Tram

Department of Physics and Astronomy,
Aarhus University,
Ny Munkegade 120,
DK-8000 Aarhus C.

E-mail: `tram@phys.au.dk`

Abstract. We calculate the thermalisation process of light, sterile neutrinos in the early Universe for a range of mixing angles and mass-differences. We show that adding a large initial lepton asymmetry suppress thermalisation significantly. In particular, a sterile neutrino that explains the anomalies seen in ground based neutrino experiments can be made compatible with cosmology. We compute the degree of thermalisation in the (1 active + 1 sterile) scenario and give the effective number of thermalised species just before BBN begins. Resonances are taken properly taken into account, since we solve the full, momentum dependent equations, knowns as the Quantum Kinetic Equations. These proceedings are partly based on the paper [1].

1. Introduction

Recent cosmological data favour additional relativistic degrees of freedom in addition to photons and the three active neutrinos [2, 3, 4]. This extra radiation is often referred to as dark radiation, and one of the prime candidates is a light, sterile neutrino. Sterile neutrinos are spin-$\frac{1}{2}$ fermions that has no standard model interactions; they are not charged under the $SU(3) \times SU(2)_L \times U(1)$ gauge group of the standard model.

Low-mass sterile neutrinos may also explain the excess $\bar{\nu}_e$ events in the LSND experiment [5, 6, 7] as well as the MiniBooNE excess events in both neutrino and antineutrino channels. Interpreted in terms of flavour oscillations, the MiniBooNE data require CP violation and thus no less than two sterile families [8, 9, 10] or additional ingredients such as non-standard interactions [11]. Very recently, the MiniBooNE collaboration published new results [12] which no longer seems to require CP violation and is fully consistent with the LSND anomaly.

A new calculation of the expected flux of $\bar{\nu}_e$ from nuclear reactors reveals a deficit in the measured flux. This deficit can also be explained by a low mass sterile neutrino [13, 14, 15].

The cosmic radiation content is usually expressed in terms of the effective number of thermally excited neutrino species, N_{eff}. Its standard value, $N_{\text{eff}} = 3.046$, slightly exceeds 3 because of e^+e^- annihilation providing residual neutrino heating [16]. The Wilkinson Microwave Anisotropy Probe (WMAP) collaboration found $N_{\text{eff}} = 4.34^{+0.86}_{-0.88}$ based on their 7-year data release and additional LSS data [17] at 1σ. Including the Sloan Digital Sky Survey (SDSS) data release 7 (DR7) halo power spectrum, [3] found $N_{\text{eff}} = 4.78^{+1.86}_{-1.79}$ at 2σ. Measurements of the CMB anisotropy on smaller scales by the ACT [18] and SPT [19] collaborations also find tentative evidence for a value of N_{eff} higher than predicted by the standard model (see also [20, 21, 22, 23, 24] for recent discussions of N_{eff}).

Content from this work may be used under the terms of the Creative Commons Attribution 3.0 licence. Any further distribution of this work must maintain attribution to the author(s) and the title of the work, journal citation and DOI.
Published under licence by IOP Publishing Ltd

2. Quantum Kinetic Equations

If we wanted to calculate the freeze-out process of neutrinos and were not interested in neutrino oscillations, we should just solve the Boltzmann equation for the neutrino distribution function. However, since we are interested in both oscillations and scattering, we must solve a Boltzmann-like equation for the density matrices [25]. This system of equations is called the Quantum Kinetic Equations (QKE).

We restrict ourselves to a system of one active and one sterile neutrino. The density matrices are Hermitian, so they have 4 independent degrees of freedom. They are conveniently expressed in terms of Pauli-matrices $\boldsymbol{\sigma}$:

$$\rho = \frac{1}{2} f_0 (P_0 + \boldsymbol{P} \cdot \boldsymbol{\sigma}), \qquad \overline{\rho} = \frac{1}{2} f_0 (\overline{P}_0 + \overline{\boldsymbol{P}} \cdot \boldsymbol{\sigma}), \tag{1}$$

where $\boldsymbol{P}$ and $\overline{\boldsymbol{P}}$ are real vectors and $f_0 = (e^x + 1)^{-1}$ is the un-normalised Fermi-Dirac distribution with zero chemical potential for comoving momentum $x = p/T$. The evolution equations for P_0 and $\boldsymbol{P}$ are given by

$$\frac{\mathrm{d}P_0}{\mathrm{d}t} = \Gamma \left[\frac{f_{\mathrm{eq}}}{f_0} - \frac{1}{2} (P_0 + P_z) \right], \tag{2a}$$

$$\frac{\mathrm{d}P_x}{\mathrm{d}t} = -V_z P_y, \tag{2b}$$

$$\frac{\mathrm{d}P_y}{\mathrm{d}t} = V_z P_x - V_x P_z, \tag{2c}$$

$$\frac{\mathrm{d}P_z}{\mathrm{d}t} = V_x P_y, \tag{2d}$$

where the potentials V_x and V_z can be found in [26, 1].

3. Lepton asymmetry

The effective lepton number $L^{(a)}$ is in general a combination of the different lepton numbers L_f. The effective lepton number in our system evolves only due to L_{ν_a}, which changes because of active-sterile oscillations. We approximate the scattering kernel by an effective equilibration rate $\Gamma = C_a G_F^2 x T^5$, where $C_e \simeq 1.27$ and $C_{\mu,\tau} = 0.92$. Unfortunately this breaks lepton number conservation, so to circumvent this problem we derive an evolution equation for L which does not explicitly depend on Γ. From the definition of L_f we find:

$$L^{(a)} = \frac{2}{8\zeta(3)} \int_0^\infty \mathrm{d}x x^2 \left(\rho_{aa} - \bar{\rho}_{aa}\right) = \frac{1}{8\zeta(3)} \int_0^\infty \mathrm{d}x x^2 f_0 \left(P_0 - \bar{P}_0 + P_z - \bar{P}_z\right) \Rightarrow \tag{3}$$

$$\frac{\mathrm{d}L}{\mathrm{d}t} = \frac{1}{8\zeta(3)} \int_0^\infty \mathrm{d}x x^2 f_0 \left(\frac{\mathrm{d}P_z}{\mathrm{d}t} - \frac{\mathrm{d}\bar{P}_z}{\mathrm{d}t}\right) = \frac{1}{8\zeta(3)} \int_0^\infty \mathrm{d}x x^2 f_0 V_x \left(P_y - \bar{P}_y\right). \tag{4}$$

The damping term D is given by $D \simeq \Gamma$ to a good approximation. This is the term responsible for destroying coherence and it is easy to see that it appears as a damping factor in the equations for the off-diagonal elements parametrised by P_x and P_y. The equilibrium distributions f_{eq} and $\bar{f}_{\mathrm{eq}}$ are the Fermi-Dirac distributions with a finite chemical potential ξ calculated from the neutrino lepton number. ξ is given by the equation

$$L_{\mathrm{eq}}^{(a)} = \frac{1}{4\zeta(3)} \int_0^\infty \mathrm{d}x\, x^2 \left[\frac{1}{1+e^{x-\xi}} - \frac{1}{1+e^{x+\xi}}\right] = \frac{1}{12\zeta(3)} \left(\pi^2 \xi + \xi^3\right). \tag{5}$$

The real root of this equation can be expressed entirely using real functions. It is given by the Chebyshev cube root, and we find

$$\xi = \frac{-2\pi}{\sqrt{3}} \sinh\left(\frac{1}{3}\text{arcsinh}\left[-\frac{18\sqrt{3}\zeta(3)}{\pi^3}L^{(a)}\right]\right). \tag{6}$$

4. Momentum grid and resonances

This system exhibits resonances similar to the well-known MSW effect. The resonance condition is $V_z = 0$ for particles and $\bar{V}_z = 0$ for anti-particles. Knowing where the resonances reside in momentum space is important, both for numerical work and for understanding the evolution of the system qualitatively. See appendix A in [1] for a careful analysis of the position of resonances.

In order to numerically solve the QKEs, we define the momentum grid in comoving coordinates ($x = p/T$). Therefore the grid becomes stationary and the partial differential equations become ordinary differential equations coupled through integrated quantities only. Using the temperature T as the evolution parameter, time derivatives, d_t, are replaced by $\rightarrow -HT\partial_T$ in the above equations, provided that the time derivative of the effective number of degrees of freedom can be ignored.

5. Results: zero initial lepton asymmetry

In Fig. 1 we show the fraction of thermalised neutrinos, δN_{eff}, for a range of mixing parameters and an initial asymmetry of $L^{(\mu)} = 0$. The top panel shows the normal hierarchy, $\delta m_s^2 > 0$, and the bottom panel the inverted hierarchy, $\delta m_s^2 < 0$. The dashed rectangle shows the range of parameters plotted in Fig. 2.

We mark with a green hexagon the best fit point of the 3+1 global analysis presented in [27], obtained from a joint analysis of Solar, reactor, and short-baseline neutrino oscillation data $(\delta m_s^2, \sin^2 2\theta_s) = (0.9\ \text{eV}^2, 0.089)$. For that point $\delta N_{\text{eff}} = 1$ in both hierarchies, i.e. complete thermalization occurs. In addition we show the parameter range preferred by CMB and large scale structure (LSS) data. The $1-2-3\sigma$ contours have been obtained interpolating the likelihood function obtained in [28] for each fixed δm_s^2 and N_{eff}. In both cases the lower left corners of parameter space where little thermalization occurs are disfavoured because of the CMB+LSS preference for extra energy density.

In the inverted hierarchy the resonance conditions are always satisfied. Therefore, we expect full thermalization for a larger region of the mass-mixing parameters than in NH, as confirmed in Fig. 1. In this case, thermalisation may proceed through resonant conversions alone.

6. Results: large initial lepton asymmetry

We now give the results for initial large lepton asymmetry. Figure 2 shows the δN_{eff} contour plot for $L^{(\mu)} = 10^{-2}$ and $\delta m_s^2 > 0$ (top panel) and $\delta m_s^2 < 0$ (bottom panel). The region with full thermalisation is now much smaller than in Fig. 1. In principle, one would expect a lepton asymmetry of the same order of magnitude as the baryon asymmetry ($\eta \simeq 10^{-10}$). However, since neutrinos are neutral particles, $L^{(a)} = 10^{-2} - 10^{-1}$ is not presently excluded [29, 30, 31] by the requirement of charge neutrality. A large lepton number can be generated by e.g. an Affleck-Dine mechanism [32] or other models that are able to produce large lepton asymmetries and small baryonic ones [33, 34]. Another interesting possibility is to grow the lepton asymmetry from some initial $L^{(a)} \sim \mathcal{O}(10^{-10})$ using active-sterile oscillations [35, 36, 37]. Solving the QKE's in IH and with an initially small but non-zero lepton number, our preliminary results point toward a final lepton number varying between 10^{-5} and 10^{-2} depending on the mixing parameters. For illustrative purposes, we choose to adopt $L^{(a)} = 10^{-2}$.

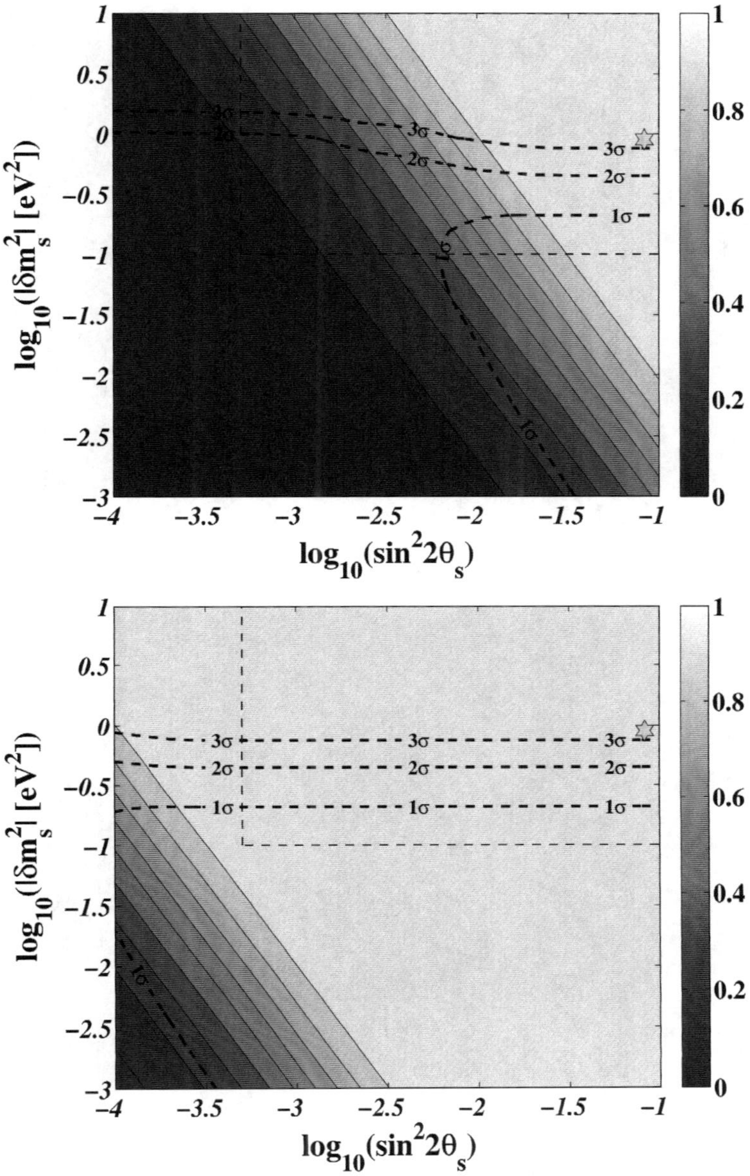

Figure 1. Iso-$\delta N_{\rm eff}$ contours in the $\sin^2 2\theta_s - \delta m_s^2$ plane for $L^{(\mu)} = 0$ and $\delta m_s^2 > 0$ (top panel) and $\delta m_s^2 < 0$ (bottom panel). The green hexagon denotes the ν_s best-fit mixing parameters as in the $3+1$ global fit in [27]: $(\delta m_s^2, \sin^2 2\theta_s) = (0.9\ {\rm eV}^2, 0.089)$. The $1-2-3\sigma$ contours denote the CMB+LSS allowed regions for ν_s with sub-eV mass as in [28].

7. Conclusions

Recent cosmological data seem to favor an excess of radiation beyond three neutrino families and photons, and light sterile neutrinos are possible candidates. The upcoming measurement of $\delta N_{\rm eff}$ by Planck will confirm or rule out the existence of such extra radiation with high precision [38, 39].

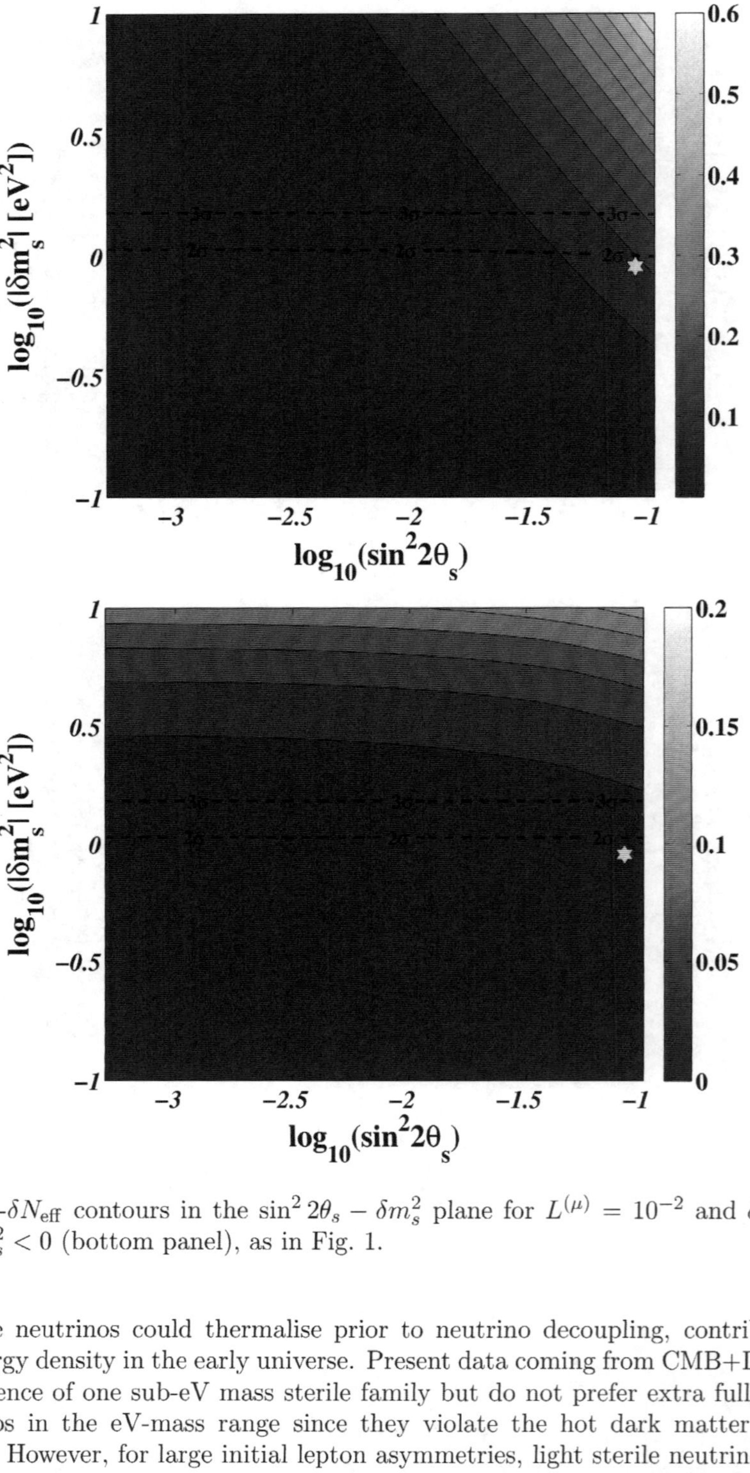

Figure 2. Iso-$\delta N_{\rm eff}$ contours in the $\sin^2 2\theta_s - \delta m_s^2$ plane for $L^{(\mu)} = 10^{-2}$ and $\delta m_s^2 > 0$ (top panel) and $\delta m_s^2 < 0$ (bottom panel), as in Fig. 1.

Light sterile neutrinos could thermalise prior to neutrino decoupling, contributing to the relativistic energy density in the early universe. Present data coming from CMB+LSS, and BBN allow the existence of one sub-eV mass sterile family but do not prefer extra fully thermalised sterile neutrinos in the eV-mass range since they violate the hot dark matter limit on the neutrino mass. However, for large initial lepton asymmetries, light sterile neutrinos are not (or only partially) thermalised for almost all the scanned parameter space. This provides a loophole for eV sterile neutrinos to be compatible with CMB+LSS constraints. For lepton asymmetries around 10^{-2} almost no thermalisation occurs for the parameters preferred by Solar, reactor and short-baseline data, and the sterile neutrinos would contribute very little to the current dark matter density.

References

[1] Hannestad S, Tamborra I and Tram T 2012 *JCAP* **1207** 025 (*Preprint* 1204.5861)
[2] Hamann J, Hannestad S, Raffelt G G and Wong Y Y Y 2007 *JCAP* **0708** 021 (*Preprint* 0705.0440)
[3] Hamann J, Hannestad S, Lesgourgues J, Rampf C and Wong Y Y Y 2010 *JCAP* **1007** 022 (*Preprint* 1003.3999)
[4] Gonzalez-Garcia M C, Maltoni M and Salvado J 2010 *JHEP* **08** 117 (*Preprint* 1006.3795)
[5] Aguilar-Arevalo A *et al.* (LSND) 2001 *Phys. Rev.* **D64** 112007 (*Preprint* hep-ex/0104049)
[6] Strumia A 2002 *Phys. Lett.* **B539** 91–101 (*Preprint* hep-ph/0201134)
[7] Gonzalez-Garcia M C and Maltoni M 2008 *Phys. Rept.* **460** 1–129 (*Preprint* 0704.1800)
[8] Aguilar-Arevalo A A *et al.* (MiniBooNE) 2009 *Phys. Rev. Lett.* **102** 101802 (*Preprint* 0812.2243)
[9] Aguilar-Arevalo A A *et al.* (MiniBooNE) 2009 *Phys. Rev. Lett.* **103** 111801 (*Preprint* 0904.1958)
[10] Karagiorgi G, Djurcic Z, Conrad J M, Shaevitz M H and Sorel M 2009 *Phys. Rev.* **D80** 073001 (*Preprint* 0906.1997)
[11] Akhmedov E and Schwetz T 2010 *JHEP* **10** 115 (*Preprint* 1007.4171)
[12] Aguilar-Arevalo A *et al.* (MiniBooNE Collaboration) 2012 (*Preprint* 1207.4809)
[13] Mention G *et al.* 2011 *Phys. Rev.* **D83** 073006 (*Preprint* 1101.2755)
[14] Huber P 2011 *Phys. Rev.* **C84** 024617 (*Preprint* 1106.0687)
[15] Kopp J, Maltoni M and Schwetz T 2011 *Phys. Rev. Lett.* **107** 091801 (*Preprint* 1103.4570)
[16] Mangano G *et al.* 2005 *Nucl. Phys.* **B729** 221–234 (*Preprint* hep-ph/0506164)
[17] Komatsu E *et al.* (WMAP) 2011 *Astrophys. J. Suppl.* **192** 18 (*Preprint* 1001.4538)
[18] Dunkley J, Hlozek R, Sievers J, Acquaviva V, Ade P *et al.* 2011 *Astrophys.J.* **739** 52 (*Preprint* 1009.0866)
[19] Keisler R, Reichardt C, Aird K, Benson B, Bleem L *et al.* 2011 *Astrophys.J.* **743** 28 (*Preprint* 1105.3182)
[20] Hamann J 2012 *JCAP* **1203** 021 (*Preprint* 1110.4271)
[21] Joudaki S 2012 (*Preprint* 1202.0005)
[22] Giusarma E, Archidiacono M, de Putter R, Melchiorri A and Mena O 2012 *Phys.Rev.* **D85** 083522 (*Preprint* 1112.4661)
[23] Nollett K M and Holder G P 2011 (*Preprint* 1112.2683)
[24] Gonzalez-Morales A X, Poltis R, Sherwin B D and Verde L 2011 (*Preprint* 1106.5052)
[25] Sigl G and Raffelt G G 1993 *Nucl. Phys.* **B406** 423–451
[26] Kainulainen K and Sorri A 2002 *JHEP* **02** 020 (*Preprint* hep-ph/0112158)
[27] Giunti C and Laveder M 2011 *Phys.Lett.* **B706** 200–207 (*Preprint* 1111.1069)
[28] Hamann J, Hannestad S, Raffelt G G, Tamborra I and Wong Y Y Y 2010 *Phys. Rev. Lett.* **105** 181301 (*Preprint* 1006.5276)
[29] Dolgov A, Hansen S, Pastor S, Petcov S, Raffelt G *et al.* 2002 *Nucl.Phys.* **B632** 363–382 (*Preprint* hep-ph/0201287)
[30] Pastor S, Pinto T and Raffelt G G 2009 *Phys.Rev.Lett.* **102** 241302 (*Preprint* 0808.3137)
[31] Castorina E, Franca U, Lattanzi M, Lesgourgues J, Mangano G *et al.* 2012 (*Preprint* 1204.2510)
[32] Kawasaki M, Takahashi F and Yamaguchi M 2002 *Phys.Rev.* **D66** 043516 (*Preprint* hep-ph/0205101)
[33] Harvey J A and Kolb E W 1981 *Phys.Rev.* **D24** 2090
[34] Dolgov A 2002 *Phys.Rept.* **370** 333–535 (*Preprint* hep-ph/0202122)
[35] Foot R, Thomson M J and Volkas R 1996 *Phys.Rev.* **D53** 5349–5353 (*Preprint* hep-ph/9509327)
[36] Barbieri R and Dolgov A 1990 *Phys.Lett.* **B237** 440
[37] Barbieri R and Dolgov A 1991 *Nucl.Phys.* **B349** 743–753
[38] Perotto L, Lesgourgues J, Hannestad S, Tu H and Wong Y Y Y 2006 *JCAP* **0610** 013 (*Preprint* astro-ph/0606227)
[39] Hamann J, Lesgourgues J and Mangano G 2008 *JCAP* **0803** 004 (*Preprint* 0712.2826)

Phase space analysis of the $F(X) - V(\phi)$ scalar field Lagrangian and scaling solutions in flat cosmology

Josue De-Santiago[1,2,3] **and Jorge L Cervantes-Cota**[2]
[1]Universidad Nacional Autónoma de México, 04510, D. F., México
[2]Depto. de Física, Instituto Nacional de Investigaciones Nucleares, México
[3]Institute of Cosmology & Gravitation, University of Portsmouth, PO1 3FX, United Kingdom
E-mail: josue@ciencias.unam.mx, jorge.cervantes@inin.gob.mx

Abstract. We review a system of autonomous differential equations developed in our previous work [1] describing a flat cosmology filled with a barotropic fluid and a scalar field with a modified kinetic term of the form $\mathcal{L} = F(X) - V(\phi)$. We analyze the critical points and summarize the conditions to obtain scaling solutions. We consider a set of transformations and show that they leave invariant the equations of motion for the systems in which the scaling solution is obtained, allowing to reduce the number of degrees of freedom.

1. Introduction

In cosmology, the scalar fields play an important role due to its broad phenomenology, which can be used in order to describe different phenomena. For that reason they have been used to model phenomena like inflation [2], dark energy [3, 4], dark matter [5], bounce scenarios [6], and unification models [7].

The canonical Lagrangian for a scalar field is given by $\mathcal{L} = X - V(\phi)$, where the first term $X = -\frac{1}{2}\partial_\mu\phi\partial^\mu\phi$ is the kinetic term, and the second is the potential that can have different functional forms depending on the model. In the last years, a generalization has been a studied in which the Lagrangian is a general function of X and ϕ. These Lagrangians were used initially in the study of inflation [8] and dark energy [9], and in a previous work we used this type of Lagrangian to unify inflation, dark matter, and dark energy [10].

One particular form of this kind of Lagrangian is the sum-separable, in which

$$\mathcal{L} = F(X) - V(\phi), \tag{1}$$

and there is a clear separation between the kinetic term $F(X)$ and the potential term $V(\phi)$. This class of Lagrangians has the advantage of being more easily studied than the more general case while still conserving some of the rich phenomenology that is not present in the canonical case. In a previous work we studied the dynamical system of this class of Lagrangians in a flat Friedmann-Lemaitre-Robertson-Walker (FLRW) cosmology filled with the field and an additional barotropic fluid [1]. This analysis allowed us to find the critical points of the system and its stability, from which we drew conclusions about the general behaviour and evolution of the system.

In the present work we will review this analysis and we will add an extra analysis for the case in which the system has non-trivial critical points. We will show that in that case there is a

Content from this work may be used under the terms of the Creative Commons Attribution 3.0 licence. Any further distribution of this work must maintain attribution to the author(s) and the title of the work, journal citation and DOI.
Published under licence by IOP Publishing Ltd

symmetry that allows the degrees of freedom of the system to reduce. This behaviour is similar to what happens in the canonical Lagrangian with exponential potential. This symmetry allows the existence of scaling solutions even when the Lagrangian is not of the form $\mathcal{L} = Xg(Xe^{\lambda\phi})$, which in the literature [11] is considered to be the most general form of Lagrangians with this behaviour.

2. Autonomous system

Along this work we will consider a flat FLRW cosmology filled with a scalar field with Lagrangian of the form (1) and a barotropic fluid with equation of state $P_m = (\gamma_m - 1)\rho_m$ where γ_m is constant. In this system, if no interaction between the field and the fluid is considered, the equations of motion correspond to the two independent Einstein equations, also called Friedmann equations and the two continuity equations as follows:

$$H^2 = \frac{1}{3M_{\rm Pl}^2}\left[2XF_X - F + V + \rho_m\right], \tag{2}$$

$$\frac{dH}{dt} = -\frac{1}{2M_{\rm Pl}^2}\left[2XF_X + \gamma_m\rho_m\right], \tag{3}$$

$$\frac{d\rho_m}{dt} = -3H\gamma_m\rho_m, \tag{4}$$

$$\frac{d}{dt}(2XF_X - F + V) = -6HXF_X, \tag{5}$$

where the subscript F_X means differentiation of F with respect to X. Here only three of the equations are independent from each other, as the system has only three degrees of freedom. We can choose an initial ϕ, $\dot{\phi}$ and ρ_m, which in turn will define the initial Hubble parameter with the equation (2) and the initial kinetic term X with the relation

$$X = \frac{1}{2}\dot{\phi}^2. \tag{6}$$

In order to obtain an autonomous system of differential equations we define as independent variable the dimensionless $dN = d\log a$, and then define the new dynamical variables

$$x = \sqrt{\frac{\rho_k}{\rho_c}}, \tag{7}$$

$$y = \sqrt{\frac{V}{\rho_c}}, \tag{8}$$

$$\sigma = -\frac{M_{\rm Pl}}{\sqrt{3\rho_k}}\frac{d\log V}{dt}, \tag{9}$$

where $\rho_k = 2XF_X - F$ is the density associated to the kinetic part of the Lagrangian, and $\rho_c = 3M_{\rm Pl}^2H^2$ is the critical density for a FLRW cosmology. The evolution equation (2) is written in terms of the new defined dynamical variables as $x^2+y^2+\Omega_m = 1$, where $\Omega_m = \rho_m/\rho_c$ is positive, which implies that both variables x and y are bounded between 0 and 1.

Using the previous relation and the evolution equations (2-5), we find the autonomous differential equations in terms of the dynamical variables as

$$\frac{dx}{dN} = \frac{3}{2}\sigma y^2 + \frac{3}{2}x\left[\gamma_k(x^2-1) + \gamma_m(1-x^2-y^2)\right], \tag{10}$$

$$\frac{dy}{dN} = \frac{3}{2}y\left[-\sigma x + \gamma_m(1-y^2) + x^2(\gamma_k - \gamma_m)\right], \tag{11}$$

$$\frac{d\sigma}{dN} = -3\sigma^2 x(\Gamma - 1) + \frac{3\sigma(2\Xi\gamma_k + \gamma_k - 2)}{2\gamma_k(2\Xi+1)}\left(\gamma_k - \frac{\sigma y^2}{x}\right), \tag{12}$$

where we required the introduction of the auxiliary variables

$$\gamma_k = \frac{\rho_k + P_k}{\rho_k} = \frac{2XF_X}{2XF_X - F}, \tag{13}$$

$$\Xi = \frac{XF_{XX}}{F_X}, \tag{14}$$

$$\Gamma = \frac{VV_{\phi\phi}}{V_\phi^2}. \tag{15}$$

These auxiliary variables depend on derivatives of the potential and kinetic terms in the Lagrangian and in general are not constant, which implies that in order to solve the system of equations (10-12) it is necessary to consider evolution equations for the extra three variables. These equations in general will depend on higher derivatives of the Lagrangian terms which in turn will have evolution equations.

In order to cut the succession of equations it is possible to demand the Lagrangian to satisfy that both γ_k and Γ be constant, which implies that

$$F(X) = AX^\eta, \tag{16}$$

and

$$V(\phi) = B(\phi - \phi_0)^n, \text{ or} \tag{17}$$

$$V(\phi) = V_0 e^{-\lambda\phi}, \tag{18}$$

where A, B, V_0, η, n, and λ are constants. In this case $\gamma_k = 2\eta/(2\eta - 1)$, $\Xi = \eta - 1$, and Γ is one if the potential is exponential or $(n-1)/n$ if it is a power-law. In this particular Lagrangian the system of equations (10-12) is closed and we can study its solutions without concern about other equations, as we will do in the next section.

3. Critical points

The critical points of the system are obtained when the equations (10-12) are equal to zero. Besides analyzing the conditions for the existence of the critical points we determine their stability. Equating only the equations (10) and (11) to zero, we obtain the critical points (a), (b), (α), (β) and (γ) summarized in table 1, where the first two are present in the canonical case and have been analized in previous works like [12, 13] and the last three are new. Another two critical points are found for a particular case of the Lagrangian, when it can be written either as a canonical Lagrangian with exponential potential (18) or as the sum of two power-law terms (16, 17), where their exponents satisfy

$$\eta = \frac{n}{2+n}. \tag{19}$$

These critical points are named (c) and (d), and written in table 1 too. As the values for x and y in these points are dependent on σ, for them to be fixed it is also necessary for the evolution equation (12) to be zero, in [1] we showed that this is the case when (19) is satisfied or the Lagrangian is a canonical with exponential potential. In the next section we will add more elements to this analysis and we will show that in fact σ is a function of x and y for these Lagrangians.

Table 1. Stability and existence of the critical points. The points labeled with Latin letters can be reduced in the canonical case to the ones already studied in the literature, the points with Greek letters are new. The points (c) and (d) exist only when (19) holds or when the Lagrangian is canonical with exponential potential.

	x	y	Existence	Stability	Ω_ϕ	γ_ϕ
(a)	0	0	Always	Unstable node for $\gamma_k < \gamma_m$ Saddle point for $\gamma_m < \gamma_k$	0	-
(α)	0	1	$\sigma = 0$	Saddle point for $\gamma_k < 0$ Stable node for $\gamma_k > 0$	1	0
(b)	1	0	Always	Unstable node for $\gamma_k > \{\gamma_m, \sigma\}$ Stable node for $\gamma_k < \{\gamma_m, \sigma\}$ Otherwise saddle point	1	γ_k
(β)	0	Arbitrary	$\sigma = 0$ $\gamma_m = 0$	Stable line for $\omega_k > -1$ Unstable otherwise	y^2	0
(γ)	Arbitrary	0	$\gamma_k = \gamma_m$	Stable line for $x\sigma > \gamma_k$ Unstable otherwise	x^2	ω_m
(c)	$\dfrac{\sigma}{\gamma_k}$	$\sqrt{1-\dfrac{\sigma^2}{\gamma_k^2}}$	$\sigma^2 > \gamma_k^2$ $\sigma\gamma_k > 0$	Saddle point for $\sigma^2 > \gamma_k\gamma_m$ Otherwise stable node	1	$\dfrac{\sigma^2}{\gamma_k}$
(d)	$\dfrac{\gamma_m}{\sigma}$	$\dfrac{\sqrt{\gamma_m(\gamma_k-\gamma_m)}}{\sigma}$	$\sigma > \sqrt{\gamma_k\gamma_m}$ $\gamma_m < \gamma_k$	Stable node for $\dfrac{\sigma^2(9\gamma_m-\gamma_k)}{8\gamma_k\gamma_m^2} < 1$ Stable spiral otherwise	$\dfrac{\gamma_k\gamma_m}{\sigma^2}$	γ_m

4. Symmetry

As we showed in the previous section, when the Lagrangian of the scalar field is the sum of two power-law terms and the relation (19) holds, two extra critical points are present in the system. In this section we will see that for this system exists a symmetry in the equations of motion (2-5). This symmetry permits the number of degrees of freedom to be reduced to only two.

The equations of motion (2-5) written for a Lagrangian with power-law terms (16, 17) and with dN as independent variable instead of dt are given by

$$H^2 = \frac{1}{3M_{\rm Pl}^2}\left[(2\eta-1)AX^\eta + B\phi^n + \rho_m\right], \tag{20}$$

$$H\frac{dH}{dN} = -\frac{1}{2M_{\rm Pl}^2}\left[2\eta AX^\eta + \gamma_m\rho_m\right], \tag{21}$$

$$\frac{d\rho_m}{dN} = -3\gamma_m\rho_m\,, \tag{22}$$

$$\frac{d}{dN}\left((2\eta-1)AX^\eta + B\phi^n\right) = -6\eta AX^\eta\,. \tag{23}$$

If we consider ϕ, X, and ρ_m as the independent variables we can apply the transformation

$$\begin{aligned} \phi &\rightarrow \xi^{2\eta}\phi\,, \\ X &\rightarrow \xi^{2n}X\,, \\ \rho_m &\rightarrow \xi^{2n\eta}\rho_m\,, \end{aligned} \tag{24}$$

where if we leave N invariant, from the definition of X

$$X = \frac{1}{2}\left(H\frac{d\phi}{dN}\right)^2, \tag{25}$$

the Hubble parameter transforms as $H \to \xi^{n-2\eta} H$. This transformation will leave invariant the equations of motion (20-23) as long as n and η satisfy the relation (19), which determines the transformation of the Hubble parameter to be $H \to \xi^{n\eta} H$.

The fact that the equations of motion are invariant under the transformations definded in (24) for arbitrary ξ means that this set of transformations is a symmetry of the system, and allows to reduce the number of degrees of freedom to two. The only thing needed is to define a set of variables that are invariant under the transformation, but x, y and σ are already invariant. Moreover, there are only two invariant degrees of freedom, which implies that one of them is dependent of the other. In fact we can see that the relation

$$\sigma = -M_{\rm Pl} n B^{1/n} \sqrt{\frac{2}{3}} (A(2\eta - 1))^{-1/2\eta} \left(\frac{x}{y}\right)^{2/n}, \tag{26}$$

holds when (19) is satisfied. If we define

$$s \equiv -M_{\rm Pl} n B^{1/n} \sqrt{\frac{2}{3}} (A(2\eta - 1))^{-1/2\eta}, \tag{27}$$

the dynamical system can be written from (10, 11) as

$$\frac{dx}{dN} = \frac{3}{2} x \left[s y \left(\frac{y}{x}\right)^{2/\gamma_k} + \gamma_k (x^2 - 1) + \gamma_m (1 - x^2 - y^2) \right], \tag{28}$$

$$\frac{dy}{dN} = -\frac{3}{2} s x^2 \left(\frac{y}{x}\right)^{2/\gamma_k} + \frac{3}{2} y \left[\gamma_m (1 - y^2) + x^2 (\gamma_k - \gamma_m) \right], \tag{29}$$

corresponding to only two equations for two variables. Here we used the fact that $\gamma_k = 2n/(n-2)$ when the relation (19) is satisfied.

4.1. Exponential potential

The behaviour stated in the this section is similar to what happens for the canonical Lagrangian with exponential potential (18), and studied in Ref. [14]. In that case a similar procedure to the one used in this section can tell us that the equations of motion are invariant under the transformations

$$\phi \to \phi + \xi, \qquad X \to e^{-\lambda \xi} X, \qquad \rho_m \to e^{-\lambda \xi} \rho_m. \tag{30}$$

The fact that the system is invariant under the transformations above, allows it to be described by only two variables as in the original work of this type of cosmological analysis [12]. In that case, unlike the relation (26), σ is a constant given by $\sqrt{2/3} M_{\rm Pl} \lambda$.

5. Phase space

The critical points found in section 3 are still valid for the 2-dimensional system (28, 29), as can be seen by equating the right-hand side of both equations to zero. The critical points (c) and (d) depend on σ that is function of x and y so we should be able to drop de dependence on σ using the relation (26). For (c) it cannot be made analytically while for (d) we obtain:

$$x_d = \frac{\sqrt{\gamma_m(\gamma_k - \gamma_m)}}{s} \left(\frac{\gamma_m}{\gamma_k - \gamma_m}\right)^{1/\gamma_k}, \qquad y_d = \frac{\gamma_k - \gamma_m}{s} \left(\frac{\gamma_m}{\gamma_k - \gamma_m}\right)^{1/\gamma_k}. \tag{31}$$

The behaviour of the critical points stated in the table 1 can be seen more graphically in the phase space plot in figure 1. There we see the phase space of a system with Lagrangian

$$\mathcal{L} = A X^2 + B \phi^{-4}, \tag{32}$$

plus a dust type barotropic fluid with $\gamma_m = 1$, where we chose $s = 4$ meaning that $AB = 4M_{\rm Pl}^4/27$. In the figure we can observe the critical points:

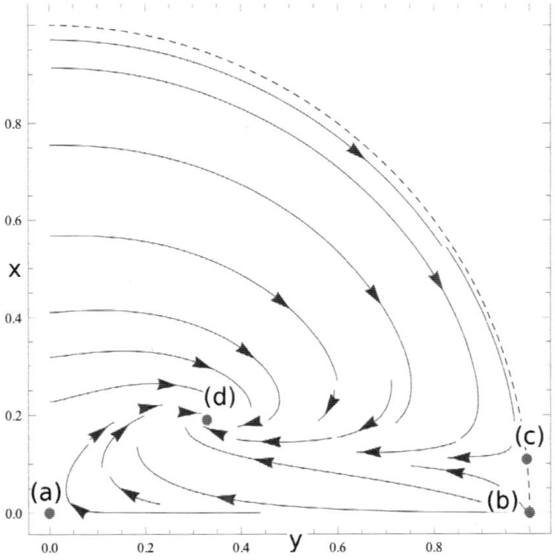

Figure 1. Phase space for the dynamical system corresponding to $\mathcal{L} = AX^2 + B\phi^{-4}$. The critical points (a), (b), (c) and (d) can be observed.

- (a) The (0,0) point that in this case is a saddle point.
- (b) The (1,0) point that in this case is an unstable node.
- (c) Corresponding to a saddle point, with domination of field density over the barotropic fluid.
- (d) The point $(3^{1/4}/4, 3^{-1/4}/4)$ corresponding to a stable spiral, and whose value can be computed from expression (31) for the values of this particular example.

The other critical points are not present in this example.

6. Conclusions

In this work we made an analysis of the dynamical system associated to a flat FLRW cosmology containing a barotropic fluid and a scalar field with Lagrangian of the form $\mathcal{L} = F(X) - V(\phi)$. We observed that this system can be described by three variables and its equations of motion, from those we defined x, y and σ and obtained their evolution equations. For Lagrangians with constant values for γ_k and Γ, the three autonomous differential equations (10 - 12) are enough to describe the system. For more general Lagrangians however, it is necessary to obtain dynamical equations for the variables γ_k, Γ, and Ξ, that in turn will include new variables in terms of higher derivatives of the Lagrangian, that will require extra equations of motion. In order to have only the minimal three equation system, in our work [1] we used power-law kinetic terms that produce γ_k constant, and power-law or exponential potentials that correspond to a constant value of Γ.

We obtained the critical points of the system, and studied their existence and stability conditions, which are summarized in table 1. The critical points (a), (b), (c), (d) can be reduced for the canonical Lagrangian to those studied in previous works [13, 15], meanwhile (α), (β) and (γ) are new.

In our previous work [1] we showed that for the existence of the critical points (c) and (d), the Lagrangian needs to be either canonical with exponential potential or the sum of two power-law terms that satisfy the relation (19). In the present work we showed that when this relation holds, the equations of motion (20 - 23) are invariant under the set of transformations defined by (24). This symmetry allows to reduce the number of degrees of freedom of the system to two. We showed that x and y are invariant under the same transformation (24) and hence are suitable to describe the system. Therefore, it was possible to obtain σ as function of x and y in equation (26). Using this equation we found the two dimensional autonomous system (28, 29).

We showed that for the two dimensional system the critical points are the same as the ones obtained in table 1, but with σ as a function of x and y. The critical point (c) can be obtained then as a function of the auxiliary variables γ_k and γ_m only, eliminating the dependence on σ, however it is not possible to do it analytically. For the critical point (d) the analytical expression can be obtained as we have done in (31).

The symmetry transformations of (24) are similar to those of (30), corresponding to canonical exponential Lagrangians and studied in [14]. Both transformations allow the systems to be treated with a two dimensional set of dynamical equations. For the canonical exponential Lagrangian the equations get reduced to (10, 11) with a constant value of σ, while for the Lagrangian whose terms satisfy (19), they get reduced to equations (28, 29).

Acknowledgments
JDS is supported by ININ Grant.

References

[1] De-Santiago J, Cervantes-Cota J L and Wands D 2013 *Phys. Rev. D* **87**(2) 023502
[2] Liddle A R and Lyth D 2000 *Cambridge University Press*
[3] Copeland E J, Sami M and Tsujikawa S 2006 *Int.J.Mod.Phys.* **D15** 1753–1936
[4] de la Macorra A and Piccinelli G 2000 *Phys.Rev.* **D61** 123503
[5] Magana J and Matos T 2012 *J.Phys.Conf.Ser.* **378** 012012
[6] Peter P and Pinto-Neto N 2008 *Phys.Rev.* **D78** 063506
[7] Bertacca D, Bartolo N and Matarrese S 2010 *Adv.Astron.* **2010** 904379
[8] Armendariz-Picon C, Damour T and Mukhanov V F 1999 *Phys. Lett.* **B458** 209–218
[9] Chiba T, Okabe T and Yamaguchi M 2000 *Phys. Rev.* **D62** 023511
[10] De-Santiago J and Cervantes-Cota J L 2011 *Phys.Rev.* **D83** 063502
[11] Piazza F and Tsujikawa S 2004 *JCAP* **0407** 004
[12] Copeland E J, Liddle A R and Wands D 1998 *Phys.Rev.* **D57** 4686–4690
[13] Gumjudpai B, Naskar T, Sami M and Tsujikawa S 2005 *JCAP* **0506** 007
[14] Holden D J and Wands D 2000 *Phys.Rev.* **D61** 043506
[15] Wands D 2009 *Adv. Sci. Lett.* **2** 194–2004

Density perturbations from curvatons revisited

Takeshi Kobayashi[1,2]

[1] Canadian Institute for Theoretical Astrophysics, University of Toronto, 60 St. George Street, Toronto, Ontario M5S 3H8, Canada

[2] Perimeter Institute for Theoretical Physics, 31 Caroline Street North, Waterloo, Ontario N2L 2Y5, Canada

E-mail: `takeshi@cita.utoronto.ca`

Abstract. We investigate density perturbations sourced by a curvaton with a generic energy potential. The key feature of a curvaton potential which deviates from a quadratic is that the curvaton experiences a non-uniform onset of its oscillation. This sources additional contributions to the resulting density perturbations, and we especially find that curvaton potentials that are flatter compared to a quadratic lead to enhancement of the linear and second-order density perturbations, while steepened potentials can generate strongly scale-dependent non-Gaussianity. As such examples, we study pseudo-Nambu-Goldstone curvatons and self-interacting curvatons. Our analyses are analytic, and thus provide a systematic framework for studying curvatons in general. The discussion in this paper are based on [1, 2].

1. Introduction

The curvaton mechanism [3, 4, 5] is an attractive mechanism for creating the primordial density perturbations of our universe. In this scenario, the curvaton field possessing large-scale field fluctuations generates the density perturbations while it oscillates about its potential minimum and comes close to dominating the universe. The mechanism also has the merit that when embedded in inflationary cosmology, it frees the inflaton from being responsible for generating the perturbations and therefore drastically relaxes constraints on inflationary model building.

In this paper, we explore density perturbations sourced by a curvaton rolling along an arbitrary energy potential. When the potential deviates from a quadratic one, the curvaton field with large-scale field fluctuations starts its oscillation at different times at different patches of the universe. Such behaviour contributes to the curvaton energy density perturbations in addition to that sourced directly from the original field fluctuations. We find that such non-trivial conversion processes of the field fluctuations into the density perturbations can lead to strong enhancement/suppression of the density perturbations of the universe, as well as for their non-Gaussianities.

It is worthwhile to investigate curvaton potentials which deviate from simple quadratic ones, both from observational and theoretical reasons: Considering a curvaton whose field fluctuations were generated during the inflationary era, in order for the curvaton to produce the red-tilted density perturbation spectrum as suggested by latest CMB observations [6] without relying on specific inflation mechanisms (such as large-field models), the curvaton needs to be located along a potential with negative curvature during inflation. This obviously suggests the curvaton potential to take non-trivial forms. Furthermore, explicit curvaton models constructed in the framework of microscopic physics can naturally possess intricate energy potentials.

Content from this work may be used under the terms of the Creative Commons Attribution 3.0 licence. Any further distribution of this work must maintain attribution to the author(s) and the title of the work, journal citation and DOI.
Published under licence by IOP Publishing Ltd

We first derive analytic expressions for the density perturbations sourced by curvatons with generic energy potentials. Then we look into pseudo-Nambu-Goldstone curvatons and self-interacting curvatons, which provide, respectively, typical examples of curvaton potentials that flatten/steepen compared to a quadratic one. We find that flat potentials can lead to enhancement of the linear and second-order density perturbations, while steepened potentials can generate strongly scale-dependent non-Gaussianity.

The results presented in the following are based on the works [1, 2], where one can find detailed discussions. (For related discussions, see also [7, 8, 9, 10, 11, 12] and references therein.)

2. Density Perturbations from Curvatons

A light curvaton acquires nearly scale-invariant field fluctuations during inflation, that are converted into the cosmological density perturbations as the curvaton oscillates and decays in the post-inflationary era. Let us first lay out analytic expressions for the density perturbations generated by a curvaton σ with a generic energy potential $V(\sigma)$. The potential is assumed to have no explicit dependence on time, and also that it is well approximated by a quadratic one around its minimum (which we set as $\sigma = 0$) so that the curvaton oscillations are sinusoidal.[1] Hence the curvaton energy density is considered to redshift similarly to nonrelativistic matter after the onset of the oscillations until when the curvaton decays into radiation, whereas we suppose the inflaton to behave as matter from the end of inflation until reheating when the inflaton decays into radiation. The energy density of the curvaton before the beginning of its oscillation is assumed to be negligibly tiny compared to the total energy of the universe, having little effect on the expansion history. Furthermore, we neglect density perturbations sourced from the inflaton.

2.1. Linear and Second Order Perturbations

Supposing the curvaton field fluctuations to be nearly Gaussian and to satisfy $\mathcal{P}_{\delta\sigma}(k) = (H|_{k=aH}/2\pi)^2$ at the time when the scale k of interest exits the horizon, then using the $\delta\mathcal{N}$-formalism, the power spectrum of the linear order density perturbations can be expressed in terms of the curvaton potential as

$$\mathcal{P}_\zeta(k_*) = \left(\frac{\partial\mathcal{N}}{\partial\sigma_*}\frac{H_*}{2\pi}\right)^2, \tag{1}$$

with

$$\frac{\partial\mathcal{N}}{\partial\sigma_*} = \frac{\hat{r}}{4+3\hat{r}}\left(1 - X(\sigma_{\rm osc})\right)^{-1}\left\{\frac{V'(\sigma_{\rm osc})}{V(\sigma_{\rm osc})} - \frac{3X(\sigma_{\rm osc})}{\sigma_{\rm osc}}\right\}\frac{V'(\sigma_{\rm osc})}{V'(\sigma_*)}. \tag{2}$$

Here, the subscript $*$ denotes values when the CMB scale k_* exits the horizon, "osc" represents values at the onset of the curvaton oscillation, and a prime denotes a derivative with respect to σ. An overdot will be used below to denote a time derivative. $\hat{r}$ is the energy density ratio between the curvaton and radiation (which originates from the inflaton) upon curvaton decay

$$\begin{aligned}\hat{r} &\equiv \left.\frac{\rho_\sigma}{\rho_r}\right|_{\rm dec} \\ &= \text{Max.}\left[\frac{V(\sigma_{\rm osc})}{3M_p^2 H_{\rm osc}^{3/2}\Gamma_\sigma^{1/2}} \times \text{Min.}\left(1, \frac{\Gamma_\phi^{1/2}}{H_{\rm osc}^{1/2}}\right), \left\{\frac{V(\sigma_{\rm osc})}{3M_p^2 H_{\rm osc}^{3/2}\Gamma_\sigma^{1/2}} \times \text{Min.}\left(1, \frac{\Gamma_\phi^{1/2}}{H_{\rm osc}^{1/2}}\right)\right\}^{4/3}\right],\end{aligned} \tag{3}$$

where the first and second terms in the Max. parentheses correspond to the curvaton being subdominant and dominant at its decay, respectively, while the Min. parentheses are due to

[1] The formulae can be generalized to cases with non-sinusoidal oscillations as well, see Appendix B of [1].

whether the onset of oscillation is after or before reheating. Γ_ϕ and Γ_σ are constants denoting respectively the decay rates of the inflaton and the curvaton. We adopt the sudden decay approximation where the scalar fields suddenly decay into radiation when $H = \Gamma$.

The function X in (2) denotes effects due to the non-uniform onset of the curvaton oscillations (which are absent for a purely quadratic curvaton potential), defined as follows:

$$X(\sigma_{\rm osc}) \equiv \frac{1}{2(c-3)} \left(\frac{\sigma_{\rm osc} V''(\sigma_{\rm osc})}{V'(\sigma_{\rm osc})} - 1 \right). \tag{4}$$

Here, c is a constant whose value is set by whether reheating (= inflaton decay, at $t_{\rm reh}$) is earlier/later than the onset of the curvaton oscillation:

$$c = \begin{cases} 5 & (t_{\rm reh} < t_{\rm osc}) \\ 9/2 & (t_{\rm reh} > t_{\rm osc}). \end{cases} \tag{5}$$

We can define the onset of the oscillation as when the time scale of the curvaton rolling becomes comparable to the Hubble time, i.e. $|\dot{\sigma}/H\sigma| = 1$. This gives the Hubble parameter at the time as

$$H_{\rm osc}^2 = \frac{V'(\sigma_{\rm osc})}{c\sigma_{\rm osc}}, \tag{6}$$

where c is given in (5). Furthermore, the curvaton field value at the onset of the oscillations $\sigma_{\rm osc}$ is obtained as a function of σ_* by solving

$$\int_{\sigma_*}^{\sigma_{\rm osc}} \frac{d\sigma}{V'} = -\frac{\mathcal{N}_*}{3H_{\rm inf}^2} - \frac{1}{2c(c-3)H_{\rm osc}^2}, \tag{7}$$

where $\mathcal{N}_*$ is the number of e-folds during inflation between the horizon exit of the CMB scale and the end of inflation, and $H_{\rm inf}$ is the inflationary Hubble scale (we are assuming a nearly constant Hubble parameter during inflation, thus $H_{\rm inf} \simeq H_*$).

Thus by combining the above expressions, one can compute the power spectrum produced by a curvaton with a generic potential $V(\sigma)$, given the parameters σ_*, Γ_ϕ, Γ_σ, $H_{\rm inf}$, and $\mathcal{N}_*$.

As for the local-type bispectrum produced by the curvaton, we parametrize its overall amplitude by the non-linearity parameter $f_{\rm NL}$ on the equilateral configuration, which can be shown to take the form

$$\begin{aligned} f_{\rm NL}(k_*) &= \frac{5}{6}\frac{\partial^2 \mathcal{N}}{\partial \sigma_*^2} \left(\frac{\partial \mathcal{N}}{\partial \sigma_*} \right)^{-2} \\ &= \frac{40(1+\hat{r})}{3\hat{r}(4+3\hat{r})} + \frac{5(4+3\hat{r})}{6\hat{r}} \left\{ \frac{V'(\sigma_{\rm osc})}{V(\sigma_{\rm osc})} - \frac{3X(\sigma_{\rm osc})}{\sigma_{\rm osc}} \right\}^{-1} \Bigg[(1 - X(\sigma_{\rm osc}))^{-1} X'(\sigma_{\rm osc}) \\ &+ \left\{ \frac{V'(\sigma_{\rm osc})}{V(\sigma_{\rm osc})} - \frac{3X(\sigma_{\rm osc})}{\sigma_{\rm osc}} \right\}^{-1} \left\{ \frac{V''(\sigma_{\rm osc})}{V(\sigma_{\rm osc})} - \frac{V'(\sigma_{\rm osc})^2}{V(\sigma_{\rm osc})^2} - \frac{3X'(\sigma_{\rm osc})}{\sigma_{\rm osc}} + \frac{3X(\sigma_{\rm osc})}{\sigma_{\rm osc}^2} \right\} \\ &\qquad + \frac{V''(\sigma_{\rm osc})}{V'(\sigma_{\rm osc})} - (1 - X(\sigma_{\rm osc})) \frac{V''(\sigma_*)}{V'(\sigma_{\rm osc})} \Bigg]. \end{aligned} \tag{8}$$

2.2. Scale-Dependence

The scale-dependence of the linear and second order perturbations are calculated by using the slow-roll approximation for the curvaton $3H_*\dot{\sigma}_* \simeq -V'(\sigma_*)$. The spectral index of the linear order perturbations is

$$n_s - 1 \equiv \frac{d}{d\ln k} \ln \mathcal{P}_\zeta \simeq 2\frac{\dot{H}_*}{H_*^2} + \frac{2}{3}\frac{V''(\sigma_*)}{H_*^2}, \tag{9}$$

and its running is

$$\alpha \equiv \frac{dn_s}{d\ln k} \simeq 2\frac{\ddot{H}_*}{H_*^3} - 4\frac{\dot{H}_*^2}{H_*^4} - \frac{4}{3}\frac{\dot{H}_*}{H_*^2}\frac{V''(\sigma_*)}{H_*^2} - \frac{2}{9}\frac{V'(\sigma_*)V'''(\sigma_*)}{H_*^4}. \tag{10}$$

In order to parametrize the scale-dependence of the non-Gaussianity, we define the spectral index of $f_{\rm NL}$ as follows:

$$\begin{aligned} n_{f_{\rm NL}} &\equiv \frac{d\ln|f_{\rm NL}|}{d\ln k} \\ &\simeq \frac{1}{f_{\rm NL}}\frac{5(4+3\hat{r})}{18\hat{r}}\left\{\frac{V'(\sigma_{\rm osc})}{V(\sigma_{\rm osc})} - \frac{3X(\sigma_{\rm osc})}{\sigma_{\rm osc}}\right\}^{-1}(1-X(\sigma_{\rm osc}))\frac{V'(\sigma_*)}{V'(\sigma_{\rm osc})}\frac{V'''(\sigma_*)}{H_*^2}. \end{aligned} \tag{11}$$

We remark that a totally scale-invariant $f_{\rm NL}$ corresponds to $n_{f_{\rm NL}} = 0$ (instead of 1).

3. Pseudo-Nambu-Goldstone Curvatons

As a simple example, let us examine the case where the curvaton is a pseudo-Nambu-Goldstone (NG) boson of a broken U(1) symmetry, possessing a potential of the form

$$V(\sigma) = \Lambda^4\left[1 - \cos\left(\frac{\sigma}{f}\right)\right]. \tag{12}$$

Here, f and Λ are mass scales. Without loss of generality, we set the field value of the curvaton at horizon exit during inflation to lie within the range $0 < \sigma_* < \pi f$, and consider its oscillation about the origin $\sigma = 0$. (For an investigation of NG curvatons located close to their potential minimum, see also [13].) Supposing that the coupling of the NG curvaton with its decay product is suppressed by the scale of symmetry breaking f, we set the curvaton decay rate as

$$\Gamma_\sigma = \frac{1}{16\pi}\frac{m^3}{f^2} = \frac{1}{16\pi}\frac{\Lambda^6}{f^5}, \tag{13}$$

where m is the mass at the minimum, i.e. $m^2 = V''(0)$.

We show the resulting density perturbations from NG curvatons in Figures 1 and 2, where the analytic estimations given by the expressions in the previous section are plotted as solid lines. We also plotted numerically computed results shown as dots. The potential parameters and the inflation/reheating scales are chosen such that the curvaton with $\sigma_* = \frac{3}{4}\pi f$ generates a density perturbation spectrum whose amplitude takes the COBE normalization value $\mathcal{P}_\zeta \approx 2.4 \times 10^{-9}$, with the spectral index lying at the central value of the WMAP7 [6] bound (for a power-law spectrum with no tensor modes) $n_s \approx 0.96$, assuming constant H during inflation. In the figures, we have fixed all the parameters except for the initial field value σ_*, and plotted the perturbations as functions of σ_*.

In the region $\sigma_* \ll \pi f$, the curvaton potential is well approximated by a quadratic, thus the density perturbations behave similarly as from quadratic curvatons. For the chosen parameters, the curvaton energy fraction at decay $\hat{r}$ is smaller than unity for $\sigma_*/\pi f \lesssim 0.5$. Therefore, as $\hat{r}$ increases with σ_*, the power spectrum also increases while $f_{\rm NL}$ decreases, which are familiar behaviors of quadratic curvatons. On the other hand, as one approaches $\sigma_*/\pi f \to 1$, it can be seen that the linear order perturbation as well as the non-Gaussianity increase. This is due to the non-uniform onset of the curvaton oscillation, and it can further be shown that the linear order perturbations blow up in the hilltop limit, accompanied by a mild increase of the non-linearity parameter $f_{\rm NL}$. This is actually a generic feature of hilltop curvatons.

NG curvatons at the hilltop can work with a wide range of inflation/reheating scales, and a detailed study of the parameter space reveals that such hilltop NG curvatons, even though it well dominates the universe before it decays, predict the non-Gaussianity to lie in the range $10 \lesssim f_{\rm NL} \lesssim 30$ [1]. (This is in strong contrast to quadratic curvatons, which produce large $f_{\rm NL}$ only when $\hat{r} < 1$.)

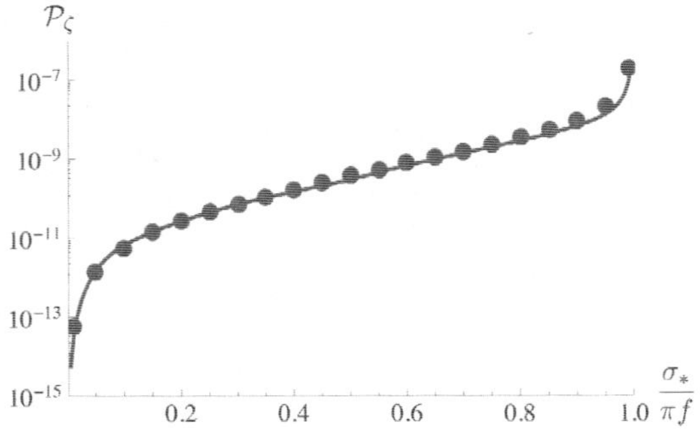

Figure 1. Power spectrum.

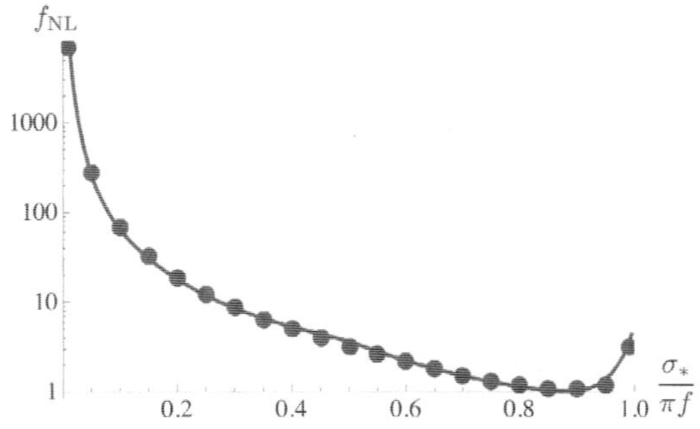

Figure 2. Non-Gaussianity.

4. Self-Interacting Curvatons

A simple example of curvaton potentials that steepen more rapidly than a quadratic is given by curvatons with self-interactions of the form

$$V(\sigma) = \Lambda^4 \left[\left(\frac{\sigma}{f}\right)^2 + \left(\frac{\sigma}{f}\right)^m \right], \tag{14}$$

where Λ and f are positive constants with mass dimension, and m is an even integer with $m > 2$. f denotes where the higher-order term becomes important, while Λ sets the overall scale of the curvaton potential. Such self-interacting curvatons do not produce a red-tilted perturbation spectrum unless relying on the time-variation of H during inflation, however, they have an interesting feature such that the resulting $f_{\rm NL}$ can be strongly scale-dependent, even when the linear order perturbations are nearly scale-invariant [2].

We plot the density perturbations produced from self-interacting curvatons with $m = 8$ in Figures 3 and 4. Here the parameters are chosen such that the power spectrum obtains the COBE normalization value at around $\sigma_*/f \approx 0.6$ (where the power spectrum obtains a peak). As in the previous section, we have fixed all parameters except for σ_*, and shown how the perturbations depend on σ_*. The blue solid lines denote the analytic calculations and the blue dots the numerically computed results. In the $n_{f_{\rm NL}}$ - $f_{\rm NL}$ plane of Figure 4, we have also shown as black dashed lines the detection limit of the Planck satellite [14], which can probe the running of the local-type $f_{\rm NL}$ if its running is large enough to satisfy $|n_{f_{\rm NL}} f_{\rm NL}| \gtrsim 5$. (See [15] for detailed analyses on the expected constraints on running non-Gaussianity.)

The $\sigma_* \ll f$ region is well-approximated by quadratic curvatons, and since $\hat{r}$ is smaller than unity throughout the plotted region, the power spectrum increases with σ_* in the small field regime due to the increase of $\hat{r}$. However for larger field values, the power spectrum starts to decrease. This is attributed to the steep self-interacting potential forcing the curvaton to roll down to small field values by the onset of the curvaton oscillation. This greatly diminishes the initial differences in the curvaton field values during inflation σ_*, thus suppresses the resulting linear order density perturbations. As a consequence, the power spectrum obtains a peak at $\sigma_*/f \approx 0.6$, and there the non-linearity parameter $f_{\rm NL}$ crosses zero and its running $n_{f_{\rm NL}}$ blows up. Therefore in the vicinity of $\sigma_*/f \approx 0.6$, a large and strongly scale-dependent $f_{\rm NL}$ can be produced, whose running is in the detectable range by upcoming CMB observations.

5. Conclusions

We have analytically explored density perturbations from a curvaton with a generic potential, and shown that non-quadratic potentials have rich phenomenology. Through studying pseudo-

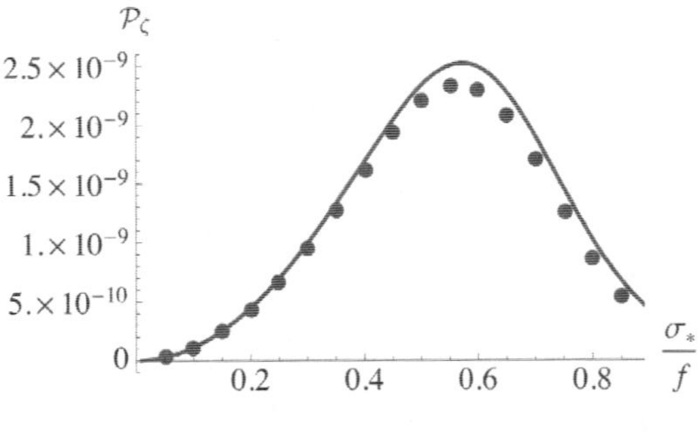

Figure 3. Power spectrum.

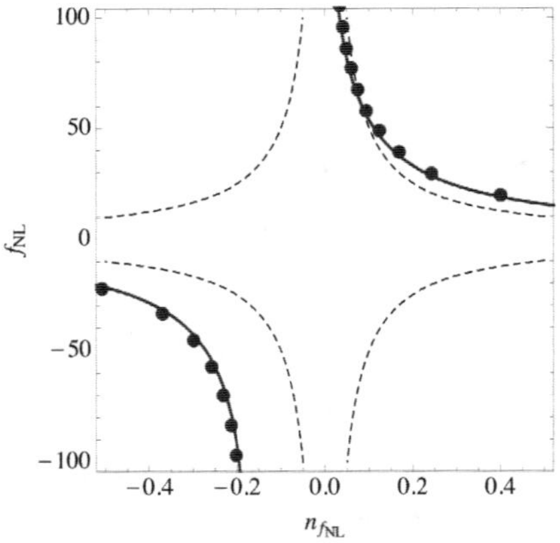

Figure 4. $n_{f_{\rm NL}}$ - $f_{\rm NL}$ plane, showing the region $0.45 \lesssim \sigma_*/f \lesssim 0.65$.

Nambu-Goldstone curvatons and self-interacting curvatons, we have seen that potentials that flatten compared to a quadratic can enhance the linear and second order density perturbations, while steepened potentials are capable of producing running non-Gaussianity. The analytic formulae presented in this paper enable us to go beyond individual case studies and give a systematic treatment of the curvaton scenario in general. This will be helpful for probing the physics of curvatons when combined with upcoming data.

References

[1] Kawasaki M, Kobayashi T and Takahashi F 2011 *Phys. Rev.* **D84** 123506 (*Preprint* `arXiv:1107.6011 [astro-ph.CO]`)
[2] Kobayashi T and Takahashi T 2012 *JCAP* **1206** 004 (*Preprint* `arXiv:1203.3011 [astro-ph.CO]`)
[3] Enqvist K and Sloth M S 2002 *Nucl. Phys.* **B626** 395 (*Preprint* `arXiv:hep-ph/0109214`)
[4] Lyth D H and Wands D 2002 *Phys. Lett.* **B524** 5 (*Preprint* `arXiv:hep-ph/0110002`)
[5] Moroi T and Takahashi T 2001 *Phys. Lett.* **B522** 215 [*Erratum-ibid.* 2002 **B539** 303] (*Preprint* `arXiv:hep-ph/0110096`)
[6] Komatsu E *et al.* (WMAP Collaboration) 2011 *Astrophys. J. Suppl.* **192** 18 (*Preprint* `arXiv:1001.4538 [astro-ph.CO]`)
[7] Enqvist K and Nurmi S 2005 *JCAP* **0510** 013 (*Preprint* `astro-ph/0508573`)
[8] Sasaki M, Valiviita J and Wands D 2006 *Phys. Rev.* **D74** 103003 (*Preprint* `arXiv:astro-ph/0607627`)
[9] Enqvist K and Takahashi T 2008 *JCAP* **0809** 012 (*Preprint* `arXiv:0807.3069 [astro-ph]`)
[10] Kawasaki M, Nakayama K, Takahashi F 2009 *JCAP* **0901** 026 (*Preprint* `arXiv:0810.1585 [hep-ph]`)
[11] Enqvist K, Nurmi S, Rigopoulos G, Taanila O and Takahashi T 2009 *JCAP* **0911** 003 (*Preprint* `arXiv:0906.3126 [astro-ph.CO]`)
[12] Byrnes C T, Enqvist K, Nurmi S and Takahashi T 2011 *JCAP* **1111** 011 (*Preprint* `arXiv:1108.2708 [astro-ph.CO]`)
[13] Dimopoulos K, Lyth D H, Notari A and Riotto A 2003 *JHEP* **0307** 053 (*Preprint* `hep-ph/0304050`)
[14] (Planck Collaboration) 2006 (*Preprint* `astro-ph/0604069`)
[15] Sefusatti E, Liguori M, Yadav A P S, Jackson M G and Pajer E 2009 *JCAP* **0912** 022 (*Preprint* `arXiv:0906.0232 [astro-ph.CO]`)

PASCOS 2012 – 18th International Symposium on Particles Strings and Cosmology IOP Publishing
Journal of Physics: Conference Series **485** (2014) 012019 doi:10.1088/1742-6596/485/1/012019

Toroidal dipole moment of the LSP in the cMSSM

L G Cabral-Rosetti[1], **M Mondragón**[2] **and E A Reyes Pérez**[2]

[1] Unidad de Ciencias Básicas, Departamento de Posgrado, Centro Interdisciplinario de Investigación y Docencia en Educación Técnica, CIIDET. Av. Universidad 282 Pte., Col. Centro, A. Postal 752, C.P. 76000, Santiago de Querétaro, Qro., México
[2] Instituto de Física, Universidad Nacional Autónoma de México, Apdo. Postal 20-364, México 01000, D.F., México

E-mail: `esteban.reyesperez@gmail.com`

Abstract. We study the toroidal dipole moment of the lightest neutralino in the constrained Minimal Supersymmetric Standard Model. The toroidal dipole moment is the only electromagnetic property of the neutralino. Since the neutralino is the LSP in many versions of the MSSM and therefore a candidate for dark matter, its characterization through its electromagnetic properties is important both for particle physics and for cosmology. We perform a scan in the parameter space of the cMSSM and find that the toroidal dipole moment is different from zero, albeit very small, in all the parameter space, and reaches a value around 10^{-3} GeV^{-2} in a particular region of the parameter space, well below experimental bounds.

1. Introduction

One of the best motivated extensions of the Standard Model (SM) is the Minimal Supersymmetric Standard Model (MSSM), since, besides giving a solution to the hierarchy problem, provides us with a good candidate for cold dark matter (CDM), namely, the lightest neutralino.

There are currently several experiments under way, and more planned for the future for direct and indirect detection of dark matter (DM). If detected, it will be necessary to discriminate between different candidates. To this end, it will be important to characterize as much as possible the different candidates. The neutralino is at present one of the best candidates for DM, and its electroweak properties can give us some insight into its nature. Because it is neutral, these properties appear only radiatively.

One of the least studied electromagnetic properties of a particle is the Toroidal Dipole Moment (TDM), which is directly related to the anapole moment. The anapole moment corresponds to a T invariant interaction, which is C and P non-invariant [1]. The electromagnetic vertex of a particle can be expressed in a multipole parametrization, including the toroidal moments, which provides a one to one correspondence between the form factors and the multipole moments.

Pospelov and ter Veldhuis have obtained an upper limit for the anapole moment of WIMPS [2], using results from the DAMA and CDMS experiments [3, 4]. In case the neutralino is the main component of dark matter, its anapole moment should comply with this limit.

In this work we calculate the TDM of the neutralino within the constrained Minimal Supersymmetric Standard Model. We do a scan in the five parameter space of the cMSSM, and compare the results with the above mentioned experimental limit.

Content from this work may be used under the terms of the Creative Commons Attribution 3.0 licence. Any further distribution of this work must maintain attribution to the author(s) and the title of the work, journal citation and DOI.
Published under licence by IOP Publishing Ltd

2. The MSSM and the neutralino as candidate for dark matter

The minimal supersymmetric extension of the Standard Model (MSSM) provides us with one of the best WIMP candidates for dark matter, namely the lighest neutralino (for reviews on SUSY see for instance [5, 6]). The MSSM requires two complex Higgs electroweak doublets to give mass to the up and down type quarks and to avoid chiral anomalies. After electroweak symmetry breaking five physical Higgs states remain: two neutral CP invariant (h^0, H^0), two charged CP invariant (H^+, H^-), and one neutral CP-odd (A^0).

The MSSM has a new discrete symmetry, R parity, defined as $R = (-1)^{3B+2S+L}$, where B and L are the baryonic and leptonic numbers respectively. This symmetry assignes a charge +1 to the SM particles and -1 to the supersymmetric partners, thus making the lightest supersymmetric particle (LSP) stable.

Supersymmetry has to be broken, or it would have already been observed. To break supersymmetry explicitly, without the reappearance of quadratic divergencies, a set of super-renormalizable terms are added to the Lagrangian, the so-called soft breaking terms. The Lagrangian for the soft breaking terms is given by

$$\mathcal{L}_{soft} = -\frac{1}{2}M_a\lambda^a\lambda^a - \frac{1}{6}A^{ijk}\phi_i\Phi_j\phi_k - \frac{1}{2}B^{ij}\phi_i\phi_j + c.c. - (m^2)^i_j\phi^{j*}\phi_i \ , \tag{1}$$

where M_a are the gaugino masses, A^{ijk} and B^{ij} are trilinear and bilinear couplings, respectively, and $(m^2)^i_j$ are scalar squared-mass terms. It is assumed that supersymmetry breaking happens in a hidden sector, which communicates to the observable one only through gravitational interactions, and that the gauge interactions unifiy. This means that at the GUT scale the soft breaking terms are "universal", i.e., the gauginos M_a have a common mass, as well as the scalars $(m^2)^i_j$ and the trilinear couplings, A^{ijk}. Requiring electroweak symmetry breaking fixes the value of B^{ij} and the absolute value of the Higgsino mixing parameter $|\mu|$. This is known as the constrained MSSM, or cMSSM, which is described by five parameters: the unified gaugino mass $m_{1/2}$, the universal scalar mass m_0, the value of the universal trilinear coupling A_0, the sign of Higgsino mass parameter μ, and the ratio of the vacuum expectation values of the two Higgses, $\tan\beta$.

After the electroweak symmetry breaking the neutral and charged states in the MSSM can mix. In the case of the neutral ones they give rise to a set of four mass eigenstates, the neutralinos. It is the lightest one of these that is the LSP and a good candidate to dark matter in many SUSY models. The lightest neutralino, in the gauge eigenstate basis, is thus a function of the neutral Higgsinos and the neutral gauginos (Wino and Bino)

$$\psi_0 = (\tilde{B}, \tilde{W}^0, \tilde{H}^0_u, \tilde{H}^0_d) \ . \tag{2}$$

The properties of the neutralinos will depend on the mixing, which in turn depends on the soft breaking parameters. Thus, the lightest neutralino can range from almost pure Bino to almost pure Higgsino.

3. Toroidal Dipole Moment

For 1/2-spin particles the most general expression for the electromagnetic vertex function, which characterizes the interaction between the particle and the electromagnetic field, is:

$$\begin{aligned}\Gamma_\mu(q) &= f_Q(q^2)\gamma_\mu + f_\mu(q^2)i\sigma_{\mu\nu}q^\nu\gamma_5 \\ &-f_E(q^2)\sigma_{\mu\nu}q^\nu + f_A(q^2)(q^2\gamma_\mu - \not{q}q_\mu)\gamma_5,\end{aligned} \tag{3}$$

where $f_Q(q^2)$, $f_\mu(q^2)$, $f_E(q^2)$ and $f_A(q^2)$ are the so called charge, magnetic dipole, electric dipole and anapole form factors, respectively; where $q_\mu = p'_\mu - p_\mu$ is the transferred 4-momentum; and

$\sigma_{\mu\nu} = (i/2)\left[\gamma_\mu, \gamma_\nu\right]$ [7, 8]. These form factors are physical observables when $q^2 \to 0$, and their combinations define the well known magnetic dipole (μ), electric dipole (d) and anapole (a) moments.

The electromagnetic properties of Majorana fermions (like the neutralino) are described by a unique form factor, the anapole, $f_A(q^2)$. This is a consequence of CPT-invariance and the C, P, T properties of $\Gamma_\mu(q^2)$ and the interaction Hamiltonian. Thus, the electromagnetic vertex function of a neutralino can be writen as

$$\Gamma_\mu(q^2) = f_A(q^2)(q^2\gamma_\mu - \not{q}q_\mu)\gamma_5. \tag{4}$$

The anapole moment was introduced by Zel'dovich to describe a T-invariant interaction that does not conserve P and C parity [1]. The anapole moment does not have a simple classical analogue, since $f_A(q^2)$ does not correspond to a multipolar distribution. A more convenient quantity to describe this interaction was proposed by V. M. Dubovik and A. A. Cheshkov [9]: the toroidal dipole moment (TDM), $\tau(q^2)$.

The TDM and the anapole moment coincide in the case of $m_i = m_f$, i.e. the incoming and outgoing particle are the same. This type of static multipole moments does not produce any external fields in vacuum but generate a free-field (gauge invariant) potential [7], which is responsible for topological effects like the Aharonov-Bohm one.

The simplest TDM model (anapole) was given by Zel'dovich as a conventional solenoid rolled up in a torus and with only one poloidal current. For such stationary solenoid, without azimuthal components for the current or the electric field, there is only one magnetic azimuthal field different from zero inside the torus.

4. One-loop calculation

The TDM of the neutralino may be defined in the one-loop approximation in the cMSSM by the Feynman diagrams shown in figs. 1 and 2, where f represents the charged fermions of the SM. Taking each fermionic family separately we obtain 94 Feynman diagrams in total: 66 corresponding to self-energy and 28 to vertex corrections.

We use *FeynCalc* to calculate the amplitude of these diagrams. Since we are only interested in the terms that contribute to the anapole form factor, we isolate the ones that have the Lorentz structure $\gamma_\mu\gamma_5$. It is important to notice here that we work in the t'Hooft-Feynman gauge ($\xi = 1$). One of the first results we obtain is that the self-energies γH^0, γh^0, γA^0 and γG^0 do not contribute to the calculation. If we call Ξ_i the coefficient that multiplies $\gamma_\mu\gamma_5$ for the *ith* diagram, then we have that

$$\sum_i \Xi_i = f_A(q^2)q^2. \tag{5}$$

To obtain the toroidal dipole moment $\tau = f_A(0)$ we use the l'Hopital rule and get

$$\tau = f_A(0) = \lim_{q^2\to 0} \frac{\sum_i \Xi_i}{q^2} = \frac{\partial \sum_i \Xi_i}{\partial q^2}\,|_{q^2\to 0}\,. \tag{6}$$

The contributions to the self-energies have two point Passarino-Veltman scalar functions of the type $B_0\left(q^2, x^2, x^2\right)$ and $B_0\left(0, x^2, x^2\right)$. Likewise, the contributions to the vertex corrections have two and three point scalar functions of the type $B_0\left(q^2, x^2, x^2\right)$, $B_0\left(M^2_{\tilde{\chi}^0_1}, y^2, x^2\right)$ and $C_0\left(q^2, M^2_{\tilde{\chi}^0_1}, M^2_{\tilde{\chi}^0_1}, x^2, x^2, y^2\right)$. In both cases x and y represent the masses of the particles in the loop.

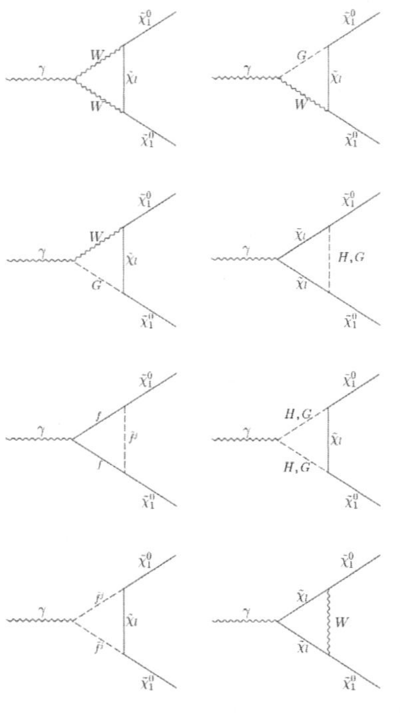

Figure 1. One-loop vertex corrections to the process $\gamma \longrightarrow \chi_1^0 \chi_1^0$.

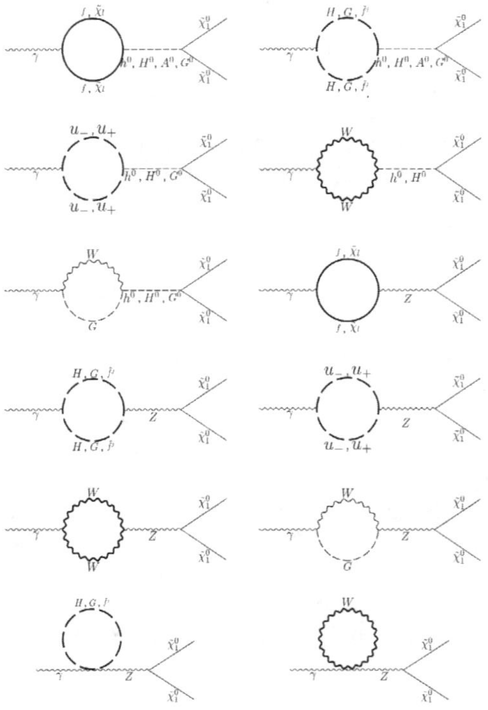

Figure 2. One-loop corrections to the self-energy for the process $\gamma \longrightarrow \chi_1^0 \chi_1^0$.

When evaluating (6), derivatives of the Passarino-Veltman functions appear. To evaluate the B_0's, as well as their derivatives, we use *LoopTools*[10]. To evaluate the C_0's and their derivatives we expande them in a power series around $q^2 = 0$.

The expression obtained for the toroidal dipole moment depends on various parameters of the MSSM, including the supersymmetric particles masses as well as the mass mixing matrix elements, the value of $\tan\beta$, and the values of the soft breaking terms. We evaluate the TDM within the cMSSM using *Suspect*[11], by fixing the value of A_0, $\tan\beta$ and signμ, and scanning over the other two parameters m_0 and $m_{1/2}$, from 0 to 1500 GeV and 250 to 1500 GeV, respectively.

Figure 3 shows the neutralino toroidal dipole moment for $\tan\beta = 10$, $\mu > 0$, and three different values of A_0, -1000, 0 and 1000 GeV (top to bottom). Comparing the three different plots, no dependence on A_0 is shown. The TDM is very low for almost every region of the parameter space scanned, with values between 10^{-5} and 10^{-8} GeV^{-2}. However the TDM increases for increasing m_0 and decreasing $m_{1/2}$, reaching values over 10^{-3} GeV^{-2} for high m_0 ($\geq$ 800 GeV) and low $m_{1/2}$ ($\leq$ 400 GeV). A similar behavior can be seen in the results for $\tan\beta = 50$, $\mu > 0$, and the same three different values of A_0, -1000, 0 and 1000 GeV (4). The TDM reaches values around 10^{-3} GeV^{-2}.

Figure 5 shows a comparison of two plots for different $\tan\beta$ but same signμ and A_0. This figure shows the dependence of the TDM on $\tan\beta$. Figure 6 shows a comparison of two plots for different values of signμ but same $\tan\beta$ and A_0. Sign$\mu > 0$ may solve the problem of the discrepancy between the measured value of $g-2$ of the muon and the one predicted by the SM. However, this does not mean negative signμ is ruled out since others mechanisms could solve this problem, and therefore sign$\mu < 0$ should be taken into consideration. This figure shows no

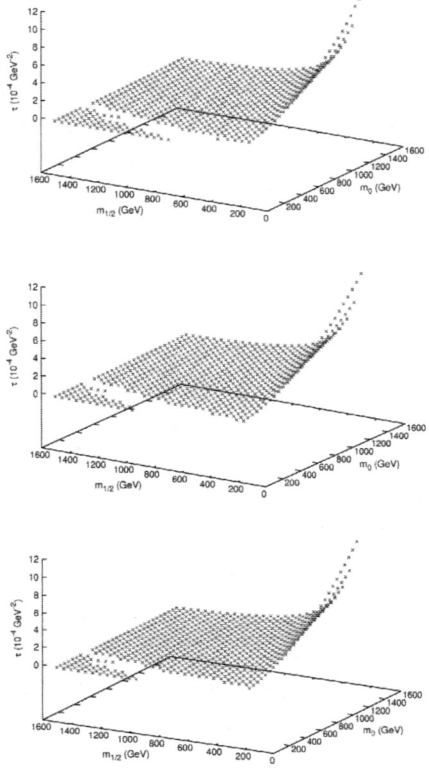

Figure 3. TDM for $\tan\beta = 10$, signμ = + y $A_0 = -1000$ (top), 0 (centre) y 1000 (bottom) GeV.

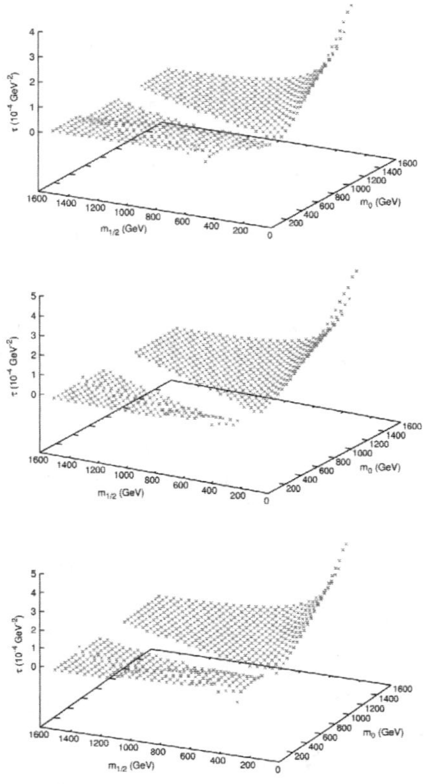

Figure 4. TDM for $\tan\beta = 50$, signμ = + y $A_0 = -1000$ (top), 0 (centre) y 1000 (bottom) GeV.

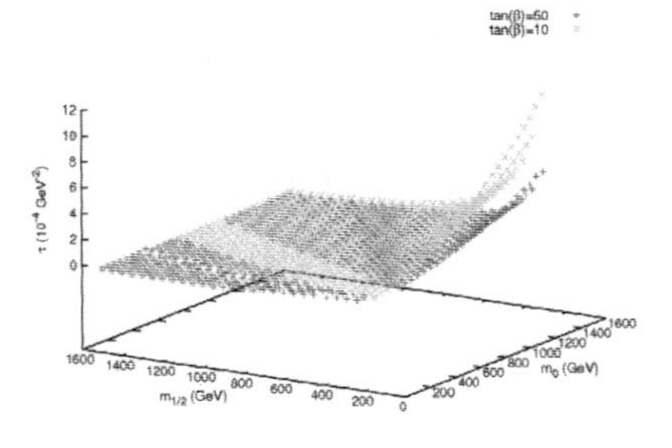

Figure 5. TDM for signμ = +, $A_0 = 0$ GeV, $\tan\beta = 10$ (green) and $\tan\beta = 50$ (red).

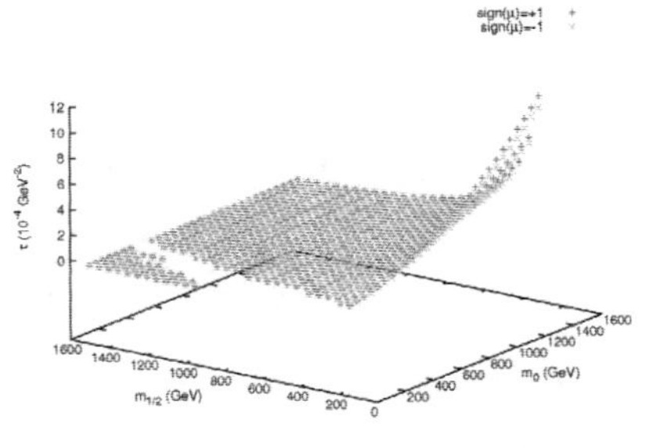

Figure 6. TDM for $\tan\beta = 10$, $A_0 = 0$ GeV, signμ = + (red) and signμ = − (green).

dependence of the TDM on signμ.

Notice that in all the plots the region for which $M_{\tilde{\chi}_1^0} = M_{\tilde{\tau}}$ is suppressed since we are not considering this possibility. This condition ($M_{\tilde{\chi}_1^0} = M_{\tilde{\tau}}$) separates the region where the neutralino $\tilde{\chi}_1^0$ is the LSP and the one where the stau $\tilde{\tau}$ is the LSP.

5. Conclusions

We calculated the only electromagnetic property of the lightest neutralino: its toroidal dipole moment. Its characterization is extremely valuable for discriminating different models which have the neutralino as dark matter candidate. We performed the calculation in the framework of the cMSSM, however a similar analysis can be performed for other models (work in progress). We found that the TDM of the neutralino is highly sensitive to m_0, $m_{1/2}$ and $\tan\beta$, but very weakly or practically non-dependent on A_0 and signμ.

All points in the parameter space we scanned give a TDM consistent with the upper limit ($\sim 10^{-2}$ GeV^{-2}) obtained by Pospelov and ter Veldhuis [2] for WIMPs interacting with heavy nuclei using data from the CDMS and DAMA experiments. However, this data can and will be improved in the next few years helping to refine the upper limit, likely ruling out some regions of the parameter space.

The TDM analysis can be used as another criteria to constrain the parameter space of a given model which has a neutralino as candidate for dark matter. Thus, according to our results, if a non-zero (around 10^{-4}-10^{-3} GeV^{-2}) TDM could be measured for the neutralino, that would indicate that the favored region of the parameter space of the cMSSM would be high m_0 (≥ 800 GeV) and low $m_{1/2}$ (≤ 400 GeV). Otherwise, other regions are compatible with a TDM lower than 10^{-5} GeV.

If combined with other criteria (such as dark matter relic density, or frequentist analysis of data from the LHC) the parameter space can be reduced even further. In fact, the combination with these other criteria favours a region with low m_0, which is compatible with a 10^{-5}-10^{-8} GeV^{-2} TDM for the neutralino. This means, among other things, that if a TDM higher than 10^{-4} GeV^{-2} is measured for a WIMP it would exclude some regions of parameter space of the cMSSM (and other more specific models), at least if the neutralino is the only component of dark matter.

Acknowledgments

We acknowledge very useful discussions with E. Ley Koo and A. Mondragón.
This work was partially supported by UNAM grants PAPIIT IN111609 and IN113412.

References

[1] Zeldovich Y 1957 *Zh. Eksp. Teor. Fiz* **33** 1531
[2] Pospelov M and ter Veldhuis T 2000 *Phys. Lett.* **B480** 181–186 (*Preprint* hep-ph/0003010)
[3] Bernabei R *et al.* 1996 *Phys. Lett.* **B389** 757–766
[4] Abusaidi R *et al.* (CDMS) 2000 *Phys. Rev. Lett.* **84** 5699–5703 (*Preprint* astro-ph/0002471)
[5] Martin S P 1997 (*Preprint* hep-ph/9709356)
[6] Aitchison I J 2005 Notes of Lectures for Graduate Students in Particle Physics Oxford, 2004 and 2005 (*Preprint* hep-ph/0505105)
[7] Dubovik V M and Kuznetsov V E 1998 *Int. J. Mod. Phys.* **A13** 5257–5278 (*Preprint* hep-ph/9606258)
[8] Bukina E N, Dubovik V M and Kuznetsov V E JINR-E4-98-95
[9] Dubovik V and Cheshkov A 1975 *Sov. J. Particles Nucl.* **5** 318–337
[10] Hahn T and Perez-Victoria M 1999 *Comput. Phys. Commun.* **118** 153–165 (*Preprint* hep-ph/9807565)
[11] Djouadi A, Kneur J L and Moultaka G 2007 *Comput. Phys. Commun.* **176** 426–455 (*Preprint* hep-ph/0211331)

Journal of Physics: Conference Series **485** (2014) 012020 doi:10.1088/1742-6596/485/1/012020

Big-bang nucleosynthesis with a long-lived CHAMP including He4 spallation process

Masato Yamanaka[1], Toshifumi Jittoh[2], Kazunori Kohri[1], Masafumi Koike[2], Joe Sato[2], Kenichi Sugai[2] and Koichi Yazaki[3,4]

[1]Theory Center, Institute of Particle and Nuclear Studies, KEK (High Energy Accelerator Research Organization), 1-1 Oho, Tsukuba 305-0801, Japan

[2]Department of Physics, Saitama University, Shimo-okubo, Sakura-ku, Saitama, 338-8570, Japan

[3]Hashimoto Mathematical Physics Laboratory , Nishina Accelerator Research Center, RIKEN, Wako, Saitama 351-0198, Japan

[4]Yukawa Institute for Theoretical Physics, Kyoto University, Kyoto 606-8502, Japan

E-mail: masato.yamanaka@kek.jp

Abstract. We propose helium-4 spallation processes induced by long-lived stau in supersymmetric standard models, and investigate an impact of the processes on light elements abundances. We show that, as long as the phase space of helium-4 spallation processes is open, they are more important than stau-catalyzed fusion and hence constrain the stau property. This talk is based on the work of ref. [1].

1. Introduction

Long-lived charged massive particles (CHAMPs) will play interesting roles in the Big-Bang Nucleosynthesis (BBN). The light nuclei will interact not only with the CHAMPs during the BBN processes [2, 3, 4, 5, 6, 7], but also with the decay products of the CHAMPs in the post-BBN era [8, 9]. The standard BBN will thus be altered, and so is the abundance of the light elements at the present time. One can thus constrain the models beyond the Standard Model by evaluating their prediction on the light elements abundance and comparing it with the current observations. We can then give stringent predictions for the forthcoming experiments and observations according to these constraints.

The Standard Model extended with supersymmetry (SUSY) is one of the models that can accommodate such long-lived CHAMPs. With the R-parity conservation, the lightest SUSY particle (LSP) is stable and become a cold dark matter. Interestingly, it can offer a long-lived CHAMPs if the LSP is the bino-like neutralino $\tilde{\chi}_1^0$. Coannihilation mechanism is required to account for the dark matter abundance in this case [10], where the LSP and the next-lightest SUSY particle (NLSP) are almost degenerate in mass. Staus, denoted by $\tilde{\tau}$ and a possible candidate of the NLSP, can acquire a long lifetime when the mass difference with the LSP is less than the mass of tau leptons. This is due to the phase space suppression of the final state that necessarily consists of three particles or more. Noting that such long-lived staus will be copious during the BBN [11, 12], we have shown in [3, 5, 7] that their presence indeed alters the prediction of the standard BBN and possibly solve the discrepancy of the lithium abundance in the Universe through the internal conversion reactions.

Content from this work may be used under the terms of the Creative Commons Attribution 3.0 licence. Any further distribution of this work must maintain attribution to the author(s) and the title of the work, journal citation and DOI.
Published under licence by IOP Publishing Ltd

In this talk, we propose new reactions

$$(\tilde{\tau}^4\mathrm{He}) \to \tilde{\chi}_1^0 + \nu_\tau + \mathrm{t} + \mathrm{n}, \tag{1a}$$

$$(\tilde{\tau}^4\mathrm{He}) \to \tilde{\chi}_1^0 + \nu_\tau + \mathrm{d} + \mathrm{n} + \mathrm{n}, \tag{1b}$$

$$(\tilde{\tau}^4\mathrm{He}) \to \tilde{\chi}_1^0 + \nu_\tau + \mathrm{p} + \mathrm{n} + \mathrm{n} + \mathrm{n}, \tag{1c}$$

in which $(\tilde{\tau}\,^4\mathrm{He})$ represents a bound state of a stau and ^{4}He nucleus. Reaction (1) is essentially a spallation of the ^{4}He nucleus, producing a triton t, a deuteron d, and neutrons n. Presence of such spallation processes has been ignored so far due to the naïve expectation that the rate of the stau-catalyzed fusion [2]

$$(\tilde{\tau}\,^4\mathrm{He}) + \mathrm{d} \to \tilde{\tau} + {}^6\mathrm{Li} \tag{2}$$

is larger than the reaction (1). Indeed, the cross section of Eq. (2) is much larger than that of $^4\mathrm{He} + \mathrm{d} \to {}^6\mathrm{Li} + \gamma$ by (6 – 7) orders of magnitude [13].

We point out that this expectation is indeed naïve; the reaction Eq. (1) is more effective than Eq. (2) as long as the spallation processes are kinematically allowed. The former reaction rapidly occurs due to the large overlap of their wave functions in a bound state. On the other hand, the latter proceeds slowly since it requires an external deuteron which is sparse at the BBN era. The overproduction of t and d is more problematic than that of ^{6}Li. This puts new constraints on the parameters of the minimal supersymmetric standard model (MSSM). Note that there is no reaction corresponding to Eq. (1) in the gravitino LSP scenario [14, 15].

2. Spallation of helium 4

The ^{4}He spallation processes of Eq. (1) is described by the Lagrangian

$$\mathcal{L} = \tilde{\tau}^* \overline{\tilde{\chi}_1^0}(g_\mathrm{L} P_\mathrm{L} + g_\mathrm{R} P_\mathrm{R})\tau + \sqrt{2} G_\mathrm{F} \nu_\tau \gamma^\mu P_\mathrm{L} \tau J_\mu + \mathrm{h.c.}, \tag{3}$$

where $G_\mathrm{F} = 1.166 \times 10^{-5}\mathrm{GeV}^{-2}$ is the Fermi coupling constant, $P_{\mathrm{L(R)}}$ represents the chiral projection operator, and J_μ is the weak current. The effective coupling constants g_L and g_R are given by

$$g_\mathrm{L} = \frac{g}{\sqrt{2}\cos\theta_\mathrm{W}} \sin\theta_\mathrm{W} \cos\theta_\tau, \; g_\mathrm{R} = \frac{\sqrt{2}g}{\cos\theta_\mathrm{W}} \sin\theta_\mathrm{W} \sin\theta_\tau \mathrm{e}^{i\gamma_\tau}, \tag{4}$$

where g is the $SU(2)_\mathrm{L}$ gauge coupling constant and θ_W is the Weinberg angle. The mass eigenstate of staus is given by the linear combination of $\tilde{\tau}_\mathrm{L}$ and $\tilde{\tau}_\mathrm{R}$, the superpartners of left-handed and right-handed tau leptons, as

$$\tilde{\tau} = \cos\theta_\tau \tilde{\tau}_\mathrm{L} + \sin\theta_\tau \mathrm{e}^{-i\gamma_\tau} \tilde{\tau}_\mathrm{R}. \tag{5}$$

Here θ_τ is the left-right mixing angle of staus and γ_τ is the CP violating phase.

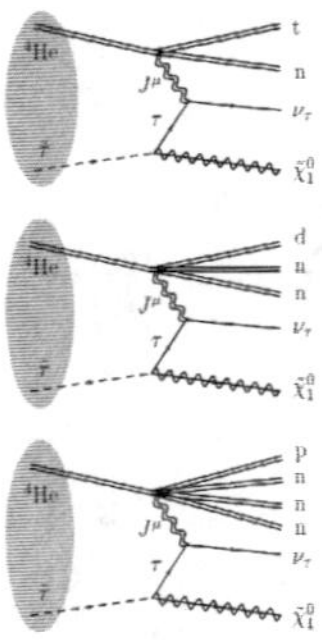

Figure 1. ^{4}He spallation processes.

2.1. $(\tilde{\tau}\,^4He) \to \tilde{\chi}_1^0 + \nu_\tau + t + n$

First we consider the process of Eq. (1a). The rate of this process is expressed as

$$\frac{1}{\tau_{\rm tn}} = \frac{1}{|\psi|^2 \cdot \sigma v_{\rm tn}}, \tag{6}$$

where $|\psi|^2$ stands for the overlap of the wave functions of the stau and the ^{4}He nucleus. We estimate the overlap by

$$|\psi|^2 = \frac{(Z\alpha m_{\rm He})^3}{\pi}, \tag{7}$$

where Z and $m_{\rm He}$ represent the atomic number and the mass of ^{4}He, respectively, and α is the fine structure constant. We assumed that the stau is pointlike particle and is much heavier than ^{4}He nucleus so that the reduced mass of the bound state is equal to the mass of ^{4}He nucleus itself. The cross section of the elementary process for this reaction is denoted by $\sigma v_{\rm tn}$ and calculated as

$$\begin{aligned}\sigma v_{\rm tn} &\equiv \sigma v\big((\tilde{\tau}^4{\rm He}) \to \tilde{\chi}_1^0 \nu_\tau {\rm tn}\big) \\ &= \frac{1}{2E_{\tilde{\tau}}} \int \frac{d^3\boldsymbol{p}_\nu}{(2\pi)^3 2E_\nu} \frac{d^3\boldsymbol{p}_{\tilde{\chi}}}{(2\pi)^3 2E_{\tilde{\chi}}} \frac{d^3\boldsymbol{q}_{\rm n}}{(2\pi)^3} \frac{d^3\boldsymbol{q}_{\rm t}}{(2\pi)^3} \\ &\quad \times \big|\mathcal{M}\big((\tilde{\tau}^4{\rm He}) \to \tilde{\chi}_1^0 \nu_\tau {\rm tn}\big)\big|^2 \\ &\quad \times (2\pi)^4 \delta^{(4)}(p_{\tilde{\tau}} + p_{\rm He} - p_\nu - q_{\rm t} - q_{\rm n}).\end{aligned} \tag{8}$$

Here p_i and E_i are the momentum and the energy of the particle species i, respectively.

We briefly show the calculation of the amplitude of this process. The amplitude is deconstructed as

$$\begin{aligned}\mathcal{M}\big((\tilde{\tau}^4{\rm He}) \to \tilde{\chi}_1^0 \nu_\tau {\rm tn}\big) &= \langle {\rm t\,n}\,\tilde{\chi}_1^0\,\nu_\tau | \mathcal{L}_{\rm int} | ^4{\rm He}\,\tilde{\tau}\rangle \\ &= \langle {\rm t\,n} | J^\mu | ^4{\rm He}\rangle\, \langle \tilde{\chi}_1^0\,\nu_\tau | j_\mu | \tilde{\tau}\rangle.\end{aligned} \tag{9}$$

The weak current J_μ consists of a vector current V_μ and an axial vector current A_μ as $J_\mu = V_\mu + g_{\rm A} A_\mu$, where $g_{\rm A}$ is the axial coupling constant. The relevant components of the currents in this reaction are V^0 and A^i $(i = 1, 2, 3)$. We take these operators as a sum of a single-nucleon operators as

$$V^0 = \sum_{a=1}^{4} \tau_a^- {\rm e}^{{\rm i}\boldsymbol{q}\cdot\boldsymbol{r}_a}, \quad A^i = \sum_{a=1}^{4} \tau_a^- \sigma_a^i {\rm e}^{{\rm i}\boldsymbol{q}\cdot\boldsymbol{r}_a}, \tag{10}$$

where $\boldsymbol{q}$ is the momentum carried by the current, $\boldsymbol{r}_a$ is the spatial coordinate of the a-th nucleon $(a \in \{1, 2, 3, 4\})$, and τ_a^- and σ_a^i denote the isospin ladder operator and the spin operator of the a-th nucleon, respectively. Each component leads to a part of hadronic matrix element:

$$\begin{aligned}\langle {\rm tn} | V^0 | ^4{\rm He}\rangle &= \sqrt{2}\mathcal{M}_{\rm tn}, \\ \langle {\rm tn} | g_{\rm A} A^+ | ^4{\rm He}\rangle &= \sqrt{2} g_{\rm A} \mathcal{M}_{\rm tn}, \\ \langle {\rm tn} | g_{\rm A} A^- | ^4{\rm He}\rangle &= -\sqrt{2} g_{\rm A} \mathcal{M}_{\rm tn}, \\ \langle {\rm tn} | g_{\rm A} A^3 | ^4{\rm He}\rangle &= -\sqrt{2} g_{\rm A} \mathcal{M}_{\rm tn},\end{aligned} \tag{11}$$

Table 1. Input values of the matter radius $R_{\rm mat}$ for d, t, and ^{4}He, the magnetic radius $R_{\rm mag}$ for p and n, nucleus mass m_X, excess energy Δ_X for the nucleus X, and each reference.

nucleus	$R_{\rm mat(mag)}$ [fm]/[GeV^{-1}]	m_X [GeV]	Δ_X [GeV]
p	0.876 / 4.439 [16]	0.9383 [21]	6.778×10^{-3} [22]
n	0.873 / 4.424 [17]	0.9396 [21]	8.071×10^{-3} [22]
d	1.966 / 9.962 [18]	1.876 [22]	1.314×10^{-2} [22]
t	1.928 / 9.770 [19]	2.809 [22]	1.495×10^{-2} [22]
^{4}He	1.49 / 7.55 [20]	3.728 [22]	2.425×10^{-3} [22]

where $A^{\pm} = (A^1 \pm iA^2)/\sqrt{2}$. Given the relevant wave functions of a ^{4}He nucleus, a triton, and a neutron (see Appendix in Ref. [1]), we obtain the hadronic matrix element as

$$\mathcal{M}_{\rm tn} = \left(\frac{128\pi}{3} \frac{a_{\rm He} a_{\rm t}^2}{(a_{\rm He}+a_{\rm t})^4} \right)^{3/4} \left\{ \exp\left[-\frac{\boldsymbol{q}_{\rm t}^2}{3a_{\rm He}} \right] - \exp\left[-\frac{\boldsymbol{q}_{\rm n}^2}{3a_{\rm He}} - \frac{(\boldsymbol{q}_{\rm t}+\boldsymbol{q}_{\rm n})^2}{6(a_{\rm He}+a_{\rm t})} \right] \right\}. \tag{12}$$

Here $\boldsymbol{q}_{\rm t}$ and $\boldsymbol{q}_{\rm n}$ are three-momenta of the triton and the neutron, respectively, and $a_{\rm He}$ and $a_{\rm t}$ are related to the mean square matter radius $R_{\rm m}$ by

$$a_{\rm He} = \frac{9}{16} \frac{1}{(R_{\rm m})^2_{\rm He}}, \quad a_{\rm t} = \frac{1}{2} \frac{1}{(R_{\rm m})^2_{\rm t}}. \tag{13}$$

We list in Table 1 input values of the matter radius for the numerical calculation in this article.

The remaining part is straightforwardly calculated to be

$$\begin{aligned} |\langle \tilde{\chi}_1^0 \nu_\tau | j_0 | \tilde{\tau} \rangle|^2 = |\langle \tilde{\chi}_1^0 \nu_\tau | j_z | \tilde{\tau} \rangle|^2 &= 4G_{\rm F}^2 |g_{\rm R}|^2 \frac{m_{\tilde{\chi}_1^0} E_\nu}{m_\tau^2}, \\ |\langle \tilde{\chi}_1^0 \nu_\tau | j_\pm | \tilde{\tau} \rangle|^2 &= 4G_{\rm F}^2 |g_{\rm R}|^2 \frac{m_{\tilde{\chi}_1^0} E_\nu}{m_\tau^2} \left(1 \mp \frac{p_\nu^z}{E_\nu} \right), \end{aligned} \tag{14}$$

where E_ν and p_ν^z are the energy and the z-component of the momentum of the tau neutrino, respectively. We assumed that the stau and the neutralino are non-relativistic. This equation includes not only all the couplings such as $G_{\rm F}$, $g_{\rm L}$, and $g_{\rm R}$, but also the effect of the virtual tau propagation in the Fig. 1. Note here that $g_{\rm L}$ coupling does not contribute. This is because the virtual tau ought to be left-handed at the weak current, and it flips its chirality during the propagation since the transferred momentum is much less than its mass.

Combining hadronic part with the other part, we obtain the squared amplitude as

$$\left| \mathcal{M}((\tilde{\tau}^4{\rm He}) \to \tilde{\chi}_1^0 \nu_\tau {\rm tn}) \right|^2 = \frac{8 m_{\tilde{\chi}_1^0} G_{\rm F}^2 |g_{\rm R}|^2}{m_\tau^2} (1+3g_A^2) E_\nu |\mathcal{M}_{\rm tn}|^2. \tag{15}$$

Integrating on the phase space of the final states, we obtain the cross section as

$$\sigma v_{\rm tn} = \frac{8}{\pi^2} \left(\frac{32}{3\pi} \right)^{3/2} g^2 \tan^2\theta_W \sin^2\theta_\tau (1+3g_A^2) G_F^2 \Delta_{\rm tn}^4 \frac{m_{\rm t} m_{\rm n}}{m_{\tilde{\tau}} m_\tau^2} \frac{a_{\rm He}^{3/2} a_{\rm t}^3}{(a_{\rm He}+a_{\rm t})^5} I_{\rm tn}, \tag{16}$$

Here $I_{\rm tn}$ is the numerically calculated factor including the information of phase space for this process. Analytic form of $I_{\rm tn}$ is shown in Ref. [1], and numerical result is depicted in Fig. 2.

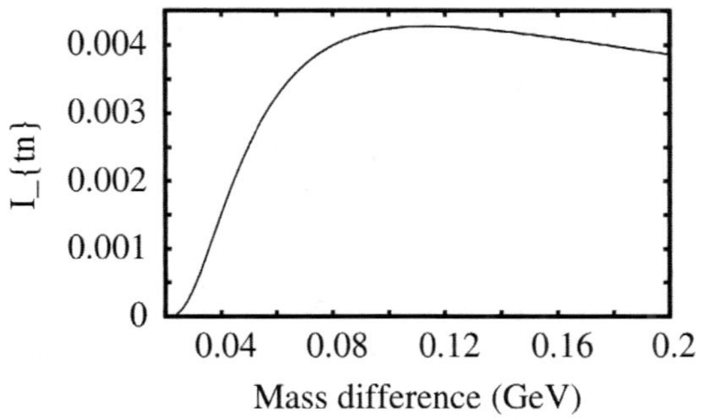

Figure 2. Factor $I_{\rm tn}$ in Eq. (16) as a function of mass difference between the stau and the neutralino. Here we took $m_{\tilde{\tau}} = 350$GeV, $\sin\theta_\tau = 0.8$, and $\gamma_\tau = 0$.

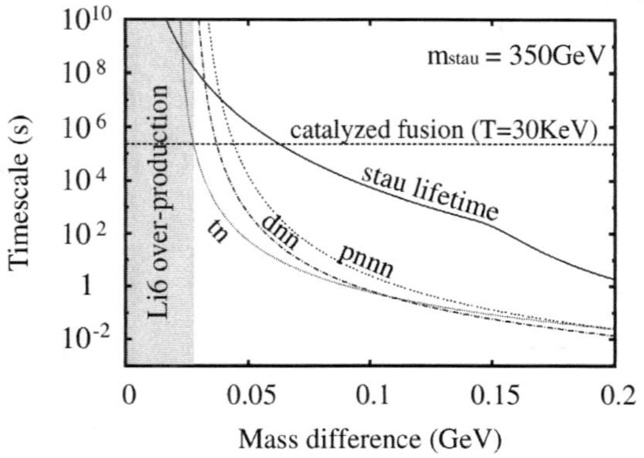

Figure 3. Timescale of spallation processes as a function of δm and the stau-catalyzed fusion at the universe temperature $T = 30$keV [13]. The lifetime of free $\tilde{\tau}$ (solid line) is also depicted. Here we took $m_{\tilde{\tau}} = 350$GeV, $\sin\theta_\tau = 0.8$, and $\gamma_\tau = 0$.

$\Delta_{\rm tn}$, $k_{\rm t}$, and $k_{\rm n}$ are defined as

$$\begin{aligned} \Delta_{\rm tn} &\equiv \delta m + \Delta_{\rm He} - \Delta_{\rm t} - \Delta_{\rm n} - E_{\rm b}, \\ k_{\rm t} &\equiv \sqrt{2m_{\rm t}\Delta_{\rm tn}}\ , \quad k_{\rm n} \equiv \sqrt{2m_{\rm n}\Delta_{\rm tn}}\ , \end{aligned} \tag{17}$$

where Δ_X is the excess energy of the nucleus X, and $E_{\rm b}$ is the binding energy of $(\tilde{\tau}\,^4{\rm He})$ system.

3. Comparing the rate of spallation reaction with that of stau-catalyzed fusion

We compare the rate of the spallation and that of the stau-catalyzed fusion. We first note that the rate of stau-catalyzed fusion strongly depends on the temperature [13], and we fix the reference temperature to be 30keV. Staus begin to form a bound state with ^{4}He at this temperature, which corresponds to cosmic time of 10^3s. Thus the bound state is formed when the lifetime of staus is longer than 10^3s.

Figure 3 shows the timescale of the spallation processes as a function of δm. The lifetime of free stau is plotted by a solid line. We took the reference values of $m_{\tilde{\tau}} = 350$GeV, $\sin\theta_\tau = 0.8$, and $\gamma_\tau = 0$. The inverted rate of the stau-catalyzed fusion at the temperature of 30keV is also shown by the horizontal dashed line. Once a bound state is formed, as long as the phase space of spallation processes are open sufficiently that is $\delta m \gtrsim 0.026$GeV, those processes dominate over other processes. There $\tilde{\tau}$ property is constrained to evade the over-production of d and/or t. For $\delta m \lesssim 0.026$GeV, the dominant process of $(\tilde{\tau}\,^4{\rm He})$ is stau-catalyzed fusion, since the free $\tilde{\tau}$ lifetime is longer than the timescale of stau-catalyzed fusion. Thus light gray region is forbidden due to the over-production of ^{6}Li.

This interpretation of Fig.2 is not much altered by varying the parameters relevant with $\tilde{\tau}$. First cross sections of spallation processes are inversely proportional to $m_{\tilde{\tau}}$, and then the timescale of each process linearly increases as $m_{\tilde{\tau}}$ increases. Thus, even when $m_{\tilde{\tau}}$ is larger than $m_{\tilde{\tau}} = 350$GeV by up to a factor of ten, the region of ^{6}Li over-production scarcely changes. Next we point out that our result depend only mildly on the left-right mixing of the stau. Indeed, cross section of the ^{4}He spallation is proportional to $\sin^2\theta_\tau$. Its order of magnitude will not change as long as the right-handed component is significant.

4. Summary

Long-lived charged massive particles provides some exotic nuclear reactions in the big bang nucleosynthesis. So it is important for understanding the property of long-lived charged massive particles to understand what type of exotic nuclear induce over-production (-destruction). Newly included in the present work is the spallation of the ^{4}He in the stau-^{4}He bound state given in Eq.(1). This process is only present in the model which predicts the long-lived charged particles due to the phase space suppression with the weakly interacting daughter particle.

We calculated the rate of the helium-4 spallation processes analytically, and compared it with that of catalyzed fusion. We found that the spallation of ^{4}He nuclei dominate over the catalyzed fusion as long as the phase space of the spallation processes are open and hence the property of long lived stau is constrained from avoiding the overproduction of a deuteron and/or a triton.

Acknowledgments

This work was supported in part by the Grant-in-Aid for the Ministry of Education, Culture, Sports, Science, and Technology, Government of Japan, No. 21111006, No. 22244030, No. 23540327(K.K.), No. 24340044 (J.S.), and No. 23740208 (M.Y.).

References

[1] Jittoh T, Kohri K, Koike M, Sato J, Sugai K, Yamanaka M and Yazaki K 2011 *Phys. Rev.* **D84** 035008 (*Preprint* `1105.1431 [hep-ph]`)
[2] Pospelov M 2007 *Phys. Rev. Lett.* **98**, 231301 (*Preprint* `hep-ph/0605215`)
[3] Jittoh T, Kohri K, Koike M, Sato J, Shimomura T and Yamanaka M 2007 *Phys. Rev.* **D76** 125023 (*Preprint* `0704.2914 [hep-ph]`)
[4] Bird C, Koopmans K and Pospelov M 2008 *Phys. Rev.* **D78** 083010 (*Preprint* `hep-ph/0703096`)
[5] Jittoh T, Kohri K, Koike M, Sato J, Shimomura T and Yamanaka M 2008 *Phys. Rev.* **D78** 055007 (*Preprint* `0805.3389 [hep-ph]`)
[6] Kamimura M, Kino Y and Hiyama E 2009 *Prog. Theor. Phys.* **121** 1059 (*Preprint* `0809.4772 [nucl-th]`)
[7] Jittoh T, Kohri K, Koike M, Sato J, Shimomura T and Yamanaka M 2010 *Phys. Rev.* **D82** 115030 (*Preprint* `1001.1217 [hep-ph]`)
[8] Kawasaki M, Kohri K and Moroi T 2005 *Phys. Lett.* **B625** 7 (*Preprint* `astro-ph/0402490`)
[9] Kawasaki M, Kohri K and Moroi T 2005 *Phys. Rev.* **D71** 083502 (*Preprint* `astro-ph/0408426`)
[10] Griest K and Seckel D 1991 *Phys. Rev.* **D43** 3191
[11] Profumo S, Sigurdson K, Ullio P and Kamionkowski M 2005 *Phys. Rev.* **D71** 023518 (*Preprint* `astro-ph/0410714`)
[12] Jittoh T, Sato J, Shimomura T and Yamanaka T 2006 *Phys. Rev.* **D73** 055009 (*Preprint* `hep-ph/0512197`)
[13] Hamaguchi K, Hatsuda T, Kamimura M, Kino Y and Yanagida T T 2007 *Phys. Lett.* **B650** 268 (*Preprint* `hep-ph/0702274`)
[14] Feng J L, Rajaraman A and Takayama F 2003 *Phys. Rev. Lett.* **91** 011302 (*Preprint* `hep-ph/0302215`)
[15] Buchmuller W, Covi L, Hamaguchi K, Ibarra A and Yanagida T 2007 *JHEP* **0703** 037 (*Preprint* `hep-ph/0702184`)
[16] Borisyuk D 2010 *Nucl. Phys.* **A843** 59 (*Preprint* `0911.4091 [hep-ph]`)
[17] Kubon G *et al.* 2002 *Phys. Lett.* **B524** 26 (*Preprint* `nucl-ex/0107016`)
[18] Wong C W 1994 *Int. J. Mod. Phys.* **E3** 821
[19] Yoshitake M *et al.*, private communication
[20] Egelhof P 2001 *Prog. Part. Nucl. Phys.* **46** 307-316
[21] Mohr P J, Taylor B N and Newell D B 2008 *Rev. Mod. Phys.* **80** 633
[22] Firestone R B 1999 *Table of Isotopes. Eighth Edition Volume I* (New York: John Wiley and Sons Inc)

Dynamical interaction between astrophysical systems and dark substructure

Alma X González-Morales [1], Octavio Valenzuela[2], L A Aguilar[3]

[1] Instituto de Ciencias Nucleares, UNAM, Apartado Postal 70-543, 04510, Ciudad Universitaria, D.F., Mexico.

[2] Instituto de Astronomia, UNAM, Apartado Postal 70-264, 04510, Ciudad Universitaria, D.F., Mexico.

[3] Instituto de Astronomia, UNAM, Apartado Postal 877, 22860 Ensenada, B.C., Mexico

E-mail: alma.gonzalez@nucleares.unam.mx

Abstract. We explore the possibility to constrain the dark matter (DM) sub-structure properties, mass and distribution, at sub-galactic scales, by considering the dynamics of some astrophysical system under the dynamical influence of such sub-structure. We present preliminary results discussing the sensitivity of the Solar System dynamics, and dSphs wide binaries abundance, to the small scale DM distribution.

1. Introduction

The Λ-CDM cosmological scenario is consistent with the large scale Universe observations; however, one of the main ingredients, the Dark Matter, has not yet been detected. One of the principle uncertainties, regarding the direct and indirect detection experiments, is the distribution of DM at the relevant scales [1]. Specifically, models based on WIMP's DM predict a large abundance of substructure at sub-galactic scales. This has been considered as a potential problem for the Λ-CDM scenario because it is not obvious that the predicted dark substructure has an observational counterpart [2], see however [3].

On the other hand, the properties of the dark matter substructure, mass and abundance, are related to the properties of the power spectrum for primordial fluctuations; the nature of the dark matter; and the dynamical evolution of the fluctuations. In the Λ-CDM model the primordial mass power spectrum is fixed by inflation, and the dynamical evolution is followed trough $N-$body simulations, so that to constrain the properties of the substructure can constitute a test for the nature of the DM particle. Once a specific DM particle is assumed, the survival of the smallest structures trough the structure formation history is by itself a topic of discussion, being the extrapolation from N -body simulations at larger scales and the semi-analytic models the most used tools used to estimate the amount of substructure that should be present today [4, 5].

It has been shown that the free streaming, and/or the kinetic decoupling, of WIMP's like DM candidates leads to a mass-power spectrum cut off at approximately $M_{min} \sim 10^{-4} - 10^{-12} M_\odot$ [6, 7], while in the case of Axion like DM the cut off could be as small as $10^{-18} - 10^{-20}\, M_\odot$ [8, 9].[1] This small scale cut-off is unreachable to study with current N-body simulations, nor with analytical models, and it is also undetectable through current observational techniques.

[1] Other DM candidates like the Warm Dark Matter are also consistent with the large scale but they predict a PS cut off at approximately the dwarf galaxies mass [10]).

Content from this work may be used under the terms of the Creative Commons Attribution 3.0 licence. Any further distribution of this work must maintain attribution to the author(s) and the title of the work, journal citation and DOI.
Published under licence by IOP Publishing Ltd

Here we explore an strategy that could allow us to set upper limits to the the existence of the dark small structure — microhalos and streams— in the Galactic halo. Since the orbital elements of planets in our Solar System are known with outstanding accuracy [11], they might be sensitive to dynamical perturbations triggered by dark matter substructure. In this work we focus on the variation of the astronomical unit, due to the presence of dark substructure and how this compares with the reported anomaly in [12]. We also use our methodology to study the stability of wide binary stars in dSph's of the Milky Way. These kind of binaries has not been detected yet, but previous works suggest that their presence could be in contradiction with the dark matter hypothesis. Here we present some preliminary results on this direction. This proposal was first presented in [13] and will be presented with more detail in [14].

2. Target tidal interactions with mini-halos and streams

The dynamics of some astrophysical systems is known with great accuracy that allows to constrain the existence of incredibly small gravitational perturbations. Here we consider the dynamical perturbations that such systems (the target) could undergo in the presence of dark substructure. Dark substructure could be in the form of mini-halos or streams, —former DM sub-halos that had been already disrupted by tidal forces or encounters—. We will treat the encounters with mini-halos, and streams, as independent of each other and separately. First, we present the equations we will use for independent single encounters, and we then explain how we follow the evolution of the target under multiple encounters.

2.1. One single encounter with the target

Our first approach is to consider spherical micro-halos: the impulsive energy injection to a target orbit during one single encounter, is

$$\frac{\langle \Delta E \rangle_{mh}}{|E_b|} = \frac{14\, G\, M_p^2}{3\, M_c\, v_0^2\, p^4} a^3, \tag{1}$$

where, the impact parameter p is the minimal distance between the target and the pertuber; M_p is the perturber mass; M_c is the mass of the central body in the target; v_0 is the relative encounter velocity; and a is the mean separation of the target system (it coincide with the semi-major axis for some systems) [15, 16]. Equation 1 is valid whenever the characteristic size of the perturber is smaller that the impact parameter (point mass approximation); when the impact parameter is much larger than the extent of target (distant tide approximation); and the encounter velocity is much larger than the internal velocity of the target (impulsive regime). We have verified that the systems under study falls in these approximations.

In our second approach, we consider the perturbers to be a sort of dark streams of cylindrical symmetry characterized by their linear mass density. The simplest model for a stream is 1-dimensional, i.e. a line of linear mass density λ_{ST}. In this case the energy injection, normalized to the binding energy of the target, is:

$$\frac{\langle \Delta E \rangle_{ST}}{|E_b|} = \frac{4\,\pi^2\, G\, \lambda_{ST}^2}{M_c\, v_0^2\, p^2 \sin^2(\theta)} a^3 \left(\cos^2(\alpha)\cos^2(\psi) + \sin^2(\psi)\right). \tag{2}$$

The angles α and ψ defines the orientation of the target orbital plane, and θ the relative direction of movement of target, both with respect to the stream. More realistic models include the finite cross section stream with constant density; or with a power law density, either with a cusp or a core in the center. These models reduces to the one dimensional when the impact parameter is larger than the cross-section of the stream. Here we will consider only the 1-dimensional case and the analysis for the other models, as well as the derivation of equation 2, will be presented in [14].

2.2. Multiple encounters during the lifetime of the target

As a result of multiple encounters, either with mini-halos or streams, the fractional energy of the target system will change. Our general approach to study the effect of this multiple encounters consists of Monte Carlo experiments: first, we draw random values for the encounter parameters (p, v_0, α, etc.), and use them to assess the effect of one single encounter trough the fractional energy, $\langle \Delta E \rangle / E_b$. We then increment the experiment timer with $\Delta t_{enc} \propto 2\,p/v_0$ (the time it takes the perturber to traverse a sphere of radius equal to the impact parameter at the encounter velocity), and then we continue sampling encounters until the time limit is reached, (the lifetime of the target), or until the fractional energy reaches the unity (i.e. the target gets unbound). We repeat this procedure a number of times, so that we can construct an ensemble of histories of the multiple encounters between the target and the perturber population. Finally, we find the median value of the total dynamical perturbations the target experimented due to the presence of the dark substructure.

In the case of encounters with mini-halos we sample the impact parameters from a nearest-neighbor distribution, with numerical density $\nu = \rho_{dm} f_s / m_{mh}$, ($f_s$ is the fraction of the mass density, ρ_{dm}, of dark matter that is in form of mini-halos of mass m_{mh}). On the other hand, if the perturbers are streams, we use the cumulative distribution shown in figure A2, and we scale the physical impact parameter using $p = \left(\sqrt{\lambda_{ST}\,\rho^* / f_s\,\rho_{dm}}\right) x$, according to the different systems under study. See Appendix A. For the velocity distribution we use a Maxwellian distribution; and a uniform distribution function for the geometric parameters, (β, α, etc.). In the next section we present a couple of examples that illustrates the applicability of the method we have described.

3. Examples of applicability

3.1. Anomalies in the Solar System

There is some evidence of a variation in the astronomical unit (AU), the increase is reported to be about $15\,cm\,yr^{-1}$ [12]. This anomaly remains unexplained,(see however [17]).

We have applied our formalism of multiple encounters with dark substructure to the Sun-Earth system, to check if this anomaly could be explained by the dynamic perturbations due to the presence of dark substructure. We assumed the canonical value for the local dark matter density, $\rho_{dm} \sim 10^{-18} M_\odot/AU^3$ $(0.3GeV/cm^3)$, and a mean encounter velocity of $v_0 = 200\,Km/s$.

In figure, 1 we show an example of the resultant probability distribution for an ensemble of Monte-Carlo experiments. Each of the runs in the ensemble corresponds to a history of encounters between the Earth-Sun system and dark streams, (with $\lambda = 10^{-6} M_\odot/AU$), for about 4 Gyr. Since the distribution is non-symmetric, we use the median value instead of the mean. In figure 2 we show the median value of the total fractional energy for different values of the perturber linear density. We see that in general these kind of perturbations are not enough to explain the reported anomaly, indicated by the dashed line. A similar trend was found when studying interactions with mini-halos.

3.2. The case of wide binaries in dSphs Galaxies

The system under study now are the wide binaries under the dynamical perturbation due to dark substructure, inside the dwarf spheroidal galaxies in the Milky Way. The survival of such kind of binary systems has been proposed as a test for the dark matter hypothesis, since it is subjected to the dynamical interaction with the dark matter, regardless it is distributed homogeneously [18], or along dark substructure [19].

Our work focuses on the dynamical interaction between such binaries and dark substructure that could be in form of mini-halos or dark streams. In [19] only considered perturbations caused by structures that produce a significant evolution of the binary as a result of one single encounter, that is the catastrophic regime. With the proposed approach in this work, we are able to test the possibility that the dynamical evolution of the binary is caused by the cumulative effect of multiple encounters with substructure that is not massive enough to be in the catastrophic regime, i.e. we consider also the diffusive regime.

We assume the following parameters for the binaries and the dark halo in which they are embedded: a

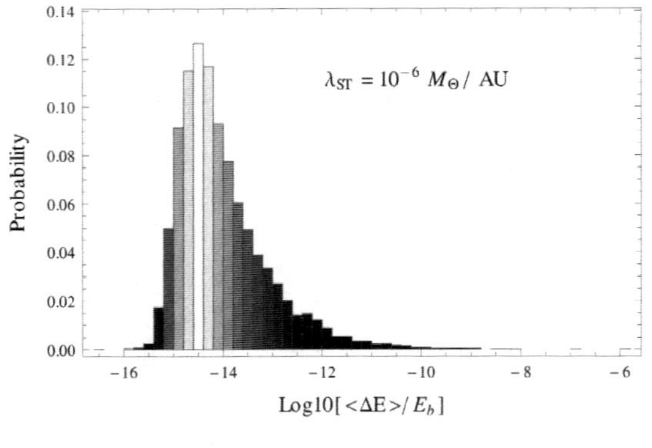

Figure 1. Probability distribution function for the Earth-Moon system binding energy change. Each run in the ensemble correspond to the result after multiple encounters with streams of linear mass density $\lambda_{ST} = 10^{-6} M_{\odot}/AU$ for about $4\,Gyr$.

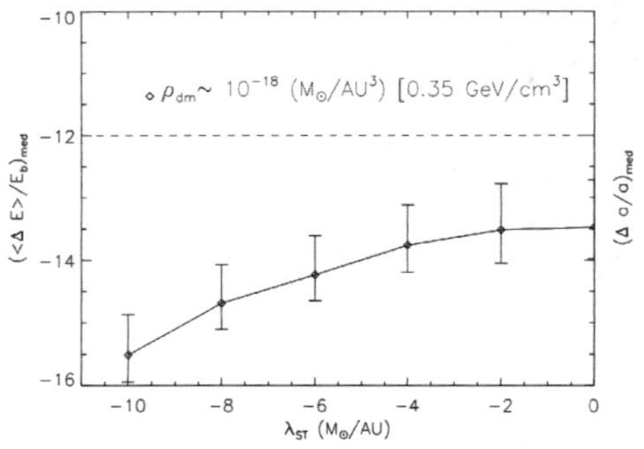

Figure 2. Median value of the Earth-Moon system binding energy change, as a result of multiple encounters with streams with linear mass density λ_{ST}. The dashed line correspond to the reported variation of the AU ($\approx 15\,cm/yr$).

fiducial binary separation of $a = 0.1pc$; a relative velocity distribution centered at $\sqrt{2}\sigma \approx 14\,Km/s$, (given that typical velocity dispersion in dSphs is about $\sigma \approx 10km/s$ [20]); a NFW density profile with virial mass $M_{vir} = 10^9 M_{\odot}$ and concentration $c = 23.1$ for the dSph (as were considered in [19]),and a radius of $r \approx 0.1\,kpc$ where the binaries could be (this radius is smaller than the typical half-light radius of the dSph's), the dark matter density at this radius is $\rho_{dm} \approx 2.2 * 10^{-17} M_{\odot}/AU^3$).

In figure 3 we show the results for dynamical interaction between the binaries and mini-halos acting as perturbers. We report the median value of the total injection energy resulted in the ensembles histories of multiple encounters, as a function of the mini-halo mass and for three different times of the total evolution. There is a minimum mass of the mini-halo that causes the total disruption of the binaries, i.e. $\delta E_{tot}/|E_b| = 1$, that depend on the time the binary is subjected to the perturbation field. In the case the evolution time is about $10Gyr$ the minimum mass needed to destroy the binaries is about $10^2 M_{\odot}$, this mass correspond to the boundary between the catastrophic and the diffusive regime set in [19], in that sense we recover their results. However, we also obtain that it is not enough to consider only the catastrophic encounters because the diffusive ones can also produce significant secular evolution of the binary system. Now, in figure 4 we show the results for the interaction of the binaries with streams like structures. Again we show the cumulative effect of the interactions for three different times of the evolution. Our findings are similar to those for mini-halos, i.e. there is a minimum value of the linear mass density of the streams at which the encounters are energetic enough to unbound the binary system, again this minimum value depends on the time the binary is under perturbation.

4. Discussion

We found that neither the perturbations due to mini-halos nor due to streams, for about the lifetime of the Earth, would be energetic enough to explain the reported anomaly in its variation. We would like to remark that this result is subjected to the mean value of the local dark matter density which is determined at larger scales (about Kilo parsecs). However, it remains open the possibility of having large over-densities at lower scales, still consistent with the mean density. Increasing the value of the dark matter density has the effect of increasing the total fractional energy given, so it would be possible to explain the anomaly in that case.

For the wide binaries, we found that its disruption could be driven by the dynamical contribution of small perturbers, as well as to the presence of dark streams like substructure. Our results strongly depends on the time the binary evolution is under the perturbation field; as the lifetime of this kind of systems is

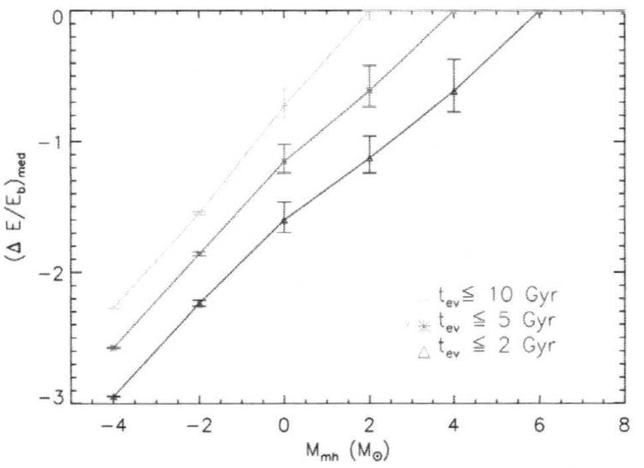

Figure 3. Median value of the binary system binding energy total change as a result of multiple encounters with mini-halos of mass m_{mh}, for three different times of evolution. The minimum mass needed to unbind the binary depends on the total evolution time.

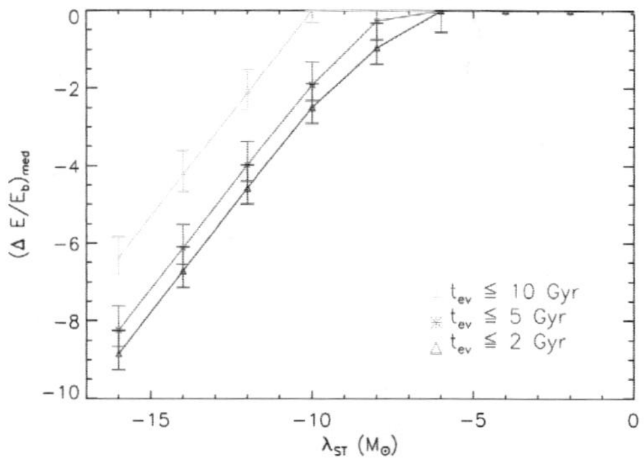

Figure 4. Median value of the binary system binding energy change as a result of multiple encounters with streams of linear mass density λ_{ST}, for three different times of evolution. Again there is a minimum linear mass density needed to unbind the binary.

uncertain thus their existence is not a sharp discriminant of the dark matter hypothesis.

In this contribution we have considered that all the dark matter is in form of mini-halos, or streams, separately. A more realistic situation would be one in which a fraction of the mass is in form of mini-halos and the rest is in form of streams. Such cases, and their implications, are considered in a work that is under revision now as well as the applicability of the proposed procedure to other astrophysical systems.

Acknowledgments
Alma wants to thanks the PASCOS's organizers, and the ICN-UNAM teaching department, for all the financial support when attending at this conference. Alma's work is supported by DGAPA-UNAM grant No. IN115311 and a CONACyT PhD fellowship.

References

[1] Green, A. M. 2010, arXiv:1004.2383
[2] Klypin, A., Kravtsov, A. V., Valenzuela, O., and Prada, F. 1999, ApJ, 522, 82
[3] Xu, D. D., Mao, S., Cooper, A. P., et al. 2010, MNRAS, 408, 1721
[4] Diemand,J. and others, *Clumps and streams in the local dark matter distribution*, *Nature* **454** (2008) 735–738.
[5] R. E. Angulo and S. D. M. White, *The Birth and Growth of Neutralino Haloes*, arXiv:0906.1730.
[6] Green,A. M., Hofmann,S. and Schwarz,D. J. *The power spectrum of SUSY - CDM on sub-galactic scales*, *Mon.Not.Roy.Astron.Soc.* **353** (2004)astro-ph/0309621.
[7] Profumo, S., K. Sigurdson, and M. Kamionkowski, *What mass are the smallest protohalos?*, *Phys.Rev.Lett.* **97** (2006) 031301,astro-ph/0603373.
[8] M. C. Johnson and M. Kamionkowski, *Dynamical and Gravitational Instability of Oscillating-Field Dark Energy and Dark Matter*, *Phys.Rev.* **D78** (2008) 063010, arXiv:0805.1748.
[9] Barranco, J. and Bernal,A., *Self-gravitating system made of axions*, *Phys.Rev.* **D83** (2011) 043525,arXiv:1001.1769.
[10] P. Colin, V. Avila-Reese, and O. Valenzuela, *Substructure and halo density profiles in a warm dark matter cosmology*, *Astrophys.J.* **542** (2000) 622–630,astro-ph/0004115
[11] E. Pitjeva and N. Pitjev, *Estimations of changes of the Sun's mass and the gravitation constant from the modern observations of planets and spacecraft*, arXiv:1108.0246.
[12] J. D. Anderson and M. M. Nieto, *Astrometric solar-system anomalies*, in *IAU Symposium* (S. A. Klioner, P. K. Seidelmann, and M. H. Soffel, eds.), vol. 261, p. 189, 2010.
[13] González-Morales, A. X., Valenzuela, O., & Aguilar, L. 2011, Revista Mexicana de Astronomia y Astrofisica Conference Series, 40, 15

[14] Gonzalez-Morales Alma X., Valenzuela, O. and Aguilar, L. A., *The Dynamics of astrophysical systems as a constraint to DM Substructure*, in preparation
[15] Binney, J. and Tremaine,S. , *Galactic Dynamics: Second Edition*. Princeton University Press, Princeton, NJ USA, 2008., 2008.
[16] L. A. Aguilar and S. D. M. White, *Tidal interactions between spherical galaxies*, *ApJ* **295** (Aug., 1985) 374–387.
[17] L. Iorio, *An Empirical Explanation of the Anomalous Increases in theAstronomical Unit and the Lunar Eccentricity*, *Astron.J.* **142** (2011) 68, [arXiv:1102.4572].
[18] X. Hernandez and W. H. Lee, *The tightening of wide binaries in dSph galaxies through dynamical friction as a test of the Dark Matter hypothesis*, arXiv:0803.1507.
[19] Penarrubia, J. *et. al.*, *Binary stars as probes of dark substructures in dwarf galaxies*, arXiv:1005.5388.
[20] M. G. Walker, M. Mateo, E. W. Olszewski, O. Y. Gnedin, X. Wang, *et. al.*, *Velocity Dispersion Profiles of Seven Dwarf Spheroidal Galaxies*, arXiv:0708.0010.

Appendix A. Distribution function of impact parameters for stream like perturbers

In the case of having streams as perturbers, the basic problem in using a nearest neighbor like distribution is that we would be linking a sample volume with the determination of the distribution of impact parameters. To break this link we will proceed along the following lines: suppose one is given a random realization of a region of 3-space where lines have been thrown at random and completely uniformly. One can then choose a collection of arbitrary and sufficiently separated points within this region, and determine the minimum impact parameter for each point. This will result in a sampling of the underlying distribution of impact parameters, completely independent of the introduction of a sample volume.
We need to relate the starting realization to a specific universe with a certain spatial mass density and linear mass density for the lines. Then we do have to introduce, now, a sample volume to measure mass density. Exploring this in the same realization, and using different sample volumes we found that the number density of filaments varies with the size of the spherical neighborhood used, but the mass density does not. Thus, the mass density is a well defined physical quantity in this procedure.

For our numerical realization we assumed a linear density of streams $\lambda^* = 1$ and the measured mass density is some value ρ^*, both given some units of mass $[m]$ and longitude $[l]$. By scaling to the physical quantities, λ_{ST} and ρ_{DM}, one finds that the physical impact parameter is given by $p = \left(\sqrt{\lambda_{ST}\rho^*/f_s\rho}\right) x AU$. For the particular realization used in this work we have $\rho^* = 1205.08$, and the other parameters are given according to the different systems under study. We have separated completely the determination of impact parameters, which is a purely geometric problem, from the mass scaling problem, which is a physical problem.

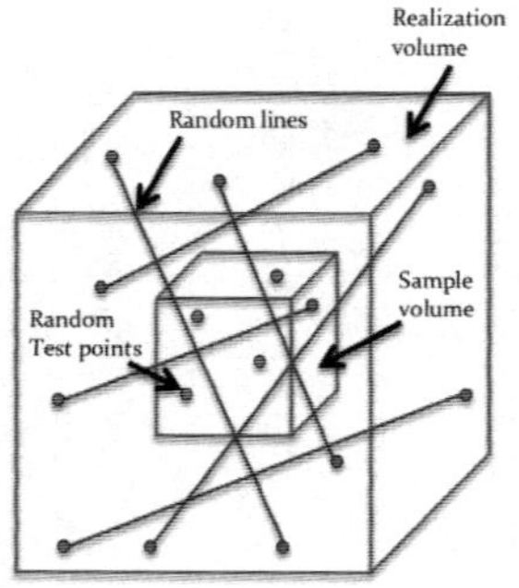

Figure A1. Illustration of the distribution of lines and points used to generate the impact parameter PDF for stream like perturbers.

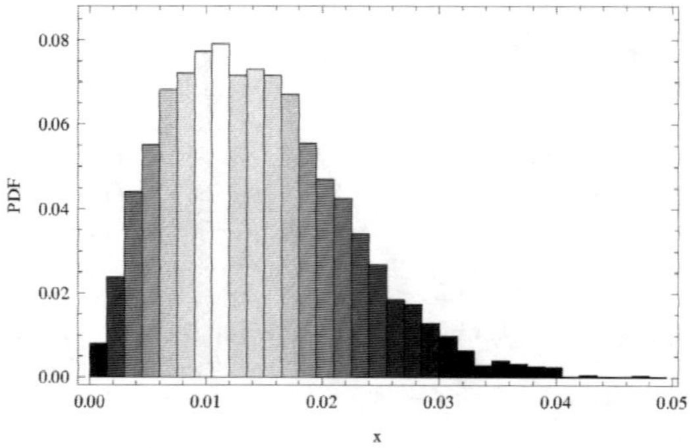

Figure A2. Probability distribution function of impact parameters for stream like perturbers. The impact parameter is scaled according to the particular physical system using: $p = \left(\sqrt{\lambda_{ST}\rho^*/f_s\rho}\right) x$

Simulating Galaxy Formation with Scalar Field Dark Matter

L A Martinez and T Matos

Departamento de Física, Centro de Investigación y de Estudios Avanzados del IPN, A.P. 14-740, 07000 México D.F., México.

E-mail: `lmedina@fis.cinvestav.mx, tmatos@fis.cinvestav.mx`

Abstract.
In this work we consider the formation of galaxies using the scalar field dark matter (SFDM) model. In this model the galactic haloes are astronomical Bose-Einstein Condensate drops of SF. This haloes are characterized by a constant-density core and are consistent with observed rotation curves of dark matter dominated galaxies, a missing feature in CDM haloes resulting from DM only simulations. We include the baryonic component to the SF haloes in a set of grid-based hydrodynamic simulations.

1. INTRODUCTION

Over the last years works on galaxy formation are gradually becoming more and more complete, incorporating a wide array of physical processes. So great progress has been made in the field,but most of this works model the formation of galaxies within the current structure formation paradigm in which the Universe is dominated by a cold dark matter component and a cosmological constant, the ΛCDM cosmological model, that has been successful in reproducing many large scale observable properties of the Universe, however, still faces many challenges from observations of galaxies on small scales. The incorporation of a new kind of DM, different from the one proposed by the CDM holds out the possibility of resolving some of these issues. An alternative scenario that has received much attention is the scalar field dark matter (SFDM) model. The main idea is simple [6], the nature of the DM is completely determined by a fundamental scalar field Φ. In this dark matter and galaxy formation scenario the DM haloes are naturally cored. Robles & Matos [10] found excellent agreement with data of rotation curves for LSB galaxies using the minimum disk hypothesis (neglecting the baryonic component).

In this work we explore the behavior of those rotation curves but now taking into account the baryonic component. We ran some simple simulations evolving the hydrodynamic equations for a gas distribution embedded in a SFDM galactic halo.

2. Basic Setup

The basic theoretical framework states that disk galaxies arises from the gravitational collapse of a rotating protogalactic cloud of gas within the gravitational potential well of the dark halo. The gas eventually settles in centrifugal equilibrium at the center of the halo potential well forming a rotationally supported gas disk. Our basic picture is a rotating distribution of gas within a spherical Dark Matter Halo. The potential is assumed to be generated by a static distribution of

Content from this work may be used under the terms of the Creative Commons Attribution 3.0 licence. Any further distribution of this work must maintain attribution to the author(s) and the title of the work, journal citation and DOI.
Published under licence by IOP Publishing Ltd

dark matter that remains fixed throughout the simulation. We focus on dark matter dominated systems so the self gravity of the gas is ignored.

2.1. Code

The simulations were performed using the latest version of the ZEUS-MP code, released by the Laboratory for Computational Astrophysics. The physics suite in this code includes gas hydrodynamics, ideal MHD, flux-limited radiation diffusion, self gravity, and multispecies advection [9]. In this work we focus in solving only the standard hydrodynamics equations, where the description of the physical state of a fluid element is specified by the following set of fluid equations relating the mass density (ρ), velocity ($\mathbf{v}$) and gas internal energy density (e).

$$\frac{D\rho}{Dt} + \rho\nabla\cdot\mathbf{v} = 0, \tag{1}$$

$$\rho\frac{D\mathbf{v}}{Dt} = -\nabla P - \rho\nabla\Phi, \tag{2}$$

$$\rho\frac{D}{Dt}\left(\frac{e}{\rho}\right) = -P\nabla\cdot\mathbf{v}, \tag{3}$$

where the Lagrangean (or comoving) derivative is given by the usual definition:

$$\frac{D}{Dt} \equiv \frac{\partial}{\partial t} + \mathbf{v}\cdot\nabla. \tag{4}$$

Spherical polar ($r;\theta;\phi$) coordinates are used. The simulations were run on a $200\times100\times100$ cell grid.

3. The Dark Matter Halo

The ΛCDM model still faces many challenges from observations of galaxies on small scales. For instance, in the ΛCDM simulations the halos present rising densities towards the central region behaving as $\rho\propto r^{-1}$ well within 1 kpc [9]. On the other hand, several observations suggest that the dynamics (rotation curves) of dwarf and LSB galaxies are more consistent with a constant central density, this is most commonly known as the cusp/core problem [3]. This and other problems motivate the proposal of alternative dark matter models to explain the observed central distribution of galaxies.

An alternative scenario that has received much attention in the last years is the scalar field dark matter (SFDM) model. The main idea is that the nature of the DM is completely determined by a fundamental scalar field Φ [6]. The SFDM model proposes that galactic haloes form by Bose-Einstein condensation of a scalar field (SF) whose boson has an ultra-light mass of the order of $m\sim10^{-22}eV$.In addition, the Compton length $\lambda_c = 2\pi\hbar/m$ associated to this boson is about $\sim$ kpc that corresponds to the dark halo-size of typical galaxies in the Universe. Thus, it has been proposed that these drops are the haloes of galaxies [8], i:e:, that haloes are huge drops of SF. Within the framework of the SFDM/BEC model exist an exact analytic solution for a static SF configuration, a finite temperature density profile which in the SFDM model represents a DM halo.

$$\rho(r) = \rho_0\frac{\sin^2(kr)}{(kr)^2}, \tag{5}$$

where the central density $\rho_0 = \rho(0)$ and k are fitting parameters. In Figure 1 we show the density profile,. it is in the log-log plot that we can see one of the main features of the

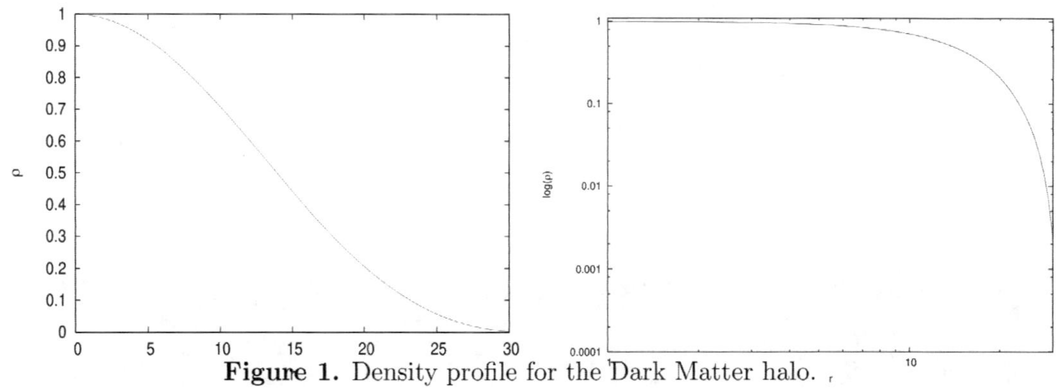

Figure 1. Density profile for the Dark Matter halo.

SFDM/BEC model, a flat core is present in the DM haloes and is explained by the model rather than just be assumed, so in this model the cusp/core problem is not even there in the first place.

From equation (5) and from Poisson's equation one can see that the potential of the DM distribution goes like

$$\phi_h(r) = \frac{GM_h}{a(r_0)}\left(\ln(r) + \frac{\sin(2kr)}{2kr} - Ci(2kr)\right), \tag{6}$$

where Ci is the cosine integral function. It is well known that a logarithmic potential maintains flat rotation curves at large radii, so one of the main features of the SFDM/BEC model is that the rotation curves are flat, which is consistent with the observations of many galaxies.

From equation (5) we obtain the accumulated mass at a radius r and the rotation curve profile

$$M(r) = \frac{2\pi\rho_0}{k^2}\left[r - \frac{\sin(2kr)}{2k}\right], \tag{7}$$

$$v^2(r) = \frac{2\pi\rho_0}{k^2}\left[1 - \frac{\sin(2kr)}{2kr}\right]. \tag{8}$$

4. Results

In this section we present the results of two simulations where the parameters, size and mass, of the DM halo remain the same but the initial distributions of gas are different. In the first configuration we have a rotating sphere of gas and for the second one we have a rotating gas disk. So the gas is embedded in a static DM halo, with the DM density distribution given in equation (5). When we say that the DM halo is static it means that its effect is treated as an external force in the momentum equation for the gas.

From equation (5) we compute the acceleration that the DM halo produces on the gas

$$-\nabla V = \frac{2\pi G\rho_0}{k^2}\left[\frac{1}{2k}\frac{\sin(2kr)}{r^2} - \frac{1}{r}\right]. \tag{9}$$

Also we imprint angular momentum to the initial distribution of gas such that it is in circular orbit in the mid-plane. Here we focus on dark matter dominated galaxies so the amount of gas that we put in the domain is such that it is the 10% of the DM halo mass.

For our first setup the initial distribution of gas is spherical. Figure 2 shows face-on slices of the gas density distribution, each plot is a color map of the gas density with the brightest colors representing the highest values, so the snapshots shows the evolution of the gas through the entire simulation. From first to second snapshot the increase in the density is appreciable, this is because the gas, initially distributed in spherical way, settles down to the mid-plane. Note that, although the initial distribution had azimuthal symmetry, a distinctive spiral pattern emerge in the center of the domain.

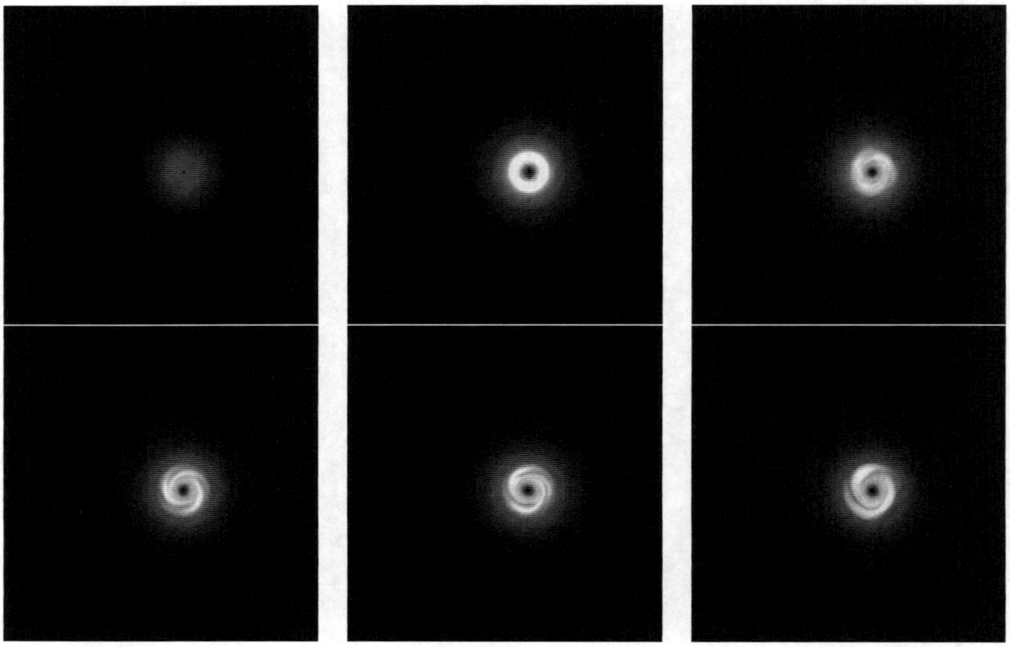

Figure 2. From left to right and from top to bottom we follow the temporal evolution of the gas density, sliced in the direction normal to the axis of rotation.

Meanwhile, Figure 3 shows edge-on slices of the gas density, here we can see the expected behavior for a rotating distribution. We begin with a spherical distribution but with the pass of time the density increase and the gas settles down in a plane perpendicular to the axis of rotation.

Figure 4 shows the temporal evolution of the circular velocity for the gas on the mid-plane, here we can see how the initial velocity drops at the right part of the domain, with the pass of time the right end of the curve rises to maintain a flat profile at the latest stages of the simulation. Clearly we can see that small oscillations appear over the profiles, and the last curve plotted in figure 4 resembles some circular velocity curves observed on LSB galaxies.

For our second setup the size and mass of the DM halo remain the same, but we start with a gas disk rotationally supported. Again, the amount of gas in the disk is such that it is the 10% of the DM halo mass. Very few changes can be seen in the shape of the gas density distribution along the simulation, so we focus on the circular velocity (figure 5).

Comparing the second plot from figure 4 and 5, which represent the simulations in their first

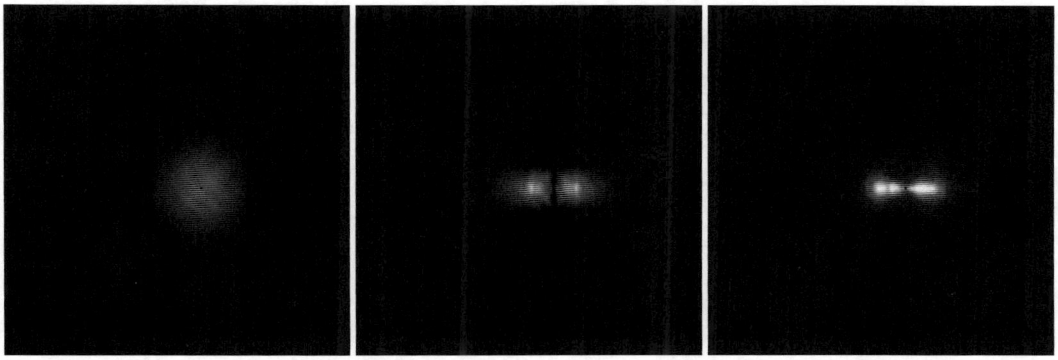

Figure 3. Temporal evolution of the gas density, sliced perpendicular to the midplane.

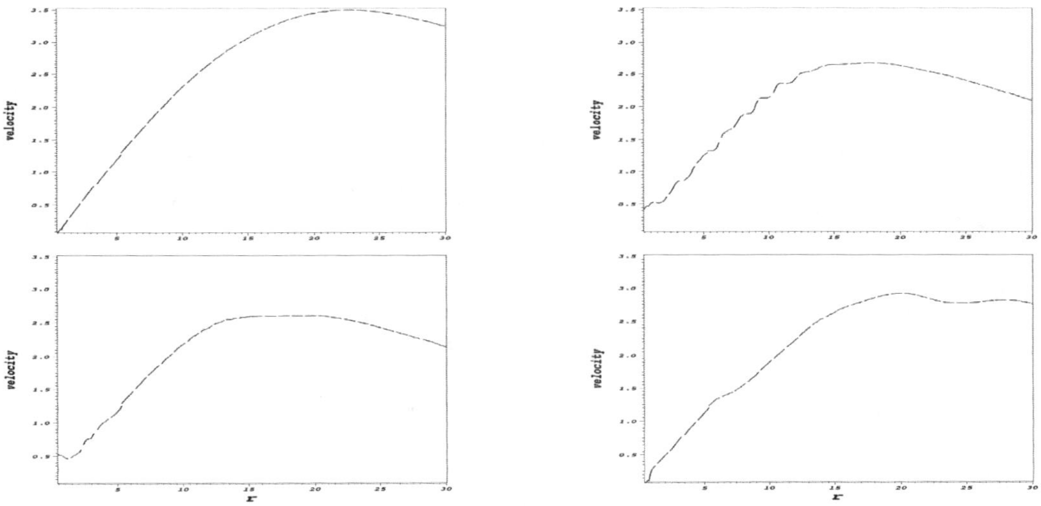

Figure 4. Temporal evolution of the gas circular velocity, taken from the midplane.

stages, we see a slightly different behavior of the curves. This is because at this stage the gas distributions are very different, it is only when the gas in the first simulation settles down to the mid-plane that the curves look very alike (see fourth plot from figure 4 and 5). This means that the final configuration in our first simulation reproduce rotation curves very similar to those seen in the simulation with the gas disk, plus the curves resemble those observed in LSB galaxies. We conclude that with this simple scenario of a rotating spherical distribution of gas embedded in a SFDM galactic halo we can reproduce some of the features seen in dark matter dominated disk galaxies.

Journal of Physics: Conference Series **485** (2014) 012022 doi:10.1088/1742-6596/485/1/012022

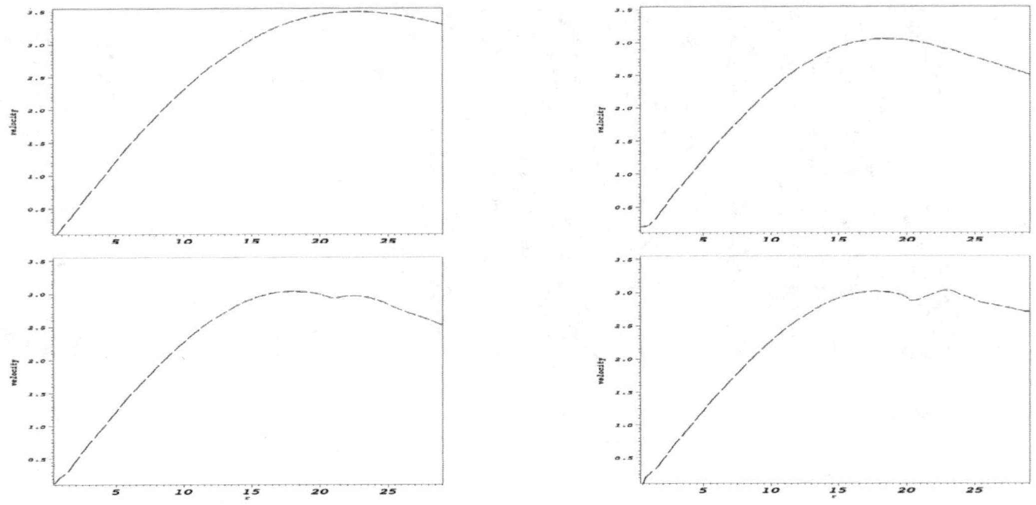

Figure 5. Temporal evolution of the gas circular velocity, taken from the midplane.

5. Conclussions

The problems that CDM faces at galactic scales had prompted the development of alternative DM models. Particularly the scalar field dark matter (SFDM) model has proved to be a serious alternative to ΛCDM. We are exploring the galaxy formation under the SFDM model by adding the baryonic component. Describing the gas with just the standard hydrodynamics equations we found that starting with a distribution of gas that eventually settles in centrifugal equilibrium, the obtained rotation curves resemble those observed in LSB galaxies.

We have performed 3D hydrodynamic simulations with simple physical processes for the gas, and although we argue that huge quantities of feedback events (star formation, supernovas) are not needed, feedback must be added to our simulations but in a way different of that used in CDM-based simulations because we are not trying to change the shape of the DM density profiles through feedback processes. In future works we will do this in an attempt to simulate realistic galaxies within the SFDM model.

References

[1] Benson A 2010 *Physics Reports* **495** 33
[2] D'Onghia E et al. 2012 arXiv:1204.0513
[3] de Blok WJG 2010 *Adv. Astron.* 2010 Article ID 789293 14 pages
[4] Governato F, Zolotov A, Pontzen A et al. 2012 *MNRAS* **422** 1231
[5] Governato F et al. 2012 *MNRAS* in press (arXiv:1202.0554v2)
[6] Guzman F S and Matos T 2000 *Class. Quantum Grav.* **17** L9-L16
[7] Kuzio de Naray R and Spekkens K 2011 *ApJ* **741** L29
[8] Matos T and Ureña L A 2001 *Phys Rev D* **63** 063506
[9] Navarro J F et al. 2010 *MNRAS* **402** 21
[10] Robles V H and Matos T 2013 *ApJ* **763** 19

Dark energy from inflation

Christophe Ringeval[1], Teruaki Suyama[2], Tomo Takahashi[3], Masahide Yamaguchi[4] and Shuichiro Yokoyama[5]

[1] Centre for Cosmology, Particle Physics and Phenomenology, Institute of Mathematics and Physics, Louvain University, 2 Chemin du Cyclotron, 1348 Louvain-la-Neuve, Belgium
[2] Research Center for the Early Universe (RESCUE), Graduate School of Science, The University of Tokyo, Tokyo 113-0033, Japan
[3] Department of Physics, Saga University, Saga 840-8502, Japan
[4] Department of Physics, Tokyo Institute of Technology, Tokyo 152-8551, Japan
[5] Department of Physics and Astrophysics, Nagoya University, Aichi 464-8602, Japan

E-mail: `christophe.ringeval@uclouvain.be`

Abstract. We show that a very light scalar field experiencing quantum fluctuations during primordial inflation can explain the current cosmic acceleration. Provided its mass does not exceed the Hubble parameter today, this field has been frozen during the cosmological ages to start dominating the universe only recently. Assuming this scenario to be correct, and using supernovae data, the model predicts the energy scale of primordial inflation to be around a few TeV and suggests that it has lasted for an extremely long period. Dark energy could therefore be a natural consequence of cosmic inflation close to the electroweak energy scale.

1. Introduction

The success of Cosmic Inflation certainly relies in its ability of providing an unified paradigm for some unaddressed questions of the standard Big-Bang model of Friedmann and Lemaître [1, 2, 3, 4]. Among those questions are the apparent fine-tuning of some cosmological parameters. For instance, the density parameter of spatial curvature is currently bounded to $|\Omega_k| < 2\%$ [5] whereas the spectral index n_{S} of a primordial power law power spectrum is close to unity, but not equal to unity [5] ($0.015 < n_{\mathrm{S}} - 1 < 0.071$ at two-sigma). Inflation provides a simple answer to those two apparently unrelated problems while addressing many others such as the adiabaticity and Gaussianity of the initial conditions. In its simpler form, the inflationary paradigm can be implemented with a self-gravitating scalar field in a potential dominated regime. At the background level, such a system triggers an accelerated expansion of the universe solving the flatness problem whereas its quantum fluctuations predicts an *almost* scale invariant power spectrum of the cosmological perturbations [6].

Another unaddressed question of the ΛCDM model concerns the present value of the cosmological constant $\Omega_{\Lambda} \simeq 0.723 \pm 0.016$, which is required to explain the current acceleration of the universe [7, 8, 9]. According to the interpretation given to the cosmological constant, such a measurement can be either viewed as a severe fine-tuning problem or a failure in our understanding of the gravitational properties of the vacuum state [10].

In this context, one may wonder whether Cosmic Inflation could also provide some insight to the current value of the cosmological constant, or more straightforwardly, to the current acceleration of the universe.

Content from this work may be used under the terms of the Creative Commons Attribution 3.0 licence. Any further distribution of this work must maintain attribution to the author(s) and the title of the work, journal citation and DOI.
Published under licence by IOP Publishing Ltd

Most alternative models of Dark Energy assume that, from still unknown symmetry reasons, the bare cosmological constant in the Einstein equation vanishes and that the current cosmic acceleration is sourced by a new dynamical degree of freedom. As a result, these models genuinely introduce extra-parameters that may need to be fine-tuned to ensure that their energy density dominates the universe only recently. This is the so-called coincidence problem. Quintessence models belong to this class as they introduce a new scalar field. However, if the potential is of the Ratra–Peebles kind, they solve the coincidence problem by exhibiting a tracking behaviour which washes out memory of the initial conditions [11, 12, 13, 14]. But still, the normalization of the potential has to be tuned to the observed cosmological constant. According to the classification of Ref. [15], these quintessence models belong to the "freezing" class as opposed the the "thawing" class. The later models are exactly as Cosmic Inflation, a field slowly rolls down a potential thereby triggering acceleration. Since the initial conditions are not erased by a tracking mechanism in the thawing models, they appear as additional and unwanted model parameters. But this is without considering the effect of Cosmic Inflation.

Inflationary effects in the freezing quintessence models have been first explored in Ref. [16] to determine how inflation affects the initial conditions of the field. As shown in Ref. [17], they have an unfortunate tendency to push the field away from the region where the tracker behavior can efficiently wash out the initial conditions. As a result, freezing quintessence on the tracker today requires that inflation lasted a low number of e-folds.

As we have shown in Ref. [18], the situation is different for the thawing models. Cosmic Inflation actually gives natural initial conditions for a thawing quintessence field which address simultaneously the coincidence problem and the apparent small value of the cosmological constant today. For this reason, thawing models within inflation have less fine-tuning than any other Dark Energy scenario and are actually predictive: the energy scale of Cosmic Inflation should be at a few TeV.

2. Storing quantum fluctuations from inflation

As an illustrative example, let us consider a thawing model which is simply a massive scalar field during inflation

$$\varphi(\boldsymbol{x},t) = \phi(t) + \delta\phi(\boldsymbol{x},t), \tag{1}$$

where ϕ is the homogeneous mode and $\delta\phi$ are its perturbations. Furthermore, assuming that inflation is almost de-Sitter with an Hubble parameter given by H_{inf}, it is straightforward to solve the Klein–Gordon equation [2, 3, 19]. The homogeneous part evolves as (in Planck units)

$$\phi(N) = \phi_{\mathrm{ini}} \exp\left(-\frac{m^2}{3H_{\mathrm{inf}}^2} N\right), \tag{2}$$

where $N \equiv \ln(a/a_{\mathrm{ini}})$ is the number of e-folds and a the scale factor. As a result, if inflation lasts long enough, the homogeneous mode reaches the minimum of its potential, as one may expect. Starting from the Bunch–Davis vacuum on sub-Hubble scales, the linear fluctuations of the field in Fourier space $\mu \equiv a\delta\phi(\boldsymbol{k},t)$ have also an exact solution as

$$\mu(\eta,k) = e^{i(\nu+1/2)\pi/2}\sqrt{\frac{\pi}{4k}}\sqrt{k\eta}\, H_\nu^{(1)}(k\eta), \quad \nu \equiv \sqrt{\frac{9}{4} - \frac{m^2}{H_{\mathrm{inf}}^2}}, \tag{3}$$

where η is the conformal time. From this expression, the primordial power spectrum of the field fluctuations after Hubble exist reads

$$\mathcal{P}_{\delta\phi} = \lim_{k\eta \ll 1} \frac{k^2}{8\pi^2}\left|\frac{\mu}{a}\right|^2 \simeq \frac{H_{\mathrm{inf}}^2}{4\pi^2}\left(\frac{k}{aH_{\mathrm{inf}}}\right)^{2m^2/(3H_{\mathrm{inf}}^2)}. \tag{4}$$

This expression shows that if the field is very light during inflation, $m \ll H_{\rm inf}$, its Fourier modes around the Hubble scale develop a scale invariant power spectrum of amplitude [20]

$$\mathcal{P}_{\delta\phi}(k) = \frac{H_{\rm inf}^2}{4\pi^2} + \cdots . \tag{5}$$

Up to a numerical factor accounting for the polarization states, this spectrum is of comparable amplitude with the one associated with the tensor modes during inflation. As a result, if inflation occurs at low energy scales, the field perturbations that re-enters the Hubble radius during the radiation and matter era are of extremely low amplitude thereby preventing the field to grow sub-Hubble inhomogeneities [21, 22, 23]. As the result, only the homogeneous part can have observable effects. As we have just discussed, its classical value vanishes but not its quantum generated variance. Indeed, going back to real space and summing only on all super-Hubble modes gives, after N e-folds of inflation

$$\left\langle \delta\phi^2 \right\rangle = \int_{a_{\rm ini} H_{\rm inf}}^{a H_{\rm inf}} \frac{\mathrm{d}^3 \boldsymbol{k}}{(2\pi)^3} \left|\delta\phi_{\boldsymbol{k}}\right|^2 = \frac{3H_{\rm inf}^4}{8\pi^2 m^2}\left[1 - \exp\left(-\frac{2m^2}{3H_{\rm inf}^2} N\right)\right] \to \frac{3H_{\rm inf}^4}{8\pi^2 m^2} . \tag{6}$$

The last limit is applicable if inflation lasts long enough, by precisely the same amount necessary to cancel the classical homogeneous mode. Let us notice that this field variance can only be realized as an homogeneous mode for an observer since we have restricted the summation on super-Hubble scales. Strictly speaking, the brackets refer to a quantum average in the de-Sitter vacuum state and, as such, the above equation is of quantum probabilistic nature. This result can be recovered and interpreted from the stochastic inflation formalism [24, 25, 26]. It shows that the field reaches an equilibrium distribution on super-Hubble scales resulting from the competition between quantum kicks that tend to push it out of the potential minimum and the classical force which drives the field towards the minimum.

For our purpose, we notice that the potential energy of such a field develops an *universal* value, i.e. it depends only on the Hubble scale during inflation

$$\langle V(\phi) \rangle = \frac{1}{2} m^2 \langle \delta\phi^2 \rangle = \frac{3H_{\rm inf}^4}{16\pi^2} . \tag{7}$$

In order to explain the current cosmic acceleration, it is necessary that the field remains frozen after inflation. This is satisfied as long as $m < H_0$, the Hubble parameter today. In other words, the field should be extremely light and as such, inflation should be extremely long. In Ref. [18], a lower limit for the total number of e-folds is estimated to be $N > 10^9$ whereas a natural number would be $N \simeq 10^{60}$. Another observable prediction of this model is therefore that the spatial curvature today should be infinitely small.

Finally, we want the field potential today to be of the same order than the energy density necessary to trigger the current acceleration. This is trivially satisfied provided

$$H_{\rm inf} \simeq (\Omega_{\Lambda})^{1/4} \sqrt{4\pi H_0 M_{\rm P}} , \tag{8}$$

where we have restored the reduced Planck mass $M_{\rm P}$ for clarity. This expression makes clear that, in our model, the apparent fine-tuning of the cosmological constant value is recast into the question of knowing the energy scale of inflation, i.e. a prediction. Plugging numbers into this expression gives

$$H_{\rm inf} \simeq 6 \times 10^{-3}\,\mathrm{eV} \quad \Rightarrow \quad E_{\rm inf} \equiv \rho_{\rm inf}^{1/4} = (3M_{\rm P}^2 H_{\rm inf}^2)^{1/4} \simeq 5\mathrm{TeV} . \tag{9}$$

In the context of inflation, the small value of $\rho_\Lambda^{1/4}$ has a natural origin. It is typical of the energy density associated with de-Sitter quantum fluctuations at TeV scale.

3. Conclusion

In Ref. [18], we have consistently taken into account the probabilistic nature associated with the theoretical predictions of our model by using Bayes statistics, and compared them with the current supernovae, BAO and CMB data [5, 9, 27]. The result is consistent with the order of magnitude considerations presented above. More quantitatively, the energy scale of inflation lies in the 95% confidence interval

$$3.8\,\mathrm{TeV} < E_{\mathrm{inf}} < 12.1\,\mathrm{TeV}, \tag{10}$$

whereas $m < 75\,\mathrm{km/s/Mpc}$ (also at 95%).

The mechanism described above actually works with any spectator field, not necessarily scalar, experiencing quantum fluctuations during inflation provided its potential is bounded from below. Indeed, in some extension of electromagnetism, gauge fields have also been shown to be able to store inflationary quantum fluctuations [28, 29]. The fact that the TeV scale seems to be universal directly stems from Eq. (8), and thus from Eq. (7). As already mentioned, this is the energy scale at which de-Sitter quantum fluctuations match the cosmological constant energy scale.

For instance, if instead of having considered a massive field, we would have assumed the potential to be $V(\phi)$, the stochastic inflation formalism gives the stationary probability distribution [26]

$$P(\phi) \propto \exp\left[-\frac{8\pi^2}{3H_{\mathrm{inf}}^4}V(\phi)\right]. \tag{11}$$

As opposed to the massive case, the field distribution is no longer Gaussian, but the potential value remains universal $\langle V(\phi)\rangle \simeq 3H_{\mathrm{inf}}^4/(8\pi^2)$ and our results still hold.

Finally, let us notice that because this model is a late-time inflationary era, triggered by the primordial one, the current acceleration of the universe will come to an end. Would our universe be filled with various uncoupled sectors having completely different mass hierarchies, one could even imagine a succession of inflationary eras, temporarily interupted with decelerating eras, and re-triggered by quantum fluctuations stored in the lightest fields. In a more pragmatic way, if the mass of the field is of comparable value with H_0 (as opposed as being much smaller), one could expect a disambiguation possible with respect to a pure cosmological constant through the redshift evolution of equation of state parameter [30].

References

[1] Guth A H 1981 *Phys. Rev.* **D23** 347–356
[2] Starobinsky A A 1980 *Phys. Lett.* **B91** 99–102
[3] Mukhanov V F and Chibisov G V 1981 *JETP Lett.* **33** 532–535
[4] Linde A D 1983 *Phys. Lett.* **B129** 177–181
[5] Komatsu E *et al.* (WMAP Collaboration) 2011 *Astrophys.J.Suppl.* **192** 18 (*Preprint* 1001.4538)
[6] Mukhanov V F, Feldman H A and Brandenberger R H 1992 *Phys. Rept.* **215** 203–333
[7] Kowalski M *et al.* (Supernova Cosmology Project) 2008 *Astrophys.J.* **686** 749–778 (*Preprint* 0804.4142)
[8] Riess A G *et al.* 2009 *Astrophys. J.* **699** 539–563 (*Preprint* 0905.0695)
[9] Amanullah R *et al.* 2010 *Astrophys. J.* **716** 712–738 (*Preprint* 1004.1711)
[10] Martin J 2012 *Comptes Rendus Physique* **13** 566–665 (*Preprint* 1205.3365)
[11] Ratra B and Peebles P J E 1988 *Phys. Rev.* **D37** 3406
[12] Zlatev I, Wang L M and Steinhardt P J 1999 *Phys. Rev. Lett.* **82** 896–899 (*Preprint* astro-ph/9807002)
[13] Brax P, Martin J and Riazuelo A 2000 *Phys.Rev.* **D62** 103505 (*Preprint* astro-ph/0005428)
[14] Martin J 2008 *Mod.Phys.Lett.* **A23** 1252–1265 (*Preprint* 0803.4076)
[15] Caldwell R R and Linder E V 2005 *Phys. Rev. Lett.* **95** 141301 (*Preprint* astro-ph/0505494)
[16] Malquarti M and Liddle A R 2002 *Phys. Rev.* **D66** 023524 (*Preprint* astro-ph/0203232)
[17] Martin J and Musso M A 2005 *Phys. Rev.* **D71** 063514 (*Preprint* astro-ph/0410190)
[18] Ringeval C, Suyama T, Takahashi T, Yamaguchi M and Yokoyama S 2010 *Phys.Rev.Lett.* **105** 121301 (*Preprint* 1006.0368)

[19] Grishchuk L P 1975 *Sov. Phys. JETP* **40** 409–415
[20] Hawking S W 1982 *Phys. Lett.* **B115** 295
[21] DeDeo S, Caldwell R and Steinhardt P J 2003 *Phys.Rev.* **D67** 103509 (*Preprint* astro-ph/0301284)
[22] Marsh D J and Ferreira P G 2010 *Phys.Rev.* **D82** 103528 (*Preprint* 1009.3501)
[23] Sefusatti E and Vernizzi F 2011 *JCAP* **1103** 047 (*Preprint* 1101.1026)
[24] Starobinsky A A 1986 *Field Theory, Quantum Gravity and Strings* (*Lecture Notes in Physics, Berlin Springer Verlag* vol 246) ed H J de Vega & N Sánchez p 107
[25] Goncharov A S, Linde A D and Mukhanov V F 1987 *Int. J. Mod. Phys.* **A2** 561–591
[26] Starobinsky A A and Yokoyama J 1994 *Phys. Rev.* **D50** 6357–6368 (*Preprint* astro-ph/9407016)
[27] Percival W J *et al.* 2010 *Mon. Not. Roy. Astron. Soc.* **401** 2148–2168 (*Preprint* 0907.1660)
[28] Beltran Jimenez J and Maroto A L 2010 *Phys.Lett.* **B686** 175–180 (*Preprint* 0903.4672)
[29] Beltran Jimenez J and Maroto A L 2011 *Mod.Phys.Lett.* **A26** 3025–3039 (*Preprint* 1112.1106)
[30] Amendola L and Mota D F 2012 (*Preprint* 1206.1225)

Production of long-lived charginos at the LHC

M G Paucar

Instituto de Física, Universidad Nacional Autónoma de México, Apdo. Postal 20-364, México 01000

E-mail: `mgerardo@fisica.unam.mx`

Abstract. We performed a study on the probable existence of long-lived charginos in the focus point region of parameter space in the framework of the MSSM with supergravity inspired soft SUSY breaking. The mass degeneracy of higgsino-like chargino and two higgsino-like neutralinos is the necessary condition for a long lifetime. In this region the chargino are long lived and may go through the detector or produce a secondary vertex when they decay inside the detector. We analyze the production and co-production cross section of long-lived charginos at LHC via electroweak interaction and found that, for light long-lived charginos, the production cross section may reach a few percent of pb.

1. Introduction

The discovery of supersymmetry is one of the main goals in the proposed experiments at the LHC.[1, 2]. However, search for supersymmetric particles at colliders usually proceeds from the assumption that all of them are relatively heavy (few hundreds of GeV) and short-lived which are created in strong and weak interactions and then decay (almost immediately) to usual Standard Model particles with additional missing energy taken away by the Lightest Supersymmetric Particle (LSP). Almost everywhere in the parameter space of the Constrained MSSM (CMSSM) (or also known as the minimal Supergravity model, mSUGRA) [3, 4], the LSP is the lightest neutralino (other, SUSY breaking scenarios lead to different experimental signatures and different LSP [5, 6]).

There are, however, some regions in mSUGRA conventions where the LSP is not the lightest neutralino $\widetilde{\chi}_1^0$, but is a stau $\widetilde{\tau}$, or a stop $\widetilde{t}$, or the first chargino $\widetilde{\chi}_1^{\pm}$ [7, 8, 9, 10]. These regions are obviously considered as forbidden ones. At the border of these regions when masses of superpartners are degenerate with the neutralino mass and hence the decay rate is suppressed, the staus, stops, and charginos become long-lived superparticles. As one departs from the border line, superpartners are slightly heavier than the neutralino and thus unstable, then they decay very fast .

Here, we explore the narrow band along the line where the radiative electroweak symmetry breaking fails. This region is consistent with the neutralino relic density constraint (the WMAP restriction on the amount of the dark matter). On the border of this region the Higgs mixing parameter μ, which is determined from the requirement of electroweak symmetry breaking via radiative corrections, tends to zero. This leads to existence of light and degenerate states: one chargino and two neutralinos, all of them being essentially higgsinos. Below we discuss their properties and possible creation at the LHC.

Content from this work may be used under the terms of the Creative Commons Attribution 3.0 licence. Any further distribution of this work must maintain attribution to the author(s) and the title of the work, journal citation and DOI.
Published under licence by IOP Publishing Ltd

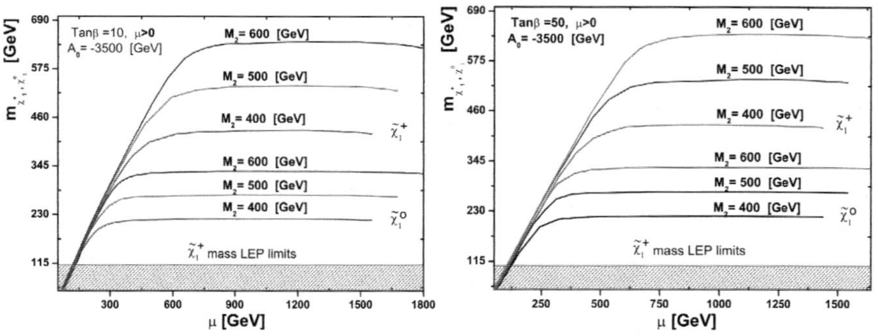

Figure 1. The masses of the lightest chargino and neutralino as functions of μ for the rest of parameters fixed. The value of M_2 is taken to be 600, 500 and 400 GeV, and $\tan\beta$=10,50, respectively. Dark (red) lower band shows the experimental limit on chargino mass.

2. Focus-point Region and Long-lived Charginos

In mSUGRA models with R-parity conservation, the only available set of parameters space is $[m_0, m_{1/2}, A_0, sign(\mu), \tan\beta]$. Fixing A_0, sign(μ), tanβ and varying m_0 and $m_{1/2}$, we can get regions in parameter space for which the cross section for $\tilde{\chi}_0^1$ LSP annihilation and co-annihilation corresponds to the dark matter density measured by WMAP. We have imposed on the mSUGRA parameter space the following constraints: i) the gauge couplings unification [11, 12, 13], ii) neutrality of the LSP [14, 15], iii) the Higgs boson and SUSY mass experimental limits [16, 17, 18], iv) radiative electroweak symmetry breaking [19, 20]. We take the sign of μ to be positive, motivated by the contribution to the anomalous magnetic moment of muon [21].

We are interested in the region of small μ (less than M_Z), which takes place in the so-called focus point region near the border line of radiative electroweak symmetry breaking. In this scenario the lightest chargino ($\tilde{\chi}_1^{\pm}$) and two lightest neutralinos ($\tilde{\chi}_{1,2}^0$) are almost degenerate and have a mass of the order of μ. The mass terms are nondiagonal and look like

$$\mathcal{L}_{Gaugino-Higgsino} = -\frac{1}{2}M_3\bar{\lambda}_a\lambda_a - \frac{1}{2}\bar{\chi}M^{(0)}\chi - \bar{\psi}M^{(c)}\psi + h.c., \tag{1}$$

where $\lambda_a, a = 1, 2, \ldots, 8$, are the Majorana gluino fields and χ, ψ are, respectively, the Majorana neutralino and Dirac chargino fields.

At the tree level the neutralino mass matrix is

$$M^{(0)} = \begin{pmatrix} M_1 & 0 & -M_Z c_\beta s_\theta & M_Z s_\beta s_\theta \\ 0 & M_2 & M_Z c_\beta c_\theta & -M_Z s_\beta c_\theta \\ -M_Z c_\beta s_\theta & M_Z c_\beta c_\theta & 0 & -\mu \\ M_Z s_\beta s_\theta & -M_Z s_\beta c_\theta & -\mu & 0 \end{pmatrix}, \tag{2}$$

where $s_\beta = \sin\beta, c_\beta = \cos\beta$ and $s_\theta = \sin\theta_W$ is the usual sinus of the Weinberg weak mixing angle ($c_W = \cos\beta$). The physical neutralino masses $m_{\tilde{\chi}_i^0}$ are obtained as eigenvalues of this matrix after diagonalization.

For charginos one has

$$M^{(c)} = \begin{pmatrix} M_2 & \sqrt{2}M_W \sin\beta \\ \sqrt{2}M_W \cos\beta & \mu \end{pmatrix}. \tag{3}$$

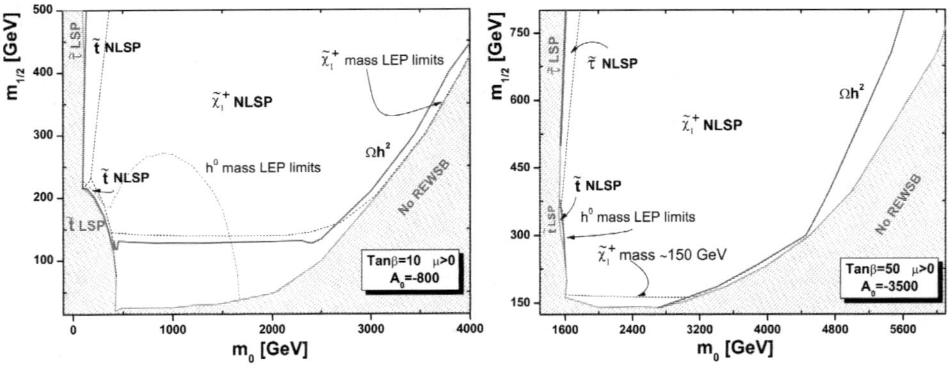

Figure 2. Focus point region of mSUGRA for $A_0 = -800, -3500$ GeV, $\mu > 0$ and $\tan\beta = 10, 50$, respectively. Dark areas show theoretically forbidden regions. Along the narrow $\Omega_{\tilde{\chi}_1^0} h^2$ band the amount of the dark matter corresponds to WMAP data $0.094 \le \Omega_{\tilde{\chi}_1^0} h^2 \le 0.125$. The focus point region is the cosmologically favored region with large m_0 and small values of μ.

This matrix has two chargino eigenstates $\tilde{\chi}^{\pm}_{1,2}$ with mass eigenvalues

$$M^2_{1,2} = \frac{1}{2}\Big[M_2^2 + \mu^2 + 2M_W^2 \mp \sqrt{(M_2^2-\mu^2)^2 + 4M_W^4\cos^2 2\beta + 4M_W^2(M_2^2+\mu^2+2M_2\mu\sin 2\beta)}\Big] \quad (4)$$

These matrices receive radiative corrections which are known in the leading order [22]. Typically they are of the order of a few per cent.

In Fig. 1, it is shown how the mass of the lightest neutralino and the mass of the lightest chargino obtained from Eqs. (2) and (4) depend on μ.

The degeneracy of masses masses $(m_{\chi_1^0}, m_{\chi_2^0}, m_{\chi_1^\pm})$ takes place for any choice of the other parameters since the tree level formulae weakly depend on them and the one-loop corrections are usually small. However, since the value of μ is not arbitrary in this approach but is taken from the requirement of electroweak symmetry breaking, one has to find the region of parameter space where it is small.

In Fig. 1, this region is just above the chargino LEP limit in the left bottom corner of the plots. One can see that masses are degenerate, and there the value of μ is of the order of 150–200 GeV depending on the value of $\tan\beta$. There is also a slight dependence on M_2 (that is on $m_{1/2}$); however, this dependence only shows how far we may go along the lines having masses degenerate. It is clearly seen that the bigger M_2 the larger values of μ are allowed. The mass of χ_2^0 is not shown, it almost coincides with the $\chi_1^\pm$ mass.

In Fig. 2, we show the projection of SUSY parameter space onto the $m_0, m_{1/2}$ plane for A=-800, -3500 GeV and $\tan\beta$=10, 50. To calculate it, we use the SuSpect v2.3.4 code [23][1]. The relic density was calculated with the help of the MicrOmegas package [24].

One can see that for small value of A_0 the Dark matter line does not go along the electroweak symmetry breaking border but deviates from it, thus not allowing the small value of μ. For large and negative A_0, on the contrary, these two lines almost coincide, the bigger the value of $\tan\beta$ the better.

[1] We have set the following input values at M_W; $\alpha_{em}^{-1} = 127.934$, $\alpha_s = 0.1172$, $m_t^{pole} = 172.7$ GeV, $m_b(m_b)^{\overline{MS}} = 4.25$ GeV, $m_\tau^{pole} = 1.777$ GeV.

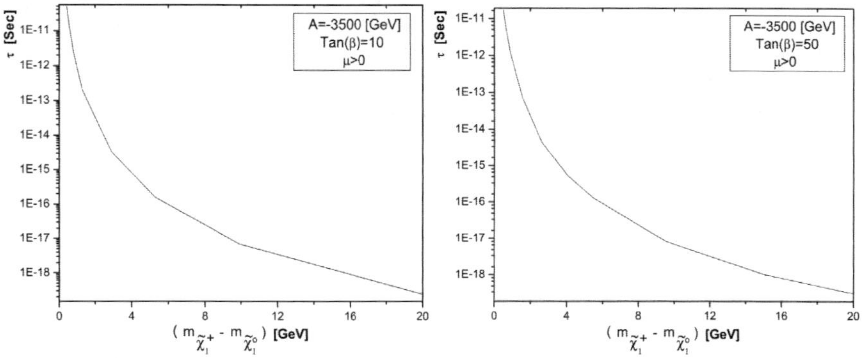

Figure 3. The lightest chargino lifetimes as a function of the mass difference between the lightest chargino (NLSP) and the lightest neutralino (LSP).

In Fig. 3, we show the lifetime of the lightest chargino as a function of the mass difference between the lightest chargino (NLSP) and the lightest neutralino (LSP). It appears that in order to get reasonable " large " lifetimes one has to go very far along the focus point region. Then keeping μ small one can get lifetimes of the order of 10^{-10} s for practically degenerate LSP and NLSP. When the mass difference increases the lifetime falls down. However, if the degeneracy is within a few GeV, charginos are long-lived.

3. Production cross-section of long-lived charginos at LHC

In the previous sections, there appeared the possibility of obtaining charginos with very long-lifetime. Now we study the production cross-section of these Long-lived sparticles at LHC and analyze the possible situation when they could go through the detector.

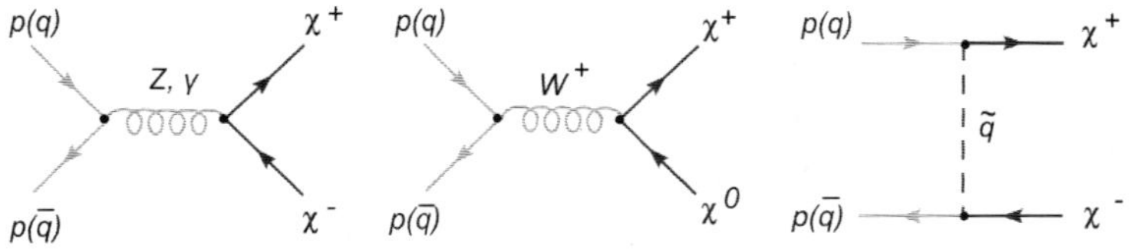

Figure 4. Main chargino production diagrams at the LHC

In Fig.4, we show the production of long-lived charginos in annihilation channel at LHC. Since three states are almost degenerate, one has also co-production which has to be taken into account. This refers also to the annihilation process that defines the amount of the Dark matter.

To calculate the production rate of long-lived chargino, we used CALCHEP code [25] which takes into account the parton distributions inside protons. For our purposes we took the MRST2002NLO parton distribution functions [26]. we choose several benchmark points in the CMSSM parameter space which are situated in the focus point region along the border of electroweak symmetry breaking (see Fig. 2). The values of μ and $\tan\beta$ are varied. As we have already mentioned for consistency with the relic density constraint the advocated scenario prefers large negative A_0 and large $\tan\beta$.

For each benchmark point we have calculated the cross-sections of chargino production and co-production with the lightest and next-to-lightest neutralino. The result of the calculations are summarized in table 1.

One can see that on average the cross-sections reach a few tenth of pb and slightly vary with the change of $\tan\beta$. The cross-sections mainly depend on μ: the bigger the value of μ the smaller the cross-section. This is due to the fact that the cross-section mostly depends on the masses determined mostly by μ and mixings.

Table 1. The cross-sections for the chargino production and co-production at the LHC in the focus point region for $\mu = 120$ GeV and $\tan\beta = 10, 30, 50$.

Process	$\tan\beta = 10$	$\tan\beta = 30$	$\tan\beta = 50$
	$pp \to \widetilde{\chi}_1^+\widetilde{\chi}_1^-$		
$\bar{u}u \to \widetilde{\chi}_1^+\widetilde{\chi}_1^-$	0.58	0.54	0.53
$u\bar{u} \to \widetilde{\chi}_1^+\widetilde{\chi}_1^-$	0.58	0.54	0.53
$\bar{d}d \to \widetilde{\chi}_1^+\widetilde{\chi}_1^-$	0.27	0.24	0.24
$d\bar{d} \to \widetilde{\chi}_1^+\widetilde{\chi}_1^-$	0.27	0.24	0.24
$\sigma_{P1}\,(pb)$	1.68	1.56	1.54
	$pp \to \widetilde{\chi}_1^+\widetilde{\chi}_1^0$		
$u\bar{d} \to \widetilde{\chi}_1^+\widetilde{\chi}_1^0$	1.08	0.99	0.98
$\bar{d}u \to \widetilde{\chi}_1^+\widetilde{\chi}_1^0$	1.08	0.99	0.98
$\sigma_{P2}\,(pb)$	2.16	1.98	1.96
	$pp \to \widetilde{\chi}_1^+\widetilde{\chi}_2^0$		
$u\bar{d} \to \widetilde{\chi}_1^+\widetilde{\chi}_2^0$	0.68	0.66	0.65
$\bar{d}u \to \widetilde{\chi}_1^+\widetilde{\chi}_2^0$	0.68	0.66	0.65
$\sigma_{P3}\,(pb)$	1.37	1.32	1.30

Table 2. The cross-sections for the chargino production and co-production at the LHC in the focus point region for $\mu = 200$ GeV and $\tan\beta = 10, 30, 50$.

Process	$\tan\beta = 10$	$\tan\beta = 30$	$\tan\beta = 50$
	$pp \to \widetilde{\chi}_1^+\widetilde{\chi}_1^-$		
$\bar{u}u \to \widetilde{\chi}_1^+\widetilde{\chi}_1^-$	0.11	0.10	0.10
$u\bar{u} \to \widetilde{\chi}_1^+\widetilde{\chi}_1^-$	0.11	0.10	0.10
$\bar{d}d \to \widetilde{\chi}_1^+\widetilde{\chi}_1^-$	0.05	0.04	0.04
$d\bar{d} \to \widetilde{\chi}_1^+\widetilde{\chi}_1^-$	0.05	0.04	0.04
$\sigma_{P1}\,(pb)$	0.32	0.28	0.28
	$pp \to \widetilde{\chi}_1^+\widetilde{\chi}_1^0$		
$u\bar{d} \to \widetilde{\chi}_1^+\widetilde{\chi}_1^0$	0.15	0.14	0.13
$\bar{d}u \to \widetilde{\chi}_1^+\widetilde{\chi}_1^0$	0.15	0.14	0.13
$\sigma_{P2}\,(pb)$	0.30	0.28	0.26
	$pp \to \widetilde{\chi}_1^+\widetilde{\chi}_2^0$		
$u\bar{d} \to \widetilde{\chi}_1^+\widetilde{\chi}_2^0$	0.12	0.12	0.12
$\bar{d}u \to \widetilde{\chi}_1^+\widetilde{\chi}_2^0$	0.12	0.12	0.12
$\sigma_{P3}\,(pb)$	0.24	0.24	0.24

4. Conclusions

We have shown that within the mSUGRA framework there exists an interesting possibility to get long-lived charginos, superpartners of the charged Higgs boson (or W-boson) which might be produced at the Large Hadron Collider. In this case the lightest chargino and the lightest neutralinos are almost fully higgsinos, the higgsino content reaching 70 - 80 % depending on a particular benchmark point. The cross-sections of chargino pair production and co-production via electroweak interaction crucially depend on the chargino mass, and for light charginos it can reach a few pb. As in the case of light stops (see ref. [9]), the light chargino NLSP scenario requires large negative values of the soft trilinear SUSY breaking parameter A_0, and large values of m_0 and $m_{1/2}$. The chargino decay would have an unusual signature and produce a noticable signal rather than pure missing energy taken away by the lightest neutralino. The two options are: charginos go through the detector, or they produce a secondary vertex when they decay inside the detector.

Acknowledgements

I would like to express my sincere thanks to Professors D.I. Kazakov and A.V. Gladyshev for their valuable discussion and constructive criticism on this work. I would like also to thank Professor M. Mondragon for her constant encouragement and help. The work of M. G. P. is supported by ICYTDF-CLAF postdoctoral fellowship. Support from PAPITT Project IN113412 is kindly acknowledged.

References

[1] Haber H E and Kane G L 1985 *Phys. Rep.* **117** 75
[2] Gladyshev A V and Kazakov D I 2007 *Phys. Atom. Nucl.* **70** 1553
[3] Hall L J, Lykken J and Weinberg S 1983 *Phys. Rev.* D **27** 2359-2378.
[4] Nath P, Arnowitt R and Chamseddine A H, 1983 *Nucl. Phys.* B **227** 121-133.
[5] Randall L and Sundrum R 1999 *Nucl. Phys.* B **557** 79-118.
[6] Dine M and Nelson A E 1993 *Phys. Rev.* D **48** 1277-1287.
[7] Gladyshev A V, Kazakov D I and Paucar M G 2005 *Mod. Phys. Lett.* A **20** 3085
[8] Paucar M G 2008 *Phys. Atom. Nucl.* **71** 695
[9] Gladyshev A V, Kazakov D I and Paucar M G 2007 Light stops in the MSSM parameter space arXiv:0704.1429v1 [hep-ph]
[10] Gladyshev A V, Kazakov D I and Paucar M G 2008 *Proc. 15th Int. Conf. on Supersymmetry and the Unification of Fundamental Interactions (Karlsruhe)* ed W de Boer and I Gebauer (Brno: Librix) Vol 1, p 338
[11] Amaldi U, de Boer W and Furstenau H 1991 *Phys. Lett.* B **260** 447
[12] Ellis J R, Kelley S and Nanopoulos D V 1991 *Phys. Lett.* B **260** 131
[13] Giunti C, Kim C W and Lee U W 1991 *Mod. Phys. Lett.* A **6** 1745
[14] Ellis J R et al. 1884 *Nucl. Phys.* B **238** 453
[15] Jungman G, Kamionkowski M and Griest K 1996 *Phys.Rep.* *267* 195
[16] LEP Higgs Working Group for Higgs boson searches, OPAL Collaboration, ALEPH Collaboration, DELPHI Collaboration and L3 Collaboration 2003 Search for the Standard Model Higgs Boson at LEP *Preprint* CERN-EP/2003-011, hep-ex/0306033
[17] Joint LEP2 SUSY Working Group 2001 Combined LEP Chargino Results up to 208 GeV `http://lepsusy.web.cern.ch/lepsusy/www/inos_moriond01/charginos_pub.html`
[18] Combined LEP Selectron/Smuon/Stau Results, 183-208 GeV `http://lepsusy.web.cern.ch/lepsusy/www/sleptons_summer02/slep_2002.html`
[19] Ibáñez L E and Ross G G 1992 *Nucl. Phys.* B **368** 3
[20] Ross G G and Roberts R G 1992 *Nucl. Phys.* B **377** 571
[21] Brown H N *et al* (Muon g-2 Collaboration) 2001 *Phys. Rev. Lett.* **86** 2227
[22] Pierce D and Papadopoulos A 1994 *Phys. Rev.* D **50** 565
[23] Djouadi A, Kneur J L and Moultaka G 2007 *Comput. Phys. Commun.* **176** 426
[24] Belanger G, Boudjema F, Pukhov A and Semenov A 2002 *Comput. Phys. Commun.* **149** 103
[25] Pukhov A 2004 Calchep 3.2: MSSM, structure functions and event generation, arXiv:hep-ph/0412191
[26] Martin A D, Roberts R G, Stirling W J and Thorne R S 2003 *Eur. Phys. J.* C **28** 455

Journal of Physics: Conference Series **485** (2014) 012025 doi:10.1088/1742-6596/485/1/012025

Axino dark matter and baryon number from Q-ball decay

Shinta Kasuya[a], **Etsuko Kawakami**[b] **and Masahiro Kawasaki**[b,c]

[a] Department of Information Sciences, Kanagawa University, Kanagawa 259-1293, Japan
[b] Institute for Cosmic Ray Research, University of Tokyo, Chiba 277-8582, Japan
[c] Institute for the Physics and Mathematics of the Universe, University of Tokyo, Chiba 277-8582, Japan

Abstract. We investigate the Q-ball decay into the axino dark matter in the gauge-mediated supersymmetry breaking. In our scenario, the Q ball decays mainly into nucleons and partially into axinos to account for the baryon asymmetry and the dark matter of the universe simultaneously. The Q ball decays well before the big bang nucleosynthesis so that it is not affected by the decay. The decay into the supersymmetric particles of the minimal supersymmetric standard model is kinematically prohibited until the very end of the decay, and we could safely make their abundances small enough for the successful big bang nucleosynthesis. We show the regions of axino model parameters which realize this scenario.

1. Introduction

The origin of baryon number asymmetry and the dark matter of the universe have been discussed for decades, but are still some of the main unsolved mysteries in cosmology. No solution can be found in the standard model (SM), and we must seek for something beyond SM. One good way is to consider supersymmetry (SUSY). In SUSY, the lightest supersymmetric particle (LSP), with R parity conservation, is stable and, in most cases, scarcely interacts with other particles. These natures make the LSP a strong candidate of the dark matter. SUSY could also explain the origin of baryon number asymmetry. The Affleck Dine (AD) baryogenesis is one of the promising mechanism [1]. The AD field, carrying the baryon number, has a large VEV during inflation and rotates in the potential after inflation, and the baryon number is created. It finally decays into quarks to become the baryon asymmetry of the universe. In the minimal supersymmetric standard model (MSSM), there exist many flat directions, which consist of squarks, sleptons and Higgs fileds, which thus carry baryon and/or lepton number. Therefore, the flat direction could be responsible for the AD field.

The very attractive feature of the AD mechanism is to provide both the baryon asymmetry and the dark matter of the universe simultaneously in the context of the Q-ball cosmology [2–12]. During the rotation, the AD condensate may fragment into nontopological solitons, Q balls. These Q balls can be the dark matter if they are stable, while the LSP dark matter could be produced from unstable Q balls. Stable Q balls form if the charge Q is large enough in the gauge-mediated SUSY breaking [2, 6]. On the other hand, Q balls are unstable in the gravity mediation producing the neutralino LSP [3, 7], the gravitino LSP [8], and the axino LSP [9], and in the gauge mediation creating the gravitino LSP if the charge is small enough [10–12].

Content from this work may be used under the terms of the Creative Commons Attribution 3.0 licence. Any further distribution of this work must maintain attribution to the author(s) and the title of the work, journal citation and DOI.
Published under licence by IOP Publishing Ltd

In this paper, we investigate a model where the Q ball decays into axino LSPs in gauge-mediated SUSY breaking. (Similar situation in the gravity mediation was investigated in Ref. [9].) The axino is a fermionic superpartner of the axion, which is introduced as a dynamical scalar field to solve the strong CP problem in quantum chromodynamics known as Peccei-Quinn (PQ) mechanism [13]. In our model, Q ball decays mainly into nucleons and partially into axinos directly in order to account for both the baryon asymmetry and the dark matter of the universe. The decay of the Q ball takes place well before the big bang nucleosynthesis (BBN) so that the decay itself does not affect the BBN. Almost throughout the process, it is kinematically prohibited to decay into the lightest supersymmetric particle (LSP) of MSSM, whose decay may destroy light elements synthesized during the BBN. However, the MSSM LSPs (MLSPs) would be produced at the very end of the Q-ball decay when the charge becomes small enough to open the kinematically allowed channel to the MLSP production [12, 14].

2. Q ball in gauge mediation

The AD field Φ is a combination of squarks, sleptons and higgs whose potential is flat in the SUSY exact limit. Because of the SUSY breaking in gauge mediation, the flat potential is lifted below the messenger scale, while it is flat above the messenger scale, $V \sim M_F^4$ [2, 15]. Here M_F is related to the F component of a gauge-singlet chiral multiplet S in the messenger sector as $M_F^4 \equiv \frac{g^2}{(4\pi)^4}\langle F_S\rangle^2$ where g is a gauge coupling constant in the standard model, and M_F is allowed in the following range: 10^3 GeV $\lesssim M_F \lesssim \frac{g^{1/2}}{4\pi}\sqrt{m_{3/2}M_{\rm P}}$, where $m_{3/2}$ and $M_{\rm P} = 2.4 \times 10^{18}$ GeV are the gravitino and the reduced Planck masses, respectively.

When the Hubble parameter becomes smaller than the curvature of the potential, the AD field begins to oscillate giving the baryon number. During the helical motion, it transforms into Q balls. The typical charge of the formed Q ball is estimated as [4] $Q = \beta\left(\frac{\phi_{\rm osc}}{M_F}\right)^4$, where $\phi_{\rm osc}$ is the field amplitude when the oscillation begins, and $\beta \simeq 6 \times 10^{-4}$ when the oscillating field has a nearly circular orbit $\epsilon = 1$ (ϵ: ellipticity of the orbit) and $\beta \simeq 6 \times 10^{-5}$ when $\epsilon \lesssim 0.1$. The charge Q is just the Φ-number, and relates to the baryon number of the Q ball as $B = bQ$, where b is the baryon number carried by a Φ particle. For example, $b = \frac{1}{3}$ for the udd direction.

3. Q-ball decay

A Q-ball decay occurs when some decay particles have the same kind of charges as the Q ball and the mass of each decay particle is less than the Q-ball mass per charge M_Q/Q. The decay rate Γ_Q has an upper bound $\Gamma_Q^{(\rm sat)}$ [16]. This saturation occurs approximately when $f_{\rm eff}\phi_Q \gtrsim \omega_Q$, where $f_{\rm eff}$ is the effective coupling constant by which the interaction is written as $\mathcal{L}_{\rm int} = f_{\rm eff}\phi\psi\bar{\psi}$. On the other hand, for the weak coupling limit such as $f_{\rm eff}\phi_Q \ll \omega_Q$, the decay rate is calculated as [16] $\Gamma_Q \simeq 12\left(\frac{f_{\rm eff}\phi_Q}{\omega_Q}\right)^2\Gamma_Q^{(\rm sat)}$. Here we are interested in the case where the Q ball decays into nucleons but not into MLSPs. It is described by the condition $bm_{\rm N} < M_Q/Q < m_{\rm MLSP}$ where m_N and $m_{\rm MLSP}$ are the nucleon and MLSP masses, respectively. The elementary process of the Q-ball decay into nucleon is squark + squark $\to$ quark + quark via gluino exchanges and this process is saturated. Next, we consider the Q-ball decay into axinos. The condition for the decay into axinos is described by $m_{\tilde{a}} < \frac{M_Q}{Q}$. The elementary process is squark $\to$ quark + axino whose effective coupling is given by [17] $f_{\rm eff}^{(\tilde{a})} = \frac{\alpha_s^2}{\sqrt{2}\pi^2}\frac{m_{\tilde{g}}}{f_a}\log\left(\frac{f_a}{m_{\tilde{g}}}\right)$, where f_a is the axion decay constant, α_s is the coupling strength for strong interaction and $m_{\tilde{g}}$ is the gluino mass. The decay may be saturated depending on the parameters f_a and Q, contrary to the case of the gravitino dark matter [12].

4. Baryon, axino and MLSP abundances from Q-ball decay

The number densities of the baryon, the axino and the MLSP are expressed in terms of the AD field number density n_ϕ as

$$n_b \simeq \epsilon b n_\phi, \tag{1}$$

$$n_{\tilde{a}} \simeq B_{\tilde{a}} n_\phi, \tag{2}$$

$$n_{\rm MLSP} \simeq \frac{Q_{\rm cr}}{Q} n_\phi, \tag{3}$$

respectively, where $B_{\tilde{a}} \equiv \frac{\Gamma_Q^{(\tilde{a})}}{\Gamma_Q^{(\rm sat)}}$ is the branching ratio and $B_{\tilde{a}} = 1$ in the saturated case. From the observations (such as the WMAP observation [18]), the ratio of the dark matter to baryon energy densities is $\rho_{\rm DM}/\rho_b \simeq 5$. Through calculations, one can see typically $\epsilon \ll 1$, so the orbit of the AD field is oblate, thus we generally set $\beta = 6 \times 10^{-5}$ below. Assuming the MLSP mass $m_{\rm MLSP} \simeq 100\,{\rm GeV}$, the MLSP abundances for the saturated and unsaturated case are obtained as

$$\left.\frac{\rho_{\rm MLSP}}{s}\right|_{\rm sat} \simeq 1.1 \times 10^{-25}\,{\rm GeV} \left(\frac{m_{\tilde{a}}}{\rm GeV}\right)^{-3} \left(\frac{T_{\rm RH}}{10^7\,{\rm GeV}}\right)^{-2} \left(\frac{T_{\rm D}}{3\,{\rm MeV}}\right)^{4}, \tag{4}$$

$$\left.\frac{\rho_{\rm MLSP}}{s}\right|_{\rm unsat} \simeq 7.8 \times 10^{-10}\,{\rm GeV} \left(\frac{m_{\tilde{a}}}{\rm GeV}\right)^{-2} \left(\frac{f_a}{10^{14}\,{\rm GeV}}\right)^{4} \left(\log\frac{f_a}{10^3\,{\rm GeV}}\right)^{-4} \times \left(\frac{T_{\rm RH}}{10^7\,{\rm GeV}}\right)^{-1} \left(\frac{T_{\rm D}}{3\,{\rm MeV}}\right)^{6}, \tag{5}$$

respectively, where $T_{\rm D}$ is the decaying temperature of Q ball.

5. Constraints on models

In this section, we see that the MLSP abundance is bounded from above by BBN constraints and bounded from below by the highest possible reheating temperature such that thermally produced axinos or gravitinos do not dominate the dark matter in the universe. We show the allowed region for the axino parameters f_a and $m_{\tilde{a}}$.

The upper limit on the MLSP abundance is given by the fact that the decay of the MLSP may affect abundances of light elements synthesized during the BBN. Here we assume the MLSP decays hadronically and neutralino as the MLSP, then adopt the BBN constraints from [19] which has the dependance of the MLSP lifetime. Comparing the lifetime of MLSP into axino [20] and gravitino [19] and using the smaller lifetime of two, we obtain the upper bound of the MLSP abundance.

On the other hand, the lower limit on the MLSP abundance comes from constraints on the reheating temperature. The constraints are due to the condition that thermally produced axinos and gravitinos cannot be the dominant component of the dark matter in the universe. We adapt the amount of the primordial abundance of the MLSP in [19]. The MLSP abundance from the Q-ball decay depends on the reheating temperature $T_{\rm RH}$ (Eqs.(4) and (5)). The upper limit of $T_{\rm RH}$ comes from the requirement that the thermally produced axinos and gravitinos should satisfy $\max(\Omega_{\tilde{a}}^{\rm TH} h^2, \Omega_{3/2}^{\rm TH} h^2) \lesssim \Omega_{\rm DM} h^2 \simeq 0.11$, where $\Omega_{\tilde{a}}^{\rm TH}$ and $\Omega_{3/2}^{\rm TH}$ respectively denote the density parameters of thermally produced axino and gravitino, and h is the Hubble constant in units of 100 km/s/Mpc. The constraint from the gravitino is written as [21],

$$T_{\rm RH} \lesssim T_{\rm RH,max}^{(3/2)} \equiv 3 \times 10^7\,{\rm GeV} \left(\frac{m_{\tilde{g}}}{500\,{\rm GeV}}\right)^{-2} \left(\frac{m_{3/2}}{\rm GeV}\right) \simeq 3 \times 10^7\,{\rm GeV} \left(\frac{m_{\tilde{a}}}{\rm GeV}\right), \tag{6}$$

where the gravitino mass is assumed to be the same as the axino mass $m_{\tilde{a}}$ On the other hand, the constraint from the axino depends on axion models. Here we consider two classes of axion models, the Kim-Shifman-Vainshtein-Zakharov (KSVZ) [22] and Dine-Fischer-Srednicki-Zhitnitskii (DFSZ) [23] models. In the KSVZ model, using the thermally produced axino density parameter [24], the constraint on $T_{\rm RH}$ is given as

$$T_{\rm RH} \lesssim T_{\rm RH,max}^{(\rm KSVZ\ \tilde{a})} \equiv 1.1 \times 10^8 \ {\rm GeV} \left(\frac{m_{\tilde{a}}}{\rm GeV}\right)^{-1} \left(\frac{f_a}{10^{14}\,{\rm GeV}}\right)^2 . \tag{7}$$

Inserting the stronger constraint of Eqs. (6) and (7) into Eqs.(4) and (5), we obtain the lowest possible abundances of the MLSP in the saturated and the unsaturated cases. On the other hand, in the DFSZ model, the axino production is dominated by the higgsino decay through the axino-Higgsino-Higgs interaction at the low reheating temperature ($T_{\rm RH} \lesssim 2 \times 10^7\,{\rm GeV}$) [25], while by scatterings due to $SU(2)_L$ coupling at the high reheating temperature ($T_{\rm RH} \gtrsim 2 \times 10^7\,{\rm GeV}$). Thus using the abundance of thermally produced axinos [26], for $T_{\rm RH} \gtrsim 2 \times 10^7\,{\rm GeV}$, we obtain the upper limit on $T_{\rm RH}$ as

$$T_{\rm RH} \lesssim T_{\rm RH,max}^{(\rm DFSZ\ \tilde{a})} \equiv 1.1 \times 10^9\,{\rm GeV} \left(\frac{m_{\tilde{a}}}{\rm GeV}\right)^{-1} \left(\frac{f_a}{10^{14}\,{\rm GeV}}\right)^2 . \tag{8}$$

On the other hand, for $T_{\rm RH} \lesssim 2 \times 10^7\,{\rm GeV}$, the condition $\Omega_{\tilde{a}}^{\rm TH} h^2 \lesssim \Omega_{\rm DM} h^2 \simeq 0.11$ only leads to the constraint on f_a and $m_{\tilde{a}}$ as

$$\left(\frac{m_{\tilde{a}}}{\rm GeV}\right) \left(\frac{f_a}{10^{14}\,{\rm GeV}}\right)^{-2} \lesssim 55, \tag{9}$$

and we must use $T_{\rm RH,max}^{(3/2)}$ in order to obtain the lowest possible abundance.

From those upper and lower constraints, we put constraints on the parameters $m_{\tilde{a}}$ and f_a. They are restricted by the condition that there exists the allowed range for the MLSP abundance; the lowest possible MLSP abundance for the KSVZ and DFSZ model must be smaller than the upper limits from the BBN. These constraints are shown in the blue solid lines in Fig. 1.

We show the allowed region for the parameters $m_{\tilde{a}}$ and f_a in both the KSVZ and the DFSZ models in Fig. 1. Q balls must be kinematically allowed to decay into axinos, which is expressed as $m_{\tilde{a}} < \frac{M_Q}{Q}$, so we obtain the upper limit on $m_{\tilde{a}}$ as the red dashed double-dotted lines. In the KSVZ model (the left panel), the upper left region is excluded by the condition $\epsilon \leq 1$ which corresponds to the black dashed dotted line, above which we cannot have enough baryon number compared to the axino dark matter. On the other hand, in the DFSZ model (the right panel), the upper left region is excluded by the condition that the higgsino decay produces too much axinos which corresponds to the black dashd dotted line, which is the main difference from the KSVZ model and considerably cuts out the allowed region compared with that in the KSVZ case. The allowed region is bounded from below for both f_a and $m_{\tilde{a}}$. The former is due to the fact that the observed neutrino cooling of SN 1987A excludes the axion decay constant smaller than 6×10^9 GeV [27], shown as the grey regions in Fig. 1. The latter comes from the constraint on the Q-ball parameter M_F such that $M_F \lesssim 0.1\sqrt{m_{3/2} M_{\rm P}} = 0.1\sqrt{m_{\tilde{a}} M_{\rm P}}$ which corresponds to the dark blue dotted lines. We find that the scenario works for rather wide range in the parameter space such that $6 \times 10^9\,{\rm GeV} \lesssim f_a \lesssim 5 \times 10^{14}\,{\rm GeV}$ and $m_{\tilde{a}} = O(0.1\,{\rm MeV}) - O({\rm GeV})$ in the KSVZ model, while the allowed region is restricted to $2 \times 10^{11}\,{\rm GeV} \lesssim f_a \lesssim 5 \times 10^{14}\,{\rm GeV}$ and $m_{\tilde{a}} = O(0.1\,{\rm MeV}) - O(10\,{\rm MeV})$ in the DFSZ model.

Using the allowed region in the parameter space (f_a, $m_{\tilde{a}}$), it is possible to check the realizations of the successful scenario for particular parameter sets of (f_a, $m_{\tilde{a}}$) in terms of Q-ball parameters, Q and M_F. Our scenario works for approximately $Q = 10^{20} - 10^{25}$ and $M_F = 10^5 - 10^7$ GeV.

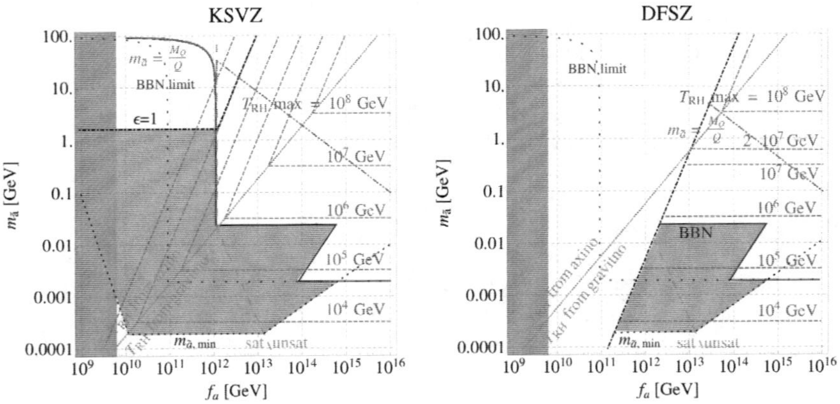

Figure 1. Allowed regions in $(f_a,\, m_{\tilde a})$ plane for the KSVZ (left) and the DFSZ (right) models are shown as blue filled regions. Below the dark red wide-dotted lines, there is no BBN constraint due to the small MLSP lifetime. The pink long-dashed lines divide the Q-ball decay into the saturated (on their left side) and unsaturated (on their right side) cases. The orange short-dashed lines distinguish which the upper limit of $T_{\rm RH}$ comes from: the thermally produced axinos (their above) or gravitinos (their below). The green dashed lines denote the iso-$T_{\rm RH,max}$ contours.

6. Summary

We have investigated the Q-ball scenario in the gauge-mediated SUSY breaking model where the Q ball decays into axinos and nucleons, providing simultaneously the dark matter and the baryon asymmetry of the universe. In our scenario, the Q ball has small enough charge to decay into nucleons and axinos, while it is large enough to kinematically forbid the decay channel to the SUSY particles in the MSSM, e.g., the lightest neutralinos. This prohibition holds almost through the decay process until the very end of the decay when the charge becomes small enough to open the channel into the MSSM LSPs, or MLSPs. We have evaluated the MLSP abundance, and imposed the condition that it should not affect the success of the BBN, which has resulted in the upper limit on the MLSP abundance. Meanwhile, the lower bound of the MLSP abundance has been obtained by the highest possible reheating temperature which comes from the condition that the thermally produced axinos and/or gravitinos are not the dominant component of the dark matter of the universe.

These conditions have led us to constrain the axino model parameters, the axino mass $m_{\tilde a}$ and the axion decay constant f_a, as well as from the condition that the axino can be produced kinematically from the Q-ball decay. The successful scenario resides in the region typically where $m_{\tilde a} = O(0.1\text{MeV}) - O(\text{GeV})$ and $6 \times 10^9\,\text{GeV} \lesssim f_a \lesssim 5 \times 10^{14}\,\text{GeV}$ in the KSVZ model, while the small f_a region ($f_a \lesssim 10^{11}\,\text{GeV}$) and the large $m_{\tilde a}$ region ($0.01\,\text{GeV} \lesssim m_{\tilde a}$) are excluded in the DFSZ model.

Acknowledgments

The work is supported by Grant-in-Aid for Scientific Research 23740206 (S.K.), 14102004 (M.K.) and 21111006 (M.K.) from the Ministry of Education, Culture, Sports, Science and Technology in Japan, and also by World Premier International Research Center Initiative (WPI Initiative), MEXT, Japan.

References

[1] Affleck I and Dine M 1985 *Nucl. Phys.* **B249** 361
[2] Kusenko A and Shaposhnikov M E 1998 *Phys. Lett.* **B 418** 46
[3] Enqvist K and McDonald J 1998 *Phys. Lett.* **B 425** 309; ibid 1999 *Nucl. Phys.* **B538** 321 ibid 1998 *Phys. Lett.***B440** 59
[4] Kasuya S and Kawasaki M 2000 *Phys. Rev.* **D61** 041301(R)
[5] Kasuya S and Kawasaki M 2000 *Phys. Rev.* **D62** 023512
[6] Kasuya S and Kawasaki M 2001 *Phys. Rev.* **D64** 123515
[7] Fujii M, Hamaguchi K 2002 *Phys. Lett.* **B525** 143
[8] Seto O 2006 *Phys. Rev.* **D73** 043509
[9] Roszkowski L and Seto O 2007 *Phys. Rev. Lett.* **98** 161304
[10] Shoemaker I M and Kusenko A 2009 *Phys. Rev.* **D80** 075021
[11] Doddato F and McDonald J 2011 *JCAP* **1106** 008; (*Preprint* 1111.2305)
[12] Kasuya S and Kawasaki M 2011 *Phys. Rev.* **D84** 123528
[13] Peccei R D and Quinn H R 1977 *Phys. Rev. Lett.* **38** 1440
[14] Kasuya S and Takahashi F 2007 *JCAP* **0711** 019
[15] de Gouvêa A, Moroi T and Murayama H 1997 *Phys. Rev.* **D56** 1281
[16] Cohen A G, Coleman S R, Georgi H and Manohar A 1986 *Nucl. Phys.* **B272** 301
[17] Covi L, Roszkowski L and Small M 2002 *JHEP* **0207** 023
[18] N. Jarosik *et al.*, Astrophys. J. Suppl. **192**, 14 (2011).
[19] Kawasaki M, Kohri K, Moroi T and Yotsuyanagi A 2008 *Phys. Rev.* **D 78** 065011
[20] Covi L, Kim B H, Kim J E and Roszkowski L 2001 *JHEP* **0105** 033
[21] Kawasaki M, Takahashi F and Yanagida T T 2006 *Phys. Rev.* **D74** 043519
[22] Kim J E 1979 *Phys. Rev. Lett.* **43** 103 M. A. Shifman, A. I. Vainshtein and V. I. Zakharov, Nucl. Phys. B **166**, 493 (1980).
[23] Dine M, Fischler W and Srednicki M 1981 *Phys. Lett.* **B104** 199; Zhitnitsky A R 1980 *Sov. J. Nucl. Phys.* **31** 260 [1980 *Yad. Fiz.* **31** 497]
[24] Brandenburg A and Steffen F D 2004 *JCAP* **0408** 008
[25] Chun E J 2011 *Phys. Rev.* **D84** 043509
[26] Choi K Y, Covi L, Kim J E and Roszkowski L 2011 em Axino Cold Dark Matter Revisited (*Preprint* 1108.2282)
[27] Kolb E W and Turner M S 1990 *The Early Universe* Addison-Wesley, Reading, MA

Discrete dark matter mechanism

E Peinado

AHEP Group, Institut de Física Corpuscular – C.S.I.C./Universitat de València
Edificio Institutos de Paterna, Apt 22085, E–46071 Valencia, Spain

E-mail: epeinado@ific.uv.es

Abstract. We present the Discrete Dark Matter mechanism. In this scenario the stability of the dark matter arises from a spontaneous breaking of a non-Abelian group (flavor symmetry) into one of its Abelian subgroups, this in the framework of a renormalizable extension of the Standard Model. Here we present the inclusion of the quarks on this scheme promoting the quarks to transform in a non-trivial way under the flavor symmetry.

1. Introduction

Non-baryonic Dark Matter (DM) is one of the most compelling problems of modern cosmology. Despite the fact that its existence is well established by cosmological and astrophysical probes, its nature remains elusive. Still, observations can constrain the properties of dark matter and give some hints about its identity. One of the fundamental requirements for a viable dark matter candidate is the stability over cosmological times.

This stability could be due to the existence of a symmetry protecting or suppresing its decay. It has been shown recently that such symmetry can be related to the flavor structure of the Standard Model [1]. The model proposed in [1] is based on an A_4 symmetry with four $SU(2)$ Higgs doublets. After the electroweak symmetry breaking, the A_4 (even permutation of four objects) group is spontaneously broken into a Z_2 subgroup which is responsible for the DM stability. The leptonic sector is also extended. It consist in four right handed neutrinos and the light neutrino masses are generated through the type-I seesaw mechanism and obey an inverted hierarchy mass spectrum with $m_{\nu_3} = 0$ and vanishing reactor angle $\theta_{13} = 0$.

We have also considered models with a different matter content for the right handed neutrinos with the same DM stability mechanism but with different neutrino phenomenology [2] or a model based on the dihedral group D_4, where some flavour changing neutral currents are present and constrain the DM sector [3]. There are some models based on flavor symmetries but with decaying DM that have also been considered, see for instance [4, 5]. For a model with non-abelian flavor symmetries with stable DM but where the DM couples to some right handed neutrinos in a similar way to our model[1] see [6].

2. Discrete Dark Matter Mechanism

The discrete dark matter (DDM) mechanism consists on the stability of the dark matter by means a residual Z_N symmetry of a spontaneously broken discrete flavor symmetry, for instance

[1] In this model the DM stability is due to an extra Z_2 where the DM and some right handed neutrinos are charged under this symmetry.

Content from this work may be used under the terms of the Creative Commons Attribution 3.0 licence. Any further distribution of this work must maintain attribution to the author(s) and the title of the work, journal citation and DOI.
Published under licence by IOP Publishing Ltd

in the original model the group of even permutations of four objects was used, A_4, and it was spontaneously broken by the electroweak symmetry breaking mechanism into a Z_2 sub group, $A_4 \to Z_2$.

The A_4 group has two generators: S, and T, which obey the properties $S^2 = T^3 = (ST)^3 = I$. S is a Z_2 generator while T is a Z_3 generator. A_4 has four irreducible representations, three singlets 1, 1′, and 1″ and one triplet. The generators of A_4 in the S-diagonal basis are

$$\begin{array}{lll} 1 & S=1 & T=1 \\ 1' & S=1 & T=\omega^2 \\ 1'' & S=1 & T=\omega \end{array} \quad 3:\; S=\begin{pmatrix} 1 & 0 & 0 \\ 0 & -1 & 0 \\ 0 & 0 & -1 \end{pmatrix} \quad T=\begin{pmatrix} 0 & 1 & 0 \\ 0 & 0 & 1 \\ 1 & 0 & 0 \end{pmatrix} \tag{1}$$

where $\omega^3 = 1$.

The model consist of an extended SM where instead of having one $SU(2)$ doublet of scalar fields, we have three of them transforming as a triplet under A_4, and four right handed neutrinos three of them transforming as a triplet and the other one as a singlet of the flavor group. In table 1 we present the relevant quantum numbers for the matter fields. The Yukawa Lagrangian

	L_e	L_μ	L_τ	l_e^c	l_μ^c	l_τ^c	N_T	N_4	H	η
$SU(2)$	2	2	2	1	1	1	1	1	2	2
A_4	1	$1'$	$1''$	1	$1''$	$1'$	3	1	1	3

Table 1. Summary of relevant model quantum numbers

of the model is given by

$$\begin{aligned} \mathcal{L} \;=\; & y_e L_e l_e^c H + y_\mu L_\mu l_\mu^c H + y_\tau L_\tau l_\tau^c H + + y_1^\nu L_e (N_T \eta)_1 + y_2^\nu L_\mu (N_T \eta)_{1''} + \\ & + y_3^\nu L_\tau (N_T \eta)_{1'} + y_4^\nu L_e N_4 H + M_1 N_T N_T + M_2 N_4 N_4 + \text{h.c.} \end{aligned} \tag{2}$$

This way H is responsible for quark and charged lepton masses, the latter automatically diagonal[2]. Neutrino masses arise from H and η. The vacumm alignment for the scalar fields [1] is

$$\langle H^0 \rangle = v_h \neq 0, \quad \langle \eta_1^0 \rangle = v_\eta \neq 0 \quad \langle \eta_{2,3}^0 \rangle = 0\,, \tag{3}$$

which means the vev alignment for the A_4 triplet of the form $\langle \eta \rangle \sim (1,0,0)$. This alignment is invariant under the S generator[3], see eq. (1), therefore the minimum of the potential breaks spontaneously A_4 into a Z_2 subgroup generated by S. All the fields in the model singlets under A_4 are even under the residual Z_2, the triplets transform as:

$$\begin{array}{lll} N_1 \to +N_1\,, & \eta_1 & \to +\eta_1 \\ N_2 \to -N_2\,, & \eta_2 & \to -\eta_2 \\ N_3 \to -N_3\,, & \eta_3 & \to -\eta_3. \end{array} \tag{4}$$

The DM candidate is the lightest particle charged under Z_2 i. e. the lightest combination of the scalars η_2 and η_3, which we will denote generically by η_{DM}. We list below all interactions of η_{DM}:

(i) Yukawa interactions

$$\eta_{DM}\, \overline{\nu}_i N_{2,3}\,, \tag{5}$$

where $i = e,\, \mu,\, \tau$.

[2] For quark mixing angles generated through higher dimension operators see reference [7].
[3] H_s is in the 1 representation of A_4 and its vev also respects the generator S.

(ii) Higgs-Vector boson couplings

$$\eta_{DM}\eta_{DM}ZZ\,,\quad \eta_{DM}\eta_{DM}WW\,,\quad \eta_{DM}\eta^{\pm}_{2,3}W^{\pm}Z\,,\quad \eta_{DM}\eta^{\pm}_{2,3}W^{\pm}\,, \eta_{DM}A_{2,3}Z\,. \tag{6}$$

(iii) Scalar interactions from the Higgs potential:

$$\begin{array}{llll} \eta_{DM}\,A_1A_2h\,, & \eta_{DM}\,A_1A_3h_1\,, & \eta_{DM}\,A_1A_2h_1\,, & \eta_{DM}\,A_1A_3h\,, \\ \eta_{DM}\,A_2A_3h_3\,, & \eta_{DM}\,h_1h_3h & \eta_{DM}\eta_{DM}hh\,, & \eta_{DM}\eta_{DM}h_1h_1\,. \end{array} \tag{7}$$

After electroweak symmetry breaking, the Higgs fields acquire vacuum (vevs) expectation values, v_h and v_η for the singlet and the first component of the triplet respectively, additional terms are obtained from those in Eq. (7) by replacing $h \to v_h$ and $h_1 \to v_\eta$. The dark matter phenomenology of this specific model has been studied in detail in [8].

3. A Discrete Dark Matter model for quarks and leptons

We were looking for a group G that contains at least two irreducible representations of dimension larger than one, namely r_a and r_b. We also require that all the components of the irreducible representation r_a transform trivially under an abelian subgroup of $G \supset Z_N$ (with $N = 2, 3$) while at least one component of the irreducible representation r_b is charged with under Z_N. The stability of the lightest component of the matter fields transforming as r_b is guaranteed by Z_N giving a potential [4] DM candidate.

The smallest group with this property we found is $\Delta(54)$, isomorphic to $(Z_3 \times Z_3) \rtimes S_3$. This group contains four triplet irreducible representations, $\mathbf{3_{1,2,3,4}}$, in addition, $\Delta(54)$ contains four different doublets $\mathbf{2_{1,2,3,4}}$ and two singlet irreducible representations, $\mathbf{1_\pm}$. The product rules for the doublets are as follows:

- The product of two equal doublets

$$\mathbf{2_k} \times \mathbf{2_k} = \mathbf{1_+} + \mathbf{1_-} + \mathbf{2_k} \tag{8}$$

- The product of two different doublets give us the other two doublets, for instance:

$$\mathbf{2_1} \times \mathbf{2_2} = \mathbf{2_3} + \mathbf{2_4} \tag{9}$$

Of the four doublets $\mathbf{2_1}$ is invariant under the $P \equiv (Z_3 \times Z_3)$ subgroup of $\Delta(54)$, while the others transform nontrivially, for example $\mathbf{2_3} \sim (\chi_1, \chi_2)$, which transforms as $\chi_1\,(\omega^2, \omega)$ and $\chi_2\,(\omega, \omega^2)$ respectively, where $\omega^3 = 1$. We can see that by taking $r_a = \mathbf{2_1}$ and $r_b = \mathbf{2_3}$ that $\Delta(54)$ is a perfect choice for our purpose.

	$\overline{L}_e$	$\overline{L}_D$	e_R	l_D	H	χ	η	Δ
$SU(2)$	2	2	1	1	2	2	2	3
$\Delta(54)$	$\mathbf{1_+}$	$\mathbf{2_1}$	$\mathbf{1_+}$	$\mathbf{2_1}$	$\mathbf{1_+}$	$\mathbf{2_1}$	$\mathbf{2_3}$	$\mathbf{2_1}$

Table 2. Lepton and higgs boson assignments of the model.

In this way we can choose the scalars in the model to belong to the $\mathbf{2_3}$ for instance, while the active scalars to belong to the $\mathbf{2_1}$ [9]. The relevant quantum numbers of the model are in

[4] Other requirements must fullfilled in order to have a viable DM candidate, such as neutrality, correct relic abundance, and consistency with constraints from DM search experiments.

	$Q_{1,2}$	Q_3	(u_R, c_R)	t_R	d_R	s_R	b_R
$SU(2)$	2	2	1	1	1	1	1
$\Delta(54)$	$\mathbf{2_1}$	$\mathbf{1_+}$	$\mathbf{2_1}$	$\mathbf{1_+}$	$\mathbf{1_-}$	$\mathbf{1_+}$	$\mathbf{1_+}$

Table 3. Quark gauge and flavour representation assignments.

Tables 2 and 3. In table 2 $L_D \equiv (L_\mu, L_\tau)$ and $l_D \equiv (\mu_R, \tau_R)$. There are five $SU(2)_L$ doublets of Higgs scalars: the H is a singlet of $\Delta(54)$, while $\eta = (\eta_1, \eta_2) \sim \mathbf{2_3}$ and $\chi = (\chi_1, \chi_2) \sim \mathbf{2_1}$ are doublets. In order to preserve a remnant P symmetry, the doublet η is not allowed to take vacuum expectation value (vev).

Lets start discussing the quark sector. In Ref. [1, 3, 8, 2] quarks were considered blind under the flavor symmetry to guarantee the stability of the DM. Consequently the generation of quark mixing angtles was difficult, see [7].

A nice feature of our current model is that with $\Delta(54)$ we can assign quarks to the singlet and doublet representations as shown in table 2, in such a way that we can fit the CKM mixing parameters.

The resulting up- and down-type quark mass matrices in our model are given by

$$M_d = \begin{pmatrix} ra_d & rb_d & rd_d \\ -a_d & b_d & d_d \\ 0 & c_d & e_d \end{pmatrix}, \quad M_u = \begin{pmatrix} ra_u & b_u & d_u \\ b_u & a_u & rd_u \\ c_u & rc_u & e_u \end{pmatrix}. \tag{10}$$

where $r = \langle\chi_2\rangle/\langle\chi_1\rangle$. Note that the Higgs fields H and χ are common to the lepton and the quark sectors and in particular the parameter r. In order to fit of all quark masses and mixings provided r lies in the range of about $0.1 < r < 0.2$ [9]. We turn to the leptonic sector, the charged leptons are basically equal to the up quarks under the flavor symmetry, so the charged lepton mass matrix is of the same form of the up quark mass matrix in Eq. (10). The neutrino masses are generated through the type-II see-saw mechanism [10]. For that we include in the model an $SU_L(2)$ Higgs triplet scalar field $\Delta \sim \mathbf{2_1}$. Regarding dark matter, note that the lightest P-charged particle in $\eta_{1,2}$ can play the role of "inert" DM [11], as it has no direct couplings to matter. The conceptual link between dark matter and neutrino phenomenology arises from the fact that the DM stabilizing symmetry is a remnant of the underlying flavor symmetry which accounts for the observed pattern of oscillations. Choosing the vev alignment $\langle\Delta\rangle \sim (1,1)$ and $\langle\chi_1\rangle \neq \langle\chi_2\rangle$, consistent with the minimization of the scalar potential one finds that

$$M_\nu \propto \begin{pmatrix} 0 & \delta & \delta \\ \delta & \alpha & 0 \\ \delta & 0 & \alpha \end{pmatrix}, \tag{11}$$

where $\delta = y_a\,\langle\Delta\rangle$, $\alpha = y_b\,\langle\Delta\rangle$ [9]. Note that this matrix has two free parameters and gives us a neutrino mass sum rule of the form $m_1^\nu + m_3^\nu = m_2^\nu$ in the complex plane, which has implications for neutrinoless double beta decay [12], as seen in Fig. (1).

We now turn to the second prediction. For simplicity, we consider in what follows only real parameters and we fix the intrinsic neutrino CP–signs [17] as $\eta = diag(-,+,+)$, where η is defined so that the CP conservation condition in the charged current weak interaction reads $U^\star = U\eta$, U being the lepton mixing matrix. The correlations we have are presented in Fig. 2.

4. Conclusions

We have extended DDM to include the quarks in the game. We have described how spontaneous breaking of a $\Delta(54)$ flavor symmetry can stabilize the dark matter by means of a residual

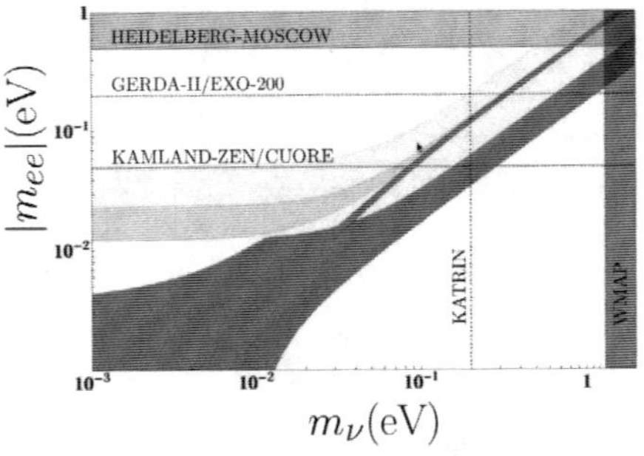

Figure 1. Effective neutrinoless double beta decay parameter m_{ee} versus the lightest neutrino mass. The thick upper and lower branches correspond the "flavor-generic" inverse (yellow) and normal (gray) hierarchy neutrino spectra, respectively. The model predictions are indicated by the green and red (darker-shaded) regions, respectively. They were obtained by taking the 3σ band on the mass squared differences. Only these sub-bands are allowed by the $\Delta(54)$ mdel. For comparison we give the current limit and future sensitivities on m_{ee} [13, 14] and m_ν [15, 16], respectively.

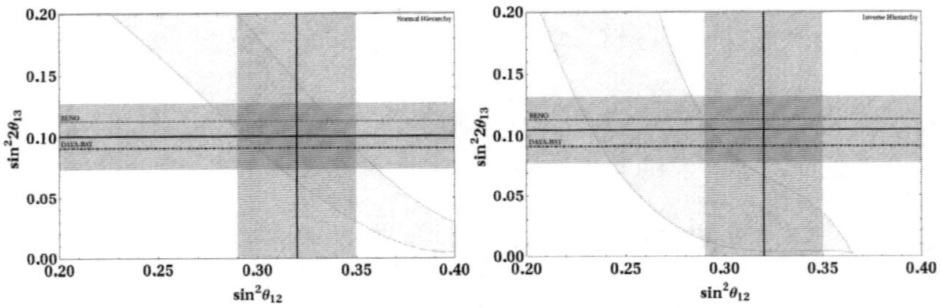

Figure 2. The left figure (right figure) is the correlation for normal hierarchy (inverse hierarchy). The shaded (yellow) curved band gives the predicted correlation between solar and reactor angles when the solar and atmospheric squared mass splittings are varied within 2σ for the normal hierarchy spectrum. The solid (black) line gives the global best fit values for θ_{12} and θ_{13}, along with the corresponding two-sigma bands, from Ref. [18]. The dashed lines correspond to the central values of the recent published reactor measurements [19, 20]. Note that θ_{23} is also within 2σ

unbroken symmetry. In our scheme left-handed leptons as well as quarks transform nontrivially under the flavor group, with neutrino masses arising from a type-II seesaw mechanism. We have found lower bounds for neutrinoless double beta decay, even in the case of normal hierarchy, as seen in Fig. 1. In addition, we have correlations between solar and reactor angles consistent with the recent Daya-Bay and RENO reactor measurements, see Fig. 2, interesting in their own right. Direct and indirect detection prospects are similar to a generic WIMP dark matter, as provided by multi-Higgs extensions of the SM, see, for example, Ref. [8].

Acknowledgments

I thank my collaborators Sofiane Boucenna, Stefano Morisi, Yusuke Shimizu and Jose W. F. Valle for the work presented here. This work was supported by the Spanish MICINN under grants FPA2008-00319/FPA, FPA2011-22975 and MULTIDARK CSD2009-00064 (Consolider-Ingenio 2010 Programme), by Prometeo/2009/091 (Generalitat Valenciana), by the EU ITN UNILHC PITN-GA-2009-237920 and by CONACyT (Mexico).

References

[1] Hirsch M, Morisi S, Peinado E and Valle J 2010 *Phys.Rev.* **D82** 116003 (*Preprint* 1007.0871)
[2] Meloni D, Morisi S and Peinado E 2011 *Phys.Lett.* **B697** 339–342 (*Preprint* 1011.1371)
[3] Meloni D, Morisi S and Peinado E 2011 *Phys.Lett.* **B703** 281–287 (*Preprint* 1104.0178)
[4] Daikoku Y, Okada H and Toma T 2010 (*Preprint* 1010.4963)
[5] Kajiyama Y and Okada H 2011 *Nucl.Phys.* **B848** 303–313 (*Preprint* 1011.5753)
[6] Eby D A and Frampton P H 2012 *Phys.Lett.* **B713** 249–254 (*Preprint* 1111.4938)
[7] de Adelhart Toorop R, Bazzocchi F and Morisi S 2012 *Nucl.Phys.* **B856** 670–681 (*Preprint* 1104.5676)
[8] Boucenna M, Hirsch M, Morisi S, Peinado E, Taoso M *et al.* 2011 *JHEP* **1105** 037 (*Preprint* 1101.2874)
[9] Boucenna M, Morisi S, Peinado E, Shimizu Y and Valle J 2012 *Phys.Rev.* **D86** 073008 (*Preprint* 1204.4733)
[10] Schechter J and Valle J 1980 *Phys.Rev.* **D22** 2227
[11] Barbieri R, Hall L J and Rychkov V S 2006 *Phys.Rev.* **D74** 015007 (*Preprint* hep-ph/0603188)
[12] Dorame L, Meloni D, Morisi S, Peinado E and Valle J 2012 *Nucl.Phys.* **B861** 259–270 (*Preprint* 1111.5614)
[13] Schwingenheuer B 2012 *J.Phys.Conf.Ser.* **375** 042007 (*Preprint* 1201.4916)
[14] Rodejohann W 2011 *Int.J.Mod.Phys.* **E20** 1833–1930 (*Preprint* 1106.1334)
[15] Osipowicz A *et al.* (KATRIN Collaboration) 2001 (*Preprint* hep-ex/0109033)
[16] Komatsu E *et al.* (WMAP Collaboration) 2011 *Astrophys.J.Suppl.* **192** 18 (*Preprint* 1001.4538)
[17] Schechter J and Valle J 1981 *Phys.Rev.* **D24** 1883–1889
[18] Schwetz T, Tortola M and Valle J 2011 *New J.Phys.* **13** 109401 (*Preprint* 1108.1376)
[19] An F *et al.* (DAYA-BAY Collaboration) 2012 *Phys.Rev.Lett.* **108** 171803 (*Preprint* 1203.1669)
[20] Ahn J *et al.* (RENO collaboration) 2012 *Phys.Rev.Lett.* **108** 191802 (*Preprint* 1204.0626)

Direct detection of vector dark matter[1]

Natsumi Nagata[1,2]

[1] Department of Physics, Nagoya University, Nagoya 464-8602, Japan
[2] Department of Physics, University of Tokyo, Tokyo 113-0033, Japan

E-mail: natsumi@eken.phys.nagoya-u.ac.jp

Abstract. We evaluate the elastic scattering cross section of vector dark matter with nucleon based on the method of effective field theory. The dark matter is assumed to behave as a vector particle under the Lorentz transformation and to interact with colored particles including quarks in the Standard Model. After formulating general formulae for the scattering cross sections, we apply them to the case of the first Kaluza-Klein photon dark matter in the minimal universal extra dimension model. The resultant cross sections are found to be larger than those calculated in previous literature.

1. Introduction

The existence of dark matter (DM) has been established by cosmological observations [2]. One of the most attractive candidates is what we call Weakly Interacting Massive Particles (WIMPs), which are stable particles with masses of the electroweak scale and weakly interact with ordinary matters. This interactions enable us to search for WIMP DM by using the scattering signal of DM with nuclei on the earth. Such kind of experiments are called the direct detection experiments of WIMP DM.

For the past years, a lot of efforts have been dedicated to the direct detection of WIMP DM, and their sensitivities have been extremely improving. The XENON100 Collaboration, for example, gives a severe constraint on the spin-independent (SI) elastic scattering cross section of WIMP DM with nucleon $\sigma_N^{\rm SI}$ ($\sigma_N^{\rm SI} < 2.0 \times 10^{-45}$ cm^2 for WIMPs with a mass of 55 GeV$/c^2$) [3]. Moreover, ton-scale detectors for the direct detection experiments are now planned and expected to have significantly improved sensitivities.

In order to study the nature of DM based on these experiments, we need to evaluate the WIMP-nucleon elastic scattering cross section precisely. In this work, we assume the WIMP DM to be a vector particle, and evaluate its cross section scattering off a nucleon. Several candidates for vector DM have been proposed in various models, and there have been a lot of previous work computing the scattering cross sections [4, 5, 6]. However, we found that in the calculations some of the leading contributions to the scattering cross section are not evaluated correctly, or in some cases completely neglected. Taking such situation into account, we study the way of evaluating the cross section systematically by using the method of effective field theory.

[1] This talk is based on the work with Junji Hisano, Koji Ishiwata, and Masato Yamanaka [1].

Content from this work may be used under the terms of the Creative Commons Attribution 3.0 licence. Any further distribution of this work must maintain attribution to the author(s) and the title of the work, journal citation and DOI.
Published under licence by IOP Publishing Ltd

2. Direct detection of vector dark matter

In this section we discuss the way of evaluating the elastic scattering cross section of vector DM with nucleon. First, we write down the effective interactions of vector DM (B_μ) with light quarks and gluon [1]:

$$\mathcal{L}^{\rm eff} = \sum_{q=u,d,s} \mathcal{L}^{\rm eff}_q + \mathcal{L}^{\rm eff}_G, \tag{1}$$

with

$$\mathcal{L}^{\rm eff}_q = f_q m_q B^\mu B_\mu \bar{q}q + \frac{d_q}{M}\epsilon_{\mu\nu\rho\sigma}B^\mu i\partial^\nu B^\rho \bar{q}\gamma^\sigma\gamma^5 q + \frac{g_q}{M^2}B^\rho i\partial^\mu i\partial^\nu B_\rho \mathcal{O}^q_{\mu\nu}, \tag{2}$$

$$\mathcal{L}^{\rm eff}_G = f_G B^\rho B_\rho G^{a\mu\nu}G^a_{\mu\nu}, \tag{3}$$

where m_q are the masses of light quarks, M is the DM mass, and $\epsilon^{\mu\nu\rho\sigma}$ is the totally antisymmetric tensor defined as $\epsilon^{0123} = +1$. The covariant derivative is defined as $D_\mu \equiv \partial_\mu + ig_s A^a_\mu T_a$, with g_s, T_a and A^a_μ being the strong coupling constant, the $\mathrm{SU}(3)_C$ generators, and the gluon fields, respectively. The gluon field strength tensor is denoted by $G^a_{\mu\nu}$, and $\mathcal{O}^q_{\mu\nu} \equiv \frac{1}{2}\bar{q}i\left(D_\mu\gamma_\nu + D_\nu\gamma_\mu - \frac{1}{2}g_{\mu\nu}\not{D}\right)q$ are the twist-2 operators of light quarks. When we write down the effective Lagrangian, we consider the fact that the scattering process is non-relativistic. The coefficients of the operators are to be determined by integrating out the heavy particles in high energy theory. The second term in Eq. (2) gives rise to the spin-dependent (SD) interaction, while the other terms yield the spin-independent (SI) interactions. We focus on the SI interactions hereafter, because the experimental constraint is much severe for the SI interactions, rather than for the SD interactions.

In order to obtain the effective coupling of the vector DM with nucleon induced by the effective Lagrangian, we need to evaluate the nucleon matrix elements of the quark and gluon operators in Eqs.(2) and (3). First, the nucleon matrix elements of the scalar-type quark operators are parametrized as

$$f_{Tq} \equiv \langle N|m_q\bar{q}q|N\rangle/m_N\ , \tag{4}$$

with $|N\rangle$ and m_N the one-particle state and the mass of nucleon, respectively. The parameters are called the mass fractions and their values are obtained from the lattice simulations [7, 8]. Second, for the quark twist-2 operators, we can use the parton distribution functions (PDFs):

$$\langle N(p)|\mathcal{O}^q_{\mu\nu}|N(p)\rangle = \frac{1}{m_N}(p_\mu p_\nu - \frac{1}{4}m_N^2 g_{\mu\nu})\left(q(2)+\bar{q}(2)\right)\ , \tag{5}$$

where $q(2)$ and $\bar{q}(2)$ are the second moments of PDFs of quark $q(x)$ and anti-quark $\bar{q}(x)$, respectively, which are defined as $q(2)+\bar{q}(2) = \int_0^1 dx\ x\ [q(x)+\bar{q}(x)]$. These values are obtained from Ref. [9]. Finally, the matrix element of gluon field strength tensor can be evaluated by using the trace anomaly of the energy-momentum tensor in QCD [10]. The resultant expression is given as

$$\langle N|G^a_{\mu\nu}G^{a\mu\nu}|N\rangle = -\frac{8\pi}{9\alpha_s}m_N f_{TG} \tag{6}$$

with $f_{TG} \equiv 1 - \sum_{q=u,d,s} f_{Tq}$. Note that the right hand side of the expression is divided by the strong coupling constant, α_s. For this reason, although the gluon contribution is induced by higher loop diagrams, it can be comparable to the quark contributions [11]. Briefly speaking, the enhancement comes from the large gluon contribution to the mass of nucleon. As a result, the SI effective coupling of vector DM with nucleon, f_N, is given as

$$f_N/m_N = \sum_{q=u,d,s} f_q f_{Tq} + \sum_{q=u,d,s,c,b}\frac{3}{4}\left(q(2)+\bar{q}(2)\right)g_q - \frac{8\pi}{9\alpha_s}f_{TG}f_G\ . \tag{7}$$

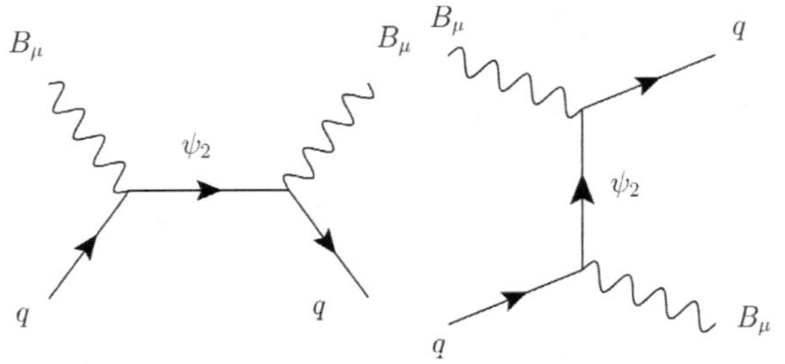

Figure 1. Tree-level diagrams of exchanging colored fermion ψ_2 to generate interaction of vector dark matter with light quarks.

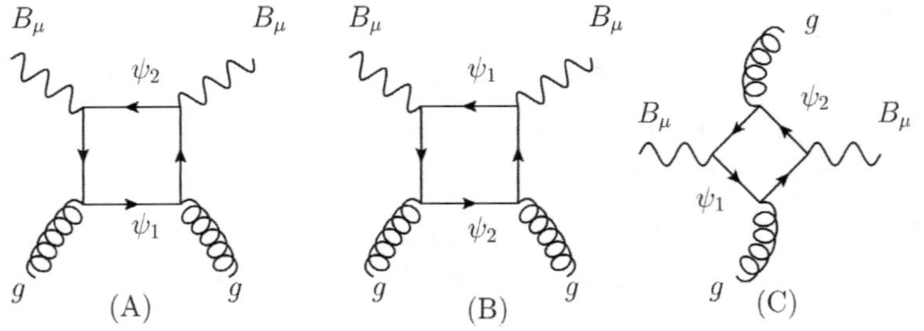

Figure 2. One-loop contributions to scalar-type effective coupling with gluon.

Using the effective coupling, we eventually obtain the SI scattering cross section of DM with nucleon:

$$\sigma_N^{(\mathrm{SI})} = \frac{1}{\pi}\left(\frac{m_N}{M+m_N}\right)^2 |f_N|^2 \,. \tag{8}$$

Now, all we have to do reduces to evaluate the coefficients of the effective operators by integrating out the heavy fields in the high-energy theories. For example, we take the case where the interaction Lagrangian of the vector DM has a generic form as

$$\mathcal{L} = \bar{\psi}_2\left(a_{\psi_2\psi_1}\gamma^\mu + b_{\psi_2\psi_1}\gamma^\mu\gamma_5\right)\psi_1 B_\mu + \mathrm{h.c.}\,, \tag{9}$$

where ψ_1 and ψ_2 are colored fermions with masses m_1 and m_2 ($m_1 < m_2$), respectively. In this case, the vector DM is scattered by light quarks at tree-level. The relevant interaction Lagrangian is given by taking $\psi_1 = q$ in Eq. (9), and the corresponding diagrams are shown in Fig. 1. After integrating out the heavy particle ψ_2, we obtain

$$f_q = \frac{a_{\psi_2 q}^2 - b_{\psi_2 q}^2}{m_q}\frac{m_2}{m_2^2 - M^2} - (a_{\psi_2 q}^2 + b_{\psi_2 q}^2)\frac{m_2^2}{2(m_2^2 - M^2)^2}, \tag{10}$$

$$g_q = -\frac{2M^2(a_{\psi_2 q}^2 + b_{\psi_2 q}^2)}{(m_2^2 - M^2)^2}. \tag{11}$$

One can easily find that the effective couplings obtained here are enhanced when the vector DM and the heavy colored fermion are degenerate in mass [12].

The effective coupling of the vector DM with gluon is induced by 1-loop diagrams illustrated in Fig. 2. In those processes, all the particles ψ_1 and ψ_2 which couple with B_μ run in the loop.

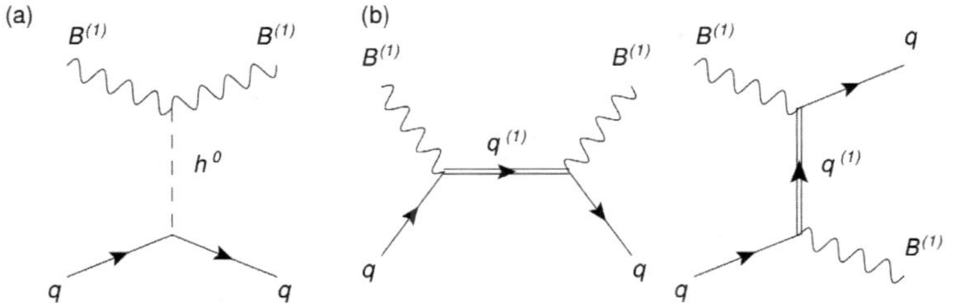

Figure 3. Tree-level diagrams for the elastic scattering of the KK photon DM $B^{(1)}$ with light quarks: (a) Higgs boson exchange contribution, and (b) KK quark exchange contributions.

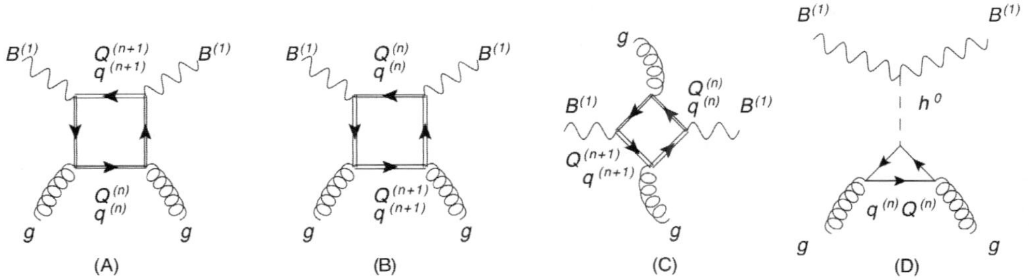

Figure 4. One-loop diagrams for the effective interaction of $B^{(1)}$ with gluon.

The resultant expressions are somewhat complicated, and thus we just quote Ref. [1] for their complete formulae as well as their derivation.

3. Application and Results

Next, we deal with a particular model for vector DM as an application. We carry out the calculation for the first Kaluza-Klein (KK) photon DM [4, 5] in the minimal universal extra dimension (MUED) model [13, 14]. In this model, an extra dimension is compactified on an $S^1/\mathbb{Z}_2$ orbifold with the compactification radius R, and all of the Standard Model (SM) particles propagate in the dimension. The lightest KK-odd particle (LKP) is prevented from decaying to the SM particles, so it becomes DM. This model has just three undetermined parameters: the radius of the extra dimension R, the mass of Higgs boson m_h, and the cutoff scale Λ.

In general, extra dimensional models give rise to the degenerate mass spectrum at tree-level, which is broken by radiative corrections. By evaluating the radiative corrections [14], one finds that the first KK photon $B^{(1)}$ is the lightest KK-odd particle, thus, becomes the DM in the Universe. Moreover, since the mass difference is induced by radiative corrections, the mass degeneracy is tight for a small cut-off scale. In such a case, although it is difficult to probe the MUED model at the LHC because of the soft QCD jets, the direct detection rate of dark matter is expected to be enhanced [12, 15], as we shall see soon later[2].

Now we evaluate the SI scattering cross section of the KK photon DM with nucleon. The effective interaction of the KK photon DM $B^{(1)}$ with light quarks is induced by the tree-level diagrams shown in Fig. 3. Here, h^0 and $q^{(1)}$ are the Higgs boson and the first KK quark, respectively. Also, there are one-loop diagrams we should evaluate for the gluon contribution. They are illustrated in Fig. 4. In these diagrams, all of the KK quarks run in the loop. Taking

[2] Indirect DM searches also might be a powerful alternative in such a case [16, 17].

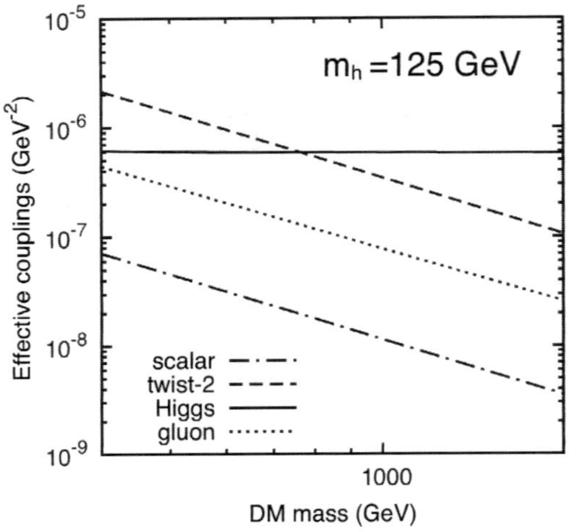

Figure 5. Each contribution in effective coupling f_N/m_N given in Eq. (7). Here we set $m_h = 125$ GeV and $(m_{1\text{st}} - M)/M = 0.1$.

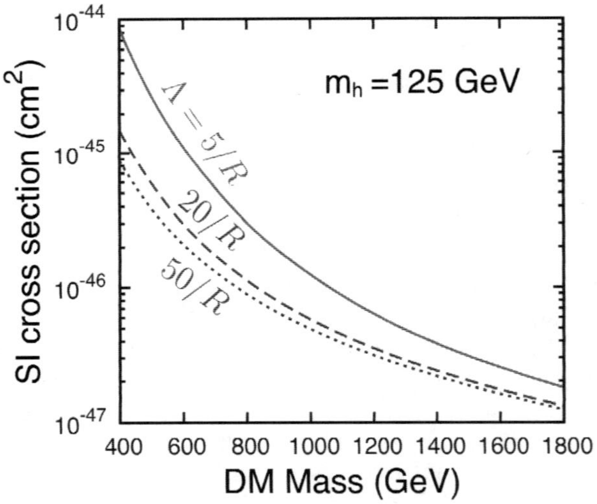

Figure 6. Spin-independent cross section with proton for $m_h = 125$ GeV. Each line corresponds to $\Lambda = 5/R$, $20/R$, and $50/R$, respectively.

all the contributions into account, we obtain the effective interactions. Their expressions are given in Ref. [1].

In Fig. 5, we plot each contribution to the SI effective coupling of DM with a proton. The solid line shows the contribution of the Higgs boson exchanging diagrams, the dashed line indicates the twist-2 type contribution, the dash-dotted line corresponds to the scalar-type contribution (except for the Higgs-exchanging contribution), and the dotted line represents the gluon contribution (except for the Higgs-exchanging contribution). Here we set the Higgs boson mass[3] equal to 125 GeV, and the mass difference between the DM and the first KK quark to be 10 % by hand. We find that all of the contributions have the same sign thus they are constructive. The twist-2 contribution is dominant when $M \lesssim 800$ GeV, while that of Higgs boson exchanging process becomes dominant above it. Moreover, although the tree-level quark contributions are dominant, it is found that the gluon contribution is not negligible at all.

By using the effective couplings obtained above, we evaluate the SI scattering cross sections. In Fig. 6, we plot the SI cross section of KK photon DM with a proton as a function of the DM mass. Here again, the Higgs boson mass is set to be 125 GeV. We take the cut-off scale $\Lambda = 5/R,\ 20/R,\ 50/R$ from top to bottom. We find that the scattering cross sections reduce as the cut-off scale is taken to be large, as expected. As a result, we obtain the SI scattering cross section which is larger than those obtained by the previous calculations. Note that the DM mass with which the thermal relic abundance is preferred by the WMAP observation [2] is around 1300 GeV [20]. It corresponds to $\sigma_N^{\rm SI} = (2.5\text{-}5)\times 10^{-47}$ cm^2, so the direct detection experiments with ton-scale detectors might be able to probe the DM in the future.

4. Conclusion

We calculate the spin independent elastic scattering cross sections of vector dark matter with nucleon based on the effective field theory. It is found that the interaction of dark matter with gluon as well as quarks yields sizable contribution to the scattering cross section, though the gluon contribution is induced at loop level. The scattering cross section of the first Kaluza-Klein photon dark matter in the MUED model turns out to be larger than those obtained by the previous calculations.

Acknowledgments

The author would like to thank Junji Hisano, Koji Ishiwata, and Masato Yamanaka for collaboration. This work is supported by Research Fellowships of the Japan Society for the Promotion of Science for Young Scientists.

References

[1] Hisano J, Ishiwata K, Nagata Nand Yamanaka M 2011 *Prog. Theor. Phys.* **126** 435 .
[2] Komatsu E *et al.* (WMAP Collaboration) 2011 *Astrophys. J. Suppl.* **192** 18 .
[3] Aprile E *et al.* (XENON100 Collaboration) 2012 (*Preprint* 1207.5988 [astro-ph.CO])
[4] Cheng H C P, Feng J L and Matchev K T 2002 *Phys. Rev. Lett.* **89** 211301
[5] Servant G and Tait T M P 2002 *New J. Phys.* **4** 99
[6] Birkedal A, Noble A, Perelstein M and Spray A 2006 *Phys. Rev.* **D74** 035002
[7] Young R D and Thomas A W 2010 *Phys. Rev.* **D81** 014503 .
[8] Ohki H *et al.* (JLQCD Collaboration) 2012 (*Preprint* arXiv:1208.4185 [hep-lat])
[9] Pumplin J, Stump D R, Huston J, Lai H L, Nadolsky P M and Tung W K 2002 *JHEP* **0207** 012
[10] Shifman M A, Vainshtein A I and Zakharov V I 1978 *Phys. Lett.* **B78** 443
[11] Hisano J, Ishiwata K and Nagata N 2010 *Phys. Rev.* **D82** 115007
[12] Hisano J, Ishiwata K and Nagata N 2011 *Phys. Lett.* **B706** 208
[13] Appelquist T, Cheng H C and Dobrescu B A 2001 *Phys. Rev.* **D64** 035002
[14] Cheng H-C, Matchev K T and Schmaltz M 2002 *Phys. Rev.* **D66** 036005
[15] Arrenberg S, Baudis L, Kong K, Matchev K T and Yoo J 2008 *Phys. Rev.* **D78** 056002
[16] Asano M, Bringmann T and Weniger C 2012 *Phys. Lett.* **B709** 128
[17] Garny M, Ibarra A, Pato M and Vogl S 2012 (*Preprint* arXiv:1207.1431 [hep-ph])
[18] Aad G *et al.* (ATLAS Collaboration) 2012 *Phys. Lett.* **B716** 1
[19] Chatrchyan S *et al.* (CMS Collaboration) *Phys. Lett.* **B716** 30
[20] Belanger G, Kakizaki M and Pukhov A 2011 *JCAP* **1102** 009

[3] Recent searches for the Standard Model Higgs boson at the LHC indicate its mass to be around 125 GeV [18, 19].

Right handed sneutrino dark matter in inverse and linear seesaw scenarios

Valentina De Romeri

AHEP Group, Instituto de Física Corpuscular – C.S.I.C./Universitat de València Edificio de Institutos de Paterna, Apartado 22085, E–46071 València, Spain

E-mail: `deromeri@ific.uv.es`

Abstract. We consider supersymmetric models in which the right–handed sneutrino is a viable WIMP dark matter candidate. These are either simple extensions of the Minimal Supersymmetric Standard Model or models with the addition of an extra U(1) group. All of them can explain small neutrino masses, through either the Inverse or the Linear Seesaw mechanism. We investigate the properties of the dark matter candidate naturally arising in these scenarios. We check for phenomenological bounds, such as correct relic abundance, consistency with direct detection cross section limits and laboratory constraints. Especially, we comment on limitations of the model space due to lepton flavour violating charged lepton decays.

1. Introduction

We analize two supersymmetric particle physics models, which contain a Right-Handed (RH) sneutrino $\tilde{\nu}_{LSP}$ as lightest supersymmetric particle (LSP) and thus as a good WIMP (weakly interactive massive particle) cold Dark Matter (CDM) candidate.
The two models we consider are:
Model I - A simple extension of the MSSM, which accounts for the *inverse seesaw* model [1], where the particle content of the SM is extended by three pairs of singlets, called ν^c and S, which form "heavy" pseudo-Dirac pairs, or for an alternative seesaw scheme that can also be realized at low-scale: the *linear seesaw*, which has been suggested as arising from a particular $SO(10)$ unified model [2, 3, 4].
The total superpotential contains then the additional terms $M_R \hat{\nu}\,\hat{S}$, (Lepton Number - LN conserving) and $\frac{1}{2}\mu_S\,\hat{S}\,\hat{S}$, (LN violating):

$$W = Y_u\,\hat{u}^c\,\hat{q}\,\hat{H}_u \;-\; Y_d\,\hat{d}^c\,\hat{q}\,\hat{H}_d \;-\; Y_e\,\hat{e}^c\,\hat{l}\,\hat{H}_d \;+\; \mu\,\hat{H}_u\,\hat{H}_d \;+\; Y_\nu\,\hat{\nu}\,\hat{l}\,\hat{H}_u \;+\; M_R\,\hat{\nu}^c\,\hat{S} \;+\; \frac{1}{2}\mu_S\,\hat{S}\,\hat{S} \qquad (1)$$

where Y_e, Y_d and Y_u are the usual MSSM Yukawa couplings for the charged leptons and the quarks. In addition there is the neutrino Yukawa coupling Y_ν.
In the linear seesaw scheme, the superpotential is

$$W = Y_u\,\hat{u}^c\,\hat{q}\,\hat{H}_u \;-\; Y_d\,\hat{d}^c\,\hat{q}\,\hat{H}_d \;-\; Y_e\,\hat{e}^c\,\hat{l}\,\hat{H}_d \;+\; \mu\,\hat{H}_u\,\hat{H}_d \;+\; Y_\nu\,\hat{\nu}\,\hat{l}\,\hat{H}_u \;+\; M_R\,\hat{\nu}^c\,\hat{S} \;+\; Y_{SL}\,\hat{S}\,\hat{l}\,\hat{H}_u \qquad (2)$$

and Lepton Number is violated by Y_{SL}.

Content from this work may be used under the terms of the Creative Commons Attribution 3.0 licence. Any further distribution of this work must maintain attribution to the author(s) and the title of the work, journal citation and DOI.
Published under licence by IOP Publishing Ltd

Model II - A left-right symmetric extension of the SM based on a variant of the basic SUSY $SO(10)$ model advocated in [2] in which an extended intermediate $U(1)_R \times U(1)_{B-L}$ stage follows a short $SU(2)_R \times U(1)_{B-L}$ phase, which can emerge, e.g., in a class of $SO(10)$ GUTs broken along the "minimal" left-right symmetric chain

$$SO(10) \to SU(3)_c \times SU(2)_L \times SU(2)_R \times U(1)_{B-L} \to SU(3)_c \times SU(2)_L \times U(1)_R \times U(1)_{B-L},$$

in which the low energy scale seesaw schemes are actually naturally realized.

The R-parity conserving superpotential is

$$W = Y_u\, \hat{u}^c\, \hat{q}\, \hat{H}_u \; - Y_d\, \hat{d}^c\, \hat{q}\, \hat{H}_d \, + Y_\nu\, \hat{\nu}^c\, \hat{l}\, \hat{H}_u \; - Y_e\, \hat{e}^c\, \hat{l}\, \hat{H}_d \, + Y_s\, \hat{\nu}^c\, \hat{\chi}_R\, \hat{S} \, + \mu\, \hat{H}_u\, \hat{H}_d \, - \mu_R\, \hat{\bar{\chi}}_R\, \hat{\chi}_R \, + \frac{1}{2}\mu_S\, \hat{S}\, \hat{S} \quad (3)$$

with the same notation as before. The main introduction here is the term featuring the Yukawa coupling Y_s, which mixes the ν^c fields with the S fields giving rise to an inverse seesaw mechanism for neutrino masses.

2. Phenomenological constraints

2.1. Neutrino masses

We fit neutrino masses according to neutrino oscillation data, i.e. with the atmospheric (atm) and solar ($\odot$) scales as in the case of normal hierarchy and the lepton mixing angles according to the most recent measurements of reactor antineutrino disappearance - Double CHOOZ, Daya Bay and RENO [5, 6, 7, 8]

2.2. Lepton flavor violation

Putting all flavor in Y_ν is actually the usual option, thus keeping Y_L , M_R and μ_S diagonal. However, this framework gives strong constraints on the sneutrino LSP coming from LFV processes notably $\mu \longrightarrow e\gamma$ and three leptons decays like $\mu \longrightarrow eee$.
Experiments set the bounds BR($\mu \longrightarrow e\gamma$) $\leq 2.4 \times 10^{-12}$ (90% C.L.) [9, 10] and BR($\mu \longrightarrow eee$) $\leq 1.0 \times 10^{-12}$ (90% C.L.) [11]. Indeed, it has been shown [12] that in models with an extended particle content, $l_i \to 3l_j$ can be more constraining than $l_i \to l_j\gamma$.
To really relax the LFV constraints, the best option is to put off diagonal terms in μ_S (in the case of the inverse seesaw) or in in Y_{SL} (linear seesaw) and to keep M_R and Y_ν diagonal

2.3. Invisible Z width constraints

Another experimental bound on MSSM neutrinos lighter than $m_Z/2$, comes from their contribution to the invisible Z–width, measured in searches at accelerators. Since in our models we have introduced also RH neutrinos ν^c, the mixing (sinθ) between the states has to be taken into account. The invisible Z–width including the mixing reads:

$$\Delta\Gamma_Z = \sin^4\theta \, \frac{\Gamma_\nu}{2} \left[1 - \left(\frac{2m_\nu}{m_Z} \right)^2 \right]^{3/2} \theta(m_Z - 2\, m_\nu). \quad (4)$$

A further contribution should in principle be considered, that is the left - right mixing of the $\tilde{\nu}_{LSP}$, which could also participate in the Z decay. However, it becomes important only when the mass of the $\tilde{\nu}_{LSP}$ is lighter than $m_Z/2$.

3. Sneutrino as Dark Matter candidate

The models considered in this paper are by construction very similar to the MSSM. In the ordinary MSSM the LSP is the neutralino in extended regions of the parameter space, which anyway provides an interesting candidate of dark matter. However our interest here is to study another dark matter candidate, which naturally arises in this simple extensions of the MSSM, which is the lightest of the supersymmetric partners of the RH neutrinos, the $\tilde{\nu}_{LSP}$. It is an interesting candidate too, since it has electroweak scale interactions, and because of the mixing between the LH and RH components, does not suffer from the problems derived from a too strong coupling to the Z boson like the LH sneutrino of the MSSM.
The existence of the RH sneutrino as LSP does not exclude the possibility of having neutralino as LSP, though; the mass of the LSP strongly depends on the new parameters added to the theory. There will then be just some regions of the parameter space where the RH sneutrino is the LSP.

3.1. Model I with inverse seesaw

The $\tilde{\nu}_{LSP}$ has a direct coupling to the Higgs, as can be seen from the term $Y_\nu \nu l H_u$ of the superpotential Eq. 1. This coupling depends on the Yukawa coupling Y_ν and on the trilinear term T_{Y_ν} and provides tree level interactions with the SM particles at the electroweak scale, thus making the $\tilde{\nu}_{LSP}$ a possible WIMP candidate. The strongest constraint on $\tilde{\nu}_{LSP}$ to be a viable CDM candidate, comes from the CMB measuraments, that is from its relic abundance (RA).
In order to perform the analysis of the sneutrino dark matter phenomenology, we have created a suitable model file for this model with `SARAH` [13]. Then, we have used `SPheno` [14] to compute the low energy spectrum through a precise mass calculation using two-loop RGEs and one-loop corrections to all masses. The calculation of the relic density of the LSP is then done with `MicrOmegas` [15] version `2.4.5` based on the `CalcHep` output of `SARAH`. To perform the scans we used a Mathematica package (`SSP`) [16].

The main annihilation channels are:

- $\tilde{\nu}_{LSP}\tilde{\nu}_{LSP} \to W^+W^-, Z^0Z^0, f\bar{f}$ via s-channel Higgs (h^0, H^0, A^0) exchange;
- $\tilde{\nu}_{LSP}\tilde{\nu}_{LSP} \to W^+W^-, Z^0Z^0$ via t- and u- channel $\tilde{l}$ exchange, and through a quartic coupling;
- $\tilde{\nu}_{LSP}\tilde{\nu}_{LSP} \to H^0H^0$ through a quartic coupling which results to be very efficient;

In Fig.1, where the $\tilde{\nu}_{LSP}$ is mainly LH, it behaves like a MSSM sneutrino. When its mass approaches half the mass of the Higgs, the "Higgs pole" is clearly visible, which is due to the opening of the annihilation channel through the Higgs boson. This channel results to be highly efficient, indeed the RA goes down to very small values of the order of 10^{-5}. Those points refer the sneutrino could not account for the totality of the CDM relic abundance, but it would still be a valid subdominant dark matter candidate. The analogue Z pole is instead suppressed, because of the dependence of the RA on Y_ν ($\propto Y_\nu^4$) and because of helicity. Another efficient channel it the quartic coupling to the Higgses, which appears when the mass of the $\tilde{\nu}_{LSP}$ is larger than the mass of the Higgs.

3.1.1. Direct Detection Direct detection of the sneutrinos would consist in detecting a signal coming from their elastic scattering with nuclei inside a detector. The direct detection rate of the DM particle depends on its mass and on the scattering cross section. The interaction, which occurs in a non relativistic limit, provided that the velocity of dark matter particles in the galactic halo is small, consists of basically two diagrams contributing at tree level, which are the t-channel exchange of a neutral Higgs or of the Z boson. This last diagram is however suppressed because of the mostly singlet nature of the $\tilde{\nu}_{LSP}$.

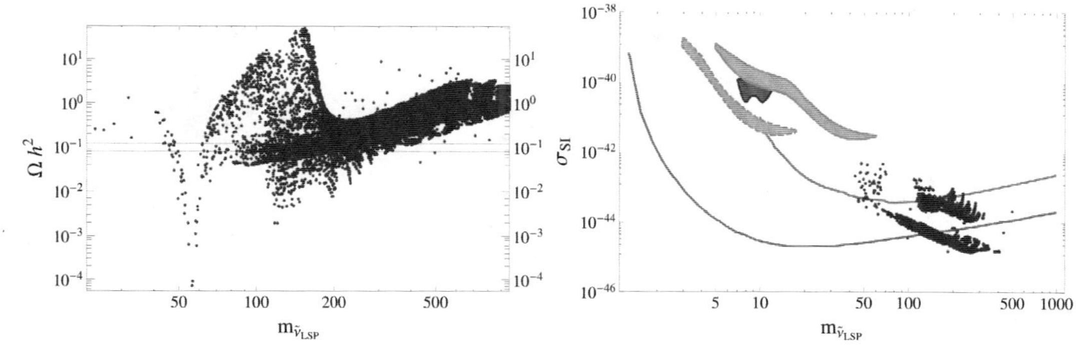

Figure 1. Fig. a (left): General scan for Model I, showing the RA versus the $\tilde{\nu}_{LSP}$ mass. Fig. b (right): Direct detection cross section for model I, and current limits from XENON100 [17] (red line), CDMS [18] (green line), DAMA (with and without channeling, orange regions) [19], and Cogent [20] (purple region).

Lepton number violation in the sneutrino mass matrix leads to a mass splitting between the real and the imaginary part of the lightest sneutrino, and the scattering via Z boson exchange occurs inelastically, through a transition from the real to the imaginary or viceversa. If the mass splitting is greater than some keV, scattering is strongly suppressed at direct detection experiments.

In Fig.1b we compare the direct detection cross section for the RH sneutrino LSP with current direct detection experiments. The major bound nowadays comes from the XENON100 experiment [17], whose best sensitivity is around 10^{-44}cm^2 for a dark matter candidate of 50 GeV. We depict the direct detection cross section versus the LSP sneutrino mass (blue points). The points are those which survive the relic abundance constraint. In the same plot, the current limits from XENON100 [17] (red line), CDMS [18] (green line), DAMA (with and without channeling, orange regions) [19], and Cogent [20] (purple region) are shown. The sneutrinos show a SI cross section $\sigma_{SI} \lesssim 10^{-42}\text{cm}^2$, and for masses $m_{\tilde{\nu}_{LSP}} \gtrsim 100$ GeV they are compatible with current limits by XENON100. However, XENON1T, whose sensibility should improve up to 10^{-46}cm^2, will test those cross sections.

3.2. Model I with linear seesaw

The phenomenology of sneutrino dark matter in the MSSM with linear seesaw is basically the same as in the previous case. Indeed, it is not possible to distinguish between the kind of seesaw mechanism through the dark matter phenomenology.

3.3. Model II

In this subsection we discuss the DM phenomenology in the minimal supersymmetric $U(1)_R \times U(1)_{B-L}$ extension of the standard model. The presence of the extra gauge boson Z' leads to important differences with respect to the previous models.

We use a version of the model where the tadpole equations for the vevs are solved analytically for $(\mu, B_\mu, \mu_R, B_{\mu_R})$ and we perform a scan over the parameters m_0, $M_{1/2}$, tanβ_R. The results are shown in Figg. 2.

In both plots the main feature clearly visible is the Z' pole. Indeed, the annihilation of the $\tilde{\nu}_{LSP}$ LSPs into SM particles via the Z' becomes efficient when the mass of the $\tilde{\nu}_{LSP}$ is close

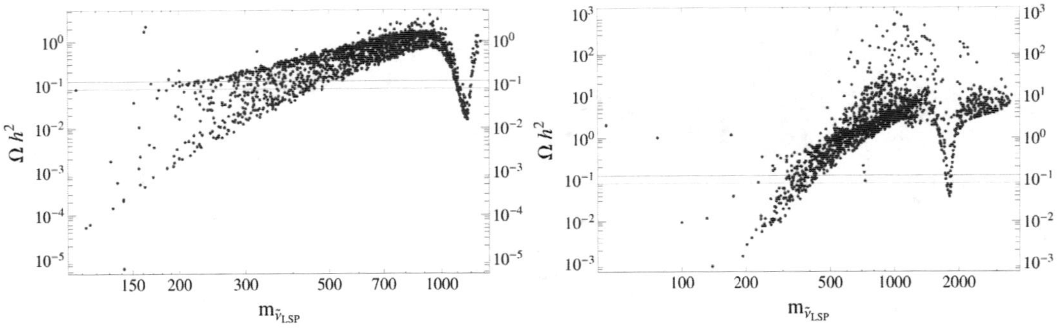

Figure 2. Scans for model II, made over the parameters $m_0 = [0, 6000]$ GeV, $M_{1/2} = [3000, 8000]$ GeV, $\tan\beta_R = [1.0, 1.3]$. The other parameters are set to the values $\tan\beta$=10, $A_0 = -4500$ GeV, $Y_S = \mathrm{diag}(0.3)$; v_R has been chosen different in the two plots, $v_R = 5.5$ TeV (left) and $v_R = 10$ TeV (right).

to half the mass of the Z'. The mass of the Z' can be calculated analytically [21] and mainly depends on the value of v_R. The ATLAS searches for a Z' set a lower limit on its mass which is 1.8 TeV, and this translates into a lower limit on $v_R \gtrsim 5$ TeV. In Fig. 2 we chose two different values of v_R: $v_R = 5.5$ TeV and $v_R = 10$ TeV, thus leading to two values of Z' and so to a shift in the position of the pole in the relic density plot. This shows that choosing a higher value of v_R we can get heavy $\tilde{\nu}_{LSP}$ DM with the correct RA.
The quartic coupling with two Higgses (h^0, h^0_{BLR} and A^0, depending on if they are kinematically allowed, according to the $\tilde{\nu}_{LSP}$ mass) is one of the most efficient annihilation channels, as before. For lower masses the annihilation via the MSSM Higgs is the most efficient, as can be noticed by the lowering of the relic density going to lower masses, expecially in the first plot, where on the left end side we are approaching the Higgs pole.

3.3.1. Direct Detection As to the direct detection cross section analysis, a bound on the v_R, and then on the mass of the Z' can be set by the DD experiments, expecially by Xenon100. This bound turns to be stronger than the one set by the collider experiments like LHC, as it is shown in Fig.3. The two clouds of points there refer to $v_R = 5.5$ TeV (black) and $v_R = 10$ TeV (blue).

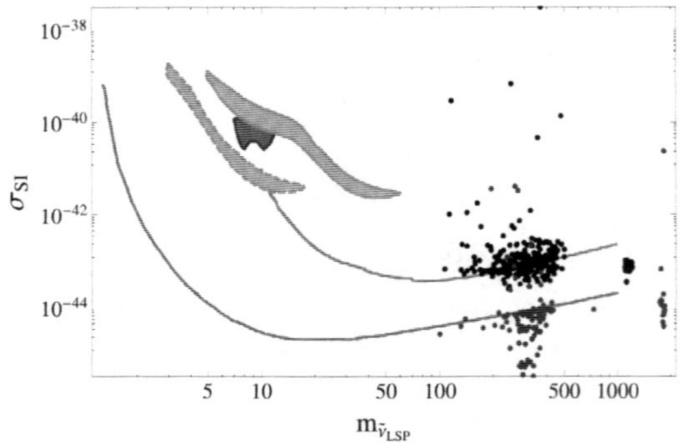

Figure 3. Direct Detection cross section limits for model II. The two clouds of points refer to $v_R = 5.5$ TeV (black) and $v_R = 10$ TeV (blue). Experimental limits as in Fig. 1b.

Acknowledgments

V.D.R. warmly thanks Martin Hirsch, with whom this work was done, for many useful comments. The work of V.D.R. is supported by the EU Network grant UNILHC PITN-GA-2009-237920. We also acknowledge support from the Spanish MICINN grants FPA2011-22975 and MULTIDARK CSD2009-00064 (Con-solider-Ingenio 2010 Programme) and by the Generalitat Valenciana grant Prometeo/2009/091.

References

[1] Mohapatra R and Valle J 1986 *Phys.Rev.* **D34** 1642
[2] Malinsky M, Romao J and Valle J 2005 *Phys.Rev.Lett.* **95** 161801 (*Preprint* hep-ph/0506296)
[3] Akhmedov E K, Lindner M, Schnapka E and Valle J 1996 *Phys.Lett.* **B368** 270–280 (*Preprint* hep-ph/9507275)
[4] Akhmedov E K, Lindner M, Schnapka E and Valle J 1996 *Phys.Rev.* **D53** 2752–2780 (*Preprint* hep-ph/9509255)
[5] Abe Y *et al.* (DOUBLE-CHOOZ Collaboration) 2012 *Phys.Rev.Lett.* **108** 131801 (*Preprint* 1112.6353)
[6] An F *et al.* (DAYA-BAY Collaboration) 2012 *Phys.Rev.Lett.* **108** 171803 (*Preprint* 1203.1669)
[7] Ahn J *et al.* (RENO collaboration) 2012 *Phys.Rev.Lett.* **108** 191802 (*Preprint* 1204.0626)
[8] Forero D, Tortola M and Valle J 2012 *Phys.Rev.* **D86** 073012 (*Preprint* 1205.4018)
[9] Nishimura K, Dey B, Aston D, Leith D, Ratcliff B *et al.* 2013 *Nucl.Instrum.Meth.* **A701** 115–126
[10] Adam J *et al.* (MEG collaboration) 2011 *Phys.Rev.Lett.* **107** 171801 (*Preprint* 1107.5547)
[11] Nakamura K *et al.* (Particle Data Group) 2010 *J.Phys.* **G37** 075021
[12] Hirsch M, Staub F and Vicente A 2012 *Phys.Rev.* **D85** 113013 (*Preprint* 1202.1825)
[13] Staub F 2008 (*Preprint* 0806.0538)
[14] Porod W 2003 *Comput.Phys.Commun.* **153** 275–315 (*Preprint* hep-ph/0301101)
[15] Belanger G, Boudjema F, Pukhov A and Semenov A 2007 *Comput.Phys.Commun.* **176** 367–382 (*Preprint* hep-ph/0607059)
[16] Staub F, Ohl T, Porod W and Speckner C 2012 *Comput.Phys.Commun.* **183** 2165–2206 (*Preprint* 1109.5147)
[17] Aprile E *et al.* (XENON100 Collaboration) 2012 (*Preprint* 1207.3458)
[18] Ahmed Z *et al.* (CDMS-II Collaboration) 2010 *Science* **327** 1619–1621 (*Preprint* 0912.3592)
[19] Bernabei R *et al.* (DAMA Collaboration, LIBRA Collaboration) 2010 *Eur.Phys.J.* **C67** 39–49 (*Preprint* 1002.1028)
[20] Aalseth C *et al.* (CoGeNT collaboration) 2011 *Phys.Rev.Lett.* **106** 131301 (*Preprint* 1002.4703)
[21] Hirsch M, Porod W, Reichert L and Staub F 2012 *Phys.Rev.* **D86** 093018 (*Preprint* 1206.3516)

$B - L$ neutralino dark matter

R J Hernández-Pinto[1,2] and A Pérez-Lorenzana[1]

[1]Departamento de Física, Centro de Investigación y de Estudios Avanzados del I.P.N., Apdo. Post. 14-740, 07000 México D.F., México.

[2]Universidad de Buenos Aires, FCEyN (1428), Pabellón 1, Ciudad Universitaria, Capital Federal, Argentina

E-mail: roger@df.uba.ar, aplorenz@fis.cinvestav.mx

Abstract. We focus on the supersymetric extension of the gauge group $SU(3)_c \times SU(2)_L \times U(1)_Y \times U(1)_{B-L}$ in which we have included the right handed neutrino superfield gauged only under $B - L$. In this context, it is possible to address the problem of neutrino masses and the DM problem can be tackled via the $B - L$ fermion sector instead of the MSSM one.

1. Introduction

Strong evidences propose the existence of a Dark Matter (DM) component in the Universe. Experiments such as Wilkinson Microwave Anisotropy Probe (WMAP) and Sloan Digital Sky Survey have determined that the dark matter density is [1]

$$\Omega_{CDM} h^2 = 0.1126^{+0.0161}_{-0.0181} \tag{1}$$

at 95% C.L. Such an accurate measurement imposes strong constraints to any theory which would attempt to bring a solution to the DM problem. In general, a Weakly Interactive Massive Particle (WIMP) would explain satisfactory this problem under certain conditions. WIMPs appear in several kind of models.

However, there is an incompatibility between DM problem and the Standard Model (SM) of particle physics; the discrepancy is simple: in the SM framework there is no candidate which could be the DM particle. Nevertheless, an sterile neutrino with mass around 1 keV could be added in the SM in order to contribute to the DM content and also in order to generate neutrino masses via the usual Dirac and Majorana mass terms [2],

$$\delta\mathcal{L} = i\bar{\nu}_R \partial_\mu \gamma^\mu \nu_R - h\sigma \bar{\nu}_R^c \nu_R - h' \bar{L}\tilde{H}\nu_R. \tag{2}$$

But, the addition by hand of these terms breaks the beauty of gauge symmetries that underly in the SM. Therefore, if Eq. (2) is considered as a consistent gauge extension of the SM, one might have a candidate for the DM problem and to neutrino masses at the same time; the issue would be the determination of which gauge symmetry would allow those additional terms in the SM Lagrangian without spoiling its own amazing predictions.

If supersymmetry is discovered, it would be a great advantage on the knowledge of the unification of the SM gauge groups, the spontaneous symmetry breaking of the electroweak sector and the solution of the hierarchy problem, among other features.

Content from this work may be used under the terms of the Creative Commons Attribution 3.0 licence. Any further distribution of this work must maintain attribution to the author(s) and the title of the work, journal citation and DOI.
Published under licence by IOP Publishing Ltd

In addition, the lightest supersymmetric particle (LSP) fits perfectly for being a DM candidate. A caveat for the Minimal Supersymmetric Standard Model (MSSM) extension is that one have to impose $R-$Parity by hand in order to avoid baryon and lepton number violating operators in the superpotential. Amazingly, imposing the discrete symmetry $U(1)_{B-L}$, operators such as,

$$W \supset \lambda^{ijk} L_i L_j \bar{e}_k + \lambda'^{ijk} L_i Q_j \bar{d}_k + \mu' L_i H_u + \lambda''^{ijk} \bar{u}_i \bar{d}_j \bar{d}_k \tag{3}$$

are no longer allowed, which is phenomenologically expected because they provoke a very fast proton decay, so far unobserved. Therefore, the supersymmetric extension of the gauge group $SU(3)_c \times SU(2)_L \times U(1)_Y \times U(1)_{B-L}$ leads naturally to a dark matter candidate, it also address the problem of neutrino masses due to the Dirac and Majorana Yukawas in the superpotential and finally, it is also attractive for collider phenomenology due to it contains inherently a Z'-type gauge boson and extra fermion fields.

2. The model

We studied a model based in the supersymmetric extension of the gauge group $\mathcal{G} = SU(3)_c \times SU(2)_L \times U(1)_Y \times U(1)_{B-L}$, where the matter superfields considered in the model are those of the MSSM, plus the right handed neutrino superfield, $\hat{N}$, and two Higgses, $\hat{\sigma}_1$ and $\hat{\sigma}_2$, assigned to the following representations of $\mathcal{G}$,

$$\hat{\bar{N}} \sim (\mathbf{1},\mathbf{1},0,1)\,, \qquad \hat{\sigma}_1 \sim (\mathbf{1},\mathbf{1},0,-2)\,, \qquad \hat{\sigma}_2 \sim (\mathbf{1},\mathbf{1},0,2)\,. \tag{4}$$

With this quantum numbers, we allow to construct the supersymmetric version of Eq. (2), and immediately $R-$parity is no longer needed. Without imposing R-parity the only contribution to the MSSM superpotential that can be constructed with the extra superfields is given by

$$\Delta W = \hat{\bar{N}} \mathbf{Y}_N^D \hat{L} \hat{H}_u + \hat{N} \mathbf{Y}_N^M \hat{N} \hat{\sigma}_1 + \mu' \hat{\sigma}_1 \hat{\sigma}_2 \;, \tag{5}$$

where $\mathbf{Y}_N^M$ and $\mathbf{Y}_N^D$, are the Yukawa matrices. One can easily realize that Eq. (5) is the supersymmetric realization of Eq. (2) which now contains the Dirac and Majorana terms plus the corresponding μ'-term.

Applying the general prescription for the soft supersymmetric breaking terms, we add the following terms to the MSSM Lagrangian,

$$\begin{aligned} -\Delta\mathcal{L}_{SB} &= \tilde{\bar{N}} \mathbf{h}_N^D \tilde{L} H_u + \tilde{N}^c \mathbf{h}_N^M \tilde{N} \sigma_1 + \tilde{N}^\dagger \mathbf{m_N^2} \tilde{N} \\ &+ m_{\sigma_1}^2 \sigma_1^\dagger \sigma_1 + m_{\sigma_2}^2 \sigma_2^\dagger \sigma_2 + B' \sigma_1 \sigma_2 + \frac{1}{2} M_{B-L} \tilde{Z}_{B-L} \tilde{Z}_{B-L} \end{aligned} \tag{6}$$

where the tilde fields are the scalar fields involved in the superpotential, and $\tilde{Z}_{B-L}$ is corresponding the $B-L$ gaugino field of the theory.

3. Neutralino sector

In the neutralino sector, one can easily check that there is no mixing between the $SU(3)_c \times SU(2)_L \times U(1)_Y$ and the $U(1)_{B-L}$ sector, due to the assignation of the $B-L$ quantum numbers. Then, we can write down the neutralino mass matrix in the basis, $\left(\tilde{\psi}^0\right)^T = (\tilde{B}^0 \;\; \tilde{W}^0 \;\; \psi_d^0 \;\; \psi_u^0 \;\; \tilde{Z}_{B-L}^0 \;\; \psi_{\sigma_1}^0 \;\; \psi_{\sigma_2}^0)$ as,

$$\mathbf{M}_{\tilde{\chi}^0} = \begin{pmatrix} \mathbf{M}_{\tilde{\chi}^0_{\mathbf{MSSM}}} & \mathbf{0} \\ \mathbf{0} & \mathbf{M}_{\tilde{\chi}^0_{\mathbf{B-L}}} \end{pmatrix} \tag{7}$$

where $\mathbf{M}_{\tilde{\chi}^0_{\mathrm{MSSM}}}$ is the usual neutralino mass matrix from the MSSM, and the $U(1)_{B-L}$ sector is,

$$\mathbf{M}_{\tilde{\chi}^0_{\mathrm{B-L}}} = \begin{pmatrix} M_{B-L} & 2\sqrt{2}g_{B-L}v's_\theta & -2\sqrt{2}g_{B-L}v'c_\theta \\ 2\sqrt{2}g_{B-L}v's_\theta & 0 & -\mu' \\ -2\sqrt{2}g_{B-L}v'c_\theta & -\mu' & 0 \end{pmatrix} \tag{8}$$

where we have defined, after $U(1)_{B-L}$ breaking,

$$\langle\sigma_i\rangle = v'_i, \quad \tan\theta = \frac{v'_1}{v'_2}, \qquad \text{and } v'^2 = v'^2_1 + v'^2_2. \tag{9}$$

It is straightforward to realize that the wave functions of the physical particles are,

$$\tilde{\chi}^0_i = N_{ij}\tilde{\psi}^0_j \qquad i,j = 1,2,3,4 \tag{10}$$

$$\tilde{\chi}^0_{B-L} = N_{ij}\tilde{\psi}^0_j \qquad i,j = 5,6,7 \tag{11}$$

where N_{ij} is now the unitary matrix that diagonalize Eq. (7). $B-L$ sector is relevant because low energy mass associated to this gauge group, M_{B-L}, turns to be very close to M_1 (under the renormalization group equations) and it would mean that the LSP is related to the $B-L$ sector, depending on the other parameters of the model. In the most general case neutralinos and neutrinos must be in the same mass matrix and masses will be defined for superposition states of neutralinos and neutrinos. Nevertheless, neutrinos are expected to be much lighter than neutralinos and therefore, we can apply a double *see saw* [3] in order to generate neutrino masses while neutralino mass matrix remains the same.

4. Neutrino sector

The mass mixing between neutrinos and neutralinos has the following structure,

$$\mathbf{M}_{\nu\tilde{\chi}^0} = \begin{pmatrix} \mathbf{0} & \mathbf{M}^D_\nu & \mathbf{\Lambda} \\ (\mathbf{M}^D_\nu)^T & \mathbf{M}^M_\nu & \mathbf{\Omega} \\ \mathbf{\Lambda}^T & \mathbf{\Omega}^T & \mathbf{M}_{\tilde{\chi}^0} \end{pmatrix} \tag{12}$$

where we have taken the basis $(\nu_L, \nu_R, \tilde{\psi}^0)$, and the elements in Eq. (12) are

$$\mathbf{M}^D_\nu = \frac{vs_\beta y_D}{\sqrt{2}}, \qquad \mathbf{M}^M_\nu = v's_\theta y_M, \tag{13}$$

$$\mathbf{\Lambda} = \begin{pmatrix} 0 & 0 & 0 & \frac{v_R y_D}{\sqrt{2}} & 0 & 0 & 0 \end{pmatrix}, \tag{14}$$

$$\mathbf{\Omega} = \begin{pmatrix} \mathbf{0}_{1\times 4} & -\sqrt{2}g_{B-L}v_R & \frac{y_M v_R}{\sqrt{2}} & 0 \end{pmatrix}, \tag{15}$$

and $\mathbf{M}_{\tilde{\chi}^0}$ is the corresponding mass matrix for neutralinos [4], the parameters are defined as, $c_\beta = \cos\beta$, $s_\beta = \sin\beta$, $\tan\beta$ corresponds to the ratio between the vacuum expectation values of the MSSM Higgses, $\tan\beta = v_u/v_d$, and $v^2 = v_u^2 + v_d^2 \approx (276\ \mathrm{GeV})^2$. Following the result of Ref. [5], we have considered that sneutrino can also acquire a vacuum expectation value labeled as $\langle\tilde{N}\rangle \equiv v_R/\sqrt{2}$. As a first approximation, we have taken the Yukawa matrices as $\mathbf{Y}^D_\nu \sim \mathrm{diag}(0,0,y_D)$ and $\mathbf{Y}^M_\nu \sim \mathrm{diag}(0,0,y_M)$.

Now, we are able to introduce a *see saw* mechanism between neutrinos and neutralinos and, in the basis (ν_L, ν_R), we find that the elements of the neutrino mass matrix are found to be,

$$[\mathbf{M}_\nu]_{11} = \frac{v_R^2 y_D^2}{4\mu\left(t_\beta - \frac{M_1 M_2 \mu c_\beta^{-2}}{m_Z^2(M_1+M_2+(M_1-M_2)c_{2\theta_W})}\right)}, \tag{16}$$

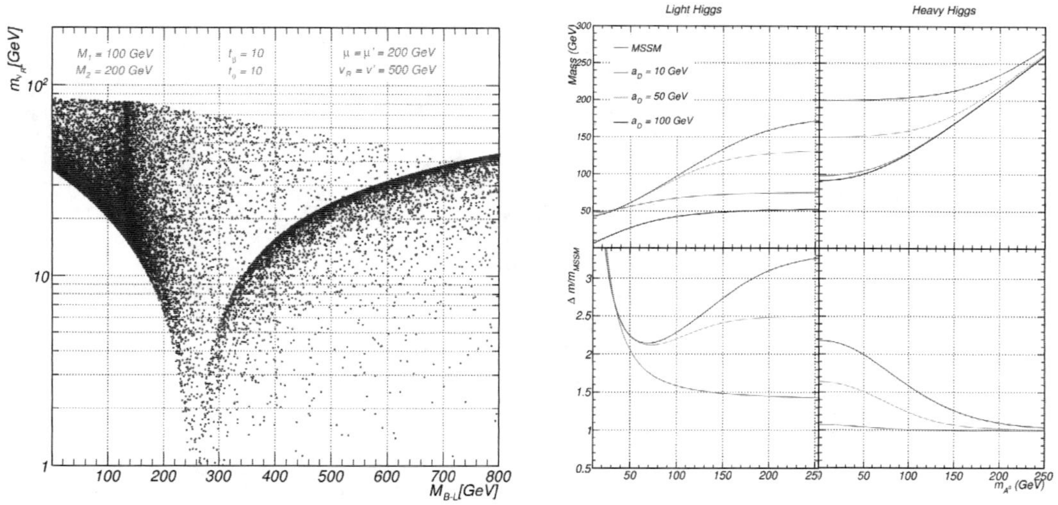

Figure 1. Left panel: Right handed neutrino mass in terms of unified $B-L$ gaugino mass for $M_1 = \frac{1}{2}M_2 = 100$ GeV, $\tan\beta = \tan\theta = 10$, $\mu = \mu' = 200$ GeV and $v_R = v' = 500$ GeV. Right panel: In the upper panels we plot the contributions to the Light and Heavy Higgs masses of the MSSM in terms of the pseudo-scalar mass for different values of a_D, where we have set the parameters to those of the left panel. In the lower panels we can observe that for $a_D = 100$ GeV, the Light Higgs mass receive contributions that makes it three times heavier than the MSSM Light Higgs for large m_{A^0}.

$$[\mathbf{M}_\nu]_{12} = \frac{v y_D s_\beta}{\sqrt{2}}, \tag{17}$$

$$[\mathbf{M}_\nu]_{22} = y_M v' s_\theta - \frac{2g_{B-L}^2 v_R^2(\mu' - \sqrt{2}y_M v' c_\theta)^2}{\mu'^2\left(M_{B-L} - 8g_{B-L}^2 s_{2\theta}\frac{v'^2}{\mu'}\right)}. \tag{18}$$

Thus, masses of right and left handed neutrinos can be computed by the standard diagonalization.

With the previous expressions, we can implement an scanning over all our parameter space. In order to find the allowed masses for the right handed neutrino, we have imposed the cosmological contraint [6], $\sum_\nu m_\nu < 2$ eV, and we have found that the allowed range for the right handed neutrino is, $m_{\nu_R} < 80$ GeV, depending on the unified $B-L$ gaugino mass parameter. A sample of the scanning can be seen on the left panel in Figure 1.

5. Higss Sector

In general, we can write the Higgs mass lagrangian as,

$$\mathcal{L} = \frac{1}{2}\Phi^T M_\Phi^2 \Phi, \qquad \text{such as,} \quad M_\Phi^2 = \begin{pmatrix} M_{B-L}^2 & M_{mix}^2 \\ (M_{mix}^2)^T & M_{MSSM}^2 \end{pmatrix}, \tag{19}$$

in the basis, $(\Phi)^T = (\widetilde{\nu}_L, \widetilde{\nu}_L^\dagger, \Sigma_1, \Sigma_2, h_u^0, h_d^0)$. Then, we have implemented the diagonalization of the mass matrix numerically. We are interested in constraining the parameters of the model, therefore, we computed the contribution to the MSSM Higgses due to the extra parameters.

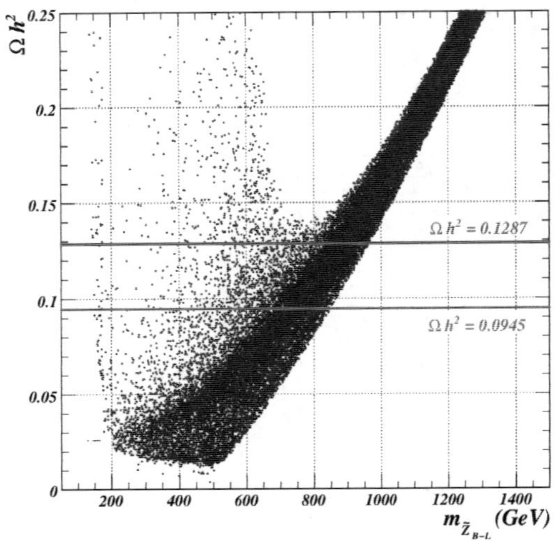

Figure 2. Relic density for the $\tilde{Z}_{B-L}$−like neutralino in terms of the mass parameter $m_{\tilde{Z}_{B-L}}$, where we have included the cosmological and the Higgs mass constraint.

Light and Heavy Higgs masses are shown in the right panel in Figure 1 in terms of the pseudo-scalar mass for different values of a_D.

We considerer a random scanning over the parameter space, and we have observed that the physical masses does not receive relevant contributions from the Yukawa parameters, the left-handed sneutrino mass nor $\tan\theta$. However, large values of a_D can avoid the limits of Higgs searches and we can see it on the left panel of Figure 1. In the lower panels we can observe that we can give large contributions to the Higgs masses by tuning the a_D parameter. Therefore, a_D will be restricted to satisfy Higgs searches in the following analysis. So, within this approximation and by considering all the constrains reviewed, we are now able to compute the corresponding relic density of the model.

6. Relic Density

In order to compute the total relic density we followed the general procedure shown in Ref. [7], thus, we needed to know which neutralino is the lightest. The solution of the renormalization group equations has shown that the $B-L$ sector could contains the lightest gaugino, $\tilde{Z}_{B-L}$; depending on the value of μ', v', and $\tan\theta$ we can make it heavier than lightest MSSM neutralino. On the other hand, we can consider in general that, the total relic density is a contribution of the MSSM and the $B-L$ sector,

$$(\Omega h^2)_{Tot} = (\Omega h^2)_{\chi_i^0} \oplus (\Omega h^2)_{\chi_{B-L}^0}. \tag{20}$$

In the following, we are going to consider that the LSP is mainly $B-L$ and $(\Omega h^2)_{\chi_i^0}$ is a small contribution to it. In this sense, we have three different scenarios. The LSP is mainly: a) $\tilde{Z}_{B-L}$-like, b) $\tilde{\sigma}_1$-like, and c) $\tilde{\sigma}_2$-like. In here, we will only show the solution for the $\tilde{Z}_{B-L}$ scenario, and the other two cases will be publish soon.

In this case, the Feynman diagrams involved are the t-channel and u-chanel of the $\tilde{Z}_{B-L}\bar{\tilde{Z}}_{B-L} \to f\bar{f}(\sigma_{1(2)}\bar{\sigma}_{1(2)})$ via exchange of an $\tilde{f}$ $(\tilde{\sigma}_{1(2)})$. We made a random scanning over the parameter space; and we set the values to be in the regions,

$$100 \text{ GeV} < m_N < 500 \text{ GeV} , \tag{21}$$

$$100 \text{ GeV} < m_{\sigma_1} < 500 \text{ GeV} , \tag{22}$$

$$m_{\sigma_2} = m_{\sigma_1} + 30 \text{ GeV}. \tag{23}$$

The solution of the relic density is shown in Figure. 2. As we can distinguish from the scanning, $\tilde{Z}_{B-L}$ masses greater than 950 GeV would overclose de universe. The correct amount of relic density can be reproduced for masses of the $\tilde{Z}_{B-L}$ around

$$200 \text{ GeV} < m_{\tilde{Z}_{B-L}} < 950 \text{ GeV}. \tag{24}$$

Nevertheless, there is still possible to overclose the universe in this range depending on the other parameters.

7. Conclusions

We have studied the phenomenological implications of the gauge group $SU(3)_c \times SU(2)_L \times U(1)_Y \times U(1)_{B-L}$. By gauging an extra superfield, the right handed neutrino, only under $B-L$, we can provide a non zero mass to neutrinos which could explain the neutrino oscillation phenomena. We have applied the double see saw mechanism induced by the mixing between neutrinos and neutralinos. We found that neutrino masses which are in agreement with cosmological constraints pushes the value of right handed neutrinos to be of the order of 1 GeV or higher. Extra neutralinos have been added to the spectra; this extra particles might help to explain the DM content of the universe and, in particular, we focused on the relic density problem. For instance, if the LSP is given by the superpartner of the extra $B-L$ gauge boson, $\tilde{Z}_{B-L}$, we can generate the correct amount of relic density if its mass resides in the range of $200 < m_{\tilde{Z}_{B-L}} < 950$ GeV.

Acknowledgments

This work was supported partially by CONACyT, México, grant 132067. RJHP work is also supported by the the Research Executive Agency (REA) of the European Union under the Grant Agreement number PITN-GA-2010-264564 (LHCPhenoNet).

References

[1] Spergel D N *et al.* (WMAP Collaboration) 2007 *Astrophys J. Suppl.* **170** 377
[2] Asaka T, Blanchet S, Shaposhnikov M 2005 *Phys. Lett.* **B631** 151
[3] Fileviez Perez P and Spinner S 2009 *Phys. Lett.* **B673** 251
[4] Haber H E and Kane G L 1985 *Phys. Rept.* **117** 75
[5] Hernández-Pinto R J and Pérez-Lorenzana A 2011 (*Preprint* arXiv:1105.0713)
[6] Lesgourgues J and Pastor S 2006 *Phys. Rept.* **429** 307
[7] Griest K, Kamionkowski M and Turner S 1990 *Phys. Rev.* **D41** 3565

A search for dark matter appearing as missing transverse energy at CMS

Tia Miceli for the CMS Collaboration
UC Davis Physics Department, 1 Shields Avenue, Davis, CA, 95616, USA
E-mail: miceli@physics.ucdavis.edu

Abstract. Results are presented of a search for dark matter production using the CMS detector at the LHC. Collisions of pp at $\sqrt{7}$ TeV totaling 5 fb^{-1} were counted for final states consisting of missing transverse energy accompanied by either a monophoton or a monojet. The data is consistent with Standard Model background, so limits are placed on both spin-dependent and spin-independent dark matter particle masses and production cross-sections. These are examined within the context of direct detection experiments where our results extend the current best limits.

1. Introduction

Proton-proton collisions from the Large Hadron Collider (LHC) totaling 5 fb^{-1} at $\sqrt{s} = 7$ TeV in which the final state contains a photon (γ) and missing transverse energy ($\not{E}_T$) or a jet and $\not{E}_T$ are used to investigate possible dark matter (DM) production. Dark matter is the dominant non-baryonic component of the universe's matter density [1]. Many searches for the DM particle (χ) rely on direct detection via a χ recoiling from a nucleus ($q\chi \rightarrow q\chi$), or DM annihilation observable in cosmic rays ($\chi\bar{\chi} \rightarrow \gamma\gamma$). At colliders however, this diagram is rotated and we look for $q\bar{q} \rightarrow \chi\bar{\chi}$. Since DM travels through the Compact Muon Solenoid detector [2] (CMS) without interaction, it appears as $\not{E}_T$ in the resulting collision fragments. To make this process visible, a γ or jet must be radiated from one of the incoming quarks. We study the monophoton and monojet final states.

Assuming a massive mediator couples with the incoming quarks and outgoing Dirac $\chi\bar{\chi}$ particles, the process can be contracted into an effective theory [3],[4]. The scale of the theory, Λ, is given by $\Lambda = M_m/\sqrt{g_\chi g_q}$, where M_m is the mediator mass, g_χ is its coupling to DM, and g_q is its coupling to quarks. This model provides a means to compare the t-channel χ-nucleon elastic scattering cross section to the s-channel DM pair production cross section. The effective operator could be either a vector operator leading to a spin-independent interaction, or an axial-vector operator leading to a spin-dependent interaction.

2. Monophoton

The monophoton data is selected from events that pass a single-photon trigger that is fully efficient within the signal region of $|\eta^\gamma| < 1.44$ (highest photon purity) and $p_T^\gamma > 145$ GeV.

The photon is identified as such by requiring the following isolation criteria. The p_T in the hadronic calorimeter (HCAL) within an annulus of $0.15 < \Delta R < 0.4$, where $\Delta R = \sqrt{(\Delta\eta)^2 + (\Delta\phi)^2}$, must be less than 2.2 GeV $+0.0025 \cdot p_T^\gamma$. The p_T in the electromagnetic

Content from this work may be used under the terms of the Creative Commons Attribution 3.0 licence. Any further distribution of this work must maintain attribution to the author(s) and the title of the work, journal citation and DOI.
Published under licence by IOP Publishing Ltd

calorimeter (ECAL) within an annulus of $0.06 < \Delta R < 0.4$ but excluding deposits within $|\Delta\eta < 0.04|$ is less than 4.2 GeV $+0.006 \cdot p_T^\gamma$. The fraction of energy in the HCAL to the energy in the ECAL within the solid cone of $\Delta R < 0.4$ must be less than 0.05. The p_T of track originating from the photon's assigned vertex within an annulus of $0.04 < \Delta R < 0.4$ but excluding deposits within $|\Delta\eta| < 0.015$ must be less than 2.0 GeV $+0.001 \cdot p_T^\gamma$. It is also required that there are no tracker hits indicative of an electron (called a pixel seed). These ignored regions in $|\Delta\eta|$ ensure high photon identification efficiency for those photons that begin to shower within the tracker. Due to the high luminosity at the LHC, there may be multiple collisions at once, which may result in an ambiguity in the assignment of a photon's origin collision vertex. To account for this, the photon is also required to pass the track isolation criteria for all reconstructed vertices, and a systematic uncertainty is included.

The $\not{E}_T$ is the magnitude of the vector sum of all reconstructed particles in the event using the particle flow algorithm [5]. The $\not{E}_T$ requirement is optimized to be greater than 130 GeV to optimize the signal strength.

Events with excessive hadronic activity are removed by requiring that there are no particle flow jets, identified with the anti-k_T algorithm with a distance parameter of 0.5 [6], within $|\eta| < 3.0$ outside of $\Delta R(jet, \gamma) = 0.5$ and $p_T > 40$ GeV and that there are no tracks outside of $\Delta R(track, \gamma) = 0.04$ with $p_T > 20$ GeV.

Backgrounds unrelated to pp collisions are removed by further cuts. Protons that stray from the beam, can shower to muons that bremsstrahlung in the ECAL. These and showers from cosmic muons are reduced by vetoing muon tracks and requiring that the time assigned to the photon is ± 3 ns of that expected of a collision particle's arrival at the ECAL. Other anomalous ECAL signals are reduced by requiring that the timing of all the deposits comprising the photon are consistent [7].

Any backgrounds unrelated to pp collisions remaining are estimated from data by examining the transverse distribution of energy in the EM cluster and the time of arrival of the signal in the channel with the largest energy deposition. Templates for anomalous signals, cosmic-ray muons, and beam halo events are fitted to a candidate sample that has no timing requirement. It is found that only the halo muons give a significant residual contribution to the in-time sample with an estimated 11.1 ± 5.6 events.

Events of $W \to e\nu$ can mimic the monophoton signal, if the electron is misreconstructed. Estimating with MC simulated events and verifying with $Z \to ee$ events in data shows that the matching of electron showers to pixel seeds has an efficiency of $\epsilon = 0.9940 \pm 0.0025$. The contribution of $W \to e\nu$ events in the candidate sample is found by scaling a control sample of electron candidates by $(1-\epsilon)/\epsilon$ yielding 3.5 ± 1.5 events.

The contamination from jets misidentified as photons is estimated by using a control sample of EM-enriched QCD events to calculate the ratio of events that pass the signal photon criteria relative to those that pass looser photon criteria but fail an isolation requirement. Since the EM-enriched sample also includes the production of direct single photons, this additional contribution to the ratio is estimated by fitting templates of energy-weighted shower widths from MC-simulated $\gamma + jets$ events to an independent QCD data sample. This contribution of true photons is subtracted from the numerator of the ratio. This corrected ratio is applied to a subset of the EM-enriched jet events that passes loose photon identification and additional monophoton event selection criteria, providing a background contribution of 11.2 ± 2.8 jet events.

The irreducible background, $Z(\to \nu\bar{\nu}) + \gamma$, as well as the $\gamma + jet$, $W(\to \ell\nu) + \gamma$, and diphoton backgrounds are estimated using Monte Carlo simulation.

A summary of the Standard Model (SM) backgrounds and the number of observed candidates are given in Table 1. Observation is consistent with the SM prediction.

Table 1. Monophoton backgrounds and observed events.

Process	Events
$Z(\nu\bar{\nu}) + \gamma$	45.3 ± 6.9
Non-collision BG	11.1 ± 5.6
$W \to e\nu$	3.5 ± 1.5
$jets$	11.2 ± 2.8
Other	4.1 ± 1.0
Total SM	75.1 ± 9.5
Data	73

3. Monojet

The monojet data is collected by jet triggers that are fully efficient within the candidate criteria that the highest p_T jet (j_1) is within $|\eta(j_1)| < 2.4$ with $p_T(j_1) > 110$ GeV.

The jet is reconstructed the same as the jets in previous section. A second jet, j_2 is allowed if $|\Delta\phi(j_1, j_2)| < 2.5$ radians to allow for accompanying initial or final state radiation while still rejecting dijet events. Furthermore, if there are additional jets with $p_T > 30$ GeV, the event is vetoed. To reduce events from W, Z, and top decays, events with isolated muons or electrons of $p_T > 10$ GeV are discarded. To be considered isolated, the lepton must be farther away from j_1 than $\Delta R = 0.3$ and the sum of the p_T of photons and hadronic particles within a cone of $\Delta R = 0.4$ divided by the lepton p_T is less than 0.2. Tau lepton decays are removed by vetoing events with isolated tracks of $p_T > 10$ GeV. An isolated track is one that requires that the sum of all tracks of $p_T > 1$ GeV within the annulus $0.02 < \Delta R < 0.3$ is less than 10% of the p_T of the track under consideration. Any instrumental backgrounds in the calorimeters are removed by requirements on neighboring channels or timing requirements.

The $\not{E}_T$ definition is the same as the previous except that muons are excluded from the calculation. Missing transverse energy value is optimized in MC to be greater than 350 GeV.

Background due to $Z(\to \nu\bar{\nu}) + jet$ is estimated from a control sample of $Z(\to \mu\bar{\mu}) + jet$ and scaled by the branching ratios, and is found to be 900 ± 94 events.

Background due to $W(\to \mu\nu)+jets$ is estimated by scaling a control sample of $W(\to \mu\nu)+jets$ by the muon reconstruction inefficiency. The background due to lost electrons and taus also use this control sample, but with a scaling produced by MC to determine the number of $W(\to e\nu) + jets$ and $W(\to \tau\nu) + jets$ and the appropriate reconstruction inefficiencies. The total $W + jets$ background is 312 ± 35 events.

The remaining backgrounds from QCD multijet events, $t\bar{t}$, and $Z(\ell\ell) + jets$ are small and are estimated from MC. The summary of monojet backgrounds and observed candidates are in Table 2. Observation agrees with the SM prediction.

4. Results

Upper limits are placed on the DM production cross sections, as a function of M_χ, assuming vector and axial-vector operators, summarized in Table 3 and Table 4. These are converted into the corresponding lower limits on the cutoff scale, Λ, also listed in Table 3 and Table 4. The Λ values are then translated into upper limits on the χ-nucleon cross sections, calculated within the effective theory framework. These are displayed in Fig. 1 and Fig. 2 as a function of M_χ. Also shown are the results from $p\bar{p}$ collider experiment CDF [8], direct detection experiments, COUPP [9], CoGeNT [10], Picasso [11], XENON100 [12], CDMS II [13, 14], and SIMPLE [15], and indirect detection experiments, IceCube [16] and Super-K [17]. Table 6 shows the 90%

Table 2. Monojet backgrounds and observed events.

Process	Events
$Z(\nu\bar{\nu}) + jets$	900 ± 94
$W + jets$	312 ± 35
$t\bar{t}$	8.5 ± 8.5
$Z(\ell\ell) + jets$	2.0 ± 2.0
Single t	1.1 ± 1.1
QCD Multijets	1.3 ± 1.3
Total SM	1225 ± 101
Data	1142

C.L. limits on Λ and the χ-nucleon cross section for the spin-dependent and spin-independent interactions.

Previously inaccessible χ masses below ~ 3.5 GeV are excluded for a χ-nucleon cross section greater than ~ 3 fb at 90% C.L. For spin-dependent scattering, the upper limits surpass all previous constraints for the mass range of 1-100 GeV. The results presented are valid for mediator masses larger than the limits on Λ, assuming unity for the couplings g_χ and g_q. The assumptions on χ interactions made in calculating the limits vary with experiments. Further, in the case of direct and indirect searches, an astrophysical model must be assumed for the density and velocity distribution of DM.

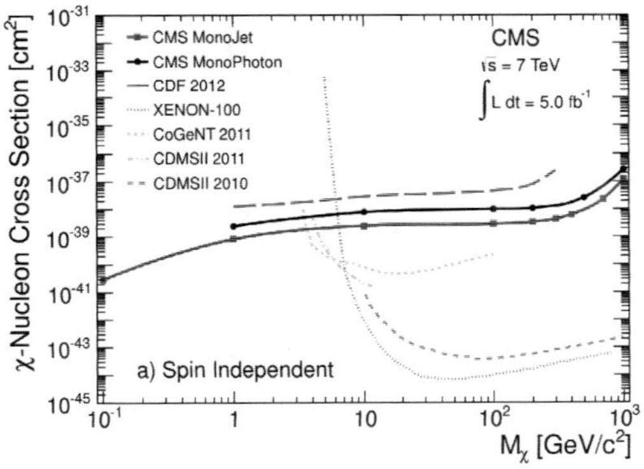

Figure 1. Comparison of the 90% CL upper limits on the dark matter-nucleon scattering cross section versus dark matter mass for the spin-independent (vector coupling) models with results from various direct detection experiments

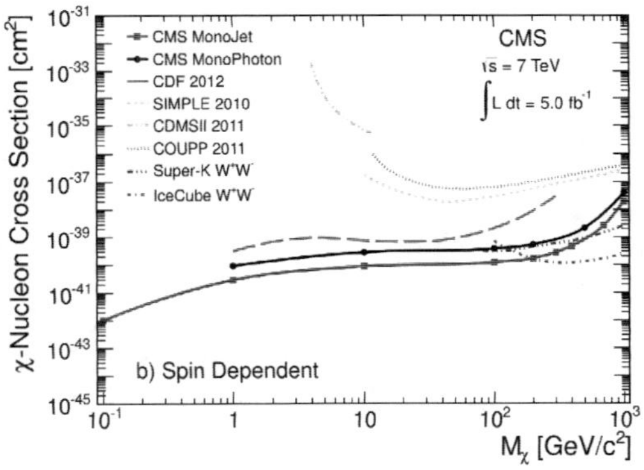

Figure 2. Comparison of the 90% CL upper limits on the dark matter-nucleon scattering cross section versus dark matter mass for the spin-dependent (axial-vector coupling) models with results from various direct detection experiments

5. Conclusion

We performed a search for new physics using 5 fb^{-1} of $\sqrt{7}$ TeV collisions in CMS for a signal of $\not{E}_T$ along with either a photon or a jet. In both cases, observation agrees with the SM, so limits were computed for the DM particle mass and it's nucleon scattering cross section in the

Table 3. Monophoton observed (expected) 90% C.L. limits on DM model parameters.

M_χ[GeV]	Vector		Axial-Vector	
	σ [fb]	Λ [GeV]	σ [fb]	Λ [GeV]
1	14.3 (14.7)	572 (568)	14.9 (15.4)	565 (561)
10	14.3 (14.7)	571 (567)	14.1 (14.5)	573 (569)
100	15.4 (15.3)	558 (558)	13.9 (14.3)	554 (550)
200	14.3 (14.7)	549 (545)	14.0 (14.5)	508 (504)
500	13.6 (14.0)	442 (439)	13.7 (14.1)	358 (356)
1000	14.1 (14.5)	246 (244)	13.9 (14.3)	172 (171)

Table 4. Monojet observed 90% C.L. limits on DM model parameters.

M_χ[GeV]	Vector		Axial-Vector	
	σ [cm^2]	Λ [GeV]	σ [cm^2]	Λ [GeV]
0.1	749	$2.90 \cdot 10^{-41}$	754	$1.03 \cdot 10^{-42}$
1	751	$8.21 \cdot 10^{-40}$	755	$2.94 \cdot 10^{-41}$
10	760	$2.47 \cdot 10^{-39}$	765	$8.79 \cdot 10^{-41}$
100	764	$2.83 \cdot 10^{-39}$	736	$1.21 \cdot 10^{-40}$
200	736	$3.31 \cdot 10^{-39}$	677	$1.70 \cdot 10^{-40}$
300	690	$4.30 \cdot 10^{-39}$	602	$2.73 \cdot 10^{-40}$
400	631	$6.15 \cdot 10^{-39}$	524	$4.74 \cdot 10^{-40}$
700	455	$2.28 \cdot 10^{-38}$	341	$2.65 \cdot 10^{-39}$
1000	302	$1.18 \cdot 10^{-37}$	206	$1.98 \cdot 10^{-38}$

regime of a high cut off scale. For spin-independent DM models, this analysis extends the limits on the DM-nucleon scattering cross section for DM particles below about 3.5 GeV, which was inaccessible to direct detection experiments. For spin-dependent models, this analysis provides the most stringent limits on the DM-nucleon scattering cross section for DM particles within the DM mass range of 1-1000 GeV.

References

[1] Gaitskell R 2004 *Annual Review of Nuclear and Particle Science* **54**
[2] CMS Collaboration (CMS) 2008 *Journal of Instrumentation* **3** S08004 URL `http://stacks.iop.org/1748-0221/3/i=08/a=S08004`
[3] Bai Y, Fox P J and Harnik R 2010 *JHEP* **12** 048 (*Preprint* 1005.3797v2)
[4] Fox P J, Harnik R, Kopp J and Tsai Y 2011 Missing energy signatures of dark matter at the lhc Tech. rep. Fermilab (*Preprint* 1109.4398)
[5] CMS Collaboration (CMS) 2009 Particle–flow event reconstruction in CMS and performance for jets, taus, and $\not{E}_T$ CMS Physics Analysis Summary CMS-PAS-PFT-09-001 CERN URL `http://cdsweb.cern.ch/record/1194487`
[6] Cacciari M, Salam G P and Soyez G 2008 *JHEP* **04** 063 (*Preprint* 0802.1189)
[7] CMS Collaboration (CMS) 2010 Electromagnetic calorimeter commissioning and first results with 7 tev data CMS Physics Analysis Summary CMS-NOTE-2010-012 CERN URL `http://cdsweb.cern.ch/record/1278160`
[8] Aaltonen T *et al.* (CDF) 2012 *Phys. Rev. Lett.* **108**(21) 211804 (*Preprint* 1203.0742)
[9] Behnke E *et al.* (COUPP) 2011 *Phys. Rev. Lett.* **106**(2) 021303 (*Preprint* 1008.3518) URL `http://link.aps.org/doi/10.1103/PhysRevLett.106.021303`
[10] Aalseth C E *et al.* (CoGeNT) 2011 *Phys. Rev. Lett.* **106**(13) 131301 (*Preprint* 1002.4703) URL `http://link.aps.org/doi/10.1103/PhysRevLett.106.131301`
[11] Archambault S *et al.* 2009 *Phys. Lett. B* **682** 185 ISSN 0370-2693 (*Preprint* 0907.0307) URL `http://www.sciencedirect.com/science/article/pii/S0370269309013525`
[12] Aprile E *et al.* (XENON100) 2011 *Phys. Rev. Lett.* **107**(13) 131302 (*Preprint* 1104.2549) URL `http://link.aps.org/doi/10.1103/PhysRevLett.107.131302`
[13] CDMS II Collaboration (CDMS II) 2010 *Science* **327** 1619 (*Preprint* `http://www.sciencemag.org/content/327/5973/1619.full.pdf`)
[14] Ahmed Z *et al.* (CDMS) 2011 *Phys. Rev. Lett.* **106**(13) 131302 (*Preprint* 1011.2482) URL `http://link.aps.org/doi/10.1103/PhysRevLett.106.131302`
[15] Felizardo M, Morlat T, Fernandes A C, Girard T A, Marques J G, Ramos A R, Auguste M, Boyer D, Cavaillou A, Sudre C, Poupeney J, Payne R F, Miley H S and Puibasset J (SIMPLE) 2010 *Phys. Rev. Lett.* **105** 211301 a more recent update can be found in `http://arxiv.org/abs/1106.3014` arXiv:1106.3014 (*Preprint* 1003.2987)
[16] Abbasi R *et al.* (IceCube) 2012 *Phys. Rev. D* **85** 042002 (*Preprint* 1112.1840)
[17] Tanaka T *et al.* (Super-Kamiokande) 2011 *Astrophys. J.* **742** 78 (*Preprint* 1108.3384)

Dark matter searches with radio observations

Marco Taoso
Department of Physics and Astronomy, University of British Columbia, Vancouver, BC V6T 1Z1, Canada
E-mail: mtaoso@phas.ubc.ca

Abstract. We compute the radio synchrotron emission induced by WIMP DM annihilations inside our galaxy and in extragalactic halos. Comparing DM fluxes with radio surveys from 22 MHz to 1420 MHz we derive bounds on the WIMP paramer space. Light WIMPs, $M_{DM} \leq 10$ GeV, with thermal annihilation cross-sections (i.e. $\langle \sigma v \rangle = 3 \times 10^{-26}$ cm^3 s^{-1}) are strongly constrained, especially those candidates annihilating mainly into leptons. We then show that DM sources could account for a significant fraction of the extragalactic-radio emission inferred by the ARCADE-2 Collaboration. We argue that future telescopes with sensitives at μJy level will be particulary suitable to search for faint DM emissions. We show that future data on source counts and angular correlations could become relevant to distinguish DM signals from astrophysical sources, such as radio loud active galactic nuclei and star forming galaxies.

1. Introduction

Indirect searches of Dark Matter (DM) are particulary promising for Weakly Interacting Massive Particles (WIMPs), which currently are the most investigated class of DM candidates. Signatures of this scenario include a multiwavelength spectrum associated to radiative emissions involving electrons and positrons generated in WIMP annihilations or decays. Here we focus our attention on the synchrotron emission produced in the interactions of electrons and positrons with galactic magnetic fields. For electrons energies below about 10 GeV and typical values of galactic magnetic fields, i.e. $\mathcal{O}(\mu G)$, the synchrotron emission falls at frequencies around and below the GHz , i.e. in the radio band. Therefore, radio observations are appropriate tools for indirect WIMP searches.

In Sec. 2 we study the galactic synchrotron emission from DM annihilations. Using current radio surveys we derive constraints on the WIMP parameter space. For light WIMP candidates, $M_{DM} \leq 10$ GeV, annihilating into charged leptons the bounds are particulary strong and disfavour thermal annihilation cross-sections (i.e. $\langle \sigma v \rangle = 3 \times 10^{-26}$ cm^3 s^{-1}) [1]. We briefly discuss the prospects for detection with future surveys.

In Sec. 3 we consider instead the extragalactic radio emission. We focus on three observables: intensity, differential number counts of sources and angular correlations. We present predictions for DM and compare with expectations for astrophysical sources and current data [1, 2].

2. Synchrotron emission from galactic DM annihilations

The goal of this section is to compare the synchrotron emission induced by DM annihilations inside our galaxy with observations (for other analysis on galactic DM radio signals see e.g. refs. [3, 4, 5, 6, 7, 8, 9, 10, 11, 12, 13]). As a benchmark case, in the following we focus on a

Content from this work may be used under the terms of the Creative Commons Attribution 3.0 licence. Any further distribution of this work must maintain attribution to the author(s) and the title of the work, journal citation and DOI.
Published under licence by IOP Publishing Ltd

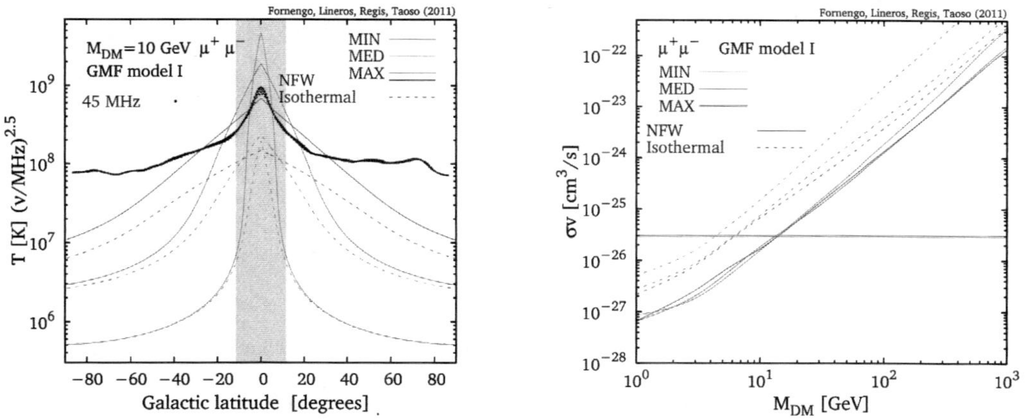

Figure 1. Left: Temperature versus galactic latitude at 45 MHz. The data corresponds to a thin strip $|l| < 3°$, with l the galactic longitude. Lines are predictions for DM models for $l = 0°$. Blue band corresponds to $|b| < 10°$ and it denotes the directions towards the galactic center region. Right: upper bounds on the annihilation cross section (σv) as a function of the DM mass for the muonic channel.

DM candidate annihilating into muon pairs. We consider two galactic DM density distributions, the popular NFW profile, which nicely fits the results of N-body simulations, and an isothermal profile which is instead used as a conservative case of a cored DM distribution.

The propagation of the electrons produced by DM annihilations is then modeled following a semi-analytical approach [14]. Electrons are assumed to be confined by the galactic magnetic field inside a propagation zone which is described by a cylinder centered at the galactic center with a radius $R_g = 20$ kpc. The number density of electrons per unit of energy $n_e(E, \mathbf{x})$ follows a transport equation which takes into account the spatial diffusion of the electrons and the energy losses due to the interactions with the galactic magnetic field and the interstellar radiation field. The parameters of the propagation model are constrained by cosmic-rays data, notably the Boron over Carbon ratio (B/C) and radioactive isotopes. In order to bracket the uncertainties on the electrons propagation we consider three propagation benchmarks, the so-called MIN, MED, and MAX models. These models are fully compatible with B/C and antiproton/proton ratio observations, describing the extreme behaviors (MIN and MAX) and best fit (MED) of these observables [15].

The structure of the Galactic Magnetic Field is still not well understood. Mainly, it is composed by a regular and a turbulent part, the last one responsible for the diffusive behaviour of the cosmic rays. Here, the total magnetic field is assumed to have a cylindrical symmetry with an exponential dependence on the distance on the galactic center and on the vertical height (more details in Ref. [1]). The normalization at the Sun position is set to 6μG, which is well consistent with the measurements.

In the left panel of Fig. 1 we compare observations and the DM emission at 45 MHz for a 10 GeV DM particle with a thermal annihilation cross section, $(\sigma v) = 3 \times 10^{-26}$ cm^3 s^{-1}. As expected the signal is enhanced at the center of the galaxy and its morphology is strongly affected by the choice of the DM density distribution and propagation model. Conservative bounds on the DM parameter space can be set comparing the DM signal with the data without attempting any astrophysical background subtraction. We employ data from five surveys from 22 MHz to 1420 MHz (see ref. [1] for more details about the method adopted to set the bounds.)

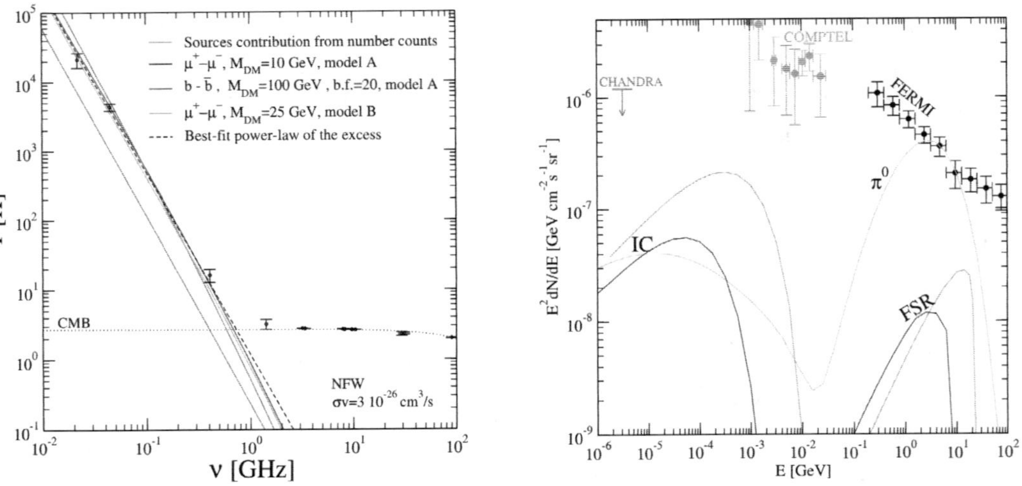

Figure 2. Left: extragalactic radio background as derived by ARCADE [16], together with three possible interpretations of the low–frequency (< 10 GHz) excess in terms of WIMP annihilations. Right: X-ray and gamma-ray fluxes for the three benchmark WIMP cases shown in left panel. The CHANDRA [17] bound in the X–ray band and the COMPTEL [18] and FERMI [19] extragalactic gamma–ray fluxes are shown.

As shown in the right panel of Fig. 1, models with DM masses $M_{DM} \leq 10$ GeV and a thermal value of the annihilation cross section are strongly constrained for a NFW profile and for for all the propagation models we have considered. We refer the reader to ref. [1] for a more complete discussion on the uncertanties affecting these bounds (see also ref.[20] for a similar work and a comparision with collider bounds).

While we have conservatively focused on the bounds which can be inferred from observations, it could be possible that the present data already contain a significant DM contribution. In Ref. [1] we have tried to single out this DM component using the 408 MHz map as a spatial template to model the galactic CR synchrotron emission in the low-frequency maps. With this method, a DM-induced component softer than the astrophysical galactic synchrotron emission would appear as an excess in the central galactic center region with an approximately spherical shape. We have concluded that present data does not support any evidence for the presence of this additional synchrotron component. However, we expect that future radio surveys, in particular with the LOFAR telescope [21], will improve coverage, angular resolution, and sensitivity in low-frequency radio maps, which may allow to disentangle a faint DM contribution.

3. DM extragalactic radio emission.

The ARCADE-2 collaboration has reported the measurement of an isotropic extragalactic radio background at frequencies below 10 GHz [16]. Surprisingly, the total contribution from the extragalactic radio sources detected in current surveys is a factor 5-6 smaller [22, 23]. Moreover, such level of cosmic radio background does not have an immediate explanation in standard astrophysical scenarios [22, 23, 24, 25]. For instance, radio supernovae, radio quiet quasars and diffuse emission from intergalactic medium and clusters have been considered, concluding that none of them can significantly contribute. Also an interpretation in terms of ordinary star-forming galaxies is strongly constrained by multi-wavelenght observations, in particular at far-infrared frequencies. The ARCADE-2 excess seems to point to the existence of a new population of numerous and faint synchrotron sources with no or faint correlated mechanisms

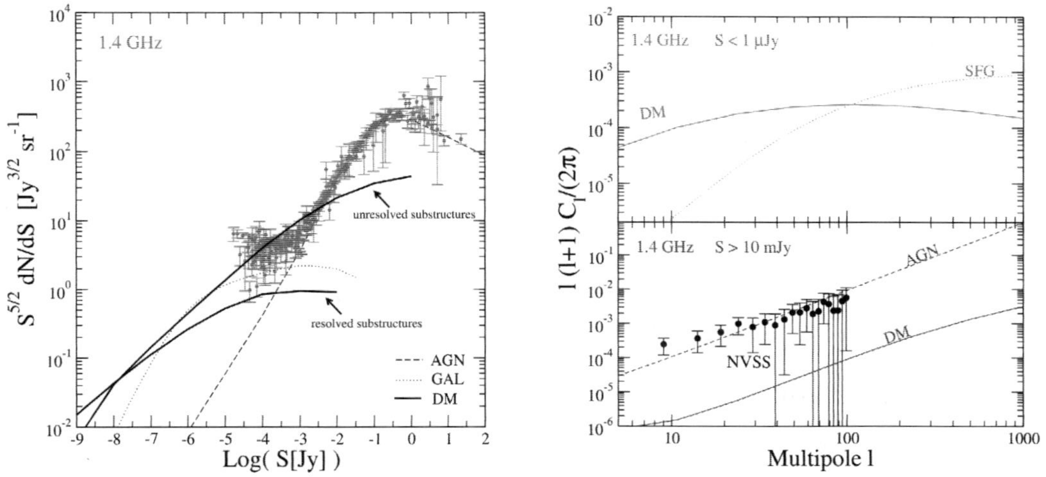

Figure 3. Left: differential number counts for AGNs (dashed line), star-forming galaxies (dotted line), and the same 10 GeV benchmark DM model (solid lines) shown in figs. 2. See ref. [2] for details. Right: angular power spectrum at 1.4 GHz of a bright ($S > 10$ mJy, lower) and a faint ($S < 1\,\mu$Jy, upper) samples for astrophysical (only dominant population is shown) and DM models. See ref. [26] for details.

at other frequencies. Here we show that WIMP annihilations in extra-galactic halos provide a possible candidate.

In the left panel of Fig. 2 we show the extragalactic radio background derived by ARCADE-2 as well as the radio signals for different DM masses, annihilation channels and assumptions on the astrophysical inputs, i.e. DM clustering and magnetic fields (see ref. [2] for more details). Leptonic channels provide an hard synchrotron spectum which can nicely match the data. In addition, for light WIMPs ($\sim$ few tens of GeV) the correct normalization of the excess can be obtained for fairly realistic assumptions on the astrophysical parameters, without the need of large DM overdensities. For these DM candidates, the X-rays and gamma-rays constraints can be satisfied, as shown in the right panel of Fig. 2. In conclusion, a viable dark matter interpretation of the radio excess requires light WIMPs annihilating mostly into electrons and muons. These results have been confirmed by the analysis in Ref.[27].

If dark matter annihilations contribute substantially to the extragalactic radio background, a population of DM halos could appear in the catalogue of sources detected by radio telescopes. In Fig. 3 we show the differential number counts of sources at 1.4 GHz as well as predictions for DM halos and two populations of astrophysical sources. While at large fluxes the counts are dominated by radio-loud AGN, DM sources dominates over astrophysical contributions at the sub μJy level. These sensitivies will be reached by the Square Kilometer Array (SKA) [28] and possibly also by its precursors, ASKAP [29] and MeerKAT [30]. In order to distinguish DM and astrophysical emissions one can also employ informations on the the angular distribution of sources. Present data on the angular power spectrum are not very relevant for DM since they refer to levels of brightness at which the number of DM sources is very low. This is shown in the lower right panel of Fig. 3. However, as discussed above, with flux thresholds at μJy level it will be possible to probe realistic WIMP scenarios and angular correlations will become important, especially at low multipoles, as shown in the upper right panel of Fig. 3.

4. Conclusions

We have argued that radio observations are particularly suitable to search for WIMPs annihilations. Present low-frequency surveys do not support any clear evidence for a galactic DM signal. However, the data can be used to set strong constraints on the WIMP parameter space which are complementary to those obtained with other indirect DM searches, like gamma-ray observations.

The isotropic extragalactic background inferred by ARCADE-2 is larger than expectations from astrophysical sources. This excess can potentially be explained by WIMPs annihilations in extragalactic halos. This scenario could be tested by the next-generation radio telescopes, in particular employing data on differential source counts and angular correlations.

Acknowledgments

The work was supported by the Institute of Particle Physics (IPP) Theory Fellowship and the Natural Sciences and Engineering Research Council (NSERC) of Canada.

References

[1] Fornengo N, Lineros R A, Regis M and Taoso M 2012 *JCAP* **1201** 005 (*Preprint* 1110.4337)
[2] Fornengo N, Lineros R, Regis M and Taoso M 2011 *Phys.Rev.Lett.* **107** 271302 (*Preprint* 1108.0569)
[3] Borriello E, Cuoco A and Miele G 2009 *Phys.Rev.* **D79** 023518 (*Preprint* 0809.2990)
[4] Delahaye T, Boehm C and Silk J 2012 *Mon.Not.Roy.Astron.Soc.Lett.* **422** L16–L20 (*Preprint* 1105.4689)
[5] Bertone G, Cirelli M, Strumia A and Taoso M 2009 *JCAP* **0903** 009 (*Preprint* 0811.3744)
[6] Boehm C, Silk J and Ensslin T 2010 (*Preprint* 1008.5175)
[7] Crocker R, Bell N, Balazs C and Jones D 2010 *Phys.Rev.* **D81** 063516 (*Preprint* 1002.0229)
[8] Regis M and Ullio P 2008 *Phys.Rev.* **D78** 043505 (*Preprint* 0802.0234)
[9] Bergstrom L, Bertone G, Bringmann T, Edsjo J and Taoso M 2009 *Phys.Rev.* **D79** 081303 (*Preprint* 0812.3895)
[10] Dobler G and Finkbeiner D P 2008 *Astrophys.J.* **680** 1222–1234 (*Preprint* 0712.1038)
[11] Cumberbatch D T, Zuntz J, Eriksen H K K and Silk J 2009 (*Preprint* 0902.0039)
[12] Linden T, Profumo S and Anderson B 2010 *Phys.Rev.* **D82** 063529 (*Preprint* 1004.3998)
[13] Laha R, Ng K C Y, Dasgupta B and Horiuchi S 2013 *Phys.Rev.* **D87** 043516 (*Preprint* 1208.5488)
[14] Maurin D, Donato F, Taillet R and Salati P 2001 *Astrophys.J.* **555** 585–596 (*Preprint* astro-ph/0101231)
[15] Donato F, Maurin D, Salati P, Barrau A, Boudoul G *et al.* 2001 *Astrophys.J.* **563** 172–184 (*Preprint* astro-ph/0103150)
[16] Fixsen D, Kogut A, Levin S, Limon M, Lubin P *et al.* 2009 (*Preprint* 0901.0555)
[17] Hickox R C and Markevitch M 2007 *Astrophys.J.* **661** L117–L121 (*Preprint* astro-ph/0702556)
[18] G Weidenspointner *et al* 2000 *American Institute of Physics Conference Series* vol 510 (AIP, eds., M. L. McConnell & J. M. Ryan) pp 467–470
[19] Abdo A *et al.* (Fermi-LAT collaboration) 2010 *Phys.Rev.Lett.* **104** 101101 (*Preprint* 1002.3603)
[20] Mambrini Y, Tytgat M H, Zaharijas G and Zaldivar B 2012 *JCAP* **1211** 038 (*Preprint* 1206.2352)
[21] http://www.lofar.org/
[22] Seiffert M, Fixsen D, Kogut A, Levin S, Limon M *et al.* 2009 (*Preprint* 0901.0559)
[23] Vernstrom T, Scott D and Wall J 2011 (*Preprint* 1102.0814)
[24] Singal J, Stawarz L, Lawrence A and Petrosian V 2010 *Mon.Not.Roy.Astron.Soc.* **409** 1172 (*Preprint* 0909.1997)
[25] Ponente P P, Ascasibar Y and Diego J M 2011 (*Preprint* 1104.3012)
[26] Fornengo N, Lineros R, Regis M and Taoso M 2012 *JCAP* **1203** 033 (*Preprint* 1112.4517)
[27] Hooper D, Belikov A V, Jeltema T E, Linden T, Profumo S *et al.* 2012 *Phys.Rev.* **D86** 103003 (*Preprint* 1203.3547)
[28] Rawlings S and Schilizzi R 2011 (*Preprint* 1105.5953)
[29] Johnston S *et al.* (ASKAP Collaboration) 2007 *PoS* **MRU** 006 (*Preprint* 0711.2103)
[30] Booth R, de Blok W, Jonas J and Fanaroff B 2009 (*Preprint* 0910.2935)

Journal of Physics: Conference Series **485** (2014) 012032 doi:10.1088/1742-6596/485/1/012032

New signals in dark matter detectors

Joachim Kopp

Fermilab, PO Box 500, Batavia, IL 60510,USA and
Max–Planck–Institut für Kernphysik, Postfach 10 39 80, 69029 Heidelberg, Germany

E-mail: jkopp@mpi-hd.mpg.de

Abstract. We investigate the scattering of solar neutrinos on electrons and nuclei in dark matter direct detection experiments. The rates of these processes are small in the Standard Model, but can be enhanced by several orders of magnitude if the neutrino sector is slightly non-minimal. This makes even the current generation of dark matter detectors very sensitive to non-standard neutrino physics. Examples discussed here are neutrino magnetic moments and toy models with a simple hidden sector containing a sterile neutrino and a light new gauge boson ("dark photon"). We discuss the expected event spectra and temporal modulation effects, as well as constraints from a variety of astrophysical, cosmological, and laboratory experiments.

1. Neutrino interactions in dark matter detectors

It is well known that solar and atmospheric neutrinos constitute an irreducible background to future dark matter searches (see for instance [1]). In particular, neutrinos can scatter on atomic nuclei or electrons, thus mimicking typical dark matter signatures. In this talk, based mostly on ref. [2], we assume that the neutrino sector is slightly non-minimal and show that, in this case, neutrino signals in dark matter detectors can be up to several orders of magnitude stronger than in the Standard Model, enough to make them highly relevant even in the current generation of experiments. (This is also discussed in refs. [3, 4].)

At the basis of our discussion lies the observation that interactions mediated by particles much lighter than the typical momentum transfers in the process become stronger at low energy. Hence, if such interactions exist in the neutrino sector, they may be very significant in dark matter detectors with their low energy thresholds of 10 keV, while dedicated neutrino detectors, whose energy thresholds are at least a few hundred keV would be insensitive to them.

We consider here two examples for light new physics in the neutrino sector: A neutrino magnetic moment (where the light mediator is the photon) and scenarios with a gauged sterile neutrino sector (where the light mediator is a new $U(1)'$ gauge boson).

1.1. Neutrino magnetic moments

Neutrino magnetic moments are described by a Lagrangian operator of the form

$$\mathcal{L}_{\mu_\nu} \supset \mu_\nu \, \bar{\nu} \sigma^{\alpha\beta} \partial_\beta A_\alpha \nu \,, \tag{1}$$

where A_α and ν denote the photon and neutrino fields, respectively, μ_ν is the neutrino magnetic moment, and as usual, $\sigma^{\alpha\beta} = \frac{i}{2}[\gamma^\alpha, \gamma^\beta]$. The Standard Model prediction for μ_ν is negligibly small [5], but in extensions of the Standard Model, it can come close to the current experimental

Content from this work may be used under the terms of the Creative Commons Attribution 3.0 licence. Any further distribution of this work must maintain attribution to the author(s) and the title of the work, journal citation and DOI.
Published under licence by IOP Publishing Ltd

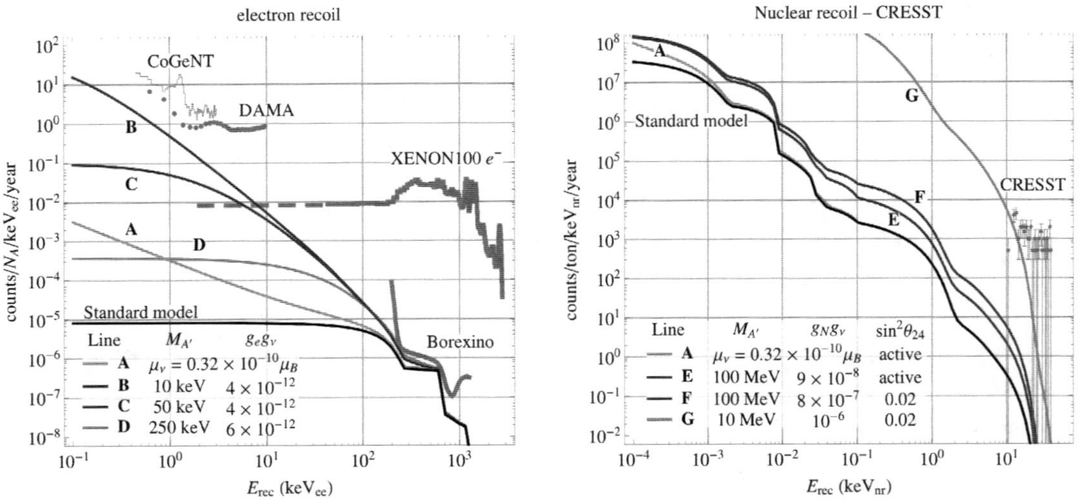

Figure 1. Neutrino–electron scattering rates (left) and neutrino–nucleus scattering rates for germanium (right) in the Standard Model (black) and in various models with extended neutrino sectors (colored lines). The curves labeled (A) illustrate the effect of a neutrino magnetic moment $\mu_\nu = 0.32 \times 10^{-10}\mu_B$; curves (B, C, D) are for a model where neutrino scattering is enhanced by exchange of a new $U(1)'$ gauge boson with mass $M_{A'}$ and couplings to neutrinos (g_ν) and electrons (g_e) or nucleons (g_N) as given in the legend. This is, for instance, realized in scenarios with a sterile neutrino charged under $U(1)'$, and coupled to Standard model particles through kinetic mixing of the $U(1)'$ gauge boson and the photon. Curve (E) shows model with active neutrino–nucleus scattering through a light A' boson (for instance a $U(1)_{B-L}$ model), and curves (F) and (G) illustrate a situation in which 2% of solar neutrinos oscillate into a "sterile" state ν_s, which is, however, charged under a $U(1)_B$ gauge group broken at a low scale [3]. We compare the model predictions to observed event rates in various experiments (red curves and data points) [7–11]. Figure taken from ref. [2].

90% CL upper limit $\mu_\nu < 0.32 \times 10^{-10}\mu_B$ (where $\mu_B = \sqrt{4\pi\alpha}/2m_e$ is the Bohr magneton, with α the electromagnetic fine structure constant and m_e the electron mass). The magnetic moment-induced neutrino–nucleus scattering cross section is [6]

$$\frac{d\sigma_\mu(\nu N \to \nu N)}{dE_{\text{rec}}} = \mu_\nu^2 \alpha Z^2 F^2(E_{\text{rec}}) \left(\frac{1}{E_{\text{rec}}} - \frac{1}{E_\nu} \right), \tag{2}$$

with the notation E_ν for the neutrino energy and E_{rec} for the nuclear recoil energy. Note that $d\sigma_\mu/dE_{\text{rec}}$ is proportional to the square of the nuclear charge Z because at low E_{rec} the scattering is coherent on all protons in the nucleus. At higher E_{rec}, loss of coherence is taken into account by the nuclear form factor $F(E_{\text{rec}})$ [2]. The neutrino–electron scattering cross section is given by eq. (2) with Z and $F(E_{\text{rec}})$ set to 1.

The differential neutrino–electron and neutrino–nucleus scattering rates in the magnetic moment scenario are shown as a function of E_{rec} in fig. 1 (curves labeled "(A)"). They exhibit the increase with $1/E_{\text{rec}}$ expected from eq. (2) at low energy, which leads to an enhancement of more than one order of magnitude at 1 keV recoil energy compared to the Standard Model.

1.2. Sterile neutrinos and a light $U(1)'$ gauge boson

An even larger enhancement of neutrino scattering rates in dark matter detectors occurs in scenarios featuring, in addition to the massless photon, another relatively light ($\ll$ 1 GeV) gauge boson ("dark photon"). Since couplings of such gauge bosons to Standard Model particles are strongly constrained [2, 12, 13], the most natural scenario of this type is one in which, in addition to the three active neutrinos, a sterile neutrino sector exists which is uncharged under the Standard Model gauge group, but couples to the new $U(1)'$ gauge force. The sterile sector can communicate with the Standard Model sector through a kinetic mixing term of the form

$$\mathcal{L}_{\text{mix}} = -\frac{1}{2}\epsilon F'_{\mu\nu}F^{\mu\nu}\,, \tag{3}$$

where $F_{\mu\nu}$ and $F'_{\mu\nu}$ are the electromagnetic and $U(1)'$ field strength tensors, respectively, and ϵ is the small mixing parameter. (An alternative possibility is a $U(1)'$ gauge boson which couples only to the sterile neutrinos and Standard Model quarks, but not leptons [3, 4].)

The sterile neutrinos can be produced in the Sun through their small mixing with the active species, and they could scatter on nuclei in a terrestrial detector with differential cross section

$$\frac{d\sigma_{A'}(\nu N \to \nu N)}{dE_{\text{rec}}} = \frac{4\pi Z^2 \alpha\alpha' \epsilon^2 m_N F^2(E_{\text{rec}})}{E_\nu^2 (M_{A'}^2 + 2E_{\text{rec}} m_N)^2}\left[2E_\nu^2 + E_{\text{rec}}^2 - 2E_{\text{rec}}E_\nu - E_{\text{rec}} m_N\right], \tag{4}$$

where α and α' are the electromagnetic and $U(1)'$ fine structure constants, m_N is the nuclear mass, $M_{A'}$ is the mass of the new gauge boson, and the other quantities are defined as in eq. (2). The expression for neutrino–electron scattering is obtained by the replacements $Z \to 1$, $F(E_{\text{rec}}) \to 1$, and $m_N \to m_e$ in eq. (4).

Possible signals of sterile neutrino scattering in dark matter detectors through new $U(1)'$ gauge bosons are shown in fig. 1, curves (B)–(G). (See plot legend for more details.) We see that these models can easily lead to neutrino signals several orders of magnitude stronger than the Standard Model prediction and thus highly relevant even to current dark matter searches: experiments like Xenon-100 can be extremely sensitive to new physics in the neutrino sector, which significantly enhances their physics case, but can also constitute a problem because often neutrino signals cannot be easily distinguished from genuine dark matter signatures.

This observation raises the question whether sterile neutrinos can explain some of the anomalous signals reported from the DAMA [9], CoGeNT [10], and CRESST [14]. As curve (B) in the left panel of fig. 1 shows, a fraction of the excess events observed in CoGeNT can indeed be accounted for if interpreted in terms of electron recoils and if very conservative assumptions are made on the sensitivity of Xenon-100 to low-energy electron recoils. The CRESST signal could also be partially explained, but note that the energy spectrum of CRESST events cannot be well reproduced, and that models like the one shown in fig. 1 right, curve (D), are constrained by the total event rate in DAMA (see fig. 4 in ref. [2]). The annual modulation signal observed in DAMA will be discussed in the next section.

2. Temporal modulation of neutrino signals

Dark matter scattering rates are expected to vary during the year because the Earth's velocity with respect to the Milky Way's dark matter halo is larger in summer than in winter [15]. However, also neutrino signals from the Sun are expected to modulate with time.

The simplest source of modulation is the varying Earth–Sun distance caused by the ellipticity of the Earth's orbit. It leads to a roughly 3% larger expected count rate in winter than in summer, which corresponds to a 180° phase shift compared to the commonly considered modulation signals from dark matter. It is, however, interesting to note, that also dark matter scattering rates can peak in winter in the higher energy parts of the recoil spectrum [15]. Thus, in a

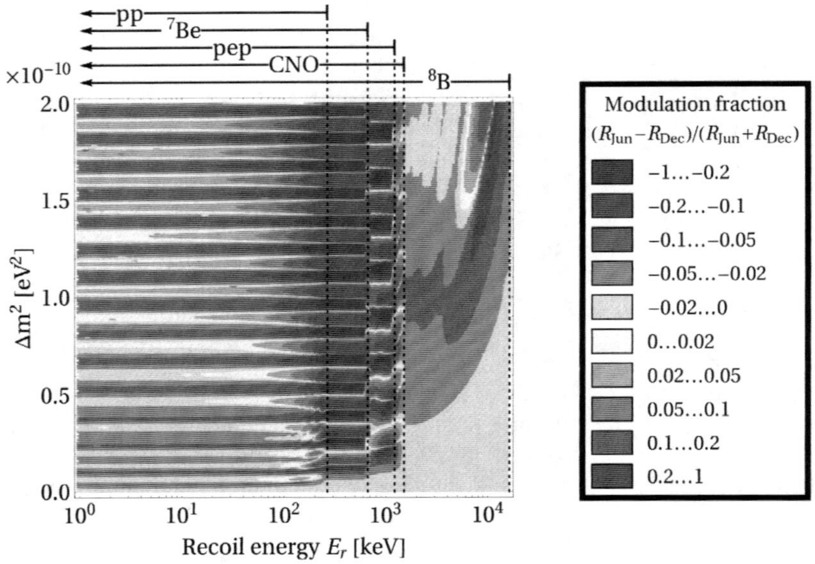

Figure 2. Relative annual modulation $(R_{\rm Jun} - R_{\rm Dec})/(R_{\rm Jun} + R_{\rm Dec})$ for neutrino–electron scattering events in the $U(1)'$ model from sec. 1.2 as a function of the recoil energy $E_{\rm rec}$ and the mass squared difference between the mostly sterile mass eigenstate and ν_2. Here $R_{\rm Jun}$ and $R_{\rm Dec}$ denote the differential count rate in events per keV at the summer and winter solstices, respectively. We have worked in a 2-flavor framework here. The vertical dashed lines indicate the cutoff energies of the different components of the solar neutrino spectrum. Figure taken from ref. [2].

detector with a high energy threshold ($\gg$ 10 keV), a dark matter modulation signal could be easily mimicked by neutrinos from the Sun.

However, depending on the details of the model, neutrino signals in dark matter detectors can also modulate with a different phase and amplitude. This happens in particular in the model discussed in sec. 1.2, where the fake dark matter signal is due to sterile neutrinos produced through oscillations (see also [3]). The sterile neutrino fraction in the solar neutrino flux can be significantly larger in summer than in winter if the oscillation length is not too different from the Earth–Sun distance, i.e. if the mass splitting Δm^2 between the mostly sterile mass eigenstate and the mostly active mass eigenstate ν_2 which dominates the solar neutrino flux, is of order 10^{-10} eV^2. We illustrate this in figure 2 by plotting the fractional annual modulation as a function of the recoil energy and of Δm^2. We see that large modulation peaking in summer can occur in large parts of the parameter space, i.e. no extreme fine-tuning is necessary to produce a dark matter-like signal. Note that, even in this case, the modulation phase would differ from the one expected from dark matter by about a month, i.e. a high statistics measurement of annual modulation would still be able to discriminate between dark matter and neutrino signals.

Other sources of temporal modulation of solar neutrino signals are Mikheyev-Smirnov-Wolfenstein type matter effects [16, 17] in the Earth (this typically requires more than one sterile neutrino), sterile neutrino absorption in the Earth, and angle-dependent detection efficiencies (see [2] for details on these modulation mechanisms).

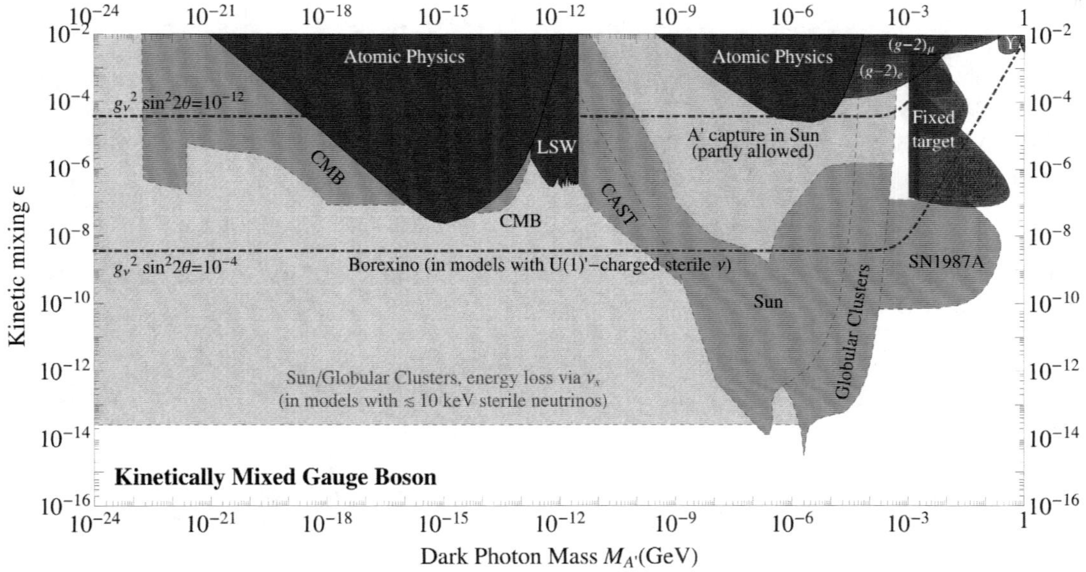

Figure 3. Constraints on the mass $M_{A'}$ and the kinetic mixing parameter ϵ (see eq. (3)) of a "dark photon". Most limits are taken from ref. [13]. See text, as well as ref. [2] and references therein for further details. Figure taken from ref. [2].

3. Constraints on dark photons and sterile neutrinos

In sec. 1.2, we have only discussed specific benchmark points for the models under consideration, but it is, of course, important to investigate the full available parameter space. Indeed, a plethora of constraints exist on light gauge bosons (dark photons) and on hidden sector particles coupled to them. Here, we discuss constraints for the specific case of a dark photon coupled to the Standard Model through a kinetic mixing term of the form (3). These constraints are summarized in fig. 3. (For constraints on other types of $U(1)'$ gauge bosons, see ref. [2] and references therein.)

At large dark photon mass $M_{A'}$, constraints come mainly from particle physics experiments, in particular fixed target (beam dump) experiments, measurements of the anomalous magnetic moment $g-2$ of the electron and the muon, and Υ decays studied at B factories [18–21]. (Note that a dark photon with parameters just below the region labeled $(g-2)_\mu$ could explain the observed discrepancy between the measured muon anomalous magnetic moment and the Standard Model prediction [22].)

For $M_{A'}$ in the eV–MeV region, very strong limits can be derived from stellar evolution: dark photons would provide a very efficient energy loss mechanism for stars and supernovae [13, 23, 24], significantly shortening their lifetime (for stars) or reducing their energy output (for supernovae). The resulting limits are especially strong if $M_{A'}$ is similar to the typical thermal energies in the star [23], so that ϵ is constrained down to 10^{-14} in this regime. Note, however, that dark photons with stronger couplings (indicated by the light blue region in fig. 3) cannot be ruled out this way because they would not be able to leave the star/supernova due to absorption [23]. Limits could still be obtained from stellar evolution constraints on anomalous heat transport inside the star [25], but for $M_{A'} \gtrsim 100$ keV, we expect these constraints to disappear as well [2, 25].

For sub-eV dark photon masses, the strongest constraints are obtained from helioscopes like

CAST [23], from "Light Shining through Walls" (LSW) experiments [26], from tests of the Coulomb law in atomic physics [27], and from distortions of the CMB spectrum [28].

In models which contain in addition to the dark photon also one or several light sterile neutrinos (like the scenario discussed in sec. 1.2), additional constraints arise from the fact that the sterile neutrino(s) would act as "minicharged particles" [13, 29]. The strongest limits in this case arise again from stellar cooling (light green region in fig. 3) and from Borexino [2, 8].

4. Summary

To summarize, we have shown that rather minimal extensions of the neutrino sector of the Standard Model can lead to very interesting signals in dark matter direct detection experiments. As examples, we have considered scenarios with neutrino magnetic moments and models with a sterile neutrino sector enjoying a new gauge symmetry broken at a scale $\ll 1$ GeV. We have shown that very large neutrino–electron and neutrino–nucleus scattering rates, well within the reach of current dark matter detectors, can occur at $\mathcal{O}(10$ keV) recoil energy. At the higher energies probed by dedicated neutrino detectors, no anomalous signals would be expected.

On the one hand, neutrino signals in dark matter detectors enhance the physics case of these experiments by allowing them to probe additional types of new physics besides dark matter. However, they can also constitute a problem because, especially for small event samples, they can be easily confused with genuine dark matter signals. For instance, we have shown that neutrino signals can exhibit temporal modulation similar to the one expected from dark matter.

References

[1] Gutlein A *et al.* 2010 *Astropart.Phys.* **34** 90–96 (*Preprint* 1003.5530)
[2] Harnik R, Kopp J and Machado P A 2012 *JCAP* **1207** 026 (*Preprint* 1202.6073)
[3] Pospelov M 2011 *Phys.Rev.* **D84** 085008 (*Preprint* 1103.3261)
[4] Pospelov M and Pradler J 2012 *Phys.Rev.* **D85** 113016 (*Preprint* 1203.0545)
[5] Beringer J *et al.* (Particle Data Group) 2012 *Phys.Rev.* **D86** 010001
[6] Vogel P and Engel J 1989 *Phys.Rev.* **D39** 3378
[7] Aprile E *et al.* 2011 *Phys.Rev.* **D83** 082001 10 pages, 12 figures (*Preprint* 1101.3866)
[8] Bellini G *et al.* (The Borexino Collaboration) 2011 *Phys.Rev.Lett.* **107** 141302 (*Preprint* 1104.1816)
[9] Bernabei R *et al.* (DAMA) 2008 *Eur. Phys. J.* **C56** 333–355 (*Preprint* 0804.2741)
[10] Aalseth C *et al.* 2011 *Phys.Rev.Lett.* **107** 141301 (*Preprint* 1106.0650)
[11] Ahmed Z *et al.* (CDMS-II Collaboration) 2010 *Phys.Rev.Lett.* (*Preprint* 1011.2482)
[12] Adelberger E *et al.* 2007 *Phys.Rev.Lett.* **98** 131104 (*Preprint* hep-ph/0611223)
[13] Jaeckel J and Ringwald A 2010 *Ann.Rev.Nucl.Part.Sci.* **60** 405–437 (*Preprint* 1002.0329)
[14] Angloher G, Bauer M, Bavykina I, Bento A, Bucci C *et al.* 2012 *Eur.Phys.J.* **C72** 1971 (*Preprint* 1109.0702)
[15] Freese K, Frieman J A and Gould A 1988 *Phys.Rev.* **D37** 3388
[16] Mikheyev S P and Smirnov A Y 1986 *Nuovo Cim.* **C9** 17–26
[17] Wolfenstein L 1978 *Phys. Rev.* **D17** 2369
[18] Bjorken J D, Essig R, Schuster P and Toro N 2009 *Phys. Rev.* **D80** 075018 (*Preprint* 0906.0580)
[19] Batell B, Pospelov M and Ritz A 2009 *Phys.Rev.* **D80** 095024 (*Preprint* 0906.5614)
[20] Essig R, Schuster P and Toro N 2009 *Phys.Rev.* **D80** 015003 (*Preprint* 0903.3941)
[21] Essig R, Harnik R, Kaplan J and Toro N 2010 *Phys.Rev.* **D82** 113008 (*Preprint* 1008.0636)
[22] Bennett G *et al.* (Muon G-2 Collaboration) 2006 *Phys.Rev.* **D73** 072003 (*Preprint* hep-ex/0602035)
[23] Redondo J 2008 *JCAP* **0807** 008 (*Preprint* 0801.1527)
[24] Dent J B, Ferrer F and Krauss L M 2012 (*Preprint* 1201.2683)
[25] Raffelt G G and Starkman G D 1989 *Phys.Rev.* **D40** 942
[26] Ahlers M, Gies H, Jaeckel J, Redondo J and Ringwald A 2008 *Phys.Rev.* **D77** 095001 (*Preprint* 0711.4991)
[27] Bartlett D and Loegl S 1988 *Phys.Rev.Lett.* **61** 2285–2287
[28] Mirizzi A, Redondo J and Sigl G 2009 *JCAP* **0903** 026 (*Preprint* 0901.0014)
[29] Davidson S, Hannestad S and Raffelt G 2000 *JHEP* **0005** 003 erratum added online, Oct/17/2000 (*Preprint* hep-ph/0001179)

Reviewing recent results from Pierre Auger Observatory

J A Chinellato for the Auger Collaboration [1]
Institute of Physics, State University of Campinas, Campinas, Brasil
E-mail: chinella@ifi.unicamp.br

[1] Full author list, in 9/6/2012:
http://www.auger.org/admin/Collaborators/author_list_alphabetical.html

Abstract. We present some results from the last three years of operation of the Pierre Auger Observatory. This is a short version of the review talk presented at the 18th International Symposium on Particles, Strings and Cosmology (PASCOS2012). The main topics are related to ultra-high energy cosmic rays and their interactions: energy spectrum, mass composition, neutrino spectrum and hadronic cross section at energies above 10^{18} eV.

1. Open questions in cosmic ray physics at energies above 10^{18} eV

A bulk of questions that were traditionally set for different cosmic-ray experiments operating below 10^{18} eV can nowadays be addressed by the Pierre Auger Observatory, due to its capability of investigating different aspects of air showers. The questions are about the topics: sources of cosmic rays at such energies, the acceleration processes taking place, the propagation along astronomical distances, the large-scale structure of the universe and the intervening magnetic fields and the mass composition. Another relevant question is: can we contribute to the knowledge of particle interactions at these otherwise inaccessible energies, of the order 450 TeV in the center-of-mass system ? We will report here topics like energy spectrum, mass composition, neutrino astrophysics and very high energy hadronic interactions.

2. The Pierre Auger Observatory

The Auger Observatory is an experimental complex that was designed to detect cosmic rays with energies above 10^{18} eV. It is situated near Malargüe, in the Province of Mendoza, Argentina at 35.1° - 35.5° S, 69.0° - 69.6° W and 1400 m asl. Its location gives a special view of the galactic center. The proposal of the Observatory aims to observe in details the development of air showers in the atmosphere, by calorimetric methods and also to observe the air shower front when it reaches ground-based detectors [1]. This is an important capability of Auger Observatory: because the design includes the use of fluorescence and ground array techniques, it is a hybrid experiment. The original design was completed in 2008, and since then many improvements have been made both to the setup and also to the analysis procedures. As a consequence, a broad view of the properties of very high energy air showers - their components and their time structure can be addressed and thus the main aspects of the primary cosmic rays can be inferred.

Content from this work may be used under the terms of the Creative Commons Attribution 3.0 licence. Any further distribution of this work must maintain attribution to the author(s) and the title of the work, journal citation and DOI.
Published under licence by IOP Publishing Ltd

3. Air shower features in the atmosphere and in ground-based modules as seen by the Auger Observatory

The fluorescence and surface detector techniques are combined, in the Observatory, to give detailed information about air showers. The experiment spans over an area of 3000 km^2 in the desertic region of Malargüe, and its aperture is about 7000 km^2 sr. Its Fluorescence Detector (FD) is an ensemble of 27 optical telescopes, grouped in 5 buildings, that can observe the longitudinal development of the shower in the atmosphere in a stereoscopic view. The shower profile is derived by detecting the fluorescence light emitted by excited nitrogen molecules - this is an established technique, used in previous experiments as Fly's Eye [2] and HiRes [3] and now in Telescope Array [4]. In the Auger Observatory, the telescopes record light profiles from showers that may be some tenths of kilometers away, over the ground array. This light suffers attenuation along the path from the production point to the telescopes, thus the detected profile must be corrected to reproduce the fluorescence production at the origin. To estimate the attenuation, which is due to molecules, aerosols and clouds, regular measurements of the atmospheric conditions are performed: UV laser shooting, radiosonde launching, optical observations and cloud measurements. They are incorporated into the event reconstruction software, so that the reconstructed fluorescence profile gives an almost calorimetric measurement of the energy of the primary particle. A typical shower profile, obtained with two telescopes, can be seen in figure 1. The duty cycle of the Fluorescence Detector is 10 - 15 % .

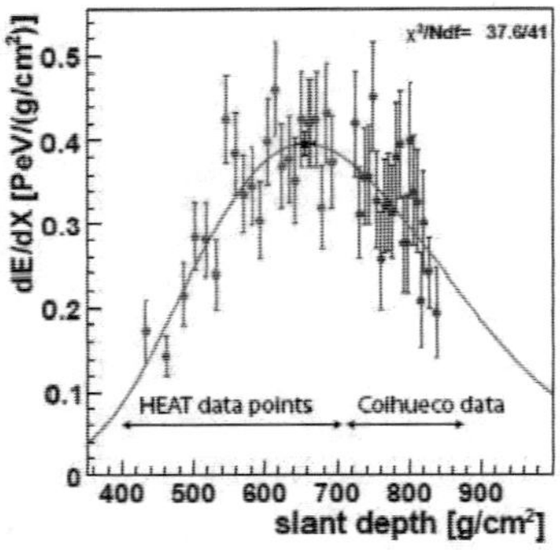

Figure 1. A typical longitudinal profile (energy deposit X depth), as seen by fluorescence telescopes. Putting together information from longitudinal and lateral (see figure at right) profiles in a event-to-event basis, a universal calibration curve is constructed.

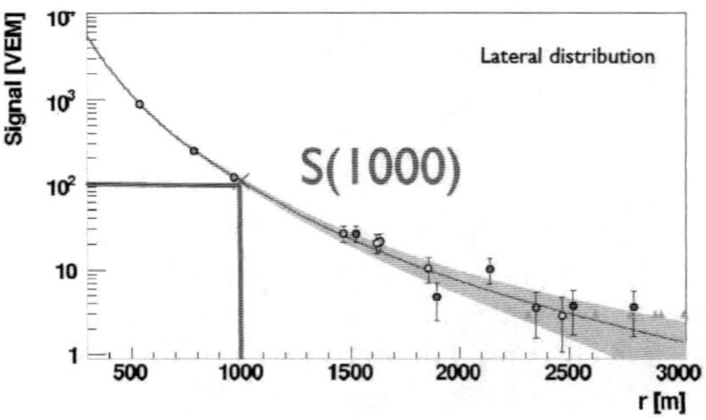

Figure 2. A typical lateral profile, as seen by the ground array (SD). VEM is the "vertical muon equivalent", the signal generated by a vertical muon at the center of the module. The signal at 1000 m from shower core is a convenient parameter to estimate shower energy.

The ground array, named Surface Detector (SD), consists of 1660 stations - water-Cherenkov tanks and their associated electronics arranged in triangular cells, where each station is 1.5 km apart from another. The Cherenkov light generated in water is collected by 3 photomultipliers and the signal is read out by 40 MHz FADC's, which give a 25 ns time scale for further shower reconstruction. The signal unit is called Vertical Muon Equivalent (VEM), the signal generated by a muon crossing the water tank vertically, at its center (figure 2). This unit is very suitable to continuously calibrating particle signals, accounting for different conditions of operation. The Surface Detector has a duty cycle near to 100%.

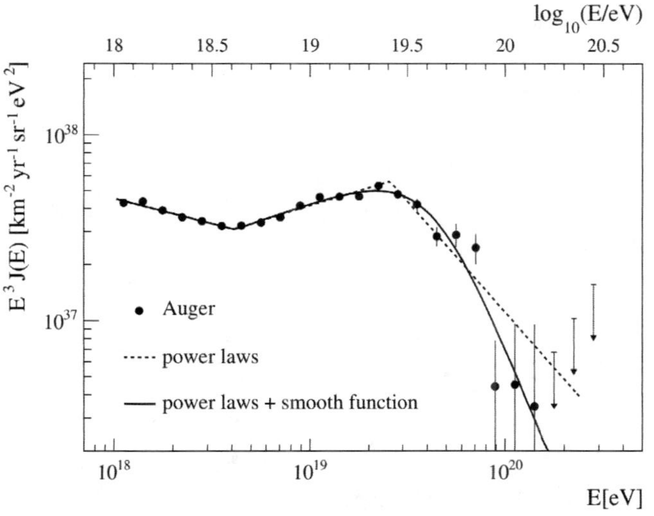

Figure 3. A high precision spectrum, obtained by combination of hybrid techniqueand SD technique. The calibration uses high-quality hybrid events, so that both inputs to this figure have the same systematic uncertainty in the energy scale. Three power laws and a smooth curve (solid) are fitted to the data. The features of this energy region are important to elucidate the well known GZK-cut quest [14].

4. Event Reconstruction

An importante piece in the event reconstruction is the lateral distribution of the signal, as recorded in the Cherenkov stations. A lateral distribution function, based on shower simulations and previous data from the Auger Observatory, is fitted to the experimental curve, in order to obtain the value of signal at 1000 m from shower core (see figure 2). A universal calibration curve using this information and those from fluorescence detector, is used to obtain the primary energy, within a systematic error of 22%, whereas the statistical error is $\sim$ 15 %. The calibration curve is then used for all showers, also for non-hybrid detection. Figure 3 shows a high-precision energy spectrum, at energies above 10^{18} eV.

The operation in hybrid mode optimizes event reconstruction: the timing information from telescope pixels together with the time and signal structure from surface stations, as cited above, gives a resolution for core location within 50 m and an angular resolution for the arrival direction of primary of 0.6 ^{o} [8]. Non hybrid events have angular resolution about 1.5^o.

5. Mass Composition

The atmospheric profile of an extensive air shower, i.e, the number of particles as a function of depth, carries important information about the mechanisms of particle production and absorption. The depth at which the shower reaches its maximum, called X_{max} in the literature gives important clue about the mass of the primary particle [6]. The change of $< X_{max} >$ per decade of energy, the so-called Elongation Rate and its shower-to-shower fluctuations, RMS (X_{max}) are sensitive to changes in mass composition with energy. These variables have different behaviors, if the primary is a proton or an iron nucleus, as predicted by different hadronic interaction models. Figure 5 shows the experimental behavior of both variables with energy, together with the predictions of the models EPOS 1.99, SIBYLL 2.1, QGSJET01C, QGSJETII [6]. We find a tendency in the composition: if the models describe correctly the interactions, when the energy gets higher, the composition departs from that expected for protons and tends to that expected for iron primary.

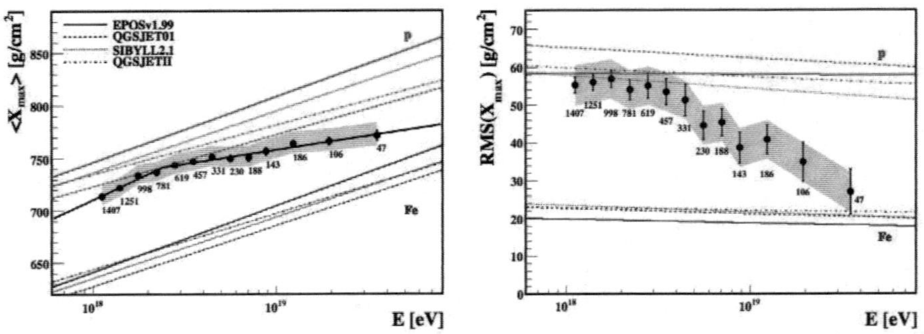

Figure 4. Behavior of $< X_{max} >$ and RMS (X_{max}) as function of primary energy and the prediction of hadronic models in the versions that are more compatible with recent experimental data: EPOS 1.99, SIBYLL 2.1, QGSJET01C, QGSJETII [6]. The numbers in each energy bin are attached to the experimental points.

6. Neutrinos

It is expected that protons with energies $10^{20} - 10^{21}$ eV would, in their interactions, produce neutrinos above the threshold of EeV telescopes [11]. Particle production scenarios like those of "top-down" and "bottom-up" [12, 13] predict very-high energy neutrinos, that could be detected by existing neutrino telescopes. The surface array of Pierre Auger Observatory can detect air showers with zenith angles from zero to 90^0 and even larger than this value - this is the case when a neutrino interacts with Earth's crust and generates a shower. Among the air showers, those almost horizontal, with zenith angles near 90° could be candidates to neutrino-initiated events, under some assumptions, that will not be discussed here. The primary neutrinos, in this case, are called down-going neutrinos, whereas those Earth-skimming are called up-going neutrinos. We can put limits at 90 percent CL per half decade of energy, for down-going (all flavours) and up-going neutrinos and compare them with other experimental results, as shown in fig 6. No candidate of neutrino induced showers were found so far, allowing us to set upper limits to their flux. Further details about neutrino initiated showers registered in Pierre Auger Observatory can be seen in ref. [15, 16].

7. Hadronic Interactions

Although cosmic ray experiments are, in general, not designed to identify single particles, like accelerators, some quantities that are measured in extensive air showers are sensitive to the non perturbative and semi-hard processes of QCD: inelastic cross-sections, multiplicities of muonic and electromagnetic components, their density per rapidity interval and others [17, 18]. The fluorescence telescopes, as they are mounted in Auger Observatory, supply enough information about the rise and fall of particle number (or energy deposited) in the atmosphere to estimate the inelastic proton-air cross section. This cross section, obtained from $E_{lab} = 10^{18}$ - $10^{18.4}$ eV showers, corresponding to about $\sqrt{s} = 57$ TeV, detected in the Auger Observatory, together with other measurements and predictions from hadronic models, is shown in figure 7. From this cross section, we can calculate, using Glauber Theory, the proton-proton cross section, that can be seen in reference [7].

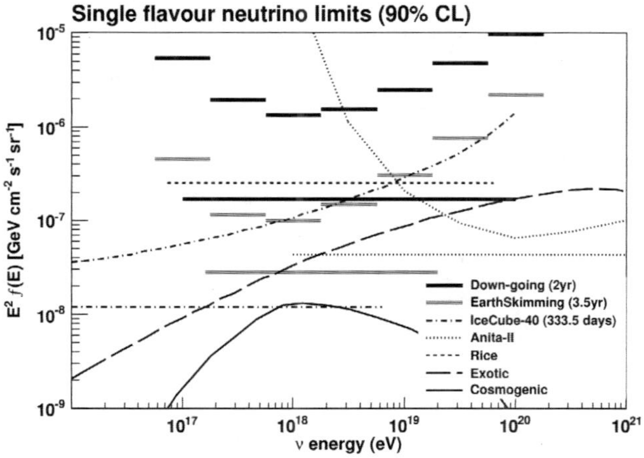

Figure 5. Ninety-percent Confidence Level limits for diffuse flux of down-going (ν) and Earth-skimming (ν_τ) neutrinos from a selected set of events as given by Pierre Auger Observatory, compared to limits from other experiments. See ref. [15] for further details.

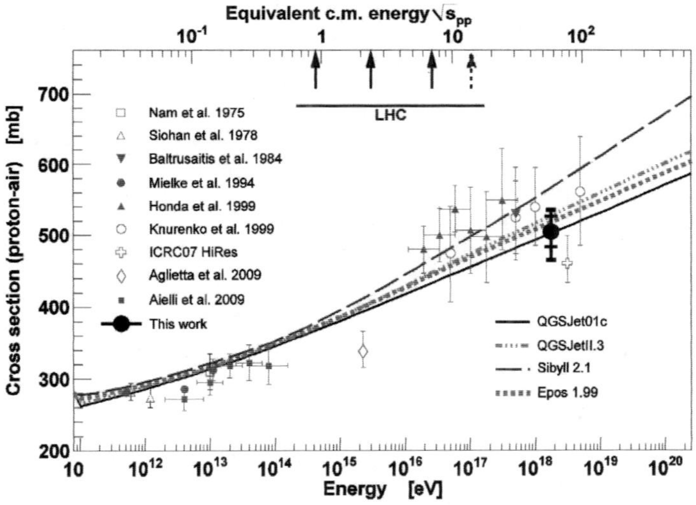

Figure 6. Inelastic proton-air cross section at $\sqrt{s} = 57$ TeV, obtained from shower longitudinal profile in Auger Observatory, together with other measurements. The predictions of hadronic models to the cross section are also shown [7].

The multiplicy of muons in a air shower is related to the type of primary particle and can also give indications of changes in the hadronic processes themselves. The Auger Collaboration has reported the observation of muon multiplicity in air showers that exceed the values predicted by hadronic interaction models [19, 20].

8. Conclusions

After the completion of the Observatory in 2008, several enhancements to the original project were added, aiming to better characterize air showers. They involve radio detection, molecular Bremsstrahlung detection, improvements in muon detection, enhancements in atmospheric monitoring, improvements in analysis procedures and others. With these enhancements, the energy spectrum will be studied with a 10^{17} eV threshold in primary energy.

The Pierre Auger Observatory has contributions to cosmic ray physics in the aspects concerning energy spectrum, photon and neutrino fluxes, hadronic interactions and others. We should stress the precision of the measurements - it is remarkable, compared to the precision of other experiments at these extremely high energies. Other outstanding features of the Observatory are, already, the large event library and also the implementation of new methods in data analysis.

Acknowledgments

The author thanks the organizers of the 18th International Symposium on Particles, Strings and Cosmology (PASCOS2012) for the oportunity of presenting recent results from the Pierre Auger Observatory. Also, the author thanks the Fundação de Amparo à Pesquisa do Estado de São Paulo (FAPESP), Brasil, for the financial support to participate in the Symposium.

References

[1] Abraham J *et al* (Auger Collaboration), 2004 Properties and Performance of the prototype instrument for the Pierre Auger Observatory *Nucl. Instr. Methods* A **523** 50-95.
[2] Baltrusaitis R M *et al* 1985 *Nucl.Instr.Meth.A* **240** 410.
[3] Abu-Zayyad T *et al* 2000 *Nucl.Instr. Meth.A* **450** 253.
[4] Kawai H *et al* 2008 *Nucl. Phys. Proc. Suppl* **175176** 221.
[5] Hillas A M 1984 *Ann. Rev. Astron. Astrophys.* **22** 425.
[6] Facal P *et al* (Pierre Auger Collaboration) The distribution of shower maxima of UHECR air showers 2011 *Proc. 32nd International Cosmic Ray Conference, Beijing* arXiv:1107.4804v1 p. 1-4
[7] Abreu P *et al* (Auger Collaboration) 2012 Measurement of the proton-air cross-section at sqrt(s) = 57 TeV with the Pierre Auger Observatory *Phys. Rev. Lett* **109** 062002
[8] Bonifazi C *et al* (Pierre Auger Collaboration) 2005 *Proceedings of the 29th International Cosmic Ray Conference* **00** 101-104, arXiv-0901.3138v1.
[9] Apel W D *et al* (Kascade-Grande Collaboration) 2010 The KASCADE-Grande experiment *Nucl. Instr. Methods A* **620** 202216
[10] Apel W D *et al* (Kascade-Grande Collaboration) 2011 Muon production height studies with the air shower experiment KASCADE-Grande *Astrop. Phys.* **34** 476485
[11] Halzen F and Hooper D, 2002 High energy neutrino astronomy: the cosmic ray connection , *Rep. Prog. Phys.* **65** 10251078.
[12] Becker J 2008 High-energy neutrinos in the context of multimessenger astrophysics *Phys. Rep* **458** 173-246.
[13] Bhattacharjee P and Sigl G 2002 Origin and propagation of extremely high-energy cosmic rays *Phys. Rep.* **327** 109-247.
[14] Settimo M *et al* (Pierre Auger Collaboration) 2012 Measurement of the cosmic ray spectrum using hybrid events of the Pierre Auger Observatory *Eur. Phys. J. Plus* **127** 87
[15] Guardincerri Y *et al* (Pierre Auger Collaboration) The Pierre Auger Observatory and ultra-high energy neutrinos: upper limits to the diffuse and point source fluxes 2011 *Proc. 32nd International Cosmic Ray Conference, Beijing* arXiv:1107.4805v1 p. 21-24.
[16] Abreu P *et al* (Pierre Auger Collaboration) 2012 Search for point-like sources of ultra-high energy neutrinos at the Pierre Auger Observatory and improved limit on the diffuse flux of tau neutrinos *Astrop. J. Letters*
[17] Ostapchenko, S 2012 Studies of Very High Energy Cosmic Rays: Status, Puzzles, and the Impact of LHC Data *Suppl. Prog. Theor. Phys* **193** 204-211.
[18] D'Enterria D *et al* 2012 The Strong Interaction at the Collider and Cosmic-Rays Frontiers *Few-Body Syst.* 53:173179.
[19] Engel R *et al* (Pierre Auger Collaboration) 2009 Test of hadronic interaction models with data from the Pierre Auger Observatory *Proceedings of the 30th International Cosmic Ray Conference* arXiv-0706.1921v1 1.
[20] Nellen L *et al* The observation of a muon deficit in simulations from data of the Pierre Auger Observatory *Proceedings for the 23rd European Cosmic Ray Symposium (ECRS 2012)* , to be published.

Cosmic-ray and gamma-ray constraints on dark matter stability

David Tran

School of Physics and Astronomy, University of Minnesota
116 Church Street SE, Minneapolis, MN 55455, USA

E-mail: tran@physics.umn.edu

Abstract. We examine different constraints on dark matter stability from cosmic-ray and gamma-ray observations and their complementarity through higher-order effects. Two-body and three-body decays of dark matter particles into charged leptons and quarks generically induce decays into monochromatic photons at the quantum level, giving rise to distinct signatures. We also present a general model-independent analysis of hadronic constraints in the mass-lifetime parameter space and compare those constraints to current and projected limits on gamma-ray lines. Furthermore, we discuss how the production of monochromatic photons can be enhanced by kinematic effects, potentially giving rise to observable lines in the gamma-ray sky.

1. Introduction

The existence of a substantial quantity of dark matter in the Universe is now established beyond reasonable doubt. Nevertheless, the physical identity of the dark matter remains mysterious. While we do know that the dark matter cannot be composed of Standard Model particles, little is known about the microscopic properties of the particles that the dark matter is made of. Among the unknown properties of the dark matter particles is their lifetime. From the observation of a population of primordial relic dark matter particles we can assume that the dark matter lifetime at the very least exceeds the age of the Universe. It is commonly assumed that the dark matter particles have infinite lifetime owing to to the effects of some stabilizing symmetry that prohibits the decay of dark matter into Standard Model or hidden-sector particles. Nevertheless, one should keep in mind that perfect dark matter stability is an unproved assumption. If the dark matter decays into Standard Model particles, one can derive from observations of cosmic-rays, photons and neutrinos much more stringent constraints on the partial decay widths, which typically exceed the age of the Universe by a factor of about 10^9. We examine two different types of constraints on unstable dark matter here, namely the hadronic constraints from measurements of the cosmic-ray antiproton spectrum as well as constraints from searches for gamma-ray lines, which may be generated at the loop-level by dark matter decay processes involving charged particles.

2. Gamma-Ray Lines from Radiative Dark Matter Decay

The observation of anomalous abundances of cosmic-ray positrons and electrons at high energies by the PAMELA [1] and Fermi LAT [2] cosmic-ray observatories has motivated many authors to study models of dark matter which self-annihilates or decays primarily or exclusively into leptons,

Content from this work may be used under the terms of the Creative Commons Attribution 3.0 licence. Any further distribution of this work must maintain attribution to the author(s) and the title of the work, journal citation and DOI.
Published under licence by IOP Publishing Ltd

thus accounting for the excess high-energy electrons without simultaneously overproducing antiprotons. We investigate here a way to test such "leptophilic" models by examining some next-to-leading order effects induced by loop diagrams. To this end, we study a toy model of unstable dark matter which at leading order decays only into charged lepton–antilepton pairs (for scalar dark matter $\phi_{\rm DM}$) or into three-body final states including lepton–antilepton pairs plus neutral fermions N (due to angular momentum conservation in the case of fermionic dark matter $\psi_{\rm DM}$). One such diagram contributing to the leptonic decay of a dark matter fermion is shown in Fig. 1.

The external charged lepton lines can be closed into a loop from which a photon is radiated. In this way, a three-body decay can induce a radiative two-body decay at the quantum level. One diagram contributing to the one-loop decay process is shown in Fig. 2. We assume here that the decay is mediated either by a charged scalar Σ or a charged vector particle V, both of which are assumed to be much heavier than the dark matter particle so that these particles cannot be produced on-shell.

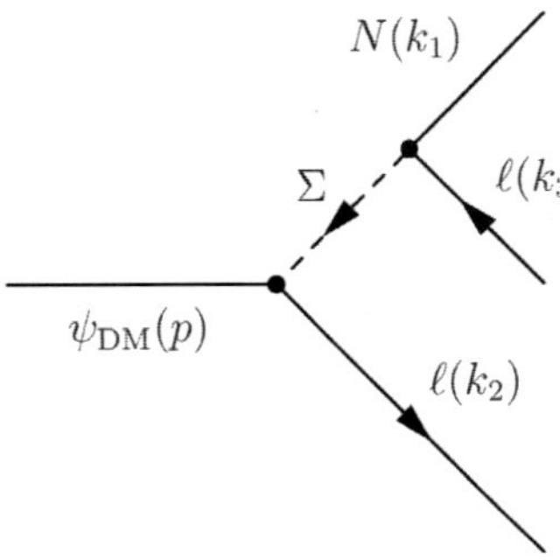

Figure 1. Tree-level diagram contributing to the three-body decay $\psi_{\rm DM} \to \ell^+\ell^-\nu$ of fermionic dark matter via a virtual scalar Σ.

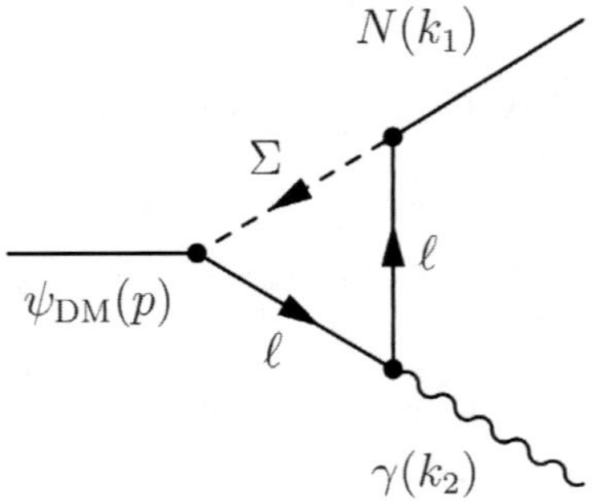

Figure 2. One-loop diagram contributing to the radiative two-body decay $\psi_{\rm DM} \to \gamma\nu$.

This subdominant two-body decay process then generates photons at a fixed energy

$$E_\gamma = \frac{m_{\psi_{\rm DM}}}{2}\left(1 - \frac{m_N^2}{m_{\psi_{\rm DM}}^2}\right), \tag{1}$$

or simply half the dark matter mass if the neutral particle N is massless.[1] For dark matter particles of roughly electroweak-scale mass these photons could conceivably show up as narrow lines in the diffuse Galactic gamma-ray emission or in the gamma-ray flux from certain astrophysical objects. Such gamma-ray lines are of great interest due to the fact that they cannot be produced by ordinary astrophysical processes. Therefore, the identification of a line in the gamma-ray spectrum would constitute a compelling signature of an underlying particle physics process.

General expressions for the tree-level and radiative decay modes of scalar and fermionic dark matter decaying leptonically via virtual scalar or vector particles can be found in [3]. For scalar dark matter we generally find that the decays into two photons are strongly chirality-suppressed by the hierarchy between lepton and dark matter masses, making the resulting lines

[1] We use the letter ν to denote a general massless neutral fermion.

Journal of Physics: Conference Series **485** (2014) 012034 doi:10.1088/1742-6596/485/1/012034

unobservable in practice. For fermionic dark matter $\psi_{\rm DM}$, the prospects of detection are more promising. It is interesting to study special cases in which the expressions for the branching ratios assume relatively simple forms. For instance, if the decay is mediated by a virtual scalar and the couplings of the dark matter are chiral, the ratio between the radiative decay $\psi_{\rm DM} \to \gamma\nu$ and the tree-level three-body decay $\psi_{\rm DM} \to \ell^+\ell^-\nu$ can be expressed in a very simple manner,

$$\frac{\Gamma(\psi_{\rm DM} \to \gamma\nu)}{\Gamma(\psi_{\rm DM} \to \ell^+\ell^-\nu)} = \frac{3\alpha_{\rm em}}{8\pi} \times R \times S, \tag{2}$$

where R and S encode the details of the various coupling constants and the ratio between the two-body and three-body kinematic factors, respectively. For generic couplings, the corresponding factor R as well as the kinematic factor S are of order one. Therefore, the relative intensity of the radiative decay mode compared to the dominant tree-level decay is determined by the numerical prefactor $3\alpha_{\rm em}/8\pi \approx 10^{-3}$. We show our results for the gamma-ray line bounds induced by the leading-order flavor-democratic leptonic decay $\psi_{\rm DM} \to \ell^+\ell^-\nu$ in the case of chiral couplings in Figs. 3 and 4 for the decay mediated by a scalar and a vector. The relative intensity of the gamma-ray line is stronger by about an order of magnitude in the case of a vector-mediated decay. We show these bounds together with the region of the parameter space that yields an explanation of the cosmic-ray anomalies observed by PAMELA and Fermi LAT, shown here in yellow and gray, respectively, indicating the 3σ and 5σ regions around the best-fit points [4]. The observational constraints shown in the plot are from Fermi LAT line searches in the Galactic halo at lower energies [5], as well as from MAGIC observations of the Perseus galaxy cluster [6], HEGRA observations of M31 [7], as well as H.E.S.S. measurements of the diffuse electron flux [8], which provides an upper limit on the gamma-ray flux. We also show our projected limits for the future Cherenkov Telescope Array (CTA). Interestingly enough, the existing Fermi LAT bounds at lower energies and the projected CTA limits at higher energies can constrain a significant portion of the parameter space relevant to dark matter interpretations of the anomalous cosmic-ray lepton abundances.

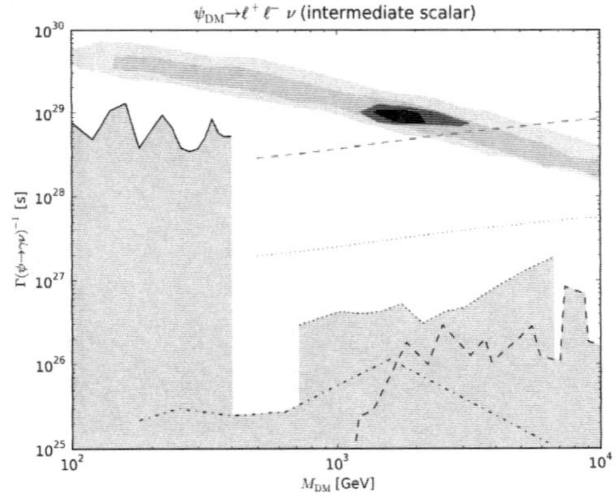

Figure 3. Gamma-ray line bounds on the radiative decay $\psi_{\rm DM} \to \gamma\nu$ induced by the tree-level decay $\psi_{\rm DM} \to \ell^+\ell^-\nu$, mediated by a virtual scalar particle.

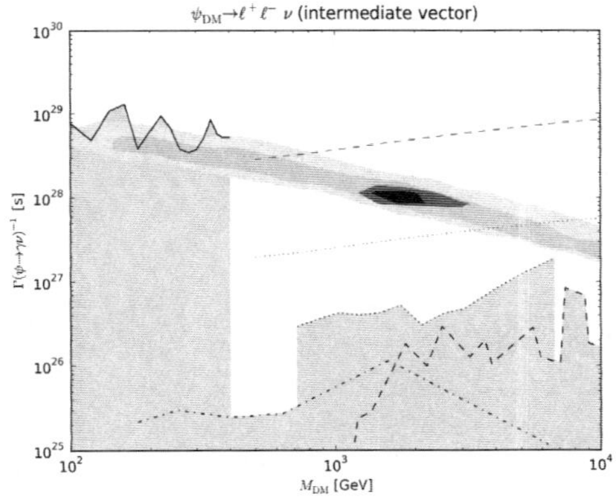

Figure 4. Gamma-ray line bounds on the radiative decay $\psi_{\rm DM} \to \gamma\nu$ induced by the tree-level decay $\psi_{\rm DM} \to \ell^+\ell^-\nu$, mediated by a virtual vector particle.

It is worth noting that if the neutral fermion N and the dark matter particle $\psi_{\rm DM}$ have opposite *CP* parities the radiative decay mode can be enhanced due to a kinematic factor

$$\mathrm{BR}(\psi_{\mathrm{DM}} \to \gamma\nu) \propto \left(1 - \frac{m_N}{m_{\mathrm{DM}}}\right)^{-2} \tag{3}$$

in the branching ratio into monochromatic photons. Therefore, as $m_N \to m_{\psi_{\mathrm{DM}}}$ the radiative branching ratio can grow dramatically. Using this kinematic enhancement factor, one can construct scenarios where the constraints from the non-observation of gamma-ray lines are stronger than the ones from leptonic cosmic rays. We show a number of benchmark points representating different scenarios in Fig. 5. Depending on the relative masses of ψ_{DM} and N the benchmark points shown in the plot can lie within the allowed or excluded regions of the parameter space.

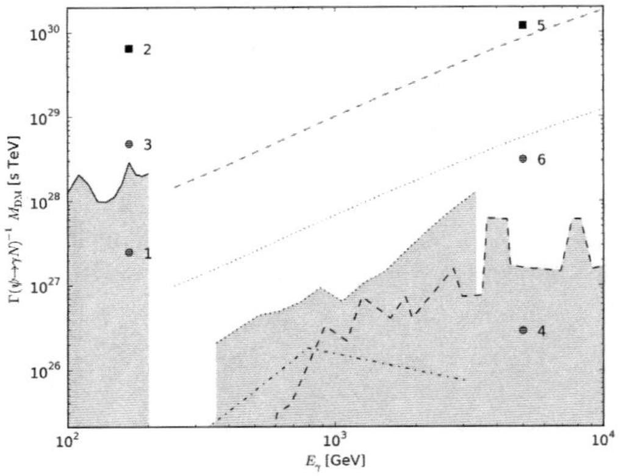

Figure 5. Benchmark points for the decay into $\psi_{\mathrm{DM}} \to \gamma N$ with a massive neutral fermion N and opposite *CP* parities of ψ_{DM} and N. Some benchmark points are excluded, allowed or borderline excluded. The constraints are rescaled compared to Figs. 3 and 4 due to the reduced gamma-ray energy available due to the non-zero mass of N.

3. Hadronic Constraints

If the dark matter decays into states involving weak gauge bosons, Higgs bosons or quarks, these particles will hadronize and generate antiprotons, which constitute a sensitive probe for the presence of primaries due to their scarcity in the cosmic radiation. The PAMELA cosmic-ray telescope has observed no deviations from the theoretically expected abundance of cosmic antiprotons if these particles are produced as secondary cosmic rays by spallation on interstellar gas. The absence of excess cosmic antiprotons over the predicted astrophysical flux from cosmic-ray spallation on interstellar gas can therefore be used to impose stringent constraints on hadronic dark matter decay modes. We utilize the recent PAMELA measurements of the antiproton-to-proton flux ratio [9] to derive limits on dark matter stability in various hadronic decay modes in a model-independent manner.

We model the propagation of charged cosmic rays in the Milky Way's magnetic halo using a stationary two-zone diffusion model with cylindrical boundary conditions [10]. This model takes into account the effects of diffusion, energy loss, convection and annihilation on interstellar gas.

Under certain assumptions, the resulting differential equation can be solved semi-analytically, with the solution depending on only a small number of parameters to be determined from cosmic-ray measurements. Using these solutions we can inject a source spectrum of antiprotons at a given rate and a spatial distribution as determined by N-body simulations. The determination of the free parameters of the transport model from cosmic-ray measurements suffers from a degeneracy between the size of the diffusion coefficient and the height of the diffusive halo. This degeneracy cannot be broken by measurements of stable secondary-to-primary cosmic-ray species which are only produced in the Galactic disk. However, since antiprotons from dark matter decay/annihilation are produced everywhere in the dark matter halo and not just in the disk, the flux of these primary antiprotons is highly sensitive to different configurations of the propagation parameters. The resulting primary antiproton fluxes can vary by as much as two orders of magnitude. Therefore, we show our constraints for the three different sets of transport parameters which correspond to minimum, medium and maximum antiproton flux.

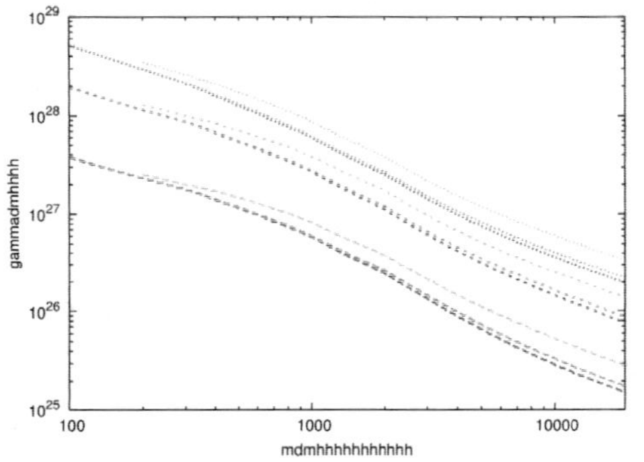

Figure 6. Lower bounds on the partial dark matter lifetime for fermionic dark matter decaying into weak gauge bosons or Higgs bosons. Red: $\psi_{\rm DM} \to W^{\pm}\ell^{\mp}$, orange: $\psi_{\rm DM} \to Z^0\nu$, green: $\psi_{\rm DM} \to h^0\nu$.

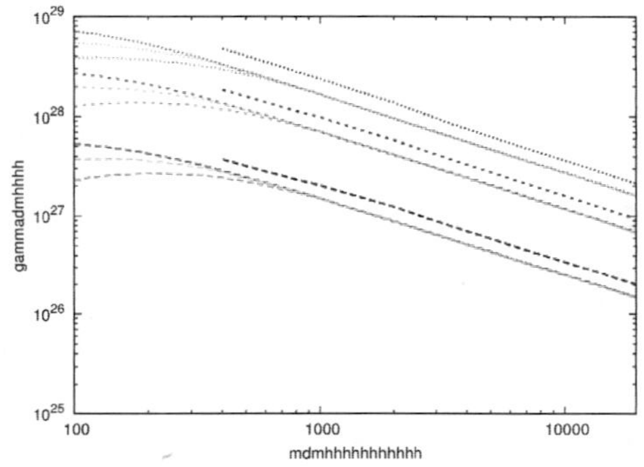

Figure 7. Lower bounds on the partial dark matter lifetime for fermionic dark matter decaying into quark–antiquark pairs and a neutral fermion. Red: $\psi_{\rm DM} \to d\bar{d}\nu$, green: $\psi_{\rm DM} \to c\bar{c}\nu$, orange: $\psi_{\rm DM} \to s\bar{s}\nu$, blue: $\psi_{\rm DM} \to t\bar{t}\nu$.

We have computed exclusion limits on various hadronic decay modes of dark matter from the requirement of not exceeding the antiproton-to-proton ratio observed by PAMELA at 95% C.L. To this end, we scanned over several orders of magnitude in the dark matter mass and lifetime. The resulting constraints on a wide range of general hadronic dark matter decay modes allowed by charge and angular momentum conservation are shown in [11]. As an example, we show in Figs. 6 and 7 the constraints on the mass–lifetime parameter space we find for the decay of a spin-1/2 dark matter particle into weak gauge bosons/Higgs bosons and into quark–antiquark pairs plus neutral fermions, respectively. Different line styles correspond to different sets of transport parameters as explained above.

Going beyond the cosmic-ray constraints, we can examine the radiative decays into monoenergetic photons in analogy to the previous section, where the photons are radiated from quark loops instead of lepton loops here. It is interesting to see how the constraints from gamma-ray line searches compare to the constraints on hadronic cosmic rays. We show our results for the cosmic-ray limits on the decay of fermionic dark matter via $\psi_{\rm DM} \to d\bar{d}\nu$ in Figs. 8 and 9 together with the gamma-ray line constraints from Fermi LAT and the projected limits for the CTA mentioned in the previous section (limits on the signal + astrophysical background flux are shown in thick lines, whereas limits on the dark matter signal alone are shown in thin lines). While the antiproton constraints are stronger as expected, the complementary constraints from the radiative next-to-leading order decays provide a model-independent alternative view of the highly model-dependent hadronic constraints. The same kinematic enhancement effect that we mentioned in the previous section can also lead to a significant relative enhancement of the strength of the radiative constraints.

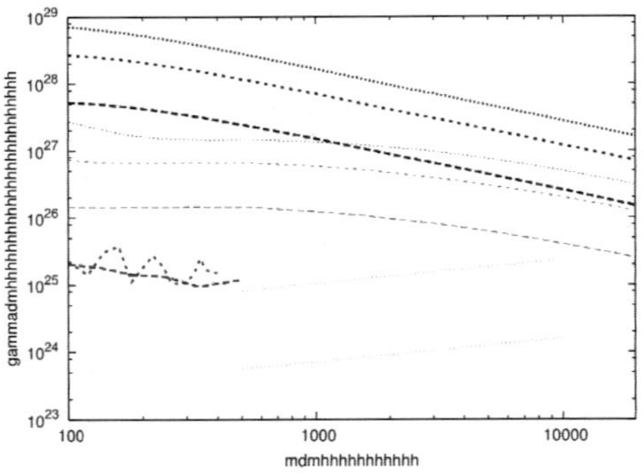

Figure 8. Comparison between hadronic constraints for the decay $\psi_{\rm DM} \to d\bar{d}\nu$ and gamma-ray line constraints from monochromatic photons radiated off of down-quark loops. Constraints shown for a decay mediated by an intermediate scalar.

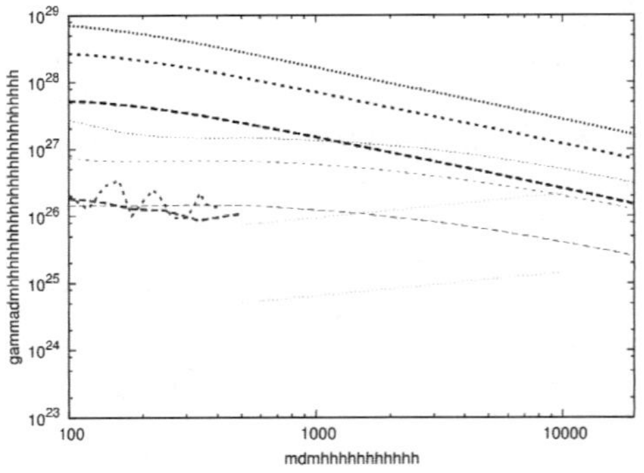

Figure 9. Comparison between hadronic constraints for the decay $\psi_{\rm DM} \to d\bar{d}\nu$ and gamma-ray line constraints from monochromatic photons radiated off of down-quark loops. Constraints shown for a decay mediated by an intermediate vector.

4. Conclusions

We have presented general constraints on dark matter stability from observations of gamma rays and cosmic-ray antiprotons. Using cosmic-ray antiproton measurements we have derived general constraints on hadronic dark matter decay modes and explored the complementarity with gamma-ray line constraints from subdominant quantum-level processes which can arise in any scenario where the dark matter decays into charged particles. We have explained that such processes can be relevant despite a general suppression by small couplings and loop factors compared to the leading-order decays.

References

[1] Adriani O *et al.* (PAMELA Collaboration) 2009 *Nature* **458** 607 (*Preprint* 0810.4995 (astro-ph])
[2] Ackermann M *et al.* (Fermi LAT Collaboration] 2010 *Phys. Rev.* **D82** 092004 (*Preprint* 1008.3999 [astro-ph.HE]).
[3] Garny M, Ibarra A, Tran D and Weniger C 2011 *JCAP* **1101** 032 (*Preprint* 1011.3786 [hep-ph])
[4] Ibarra A, Tran D and Weniger C 2010 *JCAP* **1001** 009 (*Preprint* 0906.1571 [hep-ph])
[5] Abdo A A, Ackermann M, Ajello M, Atwood W B, Baldini L, Ballet J, Barbiellini G and Bastieri D *et al.* 2010 *Phys. Rev. Lett.* **104** 091302 (*Preprint* 1001.4836 [astro-ph.HE])
[6] Aleksic J *et al.* (MAGIC Collaboration) 2010 *Astrophys. J.* **710** 634 (*Preprint* 0909.3267 [astro-ph.HE])
[7] Aharonian F A *et al.* (HEGRA Collaboration) 2003 *Astron. Astrophys.* **400** 153 (*Preprint* astro-ph/0302347).
[8] Aharonian F A *et al.* (H.E.S.S. Collaboration) 2008 *Phys. Rev. Lett.* **101** 261104 (*Preprint* 0811.3894 [astro-ph])
[9] Adriani O *et al.* (PAMELA Collaboration) 2010 *Phys. Rev. Lett.* **105** 121101 (*Preprint* 1007.0821 [astro-ph.HE])
[10] *See, for example,* Berezinskii V S, Buolanov S V, Dogiel V A, Ginzburg V L and Ptuskin V S 1990 "Astrophysics of Cosmic Rays" (Amsterdam: North–Holland)
[11] Garny M , Ibarra A and Tran D 2012 *JCAP* **1208** 025 (*Preprint* 1205.6783 [hep-ph])

The axion-photon interaction and gamma ray signals of dark matter

J. Barranco[1], A. Carrillo Monteverde[2] and D. Delepine[3]

División de Ciencias e Ingenierías, Universidad de Guanajuato, Campus León, C.P. 37150, León, Guanajuato, México.

E-mail:
[1]jbarranc@fisica.ugto.mx,[2]alcarrillo@fisica.ugto.mx,[3]delepine@fisica.ugto.mx

Abstract. We explore two scenarios where the axion-photon interaction could induce additional astrophysical gamma ray signals for the dark matter. In the first scenario, dark compact objects made of axions, named axion stars, could collide with neutron stars. The whole energy of the axion star can be dissipated in the magnetized conducting medium of the neutron star generating gamma rays. The second scenario is an indirect method for observing self-annihilating dark matter trapped in stars: Gamma rays produced by the self-annihilation of neutralinos in the interior of the Sun can be transformed into axions due to photon-axion conversion. Then, the axion will travel freely in the Sun and be converted into photons again. This process is often referred as 'shine light through walls', in this case, the wall will be the solar interior. Hence, GeV gamma rays might pass through the Sun. We may conclude that observation of GeV photons by gamma-ray observatories like HAWC, coming from the Sun, may be a signal of annihilation of neutralinos in the interior of the Sun.

1. Introduction

Last decade had witnessed the confirmation of the double dark universe: a universe where 23% of its total energy density is made of some collisionless, gravitationally attractive matter of unknown nature, named dark matter [1], plus another 74% made of a gravitationally 'repulsive' dark energy [2]. This double dark universe is complemented with a 3% of the total energy density made of the 'standard' matter that is well described by the Standard Model of Elementary Particles (SM). Now, one of the most interesting question turns out to be: what is the nature of dark matter? Although it is possible that a modification of the theory that describe the gravitation (i.e. General Relativity) could be the explanation to the dark matter problem, a more plausible explanation for the dark matter comes from the particle point of view: a weakly interactive massive particle (WIMP) arises as the most natural candidate for dark matter. Nevertheless, to say a WIMP is as generic as to say a dark matter particle. From the point of view of a particle's physicist, the WIMP could be any hypothetical particle that has a small or negligible interaction with the rest of the particles of the SM. In other words, a WIMP is a particle that is rather a cold, mostly stable and neutral particle that matches the appropriated relic density [3]. The beauty of many of the theories beyond the SM is that they naturally provide such type of WIMP candidates. Let us take as examples the axion or the lightest supersymmetric particle. The axion arises as a solution to the strong CP problem in QCD. It is the pseudo Nambu-Goldstone boson that results form the spontaneous breaking of

Content from this work may be used under the terms of the Creative Commons Attribution 3.0 licence. Any further distribution of this work must maintain attribution to the author(s) and the title of the work, journal citation and DOI.
Published under licence by IOP Publishing Ltd

the $U(1)$ Peccei-Quinn symmetry. The Peccei-Quinn symmetry was introduced to explain the apparent smallness of the strong CP violation [4]. Axion properties have been constrained from astrophysical and cosmological considerations. In particular, its mass is given in terms of the axion decay constant f_a as

$$m_a \simeq 6\mu\text{eV}\left(\frac{10^{12}\text{GeV}}{f_a}\right). \tag{1}$$

This constant f_a is related with the magnitude of the vacuum expectation value that breaks the Peccei-Quinn symmetry. Current bounds on the axion decay constant are $10^{10}\text{GeV} < f_a < 10^{12}\text{GeV}$ [5] All remaining axion couplings are inversely proportional to f_a. In particular, an axion-photon coupling is possible and it will be described by the following Lagrangian:

$$\mathcal{L} = \frac{1}{2}(\partial^\mu a \partial_\mu a - m^2 a^2) - \frac{1}{4}\frac{ca}{f_a}F_{\mu\nu}\tilde{F}^{\mu\nu} - \frac{1}{4}F_{\mu\nu}F^{\mu\nu}, \tag{2}$$

where c is a coupling constant of order unity. Concerning the lightest supersymmetric particle, in the Minimal Supersymmetric Standard Model (MSSM), the superpartners of the photon, the Z-boson and the neutral Higgs bosons are called binos ($\tilde{B}$), winos ($\tilde{W}_3$) and higgsinos ($\tilde{H}_1^0$ and $\tilde{H}_2^0$). These states mix into fermionic, majorana mass eigen-states called *neutralinos*. Those are typically labeled as $\tilde{\chi}_i^0$, $i = 1..4$, ordered with respect to their mass. The lightest of the four will be refereed as *the neutralino* and denoted simply by χ.

Both the axion and the neutralino are excellent WIMP candidates and there are current experiments looking for their direct and indirect detection signals. In the present work we sketch two scenarios where the axion-photon interaction shown in Lagrangian (2) could provide a new source of gamma rays due to the conversion of axions into photons. Hence, offering new indirect ways for dark matter detection. In the first scenario, dark compact objects made of axions, named axion stars (AS) [6], could collide with neutron stars. The whole energy of the axion star can be dissipated in the magnetized conducting medium of the neutron star generating gamma rays. A similar idea was originally proposed by Iwazaki [7] as a mechanism to generate gamma-ray-burst. Now we incorporate this idea as an indirect method for detection of axion as a dark matter candidate. The second scenario is an indirect method for observing self-annihilating dark matter trapped in stars. The usual mechanism incorporates the idea that a self-annihilating dark matters, as the case of the neutralinos, can be accumulated in the interior of the stars. The increase of the dark matter density increases the probability that two neutralinos annihilate each other into another SM particles. In particular, neutrinos or photons are possible final states of this annihilation. Since νs have a very small cross section, they can scape from the star's interior while the photons are trapped due to the strong interaction of the photons with the star's matter. The non-observation of high energy neutrinos coming from the Sun has been used to set constraints on the neutralino-nucleon cross section by the Super-Kamiokande collaboration [8]. Gamma rays produced by the self-annihilation of neutralinos in the interior of the Sun are not usually considered since photons will be trapped. But Lagrangian (2) imply the possibility that a photon can be transformed into an axion when an external magnetic field is present. Since the axion has a negligible interaction with the other SM particles, axions will travel freely in the Sun and in some cases they can be converted into photons again. This process is often referred as 'shine light through walls', in this case, the wall will be the solar interior. Hence, GeV gamma rays might pass through the Sun. We may conclude that observation of GeV gamma rays from the Sun may be a signal of annihilation of neutralinos in the interior of the Sun. In [9] it was shown that for certain range of axion parameters, the Sun is transparent to high energy photons coming form distance sources, hence, the previous mechanism for detecting neutralino annihilation into photons could be possible.

In the following sections we will complement the previous ideas with some numerical

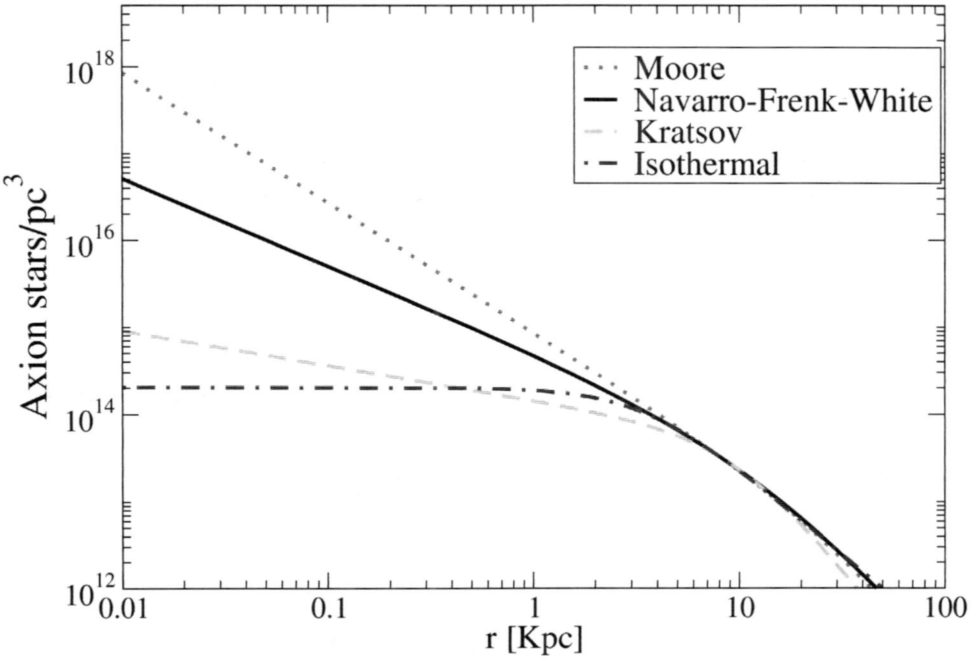

Figure 1. Number of axion stars for different dark matter profiles.

calculations in order to have an estimation on the strength of the effects. A more detailed analysis is under way and will be published elsewhere.

2. Galactic halo made of axion stars

The possibility that the galactic dark matter halo is a collisionless ensemble of scalar field mini-MACHOs is not in contradiction with the limits imposed by microlensing or gravothermal instability [10]. In fact, microlensing observations limits the dark matter mini-MACHOS have masses smaller that $10^{-7}M_\odot$, $M_\odot$ the Solar Mass. An interesting result comes from the fact that self-gravitating objects made of axions, that we will call axion-stars (AS), have the masses and radii needed to fulfill those requirements [6]. Actually, axion star masses are far from those limits. A typical AS will have a mass of $\sim 10^{-15}M_\odot$. We can estimate the number of AS in a galactic halo that follows a dark matter density profile. In Fig. 1 it is shown the expected number of axion star with a mass of $10^{-15}M_\odot$ for four different DM density profiles. The number of axion stars can be very high at the center of the galaxy. It is expected that a typical galaxy has around $\sim .1\%$ of its total stars to be neutron stars. A galaxy like the Milky-Way, there should be $\sim 10^9$ neutron stars. Both, the high number of axion stars and neutron stars, implies a high probability that AS will collide with a neutron star and dissipate their energy under the effect of the strong magnetic field of the NS. As a result, the gravitational energy of the AS is radiated away in form of low energy photons. In order to estimate the amount of energy radiated, we start with the Lagrangian (2) and express in terms of the electric and magnetic fields $\vec{E}, \vec{B}$.

$$\mathcal{L}_{a\gamma\gamma} = \frac{c\alpha}{f_a} a\vec{E}\cdot\vec{B}\,. \tag{3}$$

Then, it is possible to derive a "modified" Gauss law:

$$\partial \vec{E} = \frac{-c\alpha}{f_a} \vec{\partial} \cdot (a\vec{B}) \,. \tag{4}$$

Then, the axion field a induce a electric field $\vec{E}_a = -c\alpha a\vec{B}/f_a$, which induces an electric current $J_m = \sigma E_a$ in a magnetized conducting media with electric conductivity σ. The dissipation in the magnetized conducting media, with average σ electric conductivity can be evaluated using the Ohm's law:

$$W = \int_{AS} \sigma E_a^2 d^3x = 4\pi \frac{c^2\alpha^2 B^2 \sigma}{f_a^2} \int_0^R a(r)^2 r^2 dr \,. \tag{5}$$

Let us for the moment assume that the AS is given as in [7], and just update the values of the mass of the AS and the radius as was recently obtained in [6]. The axion star profile can be approximated as

$$a = f_a a_0 exp(-r/R_a) \,, \quad \text{with} \quad a_0 = 1.73 \times 10^6 \left(\frac{10cm}{R_a}\right)^2 \left(\frac{10^{-5}\text{eV}}{m_a}\right) . \tag{6}$$

According to [6], for $m_a = 10^{-5}$ eV, $R_a = 8.54$ meters and its mass is $M = 10^{-16} M_\odot$. Inserting this values in eq. 5, we obtain that such axion star will dissipate its whole energy

$$W \sim 10^{42} \left(\frac{\sigma}{10^{26}/s}\right) \left(\frac{M}{10^{-16} M_\odot}\right) \left(\frac{B}{10^8\text{Gauss}}\right)^2 \text{ergs/sec} \,. \tag{7}$$

That is, the energy dissipated is around $W \sim 10^9 L_\odot$, with $L_\odot$ the solar luminosity. In case there is such type of collisions, the luminous could equal the total galaxy luminosity.

3. Photon-axion conversion inside the Sun

Indirect methods for the detection of dark matter mostly consist in observing the radiation produced in dark matter annihilation. The intensity of such radiation is proportional to the annihilation rate, which depends on the dark matter density. Hence, the best places to look are the regions where dark matter density accumulates. An excellent region for such accumulation is the center of the Sun. It is commonly believed that only neutrinos can escape these dense objects. Nevertheless, here we will shown that neutralino annihilation into photons can be detected trough the photon-axion conversion in the Sun's interior. The channel we are proposing is $\chi + \chi \to \gamma$'s $\to$ 'oscillation' $\to$ *axions* $\to$ 'oscillation' $\to \gamma$'s.

The photon-axion oscillation probability is obtained by starting with the Lagrangian (2). Let us expand the photon field $A(t, x)$ in components of fixed frequency $A(x)e^{-i\omega t}$. We may assume that the variations of the magnetic field in space occurs on larger scales than the photon/axion wave-lengths. Therefore, for particles propagating in the z-axis, the operator in the equation of motion $\omega^2 + \partial_z^2$ can be expanded as $\omega^2 + \partial_z^2 \to 2\omega(\omega - i\partial_z)$ and then, the Schrödinger equation will be

$$i\partial_z \Psi = -(\omega + \mathcal{M})\Psi \,; \quad \Psi = (A_x, A_y, a) \,, \tag{8}$$

where we have defined the matrix $\mathcal{M}$ as:

$$\mathcal{M} = \begin{pmatrix} \Delta_p & 0 & \Delta_{M_x} \\ 0 & \Delta_p & \Delta_{M_y} \\ \Delta_{M_x} & \Delta_{M_y} & \Delta_m \end{pmatrix} . \tag{9}$$

and we have introduced the quantities:

$$\begin{aligned}\Delta_{M_i} &= \frac{B_i}{2f_a} = 1.755 \times 10^{-11}\left(\frac{B_i}{1\text{G}}\right)\left(\frac{10^5\text{GeV}}{f_a}\right)\text{cm}^{-1},\\ \Delta_m &= \frac{m^2}{2\omega} = 2.534 \times 10^{-11}\left(\frac{m}{10^{-3}\text{eV}}\right)\left(\frac{1\text{GeV}}{\omega}\right)\text{cm}^{-1},\\ \Delta_p &= \frac{\omega_p^2}{2\omega} = 3.494 \times 10^{-11}\left(\frac{n_e}{10^{15}\text{cm}^{-3}}\right)\left(\frac{1\text{GeV}}{\omega}\right)\text{cm}^{-1},\end{aligned} \tag{10}$$

where ω_p^2 is the plasma frequency which is given in terms of the electron density n_e as $\omega_p^2 = 4\pi\alpha n_e/m_e$. B is the magnetic field, m_e is the electron mass and α is the fine structure constant and ω is the photon's energy. The Schrödinger equation (8) is solve numerically, although it can be solved for the case of constant magnetic field B and constant electron density n_e. In this case the probability is given by:

$$P = \frac{4B^2\omega^2}{f_a^2(\omega_p^2 - m_a^2)^2 + 4B^2\omega^2}\sin^2\left(\pi\frac{z}{l_{osc}}\right), \tag{11}$$

where we have defined the oscillation length:

$$l_{osc} = \frac{4\pi\omega f_a}{\sqrt{f_a^2(\omega_p^2 - m^2)^2 + 4B^2\omega^2}}.$$

We can estimate from the probability eq. (11) the resonance condition for typical values of the electron density n_e and constant magnetic field B. In particular, we are interested in the photon energy values needed to fulfill the resonance condition for eq. (11). This can be completed if at least one oscillation length fulfill the condition $l_{osc} \simeq R_\odot$. This condition is fulfilled for photon energies given by:

$$\omega^2 < \frac{3.527^2\text{GeV}^2\left(1.38\left(\frac{n_e}{10^{15}\text{cm}^3}\right) - \left(\frac{m_a}{10^{-3}\text{eV}}\right)^2\right)^2}{64\pi^2 - 2.376 \times 10^{-9}\left(\frac{B}{\text{Gauss}}\right)^2\left(\frac{10^{10}\text{GeV}}{f_a}\right)^2}. \tag{12}$$

The electron density at the center of the Sun is as high as $n_e \sim 10^{18}/\text{cm}^3$, and condition eq. (12) gives as resonance energy of the photons to oscillate into axions to be $\omega \sim 340$ GeV, almost independent of the value of the magnetic field B. Hence, a high energy photon produced by the $\chi\chi \to \gamma\gamma$ annihilation reaction can fulfill condition (12), oscillate to axions and converted to photons again. Future Gamma-ray observatories, like HAWC [11], can be sensitive to those high energy photons, since HAWC is developed to perform a high-sensitivity synoptic survey of the sky at wavelengths between 100 GeV and 100 TeV.

4. Conclusions

We have entertained with the possibility that the axion-photon interaction given by Lagrangian (2) could be the source of new astrophysical photons. In the first scenario, collision of axion stars with neutron stars can liberate the axion star energy and to be radiated away. Such collision will liberate a luminosity as high as $10^9 L_\odot$. In the second scenario, photons produced by the neutralino-neutralino annihilation can escape the solar interior trough photon-axion conversion in a 'shine light through walls' mechanism, in this case, the wall will be the solar interior. We have estimated that the photon energy needed to fulfill the resonance condition for such oscillation to occur is around ~ 340 GeV, which for a neutralino that annihilate mono-energetically is possible for a neutralino mass of the same order, which is above the current limit for the neutralino mass.

Acknowledgments: This works has been partially supported by Conacyt Proyecto 156618, Conacyt-SNI and PIFI project of the SEP (Secretaria de Educacion Publica, Mexico) and Guanajuato University through DAIP project.

References

[1] Bertone G, Hooper D and Silk J 2005 *Phys.Rept.* **405** 279–390 (*Preprint* hep-ph/0404175)
[2] Tegmark M *et al.* (SDSS Collaboration) 2004 *Phys.Rev.* **D69** 103501 (*Preprint* astro-ph/0310723)
[3] Taoso M, Bertone G and Masiero A 2008 *JCAP* **0803** 022 (*Preprint* 0711.4996)
[4] Peccei R and Quinn H R 1977 *Phys.Rev.Lett.* **38** 1440–1443
[5] Sikivie P 2008 *Lect.Notes Phys.* **741** 19–50 (*Preprint* astro-ph/0610440)
[6] Barranco J and Bernal A 2011 *Phys.Rev.* **D83** 043525 (*Preprint* 1001.1769)
[7] Iwazaki A 1999 *Phys.Lett.* **B455** 192–196 (*Preprint* astro-ph/9903251)
[8] Desai S *et al.* (Super-Kamiokande Collaboration) 2004 *Phys.Rev.* **D70** 083523 (*Preprint* hep-ex/0404025)
[9] Fairbairn M, Rashba T and Troitsky S V 2007 *Phys.Rev.Lett.* **98** 201801 (*Preprint* astro-ph/0610844)
[10] Hernandez X, Matos T, Sussman R A and Verbin Y 2004 *Phys.Rev.* **D70** 043537 (*Preprint* astro-ph/0407245)
[11] Sinnis G (HAWC Collaboration) 2005 *AIP Conf.Proc.* **745** 234–245

Journal of Physics: Conference Series **485** (2014) 012036 doi:10.1088/1742-6596/485/1/012036

A new idea to search for charged lepton flavor violation using a muonic atom

Joe Sato
Physics Department, Saitama University, 255 Shimo-Okubo, Sakura-ku, Saitama, Saitama 338-8570, Japan

E-mail: joe@phy.saitama-u.ac.jp

Abstract. We propose a new process of $\mu^-e^- \to e^-e^-$ in a muonic atom for a quest of charged lepton flavor violation. The Coulomb attraction from the nucleus in a heavy muonic atom leads to significant enhancement in its rate, compared to $\mu^+e^- \to e^+e^-$. The upper limit of the branching ratio is estimated to be of the orders of $O(10^{-17} - 10^{-18})$ for the photonic and the four Fermi interactions from the present experimental constraints. The search for this process could serve complementarily with the other relevant processes to shed lights upon the nature of charged lepton flavor violation. This talk is based on the work [1].

1. Introduction

Charged lepton flavor violation (cLFV) is known to be one of the important rare processes to search for new physics beyond the Standard Model (SM). Various theoretical models predict sizable rates of cLFV processes, which are just below the present experimental upper limits. The on-going and future experiments for cLFV searches would reach sensitivities in the range of predictions by many theoretical models. At this moment, the cLFV searches with muons present the best limits owing to a large number of muons available for measurements [2]. Typical cLFV processes with muons include $\mu^+ \to e^+\gamma$, $\mu^+ \to e^+e^+e^-$ and μ^- - e^- conversion in a muonic atom ($\mu^-N \to e^-N$). However, after the discovery of cLFV process in future, many other different cLFV processes should be studied to shed lights upon understanding of the nature of the cLFV interactions and develop insights into new physics responsible for cLFV.

In this talk, we would like to propose a new cLFV reaction process of a bound μ^- in a muonic atom, which is

$$\mu^-e^- \to e^-e^-, \tag{1}$$

where μ^- and e^- in the initial state of Eq.(1) are the muon and the atomic 1S electron(s) bound in a Coulomb field of the nucleus in a muonic atom respectively.

This $\mu^-e^- \to e^-e^-$ process in a muonic atom has various significant advantages. First of all, this process could have not only photonic dipole interaction but also four-Fermi contact interaction, as in the processes of $\mu^+ \to e^+e^-e^-$ and $\mu^-N \to e^-N$, whereas $\mu^+ \to e^+\gamma$ has only the former. This would potentially allows us to investigate the full structure of new physics beyond the SM. Secondly, this process has a two-body final state, in which the two signal electrons are emitted almost back to back and each of them has an energy of about a half the muon mass, $m_\mu/2$. This would provide a cleaner experimental signature as well as a larger final-state phase space than $\mu^+ \to e^+e^+e^-$ decay. Also, in comparison with the $\mu^+ \to e^+\gamma$ search,

Content from this work may be used under the terms of the Creative Commons Attribution 3.0 licence. Any further distribution of this work must maintain attribution to the author(s) and the title of the work, journal citation and DOI.
Published under licence by IOP Publishing Ltd

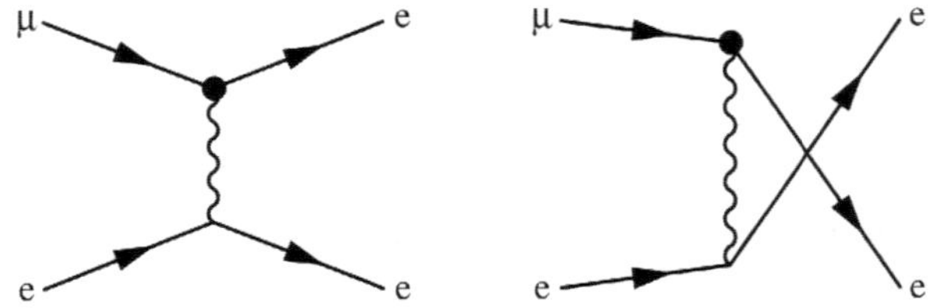

Figure 1. The process $\mu^- e^- \to e^- e^-$ induced from the photonic interactions. The black dot indicates the effective interaction that is absent from the Standard Model.

the measurement of this process would be relatively easier since no photon detection is involved. Thirdly, one can consider a similar reaction process with a muonium, such as $\mu^+ e^- \to e^+ e^-$. However, the rate of this $\mu^+ e^- \to e^+ e^-$ process can not be large because of small overlap between the μ^+ and e^- wave functions. However, in a muonic atom of atomic number Z, we can increase the overlap between the μ^- and e^- wave functions if an atom of large Z is chosen. The enhancement occurs owing to the Coulomb interaction from the nucleus which attracts the 1S state electron wave function towards the μ^- and the nucleus. The expected rate would increase by a factor of $(Z-1)^3$. For example, the rate for a lead ($Z = 82$) is 5×10^5 times that of the $\mu^+ e^- \to e^+ e^-$ reaction. However, in a muonic atom, a nuclear muon capture process occurs in addition to the normal Michel muon decay. But since a lifetime of a muonic atom changes from 2.2 μs to $\sim$ 80 ns from a hydrogen to a lead, the reduction of the number of muons of a factor of at most 20 can only be expected. Therefore, the net increase of the branching ratio would become significant for a large Z. A potential disadvantage is that the rate of the reaction process like this might not be large enough compared to rare cLFV muon decays. Therefore, in this letter we will evaluate the rate of this $\mu^- e^- \to e^- e^-$ process, and discuss its related issues.

2. The process $\mu^- e^- -> e^- + e^-$

We describe the process of $\mu^- e^- \to e^- e^-$ in a muonic atom of Eq. (1) by an effective Lagrangian at the energy scale of the muon mass m_μ. Following Ref. [2], we define

$$\begin{aligned}\mathcal{L}_{\mu^- e^- \to e^- e^-} = &-\frac{4G_F}{\sqrt{2}}\left[m_\mu A_R\, \overline{\mu_R}\sigma^{\mu\nu} e_L F_{\mu\nu} + m_\mu A_L\, \overline{\mu_L}\sigma^{\mu\nu} e_R F_{\mu\nu}\right.\\ &+ g_1(\overline{\mu_R} e_L)(\overline{e_R} e_L) + g_2(\overline{\mu_L} e_R)(\overline{e_L} e_R) + g_3(\overline{\mu_R}\gamma^\mu e_R)(\overline{e_R}\gamma_\mu e_R) + g_4(\overline{\mu_L}\gamma^\mu e_L)(\overline{e_L}\gamma_\mu e_L)\\ &\left. + g_5(\overline{\mu_R}\gamma^\mu e_R)(\overline{e_L}\gamma_\mu e_L) + g_6(\overline{\mu_L}\gamma^\mu e_L)(\overline{e_R}\gamma_\mu e_R) + (\text{H.c.})\right],\end{aligned} \tag{2}$$

where $G_F = 1.166 \times 10^{-5}\,\text{GeV}^{-2}$ is the Fermi coupling constant, and $A_{L,R}$ and g_i's ($i \in \{1, 2, \cdots, 6\}$) are dimensionless coupling constants. The first two terms in the brackets of Eq. (2) are the photonic interaction contributing to the process of Eq. (1) through the diagrams shown in Fig. 1. The remaining terms are for the direct four-Fermi contact interaction.

We estimate the branching ratio for the process of Eq. (1), which takes place in a muonic atom with an atomic number Z. The initial state has the muon and the electron in their 1S ground state of the atomic orbits. For simplicity, we ignore the three-momenta of the initial bound muon and electron. The final state has the two electrons, which, in the lowest order, can be treated as monochromatic plane waves that propagate with opposite momentum vectors. Each of the two electrons in the final state takes energy of about $m_\mu/2$, when the bound effects at the 1S state and the Coulomb interaction from the nucleus can be neglected.

3. Sensitivity

In this section, we study the upper bound for this process from the current relevant experiments.

3.1. vs. $\mu- > 3e$ with Four-Fermi-interaction dominance

We begin with the first case where the four-Fermi interaction is dominant and $A_{\rm R}$ and $A_{\rm L}$ are negligibly small, making no contribution of the photonic interactions. The remaining four-Fermi interaction allows the processes such as $\mu^-{\rm e}^- \to {\rm e}^-{\rm e}^-$ and $\mu^+ \to {\rm e}^+{\rm e}^+{\rm e}^-$. The cross section of the process of Eq. (1) is calculated to be

$$\sigma v_{\rm rel} = \frac{1}{m_\mu^2}\frac{(G_{\rm F}^2 m_\mu^2)^2}{16\pi}G\,, \tag{3}$$

where $G \equiv G_{12}+16G_{34}+4G_{56}+8G'_{14}+8G'_{23}-8G'_{56}$ with $G_{ij} \equiv |g_i|^2+|g_j|^2$ and $G'_{ij} \equiv {\rm Re}\,(g_i^* g_j)$. The transition rate is then given by

$$\Gamma(\mu^-{\rm e}^- \to {\rm e}^-{\rm e}^-) = 2\sigma v_{\rm rel}|\psi_{\rm 1S}^{(\rm e)}(0;Z-1)|^2 = m_\mu\frac{1}{8\pi}(Z-1)^3\alpha^3(G_{\rm F}^2 m_\mu^2)^2\Big(\frac{m_{\rm e}}{m_\mu}\Big)^3 G\,. \tag{4}$$

Here we took into account the facts that the 1S state can accommodate two electrons, and that the nuclear charge is shielded by the negative muon. We used the non-relativistic wave functions given by

$$\psi_{\rm 1S}^{(\rm e)}(r;Z-1) = \frac{[(Z-1)\alpha m_{\rm e}]^{3/2}}{\sqrt{\pi}}{\rm e}^{-(Z-1)\alpha m_{\rm e} r}\,, \tag{5}$$

where r is the radial coordinate, so that $\psi_{\rm 1S}^{(\rm e)}(0;Z-1) = [(Z-1)\alpha m_{\rm e}]^{3/2}/\sqrt{\pi}$. The rate of Eq. (4) is enhanced for a larger atomic number, Z, by a factor of $(Z-1)^3$, giving a notable advantage for heavy nuclei. This enhancement comes from the factor of $|\psi_{\rm 1S}^{(\rm e)}(0;Z-1)|^2$, and the large positive charge of a heavy nucleus strongly attracts the 1S wave functions of the leptons toward the nucleus position, rendering the overlap of the two wave functions large, and enhances the transition of the process of Eq. (1). We normalize the rate of Eq. (4) by the lifetime of a muonic atom, $\tilde{\tau}_\mu$, to define the branching ratio of this process as

$${\rm Br}(\mu^-{\rm e}^- \to {\rm e}^-{\rm e}^-) \equiv \tilde{\tau}_\mu\Gamma(\mu^-{\rm e}^- \to {\rm e}^-{\rm e}^-) = 24\pi(Z-1)^3\alpha^3\Big(\frac{m_{\rm e}}{m_\mu}\Big)^3\frac{\tilde{\tau}_\mu}{\tau_\mu}G\,, \tag{6}$$

The value of $\tilde{\tau}_\mu$ ranges from $\tilde{\tau}_\mu = 2.19 \times 10^{-6}\,{\rm s}$ for $^1{\rm H}$ to $\tilde{\tau}_\mu = (7-8) \times 10^{-8}\,{\rm s}$ for $^{238}{\rm U}$ as listed in Ref. [3]. This is shorter than the lifetime of free muons, $\tau_\mu = 2.197 \times 10^{-6}\,{\rm s}$ [4], which is equal to $192\pi^3/(G_{\rm F}^2 m_\mu^5)$ at the lowest order. The obtained branching ratio of Eq. (6) is to be compared with that of $\mu^+ \to {\rm e}^+{\rm e}^+{\rm e}^-$, the process that arises from the same elementary process. This branching ratio with no photonic interaction are given by [5]

$${\rm Br}(\mu^+ \to {\rm e}^+{\rm e}^+{\rm e}^-) = \frac{1}{8}(G_{12}+16G_{34}+8G_{56})\,. \tag{7}$$

The contribution from the interference among the four-Fermi interactions are not present in Eq.(7), whereas it is found to be present in Eq. (6) as the terms of G'_{ij}'s. The absence of the interferences is due to the large momenta of the final electrons. The search for the process of Eq. (1) will thereby serve complementarily with that for $\mu^+ \to {\rm e}^+{\rm e}^+{\rm e}^-$. The ratios of the two branching ratios of

$$\frac{{\rm Br}(\mu^-{\rm e}^- \to {\rm e}^-{\rm e}^-)}{{\rm Br}(\mu^+ \to {\rm e}^+{\rm e}^+{\rm e}^-)} \lesssim 192\pi(Z-1)^3\alpha^3\Big(\frac{m_{\rm e}}{m_\mu}\Big)^3\frac{\tilde{\tau}_\mu}{\tau_\mu}\,, \tag{8}$$

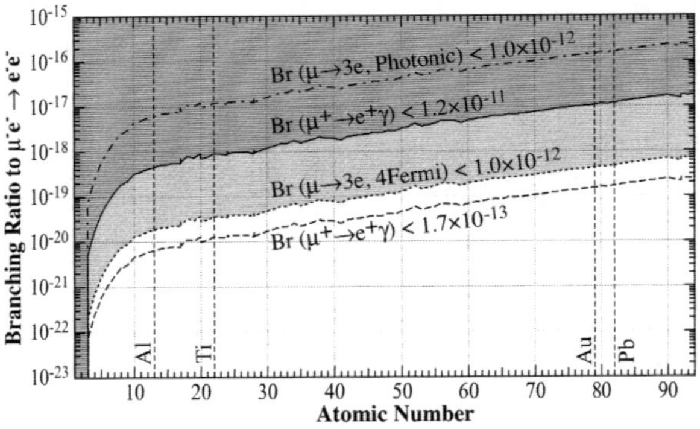

Figure 2. Limits on the branching ratio to the process of Eq. (1) imposed by the limits on $\mathrm{Br}(\mu^+ \to \mathrm{e}^+\gamma)$ and on $\mathrm{Br}(\mu^+ \to \mathrm{e}^+\mathrm{e}^+\mathrm{e}^-)$. Models without photonic interactions are excluded in the light-shaded region. Similarly, models with photonic interactions are excluded in the dark-shaded region by the current experimental limits.

is obtained by assuming $G/(G_{12} + 16G_{34} + 8G_{56}) \sim O(1)$. Hence, the upper limit on $\mathrm{Br}(\mu^-\mathrm{e}^- \to \mathrm{e}^-\mathrm{e}^-)$ is constrained by the existing limit of $\mathrm{Br}(\mu^+ \to \mathrm{e}^+\mathrm{e}^+\mathrm{e}^-) < B_{\max}$ as

$$\mathrm{Br}(\mu^-\mathrm{e}^- \to \mathrm{e}^-\mathrm{e}^-) < 192\pi(Z-1)^3\alpha^3\left(\frac{m_\mathrm{e}}{m_\mu}\right)^3\frac{\tilde{\tau}_\mu}{\tau_\mu}B_{\max}\,. \tag{9}$$

Figure 2 shows these upper limits as a function of an atomic number Z by the dotted curves. These upper limits are obtained by taking the present experimental limit of $B_{\max} = 1.0 \times 10^{-12}$ from the SINDRUM experiment [6]. The light-shaded region in Fig. 2 is excluded by the current SINDRUM limit. The reciprocal of the shown limit gives an estimation of the number of muons that is required to detect events of the process of Eq. (1). Let us take an example of the gold atom ($Z = 79$): our estimation requires a collection of $(4.21 \times 10^{-19})^{-1} = 2.38 \times 10^{18}$ muon events to improve the current limit of $B_{\max} = 1.0 \times 10^{-12}$. Assuming the detection efficiency of $O(10\%)$, the required number of muons amounts to a few times 10^{19}. Compared to this number is $O(10^{18} - 10^{19})$ of muons, which is the goal of the highly intense muon beams planned in the near future in search for the cLFV [7, 8, 9]. We thereby find that the current limit could be reachable within the capability of these muon sources. Note that relativistic effects and the binding energy of the leptons may be consequential for heavy nuclei, but are not included in our present estimation.

3.2. vs. $\mu- > 3e$, Photonic-interaction dominance

Let us now turn to the second case where the photonic interaction is present and it dominates over the four-Fermi interactions. In this case, the cross section, rate, and branching ratio of the process of Eq. (1) are calculated to be

$$\sigma v_{\mathrm{rel}} = \frac{4\alpha(G_\mathrm{F}m_\mu^2)^2}{m_\mathrm{e}^2}\left(|A_\mathrm{L}|^2 + |A_\mathrm{R}|^2\right), \tag{10}$$

$$\Gamma(\mu^-\mathrm{e}^- \to \mathrm{e}^-\mathrm{e}^-) = 2\sigma v_{\mathrm{rel}}\left|\psi_{1\mathrm{S}}^{(\mathrm{e})}(0; Z-1)\right|^2 = m_\mathrm{e}\frac{8}{\pi}(Z-1)^3\alpha^4(G_\mathrm{F}m_\mu^2)^2\left(|A_\mathrm{R}|^2 + |A_\mathrm{L}|^2\right), \tag{11}$$

and

$$\mathrm{Br}(\mu^-\mathrm{e}^- \to \mathrm{e}^-\mathrm{e}^-) = 1536\pi^2(Z-1)^3\alpha^4\left(|A_\mathrm{R}|^2 + |A_\mathrm{L}|^2\right)\frac{m_\mathrm{e}}{m_\mu}\frac{\tilde{\tau}_\mu}{\tau_\mu}, \tag{12}$$

respectively. On the other hand, the branching ratio of $\mu^+ \to e^+e^+e^-$ is given by [5]

$$\mathrm{Br}(\mu^+ \to e^+e^+e^-) = 128\pi\alpha\big(|A_R|^2 + |A_L|^2\big)\left[\log\Big(\frac{m_\mu}{m_e}\Big)^2 - \frac{11}{4}\right]. \tag{13}$$

We then have

$$\frac{\mathrm{Br}(\mu^-e^- \to e^-e^-)}{\mathrm{Br}(\mu^+ \to e^+e^+e^-)} = 12\pi(Z-1)^3\alpha^3\frac{m_e}{m_\mu}\frac{\tilde{\tau}_\mu}{\tau_\mu}\left[\log\Big(\frac{m_\mu}{m_e}\Big)^2 - \frac{11}{4}\right]^{-1}, \tag{14}$$

which we combine with the upper limit $\mathrm{Br}(\mu^+ \to e^+e^+e^-) < B_{\max}$ to yield

$$\mathrm{Br}(\mu^-e^- \to e^-e^-) < 12\pi(Z-1)^3\alpha^3\frac{m_e}{m_\mu}\frac{\tilde{\tau}_\mu}{\tau_\mu}\left[\log\Big(\frac{m_\mu}{m_e}\Big)^2 - \frac{11}{4}\right]^{-1} B_{\max}. \tag{15}$$

This upper limit is overlaid in Fig. 2 by a dash-dotted curve, according to the aforementioned SINDRUM limit of $B_{\max} = 1.0 \times 10^{-12}$.

3.3. vs. $\mu- > e\gamma$

The presence of the photonic interactions gives rise to another cLFV process $\mu^+ \to e^+\gamma$ as well, and search for this process also put a limit to $\mathrm{Br}(\mu^-e^- \to e^-e^-)$. The branching ratio to this process is given by

$$\mathrm{Br}(\mu^+ \to e^+\gamma)\frac{\Gamma(\mu^+ \to e^+\gamma)}{\Gamma(\mu^+ \to e^+\nu_e\bar{\nu}_\mu;\,\mathrm{Free})} = 384\pi^2\big(|A_R|^2 + |A_L|^2\big), \tag{16}$$

which we compare with Eq. (12) as

$$\frac{\mathrm{Br}(\mu^-e^- \to e^-e^-)}{\mathrm{Br}(\mu^+ \to e^+\gamma)} = 4(Z-1)^3\alpha^4\frac{m_e}{m_\mu}\frac{\tilde{\tau}_\mu}{\tau_\mu}. \tag{17}$$

Then the limit on $\mathrm{Br}(\mu^-e^- \to e^-e^-)$ is estimated from $\mathrm{Br}(\mu^+ \to e^+\gamma) < B_{\max}$ as

$$\mathrm{Br}(\mu^-e^- \to e^-e^-) < \frac{\mathrm{Br}(\mu^-e^- \to e^-e^-)}{\mathrm{Br}(\mu^+ \to e^+\gamma)}B_{\max} = 4(Z-1)^3\alpha^4\frac{m_e}{m_\mu}\frac{\tilde{\tau}_\mu}{\tau_\mu}B_{\max}. \tag{18}$$

A solid curve in Fig. 2 presents the upper limits given in Eq.(18) with $B_{\max} = 1.2\times10^{-11}$, which is the current upper limit from the MEGA experiment. A dashed curve in Figure 2 also shows the limits with the $B_{\max} = 1.7\times10^{-13}$, which is the goal value of the MEG experiment [11]. Even the current limit on the branching ratio to $\mu^+ \to e^+\gamma$ overwhelms the limits on $\mathrm{Br}(\mu^-e^- \to e^-e^-)$ when the photonic interaction is dominant. Accordingly, the present excluded region is above the current MEGA limit. Let us estimate, as we did earlier, the required number of muons to detect events of the process of Eq. (1), taking an example of the gold atom ($Z = 79$): an accumulation of $(1.044 \times 10^{-17})^{-1} = 9.58 \times 10^{16}$ muon events is necessary to surpass the sensitivity of MEGA, and $(1.480 \times 10^{-19})^{-1} = 6.76 \times 10^{18}$ to exceed that of the MEG goal. The required number of muons are estimated to be 10^{18} and a few of 10^{19}, respectively, assuming the $O(10\%)$ of the detection efficiency again. These are not capable now but would be possible in future with the planned highly intense muon sources.

The expected magnitudes of the branching ratio of the process of Eq. (1), which is driven by both the four-Fermi interaction and the photonic interaction, are found to be not significantly large even with the enhancement of $(Z-1)^3$. Thus, this would not be the first process to cultivate the discovery frontier of the cLFV searches. However, thanks to the enhancement of $(Z-1)^3$, this process can be accessible in future by next-generation high-intensity muon beams to produce muons of an order of $O(10^{18}\text{–}10^{19})$ per year. On the other hand, the search for $\mu^+e^- \to e^+e^-$ would not be possible with these intensities. Even higher beam intensities of $O(10^{21})$ muons per year are envisioned in the future accelerator projects such as muon colliders, neutrino factories, and Higgs factories [12].

4. Discussion

A number of critical issues should be carefully studied to consider experimental feasibility of this search. First, one should take account of potential modification of energy spectra of each of the emitted signal electrons due to the bound effects in a muonic atom. It has been known that the electron spectrum from bound muon decays in a muonic atom is strongly deformed for a large atomic number Z [13]. For the process of Eq. (1), however, it is speculated that the sum of the two signal electrons are still fixed to the initial energy of $m_\mu + m_e -$ (binding energies) due to the small kinetic energy of the recoiled nucleus. Secondly, the wave functions of the initial and final leptons should be improved to take account of the nuclear charge distribution and relativistic effects. A binding energy of the initial muon need to be also taken into account. Thirdly, the interactions of the initial leptons need detailed treatments. Various different models may predict the dependence of the rate on the initial spin state. And the interaction between the two 1S electrons in the atomic orbital of the muonic atom can be looked at. Possible contributions from 2S, 3S, $\cdots$ states should also be considered. Finally, an important experimental issue is the estimation of backgrounds, including the physical SM backgrounds from $\mu^- e^- \to e^- e^- \nu\bar{\nu}$ decay, and accidental backgrounds that are known to be detector-dependent. We will discuss these issues in our future works.

5. Summary

In summary, the new cLFV process $\mu^- e^- \to e^- e^-$ in a muonic atom is proposed. This process has the rate enhancement of $(Z-1)^3$ over the $\mu^+ e^- \to e^+ e^-$ owing to the Coulomb interaction from the nucleus in a muonic atom. This process has a final state of two electrons, which would be experimentally very clear signature. The upper limits of the branching ratio of the orders of $O(10^{-17}$–$10^{-18})$ are estimated separately for the photonic and the four Fermi interactions from the other cLFV experimental results. Once this process is observed, CP violation might be studied by comparing this process with $\mu^+ \to e^+ e^+ e^-$.

References

[1] Koike M and Kuno Y and Sato J and Yamanaka M 2010 *Phys. Rev. Lett.* **105** 121601
[2] Kuno Y and Okada Y 2001 *Rev. Mod. Phys.* **73** 151
[3] Suzuki T, Measday D F and Roalsvig J P 1997 *Phys. Rev.* **C35** 2212
[4] Amsler C*et al.* (Particle Data Group) 2008 *Phys. Lett.* **B667** 1
[5] Okada Y, Okumura K I and Shimizu Y 2000 *Phys. Rev.* **D61** 094001
[6] Bertl W H *et al.* (SINDRUM Collaboration) 1984 *Phys. Lett.* **B140** 299; *ibid.* 1985 *Nucl. Phys.* **B260** 1 Bellgardt U *et al.* (SINDRUM Collaboration) 1988 *Nucl. Phys.* **B 299** 1
[7] Carey R M *et al.* (Mu2e Collaboration) 2008 Mu2e Proposal "Proposal to Search for $\mu^- N \to e^- N$ with a Single Event Sensitivity Below 10^{-16}"
[8] Kuno Y *et al.* (COMET collaboration) 2007 J-PARC 50 GeV Proton Synchrotron Proposal P21 "A Experimental Search for Lepton Flavor Violating μ^- - e^- Conversion at Sensitivity of 10^{-16} with A Slow-Extracted Bunched Proton Beam" see http://j-parc.jp/NuclPart/pac_0801/pdf/Kuno.pdf; Cui Y G *et al.* (COMET collaboration) 2009"Conceptual Design Report for Experimental Search for Lepton Flavor Violating $\mu^- \to e^-$ Conversion at Sensitivity of 10^{-16} with a Slow-Extracted Bunched Proton Beam (COMET)" (*Preprint* KEK Report 2009-10)
[9] Kuno Y *et al.* (PRISM collaboration) 2006 J-PARC 50 GeV Proton Synchrotron LOI P20 "An Experimental Search for a μ^- - e^- Conversion at Sensitivity of the Order of 10^{-18} with a Highly Intense Muon Source: PRISM" see http://j-parc.jp/NuclPart/pac_0606/pdf/p20-Kuno.pdf.
[10] Brooks M L *et al.* (MEGA Collaboration) 1999 *Phys. Rev. Lett.* **83** 1521
[11] Baldini A *et al.* (MEG Collaboration) 2002 "The MEG experiment: search for the $\mu^+ \to e^+\gamma$ decay at PSI"
[12] Geer S 2010 "From Neutrino Factory to Muon Collider" (*Preprint* FERMILAB-CONF-10-024-APC)
[13] Watanabe R, Fukui M, Ohtsubo H and Morita M 1987 *Prog. Theor. Phys.* **78** 114

Journal of Physics: Conference Series **485** (2014) 012037 doi:10.1088/1742-6596/485/1/012037

Searching for charged lepton flavor violation: New results from the MEG experiment

Gordon Lim

Department of Physics and Astronomy, University of California, Irvine, California 92697, USA

E-mail: g.lim@uci.edu

Abstract. Discovery of charged lepton flavor violation would be unambiguous evidence of new physics beyond the Standard Model. The MEG experiment focuses on $\mu^+ \to e^+\gamma$ decay and has recently published a search result based on data collected in 2009 and 2010. No evidence of this decay mode is found in this data sample, resulting in an upper limit on the branching ratio of $2.4 \cdot 10^{-12}$ at 90% confidence level, the best limit ever achieved for this process. This result as well as the current status and prospects of the MEG experiment are reported here.

1. Introduction

Charged lepton flavor violating decays have never been observed. They are highly suppressed in the Standard Model, but many well-motivated theories beyond the Standard Model such as supersymmetric grand unified theories or theories with extra dimensions predict sizable rates up to or even higher than present experimental bounds. Observation of charged lepton flavor violating decay would, therefore, be unambiguous evidence of new physics beyond the Standard Model. The $\mu \to e\gamma$ process in particular has been used extensively to search for new physics by many experiments in the past, due to the muon's copious production rate and relatively long lifetime. Anti-muons are used because they are not captured on target nuclei. The MEG collaboration is currently searching for $\mu^+ \to e^+\gamma$ decay with unprecedented sensitivity by using innovative detector technologies and the world's most intense DC surface muon beam at the Paul Scherrer Institute (PSI) near Zurich, Switzerland. The current status, the latest results and the prospects of the MEG experiment are described in the following.

2. The MEG experiment

The $\mu^+ \to e^+\gamma$ decay mode at rest is characterized by the simultaneous emission of a back-to-back photon-positron pair in which both particles have an energy equal to half the muon rest mass (52.83 MeV). This signature has two different backgrounds: Radiative muon decays $\mu^+ \to e^+\nu_e\bar{\nu}_\mu\gamma$ (RMD) in which the neutrinos carry away little energy, and accidental coincidences (ACC) of an energetic positron from normal Michel decay $\mu^+ \to e^+\nu_e\bar{\nu}_\mu$ and a gamma-ray coming from RMD, Bremsstrahlung or positron annihilation-in-flight in the materials of the experiment. In the MEG experiment, ACC events dominate the background by more than one order of magnitude. To resolve these backgrounds, a detector with excellent spatial, temporal and energy resolution is required.

The MEG experiment [1] is being conducted at the πE5 beam line at PSI, where the world's most intense continuous surface μ^+ beam of around 10^8 μ^+/sec is available. The MEG detector

Content from this work may be used under the terms of the Creative Commons Attribution 3.0 licence. Any further distribution of this work must maintain attribution to the author(s) and the title of the work, journal citation and DOI.
Published under licence by IOP Publishing Ltd

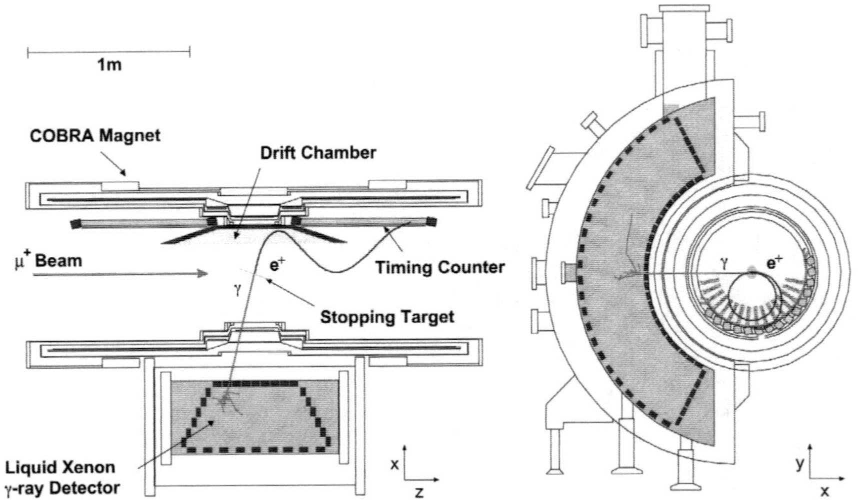

Figure 1. Top and front view of the MEG detector.

provides asymmetric coverage ($\Omega_{\rm MEG}/4\pi \simeq 11\%$) of a retractable 205 μm-thick polyethylene muon stopping target as shown in Figure 1. The detector is composed of a positron spectrometer and a photon detector. The spectrometer is used to measure the positron momentum, direction and timing. It consists of a drift chamber (DC) system flanked by two timing counters (TCs), located inside a superconducting solenoid with a gradient field along the beam axis, ranging from ~1.3 T at the center to ~0.5 T at either end. A gradient field is employed to cancel the dependence of the positron bending radius on the emission angle, and to efficiently sweep out low energy positrons from the spectrometer volume. The solenoid is filled with helium and has a fiducial thickness of 0.2 X_0. The DC system consists of 16 radially staggered double-layer modules, where each layer contains 9 axial drift cells equipped with very thin Vernier cathode pads. The counting gas in the DC modules is a 50%-50% mixture of helium and ethane. The total radiation length along positron trajectories is $\sim 2 \cdot 10^{-3} X_0$ due to the open-frame DC module design. Each TC consists of 15 axial scintillation bars read out by PMTs, and 128 semi-circular scintillation fibers read out by APDs. The photon detector, located outside of the solenoid, is a homogeneous volume (900 l) of liquid xenon (LXe) viewed by 846 UV-sensitive PMTs submerged in the LXe. It is used to reconstruct the photon energy as well as the position and time of its first interaction in the LXe. The LXe scintillation process is used because of the fast response ($\tau_{\rm decay}$ = 45 ns), the high stopping power (X_0 = 2.8 cm) and the absence of self-absorption ($\lambda_{\rm abs} > 1$ m). The cathode area coverage of the photon detector is about 40%, its total stopping power is about 14 X_0. All detector signals are individually digitized by in-house designed waveform digitizers based on the multi-GHz domino ring sampler chip (DRS) [2]. Candidate $\mu^+ \to e^+\gamma$ events are triggered by the presence of a high energy gamma-ray in the LXe detector and a hit in one of the TCs within a 20 ns window, together with an approximate back-to-back topology. Pre-scaled monitoring and calibration triggers are also recorded. The data acquisition system is based on the MIDAS software package, the online and offline data analysis are done with a common analyzer based on the framework generator ROME [3].

The MEG detector response, resolutions and stability are constantly monitored and calibrated during data acquisition using various calibration and monitoring tools. The gains and quantum-efficiencies of the photon detector PMTs are calibrated and monitored daily by a lattice of 44 LEDs and 25 point-like Am α-sources immersed in the LXe. The energy response of the photon detector is calibrated and monitored using gamma-rays from various sources. A radioactive Am/Be source and a pulsed and triggerable neutron generator are used twice per week to produce 4.4 and 9 MeV gamma-rays, respectively. A dedicated Cockcroft-Walton accelerator is also used twice per week to direct protons from the down-stream side of the detector on a lithium tetraborate target, producing monochromatic 17.6 MeV gamma-rays and a coincident gamma-ray pair of 4.4 and 11.6 MeV. Once per year negative pions are collected by the πE5 beam line and directed on a liquid hydrogen target to generate $\pi^- p$ charge exchange and radiative capture reactions that produce 54.9, 82.9 and 129 MeV gamma-rays. The spectrometer is calibrated and aligned using cosmic rays and positrons from Michel decays. The response of the spectrometer is determined using Michel positrons that make two spiral turns in the DC system. The relative timing between the photon detector and the TCs is calibrated using RMD events and coincident gamma-ray pairs generated by the Cockcroft-Walton accelerator.

3. Detector performance

The MEG experiment has been taking data since 2008. The collaboration has recently published its results based on data collected in 2009 and 2010 [1] [4]. All sub-detectors were running stably during these periods. The total number of μ^+-decays in the target in these data sets corresponds to $\sim 0.7 \cdot 10^{14}$ and $\sim 1.1 \cdot 10^{14}$, respectively. The detector resolutions, parameterized in terms of Gaussians distribution widths, are listed in Table 1. The (x, y, z) coordinate system is defined as in Figure 1. The θ and ϕ angles are the polar and azimuthal angles with respect to the beam direction. The photon detector coordinates u and v coincide with z and $r\phi$, while w is the distance from the inner wall of the LXe volume. The trigger efficiencies and the reconstruction efficiencies within the detector acceptance are also listed in Table 1.

Table 1. MEG detector performance during 2009 and 2010.

Resolution	2009	2010
γ energy	1.9% ($w > 2$ cm), 2.4% ($w < 2$ cm)	1.9% ($w > 2$ cm), 2.4% ($w < 2$ cm)
γ position	5 mm (u, v), 6 mm (w)	5 mm (u, v), 6 mm (w)
e^+ momentum	0.74% (80% core fraction)	0.74% (79% core fraction)
e^+ angles	9.4 mrad (θ), 6.7 mrad (ϕ)	11 mrad (θ), 7.2 mrad (ϕ)
e^+ vertex	1.5 mm (z), 1.1 mm (y)	2.0 mm (z), 1.1 mm (y)
γ-e^+ timing	146 ps	122 ps
Efficiency		
trigger	91%	92%
γ reconstruction	58%	59%
e^+ reconstruction	40%	34%

In 2010 a DRS upgrade resulted in an improvement in the time resolution while an increase in noise in the DC system, due to a deterioration of the high-voltage power supplies, and some unusable DC modules caused a slightly worse positron tracking performance.

[1] The 2008 data are not used due to their limited statistics and detector performance [5].

4. Analysis procedure

Several improvements to the analysis have been introduced since the presentation of the preliminary result based on the 2009 data [6]. These improvements include a new alignment technique for the DC system and the relative alignment of the photon detector and the spectrometer, a better understanding of the gradient magnetic field map, an improved spectrometer performance evaluation and an improved likelihood analysis method.

The observables used to identify $\mu^+ \to e^+\gamma$ decays are the relative time between the photon and the positron ($t_{e\gamma}$), the polar and azimuthal angles between the two particles ($\theta_{e\gamma}$, $\phi_{e\gamma}$) and their energies (E_γ, E_e). We adopt a likelihood analysis method combined with a blinding procedure: events that fall into a pre-defined blinding window that contains the signal region are saved in separate hidden data files until the analysis procedure is completely defined. The blind region is defined by $48 < E_\gamma < 58$ MeV and $|t_{e\gamma}| < 1$ ns. Events outside of this window (side-bands) are used for optimizing the analysis parameters, studying the background and, finally, constructing the probability density functions (PDFs) needed for the likelihood analysis.

A likelihood analysis is carried out for events in a portion of the blind region defined by $48 < E_\gamma < 58$ MeV, $50 < E_e < 56$ MeV, $|t_{e\gamma}| < 0.7$ ns, $|\theta_{e\gamma}| < 50$ mrad and $|\phi_{e\gamma}| < 50$ mrad. These intervals are between five and twenty sigmas wide to fully contain the signal events and also retain some background events. The best estimates of the numbers of signal, RMD and ACC events in the analysis region are obtained by maximizing the following likelihood function:

$$\begin{aligned}\mathcal{L}(N_{\rm sig}, N_{\rm RMD}, N_{ACC}) &= \frac{e^{-N}}{N_{\rm obs}!}\, e^{\frac{(N_{\rm RMD}-\langle N_{\rm RMD}\rangle)^2}{2\sigma^2_{\rm RMD}}}\, e^{\frac{(N_{\rm ACC}-\langle N_{\rm ACC}\rangle)^2}{2\sigma^2_{\rm ACC}}} \\ &\times \prod_{i=1}^{N_{\rm obs}} (N_{\rm sig}\, S(\vec{x}_i) + N_{\rm RMD}\, R(\vec{x}_i) + N_{\rm ACC}\, A(\vec{x}_i))\end{aligned}$$

where $\vec{x}_i = \{E_\gamma, E_e, t_{e\gamma}, \theta_{e\gamma}, \phi_{e\gamma}\}$ is the vector of observables for the i-th event. $N_{\rm sig}$, $N_{\rm RMD}$ and $N_{\rm ACC}$ are the fitted number of signal, RMD and ACC events, while S, R and A are their corresponding PDFs. The PDFs are constructed from the measured detector response functions and the background spectra observed in the side-bands. The dependence of the resolutions on the position of the gamma-ray conversion point and on the quality of the positron track reconstruction are taken into account in the likelihood function on an event-by-event basis. $N_{\rm obs}$ is the observed number of events, and $N = N_{\rm sig} + N_{\rm RMD} + N_{\rm ACC}$. The two Gaussian terms are the constraints on the numbers of RMD and ACC events, where $\langle N_{\rm RMD}\rangle$ and $\langle N_{\rm ACC}\rangle$ are the expected numbers of RMD and ACC events in the analysis window, while $\sigma_{\rm RMD}$ and $\sigma_{\rm ACC}$ are the corresponding uncertainties. These are estimated by extrapolating the number of RMD and ACC events observed in the side-bands, resulting in $\langle N_{\rm RMD}\rangle \pm \sigma_{\rm RMD} = 27.2 \pm 2.8\,(52.2 \pm 6.0)$ and $\langle N_{\rm ACC}\rangle \pm \sigma_{\rm ACC} = 270.9 \pm 8.3\,(610.8 \pm 12.6)$ for the 2009 (2010) data set.

The computation of the confidence interval on $N_{\rm sig}$ is based on a frequentist approach with a profile likelihood-ratio ordering [7] [8], where $N_{\rm RMD}$ and $N_{\rm ACC}$ are profiled as

$$\lambda_{\rm p}(N_{\rm sig}) = \frac{\mathcal{L}(N_{\rm sig}, \hat{\hat{N}}_{\rm RMD}(N_{\rm sig}), \hat{\hat{N}}_{\rm ACC}(N_{\rm sig}))}{\mathcal{L}(\hat{N}_{\rm sig}, \hat{N}_{\rm RMD}, \hat{N}_{\rm ACC})}$$

where the hat and double hat denote the best estimates maximizing the likelihood for floating and fixed $N_{\rm sig}$, respectively. Other, independent analysis schemes based on averaged PDFs without event-by-event information or using a Bayesian approach were also used and found to be compatible with the analysis presented here to within 10 to 20% difference in the obtained branching ratio upper limits.

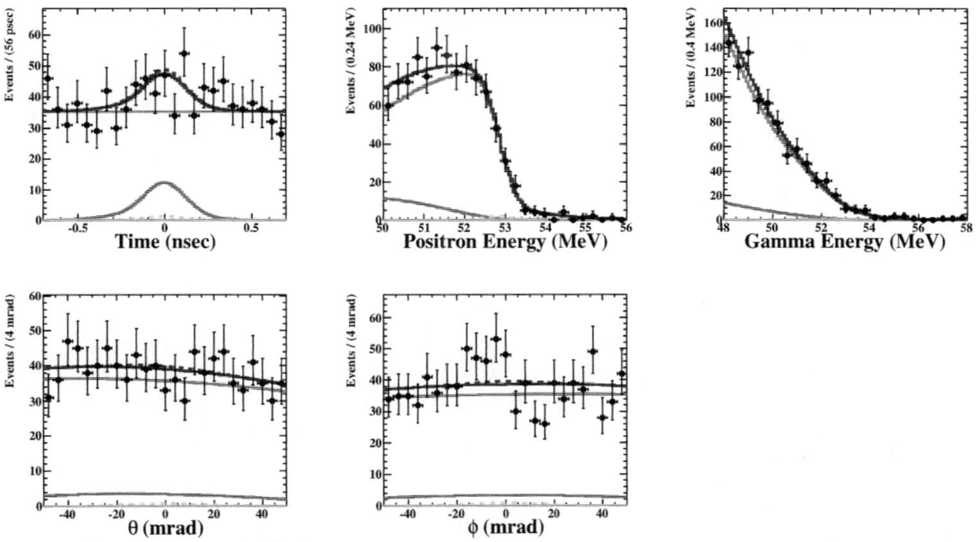

Figure 2. Distributions of the five observables $t_{e\gamma}$, E_e, E_γ, $\theta_{e\gamma}$ and $\phi_{e\gamma}$ in the analysis region. Projections of the fitted PDFs S (green), R (red), A (magenta) and their total (blue) are shown as colored lines. The dotted lines include the upper limit on N_{sig} at 90% C.L. for comparison.

In order to convert N_{sig} into a branching ratio value, the normalization relative to the Michel decay is computed by counting the number of Michel positrons that pass the same selection criteria. This is accomplished by means of a pre-scaled Michel positron trigger enabled during data acquisition, to eliminate the dependence on variations in the beam rate or detector conditions. An alternative, independent normalization is calculated using RMD events in the E_γ side-band combined with the theoretical RMD branching ratio. Both methods are in good agreement and are combined to give the normalization factor with a 7% uncertainty.

The sensitivity of the experiment with a null signal hypothesis is defined as the median of the distribution of upper limits on the branching ratio at 90% confidence level (C.L.) obtained over an ensemble of toy MC experiments, taking into account the number of RMD and ACC events as measured in the side-bands. The sensitivity of the 2009 + 2010 data sample is calculated to be 1.6×10^{-12}, and is consistent with the branching ratio upper limits derived from the side-bands demonstrating a good control over the background description.

5. Results

The distributions of the five observables in the analysis region and the projected maximum likelihood fit results are shown in Figure 2. The maximum likelihood estimate for the number of signal, RMD and ACC events in the analysis region yields $N_{\text{sig}} = -0.5^{+7.9}_{-4.7}$, $N_{\text{RMD}} = 76.5 \pm 12$ and $N_{\text{ACC}} = 882 \pm 22$, respectively, while $N_{\text{obs}} = 956$. Some 2-dimensional event distributions are shown in Figure 3.

No significant excess was found in the data set, resulting in an upper limit on the number of signal events. The corresponding upper limit on the branching ratio is $2.4 \cdot 10^{-12}$ at 90% C.L.. The systematic uncertainties of the PDF parameters and the normalization factor are taken into account in the calculation of the upper limit by varying the PDFs according to the uncertainties.

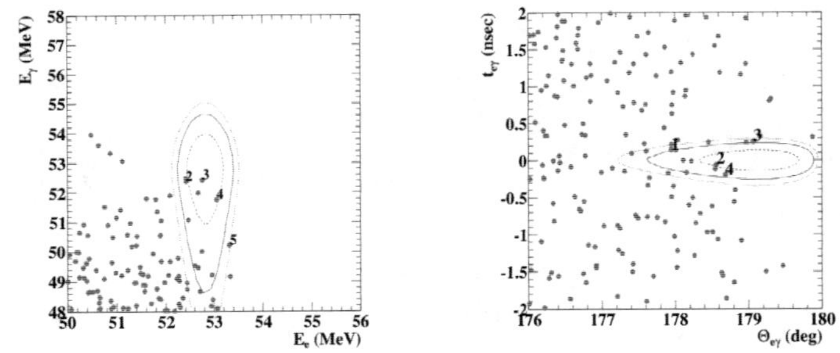

Figure 3. Event distribution in the $E_e - E_\gamma$ and the $\Theta_{e\gamma} - t_{e\gamma}$ plane, where $\Theta_{e\gamma}$ is defined as the angle between the positron and the photon. For each 2-dimensional projection, a 90% signal efficiency selection is applied on the other two observables. The $1\,\sigma$, $1.64\,\sigma$ and $2\,\sigma$ contours of the 2-dimensional projections of the signal PDF are shown as blue dashed, solid, and dotted lines. Events with the highest relative signal likelihood are numbered.

The largest contributions to the systematic uncertainty come from the uncertainties of the offsets of the relative angles, the correlations in the positron observables and the normalization. The total shift in the branching ratio upper limit amounts to about 2%.

6. Summary and outlook

The MEG experiment searches for charged lepton flavor violating $\mu^+ \to e^+\gamma$ decay with an unprecedented sensitivity. The result of the analysis on the data acquired in 2009 and 2010 is recently published. No evidence of $\mu^+ \to e^+\gamma$ decay is found in the analysis of this data sample, leading to an upper limit on the branching ratio of the $\mu^+ \to e^+\gamma$ mode of $2.4 \cdot 10^{-12}$ at 90% C.L.. This constitutes the most stringent limit on the existence of $\mu^+ \to e^+\gamma$ decay, improving the previous best limit by a factor of five [9].

During 2011, the MEG experiment collected a data sample comparable to the 2009 + 2010 statistics. Data acquisition is currently ongoing and, since the improvement in sensitivity is slowed down by the backgrounds, is planned to finish in the present detector configuration in 2013. The detector performance is expected to be similar or slightly better than in 2010, resulting in a final sensitivity in the 10^{-13} region. Various detector upgrades are currently under discussion that could improve this sensitivity by another order of magnitude in the near future. Proposed upgrades include an enlarged single volume DC or time-projection chamber, smaller photo-sensors on the inner wall of the LXe volume, the replacement of the current TC system by a lattice of small scintillating tiles, an active target and a silicon vertex tracker.

References

[1] Baldini A, Mori T *et al.* 2010 MEG Proposal to PSI, available at `http://meg.psi.ch/docs`
[2] Ritt S *et al.* 2010 *Nucl. Instr. and Meth.* **A 623** 486
[3] Cattaneo P W *et al.* 2011 *Eur. Phys. J. Plus* **126** 60
[4] Adam J *et al.* (MEG Collaboration) 2011 *Phys. Rev. Lett.* **107** 171801
[5] Adam J *et al.* (MEG Collaboration) 2010 *Nucl. Phys.* **B834** 1
[6] Sawada R for the MEG collaboration, talk presented at ICHEP2010, 35th International Conference on High Energy Physics, July 22nd-28th, 2010, Paris, France. Proceedings: PoS (ICHEP2010) 263
[7] Nakamura K *et al.* (Particle Data Group) 2010 *J. Phys.* **G37** 075021
[8] Feldman G J and Cousins R D 1998 *Phys. Rev.* **D57** 3873
[9] Brooks M L *et al.* (MEGA Collaboration) 1999 *Phys. Rev. Lett.* **83** 1521

Quantum and medium effects in (resonant) leptogenesis

Andreas Hohenegger

École Polytechnique Fédérale de Lausanne, CH-1015 Lausanne, Switzerland

E-mail: Andreas.Hohenegger@epfl.ch

Abstract. Leptogenesis offers a very attractive explanation for the origin of the baryon asymmetry of the universe. Such scenarios based on leptonic CP-violation can be realized already within minimalistic seesaw extensions of the standard model. Apart from model building issues the answer to the question of whether a given particle theory can explain the observed baryon number density depends also on the detailed statistical evolution of the asymmetry. The CP-violation within a given model leads to an asymmetry only if it is accompanied by an out-of-equilibrium evolution in the early universe. Most existing analyses employ Boltzmann-like equations (BEs) to describe it. In this context fundamental issues arise which can be addressed in the framework of non-equilibrium quantum field theory (NEQFT). Here, the relevance of quantum and medium effects for thermal leptogenesis is investigated. Within the 2PI-formalism of NEQFT, questions such as that for the justification of the particle picture arise naturally in subsequent approximations when BEs are to be derived. This specific problem is particularly important in the case of resonant leptogenesis where the relevant particle states are almost degenerate in mass. It is found that Boltzmann like equations can (only) be obtained in certain cases. But it is then possible to account for corrections due to quantum and medium effects.

1. Introduction

Leptogenesis is today the most popular mechanism for the generation of the cosmological baryon asymmetry. Its prerequisites are realized by many conceivable extensions of the standard model of particle physics (SM). In particular they are met by moderate extensions where fermionic singlets, scalar or fermionic triplet fields are added to the theory. These seesaw extensions have the additional benefit of explaining the hierarchy between known light neutrino masses and those of charged SM fermions. A lot of attention was payed in the past to the implications of variations of the models and their parameters for leptogenesis. Fundamental issues related to the non-equilibrium statistical description, in turn, have been thoroughly investigated only recently. The progress here is manly based on the application of the 2PI-formalism of NEQFT which has helped to gain a deeper understanding of leptogenesis [1, 2, 3, 4, 5, 6, 7, 8, 9, 10, 11]. Also new results based on thermal quantum field theory and semi-classical analyses brought relevant new quantitative results [12, 13, 14, 15, 16, 17]. Amongst the proposed leptogenesis scenarios the most favoured one is that of 'thermal leptogenesis' where two or more right-handed neutrino fields are added to the standard theory:

$$\mathcal{L} = \mathcal{L}_{SM} + \tfrac{1}{2}\bar{N}_i\big(i\partial\!\!\!/ - M_i\big)N_i - h_{\alpha i}\bar{\ell}_\alpha\tilde{\phi}P_R N_i - h^\dagger_{i\alpha}\bar{N}_i\tilde{\phi}^\dagger P_L \ell_\alpha\,. \tag{1}$$

Here $N_i = N_i^c$ are the heavy Majorana fields, ℓ_α are the lepton doublets, $\tilde{\phi} \equiv i\sigma_2\phi^*$ is the conjugate of the Higgs doublet, and h are the neutrino Yukawa couplings. The second ingredient is a standard cosmological evolution with high-temperature reheating. With heavy right handed neutrino masses M_i at a GUT-like scale, scenarios can comfortably be realized in which N_i stay relatively close to thermal

Content from this work may be used under the terms of the Creative Commons Attribution 3.0 licence. Any further distribution of this work must maintain attribution to the author(s) and the title of the work, journal citation and DOI.
Published under licence by IOP Publishing Ltd

equilibrium before they decay at $T \sim M_i$. The conventional analysis of this non-equilibrium process is based on cosmological 'rate equations' for the $(B-L)$-asymmetry Y_L and the abundance of the heavy Majorana neutrinos Y_{N_i} during this epoch. The coefficients for these equations ('reaction densities') are computed as Boltzmann-like collision integrals based on S-matrix amplitudes. The reaction densities for (inverse) decays, washout determine together with the *CP*-violating parameters ϵ_i the amount of asymmetry produced. In a perturbative approximation of order h^4 the graphs depicted in Fig. 1 contribute

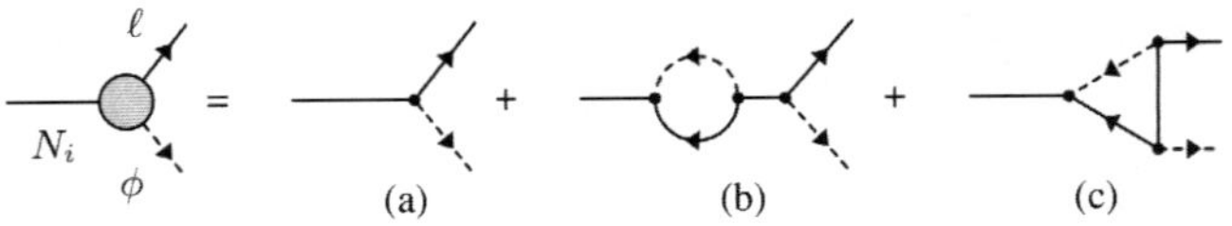

Figure 1. Tree-level (a), one-loop self-energy (b) and one-loop vertex (c) contributions to the decay of the heavy Majorana neutrino. The contributions to the *CP*-violating parameters are due to the interference terms between the tree-level and loop contributions.

to the *CP*-violating decay. At the same order the two-body scattering processes in Fig. 2 need to be taken into account. We ignore further processes involving other SM fermions and gauge interactions as they are relevant but largely independent of the present analysis. Issues (specifically the double-counting

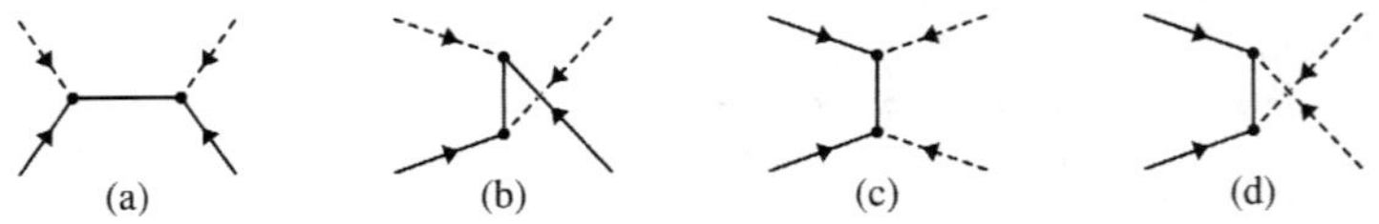

Figure 2. Contributions to the two-body scattering process $\ell\phi \leftrightarrow \bar{\ell}\bar{\phi}$ and to $\ell\ell \leftrightarrow \bar{\phi}\bar{\phi}$. Real intermediate state (RIS) contributions to the process in (a) double-count decays and inverse decays in Fig. 1. They need to be subtracted by hand to obtain consistent rate equations. Consequently decays and these two-body scattering processes have to be considered always together within the S-matrix approach.

problem) related to the insufficiency of this approach are usually fixed in an ad-hoc way to restore the compatibility with fundamental conservation laws. In equilibrium the requirement of unitarity of the S-matrix leads to well posed conditions between the amplitudes for decays and scattering which guarantee conserving equations but these are not applicable out-of-equilibrium. Also the question to which extent medium effects can be relevant arises naturally in the outlined picture of a dynamical asymmetry generation in the hot early universe.

In a variation of this simple scenario at least two of the heavy Majorana neutrinos are almost degenerate in mass $M_1 \simeq M_2$. This specific choice (which can for instance be motivated as a consequence of a radiatively broken symmetry) entails the benefit that $\epsilon_{1,2}$ can be sufficiently enhanced to lower the mass scale for the hypothetical heavy neutrino to $\sim 1\,\mathrm{TeV}$. Thermal leptogenesis can therefore be reconciled with lower reheating temperatures and the masses $M_{1,2}$ become accessible to collider experiments. The resonant contribution to the *CP*-violating parameter is maximized if the mass splitting $|M_1 - M_2|$ is comparable to the width $\Gamma_{1,2}$. In this regime peculiar issues related to the definition of quasi-particle states arise. These become evident when a first principle computation within NEQFT is performed as the spectrum of the quasi-particle excitations defining $N_{1,2}$ has to be specified in due course [3]. In the same regime oscillations between the Majorana neutrinos can become relevant and the simple description in terms of rate equations and a few reaction densities can prove to be untenable.

2. Nonequilibrium QFT

The 2PI-formalism in NEQFT represents a suitable framework for the analysis of quantum statistical processes out-of-equilibrium. It is known to yield consistent transport equations at a given truncation of a perturbative expansion and to evade the double-counting problem. Furthermore it yields equations which take medium effects into account. Upon a number of controllable approximations simple rate equations may be obtained which can be compared with conventional results. The most straight-forward access to leptogenesis is via the divergence $\mathcal{D}_\mu j^\mu_L(x)$ of the expectation value of the lepton current operator

$$j^\mu_L(x) = \big\langle \sum_{\alpha,a} \bar{\ell}^a_\alpha(x)\gamma^\mu \ell^a_\alpha(x)\big\rangle = -\sum_{\alpha,a} \operatorname{tr}\left[\gamma_\mu S^{\alpha\alpha}_{aa}(x,x)\right] .$$

The second equality uses the definition of the closed-time-path (CTP) propagator for leptons (the indices are for flavour (greek) and $SU(2)$ (latin)),

$$S^{\alpha\beta}_{ab}(x,y) \quad = \quad \langle T_\mathcal{C}\ell^a_\alpha(x)\bar{\ell}^b_\beta(y)\rangle \,. \tag{2a}$$

Similar relations hold for other propagators such as those of leptons and Higgs. For the full lepton propagator S (here in matrix notation) the 2PI-formalism provides a self-consistent Schwinger-Dyson equation

$$S^{-1}(x,y) = S_0^{-1}(x,y) - \Sigma(x,y)\,, \tag{3}$$

where the lepton self-energy Σ is obtained by functional differentiation of the 2PI effective action Γ_{2PI} with respect to S. Γ_{2PI} can be approximated by the three-loop truncation in Fig. 3(a),(b).

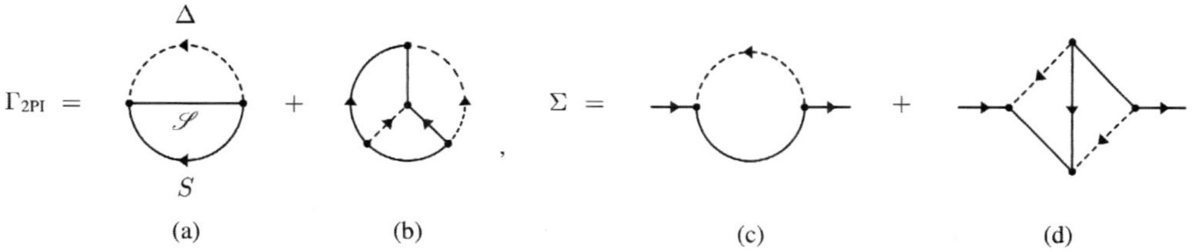

Figure 3. Two- (a) and three-loop (b) contributions to the 2PI effective action and the corresponding contributions to the lepton self-energy (c), (d). At this order all processes contributing to the S-matrix elements squared find their equivalent in the 2PI-derivation.

The divergence of the lepton current leads, upon gradient expansion and Wigner-transformation, to (the subscripts >, (<) refer to greater (lesser) components of the 'Wightman-propagators'):

$$\begin{aligned}\mathcal{D}_\mu j^\mu_\ell(t) = 2\int_0^\infty \frac{dp^0}{2\pi}\int\frac{d^3p}{(2\pi)^3}\operatorname{tr}\big\{&\big[\Sigma^{\alpha\beta}_<(t,\not p)S^{\beta\alpha}_>(t,\not p) - \Sigma^{\alpha\beta}_>(t,\not p)S^{\beta\alpha}_<(t,\not p)\big]\\ &-\big[\bar\Sigma^{\beta\alpha}_<(t,\not p)\bar S^{\alpha\beta}_>(t,\not p) - \bar\Sigma^{\beta\alpha}_>(t,\not p)\bar S^{\alpha\beta}_<(t,\not p)\big]\big\}\,.\end{aligned} \tag{4}$$

The lepton self-energies are functionals of the Majorana-neutrino- and Higgs-propagators $\mathscr{S}$ and Δ as depicted in Fig. 3(c),(d). These are governed by their respective Schwinger-Dyson equations similar to Eq. (3). Further simplification of the evolution equation for the lepton current can be achieved if their solution yields additional analytic information about the propagators. In particular, if the spectral dependence of the propagators is that of a sharply peaked on-shell state, as is the case for leptons and Higgs. It is then valid to introduce a 'quasi-particle approximation' and the degrees of freedom in the propagators S, Δ are reduced to those of one-particle distribution functions f_ℓ and f_ϕ which depend only on space, time and the three-momenta. The approximation of the Majorana neutrino propagator $\mathscr{S}$ is more intricate because it can deviate from equilibrium and the finite spectral width can be relevant.

The vertex contribution to the *CP*-violating parameter arises as cut of the 2PI graph in Fig. 3(a). This diagram yields also the $s \times u$ contributions to $\ell\phi \leftrightarrow \bar{\ell}\bar{\phi}$ as well as the $u \times t$ contribution to $\ell\ell \leftrightarrow \bar{\phi}\bar{\phi}$. To obtain the self-energy contribution, the off-diagonal elements of the Majorana neutrino propagator have to be taken into account. Furthermore an 'extended quasi-particle approximation' needs to be employed in order to obtain the $s \times s$ and $u \times u$ contributions to $\ell\phi \leftrightarrow \bar{\ell}\bar{\phi}$ as well as $u \times u$ and $t \times t$ contributions to $\ell\ell \leftrightarrow \bar{\phi}\bar{\phi}$ scattering from the setting-sun graph in Fig. 3. To a large extent, the answer to the question if conventional Boltzmann-like equations are applicable for leptogenesis depends on whether such quasi-particle approximations for the Majorana neutrino propagator are justified.

3. Non-resonant Leptogenesis

If the heavy Majorana neutrino masses are hierarchical and deviations from thermal equilibrium not too large, the quasi-particle approximation (and a 'Kadanoff-Baym' ansatz which relates the statistical dependence of the propagators to the distribution functions) can be applied. In a homogeneous and isotropic FRW-universe, it is then possible to reduce the quantum kinetic equations to a system of rate equations for the abundances which may be compared to conventional results:

$$\frac{s\mathcal{H}}{z}\frac{dY_L}{dz} = \sum_i \langle \epsilon_i \gamma^D_{N_i}\rangle \left(\frac{Y_{N_i}}{Y^{eq}_{N_i}} - 1\right) - \frac{Y_L}{2Y^{eq}_\ell}(1 + c_{\phi\ell})c_\ell\Big(\sum_i \langle \gamma^W_{N_i}\rangle + 4\langle \gamma'^{\ell\phi}_{\bar{\ell}\bar{\phi}}\rangle + 4\langle \gamma^{\bar{\ell}\bar{\ell}}_{\phi\phi}\rangle\Big), \tag{5a}$$

$$\frac{s\mathcal{H}}{z}\frac{dY_{N_i}}{dz} = -\langle \gamma^D_{N_i}\rangle \left(\frac{Y_{N_i}}{Y^{eq}_{N_i}} - 1\right). \tag{5b}$$

Here we introduced decay reaction densities $\langle \epsilon_i \gamma^D_{N_i}\rangle$, $\langle \gamma^W_{N_i}\rangle$ by defining the averages

$$\langle X\gamma^D_{N_i}\rangle \equiv \int d\Pi^{\ell N_i \phi}_{pqk}\,(2\pi)^4\delta(q-k-p)X\,|\mathcal{M}|^2_{N_i}\,f^{eq}_{N_i} f_{\ell\phi}\,,\quad \langle X\gamma^W_{N_i}\rangle \equiv \langle X(1-f^{eq}_{N_i})\gamma^D_{N_i}\rangle\,, \tag{6}$$

where $f_{\ell\phi} \equiv (1 - f^{eq}_\ell + f^{eq}_\phi)$, and $d\Pi^{\{i\}}_{\{p_i\}}$ represents the complete phase space element. Similar for scattering:

$$\langle X\gamma^{ab}_{ij}\rangle \equiv \int d\Pi^{abij}_{kpqr}(2\pi)^4\delta(k+p-q-r)X\,|\mathcal{M}|^2_{ab\leftrightarrow ij}\,(1 \pm f^{eq}_a)(1 \pm f^{eq}_b) f^{eq}_i f^{eq}_j\,. \tag{7}$$

The amplitudes squared obtained here agree with those obtained in S-matrix theory but medium effects are implied by the quantum statistical factors and the term $f_{\ell\phi}$. Further such contributions show up in the (momentum dependent) *CP*-violating parameters. The contributions to $\epsilon_i = \epsilon^S_i + \epsilon^V_i$ read

$$\epsilon^S_i(q,p) = -\sum_{j\neq i} \frac{\mathrm{Im}\{(h^\dagger h)^2_{ij}\}}{(h^\dagger h)_{ii}(h^\dagger h)_{jj}} \frac{M_i\Gamma_j}{M_j^2 - M_i^2} \int \frac{d\Omega}{4\pi}(1-\cos\theta) f_{\ell\phi}\,, \tag{8a}$$

$$\epsilon^V_i(q,p) = -\sum_{j\neq i} \frac{\mathrm{Im}\{(h^\dagger h)^2_{ij}\}}{(h^\dagger h)_{ii}(h^\dagger h)_{jj}} \frac{\Gamma_j}{M_i} \int \frac{d\Omega}{4\pi} \frac{1-\cos\theta}{1+\cos\theta + 2M_j^2/M_i^2} f_{\ell\phi}\,, \tag{8b}$$

where the same dependence on the distributions, $f_{\ell\phi}$ enters. The reaction density $\langle \gamma'^{\ell\phi}_{\bar{\ell}\bar{\phi}}\rangle$ has to be compared with the RIS-subtracted two-body scattering rate. In the narrow width limit:

$$\langle \gamma'^{\ell\phi}_{\bar{\ell}\bar{\phi}}\rangle \equiv \left\langle \Big[1 - \sum_i \frac{\pi\,|\mathcal{M}|^2_{N_i}}{4\,|\mathcal{M}|^2_{\ell\phi\leftrightarrow\bar{\ell}\bar{\phi}} M_i\Gamma_i}\delta(s - M_i^2)\Big]\gamma^{\ell\phi}_{\bar{\ell}\bar{\phi}}\right\rangle.$$

A peculiar medium effect becomes relevant if, at high temperatures, the thermal Higgs mass exceeds that of the heavy Majorana neutrinos $m_\ell + m_\phi > M_i$. In this case the Majorana neutrino decay

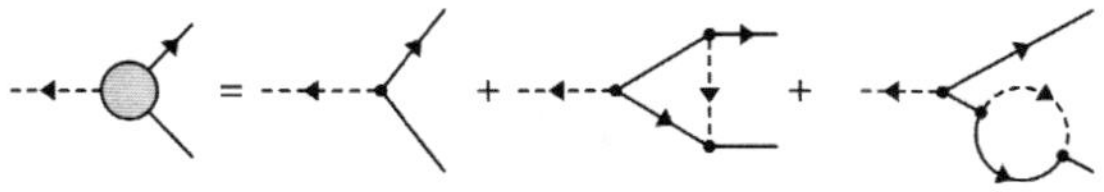

Figure 4. Contributions to the Higgs decay into Majorana neutrinos and leptons. The process violates *CP* due to the purely thermal loop contributions.

becomes impossible but decays of the Higgs into a Majorana neutrino and a lepton can become allowed instead. This process violates *CP* due to the purely thermal one-loop contributions in Fig. 4. To obtain quantitative results for thermal corrections to the *CP*-violating parameters, which become relevant at high temperatures, it needs to be taken into account. The main difference compared to Majorana neutrino decay consist in the replacement of $f_{\ell\phi} = (1 - f_\ell^{eq} + f_\phi^{eq}) \to (f_\ell^{eq} + f_\phi^{eq})$. The simple picture of temperature dependent thermal masses is altered by the modified dispersion relations obtained for leptons in the framework of a hard thermal loop resummation [18, 13] or if the collinear enhancement effects described in [15, 16, 17] are taken into account. At the present order (in μ_ℓ/T, Δf_{N_i} and ϵ_i) Eqs. (5) can also be obtained within thermal quantum field theory if the RIS subtraction is generalized to include quantum statistical factors and to compensate the double-counting related to Higgs-decays. The *CP*-violating parameters (i.e. the amplitudes that enter the Boltzmann equations) can be computed using retarded/advanced products in the real time formalism [12] or in the imaginary time formalism [13, 14]. Differences appear at higher order in the expansion in these quantities [5]. Further differences arise when cuts through the internal Majorana neutrino line [10] or flavour effects are to be taken into account [9].

4. Resonant Case

According to established reasoning it is sufficient to modify the *CP*-violating parameters such that the divergence, also present in Eq. (8a), for $M_j \to M_i$ is regulated by the finite width of the Majorana neutrinos:

$$\epsilon_i^{S,vac} = -\sum_{j\neq i} \frac{\mathrm{Im}\{(h^\dagger h)_{ij}^2\}}{(h^\dagger h)_{ii}(h^\dagger h)_{jj}} \frac{R}{R^2 + A^2}, \quad \text{with } R \equiv \frac{M_j^2 - M_i^2}{M_i \Gamma_j}, \tag{9}$$

with (e.g.) $A = M_i/M_j$. However, as outlined in Sec. 2, the applicability of simple Boltzmann-like equations depends on the validity of the quasi-particle approximation. Staying within the quasi-particle picture, an improved expression for the *CP*-violating parameters may be derived if there is only a mild degeneracy of the Majorana neutrino masses [19]:

$$\epsilon_i^S(q,p) = -\sum_{j\neq i} \frac{\mathrm{Im}\,\{(h^\dagger h)_{ij}^2\}}{(h^\dagger h)_{ii}(h^\dagger h)_{jj}} \frac{R}{R^2 + A^2} \frac{p\,L_\rho}{p\,q}, \tag{10}$$

with $A = qL_\rho/(M_i M_j)$. It takes into account a 'collisional broadening' of the Majorana neutrino propagator. If the two states get so close that their spectral functions have significant overlap, the approximation by two δ-like peaks becomes impossible and the dynamics of the flavour off-diagonals of the Majorana neutrino propagator $\mathscr{S}$ cannot be reduced to that of its diagonal elements. It is possible to find analytic solutions for $\mathscr{S}$ if one assumes that the expansion and washout effects are negligible [20]. This leads to an immediate result for the time dependence of the lepton asymmetry $n_L = j_0$ in the degenerate limit [21]:

$$n_L(t) = -\frac{\mathrm{Im}\{(h^\dagger h)^2_{12}\}}{8\pi} \frac{M_1 M_2 \Delta M^2_{12}}{(\Delta M^2_{12})^2 + (M_1\Gamma_1 - M_2\Gamma_2)^2}$$
$$\times 2 \int d\Pi^{N_{1,2}\ell\phi}_{qpk} (2\pi)^3 \delta(\mathbf{q}-\mathbf{p}-\mathbf{k}) \frac{\Gamma_{\ell\phi}}{(E_{N_{1,2}} - p - k)^2 + \Gamma^2_{\ell\phi}/4} 4pq\, f_{\ell\phi} f^{eq}_N$$
$$\times \left[\sum_{i=1,2} \frac{1 - e^{-M_i\Gamma_i/E_{N_i} t}}{M_i\Gamma_i/E_{N_i}} - 4\,\mathrm{Re}\left(\frac{1 - e^{-i(E_{N_1} - E_{N_2})t} e^{-\Gamma t/4}}{\Gamma/2 + 2i(E_{N_1} - E_{N_2})} \right) \right], \qquad (11)$$

where $\Delta M^2_{ij} = (M^2_i - M^2_j)$. Γ is the average of the Majorana neutrino decay rates, see [21] for details. The energy conserving Dirac-delta is here replaced by a Breit-Wigner curve dependent on the sum of the thermal widths of leptons and Higgs $\Gamma_{\ell\phi}$. The additional square bracket in Eq. (11) is related to coherent flavour transitions between the Majorana neutrino states N_1 and N_2. As advertised above, this term can become relevant if the size of the mass splitting $|\Delta M| = |M_i - M_j|$ is of the same order as the width Γ. Under similar assumptions density matrix like equations have been obtained in [22], as well based on analytic solutions for $\mathscr{S}$. They describe the oscillations between the degenerate states and how they source the *CP*-asymmetry. Note that although the relative size of the thermal corrections to the Majorana neutrino masses (not described here) is generically small it can lead to peculiar effects in the degenerate limit if they are of the same order as the mass splitting, so that their effect should be taken into account in such cases. Oscillatory effects are particularly important if the asymmetry production is mainly due to the mechanism of leptogenesis via neutrino oscillations [23].

The application of NEQFT methods to leptogenesis as it evolved in recent years put leptogenesis on a solid footing. The employed 2PI-formalism represents an adequate basis for the description of this strikingly elegant link between the fundamental micro-physics of *CP*-violation and the macroscopic dominance of baryons in the universe. The treatment confirmed conventional results in the limit of zero temperature and led to a corrected account of medium effects otherwise. Beyond the simplest scenarios where the 'Boltzmann-limit' is applicable it can lead to more elaborate kinetic equations which govern the generation of an asymmetry. These increasingly challenge numerical methods as they are in the class of integro-differential equations and further reduction is not always possible.

Acknowledgements
I like to thank my collaborators T. Frossard, M. Garny and A. Kartavtsev.

References

[1] Hohenegger A, Kartavtsev A and Lindner M 2008 *Phys. Rev. D* **78** 085027 (*Preprint* 0807.4551)
[2] Garny M, Hohenegger A, Kartavtsev A and Lindner M 2009 *Phys. Rev. D* **80** 125027 (*Preprint* 0909.1559)
[3] Garny M, Hohenegger A, Kartavtsev A and Lindner M 2010 *Phys. Rev. D* **81** 085027 (*Preprint* 0911.4122)
[4] Garny M, Hohenegger A and Kartavtsev A 2010 (*Preprint* 1005.5385)
[5] Frossard T, Garny M, Hohenegger A, Kartavtsev A and Mitrouskas D 2012 in preparation
[6] Hohenegger A 2010 *Journal of Physics: Conference Series* **259** 012074
[7] Anisimov A, Buchmüller W, Drewes M and Mendizabal S 2010 *Phys. Rev. Lett.* **104** 121102 (*Preprint* 1001.3856)
[8] Beneke M, Garbrecht B, Herranen M and Schwaller P 2010 *Nuclear Physics B* **838** 1–27 (*Preprint* 1002.1326)
[9] Beneke M, Garbrecht B, Fidler C, Herranen M and Schwaller P 2011 *Nucl.Phys.* **B843** 177–212 (*Preprint* 1007.4783)
[10] Garbrecht B 2011 *Nucl.Phys.* **B847** 350–366 (*Preprint* 1011.3122)
[11] De Simone A and Riotto A 2007 *JCAP* **0708** 002 (*Preprint* hep-ph/0703175)
[12] Garny M, Hohenegger A and Kartavtsev A 2010 *Phys. Rev. D* **81** 085028 (*Preprint* 1002.0331)
[13] Kiessig C and Plumacher M 2012 *JCAP* **1207** 014 (*Preprint* 1111.1231)
[14] Kiessig C and Plumacher M 2012 *JCAP* **1209** 012 (*Preprint* 1111.1235)
[15] Anisimov A, Besak D and Bodeker D 2011 *JCAP* **1103** 042 (*Preprint* 1012.3784)
[16] Laine M and Schroder Y 2012 *JHEP* **1202** 068 (*Preprint* 1112.1205)
[17] Besak D and Bodeker D 2012 *JCAP* **1203** 029 (*Preprint* 1202.1288)

[18] Giudice G F, Notari A, Raidal M, Riotto A and Strumia A 2004 *Nucl. Phys. B* **685** 89–149 (*Preprint* hep-ph/0310123)
[19] Frossard T, Garny M, Hohenegger A and Kartavtsev A 2012 in preparation
[20] Anisimov A, Buchmuller W, Drewes M and Mendizabal S 2011 *Annals Phys.* **326** 1998–2038 (*Preprint* 1012.5821)
[21] Garny M, Kartavtsev A and Hohenegger A 2011 (*Preprint* 1112.6428)
[22] Garbrecht B and Herranen M 2012 *Nucl.Phys.* **B861** 17–52 (*Preprint* 1112.5954)
[23] Canetti L, Drewes M, Frossard T and Shaposhnikov M 2012 (*Preprint* 1208.4607)

Journal of Physics: Conference Series **485** (2014) 012039 doi:10.1088/1742-6596/485/1/012039

Triplet seesaw model: from inflation to asymmetric dark matter and leptogenesis

Chiara Arina

Institut für Theoretische Teilchenphysik und Kosmologie, RWTH Aachen, 52056 Aachen, Germany

E-mail: `chiara.arina@physik.rwth-aachen.de`

Abstract. The nature of dark matter (DM) particles and the mechanism that provides their measured relic abundance are currently unknown. Likewise, the nature of the inflaton is unknown as well. We investigate the triplet seesaw model in an unified picture. At high energy scale, we consider Higgs inflation driven by an admixture of standard model and triplet Higgs fields, both coupled non-minimally to gravity. At intermediate and low energies we investigate vector like fermion doublet DM candidates with a charge asymmetry in the dark sector, which is generated by the same mechanism that provides the baryon asymmetry, namely baryogenesis-via-leptogenesis induced by the decay of scalar triplets. At the same time the model gives rise to neutrino masses in the ballpark of oscillation experiments via type-II seesaw. We then apply Bayesian statistics to infer the model parameters giving rise to the observed baryon asymmetry and DM density, compatibly with inflationary and DM direct detection constraints, updated with the CRESST-II excess, the new XENON100 data release and KIMS exclusion limit.

1. Introduction

A widely accepted theory describing the early universe is inflation and many realization of it have been modeled (see *e.g.* [1, 2] for a review). Among them, the Higgs boson has been suggested as inflaton [3] by means of a non-minimal coupling to gravity ξ_H: inflation will occur at the unitarity scale $M_{\rm pl}/\xi_H \simeq 10^{14}$ GeV. The observation of a new boson at LHC by CMS [4] and ATLAS [5] points towards a Higgs-like particle with a mass of 125 GeV. At this mass value the Higgs potential is likely in a metastable vacuum [6] around 10^9 GeV. A viable way of curing the potential instability and restoring perturbative unitarity is to introduce new physics at that scale, *i.e.* a triplet scalar with a mass $M_\Delta = 10^8$ GeV, as it has been discussed in [7]. The triplet mixes with the Higgs and has a non-minimal coupling to the Ricci scalar R, ξ_Δ, such that both scalars behave as inflaton: the potential will be flat enough to produce 60 e-folds from slow-roll needed to solve the horizon problem. At the end of inflation the early universe enters the radiation era: within the thermal bath the visible and dark matter are created in the measured amount [8]: $\Omega_{\rm B} \simeq 0.04$ and $\Omega_{\rm DM} \simeq 0.23$. We suppose that the DM is asymmetric as the baryonic matter, instead its relic abundance being due to the usual freeze-out mechanism. Both asymmetries are linked together and induced by triplet decay via leptogenesis, as detailed in [9, 7]. The DM candidate is part of a doublet under $SU(2)_L$, either scalar or fermion, and leads to distinct signatures in DM direct detection experiments: its scattering is inelastic [10] and accounts for the DAMA/LIBRA signal [11]. We investigate if it may account for the CRESST-II

Content from this work may be used under the terms of the Creative Commons Attribution 3.0 licence. Any further distribution of this work must maintain attribution to the author(s) and the title of the work, journal citation and DOI.
Published under licence by IOP Publishing Ltd

excess [12] as well, while being compatible with the most recent upper bound by XENON100 [13] and by KIMS [14].

2. The model and the DM candidate

The standard model (SM) is extended by introducing a scalar triplet Δ with charges $(3, 2)$ under the gauge group $SU(2)_L \times U(1)_Y$. The scalar potential involving the Higgs field as well is:

$$\begin{aligned} V_J(\Delta, H) &= M_\Delta^2 \Delta^\dagger \Delta + \frac{\lambda_\Delta}{2}(\Delta^\dagger\Delta)^2 - M_H^2 H^\dagger H + \frac{\lambda_H}{2}(H^\dagger H)^2 \\ &+ \lambda_{\Delta H} H^\dagger H \Delta^\dagger \Delta + \frac{1}{\sqrt{2}}\left[\mu_H \Delta^\dagger H H + \text{h.c.}\right] . \end{aligned} \tag{1}$$

The bilinear couplings of the triplet with the leptons are:

$$-\mathcal{L} \supset +\frac{1}{\sqrt{2}}\left[f_H \Delta^\dagger H H + f_L \Delta L L + \text{h.c.}\right] , \tag{2}$$

with the redefinition $f_H = \mu_H/M_\Delta$. These terms give rise to neutrino masses via type-II seesaw [15], $M_\nu \propto f_H f_L(-v^2)/M_\Delta$, where v is the Higgs vev (vacuum expectation value). In order to have Majorana masses in the range indicated by neutrino oscillations, $f_L \simeq 10^{-9}$, because M_Δ is at the scale of 10^8 GeV, slightly below the metastability one.

We extend the Lagrangian with the DM candidate given by the neutral component of $\chi \equiv (\chi^+ \chi^0)^T$ for scalars and of vector-like doublet $\psi \equiv (\psi_{\text{DM}}, \psi_-)$ for fermions:

$$-\mathcal{L}_{\text{DM}} \supset \begin{cases} M_\chi^2 \chi^\dagger \chi + \lambda_\chi (\chi^\dagger \chi)^2 + \left[\mu_\chi \Delta^\dagger \chi\chi + \text{h.c.}\right] + & \text{scalar DM} \\ +\lambda_3 |H|^2 |\chi|^2 + \lambda_4 |H^\dagger \chi|^2 + \frac{\lambda_5}{2}\left[(H^\dagger\chi)^2 + \text{h.c.}\right] , & \\ \overline{\psi} i\gamma^\mu \mathcal{D}_\mu \psi + M_D \overline{\psi}\psi + \frac{1}{\sqrt{2}}\left[f_\psi \Delta \psi\psi + \text{h.c.}\right] . & \text{fermionic DM} \end{cases} \tag{3}$$

The doublets are odd under a Z_2 symmetry making the DM particle stable, besides being inert, since there are no coupling with the SM fermions. In the case of scalar DM, the Z_2 symmetry is elevated to global $U(1)$ Peccei-Quinn symmetry for $\lambda_5 \to 0$, that is null DM number violating coupling. For $\lambda_5 \simeq 10^{-7}$ a small mass splitting, $\Delta m \equiv \delta$, is generated between the odd and even neutral states, the lightest one being the DM particle, making the interaction with Z boson off-diagonal. As for the fermionic DM candidate, in the same way a Majorana mass is generated for neutrino masses, the interaction with the triplet scalar generates a Majorana mass for ψ_{DM}, $m = f_H f_\psi \frac{-v^2}{M_\Delta}$, via the low scale dimension five operator $\mathcal{O}_5 = \psi\psi H H$, leading to an off diagonal coupling between ψ_{DM} and the Z boson as well.

In case of a scalar DM χ_0 any asymmetry in the DM sector gets washed out below electroweak (EW) phase transition by fast oscillations between χ_0 and its complex conjugate field $\overline{\chi_0}$. This sets a limit of the mass scale of χ_0 to be $M_{\chi_0} \equiv m_{\text{DM}} \geq 2$ TeV, so that the DM freezes out before oscillations begin to occur.

The DM scattering off nuclei is mediated by the Z boson, hence inelastic. The predictions in the $\{m_{\text{DM}}, \delta\}$-plane are given in figure 1 (see caption for the labeling), following the Bayesian inference procedure explained in [9]. Scalar DM in the mass range above 2 TeV is excluded as explanation of DAMA annual modulation and CRESST-II excess by XENON100 and KIMS. We therefore do not consider any further the scalar case for asymmetric DM. Contrary fermionic DM is compatible at 99% C.L. with all exclusion bounds in the mass range 50 $\to$ 350 GeV. This candidate is compatible as well with the CRESST commissioning run on W [16, 7]. Notice that at those masses the regions that explain DAMA and CRESST-II overlap. For the fermionic candidate to have a mass splitting $\delta \simeq 150$ keV we require $f_\psi \geq 10^{-4}$. This hierarchy between f_L and f_ψ, and consequently in the Majorana masses will fix the phenomenology of leptogenesis and DM asymmetry generation, as it will be discussed in 4.

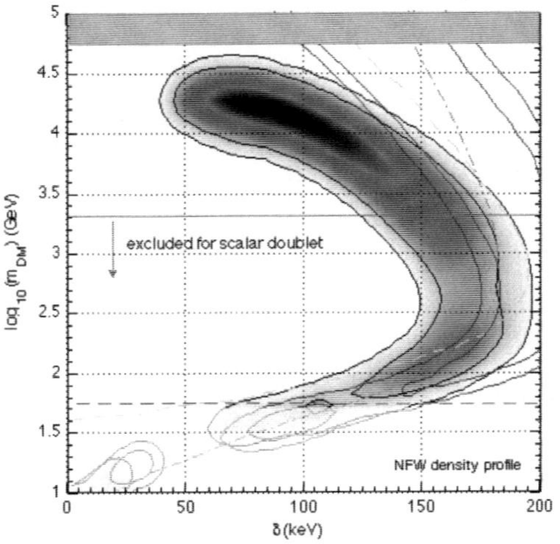

Figure 1. 2D credible regions in the plane $\{\delta - m_{\rm DM}\}$: individual experimental bounds and regions assuming the NFW DM density profile and marginalizing over the astrophysical uncertainties, combined in a single plot. For DAMA (shaded) and CRESST-II (solid blue) we show the 90% and 99% contours. The 90_S% contours are given by respectively the pink dot-dash curve for XENON100 ($\Delta\chi^2 = 3.1$) and the dashed green line for KIMS ($\Delta\chi^2 = 4.6$). The region below $m_{\rm DM} = 2$ TeV is excluded by $\chi_0 - \bar{\chi}_0$ oscillation (horizontal red solid line). The light gray line is the Z invisible decay width bound. The orange/dark gray region is excluded by unitarity bound.

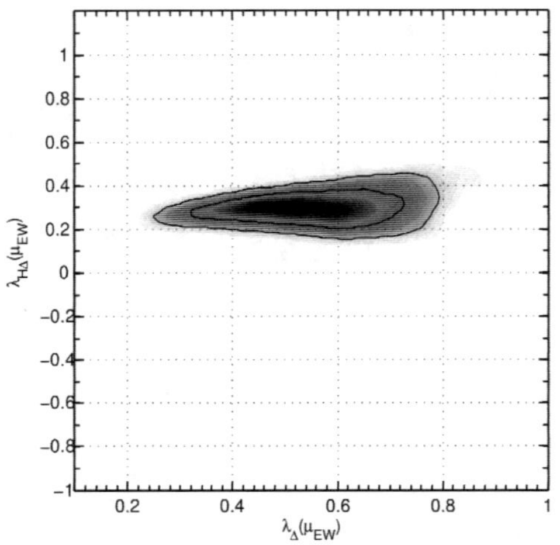

Figure 2. 2D marginal posterior in the $\{\lambda_\Delta, \lambda_{H\Delta}\}$-plane for mixed inflation at the EW renormalization scale $\mu_{\rm EW}$. The black solid lines enclose the 68% and 95% credible regions. All of other parameters entering in the renormalization group equations, specifically ξ_H, ξ_Δ, μ_H/M_Δ, f_H and f_ψ, have been marginalized over. The triplet mass is fixed at $M_\Delta = 10^8$ GeV, while the Higgs mass is set at $M_H = 125$ GeV. Predictions for the pure Higgs and pure triplet scenarios are very similar. See for details [7] as well as for the Markov Chain Monte Carlo (MCMC) setup.

3. Inflation from triplet scalar and Higgs boson

By adding non-minimal couplings to gravity for the Higgs and triplet, the action in the Jordan frame is given by:

$$S_J = \int d^4x \sqrt{-g} \left[\frac{R}{2} + \left(\xi_H H^\dagger H + \xi_\Delta \Delta^\dagger \Delta + c.c. \right) R - |\mathcal{D}_\mu H|^2 - |\mathcal{D}_\mu \Delta|^2 - V_J(H, \Delta) \right], \quad (4)$$

with the scalar content of the model described by the potential in equation 1. In order to have the usual interaction for gravity, it is usual to perform a conformal transformation from the Jordan frame to the Einstein frame, where the matter content ends up with non standard kinetic terms. The Higgs and triplet fields are defined in the unitarity gauge, for a total of three degree of freedom h, δ and θ that can be redefined in the large field limit as $r = \delta/h, \varphi = \sqrt{3/2}\log\left(1 + \xi_h h^2 + \xi_\Delta \delta^2\right)$ and θ. During inflation the mass value for the field r is always large compared to the Hubble parameter, hence this heavy field sets quickly in a minimum of the potential: we are left with an effective theory for the light inflatons φ and θ. Note that with respect to the quadratic term in the scalar potential the mass terms are negligible as far as $M_\Delta \leq 10^{12}$ GeV, while the μ_H term becomes negligible for $\mu_H < 10^{10}$ GeV. Hence only the quartic term contributes to inflation and is θ independent, namely this model is equivalent

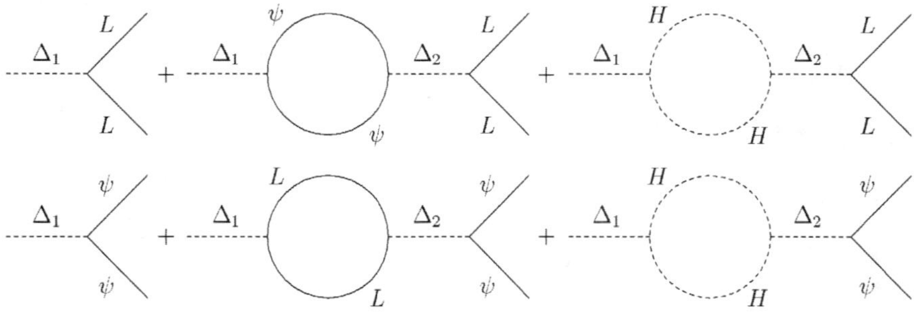

Figure 3. The CP asymmetry in the leptonic and dark sectors, first and second raw respectively, generated by the interference of tree and one-loop self energy correction diagrams of the scalar triplet Δ_1 and Δ_2.

to Higgs inflation:

$$V(\varphi) = V_0 \left(1 - e^{-2\varphi/\sqrt{6}}\right)^2 , \tag{5}$$

with

$$V_{\varphi-\text{indep}} = \begin{cases} V_0^{(\text{mixed})} = \frac{\lambda_\Delta \lambda_H - \lambda_{H\Delta}^2}{8\,(\lambda_\Delta\, \xi_H^2 + \lambda_H\, \xi_\Delta^2 - 2\lambda_{H\Delta}\xi_\Delta\xi_H)} , \\ V_0^{(H)} = \frac{\lambda_H}{8\xi_H^2} , \\ V_0^{(\Delta)} = \frac{\lambda_\Delta}{8\xi_\Delta^2} . \end{cases} \tag{6}$$

The three different V_0 arise from different minimum in which r sits: depending on its value, the inflaton is an admixture of Higgs and triplet or a pure Higgs or a pure triplet. In figure 2 we show as example of the model parameter space for a mixed inflaton, constrained by matching the amplitude of the power spectrum with the WMAP data. The parameters are fixed at EW and run up to unitarity scale using the appropriate renormalization group equations, with the inclusion of the triplet and DM contributions. Since the triplet is at a scale of 10^8 GeV, the inflationary constraints loosely bound the EW parameters, while λ_Δ or $\lambda_{H\Delta}$ are unaccessible at the low energy of the Large Hadron Collider (LHC).

4. Generation of asymmetries in the visible and dark sectors

From equations 2 and 3, note that there are three channels for triplet decay: $\Delta \to HH$, $\Delta \to LL$ and $\Delta \to \psi\psi$. Since the couplings are complex, the quasi-equilibrium decay of Δ via these channels produce asymmetries in lepton and DM sectors, as described in figure 3, with the requirement of having at least two scalar triplet Δ_1 and Δ_2. When the total decay rate of the lightest of the triplet mass eigenstate (ζ_1) fails to compete with the Hubble expansion rate of the universe, then ζ_1 decays and produces the asymmetries. These are proportional to the CP asymmetries ϵ_i (with $i = DM, H, L$), to the triplet yield $X_{\zeta_1} = n_{\zeta_1}/s$ (s is the entropy density) and to the efficiency factor η_i, that account for the washout effect on the asymmetries due to the number violating processes involving ψ, L and H. At a temperature above EW phase transition a part of the lepton asymmetry gets converted to the baryon asymmetry via the $SU(2)_L$ sphaleron processes. We solve the relevant Boltzmann equations for the quasi-equilibrium decays of ζ_1 and require that the relic abundance of the visible and dark matter are satisfied:

$$\frac{\Omega_{\text{DM}}}{\Omega_B} = \frac{1}{0.55} \frac{m_{\text{DM}}}{m_p} \frac{\epsilon_{\text{DM}}}{\epsilon_L} \frac{\eta_{\text{DM}}}{\eta_L} , \tag{7}$$

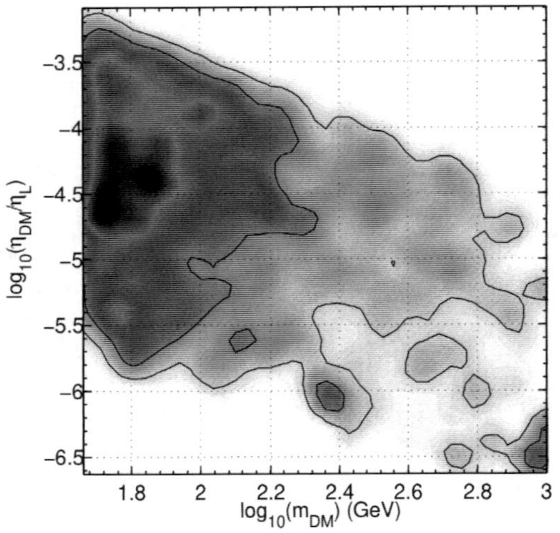

Figure 4. 2D credible regions at 68% and 95% C.L. in the $\{m_{\rm DM}, \eta_{\rm DM}/\eta_L\}$-plane. All of other parameters have been marginalized over.

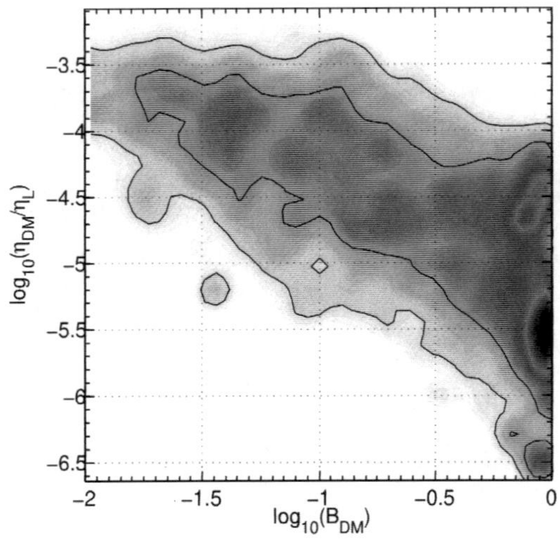

Figure 5. Same as left in the $\{B_{\rm DM}, \eta_{\rm DM}/\eta_L\}$-plane. For details about the MCMC sampling we refer to [7].

with m_p being the proton mass, as well as the asymmetry in the baryon sector matches the measured ones:

$$\frac{n_B}{n_\gamma} \sim 6.15 \times 10^{-10} = 7.02 \times 0.55 \times Y_L \,, \tag{8}$$

where Y_L is the yield for the leptons sector, defined as $Y_L = \eta_L \epsilon_L X_\varsigma$. The sampling over the parameter space of the model, given by the branching ratios $B_L, B_{\rm DM}$, the CP asymmetries ϵ_L and $\epsilon_{\rm DM}$ and the dark matter mass $m_{\rm DM}$ is performed via MCMC methods.

An overview of the results is shown in figure 4 and figure 5, where the 2D posterior pdfs are plotted for different parameters of the model. From figure 4 we see that all the mass range from 45 GeV up to 1 TeV can lead to successful leptogenesis and to an asymmetric dark matter candidate, namely satisfies equations 8 and 7. Note from the posterior pdf that the most favored region is for low mass candidates, even though there are candidates viable up to 1 TeV with smaller statistical significance. For a DM mass up to around 150 GeV, the preferred values of the ratio $\eta_{\rm DM}/\eta_L$ are of $\mathcal{O}(10^{-4} - 10^{-5})$, which compensate the large CP asymmetry ratio $\epsilon_{\rm DM}/\epsilon_L$. For DM masses around $\mathcal{O}$(TeV), $\epsilon_{\rm DM}/\epsilon_L$ is even larger for the preferred values of $\eta_{\rm DM}/\eta_L$, which decreases down to 10^{-6}. For a triplet mass of 10^8 GeV the important quantities which drive Boltzmann equations are the branching ratios. In figure 5 we show the correlation of $\eta_{\rm DM}/\eta_L$ versus $B_{\rm DM}$ with the 68% and 95 % credible regions. We see that the largest efficiency ratio $\eta_{\rm DM}/\eta_L$ is preferred when $B_{\rm DM} \to 1$, which implies a small $B_{\rm L} \to 10^{-3} - 10^{-4}$. This is because of the required hierarchy between the sub-eV neutrino mass and the Majorana mass splitting between the DM mass eigenstates, mentioned in section 2. In all cases the slow channel that builds and conserves the asymmetry is the leptonic one. The fast channel is either the Higgs, to compensate the neutrino mass in the total rate (see [17, 18] for accurate discussion) or the DM one. Since the DM channel is not related to the neutrino mass via the total rate, its branching ratio can assume different values all along the DM mass range. The small values for the efficiency ratio are compensated by the large CP asymmetry ratio, as confirmed in figure 5. We note that without this constraint on the hierarchy between Majorana masses the behavior for the CP asymmetry ratio and for the efficiency factor ratio would be the opposite ones, as shown in [9].

5. Conclusions

In these Proceedings we have discussed an extension of the SM with two heavy triplet scalars whose partial decay to SM leptons and inert doublet scalars (χ) or vector like fermions (ψ), could explain a common origin of asymmetric dark matter and visible matter through the leptogenesis mechanism. The induced vev of the triplets give rise to neutrino masses, as required by the oscillation experiments, via the type-II seesaw mechanism.

Regarding the DM phenomenology, odd under a Z_2 symmetry and thus stable, for the scalar candidate fast oscillations between χ_0 and $\overline{\chi_0}$ strongly deplete the asymmetry below EW symmetry breaking scale. Therefore, the survival of the asymmetry leads to a lower bound on the DM mass of 2 TeV so that the DM freezes out before it begins to oscillate. Both scalar and fermionic candidate scatters inelastically off nuclei in underground detectors. By performing a Bayesian analysis we found that an asymmetric scalars of mass larger than 2 TeV is excluded at 90% C.L. by the XENON100 data while an asymmetric fermionic DM of mass $\mathcal{O}(100)$ GeV is suitable to explain DAMA annual modulation signal within the 99% C.L. given by the XENON100 experiment. For the first time we propose it as explanation of the CRESST-II excess as well.

The indication at the LHC of a SM like Higgs boson with mass around 125-126 GeV suggests that the SM vacuum might be metastable. The scalar triplet extension of the SM evades as well the possibility of having a vacuum instability at least up to the unitarity scale. We introduced non-minimal couplings to gravity for both scalar triplet and the SM Higgs. In presence of these couplings the scalar triplet, mixed with the SM Higgs, drives inflation in the early Universe. We showed that the extended scalar potential gives rise to slow-roll single field inflation, once the heavy field is stabilized at a minimum of the potential. In general the inflaton is an admixture of triplet scalar and the SM Higgs or given by a pure Higgs or triplet field. Unfortunately it is not possible to measure the quartic couplings of the triplet at the LHC, because of its large mass. Hence it is not possible to distinguish between the three type of inflationary pictures and standard Higgs inflation. We leave for future work numerical treatment of Higgs inflation and its variant as well as future investigation on the triplet model related to Higgs properties, as soon as more data at LHC will be collected.

Acknowledgments

We thank Jinn-Ouk Gong for useful discussion. This work acknowledges use of the COSMO computing resource at CP3 of Louvain University.

References

[1] Lyth D H and Riotto A 1999 *Phys.Rept.* **314** 1–146 (*Preprint* hep-ph/9807278)
[2] Martin J and Ringeval C 2006 *JCAP* **0608** 009 (*Preprint* astro-ph/0605367)
[3] Bezrukov F and Shaposhnikov M 2008 *Phys.Lett.* **B659** 703–706 (*Preprint* 0710.3755)
[4] Chatrchyan S *et al.* (CMS Collaboration) 2012 *Phys.Lett.B* (*Preprint* 1207.7235)
[5] Aad G *et al.* (ATLAS Collaboration) 2012 (*Preprint* 1207.7214)
[6] Degrassi G, Di Vita S, Elias-Miro J, Espinosa J R, Giudice G F *et al.* 2012 (*Preprint* 1205.6497)
[7] Arina C, Gong J O and Sahu N 2012 (*Preprint* 1206.0009)
[8] Komatsu E *et al.* (WMAP Collaboration) 2011 *Astrophys.J.Suppl.* **192** 18 (*Preprint* 1001.4538)
[9] Arina C and Sahu N 2012 *Nucl.Phys.* **B854** 666–699 (*Preprint* 1108.3967)
[10] Tucker-Smith D and Weiner N 2001 *Phys. Rev.* **D64** 043502 (*Preprint* hep-ph/0101138)
[11] Bernabei R, Belli P, Cappella F, Cerulli R, Dai C *et al.* 2010 *Eur.Phys.J.* **C67** 39–49 (*Preprint* 1002.1028)
[12] Angloher G, Bauer M, Bavykina I, Bento A, Bucci C *et al.* 2012 *Eur.Phys.J.* **C72** 1971 (*Preprint* 1109.0702)
[13] Aprile E *et al.* (XENON100 Collaboration) 2012 (*Preprint* 1207.5988)
[14] Kim S, Bhang H, Choi J, Kang W, Kim B *et al.* 2012 *Phys.Rev.Lett.* **108** 181301 (*Preprint* 1204.2646)
[15] Ma E 2006 *Phys.Rev.* **D73** 077301 (*Preprint* hep-ph/0601225)
[16] Angloher G *et al.* 2008 (*Preprint* 0809.1829)
[17] Hambye T, Raidal M and Strumia A 2006 *Phys.Lett.* **B632** 667–674 (*Preprint* hep-ph/0510008)
[18] Chun E J and Scopel S 2007 *Phys.Rev.* **D75** 023508 (*Preprint* hep-ph/0609259)

Journal of Physics: Conference Series **485** (2014) 012040 doi:10.1088/1742-6596/485/1/012040

Leptogenesis in an S_3 model

Arturo Álvarez Cruz and Myriam Mondragón
Instituto de Física, Universidad Nacional Autónoma de México,
Apdo. Postal 20-360, 04515 México D.F., México
E-mail: aalvarez@fciencias.unam.mx

Abstract. We study the possibility of having leptogenesis in an S_3 extension of the Standard Model with three Higgs doublets and three right-handed neutrinos. In the leptonic sector we introduce also a Z_2 symmetry to reduce further the number of parameters. The characteristics of the model allow in a very natural way for resonant leptogenesis, which is highly dependent on the values of the Dirac and Majorana phases. The associated baryogenesis is also maximal when the neutrino masses are almost degenerate.

1. Introduction

It is well known that there are more baryons than antibaryons in the Universe. Big Bang Nucleosynthesis is a solid and consistent model of the creation of the nuclei in the early Universe, and predicts the following baryonic density

$$\eta = \frac{\eta_b - \eta_{\bar{b}}}{\eta_\gamma} = \eta = (2.6 - 6.2) \times 10^{-10}. \tag{1}$$

Measurements of the cosmic background radiation [1–3] show a density of

$$\eta = (6.1 \pm 0.3) \times 10^{-10} \ ,$$

in full agreement with the baryon density obtained from nucleosynthesis [2, 4]. This baryonic density can be express in terms of the entropy of the Universe,

$$Y_B \equiv \frac{\eta_b - \eta_{\bar{b}}}{s}|_0 = (8.77 \pm 0.24) \times 10^{11} \ . \tag{2}$$

Baryogenesis, which comprises all the processes that could lead to an abundance of matter over anti-matter, cannot be achieved with the amount of CP violation present in the Standard Model (SM). For this reason it is necessary to consider models beyond the SM that include new ways to increase either the production of baryons or the amount of CP violation, or both. An increasingly popular idea to achieve this is leptogenesis. Leptogenesis is a mechanism by which a leptonic asymmetry is generated, and then by sphaleron processes, this becomes a baryon asymmetry [5]. Several things are needed for the occurrence of leptogenesis: heavy right-handed neutrinos, that the neutrinos be of Majorana nature, and a possible decay of the right-handed neutrinos to the left-handed ones.

There have been many attempts to extend the SM [6, 7], one way to do it is through flavour symmetries, that relate the three families of particles known. Among the smallest more common

Content from this work may be used under the terms of the Creative Commons Attribution 3.0 licence. Any further distribution of this work must maintain attribution to the author(s) and the title of the work, journal citation and DOI.
Published under licence by IOP Publishing Ltd

groups to relate the three generations of particles are A_4, Q_6, S_4, and S_3. The three Higgs S_3 model [8–10] is an extension of the SM in which a symmetry representing invariance under the permutation of three objects is imposed. This model has, besides the quark and lepton particle content of the SM, three Higgs electroweak doublets and three right handed majorana neutrinos. The right handed neutrinos are related to the left ones through the see-saw mechanism (type I).

2. The S_3 model with three Higgs doublets

In the Standard Model analogous fermions in different generations have identical couplings to all gauge bosons of the strong, weak and electromagnetic interactions [11]. Prior to the electroweak symmetry breaking, the Lagrangian is chiral and invariant regarding to permutations of the left and right fermionic fields. The group S_3 consists of the six possible permutations of three objects (f_1, f_2, f_3), and is the smallest, discrete, non-abelian group. Its irreducible representations (irreps) are the two-dimensional one, and two singlet irreps, a symmetric and an anti-symmetric one. We will only consider the doublet and the symmetric singlet here, given by

$$\mathbf{f}_S = \frac{1}{\sqrt{3}}(f_1 + f_2 + f_3)\ ,\ \mathbf{f}_D^T = (\ \frac{1}{\sqrt{2}}(f_1 - f_2), \frac{1}{\sqrt{6}}(f_1 + f_2 - 2f_3)\ ).$$

The direct product of two doublets $p_D^T = (p_{D1}, p_{D2})$ and $q_D^T = (q_{D1}, q_{D2})$ may be decomposed into the direct sum of two singlets r_s and $r_{s'}$, and one doublet r_D^T. In this way we can associate the known particles to doublets or to singlets with the following rules: we accommodate the first two families in a doublet flavour representation, and the third one in the symmetric singlet one. We follow this same pattern for the Higgs fields and the right handed neutrinos. The quark and Higgs fields are denoted as

$$Q^T = (u_L, d_L), u_r, d_r, \quad L^T = (\nu_L, e_L), e_R, \nu_R \quad \text{and} \quad H. \tag{3}$$

The doublets carry capital subindices I and J, which run from 1 to 2, and the singlets are denoted by Q_3, u_{3R}, d_{3R}, L_3, e_{3R}, ${}_{3R}$ and H_S. Notice that the subscript 3 denotes the singlet representation and not the third generation. The most general renormalizable Yukawa interactions of this model are given by

$$L_Y = L_{Y_D} + L_{Y_U} + L_{Y_E} + L_{Y_\nu} + L_{Y_M} \tag{4}$$

where

$$\begin{aligned} L_{Y_\nu} = &- Y_1^\nu \overline{L_I}(i\sigma_2 H_S^* \nu_{IR}) - Y_3^\nu \overline{L_3}(i\sigma_2 H_S^* \nu_{3R}) \\ &- Y_2^\nu [\overline{L_I}\kappa_{IJ}(i\sigma_2 H_1^* \nu_{JR}) - \overline{L_I}\eta_{IJ}(i\sigma_2 H_2^* \nu_{JR})] \\ &- Y_4^\nu \overline{L_3}(i\sigma_2 H_I^* \nu_{IR}) - Y_5^\nu \overline{L_I}(i\sigma_2 H_I^* \nu_3 R) + h.c., \end{aligned} \tag{5}$$

with

$$\kappa = \begin{pmatrix} 0 & 1 \\ 1 & 0 \end{pmatrix} \qquad \eta = \begin{pmatrix} 1 & 0 \\ 0 & -1 \end{pmatrix}.$$

The Majorana mass terms for the right-handed neutrinos are given by

$$L_M = -M_1 \nu_{IR}^T C \nu_{IR} - M_3 \nu_{3R}^T C \nu_{3R}. \tag{6}$$

Due to the presence of three Higgs fields, the Higgs potential $V_H(H_S, H_D)$ is more complicated than that of the Standard Model. We assume that all the vacuum expectation values (vev's) of the Higgs fields are real and that $\langle H_1 \rangle = \langle H_2 \rangle$, which implies that besides the S_3 symmetry, it has a permutational symmetry $Z_2 : H_1 \leftrightarrow H_2$, which is not a subgroup of the flavour group

S_3 [12]. We will assume that the vacuum respects this accidental Z_2 symmetry. With these assumptions, the Yukawa interactions yield mass matrices, for all fermions in the theory, of the general form [8]

$$M = \begin{pmatrix} \mu_1 + \mu_2 & \mu_2 & \mu_5 \\ \mu_2 & \mu_1 - \mu_2 & \mu_5 \\ \mu_4 & \mu_4 & \mu_3 \end{pmatrix} . \quad (7)$$

The Majorana mass for the left handed neutrinos ν_L is generated by the see-saw mechanism. The corresponding mass matrix is given by

$$M = M_{\nu D} M^{-1} (M_{\nu D})^T, \quad (8)$$

where $M = \text{diag}(M_1, M_1, M_3)$. In principle, all entries in the mass matrices can be complex since there is no restriction coming from the flavour symmetry S_3 . The mass matrices are diagonalized by bi-unitary transformations as

$$U^\dagger_{d(u,e)L} M_{d(u,e)} U_{d(u,e)R} = \text{diag}(m_{d(u,e)}, m_{s(u,e)}, m_{b(u,e)})$$

$$U^T_\nu M_\nu U_\nu = \text{diag}(m_{\nu 1}, m_{\nu 2}, m_{\nu 3}) \ .$$

The entries in the diagonal matrices may be complex, so the physical masses are their absolute values. The mixing matrices are, $V_{CKM} = U^\dagger_{uL} U_{dL}$, $V_{PMNS} = U_{eL} U_\nu K$, where K is the diagonal matrix of the Majorana phase factors.

The theoretical mixing matrix, V_{PMNS}, is given by

$$V^{th}_{PMNS} = \begin{pmatrix} O_{11}\cos\eta + O_{31}\sin\eta e^{i\delta} & O_{11}\sin\eta - O_{31}\cos\eta e^{i\delta} & -O_{21} \\ -O_{12}\cos\eta + O_{32}\sin\eta e^{i\delta} & -O_{12}\sin\eta - O_{32}\cos\eta e^{i\delta} & O_{22} \\ O_{13}\cos\eta - O_{33}\sin\eta e^{i\delta} & O_{13}\sin\eta + O_{33}\cos\eta e^{i\delta} & O_{23} \end{pmatrix} \times K \ , \quad (9)$$

with

$$\sin^2\eta = \frac{m_{\nu 3} - m_{\nu 1}}{m_{\nu 2} - m_{\nu 1}} \ , \quad \cos^2\eta = \frac{m_{\nu 2} - m_{\nu 3}}{m_{\nu 2} - m_{\nu 1}}, \quad \delta = \delta_\nu - \delta_e.$$

Neglecting the smalls terms proportional to O_{11} and O^2_{11} we get

$$\tan^2\theta_{12} = \frac{(\Delta m^2_{12} + \Delta m^2_{13} + |m_{\nu 3}|^2 cos^2\phi_{nu})^{1/2} - |m_{\nu 3}||cos\phi_{nu}|}{(\Delta m^2_{13} + |m_{\nu 3}|^2 cos^2\phi_{nu})^{1/2} + |m_{\nu 3}||cos\phi_{nu}|} \ . \quad (10)$$

From this expression, we may readily derive expressions for neutrinos masses in terms of $\tan\theta_{12}, \phi_\nu$ and the differences of the squared masses. The sum of the neutrino masses is

$$\sum_{i=1}^{3} |m_{\nu i}| \approx \frac{\Delta m^2_{13}}{2\cos\phi_\nu \tan\theta_{12}} (1 + 2\sqrt{1 + 2\tan^2\theta_{12}(2\cos^2\phi_\nu 1) + \tan^4\theta_{12}\tan^2\theta_{12}}). \quad (11)$$

The most restrictive cosmological upper bound for this sum is

$$\sum |m_\nu| \leq 0.17eV \ . \quad (12)$$

The neutrino masses $|m_{\nu i}|$ assume their minimal values when $\cos\phi_\nu = 1$. When $\cos\phi_\nu$ takes values in the range $0.55 \leq \cos\phi \leq 1$, the neutrino masses change very slowly with $\cos\phi_\nu$. In the absence of experimental information we will assume that the Majorana phase ϕ_ν vanishes. Hence, setting $\phi_\nu = 0$ in our formula, we find

$$m_{\nu 1} = 0.052 \ eV, \quad m_{\nu 2} = 0.053 \ eV, \quad m_{\nu 3} = 0.019 \ eV. \quad (13)$$

The computed sum of the neutrino masses is

$$(\sum_{i=1}^{3} |m_{\nu i}|)^{th} = 0.13\ eV, \tag{14}$$

below the cosmological upper bound given in eq. (12), as expected, since we used the cosmological bound to fix the bound on $\cos\phi_\nu$. The most restrictive direct neutrino measurement involving electron type neutrinos, is based on fitting the shape of the beta spectrum. In such measurement, the quantity

$$\bar{m}_{\nu e} = \sqrt{\sum_i |V_{ei}|^2 m_{\nu i}}$$

is determined or constrained. A very restrictive upper bound for this sum is obtained from nucleosynthesis processes [13, 14]

$$(m_{\nu e})^{exp} < 0.37\ eV\ . \tag{15}$$

We find

$$(m_{\nu e})^{th} = 0.053\ eV\ , \tag{16}$$

again, well below the experimental upper bound given in eq. (15)

3. Leptogenesis in an S_3 model

The Yukawa couplings of the neutrinos allow the decay of the right-handed neutrinos to the left-handed ones $-Y_1^\nu \overline{L_I}(i\sigma_2 H_S^* \nu_{IR})$.
The asymmetry is defined as

$$\epsilon_1 = \frac{\sum_\alpha \Gamma(N_1 \to \ell_\alpha H) - \Gamma(N_1 \to \overline{\ell}_\alpha \overline{H})}{\sum_\alpha \Gamma(N_1 \to \ell_\alpha H) + \Gamma(N_1 \to \overline{\ell}_\alpha \overline{H})}. \tag{17}$$

where Γ is the decay rate, and N_1 is a singlet neutrino. The possible decays up to tree level are shown in fig. 1.

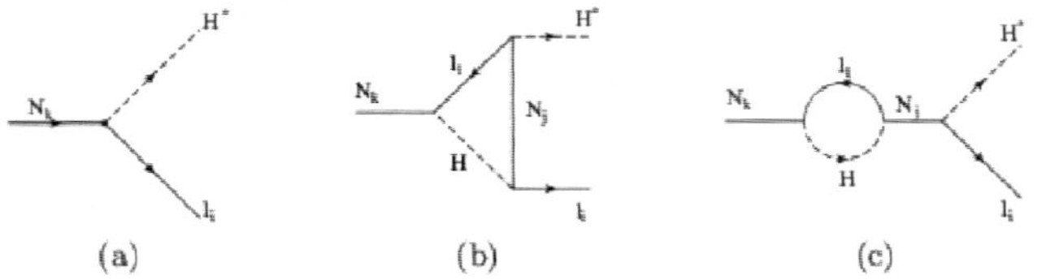

Figure 1. Feynman diagrams up to second order: a) tree level b) second order c) self energy diagram

The asymmetry generated by these decays is,

$$\epsilon \simeq -\frac{3}{8\pi}\frac{1}{(h_\nu h_\nu^\dagger)}\sum_{i=2,3} \mathrm{Im}\{(h_\nu h_\nu^\dagger)^2_{1i}\}[f(\frac{M_i^2}{M_1^2}) + g(\frac{M_i^2}{M_1^2})]\ , \tag{18}$$

and the self interactions are

$$f(x) = \sqrt{x}[1-(1+x)\ln(\frac{1+x}{x})]\ , \quad g(x) = \frac{\sqrt{x}}{1-x}.$$

This function depends strongly on the hierarchy of light neutrino masses. It can lead to a strong enhancement of the CP asymmetries if the masses M_2 and M_3 are nearly degenerate. The asymmetry for the S_3 model is given by

$$\epsilon = \frac{\mathrm{Im}[e^{2i\delta^*} M_2 m_3 \frac{\sqrt{M_2(m_2-m_3)(m_3-m_1)}}{\sqrt{m_3}}](f[\frac{M_3^2}{M_1^2}] + g[\frac{M_3^2}{M_1^2}])}{8\pi|M_2 m_3|} \tag{19}$$

The relation between the lepton and baryon asymmetry is given through the sphaleron process

$$Y_B = aY_{B-L} = \frac{a}{a-1}Y_L\ , \tag{20}$$

where $a = (8N_f = 4N_H)/(22N_f + 13N_H)$ and $Y_L \equiv \frac{n_L - n_{\bar{L}}}{s}$. We can express the lepton asymmetry in terms of the CP asymmetry

$$Y_L = \kappa\frac{\epsilon_i}{g*}\ ,$$

where g is 110, the number of relativistic degrees of freedom, κ is obtained from solving the Boltzmann equations, and it can be reparametrized in terms of K, defined as the ratio of Γ_1, the tree-level decay width of N_1 to H the Hubble parameter at temperature M_1. Thus, $\kappa = \Gamma_1/H < 1$ describes processes out of thermal equilibrium and $\kappa < 1$ describes washout effects

$$\kappa \approx \frac{0.3}{K(\ln(K))^{0.6}} \quad \text{for} \quad 10 < K < 10^6$$

$$\kappa \approx \frac{1}{2\sqrt{K^2+9}} \quad \text{for} \quad 0 < K < 10.$$

The decay width of N_1 by the Yukawa interaction at tree level and Hubble H parameter in terms of the temperature T and Planck scale M_{Pl} are

$$\Gamma_1 = (m_D^\dagger m_D)_{11} M_1/(8\pi v^2) \quad \text{and} \quad H = 1.66 g^{*1/2} T^2/M_{Pl}\ ,$$

respectively. At temperature $T = M_1$ the ratio K is

$$K = \frac{M_{Pl}}{1.66\sqrt{g*}(8\pi v^2)} \frac{(m_D^\dagger m_D)_{11}}{M_1} \tag{21}$$

The value of the baryon asymmetry on the S_3 model has a dependence on ϕ_ν and on the real masses of the neutrinos $m_1, M_1, m_2, M_2, m_3, M_3$, where the masses of the right-handed neutrinos are considered real.

We can calculate the dependence of the baryon asymmetry on the phases δ and ϕ_ν independently of the masses. Knowing the value of the masses would give the actual value of the baryon asymmetry. The maximum value is found on $\delta = 3/4\pi$ and $\phi = \pi$, this values are the same that make the best fit to the neutrino masses. As fig. 3 shows, leptogenesis in this model is highly dependent on values of the phases.

The value of the baryon asymmetry is determined by the masses of the light neutrinos and the difference of the masses of the right-handed neutrinos, the see-saw mechanism relates the masses of the right handed neutrinos making the right-handed neutrino masses larger than 10^{12} GeV. The maximal contribution to baryogensis occurs in the case where the three masses are nearly degenerate, basically almost equal.

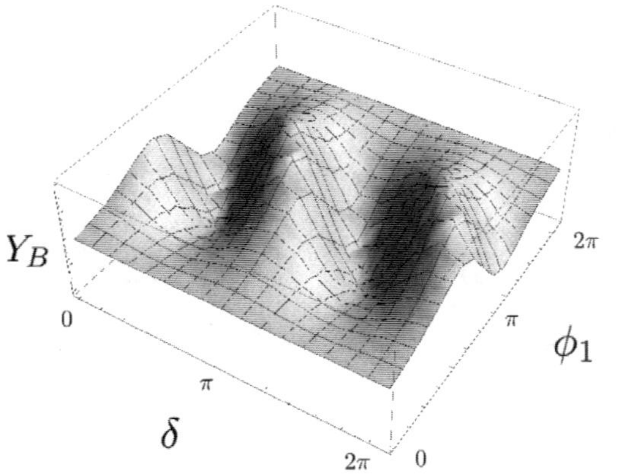

Figure 2. Baryon asymmetry in the S_3 flavour symmetry model. It can be seen the strong dependence not only on the Dirac phase δ, but also on the neutrino phase ϕ_ν.

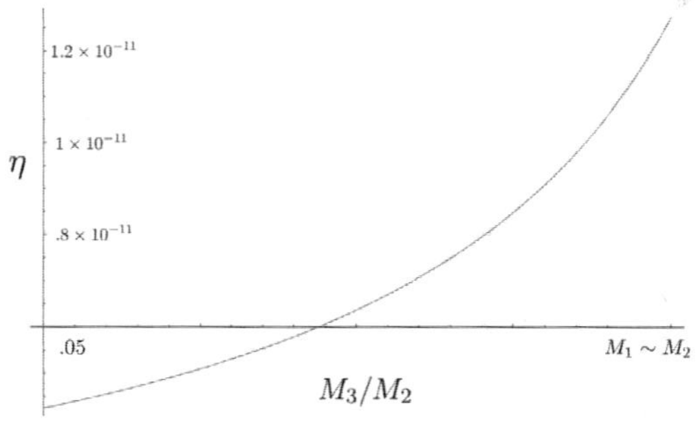

Figure 3. Baryogenesis dependence on the right-handed neutrino masses in the S_3 flavour symmetry. For leptogenesis to contribute to the explanation of the baryon asymmetry, an almost degenerate heavy ($M \sim 10^{12}$ GeV) right-handed neutrinos mass is required.

4. Conclusions

The S_3 extension of the SM with three Higgs doublets allows a definite structure of the Yukawa couplings, and we have studied its consequences regarding leptogenesis. The theoretical predictions obtained from the $S3 \otimes Z2$ symmetry are consistent with the experimental observations made to this time and the model allows leptogenesis as a mechanism to solve in part the baryonic asymmetry.

Acknowledgments

This work was supported by a PAPIIT-UNAM grant IN113412.

References

[1] Steigman G 1976 *Ann. Rev. Astron. Astrophys.* **14** 339–372
[2] Trodden M and Carroll S M 2004 (*Preprint* astro-ph/0401547)
[3] Bennett C *et al.* (WMAP) 2003 *Astrophys. J. Suppl.* **148** 97 (*Preprint* astro-ph/0302208)
[4] Lyth D H and Liddle A R 2009 *The primordial density perturbation: cosmology, inflation and the origin of structure; rev. version* (Cambridge: Cambridge Univ. Press)
[5] Fukugita M and Yanagida T 1986 *Phys.Lett.* **B174** 45
[6] Georgi H and Glashow S L 1974 *Phys. Rev. Lett.* **32** 438–441
[7] Georgi H In *Coral Gables 1975, Proceedings, Theories and Experiments In High Energy Physics*, New York 1975, 329-339
[8] Kubo J, Mondragon A, Mondragon M and Rodriguez-Jauregui E 2003 *Prog. Theor. Phys.* **109** 795–807 (*Preprint* hep-ph/0302196)
[9] Mondragon A, Mondragon M and Peinado E 2008 *AIP Conf. Proc.* **1026** 164–169 (*Preprint* 0712.2488)
[10] Felix O, Mondragon A, Mondragon M and Peinado E 2007 *AIP Conf. Proc.* **917** 383–389 (*Preprint* hep-ph/0610061)
[11] Mondragon A, Mondragon M and Peinado E 2008 *J. Phys.* **A41** 304035 (*Preprint* 0712.1799)
[12] Pakvasa S and Sugawara H 1978 *Phys.Lett.* **B73** 61
[13] Kolb E W and Turner M S 1991 *Phys.Rev.Lett.* **67** 5–8
[14] Fields B D, Kainulainen K and Olive K A 1997 *Astropart.Phys.* **6** 169–186 (*Preprint* hep-ph/9512321)

Tests of Lorentz and CPT violation with MiniBooNE neutrino oscillation excesses

Teppei Katori
Laboratory for Nuclear Science, Massachusetts Institute of Technology, Cambridge, MA 02139, U.S.A.
E-mail: katori@mit.edu

Abstract. Lorentz and CPT symmetry violaton is a predicted phenomenon of Planck-scale physics. Various types of data are analyzed to search for Lorentz violation under the Standard-Model Extension (SME) framework, including neutrino oscillation data. MiniBooNE is a short-baseline neutrino oscillation experiment at Fermilab. The measured excesses from MiniBooNE cannot be reconciled within the neutrino Standard Model (νSM); thus it might be a signal of new physics, such as Lorentz violation. We have analyzed the sidereal time dependence of MiniBooNE data for signals of the possible sidereal time dependence of the ocillation signals. we find that the ν_e appearance data prefer a sidereal time-independent solution, and the $\bar{\nu}_e$ appearance data slightly prefer a sidereal time-dependent solution, however, the statistical significance is not high to claim the discovery. Limits of order 10^{-20} GeV are placed on combinations of SME coefficients.

1. Introduction

Every fundamental symmetry needs to be tested, including Lorentz symmetry. The breakdown of Lorentz invariance naturally arises in different scenarios of physics at the Planck scale. For this reason the expected scale of Lorentz-violating phenomena is more than the Planck mass $m_P \simeq 10^{19}$ GeV, or in other words, Lorentz violation is expected to be suppressed until at least $\simeq 10^{-19}$ GeV in our energy scale. Lorentz symmetry is a fundamental symmetry both in quantum field theory and general relativity; the consequence of its violation would be enormous, and it seems it is impossible to establish a self-consistent theory with Lorentz violation. However, Lorentz violation can be incorporated into existing theories by spontaneous breaking [1]. In this way, Lorentz-violating terms in the Lagrangian do not conflict with the Standard Model (SM).

2. What is Lorentz violation?

We introduce Lorentz violation as coupling terms between ordinary SM fields and background fields in the universe. They are Lorentz scalars of the coordinate transformation; however since background fields are fixed in space, motion of the SM particles generates direction-dependent physics.

The situation is illustrated in Figure 1 [2]. The top cartoon (Fig. 1a) shows our setting: a SM particle is moving in two-dimensional space, from bottom to top, as seen by the local observer (Einstein). The space is filled with a hypothetical Lorentz-violating background field, a^μ (depicted by arrows). There are two ways to move this particle from left to right for the local observer: Particle Lorentz transformation and Observer Lorentz transformation.

Content from this work may be used under the terms of the Creative Commons Attribution 3.0 licence. Any further distribution of this work must maintain attribution to the author(s) and the title of the work, journal citation and DOI.
Published under licence by IOP Publishing Ltd

2.1. Particle Lorentz transformation

The first one is Particle Lorentz transformation, where the motion of a SM particle is actively transformed in the fixed coordinate system (Fig. 1b). Since the background field is unchanged, as a consequence, a coupling between the SM particle and the background field is not preserved; therefore, one can see Lorentz violation. In other words, Lorentz violation generally means the Particle Lorentz violation, and it implies direction-dependent physics of SM particles in the fixed coordinate space.

2.2. Observer Lorentz transformation

The second way is Observer Lorentz transformation, where the coordinate is inversely transformed (Fig. 1c). In this cartoon, if Einstein (the local observer) turns his neck 90° counterclockwise, the SM particle moves from left to right for the local observer. However, the background field is also transformed to the new coordinate system leaving the coupling with the SM particle unchanged. One cannot generate Lorentz violation by Observer Lorentz transformation, because Observer Lorentz transformation only corresponds to a coordinate transformation. In other words, coordinate transformations preserve the Lorentz-violating effect, and every observers agree with the same Lorentz-violating effect.

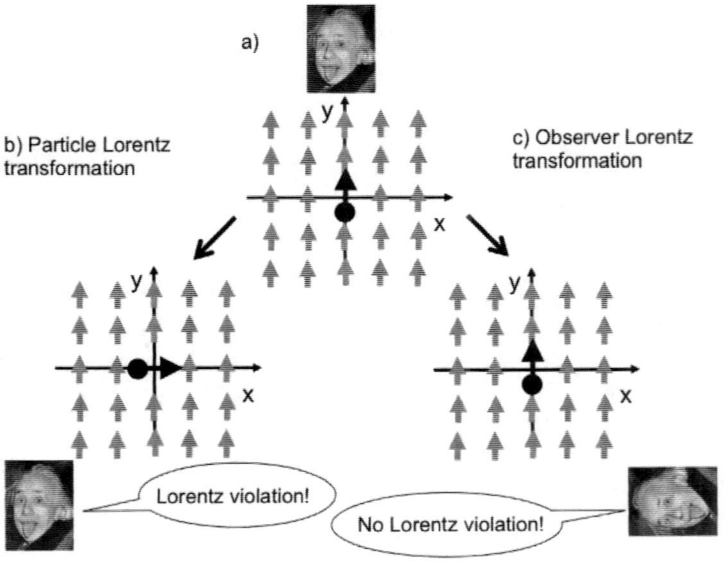

Figure 1. An illustration of Particle Lorentz transformation and Observer Lorentz transformation.

3. The MiniBooNE experiment

The MiniBooNE experiment is a short-baseline neutrino oscillation experiment at Fermilab, USA (2002-2012). Its primary goal is to find $\nu_\mu \to \nu_e$ ($\bar{\nu}_\mu \to \bar{\nu}_e$) oscillations with an ~800 (600) MeV neutrino (antineutrino) beam with an ~500 m baseline. Figure 2 shows the overview of the MiniBooNE experiment [2].

3.1. Booster neutrino beamline (BNB)

MiniBooNE uses neutrinos (antineutrinos) from the Booster neutrino beamline (BNB) [3], which is illustrated in Figure 2, bottom. The 8 GeV protons, the "primary" beam, are extracted from the Booster and steered to collide with the beryllium target in the magnetic focusing horn. The collision of protons and the target makes a shower of mesons, the "secondary" beam; and the toroidal field created by the horn focuses π^+ (π^-) for neutrino (antineutrino) mode.

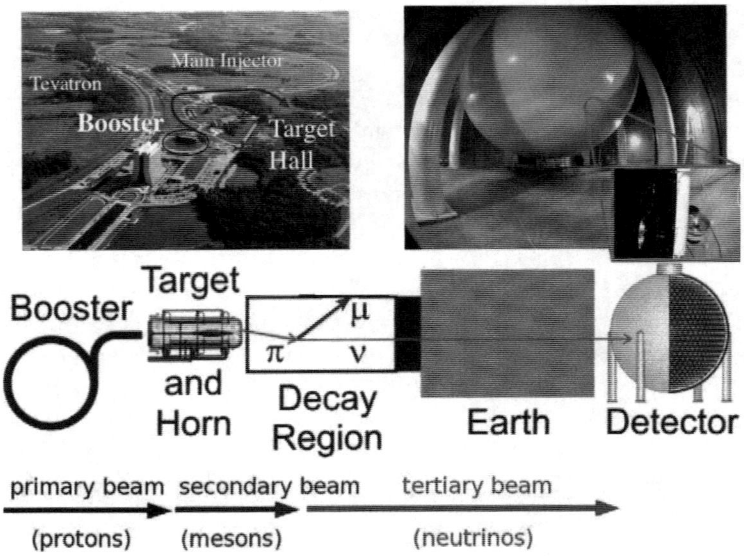

Figure 2. An overview of the MiniBooNE experiment. The top left picture shows the Fermilab site, including the Booster. The top right picture is the tank of the MiniBooNE detector. The bottom cartoon shows the sketch of the Booster Neutrino Beamline (BNB).

3.2. MiniBooNE detector

The MiniBooNE detector is located 541 m to the north of the target [4]. The detector is a 12.2 m diameter spherical Cherenkov detector, filled with mineral oil. An optically separated inner black region is covered with 1,280 8-inch PMTs, and an outer white region has 240 8-inch PMTs which act as a veto (Fig. 2, top right).

3.3. Events in detector

The time and charge information of the Cherenkov ring from the charged particle is used to estimate particle type, energy, and direction [5]. The most important reaction for the oscillation analysis is the charged current quasielastic (CCQE) interaction [6], which is characterized by one outgoing charged lepton (at the BNB energy, protons seldom exceed a Cherenkov threshold of $\sim$350 MeV kinetic energy). If a charged lepton is detected from the CCQE interaction, one can reconstruct the neutrino energy, E_ν^{QE}, by assuming the target nucleon is at rest and the interaction type is truly CCQE [6] (QE assumption).

$$E_\nu^{QE} = \frac{2(M_n - B)E_l - ((M_n - B)^2 + m_l^2 - M_p^2)}{2 \cdot [(M_n - B) - E_l + \sqrt{E_l^2 - m_l^2}\cos\theta_l]}. \qquad (1)$$

Here, M_n, M_p, and m_l are the neutron, proton, and lepton masses, E_l is the total lepton energy, θ_l is the lepton scattering angle, and B is the binding energy of carbon. Ability to reconstruct neutrino energy is essential for neutrino oscillation physics, since neutrino oscillations are function of neutrino energy.

3.4. Oscillation analysis

The signature of the $\nu_\mu \to \nu_e$ ($\bar{\nu}_\mu \to \bar{\nu}_e$) oscillation is the single, isolated electron-like Cherenkov ring produced by the CCQE interaction.

$$\nu_\mu \overset{oscillation}{\longrightarrow} \nu_e + n \to e^- + p\ ,$$
$$\bar{\nu}_\mu \overset{oscillation}{\longrightarrow} \bar{\nu}_e + p \to e^+ + n\ .$$

Note, since MiniBooNE was not magnetized, electrons and positrons were not distinguished. Thus, the analysis of neutrino mode and antineutrino mode is reasonably parallel, except for some differences in handling backgrounds.

There are two backgrounds which contribute equally to our signals. The first class is the "misID", and this is dominated by a single gamma ray from the neutral current channels, such as radiative Δ decay and π° production, where only one gamma ray is detected. Another major background is the intrinsic background, namely ν_e ($\bar{\nu}_e$) as beam contamination.

3.5. MiniBooNE oscillation analysis results

Figure 3 and 4 show the ocillation candidate events after all cuts. The data points with errors are shown together with predicted backgrounds. For neutrino mode data analysis (Fig. 3), an excess of ν_e candidate events in the low-energy region (200< E_ν^{QE}(MeV)<475) was observed [7]. For the antineutrino mode analysis (Fig. 4), an excess is observe in the entire oscillation analysis region (200< E_ν^{QE}(MeV)<1300) [8] [1]. The νSM does not predict these excesses, which therefore has the potential to be new physics.

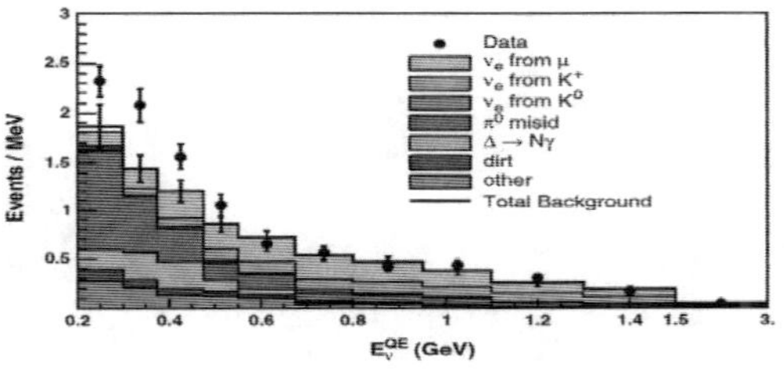

Figure 3. MiniBooNE neutrino mode ν_e appearance search result.

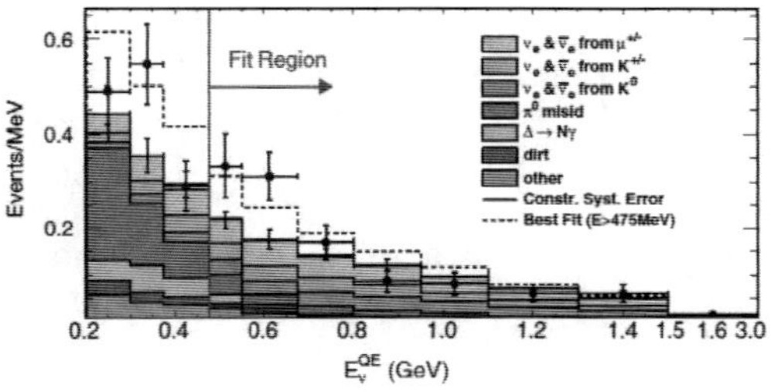

Figure 4. MiniBooNE antineutrino mode $\bar{\nu}_e$ appearance search result.

4. Lorentz-violating neutrino oscillation analysis in MiniBooNE

4.1. Standard-Model Extension (SME)

For the general search for Lorentz violation, the Standard-Model Extension (SME) [10] is widely used by the community. Various types of data are analyzed under SME [11, 12], including neutrino oscillation data [13, 14, 15]. The SME is a self-consistent effective field theory including Particle Lorentz violation. In principle, SME is an infinite series of all types of interactions, but many analyses are limited to the renormalizable sector, called the minimal SME. The neutrino sector minimal SME Lagrangian can be written with eight additional couplings with neutrinos and background fields, on top of SM vacuum Lagrangian [16].

By assuming the baseline is short enough compared to the oscillation length, under the minimal SME, the $\nu_\mu - \nu_e$ oscillation probability can be written as follows [17],

$$P_{\nu_\mu \to \nu_e} \simeq \frac{L^2}{(\hbar c)^2} |\, (\mathcal{C})_{e\mu} + (\mathcal{A}_s)_{e\mu} \sin \omega_\oplus T_\oplus + (\mathcal{A}_c)_{e\mu} \cos \omega_\oplus T_\oplus + (\mathcal{B}_s)_{e\mu} \sin 2\omega_\oplus T_\oplus + (\mathcal{B}_c)_{e\mu} \cos 2\omega_\oplus T_\oplus \,|^2. \quad (2)$$

Here, $\omega_\oplus$ stands for the sidereal time angular frequency ($\omega_\oplus = \frac{2\pi}{86164.1}$ rad/s), as opposed to the solar time angular frequency ($\omega_\odot = \frac{2\pi}{86400.0}$ rad/s). The neutrino oscillation is described by the

[1] The time of this analysis, only the half of antineutrino mode data was available. The full statistics antineutrino result is available recently [9].

function of the sidereal time $T_\oplus$, with the sidereal time independent amplitude $(\mathcal{C})_{e\mu}$, and the sidereal time dependent amplitudes, $(\mathcal{A}_s)_{e\mu}$, $(\mathcal{A}_c)_{e\mu}$, $(\mathcal{B}_s)_{e\mu}$, and $(\mathcal{B}_c)_{e\mu}$. Thus, an analysis of Lorentz and CPT violation in neutrino oscillation data involves fitting the data with Eq. 2 to find nonzero sidereal time dependent amplitudes. These five parameters are the combinations of fourteen SME coefficients [17]. Therefore, this analysis at most find five combinations of fourteen SME coefficients by the fit.

Neither the neutrino nor the antineutrino mode data allow us to extract errors if we fit all five parameters at once (Eq. 2), due to the high correlation of parameters. Therefore, we focus on "three"-parameter fit, by turning off $(\mathcal{B}_s)_{e\mu}$ and $(\mathcal{B}_c)_{e\mu}$, to discuss errors and limits. Since the full five-parameter fit is quantitatively similar to the three-parameter fit, we will focus the discussion of these results on the three-parameter fit.

5. Results

This section describes the results of the fits.

5.1. Fit result of neutrino mode data

Figure 5 shows the neutrino mode low energy region fit results [18]. The solid and dash-dotted lines are the best fit curves from three- and five-parameter fit. The dotted line is the flat solution. Since the fit is dominated by the $(\mathcal{C})_{e\mu}$ parameter (sidereal time-independent amplitude), both three- and five-parameter fit solutions have small time-dependent amplitudes and look like the flat solution. The details of the fit results were tested by using a fake data set. We constructed a fake data set with signal to evaluate errors of fit parameters of three-parameter fit ($(\mathcal{C})_{e\mu}$, $(\mathcal{A}_s)_{e\mu}$, and $(\mathcal{A}_c)_{e\mu}$). Then we constructed a fake data set without signal, to evaluate the compatibility with flat solution over three parameter fit solution by $\Delta\chi^2$ method. It turns out data is compatible with flat solution over a 26.9%, and it concludes ν_e candidate data is consistent with flat.

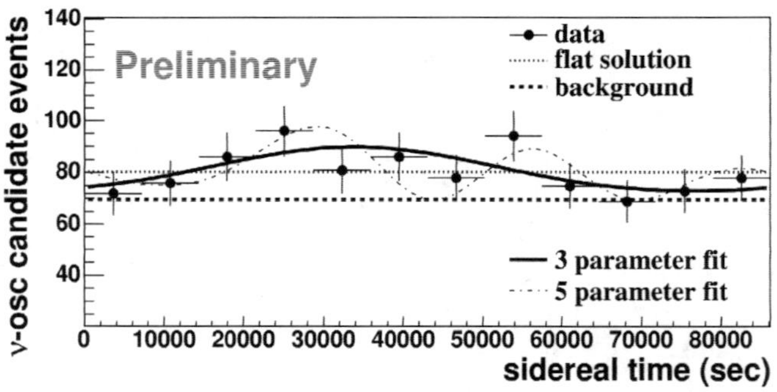

Figure 5. The fit results for the neutrino mode low energy region. The plot shows the curves corresponding to the flat solution (dotted), three-parameter fit (solid), five-parameter fit (dash-dotted), and fitted background (dashed), together with binned data (solid marker).

5.2. Fit result of antineutrino mode data

Figure 6 shows the fit result in antineutrino mode, which is analogous with Figure 5 [18]. The fit result is more curious in antineutrino mode, because now the $(\mathcal{C})_{e\mu}$ parameter no longer significantly deviates from zero, but the fit favors a nonzero solution for the $(\mathcal{A}_s)_{e\mu}$ and $(\mathcal{A}_c)_{e\mu}$ parameters. The fit solutions look more different from the flat distribution. We again constructed a fake data set to find the significance of this solution, and it turns out that the compatibility with the flat solution is now only 3.0%. Although this is interesting, the significance is not high enough to claim discovery.

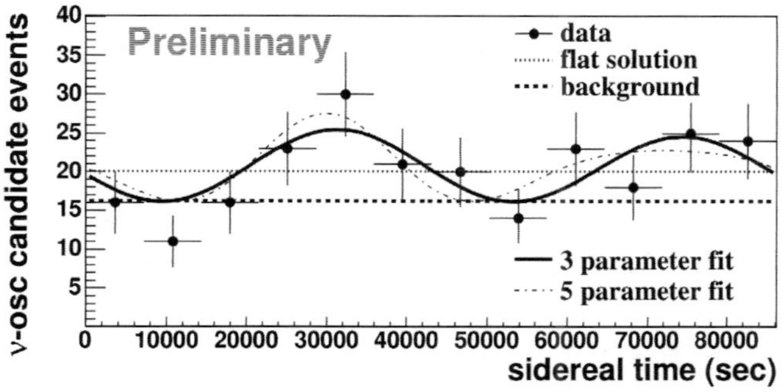

Figure 6. The fit results for the antineutrino mode combined region. Notations are the same as Fig. 5. The fitted background is shown as a dashed line, and the best fit for the background is 0.97 (3% lower than the default prediction).

Conclusion

Lorentz and CPT violation is a predicted signal at the Planck scale, and there are worldwide efforts to search for it. Neutrino oscillation is a natural interferometer, and the sensitivity to Lorentz violation is comparable with precise optical experiments. The MiniBooNE short-baseline neutrino oscillation experiment at Fermilab observed a ν_e ($\bar{\nu}_e$) candidate event excess from a ν_μ ($\bar{\nu}_\mu$) beam. These excesses are not understood by νSM; therefore, they might be the first signal of new physics. The analysis found that neutrino mode data prefer a sidereal time independent solution, and the data is compatible with a flat distribution over 26.9%. The antineutrino mode prefers a sidereal time dependent solution, and the data is compatible with a flat solution only 3.0%, making this solution very interesting; however, the statistical significance is not high enough to claim as evidence. Since the data set we used for this analysis is about ~50% of the total antineutrino mode data set, reanalysis including full data set may increase the significance.

Acknowledgments

I thank PASCOS 2012 organizer to invite me to the conference. This work is supported by NSF PHYS-084784.

References

[1] For example, see, Kostelecký V A and Samuel S 1989 *Phys. Rev.* **D39** 683
[2] Katori T (MiniBooNE Collaboration) 2012 *Mod. Phys. Lett.* **A27** 1230024
[3] Aguilar-Arevalo A A *et al.* 2009 *Phys. Rev.* **D79** 072002
[4] Aguilar-Arevalo A A *et al.* 2009 *Nucl. Instrum. Meth.* **A599** 28
[5] Patterson R B *et al.* 2009 *Nucl. Instrum. Meth.* **A608** 206
[6] Aguilar-Arevalo A A *et al.* 2008 *Phys. Rev. Lett.* **100** 032301; *ibid. Phys. Rev.* **D81** 092005
[7] Aguilar-Arevalo A A *et al.* 2009 *Phys. Rev. Lett.* **102** 101802
[8] Aguilar-Arevalo A A *et al.* 2010 *Phys. Rev. Lett.* **105** 181801
[9] Aguilar-Arevalo A A *et al.* 2012 (*Preprint* `arXiv:1207.4809v2 [hep-ex]`)
[10] Colladay D and Kostelecký V A 1997 *Phys. Rev.* **D55** 6760; *ibid.* 1998 *Phys. Rev.* **D58** 116002; Kostelecký V A 2004 *Phys. Rev.* **D69** 105009
[11] For recent reviews see, for example, *CPT and Lorentz Symmetry V*, edited by Kostelecký V A 2011 (World Scientific, Singapore)
[12] Kostelecký V A and Russell N 2011 *Rev. Mod. Phys.* **83** 11
[13] Auerbach *et al.* L B 2005 *Phys. Rev.* **D72** 076004
[14] Adamson P *et al.* 2008 *Phys. Rev. Lett.* **101** 151601; *ibid.* 2010 *Phys. Rev. Lett.* **105** 151601; *ibid.* 2012 *Phys. Rev.* **D85** 031101
[15] Abbasi R *et al.*, 2010 *Phys. Rev.* **D82** 112003
[16] Kostelecký V A and Mewes M 2004 *Phys. Rev.* **D69** 016005
[17] Kostelecký V A and Mewes M 2004 *Phys. Rev.* **D70** 076002
[18] Aguilar-Arevalo A A *et al.* 2011 (*Preprint* `arXiv:1109.3480v2 [hep-ex]`)

The MAJORANA DEMONSTRATOR: Progress towards showing the feasibility of a tonne-scale ^{76}Ge neutrinoless double-beta decay experiment

P Finnerty[1,2], E Aguayo[3], M Amman[4], F T Avignone III[5,6], A S Barabash[7], P J Barton[8], J R Beene[6], F E Bertrand[6], M Boswell[9], V Brudanin[10], M Busch[11,2], Y-D Chan[8], C D Christofferson[12], J I Collar[13], D C Combs[14,2], R J Cooper[6], J A Detwiler[8], P J Doe[15], Yu Efremenko[16], V Egorov[10], H Ejiri[17], S R Elliott[9], J Esterline[11,2], J E Fast[3], N Fields[13], F M Fraenkle[1,2], A Galindo-Uribarri[6], V M Gehman[9], G K Giovanetti[1,2], M P Green[1,2], V E Guiseppe[18], K Gusey[10], A L Hallin[19], R Hazama[17], R Henning[1,2], E W Hoppe[3], M Horton[12], S Howard[12], M A Howe[1,2], R A Johnson[15], K J Keeter[20], M F Kidd[9], A Knecht[15], O Kochetov[10], S I Konovalov[7], R T Kouzes[3], B D LaFerriere[3], J Leon[15], L E Leviner[14,2], J C Loach[8], P N Luke[4], S MacMullin[1,2], M G Marino[15], R D Martin[8], J H Merriman[3], M L Miller[15], L Mizouni[5,3], M Nomachi[17], J L Orrell[3], N R Overman[3], G Perumpilly[18], D G Phillips II[14,2], A W P Poon[8], D C Radford[6], K Rielage[9], R G H Robertson[15], M C Ronquest[9], A G Schubert[15], T Shima[17], M Shirchenko[10], K J Snavely[1,2], D Steele[9], J Strain[1,2], V Timkin[10], W Tornow[11,2], R L Varner[6], K Vetter[8,21], K Vorren[1,2], J F Wilkerson[1,2,6], E Yakushev[10], H Yaver[4], A R Young[14,2], C-H Yu[6] and V Yumatov[7]
The MAJORANA Collaboration

[1]Department of Physics and Astronomy, University of North Carolina, Chapel Hill, NC, USA
[2]Triangle Universities Nuclear Laboratory, Durham, NC, USA
[3]Pacific Northwest National Laboratory, Richland, WA, USA
[4]Engineering Division, Lawrence Berkeley National Laboratory, Berkeley, CA, USA
[5]Department of Physics and Astronomy, University of South Carolina, Columbia, SC, USA
[6]Oak Ridge National Laboratory, Oak Ridge, TN, USA
[7]Institute for Theoretical and Experimental Physics, Moscow, Russia
[8]Nuclear Science Division, Lawrence Berkeley National Laboratory, Berkeley, CA, USA
[9]Los Alamos National Laboratory, Los Alamos, NM, USA
[10]Joint Institute for Nuclear Research, Dubna, Russia
[11]Department of Physics, Duke University, Durham, NC, USA
[12]South Dakota School of Mines and Technology, Rapid City, SD, USA
[13]Department of Physics, University of Chicago, Chicago, IL, USA
[14]Department of Physics, North Carolina State University, Raleigh, NC, USA
[15]Center for Experimental Nuclear Physics and Astrophysics and Department of Physics, University of Washington, Seattle, WA, USA
[16]Department of Physics and Astronomy, University of Tennessee, Knoxville, TN, USA

[21] Alternate Address: Department of Nuclear Engineering, University of California, Berkeley, CA, USA

Content from this work may be used under the terms of the Creative Commons Attribution 3.0 licence. Any further distribution of this work must maintain attribution to the author(s) and the title of the work, journal citation and DOI.
Published under licence by IOP Publishing Ltd

[17]Research Center for Nuclear Physics and Department of Physics, Osaka University, Ibaraki, Osaka, Japan
[18]Department of Physics, University of South Dakota, Vermillion, SD, USA
[19]Centre for Particle Physics, University of Alberta, Edmonton, AB, Canada
[20]Department of Physics, Black Hills State University, Spearfish, SD, USA

E-mail: paddy@unc.edu

Abstract. The MAJORANA DEMONSTRATOR will search for the neutrinoless double-beta decay ($0\nu\beta\beta$) of the ^{76}Ge isotope with a mixed array of enriched and natural germanium detectors. The observation of this rare decay would indicate the neutrino is its own anti-particle, demonstrate that lepton number is not conserved, and provide information on the absolute mass-scale of the neutrino. The DEMONSTRATOR is being assembled at the 4850 foot level of the Sanford Underground Research Facility in Lead, South Dakota. The array will be contained in a low-background environment and surrounded by passive and active shielding. The goals for the DEMONSTRATOR are: demonstrating a background rate less than 3 t^{-1} y^{-1} in the 4 keV region of interest (ROI) surrounding the 2039 keV ^{76}Ge endpoint energy; establishing the technology required to build a tonne-scale germanium based double-beta decay experiment; testing the recent claim of observation of $0\nu\beta\beta$ [1]; and performing a direct search for light WIMPs (3-10 GeV/c^2).

1. Introduction

The MAJORANA collaboration [2, 3, 4] will search for the neutrinoless double-beta decay ($0\nu\beta\beta$) of ^{76}Ge. The observation of this rare decay would indicate the neutrino is its own anti-particle, demonstrate that lepton number is not conserved, and provide information on the absolute mass-scale of the neutrino (see Ref. [5] for a recent review of $0\nu\beta\beta$ theory). Reaching the neutrino mass-scale associated with the inverted mass hierarchy, $20-50$ meV, will require a half-life sensitivity on the order of 10^{27} y. This corresponds to a signal of a few counts or less per tonne-year in the $0\nu\beta\beta$ peak (2039 keV for ^{76}Ge). To observe such a rare signal, one will need to construct tonne-scale detectors with backgrounds in the region of interest at or below $\sim$1 t^{-1} y^{-1}. The MAJORANA collaboration is constructing the DEMONSTRATOR, an array of high-purity germanium (HPGe) detectors at the 4850 foot level of the Sanford Underground Research Facility (SURF) in Lead, South Dakota. The DEMONSTRATOR will consist of a mixture of natural (10-15 kg) and >86% enriched ^{76}Ge (up to 30 kg) HPGe detectors in two low-background cryostats. Each cryostat will contain seven closely-packed stacks, called strings, and each string will have up to five crystals (see Figs. 1 and 2). The DEMONSTRATOR aims to:

(i) demonstrate a background rate less than 3 t^{-1} y^{-1} in the 4 keV region of interest (ROI) surrounding the 2039 keV ^{76}Ge endpoint energy;

(ii) establish the technology required to build a tonne-scale germanium based $0\nu\beta\beta$ experiment;

(iii) test the recent claim of observation of $0\nu\beta\beta$ [1];

(iv) and perform a direct search for light WIMPs (3-10 GeV/c^2).

The MAJORANA and GERDA [6] collaborations are working together to prepare for a single tonne-scale ^{76}Ge experiment that will combine the best technical features of both experiments. There are two main differences between the experimental techniques employed by MAJORANA and GERDA. First, the DEMONSTRATOR array will be deployed in a custom vacuum cryostat, whereas GERDA is submerging theirs in liquid argon. Secondly, the DEMONSTRATOR will use a compact shield with lead, oxygen-free copper, electroformed copper, and scintillator paddles, whereas GERDA is using the liquid argon and high-purity water as a shield.

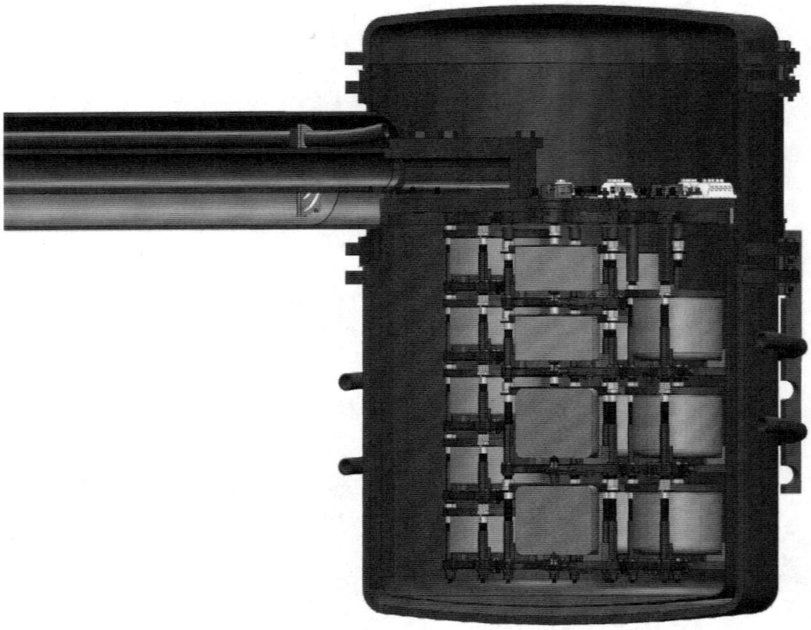

Figure 1. A cross sectional view of a MAJORANA DEMONSTRATOR cryostat. The strings within the cryostat hold a mixture of natural (smaller/light blue) and enriched (larger/dark blue) germanium detectors (Color online).

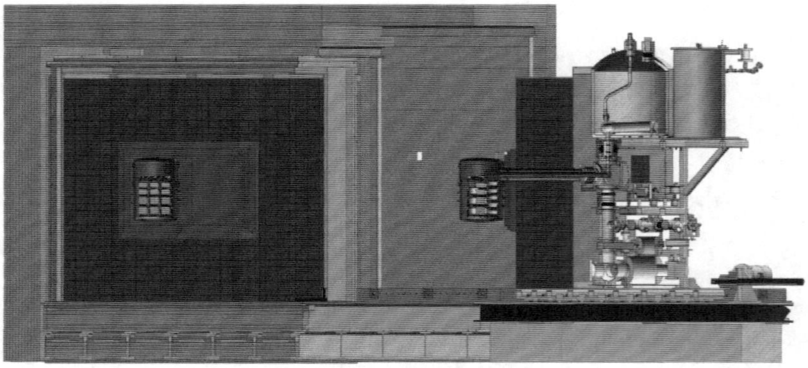

Figure 2. The MAJORANA DEMONSTRATOR is shown here with both active and passive shielding in place. One cryostat is in place inside the shield while the other is being positioned for insertion. For scale, the inner copper shield is 20" high and 30" in length (Color online).

2. Detector Technology

The MAJORANA collaboration will use p-type point contact (PPC) HPGe detectors. These detectors [7, 8] have been demonstrated to provide both exceptional energy resolution (<2.0 keV FWHM at 1332 keV, <4.0 keV FWHM at 2039 keV) and low-energy thresholds (~500 eV) [8, 9], see e.g. Figure 1 in Ref. [9]. Several successful prototypes have been commercially produced and successfully operated in an underground environment [9, 10, 11, 12, 13]. The PPC detectors used in the DEMONSTRATOR will each have a mass of 0.6-1.0 kg.

3. Background Mitigation Techniques

One of the technical goals of the DEMONSTRATOR is to show that a 1 t^{-1} y^{-1} ROI^{-1} background rate is achievable for a tonne-scale experiment; this scales to 3 t^{-1} y^{-1} ROI^{-1} background rate for the DEMONSTRATOR, which is ~100 times lower than previous Ge experiments. One of the primary methods for achieving this rate is to deploy the detectors inside two independent cryostats that minimize the amount of interstitial material (see Figs. 1 and 2). In addition, the materials used to fabricate the DEMONSTRATOR have been screened and selected based on strict radiopurity requirements. The main structural material closest to the detectors within the DEMONSTRATOR is electroformed copper due to its intrinsically low-background and excellent physical properties. Electroformed copper is made by electroplating from high-purity commercial oxygen-free copper nuggets onto stainless steel mandrels. The electroforming process greatly reduces the radioactivity due to U, Th and cosmogenically-produced ^{60}Co. Electroformed copper is being used to fabricate both of the cryostats for the DEMONSTRATOR. A prototype cryostat has been built from commercial oxygen-free copper and is currently being tested. The successful performance of the prototype cryostat will give the MAJORANA collaboration confidence in the future performance of both electroformed copper cryostats.

Further background reduction in the DEMONSTRATOR is achieved using the technological features of the PPC detectors. A $0\nu\beta\beta$ event will deposit all of its energy within a ~1 mm^3 region inside the PPC detector. These are called single-site events (SSE). In contrast, gamma-rays from radioactive contaminants of sufficient energy to affect the 2039 keV ROI will typically Compton scatter several times with a scattering length of ~1 cm. These are called multi-site events (MSE). The DEMONSTRATOR will reduce Compton scattered gamma-ray backgrounds by implementing sophisticated pulse shape analysis (PSA) techniques to separate SSE from MSE within a single Ge crystal. In addition to signal PSA, background events that deposit energy in more than one detector can be removed with an anti-coincidence cut.

Cosmogenically-produced isotopes represent another source of background. Cosmic rays can create ^{60}Co in copper components as well as ^{68}Ge, ^{60}Co and tritium within the detectors themselves. In an effort to minimize these backgrounds, the surface exposure of detectors and copper components will be minimized. Also, all copper components within the cryostats are being electroformed underground to reduce the amount of ^{60}Co activation. Reducing surface exposure is the only option for most cosmogenically-produced isotopes. However, the background due to ^{68}Ge is a special case. ^{68}Ge decays to ^{68}Ga which has a half-life of 67.71 minutes. The low-thresholds and excellent energy resolution of PPC detectors allow for a single-site time-correlation (SSTC) cut, which looks forward and backward in time from the current event in the ROI to search for signatures of parent or daughter isotopes [14, 15]. The DEMONSTRATOR will implement a SSTC cut for the ^{68}Ge-^{68}Ga coincident decay for five ^{68}Ga half-lives following the 10 keV K-shell de-excitations. The SSTC method can also be used to reduce backgrounds due to ^{208}Tl ($T_{1/2}$ = 3.05 min) and ^{214}Bi ($T_{1/2}$ = 19.9 min) in the Ge crystals and the inner mount. The SSTC and SSE/MSE cuts will both be used in order to maximize background reduction.

4. Background Modeling Tools

The MAJORANA collaboration uses a Monte Carlo simulation framework, MaGe [16], which is co-developed by the MAJORANA and GERDA collaborations, to estimate background contributions to the ROI. MaGe is based on the GEANT4 [17, 18] and ROOT [19] packages and provides a common set of physics processes, event generators, and various management classes. MaGe also provides a user-interface for output classes and detector geometries. The radioactivities of all components within the DEMONSTRATOR have been measured. The results of these measurements are used with MaGe to develop a comprehensive background model for the DEMONSTRATOR.

5. Dark Matter

The low-energy performance of PPC detectors, due to their low-capacitance point-contact design, makes them suitable for dark matter (DM) searches. The direct detection of DM, which makes up ~23% of the Universe, remains an active area of research. Weakly interacting massive particles (WIMPs), a generic class of potential DM candidates, are widely regarded as the most promising candidate [20]. Recent direct searches for WIMPs have hinted towards a low-mass WIMP at ~7 GeV/c^2 [9, 11, 12, 13, 21]. The favored ~7 GeV/c^2 WIMP is also consistent with the theoretical model proposed by Hooper and Goodenough [22] to explain the excess flux of gamma-rays from the Galactic center. With a 500 eV threshold, the sensitivity of the DEMONSTRATOR to a 7 GeV/c^2 WIMP is expected to reach 10^{-43} cm^2 (normalized to nucleon) allowing it to test these recent claims [10, 13, 23].

As a part of the research and development (R&D) efforts for the DEMONSTRATOR, the MAJORANA collaboration has deployed a custom PPC detector (MALBEK) in the Kimballton Underground Research Facility (KURF) in Ripplemead, Virginia, at a depth of 1450 meters water equivalent [24]. MALBEK is a 465 g natural Ge crystal housed in a low-background cryostat and shield. The data acquisition (DAQ) and slow-control system for MALBEK was built around ORCA, an object oriented data acquisition that provides an easy to use, graphical interface for manipulation of experimental hardware and data streams [25]. The goals for this R&D effort are to:

(i) test a MAJORANA DEMONSTRATOR-like DAQ;
(ii) validate the MAJORANA DEMONSTRATOR background model;
(iii) systematically characterize the low-energy spectrum;
(iv) and perform a light WIMP search.

6. DEMONSTRATOR Implementation

The MAJORANA DEMONSTRATOR prototype cryostat is expected to be completed by late 2012 and will contain two strings of natural Ge PPC crystals. The electroformed cryostats, Cryostat 1 and Cryostat 2, are being fabricated in a phased approach. Cryostat 1 will contain seven strings of Ge crystals, four holding natural Ge crystals and three containing Ge crystals enriched to greater than 86% ^{76}Ge. Cryostat 1 is expected to be completed towards the end of 2013. Cryostat 2 will contain seven strings of Ge crystals, all of which enriched to greater than 86% ^{76}Ge. Cryostat 2 is expected to be completed by the end of 2014. The MAJORANA DEMONSTRATOR will search for $0\nu\beta\beta$ and low-mass WIMPs once fabrication is completed at SURF. In addition, the MAJORANA DEMONSTRATOR should be able to verify or refute the recent observational claim of $0\nu\beta\beta$ [1] within 2-3 years of commissioning Cryostat 1.

Acknowledgments

We acknowledge support from the Office of Nuclear Physics in the DOE Office of Science under grant numbers DE-AC02-05CH11231, DE-FG02-97ER41041, DE-FG02-97ER41033, DE-FG02-97ER4104, DE-FG02-97ER41042, DE-SCOO05054, DE-FG02-10ER41715, and DE-FG02-

97ER41020. We acknowledge support from the Particle and Nuclear Astrophysics Program of the National Science Foundation through grant numbers PHY-0919270, PHY-1003940, 0855314, and 1003399. We gratefully acknowledge support from the Russian Federal Agency for Atomic Energy. We gratefully acknowledge the support of the U.S. Department of Energy through the LANL/LDRD Program. N. Fields is supported by the DOE/NNSA SSGF program. G. K. Giovanetti is supported by the Department of Energy Office of Science Graduate Fellowship Program (DOE SCGF), made possible in part by the American Recovery and Reinvestment Act of 2009, administered by ORISE-ORAU under contract no. DE-AC05-06OR23100.

References

[1] Klapdor-Kleingrothaus H V and Krivosheina I V 2006 *Mod. Phys. Lett. A* **21** 1547 – 1566
[2] Aguayo E *et al.* 2011 *AIP Conf. Proc.* **1417** 95–99
[3] Wilkerson J F *et al.* 2012 *J. Phys. Conf. Ser.* **375** 042010
[4] Schubert A *et al.* (MAJORANA Collaboration) 2012 *AIP Conf. Proc.* **1441** 480–482
[5] Vergados J D, Ejiri H and Simkovic F 2012 *Rep. Prog. Phys.* **75** 106301
[6] Schönert S *et al.* (GERDA Collaboration) 2005 *Nucl. Phys. B (Proc. Suppl.)* **145** 242 – 245
[7] Luke P, Goulding F, Madden N and Pehl R 1989 *IEEE Trans. Nucl. Sci.* **36** 926
[8] Barbeau P S, Collar J I and Tench O 2007 *J. Cosm. Astro. Phys.* **0709** 009
[9] Aalseth C E *et al.* (CoGeNT Collaboration) 2008 *Phys. Rev. Lett.* **101** 251301
[10] Barbeau P S 2009 *Neutrino and Astroparticle Physics with P-Type Point Contact High Purity Germanium Detectors* Ph.D. thesis University of Chicago
[11] Aalseth C E *et al.* (CoGeNT Collaboration) 2011 *Phys. Rev. Lett.* **106**(13) 131301
[12] Aalseth C E *et al.* (CoGeNT Collaboration) 2011 *Phys. Rev. Lett.* **107**(14) 141301
[13] Marino M G 2010 *Dark matter physics with p-type point-contact germanium detectors: extending the physics reach of the* MAJORANA *experiment* Ph.D. thesis University of Washington
[14] Ejiri H 2004 *Czech. J. Phys.* **54** B299
[15] Ejiri H 2005 *J. Phys. Soc. Japan* **74** 2101
[16] Boswell M *et al.* 2011 *IEEE Trans. Nucl. Sci.* **58** 1212 –1220
[17] Agostinelli S *et al.* 2003 *Nucl. Inst. & Meth. A* **506** 250 – 303
[18] Allison J *et al.* 2006 *IEEE Trans. Nucl. Sci.* **53** 270 –278
[19] Brun R and Rademakers F 1997 *Nucl. Inst. & Meth. A* **389** 81–86 URL `http://root.cern.ch/`
[20] Beringer J *et al.* (Particle Data Group) 2012 *Phys. Rev. D* **86**
[21] Angloher G, Bauer M, Bavykina I, Bento A, Bucci C *et al.* 2011 ArXiv:astro-ph/1109.0702
[22] Hooper D and Goodenough L 2011 *Phys. Lett. B* **697** 412 – 428
[23] Giovanetti G K *et al.* 2012 *J. Phys. Conf. Ser.* **375** 012014
[24] Finnerty P, Collar J, Giovanetti G, Henning R, Marino M, Schubert A and Wilkerson J 2010 Astroparticle physics with a customized low-background broad energy germanium detector *Nuclear Science Symposium Conference Record (NSS/MIC), 2010 IEEE* pp 671 –673
[25] Howe M, Cox G, Harvey P, McGirt F, Rielage K, Wilkerson J and Wouters J 2004 *IEEE Trans. Nucl. Sci.* **51** 878 – 883

Journal of Physics: Conference Series **485** (2014) 012043 doi:10.1088/1742-6596/485/1/012043

Constraining $0\nu\beta\beta$

Luis Dorame

Instituto de Fisica Corpuscular - C.S.I.C./Universitat de Valencia// Edificio Institutos de Paterna, Apt 22086, E-46071 Valencia, Spain

E-mail: `dorame@ific.uv.es`

Abstract. A class of discrete flavor-symmetry-based models predicts constrained neutrino mass matrix schemes that lead to specific neutrino mass sum-rules (MSR). We show how these theories may constrain the absolute scale of neutrino mass, leading in most of the cases to a lower bound on the neutrinoless double beta decay effective amplitude.

1. Introduction

The discovery of neutrino oscillations implies non-vanishing neutrino masses and mixing providing one of the most solid indications for physics beyond the Standard Model. The fact that neutrinos have very tiny masses, in contrast to charged leptons and quarks, and that two of the mixing angles are large, is among the deepest theoretical puzzles in particle physics. Since neutrinos carry no electric charge, they are expected on general grounds to be Majorana particles [1], leading to the existence of lepton number violating processes. This intriguing possibility will be hopefully confirmed by the observation of neutrinoless double beta decay ($0\nu\beta\beta$) processes [2, 3]. Indeed, upcoming $0\nu\beta\beta$ experiments are expected to improve the sensitivity by up to about one order of magnitude.

The observed pattern of lepton mixing angles suggests the existence of an underlying flavor symmetry of some sort, either an Abelian [4] symmetry or a non-Abelian one [5]. In the former case one typically obtains texture zeros for the mass matrices but is unable to predict mixing angles. In contrast, non-Abelian flavor symmetries are potentially more powerful, allowing also in principle for mixing angle predictions. As an example, several realizations of non-Abelian discrete flavor symmetry schemes lead to an effective neutrino mass matrix which corresponds to a numerical (parameter-free) prediction for lepton mixing. A popular example of such neutrino mass matrix M^ν is the tri-bimaximal (TBM) [6] type, characterized by

$$M^\nu = M_{TBM} \equiv \begin{pmatrix} x & y & y \\ y & x+z & y-z \\ y & y-z & x+z \end{pmatrix}, \tag{1}$$

which depends only on three complex parameters x, y and z. In the mass eigentstate basis the three complex parameters x, y and z would correspond to three neutrino mass parameters plus two Majorana CP phases [1, 7], as the Dirac phase disappears since $\theta_{13} = 0$. Many such schemes are characterized by a specific (complex) relation among the parameters x, y and z (see [8] for a complete list of models that predicts such relationships), leaving only two free complex parameters, further reducing the number of independent model parameters describing the lepton

Content from this work may be used under the terms of the Creative Commons Attribution 3.0 licence. Any further distribution of this work must maintain attribution to the author(s) and the title of the work, journal citation and DOI.
Published under licence by IOP Publishing Ltd

sector. In the mass basis these correspond to only two independent neutrino mass eigenvalues (the other follows from the existence of a neutrino mass sum rule), plus two Majorana CP phases (as mentioned, the Dirac phase is unphysical).

2. Mass relations

In this section we focus on a general sub-class of mass matrices leading to a numerical (parameter-free) prediction for the lepton mixing matrix, where the following types of mass relations hold:

$$A) \quad \chi\, m_2^\nu + \xi\, m_3^\nu = m_1^\nu, \tag{2}$$

$$B) \quad \frac{\chi}{m_2^\nu} + \frac{\xi}{m_3^\nu} = \frac{1}{m_1^\nu}, \tag{3}$$

$$C) \quad \chi\sqrt{m_2^\nu} + \xi\sqrt{m_3^\nu} = \sqrt{m_1^\nu}\,. \tag{4}$$

Here $m_i^\nu = m_i^0$ denote neutrino mass eigenvalues, up to a Majorana phase factor, while χ and ξ are free parameters which specify the model, taken to be positive without loss of generality. For the sake of completeness, we also consider a fourth mass relation,

$$D) \quad \frac{\chi}{\sqrt{m_2^\nu}} + \frac{\xi}{\sqrt{m_3^\nu}} = \frac{1}{\sqrt{m_1^\nu}}\,. \tag{5}$$

.

These kind of mass sum-rule can be obtained in the scenario of inverse seesaw mechanism [9, 10, 11], which arises when introducing a fermion singlet S with opposite lepton number with respect to the right-handed neutrinos, so that the effective light neutrino mass matrix is $m_\nu = m_D \frac{1}{M}\mu\frac{1}{M}m_D^T$. Assuming m_D and μ to be proportional to the identity matrix and $M \sim M_{TBM}$, it is straightforward to show that we can obtain the mass sum-rule of type *(D)*.

3. Lower bound for neutrinoless double-β decay

Let us first consider the amplitude for neutrinoless double-β decay within a flavor-generic model. One can plot the effective neutrino mass parameter $|m_{ee}|$ determining the $0\nu\beta\beta$ decay amplitude, as a function of the lightest neutrino mass. As is well-known, by varying the neutrino oscillation parameters Δm^2_{atm}, Δm^2_{sol}, θ_{12}, θ_{13}, θ_{23} in their allowed ranges one obtains two types of relatively broad bands in the $(|m_{ee}|, m^\nu_{light})$ plane corresponding to normal and inverse hierarchy spectra.

In this "generic" case there is a lower bound on the neutrinoless double-β decay effective mass parameter $|m_{ee}|$ only in the case of inverse mass hierarchy: due to the possibility of destructive interference among the light neutrinos from the effect of having opposite CP signs or due to the effect of majorana phases, no lower bound can be established for the case of normal hierarchy [12, 13, 14].

Let us now turn to the case where MSR relations like *(A)*,*(B)*,*(C)* and *(D)* hold. As discussed above these can be obtained in flavour models where the neutrino mass matrix only depends on two independent free parameters, so that the resulting mixing angles are fixed, like for example for the tri-bimaximal or bimaximal mixing patterns.

For definiteness here we focus on the case where the rotation in the neutrino sector is of tri-bimaximal form. Corrections from higher dimensional operators and/or from the charged lepton sector can yield $\theta_{13} \neq 0$.

Hence we retain the TBM approximation as a useful starting point to obtain our MSR relations. However, when evaluating a lower bound on the effective neutrino mass parameter $|m_{ee}|$ determining the neutrinoless double-β decay amplitude, we include explicitly the effects of non-vanishing θ_{13}. We do this by taking the values at 3 σ determined in Ref. [15]. Such a lower bound can be obtained from the following procedure.

We first consider that the neutrino masses are complex parameters, where the two Majorana phases are encoded in m_2^ν and m_3^ν, i.e.

$$m_1^\nu = m_1^0, \tag{6}$$
$$m_2^\nu = m_2^0 e^{i\alpha}, \tag{7}$$
$$m_3^\nu = m_3^0 e^{i\beta}. \tag{8}$$

As shown in Fig. 1, the neutrino mass sum-rule can then be interpreted geometrically as a triangle in the complex plane, whose area provides a measure of Majorana CP violation [1]. Each

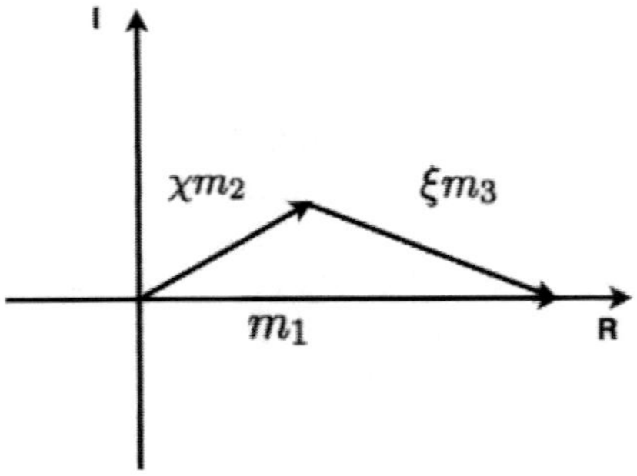

Figure 1. Neutrino mass sum-rule triangle in the complex plane.

of the above equations (6), (7) and (8) can be split into two independent equations for the real and imaginary parts.

For simplicity let us start from the idealized case where the neutrino oscillation parameters Δm^2_{atm}, Δm^2_{sol}, θ_{12}, θ_{13}, θ_{23} are perfectly well-measured quantities. In this case one can extract the two Majorana phases α and β as functions of the base of the triangle, which is determined by m_1^0 in case of normal hierarchy (NH) or by m_3^0 in case of inverted hierarchy (IH), as well as the parameters χ and ξ labeling the particular model under consideration.

These relations obtained can then be inserted into the general expression of $|m_{ee}|$:

$$|m_{ee}| = \left| c_{12}^2 c_{13}^2 m_1 + s_{12}^2 c_{13}^2 e^{i\alpha} m_2 + s_{13}^2 e^{i\beta} m_3 \right| . \tag{9}$$

For each (χ, ξ) model this effective mass parameter depends on a single parameter, namely the length of the triangle base, which gives a measure of the abolute scale of neutrino mass.

For instance, for case *(A)* this procedure gives:

$$\xi \cos\alpha \, m_2^0 + \chi \cos\beta \, m_3^0 = m_1^0, \quad \xi \sin\alpha \, m_2^0 + \chi \sin\beta \, m_3^0 = 0\,, \tag{10}$$

[1] As the area shrinks to zero one obtains the CP-conserving limits corresponding to the four independent choices of CP sign [12, 13].

so that the Majorana CP phases are determined as:

$$\begin{aligned} \cos\alpha &= \frac{m_1^2 - \chi^2\left(\Delta m_{atm}^2 + m_1^2\right) + \xi^2\left(\Delta m_{sol}^2 + m_1^2\right)}{2m_1\xi\sqrt{\Delta m_{sol}^2 + m_1^2}} \\ \cos\beta &= \frac{m_1^2 + \chi^2\left(m_1^2 + \Delta m_{atm}^2\right) - \xi^2\left(m_1^2 + \Delta m_{sol}^2\right)}{2m_1\chi\sqrt{\Delta m_{atm}^2 + m_1^2}} . \end{aligned} \tag{11}$$

The lower bound for the lightest neutrino mass can be obtained from our MSR, using the triangle inequality in the complex plane as suggested in [16], see Fig.(1) for a schematic view.

We must first select the biggest side of the triangle; calling them x_1, x_2 and x_3, then the triangle inequality $|x_i| \leq |x_j| + |x_k|$ must be fulfilled, where $|x_i| \equiv Max(|x_1|, |x_2|, |x_3|)$ and $i \neq j \neq k$ [2]. In case *(A)* and assuming NH for the neutrino mass spectrum, we always have $(\chi m_2^0, \xi m_3^0) > m_1^0$ and the largest side of the triangle can be either χm_2^0 or ξm_3^0; so we must consider separately these two cases. After rewriting two masses in terms of the two squared mass differences, we can obtain a lower limit for the lightest neutrino mass from the triangle inequality $|x_i| \leq |x_j| + |x_k|$. For the other cases we follow the same procedure. The lower bound on the lightest neutrino mass obtained in this way is then used to estimate the lower bound for $|m_{ee}|$ from the general expression in eq. (9). Notice that, although we focus here on TBM schemes, some of the MSR considered in our analysis can also be derived using bimaximal mixing as a starting point.

Classification

χ, ξ	A–NH	A–IH	Ref.	B–NH	B – IH	Ref.	C– NH	C –IH	Ref.	D –NH	D– IH	Ref.
1,1	0.010	0.044	†	0.008	0.036	†	0.006	0.029	-	0.005	0.008	-
1,2	⋆	0.046	-	0.008	0.027	-	⋆	0.014	-	0.004	0.026	†
1,3	⋆	0.011	-	0.030	0.005	-	⋆	0.014	-	0.018	0.025	-
2,1	0.006	⋆	†	0.006	0.007	†	0.000	⋆	†	⋆	0.007	-
2,2	0.019	0.026	-	0.023	0.008	-	0.017	⋆	-	0.003	0.015	-
2,3	⋆	0.046	†	0.007	0.008	-	⋆	0.031	-	0.005	0.026	-
3,1	0.004	⋆	-	0.004	0.008	-	⋆	⋆	-	⋆	⋆	-
3,2	0.011	⋆	-	0.004	0.021	-	0.000	⋆	-	⋆	0.007	-
3,3	0.023	0.061	-	0.029	0.031	-	0.011	0.019	-	0.018	0.016	-

Table 1. Minimal values for the effective $0\nu\beta\beta$ decay mass parameter $|m_{ee}|$, in eV, see text for details.

The results obtained from the procedure discussed above are summarized in Tab.1, where we report the lower limits of $|m_{ee}|$ corresponding to different integer choices of (χ, ξ) between 1 and 3 and for each of the four MSR considered in eqs. (2)-(5), for both normal and inverted hierarchies. The cases already discussed in the literature are indicated by a †, see [8] for a complete list of references. The entries denoted with the symbol ($\star$) represent situations that do not satisfy the inequality for any value of the lightest neutrino mass m_{light}. Cases marked by a (-) correspond to models which, as far as we can tell, have not been considered. Some comments

[2] Notice that there are three inequalities of the type $|x_i| \leq |x_j| + |x_k|$ obtained by permuting the three indices i, j and k, but only one of these constrains the lightest neutrino mass.

are in order. First let us consider the effect of a possible non-zero effect of θ_{13}. In Fig 2 we show the prediction for $|m_{ee}|$ as function of m_{light} obtained from the MSR $\frac{1}{\sqrt{m_2}} + \frac{2}{\sqrt{m_3}} = \frac{1}{\sqrt{m_1}}$. The red band corresponds to the normal hirarchy case and the green band the inverse hierarchy case, varying the values of θ_{13}, θ_{23} and θ_{12} within their 3σ C.L. interval. By looking at Fig. 2

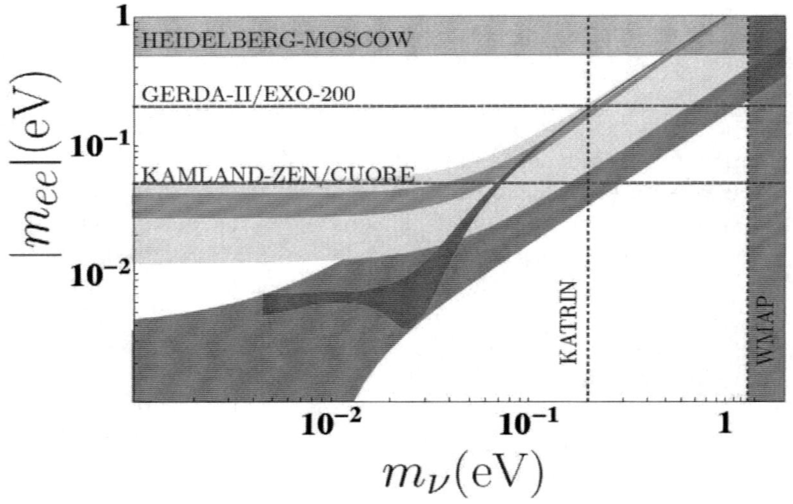

Figure 2. $\langle |m_{ee}| \rangle$ as a function of the lightest neutrino mass corresponding to the mass sum-rule $\frac{1}{\sqrt{m_2}} + \frac{2}{\sqrt{m_3}} = \frac{1}{\sqrt{m_1}}$ [11]. The gray and yellow bands are the allowed range for $\langle |m_{ee}| \rangle$ for the normal hierarchy (NH) and inverse hierarchy (IH) respectively. The red (NH) and green(IH) are the predicted values for the MSR varying the values of θ_{13}, θ_{23} and θ_{12} to 3σ C.L. range.

one sees that, indeed, the $0\nu\beta\beta$ lower bound is sensitive to the value of θ_{13}.

One also finds that, as expected on general gounds, all inverse hierarchy schemes corresponding to various choices of (χ, ξ) within sum-rules A-D have a lower bound for the parameter $|m_{ee}|$. However, the numerical value obtained depends on the MSR scheme, signalling that not all values within the corresponding band in Fig. 2 are covered.

On the other hand, even though normal hierarchy models do not lead to a lower bound on the $0\nu\beta\beta$ amplitude due to the possibility of destructive interference amongst the light neutrinos, one finds that the possibility of full cancellation is precluded for all schemes in the table, except for the (2,1) case considered in Ref. [17] and the (3,2) scheme, both of which correspond to MSR of type (C). All other NH MSR schemes considered here imply a minimum value for the $0\nu\beta\beta$ decay amplitude [3].

In particular, the maximal value we have found for the lower bound on $|m_{ee}|$ is $|m_{ee}| = 0.061\,\text{eV}$, obtained in correspondence with the set of values $(\chi, \xi) = (3, 3)$ for the case *(A)* in IH. Such a value for $|m_{ee}|$ lies within the sensitivity of upcoming experiments; hence it would be interesting, from the model building point of view, to find from first principles a flavor-symmetry-based model predicting such a mass relation.

Mass sum rules are classified in four different categories, some have already been considered in the literature. For each case, we have first extracted the allowed numerical values of $|m_{ee}|$, for both mass orderings of the neutrino mass eigenstates and we have then given the behaviour

[3] Of course some of the bounds are phenomenologically less interesting since they fall outside realistic sensitivities of coming experiments.

of $|m_{ee}|$ as a function of the lightest neutrino mass. Although our MSR schemes were obtained within the TBM anzatz, we have computed the possible values of $|m_{ee}|$ considering all the neutrino parameters (including a non-vanishing θ_{13}) within their 3σ allowed ranges. In most MSR schemes one finds a lower bound for the $0\nu\beta\beta$ amplitude, even for NH spectra. We find that the most favourable case (large lower bound) corresponds to a sum-rule of type (A) obtained in correspondence of the set of values $(\chi, \xi) = (3, 3)$, $|m_{ee}| = 0.061\,\mathrm{eV}$.

Acknowledgments

This talk was based on work in collaboration with Eduardo Peinado, Stefano Morisi, Davide Meloni and J.W.F. Valle. LD is supported by a CSIC JAE predoctoral fellowship, by the Spanish MEC under grants FPA2008-00319/FPA, FPA2011-22975 and MULTI-DARK CSD2009-00064 (Consolider-Ingenio 2010 Programme), by Prometeo/2009/091 (Generalitat Valenciana) and by the EU ITN UNILHC PITN-GA-2009-237920.

References

[1] Schechter J and Valle J. W. F. 1980 *Phys. Rev.* **D22**, 2227
[2] Schechter J and Valle J W F 1982 *Phys. Rev.*D **D25**, 2951
[3] Duerr M, Lindner M and Merle A 2011 *J. High Energy Phys.* JHEP 1106(2011)091 (*Preprint* hep-ph/1105.0901)
[4] Froggatt C D and Nielsen H B 1979 *Nucl. Phys.* **B147** 277
[5] Ishimori H *et al.* 2010 *Prog. Theor. Phys. Suppl.* **183** 1 (*Preprint* hep-ph/1003.3552)
[6] Harrison P F, Perkins D H and Scott D H 2002 *Phys. Lett.* **B530** 167
[7] Rodejohann W and Valle J 2011 *Phys.Rev.* D **84** 073011 (*Preprint* hep-ph/1108.3484)
[8] Dorame L, Meloni D, Morisi S, Peinado E and Valle J W F 2012 *Nucl. Phys.* B **861** 259 (*Preprint* hep-ph/1111.5614)
[9] Mohapatra R N and Valle J W F 1986 *Phys. Rev.* **D34** 1642
[10] Gonzalez-Garcia M C and Valle J W F 1989 *Phys. Lett.* **B216** 360
[11] Dorame L, Morisi S, Peinado E, Valle J W F and Rojas A D 2012 *Phys. Rev.* D **86** 056001 (*Preprint* hep-ph/1203.0155)
[12] Wolfenstein L 1981 *Phys. Lett.* **B107** 77
[13] Schechter J and Valle J W F 1981 *Phys. Rev.* **D24** 1883; erratum ibid 1982 **D25** 283
[14] Valle J W F 1983 *Phys. Rev.* **D27** 1672
[15] Schwetz T, Tortola M and Valle J W F 2011 *New J. Phys.* **13** 063004
Schwetz T, Tortola M and Valle J W F 2011 *New J. Phys.* **13** 109401
[16] Barry J and Rodejohann W 2011 *Nucl. Phys.* **B842** 33
[17] Hirsch M, Morisi S and Valle J W F 2008 *Phys. Rev.* **D78** 093007

PASCOS 2012 – 18th International Symposium on Particles Strings and Cosmology IOP Publishing
Journal of Physics: Conference Series **485** (2014) 012044 doi:10.1088/1742-6596/485/1/012044

Testing the Standard Model and beyond with the LENA proposal

Estela A Garcés[1], O G Miranda[2], M Tortola [3] and JWF Valle[3]

[1] Departamento de Física, Facultad de Estudios Superiores Cuautitlán, UNAM, CP 54770, Edo de México, Mexico

[2]Departamento de Física, Centro de Investigación y de Estudios Avanzados del IPN, Apdo. Postal 14-740 07000 México, D F, Mexico

[3]AHEP Group, Institut de Física Corpuscular – C.S.I.C./Universitat de València Edificio Institutos de Paterna, Apt. 22085, E–46071 Valencia, Spain

E-mail: egarces@fis.cinvestav.mx

Abstract. We discuss the possibility of a precision measurement of the electroweak mixing angle and a probe for new physics in the leptonic process of neutrino electron scattering. In the new physics schemes we explore the case of non standard neutrino interactions (NSI). The LENA proposal, currently under discussion, considers a large detector and the use of an artificial, ^{51}Cr, radioactive neutrino source with of 5 MCi intensity. We also discuss the possible use of the solar neutrino flux as a neutrino source, in particular the case of the ^{7}Be emission line.

1. Introduction

The evidence of neutrino oscillations reveal the implication of massive neutrinos, this fact can be acknowledged as the first evidence of new physics beyond the Standard Model [1, 2]. The Standard Model of electroweak interactions has been extremely successful, experiments confirm its predictions to a very high degree of accuracy. Yet many open questions remain, especially in the neutrino sector. The clear knowledge of neutrino properties will play a very important role to show us the right physical laws beyond the Standard Model.

There are several proposals of large future neutrino experiments. Large sized detectors would combine the technological advantages of previous oscillation experiments; characteristics such as low energy threshold and direction determination will be a key feature in liquid scintillators, water Cerenkov and liquid Argon detectors. LENA [3] is one among other multipurpose experiments in R&D phase that would extend our current knowledge of the neutrino oscillation phenomenon and can also test physics beyond the Standard Model [4, 5]. In this talk we study in detail the LENA proposal as a tool to search for new physics [6].

The weak mixing angle is a very important parameter in the Standard Model, while it has been measured with high accuracy only at high energies, at low energies it is still a challenge to achieve similar accuracy. In neutrino nucleon scattering there is the controversial result of NuTeV [7], reporting an almost 3 σ deviation from the value determined in global precision electroweak fits.

When systematic uncertainties are recalculated in the NuTeV data an error for $\sin^2\theta_W$ of $1-5\%$ is found [8, 9]. In neutrino electron scattering the accuracy attained by reactor experiments is

Content from this work may be used under the terms of the Creative Commons Attribution 3.0 licence. Any further distribution of this work must maintain attribution to the author(s) and the title of the work, journal citation and DOI.
Published under licence by IOP Publishing Ltd

over 10%, the TEXONO collaboration measured $\sin^2\theta_W = 0.251 \pm 0.031(\text{stat}) \pm 0.024(\text{sys})$ [10]. The study of low energy neutrino electron scattering could be also an excellent probe for various kinds of new physics, such as non-standard interactions (NSI) potentially associated to the mechanism of neutrino mass generation [11, 12, 13] and/or new gauge bosons [14, 15, 16].

2. LENA potential with an artificial ^{51}Cr neutrino source

The LENA proposal has considered the use of an artificial radioactive neutrino source for oscillometry tests. Here we study, as an application of that kind of source: (A) the precise determination of the electroweak mixing by measuring the neutrino electron differential cross section and (B) the sensitivity to new physics such as NSI and/or additional neutral gauge bosons.

2.1. LENA sensitivity to the electroweak mixing angle with an artificial source

The $\nu_e e$ differential cross section in the Standard Model is given as

$$\frac{d\sigma}{dT} = \frac{2G_F m_e}{\pi}[g_L^2 + g_R^2(1-\frac{T}{E_\nu})^2 - g_L g_R \frac{m_e T}{E_\nu^2}], \tag{1}$$

with G_F the Fermi constant, m_e the electron mass, T the kinetic energy of the recoil electron and E_ν the neutrino energy.

The coupling constants at tree level are expressed as $g_L = \frac{1}{2} + \sin^2\theta_W$ and $g_R = \sin^2\theta_W$. Radiative corrections to the $\nu_e e$ process give a modification to these coupling constants of around 2% [17] and will be included in all our calculations.

We consider a ^{51}Cr source of 5 MCi intensity with a monochromatic neutrino line at 747 keV as the one in the LENA proposal [3], in the deployment time of the source a signal of 1.9×10^5 neutrino events is expected. The number of events per energy bin is calculated as

$$N_i = n_e \phi_{Cr} \Delta t \int_{T_i}^{T_{i+1}} \int \frac{d\sigma}{dT} R(T,T')dT'dT, \tag{2}$$

with n_e the number of electron targets, ϕ_{Cr} the neutrino flux and Δt the the half-life of the radioactive source. The resolution function $R(T,T')$ accounts for the distribution of the measured recoil electron energy, T, around the true energy T',

$$R(T,T') = \frac{1}{\sigma\sqrt{2\pi}} \exp[-\frac{(T-T')^2}{(2\sigma^2)}], \tag{3}$$

with the energy resolution, $\sigma = 0.075\sqrt{T/MeV}$.

If we assume that the detector will measure the SM prediction, Fig. 1, an idealistic χ^2 analysis can be performed for different given errors for the data to obtain an expected sensitivity to $\sin^2\theta_W$. We use the χ^2 function with the form

$$\chi^2 = \sum_i \frac{(N_i^{\text{theo}} - N_i^{\text{exp}})^2}{\sigma_i^2}, \tag{4}$$

N_i^{theo} is the expected number of events for different values of $\sin^2\theta_W$ for a given bin i, N_i^{exp} is the 'experimental' value given by the expected number of events for the SM prediction and σ_i^2 is the error per bin. We have assigned different values to the error σ_i^2. The results of the statistical analysis are shown in Fig. 2, for three different values of the experimental error (0.5, 1 and 5 %). In the left panel we show a less optimistic scenario where the number of events is not divided in energy bins but is integrated over the whole energy range. The right panel shows

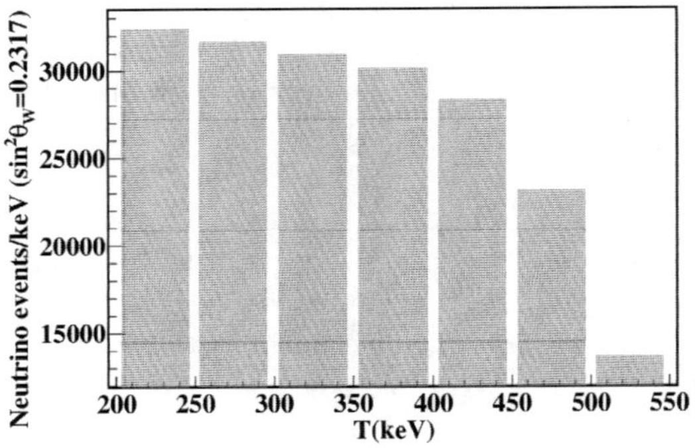

Figure 1. LENA spectrum obtained from Eq. 2, assuming the SM value of $\sin^2\theta_W = 0.2313$.

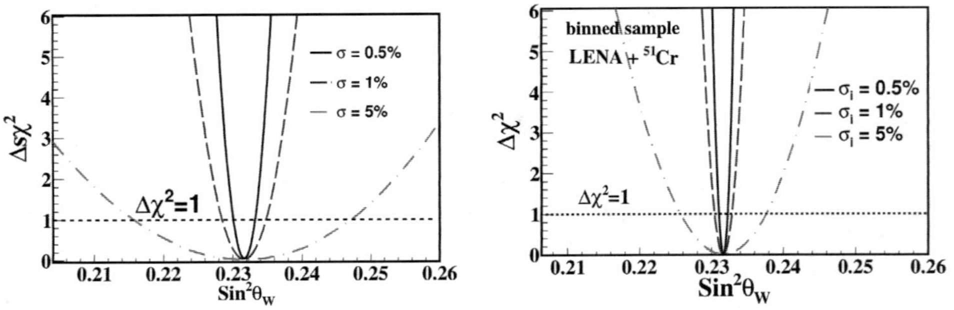

Figure 2. Expected sensitivity to the electroweak mixing angle of a ^{51}Cr neutrino source with the LENA detector. We show the result of a simulated χ^2 analysis of the total number of events with a given total 'error' of 0.5, 1 and 5% taking the whole energy range (left panel) or dividing the sample in bins(right panel).

the case where we divide the energy range in seven bins of 50 keV width. If we compare this analysis with the NuTeV results [8, 9] we find that an experimental error of the order of 1.-2.5 % would be required in order to improve their determination of the electroweak mixing angle.

A total experimental error in the LENA detector of the order of 2 % would suffice to improve the present sensitivity on the electroweak mixing angle given from the more precise calculations of the NuTeV measurements, as in the case of the two recent evaluations of the systematical errors of the experiment [8, 9].

2.2. LENA sensitivity to new physics, NSI and SM gauge structure

We first consider the sensitivity to the non standard neutrino interaction (NSI) parameters that could be generically associated to the generation of neutrino mass through a low-scale seesaw mechanism or through scalar boson mediation [11, 13]. A generic effective four-fermion NSI Lagrangian is given as

$$-\mathcal{L}^{eff}_{\rm NSI} = \varepsilon^{fP}_{\alpha\beta} 2\sqrt{2} G_F (\bar{\nu}_\alpha \gamma_\rho L \nu_\beta)(\bar{f}\gamma^\rho P f)\,. \tag{5}$$

Here G_F is the Fermi constant and $\varepsilon^{fP}_{\alpha\beta}$ parametrize the strength of the NSI. This term must be added to the Standard Model Lagrangian. For laboratory experiments f is a first generation SM fermion (e, u or d). The chiral projectors P denote $\{L, R = (1 \pm \gamma^5)/2\}$, while α and β denote the three neutrino flavors: e, μ and τ. We can obtain restrictions on the strength of the NSI

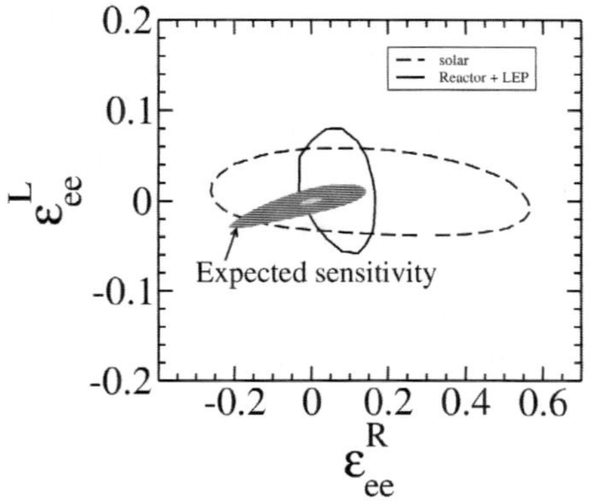

Figure 3. Expected LENA sensitivity at 90% CL to non-universal NSI using a [51] Cr neutrino source. The shaded areas correspond to a binned data sample divided in seven bins of 50 keV each and an 'error' per bin of either 1 (grey inner region) or 5% (magenta outer region). For comparison we show current limits to these parameters from an analysis coming from solar and KamLAND neutrino data [22] (dashed line) as well as from an analysis to the LEP and reactor data [23] (solid line).

parameters. To illustrate the physics potential of LENA to this type of new physics we focus on the sensitivity to non universal NSI parameters for the interaction of neutrinos with electrons, a similar analysis could be done for the flavor changing NSI parameters.

The differential cross section for neutrino electron scattering is modified in presence of the new interactions. The coupling constants for the Eq. (1) will be modified to $g_R \to g_R + \varepsilon_{ee}^R$ and $g_L \to g_L + \varepsilon_{ee}^L$.

After performing an idealistic statistical analysis, similar as the one explained in the previous section, we find that for a 5% error in the measured event number per bin the constraint on ε_{ee}^L would be below a few percent, while for the case of a 1% error the constraint on this parameter will lie below the percent level, in Fig. 3 we compare the estimated sensitivity to the NSI parameters attainable by LENA with previous limits from reactor and solar neutrino data.

This limits can be put in the context of specific extensions to the SM, some of them as those arising from a E_6 gauge group [14, 15, 16] involve an additional gauge boson. In that case the non universal NSI parameters will be expressed as [15]

$$\begin{aligned} \varepsilon_{ee}^L &= 2\gamma \sin^2\theta_W \rho_{\nu e}^{NC} \left(\frac{3\cos\beta}{2\sqrt{6}} + \frac{\sin\beta}{3}\sqrt{\frac{5}{8}} \right)^2 \\ \varepsilon_{ee}^R &= 2\gamma \sin^2\theta_W \rho_{\nu e}^{NC} \left(\frac{\cos\beta}{2\sqrt{6}} - \frac{\sin\beta}{3}\sqrt{\frac{5}{8}} \right), \left(\frac{3\cos\beta}{\sqrt{24}} + \frac{\sin\beta}{3}\sqrt{\frac{5}{8}} \right) \end{aligned} \quad (6)$$

where $\gamma = \frac{M_{Z'}{}^2}{M_Z{}^2}$, with M_Z the mass of the Z boson, and M_Z' the mass of the extra gauge boson. As an example we take the χ model ($\cos\beta = 1$) discussed in [15]. After the idealistic statistical analysis we find that it would be possible to put a lower bound in $M_{Z'}$ from 360 GeV to 1.1 TeV if the total error is assumed to be from 0.5% to 5%. This limit would be complementary to those expected at the Large Hadron Collider [18]. Never the less in specific models with different couplings for leptons and quarks, as in the case of leptophilic scenarios [19, 20] the LENA sensitivities would be dominant. The LENA proposal also has potential to restrict models with trilinear R -parity violating couplings [21].

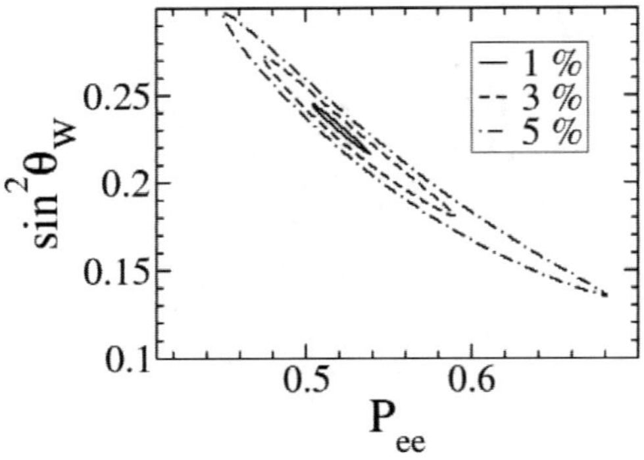

Figure 4. Expected sensitivity (at 68.27% CL) of the LENA detector to the solar neutrino Beryllium signal, assuming a 50 keV recoil energy binning and an error per bin of 1, 3 and 5%.

3. LENA potential with solar neutrinos

A similar analysis can be done considering solar neutrinos collected by the LENA detector. We take as an example the case of the monoenergetic line of ^{7}Be neutrinos of $E_\nu = 0.862$ MeV. LENA will be capable to detect this part of the solar spectrum in a similar fashion as the Borexino experiment [24].

To estimate the sensitivity of LENA to $\sin^2\theta_\mathrm{W}$ with solar neutrinos we need to take into account the dependence of the signal on the survival probability, P_{ee}. With $\sin^2\theta_\mathrm{W}$ and P_{ee} as free parameters in the statistical analysis we have, for the number of events per bin,

$$N_i = n_e \phi_{Cr} \Delta t \int_{T_i}^{T_{i+1}} \int \left(P_{ee} \frac{d\sigma^{\nu_e e}}{dT} + (1 - P_{ee}) \frac{d\sigma^{\nu_{\mu,\tau} e}}{dT} \right) R(T, T') dT' dT. \tag{7}$$

The differential cross section for the electron-neutrino is given by Eq. (1). For muon or tau neutrinos one has the coupling constants: $g_L^{\nu_{\mu,\tau}} = -\frac{1}{2} + \sin^2\theta_\mathrm{W}$, and $g_R^{\nu_{\mu,\tau}} = \sin^2\theta_\mathrm{W}$.

In Fig. 4 is shown the result of the estimated sensitivity to $\sin^2\theta_\mathrm{W}$. We find that a better determination of the Beryllium spectrum is needed in order to have a precise measurement of $\sin^2\theta_\mathrm{W}$. However, in the optimistic case of an error in the number of events per bin of 1 %, there would be a sensitivity to the electroweak mixing parameter of the order ~ 2 %.

4. Conclusions

We showed that by using a Chromium neutrino source LENA could provide a precise measurement of the electroweak mixing angle in a region of energy that is not easy to study with other experiments. LENA can also be applied to probe physics beyond the Standard Model. Here we discussed the case of non universal non standard neutrino interactions, and the possible existence of new electroweak neutral gauge bosons.

The accuracy in the determination of the electroweak mixing parameter expected in LENA lies below the percent level for the most optimistic expectations on the systematical error. These results indicate that LENA holds very good prospects.

Acknowledgments

This work was partially supported by CONACyT grant 166639 (Mexico) and postdoctoral DGAPA-UNAM grant.

References

[1] Nakamura K *et al.* (Particle Data Group) 2010 *J.Phys.* **G37** 075021

[2] Schwetz T, Tortola M and Valle J 2011 *New J.Phys.* **13** 109401 (*Preprint* 1108.1376)
[3] Wurm M *et al.* (LENA Collaboration) 2012 *Astropart.Phys.* **35** 685–732 (*Preprint* 1104.5620)
[4] Alonso J, Avignone F, Barletta W, Barlow R, Baumgartner H *et al.* 2010 (*Preprint* 1006.0260)
[5] de Bellefon A, Bouchez J, Busto J, Campagne J E, Cavata C *et al.* 2006 (*Preprint* hep-ex/0607026)
[6] Garces E, Miranda O, Tortola M and Valle J 2012 *Phys.Rev.* **D85** 073006 (*Preprint* 1112.3633)
[7] Zeller G *et al.* (NuTeV Collaboration) 2002 *Phys.Rev.Lett.* **88** 091802 (*Preprint* hep-ex/0110059)
[8] Bentz W, Cloet I, Londergan J and Thomas A 2010 *Phys.Lett.* **B693** 462–466 (*Preprint* 0908.3198)
[9] Ball R D *et al.* (NNPDF Collaboration) 2009 *Nucl.Phys.* **B823** 195–233 (*Preprint* 0906.1958)
[10] Deniz M *et al.* (TEXONO Collaboration) 2010 *Phys.Rev.* **D81** 072001 (*Preprint* 0911.1597)
[11] Valle J W F
[12] Schechter J and Valle J 1980 *Phys.Rev.* **D22** 2227
[13] Maltoni M, Schwetz T, Tortola M A and Valle J W F 2004 *New J. Phys.* **6** 122 (*Preprint* hep-ph/0405172)
[14] Mohapatra R N and Senjanovic G 1981 *Phys.Rev.* **D23** 165
[15] Valle J 1987 *Phys.Lett.* **B196** 157
[16] Malinsky M, Romao J and Valle J 2005 *Phys.Rev.Lett.* **95** 161801 (*Preprint* hep-ph/0506296)
[17] Bahcall J N, Kamionkowski M and Sirlin A 1995 *Phys.Rev.* **D51** 6146–6158 (*Preprint* astro-ph/9502003)
[18] Aad G *et al.* (ATLAS Collaboration) 2011 *Phys.Lett.* **B700** 163–180 (*Preprint* 1103.6218)
[19] Fox P J and Poppitz E 2009 *Phys.Rev.* **D79** 083528 (*Preprint* 0811.0399)
[20] Ko P and Omura Y 2011 *Phys.Lett.* **B701** 363–366 (*Preprint* 1012.4679)
[21] Sessolo E, Tahir F and McKay D 2009 *Phys.Rev.* **D79** 115010 (*Preprint* 0903.0118)
[22] Bolanos A, Miranda O, Palazzo A, Tortola M and Valle J 2009 *Phys.Rev.* **D79** 113012 (*Preprint* 0812.4417)
[23] Barranco J, Miranda O, Moura C and Valle J 2008 *Phys.Rev.* **D77** 093014 (*Preprint* 0711.0698)
[24] Bellini G, Benziger J, Bick D, Bonetti S, Bonfini G *et al.* 2011 *Phys.Rev.Lett.* **107** 141302 (*Preprint* 1104.1816)

A new neutrino mass sum-rule from inverse seesaw

Alma D Rojas

Facultad de Ciencias, CUICBAS, Universidad de Colima, Colima, México

E-mail: alma.drp@gmail.com

Abstract. We propose a new model based on the S_4 flavor symmetry that leads to a new neutrino mass sum-rule and discuss how to generate a nonzero value for the reactor angle θ_{13} indicated by recent experiments, and the resulting correlation with the solar angle θ_{12}. The model implies a lower bound on the effective neutrinoless double beta mass parameter, even for normal hierarchy neutrinos.

1. Introduction

Although nonvanishing neutrino masses have been confirmed by the discovery of neutrino oscillations, their nature, if they are Dirac or Majorana particles, is still an unanswered question. While there is no way to probe the Dirac nature of neutrinos, a confirmation of the Majorana nature would be the observation of neutrinoless double beta decay ($0\nu2\beta$) [1]. The mass matrix characterizing Majorana neutrinos is a symmetric mass matrix whose parameters are restricted by the experimental data: the neutrino oscillation parameters as well as the limits on the $0\nu2\beta$ effective mass parameter [2, 3].

The general neutrino mixing matrix can be parametrized in different equivalent ways [4, 5, 6]. The Tribimaximal Mixing Matrix (TBM) [7] is a particular *ansatz* of the mixing matrix in which the θ_{13} angle has a zero value. However, the recent experiments [8, 9, 10, 11] indicated a non-zero value of the θ_{13} angle, but taking into account that the θ_{13} value can receive corrections from charged lepton diagonalization and/or from renormalization effects, the TBM can still be used as a good first aproximation.

In Ref. [12] is pointed out that a two-parameter neutrino mass matrix implying a particular mixing matrix form can be obtained from several flavor models based in non-Abelian discrete symmetries. There it was noted that in these models only the following mass relations can be obtained,

$$\chi m_2^\nu + \xi m_3^\nu = m_1^\nu, \tag{1}$$

$$\frac{\chi}{m_2^\nu} + \frac{\xi}{m_3^\nu} = \frac{1}{m_1^\nu}, \tag{2}$$

$$\chi\sqrt{m_2^\nu} + \xi\sqrt{m_3^\nu} = \sqrt{m_1^\nu}, \tag{3}$$

$$\frac{\chi}{\sqrt{m_2^\nu}} + \frac{\xi}{\sqrt{m_3^\nu}} = \frac{1}{\sqrt{m_1^\nu}}, \tag{4}$$

where χ and ξ are free parameters that characterize each specific model. Also, a classification of all models predicting TBM mixing which generate mass relations similar to the first three is

Content from this work may be used under the terms of the Creative Commons Attribution 3.0 licence. Any further distribution of this work must maintain attribution to the author(s) and the title of the work, journal citation and DOI.
Published under licence by IOP Publishing Ltd

presented there, but the last case correspond to a completely new case. In this work we present a model implementing the inverse seesaw mechanism [13, 14] as well as a non-Abelian flavor symmetry [15], along the lines of Ref. [16], but adopting S_4, instead of A_4.

The characteristic which distinguish the inverse seesaw scheme from other schemes is that it allows a low-scale seesaw scheme [17] with naturally light neutrinos. The particle content of the Standard Model (SM) is extended by adding a pair of two component gauge singlet leptons, ν_i^c and S_i, with i running over the three generations. The fermion singlets S_i have opposite lepton number with respect to that of the three singlets ν_i^c associated to the "right-handed" neutrinos.

In the ν, ν^c, S basis the 9×9 neutral lepton mass matrix M_ν has the form:

$$M_\nu = \begin{pmatrix} 0 & m_D^T & 0 \\ m_D & 0 & M^T \\ 0 & M & \mu \end{pmatrix}, \tag{5}$$

where m_D and M are arbitrary 3×3 complex matrices, while μ is symmetric due to the Pauli principle.

Following the seesaw diagonalization method in [18] one obtains the effective light neutrino mass matrix $m_\nu \sim m_D^T M^{T-1} \mu M^{-1} m_D$, with the entry μ being very small.

If m_D and μ are both proportional to the identity, and

$$M \sim M_{TBM} = \begin{pmatrix} x & y & y \\ y & x+z & y-z \\ y & y-z & x+z \end{pmatrix}, \tag{6}$$

in the basis where the charged lepton mass matrix is diagonal, then there is a specific (complex) relation among the parameters x, y and z [19], leaving only two free complex parameters, and we obtain the mass sum-rule in Eq. (4).

Our model is presented in the next section, and the predictions regarding the lower bound on the $0\nu2\beta$ amplitude are presented in section III, where we also discuss possible departures from tribimaximality, including a finite θ_{13} value.

2. The model

We follow Table I given in Ref. [16], where some possible schemes realizing the TBM pattern are summarized for the inverse seesaw case. From these, adopting the S_4 flavor symmetry instead of A_4, we will implement case 2)

$$M_D \propto \mathcal{I} \qquad \mu \propto \mathcal{I}, \qquad M \propto \begin{pmatrix} A & 0 & 0 \\ 0 & B & C \\ 0 & C & B \end{pmatrix}. \tag{7}$$

In order to obtain the S_4-based inverse seesaw model we assign the charge matter fields as in Table 1 (see [20] for S_4 multiplication rules). To generate the desired mass matrix structures five flavon fields, ϕ_ν, ϕ'_ν, ϕ_l, ϕ'_l, ϕ''_l are introduced, and the extra symmetries Z_3 and Z_2. The introduction of a scalar field σ is required to break lepton number through the coupling with S_iS_j terms, and when this field acquire its vacuum expectation value (VEV) it will be responsible of the mass terms. In order to keep the renormalizability of the Lagrangian we add Frogatt-Nielsen fermion, singlet under the weak $SU(2)$ gauge group, χ and its conjugate χ^c [21, 22, 23, 24]. The quantum numbers under these extra symmetries are shown in Table 2.

The renormalizable Lagrangian relevant for neutrinos is

$$\mathcal{L}_\nu = Y_{D_{ij}} \overline{L}_i \nu_{R_j} h + Y^k_{\nu ij} \nu_{R_i} S_j \phi_{\nu_k} + Y'_{\nu ij} \nu_{R_i} S_j \phi'_\nu + \mu_{ij} S_i S_j \sigma, \tag{8}$$

Table 1. Fields and transformation properties under $SU(2)$, the S_4 flavor symmetry, and global lepton number $U_l(1)$

	$\overline{L}$	ν_R	l_R	h	S	ϕ_ν	ϕ'_ν	ϕ_l	ϕ'_l	ϕ''_l	σ	χ	χ^c
$SU(2)$	2	1	1	2	1	1	1	1	1	1	1	1	1
S_4	3_1	3_1	3_1	1_1	3_1	3_1	1_1	3_1	3_2	1_1	1_1	3_1	3_1
$U_l(1)$	-1	1	1	0	-1	0	0	0	0	0	2	1	-1

Table 2. Fields and their transformation properties under the Z_3, and Z_2 flavor symmetries

	$\overline{L}$	ν_R	l_R	h	S	ϕ_ν	ϕ'_ν	ϕ_l	ϕ'_l	ϕ''_l	σ	χ	χ^c
Z_3	ω^2	ω	1	1	1	ω^2	ω^2	ω	ω	ω	1	ω	ω^2
Z_2	+	+	+	+	−	−	−	+	+	+	+	+	+

while the renormalizable Yukawa terms involving the messenger fields are

$$\mathcal{L}_\chi = M_\chi \chi\chi^c + \bar{L}h\chi + \chi^c l_R\phi_l + \chi^c l_R\phi'_l + \chi^c l_R\phi''_l. \tag{9}$$

The effective Lagrangian for charged leptons, after integrating out the messenger fields χ, takes the form

$$\mathcal{L}_l = \frac{y_l}{\Lambda}(\bar{L}l_R)h\phi_l + \frac{y'_l}{\Lambda}(\bar{L}l_R)h\phi'_l + \frac{y''_l}{\Lambda}(\bar{L}l_R)h\phi''_l, \tag{10}$$

being Λ the effective scale. This effective Lagrangian is responsible for charged lepton mass generation.

To obtain the desired neutrino mixing matrix the flavon fields must have the following alignments:

$$\langle\phi_\nu\rangle = v_\nu(1,0,0), \quad \langle\phi_l\rangle = v_l(1,1,1), \quad \langle\phi'_l\rangle = v'_l(1,1,1), \tag{11}$$

where we define $\langle\phi'_\nu\rangle = v'_\nu$, $\langle\phi''_l\rangle = v''_l$, $\langle\sigma\rangle = v_\sigma$ and $\langle h\rangle = v$. We do not write here explicitly the potential (see [25] for more details) but we have verified that it is possible to find parameters for which these alignments provide a minimum of the potential. With the previous alignments the mass matrices take the form

$$(\mu) = \begin{pmatrix} \mu v_\sigma & 0 & 0 \\ 0 & \mu v_\sigma & 0 \\ 0 & 0 & \mu v_\sigma \end{pmatrix}, \quad M_D = \begin{pmatrix} Y_D v & 0 & 0 \\ 0 & Y_D v & 0 \\ 0 & 0 & Y_D v \end{pmatrix}, \quad M = \begin{pmatrix} Y'_\nu v'_\nu & 0 & 0 \\ 0 & Y'_\nu v'_\nu & Y_\nu v_\nu \\ 0 & Y_\nu v_\nu & Y'_\nu v'_\nu \end{pmatrix}, \tag{12}$$

and

$$M_l = \begin{pmatrix} y''_l v''_l & y_l v_l - y'_l v'_l & y_l v_l + y'_l v'_l \\ y_l v_l + y'_l v'_l & y''_l v''_l & y_l v_l - y'_l v'_l \\ y_l v_l - y'_l v'_l & y_l v_l + y'_l v'_l & y''_l v''_l \end{pmatrix} \frac{v}{\Lambda}. \tag{13}$$

This matrix M_l is diagonalized by the "magic" matrix" U_ω[16, 25]. On the other hand, after diagonalization, the light neutrino mass matrix takes the form

$$M_\nu = \begin{pmatrix} \frac{1}{a^2} & 0 & 0 \\ 0 & \frac{a^2+b^2}{(b^2-a^2)^2} & -\frac{2ab}{(b^2-a^2)^2} \\ 0 & -\frac{2ab}{(b^2-a^2)^2} & \frac{a^2+b^2}{(b^2-a^2)^2} \end{pmatrix} \tag{14}$$

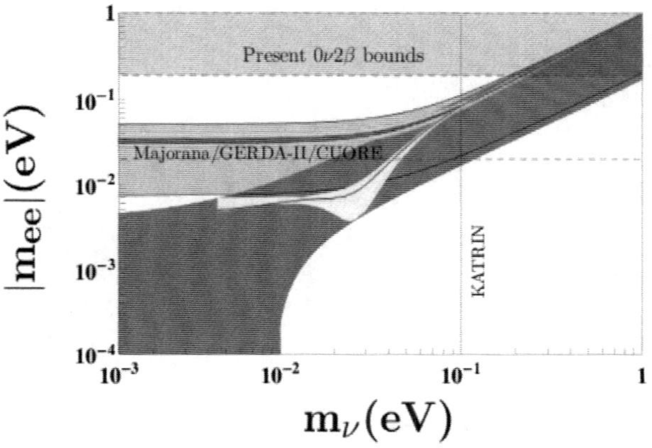

Figure 1. $|m_{ee}|$ as a function of the lightest neutrino mass corresponding to the mass sum-rule in Eq. (4). The bands in gray and blue correspond to generic normal and inverse hierarchy regions, while the yellow and green bands correspond to our flavor prediction varying the values of oscillation parameters in their 3σ C.L. range. The thin red bands correspond to the TBM limit. The upper band in lavender corresponds to the present bounds on $0\nu 2\beta$.

where $a = Y_\nu' v_\nu'/(\sqrt{\mu v_\sigma} Y_D v)$ and $b = Y_\nu v_\nu/(\sqrt{\mu v_\sigma} Y_D v)$. In the basis where charged lepton mass matrix is diagonal, the light neutrino mass matrix is diagonalized by the TBM form, and the corresponding eigenvalues are given by $m_1 = 1/(a+b)^2$, $m_2 = 1/(a-b)^2$ and $m_3 = 1/a^2$, and with these eigenvalues we obtain the desired neutrino mass sum-rule $\frac{1}{\sqrt{m_1}} = \frac{2}{\sqrt{m_3}} - \frac{1}{\sqrt{m_2}}$.

3. Phenomenology

3.1. Neutrinoless double beta decay

We can write the general expression of the mass parameter $|m_{ee}|$ which determines the $0\nu 2\beta$ decay amplitude as

$$|m_{ee}| = \left|\sum_j U_{ej}^2\, m_j\right| = \begin{cases} \left|c_{12}^2 c_{13}^2\, m_1 + s_{12}^2 c_{13}^2\, m_2\, e^{\frac{1}{2}i\alpha_{21}} + s_{13}^2\, m_3\, e^{\frac{1}{2}i(\alpha_{31}-2\delta)}\right| & \text{(PDG [4])}, \\ \left|c_{12}^2 c_{13}^2\, m_1 + s_{12}^2 c_{13}^2\, m_2\, e^{2i\phi_{12}} + s_{13}^2\, m_3\, e^{2i\phi_{13}}\right| & \text{(symmetrical[5, 6])}. \end{cases} \tag{15}$$

where $c_{ij} = \cos\theta_{ij}$ and $s_{ij} = \sin\theta_{ij}$, m_i, $i = 1, 2, 3$, are the neutrino masses, and ϕ_{12} and ϕ_{13} are the two Majorana phases.

The parameter $|m_{ee}|$ can be plotted in terms of the lightest neutrino mass, varying the neutrino oscillation parameters in their allowed range, and depending on which is the lightest neutrino one can have normal or inverse hierarchy, having a lower bound in the latter case. However, as was noted in Ref. [12], the neutrino mass sum-rule can be interpreted geometrically as a triangle in the complex plane, giving its area a measure of the Majorana CP violation, see Ref. [12] for details.

In the present scheme, as a result, there is a lower bound on $|m_{ee}|$ even in the case of normal hierarchy, and since the allowed ranges for normal and inverse hierarchy are much more constrained than in the generic case, it becomes possible to distinguish the neutrino mass hierarchy even for lighter neutrinos.

3.2. Quark sector

For the quark sector, we will only mention here that its possible to fit quark masses and mixing assigning charges as in Table 3 and adding flavons $\phi_{D,S}$ in doublet and singlet representations of the S_4. As in the charged lepton sector the dimension five operators can be given in terms of renormalizable interaction by introducing suitable messenger fields, and the alignment required for the VEV of ϕ_D is $\langle\phi_D\rangle \sim (-\sqrt{3}, 1)$ (see [25] for further details).

Table 3. Quark sector and their transformation properties under the Z_3, and Z_2 flavor symmetries

	$\overline{Q}_D$	$\overline{Q}_S$	u_{R_D}	u_{R_S}	d_{R_D}	d_{R_S}	ϕ_D	ϕ_S
$SU(2)$	2	2	1	1	1	1	1	1
S_4	2	1_1	2	1_1	2	1_1	2	1_1
Z_3	ω	ω	ω^2	ω^2	ω	ω	ω	ω
Z_2	+	+	−	−	−	−	−	−

3.3. Finite θ_{13} value

Although the model leads to the TBM pattern we can obtain corrections from the charged lepton sector by coupling an extra S_4-doublet flavon, inducing in this way a nonzero values of θ_{13} as recently suggested by experiments [8, 9, 10, 11].

For instance, consider a flavon scalar doublet under S_4, $\phi \sim \mathbf{2}$, transforming as $(\omega, +)$ under $Z_3 \times Z_2$. In the Lagrangian we must then include the term $(\bar{L} l_R) h\phi$, which is a dimension five operator that can be obtained from a renormalizable Lagrangian by means of the messenger fields χ, χ^c.

Assuming that ϕ acquires VEV $\langle\phi\rangle = (u_1, u_2)$, a natural vacuum alignment, consistent with the previous alignments, is $u_1 = -\sqrt{3}u_2$. Then, one finds that the contribution from this term to the charged lepton mass matrix is

$$\delta M_l = \begin{pmatrix} -\sqrt{\frac{2}{3}}vu_2 & 0 & 0 \\ 0 & \sqrt{\frac{1}{2}}vu_1 + \sqrt{\frac{1}{6}}vu_2 & 0 \\ 0 & 0 & -\sqrt{\frac{1}{2}}vu_1 + \sqrt{\frac{1}{6}}vu_2 \end{pmatrix}, \quad (16)$$

which modifies the diagonal entries in the charged lepton mass matrix, M_l, so that the total $M_l + \delta M_l$ is no longer diagonalized by U_ω. In this way one can induce a potentially large value for θ_{13} and also potential departures of the solar and atmospheric angles from their TBM values. Besides, one finds relations among these three neutrino mixing angles. The most interesting of these is the correlation involving the solar and reactor angles, as illustrated in Fig. 2. The

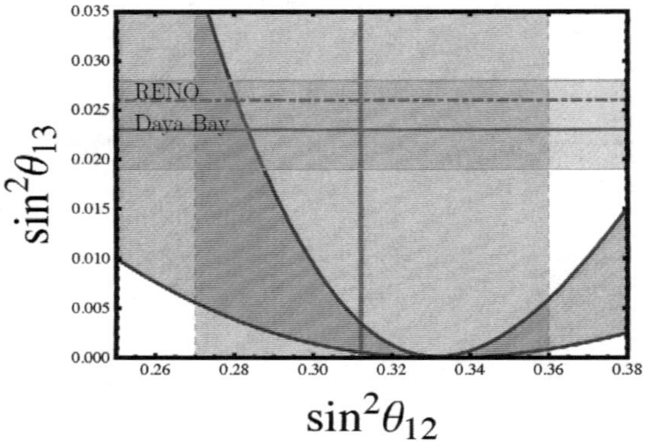

Figure 2. Correlations between reactor and solar neutrino mixing angles. See text for explanations.

horizontal green band represents the 2σ Daya Bay measurement [8], and its central value is indicated by the horizontal red line, while the horizontal dot-dashed line indicates the central value of the recent RENO measurement [11]. On the other hand, the vertical band, delimited by dotted lines, corresponds to the 2σ region for $\sin^2\theta_{13}$ found in the global analysis in Ref. [26], and the vertical line corresponds to the central value. The region in lavender shows the correlation between the reactor and solar angles. We observe that the deviation of θ_{13} from zero can be substantial provided the departure of θ_{12} from its TBM value is also large. Moreover, the model is consistent with the measurements of the two recent reactor experiments, only if the solar angle lies substantially below the TBM prediction (at 2σ).

Acknowledgments

I thank to my collaborators L. Dorame, S. Morisi, E. Peinado and J. W. F. Valle for the work presented here. I would like also thank to the IFIC-CSIC by its hospitality while this work was carried out and to the Fundación Carolina by its support.

References

[1] Schechter J and Valle J Wen F 1982 *Phys. Rev.* D **25** 2951; Duerr M, Lindner M and Merle A 2011 On the Quantitative Impact of the Schechter-Valle Theorem *J. High Energy Phys.* JHEP 1106(2011)091 (*Preprint* hep-ph/1105.0901)
[2] Rodejohann W 2011 *Int.J.Mod.Phys.* E **20** 1833 (*Preprint* hep-ph/1106.1334)
[3] Barabash A 75 years of double beta decay: yesterday, today and tomorrow (*Preprint* nucl-ex/1101.4502v1)
[4] Nakamura K *et al* 2010 *Journal of Physics G: Nuclear and Particle Physics* **37** 075021
[5] Schechter J and Valle J. W. F. 1980 *Phys. Rev.* D **22**, 2227
[6] Rodejohann W and Valle J 2011 *Phys.Rev.* D **84** 073011 (*Preprint* hep-ph/1108.3484)
[7] Harrison P F, Perkins D H and Scott D H 2002 *Phys. Lett.* B **530** 167
[8] An F P *et al* (DAYA-BAY Collaboration) 2012 *Phys. Rev. Lett.* 108 171803 (*Preprint* hep-ex/1203.1669)
[9] Abe K *et al* (T2K Collaboration) 2011 *Phys. Rev. Lett.* **107** 041801
[10] Abe Y *et al* (DOUBLE-CHOOZ Collaboration) 2012 *Phys. Rev. Lett.* 108 131801 (*Preprint* hep-ph/1112.6353)
[11] Ahn J K *et al* (RENO Collaboration) *Phys. Rev. Lett. 108, 131801* (*Preprint* hep-ex/1204.0626)
[12] Dorame L, Meloni D, Morisi S, Peinado E and Valle J W F 2012 *Nucl. Phys.* B **861** 259 (*Preprint* hep-ph/1111.5614)
[13] Mohapatra R N and Valle J W F 1986 *Phys. Rev.* D **34** 1642
[14] Gonzalez-Garcia M C and Valle J W F 1989 *Phys. Lett.* B **216** 360
[15] Hirsch M *et al* (*Preprint* hep-ph/1201.5525)
Ishimori H *et al.* 2010 *Prog. Theor. Phys. Suppl.* **183** 1 (*Preprint* hep-ph/1003.3552)
[16] Hirsch M, Morisi S and Valle J W F 2009 *Phys. Lett.* B **679** 454 (*Preprint* hep-ph/0905.3056)
[17] Valle J W F 2006 *J. Phys.*: Conf. Series **53** 473-505 (*Preprint* hep-ph/0608101)
[18] Schechter J and Valle J W F 1982 *Phys. Rev.* D **25** 774
[19] Ma E 2006 *Phys. Rev.* D **73** 057304
[20] Hagedorn C, Lindner M and Mohapatra R 2006 S4 Flavor Symmetry and Fermion Masses: Towards a Grand Unified theory of Flavor *J. High Energy Phys.* JHEP0606(2006)042 (*Preprint* hep-ph/0602244)
[21] Froggatt C D and Nielsen H B 1979 *Nucl. Phys.* B **147** 277
[22] Barbieri R, Hall L J, Raby S and Romanino A 1997 *Nucl.Phys.* B **493** 3 (*Preprint* hep-ph/9610449)
[23] Varzielas I de M and Ross G G 2006 *Nucl.Phys.* B **733** 31 (*Preprint* hep-ph/0507176)
[24] King S F and Malinsky M 2006 "Towards a Complete Theory of Fermion Masses and Mixings with SO(3) Family Symmetry and 5d SO(10) Unification" *J. High Energy Phys.* JHEP0611(2006)071
[25] Dorame L, Morisi S, Peinado E, Valle J W F and Rojas A D 2012 *Phys. Rev.* D **86** 056001 (*Preprint* hep-ph/1203.0155)
[26] Schwetz T, Tortola M and Valle J W F 2011 *New J. Phys.* **13** 063004; Schwetz T, Tortola M and Valle J W F 2011 *New J. Phys.* **13** 109401

PASCOS 2012 – 18th International Symposium on Particles Strings and Cosmology IOP Publishing
Journal of Physics: Conference Series **485** (2014) 012046 doi:10.1088/1742-6596/485/1/012046

S_3 flavour symmetry and the reactor mixing angle

O Felix-Beltran[1,a], **F González Canales**[1,b], **A Mondragón**[2,c] **and M Mondragón**[2,d]

[1]Facultad de Ciencias de la Electrónica, Benemérita Universidad Autónoma de Puebla, Apdo. Postal 157, 72570, Puebla, Pue., México
[2]Instituto de Física, Universidad Nacional Autónoma de México, 04510, México D.F., México

E-mail: [a]olga_flix@ece.buap.mx, [b]ffelix@fisica.unam.mx, [c]mondra@fisica.unam.mx, [d]myriam@fisica.unam.mx

Abstract. In the framework of a minimal S_3 symmetric extension of the Standard Model, the mass matrices of all fermions take the same generic form with two texture zeroes. The neutrino mixing matrix, V_{PMNS}, is computed and exact, explicit analytical expressions for the neutrino mixing angles as functions of the masses of neutrinos and charged leptons are obtained in good agreement with the latest experimental data. We also compute the branching ratios of some selected flavour-changing neutral current (FCNC) processes, as well as the contribution of the exchange of neutral flavour-changing scalars to the anomaly of the magnetic moment of the muon, as functions of the masses of charged leptons and the neutral Higgs bosons.

1. Introduction

In the last ten years, flavour mixing research in the leptonic sector progressed from the fascinating exploratory stage of discovery to the more advanced stage of qualitative precision measurements of the observables in neutrino flavour oscillations. A global fit of the observables to the current experimental data in neutrino oscillations of solar, atmospheric, reactor and accelerator neutrinos yields the following results:

a) the differences of the squared masses, to 1σ, are [1]:

$$\begin{aligned} &\Delta m_{31}^2 = 2.47^{+0.069}_{-0.067} \times 10^{-3}\ \mathrm{eV}^2 \text{ (Normal hierarchy)}, \quad \Delta m_{21}^2 = 7.50 \pm 0.185 \times 10^{-5}\ \mathrm{eV}^2, \\ &\Delta m_{32}^2 = -2.43^{+0.042}_{-0.065} \times 10^{-3}\ \mathrm{eV}^2 \text{ (Inverted hierarchy)}, \end{aligned} \tag{1}$$

b) the magnitudes of the leptonic mixing matrix ($PMNS$) elements, to 3σ, are [1]:

$$|\mathbf{U}_{PMNS}| = \begin{pmatrix} 0.795 \leftrightarrow 0.846 & 0.513 \leftrightarrow 0.585 & 0.126 \leftrightarrow 0.178 \\ 0.205 \leftrightarrow 0.543 & 0.416 \leftrightarrow 0.730 & 0.579 \leftrightarrow 0.808 \\ 0.215 \leftrightarrow 0.548 & 0.409 \leftrightarrow 0.725 & 0.567 \leftrightarrow 0.800 \end{pmatrix}. \tag{2}$$

Specially worth of notice is the recent measurement of the reactor mixing angle $\theta_{13}^{l^{exp}}$, in the following range of values

$$7.2^\circ \lesssim \theta_{13}^{l^{exp}} \lesssim 10.25^\circ, \tag{3}$$

which can discriminate between different possible flavour symmetry groups as realized in neutrino mixing. However, there are some still unsolved problems, such as: i) the absolute value of the

Content from this work may be used under the terms of the Creative Commons Attribution 3.0 licence. Any further distribution of this work must maintain attribution to the author(s) and the title of the work, journal citation and DOI.
Published under licence by IOP Publishing Ltd

neutrino masses; *ii*) the neutrino mass hierarchy. *iii*) the experimental detection of CP violation in the neutrino oscillation and *iv*) the Dirac o Majorana nature of the neutrinos.

The experimentally determined non vanishing and rather sizeable value of the reactor mixing angle θ_{13}^{lexp} [2, 3, 4, 5, 6] opened a gate to the experimental and theoretically important research on CP violation in the leptonic sector as well as the determination of the neutrino masses hierarchy.

2. The S_3-invariant Extension of the Standard Model

A convenient theoretical framework for a unified interpretation of the experimental data on masses and mixing of quarks and leptons is the minimal S_3-invariant extension of the Standard Model of J. Kubo, A. Mondragón, M. Mondragón and E. Rodriguez Jauregui [7]. For a recent review see [8, 9]. In this model the concept of flavour and generations is extended to the Higgs sector by introducing in the theory three Higgs fields which are $SU(2)$ doublets, so that, all matter fields, quark, lepton and Higgs fields, have three families. The permutational group S_3 is introduced in the theory as a global flavour symmetry unbroken at the Fermi scale. Then, quark, lepton and Higgs fields are assigned to singlets and doublets of a reducible representation $\mathbf{2}\oplus\mathbf{1}_s$ of the flavour group S_3. This multi-Higgs model has tree level flavour changing neutral currents whose exchange may give rise to lepton flavour violating processes and may also contribute to the anomalous magnetic moment of the muon [7, 8]. An effective test of the phenomenological success of the model is obtained by verifying that all flavour changing neutral current processes and the magnetic anomaly of the muon, computed in the S_3–Invariant extended form of the Standard Model, agree with the experimental values [8, 10, 11].

The most general renormalizable Yukawa interactions for leptons are given by $\mathcal{L}_Y = \mathcal{L}_{Y_E} + \mathcal{L}_{Y_\nu}$, where

$$\begin{aligned}
\mathcal{L}_{Y_E} = \; & -Y_1^e \overline{L}_I H_S e_{IR} - Y_3^e \overline{L}_3 H_S e_{3R} - Y_2^e \left[\, \overline{L}_I \kappa_{IJ} H_1 e_{JR} + \overline{L}_I \eta_{IJ} H_2 e_{JR} \,\right] \\
& -Y_4^e \overline{L}_3 H_I e_{IR} - Y_5^e \overline{L}_I H_I e_{3R} + \text{ h. c. }, \\
\mathcal{L}_{Y_\nu} = \; & -Y_1^\nu \overline{L}_I \,(i\sigma_2)\, H_S^* \nu_{IR} - Y_3^\nu \overline{L}_3 \,(i\sigma_2)\, H_S^* \nu_{3R} - Y_4^\nu \overline{L}_3 \,(i\sigma_2)\, H_I^* \nu_{IR} \\
& -Y_2^\nu \left[\, \overline{L}_I \kappa_{IJ} \,(i\sigma_2)\, H_1^* \nu_{JR} + \overline{L}_I \eta_{IJ} \,(i\sigma_2)\, H_2^* \nu_{JR} \,\right] - Y_5^\nu \overline{L}_I \,(i\sigma_2)\, H_I^* \nu_{3R} + \text{ h. c. }.
\end{aligned} \tag{4}$$

In this expressions $I, J = 1, 2$, the singlets are denoted by Q_3, u_{3R}, d_{3R}, L_3, e_{3R}, ν_{3R} and H_S, and

$$\kappa = \begin{pmatrix} 0 & 1 \\ 1 & 0 \end{pmatrix}, \qquad \eta = \begin{pmatrix} 1 & 0 \\ 0 & -1 \end{pmatrix}. \tag{5}$$

Then, the Yukawa interactions, eq. (4), yield the following generic form for the mass matrices of all Dirac fermions in the theory [7],

$$\mathbf{M} = \begin{pmatrix} \mu_1 + \mu_2 & \mu_2 & \mu_5 \\ \mu_2 & \mu_1 - \mu_2 & \mu_5 \\ \mu_4 & \mu_4 & \mu_3 \end{pmatrix}, \tag{6}$$

where the μ's are, in general, complex since there is no restriction coming from flavour symmetry. For details see [7, 8]. Furthermore, we add to the Lagrangian the Majorana mass terms for the right-handed neutrinos, $\mathcal{L}_M = -\nu_R^T \mathbf{C} \mathbf{M}_{\nu_R} \nu_R$, where $\mathbf{M}_{\nu_R} = \text{diag}\{M_1, M_2, M_3\}$ is the mass matrix for the right-handed neutrinos and $\mathbf{C}$ is the charge conjugation matrix [8, 12]. The small masses of left-handed Majorana neutrinos are obtained through the type I seesaw mechanism: $\mathbf{M}_\nu = \mathbf{M}_{\nu_D} \mathbf{M}_{\nu_R}^{-1} (\mathbf{M}_{\nu_D})^T$, where $\mathbf{M}_{\nu_D}$ are the Dirac neutrino mass matrices.

3. The leptonic mass matrices and Z_2 symmetry

The number of parameters in the lepton sector may be further reduced by means of an Abelian Z_2 symmetry, which forbids the following Yukawa couplings [7] $Y_1^e = Y_3^e = Y_1^\nu = Y_5^\nu = 0$.

Therefore, the mass matrices of the charged leptons and Dirac neutrinos take the form [7, 8]

$$\mathbf{M}_e = m_\tau \begin{pmatrix} \tilde{\mu}_2 & \tilde{\mu}_2 & \tilde{\mu}_5 \\ \tilde{\mu}_2 & -\tilde{\mu}_2 & \tilde{\mu}_5 \\ \tilde{\mu}_4 & \tilde{\mu}_4 & 0 \end{pmatrix} \quad \text{and} \quad \mathbf{M}_{\nu_D} = \begin{pmatrix} \mu_2^\nu & \mu_2^\nu & 0 \\ \mu_2^\nu & -\mu_2^\nu & 0 \\ \mu_4^\nu & \mu_4^\nu & \mu_3^\nu \end{pmatrix}. \tag{7}$$

respectively. The unitary matrix $\mathbf{U}_{eL}$ that diagonalizes $\mathbf{M}_e$ and enters in the definition of the neutrino mixing matrix $\mathbf{U}_{PMNS}$ may be written in polar form as

$$\mathbf{U}_{eL} = \mathbf{P}_e \mathbf{O}_{eL}, \tag{8}$$

where $\mathbf{P}_e = \mathrm{diag}\left(1,\ 1,\ e^{i\delta_e}\right)$ and the explicit form of orthogonal matrix $\mathbf{O}_{eL}$ (exact to order 10^{-9} in units of the τ mass) is

$$\begin{pmatrix} \frac{1}{\sqrt{2}} x \frac{\left(1+2\widetilde{m}_\mu^2+4x^2+\widetilde{m}_\mu^4+2\widetilde{m}_e^2\right)}{\sqrt{1+\widetilde{m}_\mu^2+5x^2-\widetilde{m}_\mu^4-\widetilde{m}_\mu^6+\widetilde{m}_e^2+12x^4}} & -\frac{1}{\sqrt{2}} \frac{\left(1-2\widetilde{m}_\mu^2+\widetilde{m}_\mu^4-2\widetilde{m}_e^2\right)}{\sqrt{1-4\widetilde{m}_\mu^2+x^2+6\widetilde{m}_\mu^4-4\widetilde{m}_\mu^6-5\widetilde{m}_e^2}} & \frac{1}{\sqrt{2}} \\ -\frac{1}{\sqrt{2}} x \frac{\left(1+4x^2-\widetilde{m}_\mu^4-2\widetilde{m}_e^2\right)}{\sqrt{1+\widetilde{m}_\mu^2+5x^2-\widetilde{m}_\mu^4-\widetilde{m}_\mu^6+\widetilde{m}_e^2+12x^4}} & \frac{1}{\sqrt{2}} \frac{\left(1-2\widetilde{m}_\mu^2+\widetilde{m}_\mu^4\right)}{\sqrt{1-4\widetilde{m}_\mu^2+x^2+6\widetilde{m}_\mu^4-4\widetilde{m}_\mu^6-5\widetilde{m}_e^2}} & \frac{1}{\sqrt{2}} \\ -\frac{\sqrt{1+2x^2-\widetilde{m}_\mu^2-\widetilde{m}_e^2}\left(1+\widetilde{m}_\mu^2+x^2-2\widetilde{m}_e^2\right)}{\sqrt{1+\widetilde{m}_\mu^2+5x^2-\widetilde{m}_\mu^4-\widetilde{m}_\mu^6+\widetilde{m}_e^2+12x^4}} & -x \frac{\left(1+x^2-\widetilde{m}_\mu^2-2\widetilde{m}_e^2\right)\sqrt{1+2x^2-\widetilde{m}_\mu^2-\widetilde{m}_e^2}}{\sqrt{1-4\widetilde{m}_\mu^2+x^2+6\widetilde{m}_\mu^4-4\widetilde{m}_\mu^6-5\widetilde{m}_e^2}} & \frac{\sqrt{1+x^2}\widetilde{m}_e\widetilde{m}_\mu}{\sqrt{1+x^2-\widetilde{m}_\mu^2}} \end{pmatrix}, \tag{9}$$

where $\widetilde{m}_\mu = m_\mu/m_\tau$, $\widetilde{m}_e = m_e/m_\tau$ and $x = m_e/m_\mu$.

3.1. The mass matrix of the neutrinos

In the leptonic sector of this extension of Standard Model the mass matrix of the left-handed Majorana neutrinos, $\mathbf{M}_\nu$, is generated by the type I seesaw mechanism. The mass matrix $\mathbf{M}_\nu$ takes the form [12, 8]:

$$\mathbf{M}_\nu = \begin{pmatrix} \frac{2\left(\mu_2^\nu\right)^2}{\overline{M}} & \lambda\frac{2\left(\mu_2^\nu\right)^2}{\overline{M}} & \frac{2\mu_2^\nu\mu_4^\nu}{\overline{M}} \\ \lambda\frac{2\left(\mu_2^\nu\right)^2}{\overline{M}} & \frac{2\left(\mu_2^\nu\right)^2}{\overline{M}} & \lambda\frac{2\mu_2^\nu\mu_4^\nu}{\overline{M}} \\ \frac{2\mu_2^\nu\mu_4^\nu}{\overline{M}} & \lambda\frac{2\mu_2^\nu\mu_4^\nu}{\overline{M}} & \frac{2\left(\mu_4^\nu\right)^2}{\overline{M}} + \frac{\left(\mu_3^\nu\right)^2}{M_3} \end{pmatrix}, \text{ with } \quad \begin{array}{l} \lambda = \left(\frac{M_2-M_1}{M_1+M_2}\right), \\ \overline{M} = 2\frac{M_1 M_2}{M_2+M_1}. \end{array} \tag{10}$$

M_i $(i = 1, 2, 3)$ are the masses of right-handed neutrinos, in this expression the μ_2, μ_3 and μ_4 are complex parameters that come from the mass matrix of the Dirac neutrinos [12, 8].

The identity of the leptons is encoded in the mass matrices $\mathbf{M}_e$ and $\mathbf{M}_\nu$, for charged leptons and left-handed neutrinos respectively. Nevertheless, the form or texture of these matrices is basis dependent, since given any pair $\mathbf{M}_e$, $\mathbf{M}_\nu$ one can obtain other pairs of equivalent matrices through a unitary rotation, without affecting the physics. On other hand, the phase factors may be factored out of $\mathbf{M}_\nu$ if $\phi_4 = \arg\{\mu_4^\nu\}$ and $\phi_3 = \arg\{\mu_3^\nu\}$ satisfy the relation $\phi_4 = \phi_3$ [12]. Hence, the mass matrix of the left-handed Majorana neutrinos can be written as [8, 12]:

$$\mathbf{M}_\nu = \mathbf{Q}^\nu\, \mathcal{U}_{\frac{\pi}{4}} \left(\mu_0 \mathbf{I}_{3\times 3} + \widehat{\mathbf{M}}\right) \mathcal{U}_{\frac{\pi}{4}}^\dagger\, \mathbf{Q}^\nu, \tag{11}$$

where $\mathbf{Q}^\nu = e^{i\phi_2}\mathrm{diag}\left\{1, 1, e^{i\delta_\nu}\right\}$ with $\delta_\nu = \phi_4 - \phi_2 = \arg\{\mu_4^\nu\} - \arg\{\mu_2^\nu\}$, $\mu_0 = \xi - d$,

$$\mathcal{U}_{\frac{\pi}{4}} = \begin{pmatrix} \frac{1}{\sqrt{2}} & 0 & \frac{1}{\sqrt{2}} \\ -\frac{1}{\sqrt{2}} & 0 & \frac{1}{\sqrt{2}} \\ 0 & 1 & 0 \end{pmatrix} \quad \text{and} \quad \widehat{\mathbf{M}} = \begin{pmatrix} 0 & a & 0 \\ a & b & c \\ 0 & c & 2d \end{pmatrix}. \tag{12}$$

Here, $\xi = 2\left|\mu_2\right|^2 \left|\overline{M}\right|^{-1}$, $a = \sqrt{2}\left|\mu_2^\nu\right|\left|\mu_4^\nu\right| \left|\overline{M}\right|^{-1}(1-|\lambda|)$, $b \equiv 2\left|\mu_4^\nu\right|^2 \left|\overline{M}\right|^{-1} + \left|\mu_3^\nu\right|^2 M_3^{-1} - \xi + d$, $c = \sqrt{2}\left|\mu_2^\nu\right|\left|\mu_4^\nu\right| \left|\overline{M}\right|^{-1}(1+|\lambda|)$ and $d \equiv 2|\lambda|\left|\mu_2^\nu\right|^2 \left|\overline{M}\right|^{-1}$. As mentioned before, the diagonalization of $\mathbf{M}_\nu$ is reduced to the diagonalization of the real symmetric matrix $\widehat{\mathbf{M}}$, which is a matrix with two texture zeroes of class I [13]. Hence, the matrix $\mathbf{M}_\nu$ is diagonalized by a unitary matrix of the following form

$$\mathbf{U}_\nu = \mathbf{Q}^\nu \mathcal{U}_{\frac{\pi}{4}} \mathbf{O}_\nu^{N[I]}. \tag{13}$$

In the literature, these similarity transformations are also known as weak basis transformations, since they leave invariant the gauge currents [14].

The orthogonal matrix $\mathbf{O}_\nu^{N[I]}$ reparametrized in terms of the neutrino masses is given by [8, 12]

$$\begin{pmatrix} \sqrt{\frac{[-1](m_{\nu_3}-\mu_0)(m_{\nu_2}-\mu_0)f_1}{\mathcal{D}_1^{N[I]}}} & \sqrt{\frac{(m_{\nu_{3[1]}}-\mu_0)(\mu_0-m_{\nu_{1[3]}})f_2^{N[I]}}{\mathcal{D}_2^{N[I]}}} & -\sqrt{\frac{[-1](\mu_0-m_{\nu_1})(m_{\nu_2}-\mu_0)f_3^{N[I]}}{\mathcal{D}_3^{N[I]}}} \\ \sqrt{\frac{[-1]2d(\mu_0-m_{\nu_1})f_1}{\mathcal{D}_1^{N[I]}}} & \sqrt{\frac{2d(m_{\nu_2}-\mu_0)f_2^{N[I]}}{\mathcal{D}_2^{N[I]}}} & \sqrt{\frac{[-1]2d(m_{\nu_3}-\mu_0)f_3^{N[I]}}{\mathcal{D}_3^{N[I]}}} \\ -\sqrt{\frac{[-1](\mu_0-m_{\nu_1})f_2^{N[I]}f_3^{N[I]}}{\mathcal{D}_1^{N[I]}}} & \sqrt{\frac{(m_{\nu_2}-\mu_0)f_1 f_3^{N[I]}}{\mathcal{D}_2^{N[I]}}} & -\sqrt{\frac{[-1](m_{\nu_3}-\mu_0)f_1 f_2^{N[I]}}{\mathcal{D}_3^{N[I]}}} \end{pmatrix}, \tag{14}$$

where: $f_1 = (2d+\mu_0-m_{\nu_1})$, $f_2^{N[I]} = [-1](2d+\mu_0-m_{\nu_2})$, $f_3^{N[I]} = [-1](m_{\nu_3}-\mu_0-2d)$, $\mathcal{D}_1^{N[I]} = [-1]2d(m_{\nu_2}-m_{\nu_1})(m_{\nu_3}-m_{\nu_1})$, $\mathcal{D}_2^{N[I]} = [-1]2d(m_{\nu_2}-m_{\nu_1})(m_{\nu_3}-m_{\nu_2})$, and $\mathcal{D}_3^{N[I]} = 2d\left(m_{\nu_{3[1]}}-m_{\nu_{1[3]}}\right)\left(m_{\nu_{3[2]}}-m_{\nu_{2[3]}}\right)$. The allowed values for the parameters μ_0 and $2d+\mu_0$ are in the following ranges: $m_{\nu_{2[1]}} > \mu_0 > m_{\nu_{1[3]}}$ and $m_{\nu_{3[2]}} > 2d+\mu_0 > m_{\nu_{2[1]}}$. The superscripts N and I denote the normal and inverted hierarchies respectively.

4. The leptonic mixing matrix *PMNS*

The leptonic mixing matrix $PMNS$ is the product $\mathbf{U}_{eL}^\dagger \mathbf{U}_\nu \mathbf{K}$, where $\mathbf{K}$ is the diagonal matrix of the Majorana phase factors, defined by $\mathbf{K} = \mathrm{diag}\left(1, e^{i\alpha}, e^{i\beta}\right)$ [15]. In this case, the theoretical expression for the lepton mixing matrix, $\mathbf{U}_{PMNS}^{th}$, is:

$$\begin{pmatrix} \frac{\tilde{m}_e}{\tilde{m}_\mu}O_{11}^{N[I]} - O_{21}^{N[I]}e^{i\delta_l} & \left(\frac{\tilde{m}_e}{\tilde{m}_\mu}O_{12}^{N[I]} - O_{22}^{N[I]}e^{i\delta_l}\right)e^{i\alpha} & \left(\frac{\tilde{m}_e}{\tilde{m}_\mu}O_{13}^{N[I]} - O_{23}^{N[I]}e^{i\delta_l}\right)e^{i\beta} \\ -O_{11}^{N[I]} - \frac{\tilde{m}_e}{\tilde{m}_\mu}O_{21}^{N[I]}e^{i\delta_l} & \left(-O_{12}^{N[I]} - \frac{\tilde{m}_e}{\tilde{m}_\mu}O_{22}^{N[I]}e^{i\delta_l}\right)e^{i\alpha} & \left(-O_{13}^{N[I]} - \frac{\tilde{m}_e}{\tilde{m}_\mu}O_{23}^{N[I]}e^{i\delta_l}\right)e^{i\beta} \\ O_{31}^{N[I]} & O_{32}^{N[I]}e^{i\alpha} & O_{33}^{N[I]}e^{i\beta} \end{pmatrix}, \tag{15}$$

where the elements $O_{ij}^{N[I]}$ $(i,j=1,2,3)$ are given in eq. (14) and the phase $\delta_l = \delta_\nu - \delta_e$ is associated with the CP violation in the leptonic sector.

Now, in the case of an inverted neutrino mass hierarchy $(m_{\nu_2} > m_{\nu_1} > m_{\nu_3})$, a numerical fit of eq. (15) to the magnitude of the entries of the matrix U_{PMNS} given in [1] yields the numerical values for the solar and reactor mixing angle as:

$$\sin^2\theta_{12}^l \approx 0.307 \longrightarrow \theta_{12}^l \approx 33.7^\circ, \quad \sin^2\theta_{13}^l \approx 0.028 \longrightarrow \theta_{13}^l \approx 9.6^\circ, \tag{16}$$

with the following values for the neutrino masses $m_{\nu_2} = 0.088$ eV, $m_{\nu_1} = 0.087$ eV and $m_{\nu_3} = 0.073$ eV, and the parameter values $\delta_l = 75^\circ$, $\mu_0 = 0.0727$ eV and $d = 7.34 \times 10^{-3}$ eV.

5. Flavour Changing Neutral Currents (FCNC)

The Yukawa couplings that enter in the computation of the flavour changing neutral currents are defined in the mass basis, according to $\tilde{Y}_m^{EI} = \mathbf{U}_{eL}^\dagger Y_w^{EI}\mathbf{U}_{eR}$, where $\mathbf{U}_{eL}$ and $\mathbf{U}_{eR}$ are the matrices that diagonalize the charged lepton mass matrix. We obtain [10]

$$\tilde{Y}_m^{E1} \approx \frac{m_\tau}{v_1}\begin{pmatrix} 2\tilde{m}_e & -\frac{1}{2}\tilde{m}_e & \frac{1}{2}x \\ -\tilde{m}_\mu & \frac{1}{2}\tilde{m}_\mu & -\frac{1}{2} \\ \frac{1}{2}\tilde{m}_\mu x^2 & -\frac{1}{2}\tilde{m}_\mu & \frac{1}{2} \end{pmatrix}_m \quad \text{and} \quad \tilde{Y}_m^{E2} \approx \frac{m_\tau}{v_2}\begin{pmatrix} -\tilde{m}_e & \frac{1}{2}\tilde{m}_e & -\frac{1}{2}x \\ \tilde{m}_\mu & \frac{1}{2}\tilde{m}_\mu & \frac{1}{2} \\ -\frac{1}{2}\tilde{m}_\mu x^2 & \frac{1}{2}\tilde{m}_\mu & \frac{1}{2} \end{pmatrix}_m . \tag{17}$$

All the non-diagonal elements are responsible for tree-level FCNC processes. For example, the amplitude of the flavour violating process $\mu \to 3e$, is proportional to $\tilde{Y}_{\mu e}^E \tilde{Y}_{ee}^E$ [16]. Then, in the leading order of perturbation theory, the leptonic branching ratio for this process

$$Br(\mu \to 3e) \approx 2(2+\tan^2\beta)^2 \left(\frac{m_e m_\mu}{m_\tau^2}\right)^2 \left(\frac{m_\tau}{M_H}\right)^4 , \tag{18}$$

taking for $M_H \approx 120\ GeV$ and $\tan\beta = 1$ we obtain $Br(\mu \to 3e) = 2.53 \times 10^{-16}$, well below the experimental upper bound for this process, which is 1×10^{-12} [17]. Similar computations give the numerical estimates of the branching ratios for some other flavour violating processes in the leptonic sector. These results and the corresponding experimental upper bounds are shown in following table

FCNC processes	Theoretical BR	Experimental upper bound BR
$\tau \to 3\mu$	8.43×10^{-14}	2.0×10^{-7} [18]
$\tau \to \mu e^+ e^-$	3.15×10^{-17}	2.7×10^{-7} [18]
$\tau \to \mu\gamma$	9.24×10^{-15}	6.8×10^{-8} [19]
$\tau \to e\gamma$	5.22×10^{-16}	1.1×10^{-11} [20]
$\mu \to 3e$	2.53×10^{-16}	1.0×10^{-12} [17]
$\mu \to e\gamma$	2.42×10^{-20}	1.2×10^{-11} [21]

In all cases considered, the theoretical estimations made in the framework of the minimal S_3-invariant extension of the SM are well below the corresponding experimental upper bounds [10].

6. Muon anomalous magnetic moment

In the minimal S_3-invariant extension of the Standard Model that we are considering here, we have three Higgs $SU(2)$ doublets, one in the singlet and the other two in the doublet representations of the S_3 flavour group. The Z_2 symmetry decouples the charged leptons from the Higgs boson in the S_3 singlet representation. Therefore, in the leading order of perturbation theory there are only two neutral scalars and two neutral pseudoscalars whose exchange will contribute to the anomalous magnetic moment of the muon. Since the heavier generations have larger flavour-changing couplings, the largest contribution comes from the heaviest charged leptons coupled to the lightest of the neutral Higgs bosons. A straightforward computation gives

$$\delta a_\mu^{(H)} = \frac{Y_{\mu\tau} Y_{\tau\mu}}{16\pi^2} \frac{m_\mu m_\tau}{M_H^2}\left(log\left(\frac{M_H^2}{m_\tau^2}\right) - \frac{3}{2}\right) = \frac{m_\tau^2}{(246\ GeV)^2} \frac{(2+\tan^2\beta)}{32\pi^2} \frac{m_\mu^2}{M_H^2}\left(log\left(\frac{M_H^2}{m_\tau^2}\right) - \frac{3}{2}\right). \tag{19}$$

Taking again $M_H = 120\ GeV$ and the upper bound for $\tan\beta = 14$ gives an estimate of the largest possible contribution of the FCNC to the anomaly of the muon's magnetic moment $\delta a_\mu^{(H)} \approx 1.7 \times 10^{-10}$. This number should be compared with the difference between the experimental value and the Standard Model prediction for the anomaly of the muon's magnetic moment [22]

$$\Delta a_\mu = a_\mu^{exp} - a_\mu^{SM} = (28.7 \pm 9.1) \times 10^{-10} \quad \Longrightarrow \quad \frac{\delta a_\mu^{(H)}}{\Delta a_\mu} \approx 0.06. \tag{20}$$

Hence, the contribution of the flavour changing neutral currents to the anomaly of the magnetic moment of the muon is smaller than or of the order of 6% of the discrepancy between the experimental value and the Standard Model prediction.

7. Conclusions

The minimal S_3-invariant extension of the Standard Model permits the calculations of mass and mixing matrices for quark and leptons in a unified way, since the mass matrices of all fermions are brought to the same generic form with two texture zeroes by means of a similarity transformation. In this model we computed the reactor mixing angle, θ_{13}, as function of the masses of the leptons. The numerical values of the neutrino masses extracted from the fit are $m_{\nu_2} = 0.088$ eV, $m_{\nu_1} = 0.087$ eV and $m_{\nu_3} = 0.073$ eV, which in this analysis have an inverted mass hierarchy. The numerical value obtained for θ_{13} is $\theta^l_{13} \approx 9.6°$ in good agreement with the latest analysis of the experimental data on neutrino oscillations [2, 3, 4, 5, 6]. Also, with the help of the Yukawa matrices, we computed the branching ratios of a number of FCNC processes and found that the branching ratios of all FCNC processes considered here are strongly suppressed by powers of the small mass ratios m_e/m_τ and m_μ/m_τ, and by the ratio $\left(m_\tau/M_{H_{1,2}}\right)^4$, where $M_{H_{1,2}}$ is the mass of the neutral Higgs bosons in the S_3-doublet. Taking for $M_{H_{1,2}}$ a very conservative value ($M_{H_{1,2}} \approx 120\ GeV$), we found that the numerical values of the branching ratios of the FCNC in the leptonic sector are well below the corresponding experimental upper bounds by many orders of magnitude. Finally, the contribution of the flavour changing neutral currents to the anomalous magnetic moment of the muon is small but non-negligible, and it is compatible with the best, state of the art measurements and theoretical computations.

Acknowledgements

This work was partially supported by PAPIIT IN113712 and CONACyT-132059. F. González Canales acknowledges the financial support received from PROMEP through a postdoctoral scholarship under contract /103.5/13/4366.

References

[1] Gonzalez-Garcia M, Maltoni M, Salvado J and Schwetz T 2012 (*Preprint* 1209.3023)
[2] Abe K *et al.* (T2K Collaboration) 2011 *Phys. Rev. Lett.* **107**(4) 041801
[3] Abe Y *et al.* (DOUBLE-CHOOZ) 2011 *arXiv:1112.6353 [hep-ex]* (*Preprint* 1112.6353)
[4] Adamson P *et al.* (MINOS Collaboration) 2011 *Phys. Rev. Lett.* **107**(18) 181802
[5] An F P *et al.* (DAYA-BAY) 2012 *arXiv:1203.1669 [hep-ex]* (*Preprint* 1203.1669)
[6] Ahn J *et al.* (RENO collaboration) 2012 *arXiv:1204.0626 [hep-ex]* (*Preprint* 1204.0626)
[7] Kubo J, Mondragon A, Mondragon M and Rodriguez-Jauregui E 2003 *Prog. Theor. Phys.* **109** 795–807
[8] Gonzalez Canales F, Mondragon A and Mondragon M 2012 (*Preprint* 1205.4755)
[9] González Canales F, Mondragon A, Salazar U S and Velasco-Sevilla L 2012 (*Preprint* 1210.0288)
[10] Mondragon A, Mondragon M and Peinado E 2007 *Phys. Rev.* **D76** 076003 (*Preprint* 0706.0354)
[11] Mondragon A, Mondragon M and Peinado E 2008 *J. Phys.* **A41** 304035 (*Preprint* 0712.1799)
[12] González Canales F and Mondragón A 2012 *Journal of Physics: Conference Series* **378** 012014
[13] Gonzalez Canales F and Mondragon A 2011 *J. Phys. Conf. Ser.* **287** 012015 (*Preprint* 1101.3807)
[14] Branco G C, Emmanuel-Costa D, Gonzalez Felipe R and Serodio H 2009 *Phys. Lett.* **B670** 340–349
[15] Beringer J *et al.* (Particle Data Group) 2012 *Phys.Rev.* **D86** 010001
[16] Sher M and Yuan Y 1991 *Phys. Rev.* **D44** 1461–1472
[17] Bellgardt U *et al.* (SINDRUM) 1988 *Nucl. Phys.* **B299** 1
[18] Aubert B *et al.* (BABAR) 2004 *Phys. Rev. Lett.* **92** 121801 (*Preprint* hep-ex/0312027)
[19] Aubert B *et al.* (BABAR) 2005 *Phys. Rev. Lett.* **95** 041802 (*Preprint* hep-ex/0502032)
[20] Aubert B *et al.* (BABAR) 2006 *Phys. Rev. Lett.* **96** 041801 (*Preprint* hep-ex/0508012)
[21] Brooks M L *et al.* (MEGA) 1999 *Phys. Rev. Lett.* **83** 1521–1524 (*Preprint* hep-ex/9905013)
[22] Jegerlehner F 2007 *Acta Phys. Polon.* **B38** 3021 (*Preprint* hep-ph/0703125)

Journal of Physics: Conference Series **485** (2014) 012047 doi:10.1088/1742-6596/485/1/012047

How to construct self/anti-self charge conjugate states?

V V Dvoeglazov

UAF, Universidad de Zacatecas, Ap. Postal 636, Suc. 3 Cruces, Zacatecas 98068 Zac., México

E-mail: valeri@fisica.uaz.edu.mx

Abstract. We construct self/anti-self charge conjugate (Majorana-like) states for the $(1/2,0)\oplus(0,1/2)$ representation of the Lorentz group, and their analogs for higher spins within the quantum field theory. The problem of the basis rotations and that of the selection of phases in the Dirac-like and Majorana-like field operators are considered. The discrete symmetries properties (P, C, T) are studied. The corresponding dynamical equations are presented. In the $(1/2,0)\oplus(0,1/2)$ representation they obey the Dirac-like equation with eight components, which has been first introduced by Markov. Thus, the Fock space for corresponding quantum fields is doubled (as shown by Ziino). The particular attention has been paid to the questions of chirality and helicity (two concepts which are frequently confused in the literature) for Dirac and Majorana states. We further review several experimental consequences which follow from the previous works of M. Kirchbach *et al.* on neutrinoless double beta decay, and G. J. Ni *et al.* on meson lifetimes.

1. The Majorana-like Spinors

Various authors introduced *self/anti-self* charge-conjugate 4-spinors (including in the momentum representation), see, e. g., [1, 2, 3, 4]. Later, these spinors have been studied in Refs. [5, 6, 7, 8, 9]. The authros found corresponding dynamical equations, gauge transformations and other specific features of them. On using $C = -e^{i\theta}\gamma^2\mathcal{K}$, the anti-linear operator of charge conjugation ($\mathcal{K}$ is the complex conjugation operator) we define the *self/anti-self* charge-conjugate 4-spinors in the momentum space $C\lambda^{S,A}(\mathbf{p}) = \pm\lambda^{S,A}(\mathbf{p})$, $C\rho^{S,A}(\mathbf{p}) = \pm\rho^{S,A}(\mathbf{p})$. Such definitions of 4-spinors differ, of course, from the original Majorana definition in the x-representation $\nu(x) = \frac{1}{\sqrt{2}}(\Psi_D(x) + \Psi_D^c(x))$, $C\nu(x) = \nu(x)$ that represents the positive real $C-$ parity only. However, see [8], "for imaginary C parities, the neutrino mass can drop out from the single β decay trace and reappear in $0\nu\beta\beta$, a curious and in principle experimentally testable signature for a non-trivial impact of Majorana framework in experiments with polarized sources."

The rest λ and ρ spinors can be defined in analogous way with the Dirac spinors:

$$\lambda_\uparrow^S(\mathbf{0}) = \sqrt{\frac{m}{2}}\begin{pmatrix}0\\ i\\ 1\\ 0\end{pmatrix}, \lambda_\downarrow^S(\mathbf{0}) = \sqrt{\frac{m}{2}}\begin{pmatrix}-i\\ 0\\ 0\\ 1\end{pmatrix}, \lambda_\uparrow^A(\mathbf{0}) = \sqrt{\frac{m}{2}}\begin{pmatrix}0\\ -i\\ 1\\ 0\end{pmatrix}, \lambda_\downarrow^A(\mathbf{0}) = \sqrt{\frac{m}{2}}\begin{pmatrix}i\\ 0\\ 0\\ 1\end{pmatrix}, \tag{1}$$

$$\rho_{\uparrow\downarrow}^S(\mathbf{0}) = \mp i\lambda_{\downarrow\uparrow}^A(\mathbf{0}),\, \rho_{\uparrow\downarrow}^A(\mathbf{0}) = \pm i\lambda_{\downarrow\uparrow}^S(\mathbf{0})\,. \tag{2}$$

Content from this work may be used under the terms of the Creative Commons Attribution 3.0 licence. Any further distribution of this work must maintain attribution to the author(s) and the title of the work, journal citation and DOI.
Published under licence by IOP Publishing Ltd

Thus, in this basis with the appropriate normalization ("mass dimension") the explicit forms of the 4-spinors of the second kind $\lambda^{S,A}_{\uparrow\downarrow}(\mathbf{p})$ and $\rho^{S,A}_{\uparrow\downarrow}(\mathbf{p})$ can be found (ϕ_L, ϕ_R can be boosted with $\Lambda_{L,R}$ matrices). As claimed in [4], λ and ρ 4-spinors are *not* the eigenspinors of the helicity. Moreover, λ and ρ are *not* the eigenspinors of the parity, as opposed to the Dirac case ($P = \gamma^0 R$, where $R = (\mathbf{x} \to -\mathbf{x})$). The indices $\uparrow\downarrow$ should be referred to the chiral helicity quantum number introduced in the 60s, $\eta = -\gamma^5 h$, Ref. [10]. The normalizations of the spinors $\lambda^{S,A}_{\uparrow\downarrow}(\mathbf{p})$ and $\rho^{S,A}_{\uparrow\downarrow}(\mathbf{p})$ have been given in the previous works.

The dynamical coordinate-space equations are:[1]

$$i\gamma^\mu \partial_\mu \lambda^S(x) - m\rho^A(x) = 0\,,\; i\gamma^\mu \partial_\mu \rho^A(x) - m\lambda^S(x) = 0\,, \tag{3}$$

$$i\gamma^\mu \partial_\mu \lambda^A(x) + m\rho^S(x) = 0\,,\; i\gamma^\mu \partial_\mu \rho^S(x) + m\lambda^A(x) = 0\,. \tag{4}$$

These are *not* the Dirac equation. However, they can be written in the 8-component form. One can also re-write the equations into the two-component form. Thus, one obtains the equations of Ref. [11] equations. Similar formulations have been presented by M. Markov [12], and by A. Barut and G. Ziino [3]. The group-theoretical basis for such doubling has been given in the papers by Gelfand, Tsetlin and Sokolik [13], who first presented the theory in the 2-dimensional representation of the inversion group in 1956 (later called as "the Bargmann-Wightman-Wigner-type quantum field theory" in 1993). The Lagrangian was found. The connection with the Dirac spinors has been found [6, 8]. We can see that the two sets are connected by the unitary transformations, and this represents itself the rotation of the spin-parity basis.

It was shown [6] that the covariant derivative (and, hence, the interaction) can be introduced in this construct in the following way $\partial_\mu \to \nabla_\mu = \partial_\mu - ig\mathrm{L}^5 A_\mu$, where $\mathrm{L}^5 = \mathrm{diag}(\gamma^5 \quad -\gamma^5)$, the 8×8 matrix. In other words, with respect to the transformations

$$\lambda'(x) \to (\cos\alpha - i\gamma^5 \sin\alpha)\lambda(x)\,,\; \overline{\lambda}'(x) \to \overline{\lambda}(x)(\cos\alpha - i\gamma^5 \sin\alpha)\,, \tag{5}$$

$$\rho'(x) \to (\cos\alpha + i\gamma^5 \sin\alpha)\rho(x)\,,\; \overline{\rho}'(x) \to \overline{\rho}(x)(\cos\alpha + i\gamma^5 \sin\alpha) \tag{6}$$

the spinors retain their properties to be self/anti-self charge conjugate spinors and the proposed Lagrangian [6] remains to be invariant. This tells us that while self/anti-self charge conjugate states have zero eigenvalues of the ordinary (scalar) charge operator but they can possess the axial charge (cf. with the discussion of [3] and the old idea of R. E. Marshak – they claimed the same). Next, due to the fact that the transformations

$$\lambda'_S(\mathbf{p}) = \begin{pmatrix} \Xi & 0 \\ 0 & \Xi \end{pmatrix} \lambda_S(\mathbf{p}) \equiv \lambda^*_A(\mathbf{p}), \lambda''_S(\mathbf{p}) = \begin{pmatrix} i\Xi & 0 \\ 0 & -i\Xi \end{pmatrix} \lambda_S(\mathbf{p}) \equiv -i\lambda^*_S(\mathbf{p}), \tag{7}$$

$$\lambda'''_S(\mathbf{p}) = \begin{pmatrix} 0 & i\Xi \\ i\Xi & 0 \end{pmatrix} \lambda_S(\mathbf{p}) \equiv i\gamma^0\lambda^*_A(\mathbf{p}),\; \lambda^{IV}_S(\mathbf{p}) = \begin{pmatrix} 0 & \Xi \\ -\Xi & 0 \end{pmatrix} \lambda_S(\mathbf{p}) \equiv \gamma^0\lambda^*_S(\mathbf{p}) \tag{8}$$

with the 2×2 matrix Ξ defined in Ref. [4], $\Xi\Lambda_{R,L}(\mathbf{p} \leftarrow \mathbf{0})\Xi^{-1} = \Lambda^*_{R,L}(\mathbf{p} \leftarrow \mathbf{0})$, and corresponding transformations for λ^A, do *not* change the properties of bispinors to be in the self/anti-self charge-conjugate spaces, the Majorana-like field operator ($b^\dagger \equiv a^\dagger$) admits additional phase (and, in general, normalization) transformations $\nu^{ML\,\prime}(x^\mu) = [c_0 + i(\tau \cdot \mathbf{c})]\, \nu^{ML\,\dagger}(x^\mu)$, where c_α are arbitrary parameters. The τ matrices are defined over the field of 2×2 matrices. One can parametrize $c_0 = \cos\phi$ and $\mathbf{c} = \mathbf{n}\sin\phi$ and, thus, define the $SU(2)$ group of phase transformations. One can select the Lagrangian which is composed from the both field operators (with λ spinors and ρ spinors) and which remains to be invariant with respect to this kind of

[1] Of course, the signs at the mass terms depend on how do we associate the positive- or negative- frequency solutions with λ and ρ.

transformations. The conclusion is: it is permitted the non-Abelian construct which is based on the spinors of the Lorentz group only (cf. with the old ideas of T. W. Kibble and R. Utiyama) .

The Dirac-like and Majorana-like field operators can be built from both $\lambda^{S,A}(\mathbf{p})$ and $\rho^{S,A}(\mathbf{p})$, or their combinations. It is interesting to note that $[\nu^{ML}(x^\mu) \pm C\nu^{ML^\dagger}(x^\mu)]/2$ lead naturally to the Ziino-Barut scheme of massive chiral fields, Ref. [3], if the former are composed from $\lambda^{S,A}$ spinors. Recently, the interest to these models raised again [9, 15].

2. Chirality and Helicity

Ahluwalia [4] claimed "Incompatibility of Self-Charge Conjugation with Helicity Eignestates and Gauge Interactions". I showed that the gauge interactions of λ and ρ 4-spinors are different. As for the self/anti-self charge-conjugate states and their relations to helicity eigenstates the question is much more difficult, see below. Either we should accept that the rotations would have physical significance, or, due to some reasons, we should not apply the equivalence transformation to the discrete symmetry operators. As far as I understood his paper [4], the latter standpoint is precisely his standpoint. He claimed [4]: "Just as the operator of parity in the $(j,0)\oplus(0,j)$ representation space is independent of which wave equation is under study, similarly the operations of charge conjugation and time reversal do not depend on a specific wave equation. Within the context of the logical framework of the present paper, without this being true we would not even know how to define self-/anti self conjugate $(j,0)\oplus(0,j)$ spinors."

Z.-Q. Shi and G. J. Ni promote a very extreme standpoint. Namely, "the spin states, the helicity states and the chirality states of fermions in Relativistic Quantum Mechanics ... are entirely different: a spin state is helicity degenerate; a helicity state can be expanded as linear combination of the chirality states; the polarization of fermions in flight must be described by the helicity states" (see also their Conclusion section in [16]). In fact, they showed experimental consequences of their statement: "the lifetime of RH polarized fermions is always greater than of LH ones with the same speed in flight". However, we showed that the helicity, chiral helicity and chirality operators are connected by the unitary transformations (see below). Do rotations have physical significance in their opinion?

Let us analize the above statements. The helicity operator is $\hat{h} = \frac{1}{2}\begin{pmatrix} (\sigma\cdot\hat{\mathbf{p}}) & 0 \\ 0 & (\sigma\cdot\hat{\mathbf{p}})\end{pmatrix}$. However, we can do the equivalence transformation of the helicity $\hat{h}$-operator by the unitary matrix. It is known [18] that one can $\mathcal{U}_1(\sigma\cdot\mathbf{a})\mathcal{U}_1^{-1} = \sigma_3|\mathbf{a}|$. In the case of the momentum vector $(\mathbf{n}\equiv\hat{\mathbf{p}} = \mathbf{p}/|\mathbf{p}|)$, one has

$$\mathcal{U}_1 = \begin{pmatrix} 1 & p_l/(p+p_3) \\ -p_r/(p+p_3) & 1 \end{pmatrix},\ U_1 = \begin{pmatrix} \mathcal{U}_1 & 0 \\ 0 & \mathcal{U}_1 \end{pmatrix},\ U_1\hat{h}U_1^{-1} = |\frac{\mathbf{n}}{2}|\begin{pmatrix} \sigma_3 & 0 \\ 0 & \sigma_3 \end{pmatrix}. \quad (9)$$

Then, applying other unitary matrix U_3:

$$\begin{pmatrix} 1&0&0&0\\0&0&1&0\\0&1&0&0\\0&0&0&1 \end{pmatrix}\begin{pmatrix} \sigma_3 & 0 \\ 0 & \sigma_3 \end{pmatrix}\begin{pmatrix} 1&0&0&0\\0&0&1&0\\0&1&0&0\\0&0&0&1 \end{pmatrix} = \begin{pmatrix} 1&0&0&0\\0&1&0&0\\0&0&-1&0\\0&0&0&-1 \end{pmatrix}. \quad (10)$$

we transform to the basis, where helicity is equal (within the factor $\frac{1}{2}$) to γ^5, the chirality operator.

The author of [10] and others introduced the *chiral* helicity $\eta = -\gamma_5 h$, which is equal (within the sign and the factor $\frac{1}{2}$) to the well-known matrix α multiplied by $\mathbf{n}$. Again, $U_1(\alpha\cdot\mathbf{n})U_1^{-1} = \alpha_3|\mathbf{n}|$, with the same matrix U_1. And applying the second unitary transformation:

$$U_2\alpha_3U_2^\dagger = \begin{pmatrix} 1&0&0&0\\0&0&0&1\\0&0&1&0\\0&1&0&0 \end{pmatrix}\alpha_3\begin{pmatrix} 1&0&0&0\\0&0&0&1\\0&0&1&0\\0&1&0&0 \end{pmatrix} = \begin{pmatrix} 1&0&0&0\\0&1&0&0\\0&0&-1&0\\0&0&0&-1 \end{pmatrix} \quad (11)$$

we again come to the γ_5 matrix. The determinats are: $DetU_1 = 1 \neq 0$, $DetU_{2,3} = -1 \neq 0$. Thus, helicity, chirality and chiral helicity are connected by the unitary transformations.

It is *not* surprising to have such a situation because the different helicity 2-spinors can be also connected *not only* by the anti-linear transformation [14, 4] $\xi_h = (-1)^{1/2+h} e^{i\alpha_h} \Theta_{[1/2]} \mathcal{K} \xi_{-h}$, but the unitary transformation too. For instance, when we parametrize the 2-spinors as in [19] we obtain

$$\xi_\downarrow = U\xi_\uparrow = e^{i(\beta-\alpha)} \begin{pmatrix} 0 & e^{-i\phi} \\ -e^{i\phi} & 0 \end{pmatrix} \xi_\uparrow,\ \xi_\uparrow = U^\dagger \xi_\downarrow = e^{i(\alpha-\beta)} \begin{pmatrix} 0 & -e^{-i\phi} \\ e^{i\phi} & 0 \end{pmatrix} \xi_\downarrow\,. \tag{12}$$

To say that the 4-spinor is the eigenspinor of the *chiral helicity*, and, at the same time, it is *not!* the eigenspinor of the helicity operator (and that the physical results would depend on this) signifies the same as to say that rotations have physical significance on the fundamental level.

3. Charge Conjugation and Parity for $S = 1$

Several formalisms have been used for higher spin fields, e. g., [20, 21]. The $2(2S+1)$ formalism gives the equations which are in some sense on an equal footing with the Dirac equation. For instance, for the spin-1 field the equation is $[\gamma_{\mu\nu} p_\mu p_\nu - m^2]\Psi(x) = 0$, with the $\gamma_{\mu\nu}$ being the 6x6 covariantly-defined matrices. However, it was argued later that the signs before the mass terms should be opposite for charged particles of positive- and negative- frequencies [22, 23]: $[\gamma_{\mu\nu} p_\mu p_\nu - (\frac{i\partial/\partial t}{E}) m^2]\Psi(x) = 0$. Hence, Ahluwalia *et al.* write: "The charge conjugation operation C must be carried through with a little greater care for bosons than for fermions within [this] framework because of $\wp_{u,v} = \pm 1$ factor in the mass term. For the $(1,0)\oplus(0,1)$ case, at the classical level we want

$$C : \left(\gamma_{\mu\nu}\, D_+^\mu\, D_+^\nu \,+\, m^2\right)\, u(x) = 0 \rightarrow \left(\gamma_{\mu\nu}\, D_-^\mu\, D_-^\nu \,-\, m^2\right)\, v(x) = 0, \tag{13}$$

where the local $U(1)$ gauge covariant derivatives are defined as: $D_+^\mu = \partial^\mu + i\,q\,A^\mu(x)$, $D_-^\mu = \partial^\mu - i\,q\,A^\mu(x)$", Ref. [23].

"These results read [Ref. [4]]:

$$S^c_{[1]} = e^{i\vartheta^c_{[1]}} \begin{pmatrix} 0 & \Theta_{[1]} \\ -\Theta_{[1]} & 0 \end{pmatrix} \mathcal{K} \equiv C_{[1]}\,\mathcal{K}\,,\ S^s_{[1]} = e^{i\vartheta^s_{[1]}} \begin{pmatrix} 0 & 1_3 \\ 1_3 & 0 \end{pmatrix} = e^{i\vartheta^s_{[1]}}\,\gamma_{00}\,. \tag{14}$$

Note that neither $S^c_{[1/2]}$ nor $S^c_{[1]}$ are unitary (or even linear)." $\Theta_{[1]}$ is the 3x3 representation of the $\Theta_{[1/2]} = -i\sigma_2$.

"For spin-1 ... the requirement of self/anti-self charge conjugacy *cannot* be satisfied. That is, there does not exist a ζ [the phase factors between right- and left- 3-"spinors"] that can satisfy the spin-1 ... requirement" $S^c_{[1]}\,\lambda(p^\mu) = \pm\,\lambda(p^\mu)$, $S^c_{[1]}\,\rho(p^\mu) = \pm\,\rho(p^\mu)$" (?). This is due to the fact that $C^2 = -1$ within this definition of the charge conjugation operator. "We find, however, that the requirement of self/anti-self conjugacy under charge conjugation can be replaced by the requirement of self/anti-self conjugacy under the operation of $\Gamma^5\, S^c_{[1]}$ [precisely, which was used by Weinberg in Ref. [21] due to the different choice of the equation for the negative-frequency 6-"bispinors"], where Γ^5 is the *chirality* operator for the $(1,\,0)\oplus(0,\,1)$ representation space...

The requirement $\left[\Gamma^5\, S^c_{[1]}\right]\,\lambda(p^\mu) = \pm\,\lambda(p^\mu)$, $\left[\Gamma^5\, S^c_{[1]}\right]\,\rho(p^\mu) = \pm\,\rho(p^\mu)$ determines $\zeta^S_\lambda = +1 = \zeta^S_\rho$ for the self $\left[\Gamma^5\, S^c_{[1]}\right]$-conjugate spinors $\lambda^S(p^\mu)$ and $\rho^S(p^\mu)$; and $\zeta^A_\lambda = -1 = \zeta^A_\rho$ for the anti-self $\left[\Gamma^5\, S^c_{[1]}\right]$-conjugate spinors $\lambda^A(p^\mu)$ and $\rho^A(p^\mu)$".

The covariant equations for $\lambda-$ and $\rho-$ objects in the $(1,0)\oplus(0,1)$ representation have been obtained in Ref. [6] under the certain choice of the phase factors in the definition of left- and right- 3-objects.

On the quantum-field level we have to introduce the unitary operators for the charge conjugation and the parity in the Fock space $U^c_{[S]}\Psi_{[S]}(x^\mu)(U^c_{[S]})^{-1} = C_{[S]}\Psi^\dagger_{[S]}(x^\mu)$, $U^s_{[S]}\Psi_{[S]}(x^\mu)(U^s_{[S]})^{-1} = \gamma^0\Psi_{[S]}(x'^\mu)$. For the spin $S = 1/2$ they can be find in the well-known textbooks [25].

R. da Rocha *et al.* write [15]: "Now let one denote the eigenspinors of the Dirac operator for particles and antiparticles respectively by $u_\pm(\mathbf{p})$ and $v_\pm(\mathbf{p})$. The subindex $\pm$ regards the eigenvalues of the helicity operator $(\sigma\cdot\widehat{\mathbf{p}})$. The parity operator acts as $Pu_\pm(\mathbf{p}) = +\,u_\pm(\mathbf{p})$, $Pv_\pm(\mathbf{p}) = -\,v_\pm(\mathbf{p})$, which implies that $P^2 = 1$ in this case. The action of C on these spinors is given [in textbooks [25]], which implies that $\{C,P\} = 0$, [anticommutator]. On the another hand the parity operator P acts on ELKO by $P\lambda^S_{\mp,\pm}(\mathbf{p}) = \pm\, i\,\lambda^A_{\pm,\mp}(\mathbf{p})$, $P\lambda^A_{\mp,\pm}(\mathbf{p}) = \mp\, i\,\lambda^S_{\pm,\mp}(\mathbf{p})$, and it follows that $[C,P] = 0$ [when acting on the Majorana-like states]." In the previous works of the 50s-60s, Ref. [26] it is this case which has been attributed to the $Q = 0$ eigenvalues (the truly neutral particles). You may compare these results with those of Refs. [4, 7, 27], where the same statements have been done on the quantum-field level even at the earlier time comparing with [15]. The notation for the 4-spinors used in the cited papers is a bit different. The acronym "ELKO" is (**almost**) the synonym for the self/anti-self charge conjugated states (the Majorana-like spinors). So, why the difference appeared in the da Rocha formulas compared to my previous results on the classical level? In my papers, see, e.g., Ref. [6, 7, 27], I presented the explicit forms of the $\lambda-$ and $\rho-$ 2-spinors in the basis $\hat{S}_3\xi(\mathbf{0}) = \pm\frac{1}{2}\xi(\mathbf{0})$. The corresponding properties with respect to the parity (on the classical level) are different: $\gamma^0\lambda^S_{\uparrow\downarrow}(p^{\mu'}) = \pm i\lambda^S_{\downarrow\uparrow}(p^\mu)$, $\gamma^0\lambda^A_{\uparrow\downarrow}(p^{\mu'}) = \mp i\lambda^A_{\downarrow\uparrow}(p^\mu)$.

It is easy to find the correspondence between "the new notation", Refs. [30, 15] and the previous one. Namely, $\lambda^{S,A}_\uparrow \to \lambda^{S,A}_{-,+}$, $\lambda^{S,A}_\downarrow \to \lambda^{S,A}_{+,-}$. However, the difference is also in the choice of the basis for the 2-spinors (!). As in Ref. [24], Ahluwalia, Grumiller and da Rocha have chosen the well-known helicity basis (cf. [28, 19]). In my work of 2002 (published in 2004) I have shown that the helicity-basis 4-spinors satisfies the same Dirac equation, the parity matrix can be defined in the similar fashion as in the spinorial basis (according to the Itzykson-Zuber textbook [25]), but the helicity-basis 4-spinors are *not* the eigenspinors of the parity (in full accordance with the claims made in the 4th volume of the Landau course of theoretical physics and with the fact that $[\hat{h},\hat{P}]_+ = 0$, Ref. [29]). In this basis, the parity transformation ($\theta\to\pi-\theta$, $\phi\to\pi+\phi$) lead to the properties:

$$R\phi^-_L(\mathbf{0}) = -ie^{i(\theta_2-\theta_1)}\phi^+_L(\mathbf{0})\,,\; R\phi^+_L(\mathbf{0}) = -ie^{i(\theta_1-\theta_2)}\phi^-_L(\mathbf{0}), \tag{15}$$

$$R\Theta(\phi^-_L(\mathbf{0}))^* = -ie^{-2i\theta_2}\phi^-_L(\mathbf{0})\,,\; R\Theta(\phi^+_L(\mathbf{0}))^* = +ie^{-2i\theta_1}\phi^+_L(\mathbf{0}). \tag{16}$$

This opposes to the spinorial basis, where, for instance: $R\phi^\pm_L(\mathbf{0}) = \phi^\pm_L(\mathbf{0})$. Further calculations are straightforward, and, indeed, they can lead to $[C,P]_- = 0$ when acting on the "ELKO" states, due to $[C,\gamma^5]_+ = 0$.

In the $(1,0)\oplus(0,1)$ representation the situation is similar. If we would like to extend the Nigam-Foldy conclusion, Ref. [26] (about $[C,P]_- = 0$ corresponds to the neutral particles even in the higher spin case (?)) then we should use the helicity basis on the classical level. However, on the level of the quantum-field theory (the "secondary" quantization) the situation is self-consistent. As shown in 1997, Ref. [7, 27], we can obtain easily **both** cases (commutation and anti-commutation) on using $\lambda^{S,A}$ 4-spinors, which have been used earlier (in the basis *column*(1 0) *column*(0 1)).

4. Conclusions

We presented a review of the formalism for the momentum-space Majorana-like particles in the $(S,0)\oplus(0,S)$ representation of the Lorentz Group. The λ and ρ 4-spinors satisfy the 8-component analogue of the Dirac equation. Besides, they have different gauge transformations comparing to the usual Dirac 4-spinors. Their helicity, chirality and chiral helicity properties have been investigated in detail. These operators are connected by the given unitary transformations. At the same time, we showed that the Majorana-like 4-spinors can be obtained by the rotation of the spin-parity basis. Meanwhile, several authors have claimed that the physical results would be different on using calculations with these Majorana-like spinors. Thus, the $(S,0)\oplus(0,S)$ representation space (even in the case of $S=1/2$) has additional mathematical structures leading to deep physical consequences, which have not yet been explored before. However, several claims made by other researchers concerning with chirality, helicity, chiral helicity should not be considered to be true until the time when experiments confirm them. Usually, it is considered that the rotations (unitary transformations) have *no* physical consequences on the level of the Lorentz-covariant theories. Next, we discussed the $[C,P]_\pm=0$ dilemma for neutral and charged particles on using the analysis of the basis rotations and phases.

References

[1] Majorana E 1937 *Nuovo Cimento* **14** 171
[2] Bilenky S M and Pontekorvo B M 1978 *Phys. Repts.* **42** 224
[3] Barut A and Ziino G 1993 *Mod. Phys. Lett.* **A8** 1099; Ziino G 1996 *Int J. Mod. Phys.* **A11** 2081
[4] Ahluwalia D V 1996 *Int. J. Mod. Phys.* **A11** 1855; 1994 Incompatibility of Self-Charge Conjugation with Helicity Eignestates and Gauge Interactions. *Preprint* LANL UR-94-1252 (Los Alamos)
[5] Lounesto P 2002 *Clifford Algebras and Spinors.* (Cambridge: Cambridge University Press), Ch. 11 and 12
[6] Dvoeglazov V V 1995 *Int. J. Theor. Phys.* **34** 2467; 1995 *Nuovo Cim.* **108A** 1467, 1997 *Hadronic J.* **20** 435; 1998 *Acta Phys. Polon.* **B29** 619
[7] Dvoeglazov V V 1997 *Mod. Phys. Lett.* **A12** 2741
[8] Kirchbach M, Compean C and Noriega L 2004 *Eur. Phys. J.* **A22** 149
[9] da Rocha R and Rodrigues W, Jr. 2006 *Mod. Phys. Lett.* **A21** 65, *Preprint* math-ph/0506075
[10] Sen Gupta N D 1967 *Nucl. Phys.* **B4** 147
[11] Feynman R P and Gell-Mann M 1958 *Phys. Rev.* **109** 193
[12] Markov M 1937 *ZhETF* **7** 579: ibid. 603; 1964 *Nucl. Phys.* **55** 130
[13] Gelfand I M and Tsetlin M L 1956 *ZhETF* **31** 1107; Sokolik G A 1957 *ZhETF* **33** 1515
[14] Ryder L H 1985 *Quantum Field Theory.* (Cambridge: Cambridge University Press)
[15] da Rocha R, Bernardini A E and Hoff da Silva J M 2011 *JHEP* **1104** 110
[16] Shi Z Q and Ni G J 2002 *Chin. Phys. Lett.* **19** 1427; Shi Z Q 2011 *Preprint* arXiv:1101.0481v1
[17] Dvoeglazov V V 1998 *Int. J. Theor. Phys.* **37** 1915
[18] Berg R A 1966 *Nuovo Cim.* **42** 148
[19] Dvoeglazov V V 1997 *Fizika* **B6** 111
[20] Bargmann V and Wigner E P 1948 *Proc. Nat. Acad. Sci. (USA)* **34** 211
[21] Weinberg S 1964 *Phys. Rev.* **133** B1318; ibid. 1964 **134** B882
[22] Weaver D L, Hammer C L and Good R H, Jr. 1964 *Phys. Rev.* **B135** 241; Sankaranarayanan A and Good R H, jr. 1965 *Nuovo Cim.* **36** 1303
[23] Ahluwalia D V, Johnson M B and Goldman T 1993 *Phys. Lett.* **B316** 102
[24] Dvoeglazov V V 2004 *Int. J. Theor. Phys.* **43** 1287
[25] Itzykson C and Zuber J.-B. 1980 *Quantum Field Theory.* (McGraw-Hill Book Co.)
[26] Nigam B and Foldy L L 1956 *Phys. Rev.* **102** 1410
[27] Dvoeglazov V V 2012 *J. Phys. Conf. Ser.* **343** 012033
[28] Varshalovich D A *et al.* 1988 *Quantum Theory of Angular Momentum* (Singapore: World Scientific)
[29] Berestetskii V B *et al.* 1982 *Quantum Electrodynamics.* 2nd Edition (Butterworth-Heinemann)
[30] Ahluwalia D V and Grumiller D 2005 *Phys. Rev.* **D72** 067701

Sum rules for tetraquark decay in broken SU(3) symmetry

V Gupta,[1] G Sánchez-Colón[1] and S Rajpoot[2]

[1]Departamento de Física Aplicada,
Centro de Investigación y de Estudios Avanzados del IPN,
Unidad Mérida, A.P. 73, Cordemex, Mérida, Yucatán, 97310, Mexico
[2]Department of Physics & Astronomy,
California State University, Long Beach, CA 90840, USA

E-mail: virendra@mda.cinvestav.mx, gsanchez@mda.cinvestav.mx, subhash.rajpoot@csulb.edu

Abstract.
We give sum rules for tetraquark decay coupling constants, taking into account the $SU(3)$ symmetry breaking interactions to first order.

1. Introduction

Recently [1], the Belle Collaboration has found the decay of heavy mesons of masses 10.61 GeV and 10.65 GeV into $\Upsilon + \pi^{\pm}$, refered to as $Z_b(10610)$ and $Z_b(10650)$, respectively. In this note we suggest that these new states are tetraquarks made of $(b\bar{b}\, q\bar{q})$ with $q = u, d, s$. We workout the consequences of broken SU(3) symmetry to derive sum rules for tetraquark decay coupling constants.

2. Notation

Let us consider a general tetraquark $T(Q\bar{Q}\, M(q\bar{q}))$ with Q ($= b$ or c) quark a $SU(3)$ flavour singlet and $M(q\bar{q})$ made of the light quarks q ($= u, d,$ and s) a $SU(3)$ flavour octet. $M(q\bar{q})$ could be the 0^- pseudoscalar octet ($\pi^{\pm}$, $K^{\pm}$, etc.) or the 1^- vector octet ($\rho^{\pm}$, $K^{*\pm}$, etc.).

The natural decay mode of the tetraquark $T(Q\bar{Q}\, M(q\bar{q}))$ would be

$$T(Q\bar{Q}\, M(q\bar{q})) \rightarrow H(Q\bar{Q}) + M(q\bar{q}), \tag{1}$$

where $H(Q\bar{Q})$ is a heavy meson since $Q = b$ or c. Furthermore, one expects that $H(Q\bar{Q})$ and $M(q\bar{q})$ in the final state have the same J^P they have in the tetraquark. This is the simplest possibility.

3. Sum rules

If the decay interaction is flavour $SU(3)$ singlet, then all the eight decays will have the same coupling constant G_0. However if one considers a first order breaking interaction transforming like λ_8 (or T_3^3) which conserves isospin invariance then this will give rise to two $SU(3)$ breaking coupling constants, G_F and G_D, which arise from the antisymmetric and symmetric octets

Content from this work may be used under the terms of the Creative Commons Attribution 3.0 licence. Any further distribution of this work must maintain attribution to the author(s) and the title of the work, journal citation and DOI.
Published under licence by IOP Publishing Ltd

Journal of Physics: Conference Series **485** (2014) 012048 doi:10.1088/1742-6596/485/1/012048

resulting from the breaking interaction and the $M(q\bar{q})$ octet in the final state. Thus, the eight decays will be given in terms of three coupling constants. However, since isospin invariance is present, there will be four independent coupling constants namely, when the $M(q\bar{q})$ meson is (K^+, K^0), ($\bar{K}^0$, K^0_-), (π^+, π^0, π^-), and $s\bar{s}$. These four coupling constants will be linear combinations of G_0, G_F, and G_D, and hence, we expect a sume rule among the four coupling constants of the observable decays [2, 3, 4].

a) Since the $SU(3)$ calculation is analogous to the calculation of the meson masses with λ_8 breaking, one obtains a coupling constants sum rule analogous to the Gell-Mann-Okubo mass formula for mesons [5, 6]. Thus, if our interpretation of the observed particles as tetraquarks is right we expect,

$$\begin{aligned} 2\,G\left[T(Q\bar{Q}\,K) \to H(Q\bar{Q}) + K\right] + 2\,G\left[T(Q\bar{Q}\,K) \to H(Q\bar{Q}) + \bar{K}\right] = \\ G\left[T(Q\bar{Q}\,\pi) \to H(Q\bar{Q}) + \pi\right] + 3\,G\left[T(Q\bar{Q}\,s\bar{s}) \to H(Q\bar{Q}) + s\bar{s}\right]. \end{aligned} \tag{2}$$

Here $K = K^+$ or K^0, $\bar{K} = \bar{K}^0$ or K^-, $\pi = \pi^+, \pi^-$, or π^0. The heavy quark Q is a $SU(3)$ singlet and could be either b or c quark.

b) The above formula for decay coupling constants is also valid for tetraquarks $T(Q\bar{Q}'\,M(q\bar{q}))$ where $Q = b$ (or c) with $Q' = c$ (or b). As noted earlier, the light quark octet $M(q\bar{q})$ could be the vector octet (ρ, K^*, etc.). The sum rule would be valid with the obvious change $\pi \to \rho$, $K \to K^*$, etc.

The sum rules contain $\eta = s\bar{s}$. However, as is well known, both for pseudoscalar and vector mesons, $s\bar{s}$ is a combination of the $SU(3)$ singlet η_1 and the eigth component η_8. This mixing must be taken into account in testing our coupling constants sume rule. Specifically, in terms of the SU(3) flavor states η_1 and η_8,

$$s\bar{s} = \frac{1}{\sqrt{3}}\eta_1 - \frac{2}{\sqrt{6}}\eta_8, \tag{3}$$

where

$$\eta_1 = \frac{1}{\sqrt{3}}(u\bar{u} + d\bar{d} + s\bar{s}) \quad \text{and} \quad \eta_8 = \frac{1}{\sqrt{6}}(u\bar{u} + d\bar{d} - 2s\bar{s}). \tag{4}$$

For $J^P = 0^-$, $\eta = \eta_8 \cos\theta - \eta_1 \sin\theta$ and $\eta'(958) = \eta_8 \sin\theta + \eta_1 \cos\theta$, with $\theta = -11.5°$. For $J^P = 1^-$, $\eta \to \phi(1020)$ and $\eta'(958) \to \omega(782)$, with $\theta = 38.7°$ [7].

c) In case there are baryonic pentaquarks of the form $P(Q\bar{Q}\,B)$ where B is usual baryon octet then our sum rule will be valid for the decay coupling constants of

$$P(Q\bar{Q}\,B) \to H(Q\bar{Q}) + B, \tag{5}$$

with appropiate changes, namely,

$$\begin{aligned} 2\,G\left[P(Q\bar{Q}\,N) \to H(Q\bar{Q}) + N\right] + 2\,G\left[P(Q\bar{Q}\,\Xi) \to H(Q\bar{Q}) + \Xi\right] = \\ G\left[P(Q\bar{Q}\,\Sigma) \to H(Q\bar{Q}) + \Sigma\right] + 3\,G\left[P(Q\bar{Q}\,\Lambda) \to H(Q\bar{Q}) + \Lambda\right]. \end{aligned} \tag{6}$$

The above sum rule with appropriate changes would be valid for pentaquarks $P(Q\bar{Q}'\,B)$, where $Q = b$ (or c) with $Q' = c$ (or b).

4. Conclusions

The above sum rules were obtained by taking into account the $SU(3)$ symmetry breaking only up to the first order. At present we do not have data to extract the coupling constants involved in the sum rules. Hopefully, we will have sufficient data in the near future to enable us to check the sum rules given here.

Acknowledgments
V. Gupta and G. Sánchez-Colón would like to thank CONACyT (México) for partial support. The work of S. Rajpoot was supported by DOE Grant #: DE-FG02-10ER41693.

References
[1] Bondar A *et al.* (Belle Collaboration) 2012 *Phys. Rev. Lett.* **108** 122001
[2] 1964 *Phys. Rev.* **135** 1442
[3] Gupta V and Singh V 1964 *Phys. Rev.* **136** 782
[4] Gupta V 1966 *Il Nvo. Cim.* **42** 737
[5] Gell-Mann M 1961 *"The Eightfold Way: A Theory of Strong Interaction Symmetry"* California Institute of Technology Synchrotron Laboratory Report CTSL-20 (1961), unpublished.
[6] Okubo S 1962 *Prog. Theor. Phys.* **27** 949
[7] Nakamura K et al. (Particle Data Group) 2010 *J. Phys.* **G37** 075021

PASCOS 2012 – 18th International Symposium on Particles Strings and Cosmology IOP Publishing
Journal of Physics: Conference Series **485** (2014) 012049 doi:10.1088/1742-6596/485/1/012049

Non-Abelian tensor multiplet in four dimensions

Hitoshi Nishino[†] **and Subhash Rajpoot**[‡,⋆]

Department of Physics and Astronomy, California State University, 1250 Bellflower Blvd., Long Beach, CA 90840, USA.

E-mail: [†] H.Nishino@csulb.edu
E-mail: [‡] Subhash.Rajpoot@csulb.edu, ⋆ Conference Speaker

Abstract. The long-standing problem with a non-Abelian tensor with non-trivial consistent couplings in four dimensions has been solved. The key technique is double-fold: (1) Adding extra Chern-Simons terms for the field strength of non-Abelian tensor, and (2) employing a compensator mechanism. We generalize this mechanism to supersymmetric system. Our system has three multiplets: (i) The usual non-Abelian vector multiplet (VM) $(A_\mu{}^I, \lambda^I)$, (ii) A non-Abelian tensor multiplet (TM) $(B_{\mu\nu}{}^I, \chi^I, \varphi^I)$, and (iii) A compensator vector multiplet (CVM) $(C_\mu{}^I, \rho^I)$. The indices $I, J, \cdots$ are for the adjoint representation of a non-Abelian group G. All of our fields are propagating with kinetic terms. The $C_\mu{}^I$-field plays the role of a compensator absorbed into the longitudinal component of $B_{\mu\nu}{}^I$. We give both the component lagrangian and a corresponding superspace reformulation, reconfirming the total consistency of the system.

1. The Conventional Problem with Non-Abelian Tensors

First, the conventional problem with non-Abelian tensors will be discussed. A solution will be presented. Also presented will be a supersymmetrized physical non-Abelian field with consistent interactions [1], both in component language and superfield language.

The long-standing problem with a non-Abelian tensor is described as follows. Let I be the adjoint index of a non-Abelian group G gauged by the a non-Abelian vector field $A_\mu{}^I$, minimally coupled to the antisymmetric tensor $B_{\mu\nu}{}^I$ with the coupling constant g. Consider the conventional field strength[3]

$$G^{(0)}_{\mu\nu\rho}{}^I \equiv +3D_{[\mu}B_{\nu\rho]}{}^I \equiv +3(\partial_{[\mu}B_{\nu\rho]}{}^I + gf^{IJK}A_{[\mu}{}^J B_{\nu\rho]}{}^K) \ , \tag{1.1}$$

where D_μ is the usual gauge-covariant derivative with the structure constant f^{IJK} of the group G. Consider an action $I_0 \equiv \int d^4x\, \mathcal{L}_0$ with the lagrangian[4]

$$\mathcal{L}_0 \equiv -\frac{1}{12}(G^{(0)}_{\mu\nu\rho}{}^I)^2 - \frac{1}{4}(F_{\mu\nu}{}^I)^2 \ , \tag{1.2}$$

with $F_{\mu\nu}{}^I \equiv 2\partial_{[\mu}A_{\nu]}{}^I + gf^{IJK}A_\mu{}^J A_\nu{}^K$. Obviously, the B-field equation is[5]

$$\frac{\delta\mathcal{L}_0}{\delta B_{\mu\nu}{}^I} = +\frac{1}{2}D_\rho G^{(0)\mu\nu\rho I} \doteq 0 \ . \tag{1.3}$$

[3] The formulation in this section is the so-called 'tensor hierarchy' [2, 3, 4, 5], but we re-formulate their general expressions in terms of our objectives here.
[4] We use the signature $(-,+,+,+)$ for four dimensions (4D) as in [1].
[5] The symbol $\doteq$ stands for a field equation, to be distinguished from an algebraic identity. We also use the symbol $\stackrel{?}{=}$ for an equality under question.

Content from this work may be used under the terms of the Creative Commons Attribution 3.0 licence. Any further distribution of this work must maintain attribution to the author(s) and the title of the work, journal citation and DOI.
Published under licence by IOP Publishing Ltd

The problem is that the divergence of this B-field equation does *not* vanish:

$$0 \stackrel{?}{=} D_\nu \left(\frac{\delta \mathcal{L}_0}{\delta B_{\mu\nu}{}^I} \right) = +\frac{1}{4} g f^{IJK} F_{\nu\rho}{}^J G^{(0)\mu\nu\rho K} \neq 0 \ , \tag{1.4}$$

unless $F_{\mu\nu}{}^I$ or $G^{(0)}_{\mu\nu\rho}{}^I$ vanishes trivially. This inconsistency arises already at the *classical* level. This is also one of the reasons, why topological formulations with vanishing field strengths $F_{\mu\nu}{}^I \doteq 0$, $G^{(0)}_{\mu\nu\rho}{}^I \doteq 0$ such as in [6] are easier to formulate for non-Abelian tensors.

An additional problem is related to the so-called local tensorial gauge transformation of the B-field:

$$\delta_\beta B_{\mu\nu}{}^I = +D_{[\mu}\beta_{\nu]}{}^I - D_{[\nu}\beta_{\mu]}{}^I \ , \tag{1.5}$$

because the field strength $G_{\mu\nu}{}^I$ is *not* invariant under δ_β:

$$\delta_\beta G^{(0)}_{\mu\nu\rho}{}^I = +3 g f^{IJK} F_{[\mu\nu}{}^J \beta_{\rho]}{}^K \neq 0 \ . \tag{1.6}$$

This further implies the *non*-invariance of the action: $\delta_\beta I_0 \neq 0$. These two problems are mutually related, because the non-vanishing of (1.4) is also re-casted into $\delta_\beta I_0 \neq 0$.

2. Solution to the Problem

The solution to the problem above is to introduce a non-trivial Chern-Simons (CS) term into the G-field strength:

$$\begin{aligned} G_{\mu\nu\rho}{}^I &\equiv +3(\partial_{[\mu} B_{\nu\rho]}{}^I + g f^{IJK} A_{[\mu}{}^J B_{\nu\rho]}{}^K) - 3 f^{IJK} C_{[\mu}{}^J F_{\nu\rho]}{}^K \\ &= +3 D_{[\mu} B_{\nu\rho]}{}^I - 3 f^{IJK} C_{[\mu}{}^J F_{\nu\rho]}{}^K \equiv +G^{(0)}_{\mu\nu\rho}{}^I - 3 f^{IJK} C_{[\mu}{}^J F_{\nu\rho]}{}^K \ , \end{aligned} \tag{2.1}$$

where $C_\mu{}^I$ is a 'compensator' vector field, also carrying the adjoint index. The field strength for C is defined by

$$H_{\mu\nu}{}^I \equiv +D_{[\mu} C_{\nu]}{}^I - D_{[\nu} C_{\mu]}{}^I + g B_{\mu\nu}{}^I \ . \tag{2.2}$$

Now these field strengths G and H are *invariant* under the δ_β-transformation

$$\delta_\beta B_{\mu\nu}{}^I = + D_{[\mu}\beta_{\nu]}{}^I - D_{[\nu}\beta_{\mu]}{}^I \tag{2.3a}$$

$$\delta_\beta C_\mu{}^I = - g \beta_\mu{}^I \ , \tag{2.3b}$$

which is the 'proper' gauge transformation for $B_{\mu\nu}{}^I$, and δ_γ-transformation

$$\delta_\gamma B_{\mu\nu}{}^I = - f^{IJK} F_{\mu\nu}{}^J \gamma^K \ , \tag{2.4a}$$

$$\delta_\gamma C_\mu{}^I = D_\mu \gamma^I \ . \tag{2.4b}$$

which is the 'proper' gauge transformation for $C_\mu{}^I$. As (2.3b) shows, $C_\mu{}^I$ is a compensator field for the δ_β-transformation.

The role played by the $C \wedge F$-term in (2.1) is to cancel the unwanted term in (1.6). The C-field itself should have its own 'gauge' transformation as the covariant gradient (2.4b). The contribution of $\delta_\gamma(2D_{[\mu}C_{\nu]}{}^I)$ in (2.2) is cancelled by the contribution of $\delta_\gamma(gB_{\mu\nu}{}^I)$. In other words, we have the total invariances

$$\delta_\beta(G_{\mu\nu\rho}{}^I, H_{\mu\nu}{}^I) = (0, 0) \ , \quad \delta_\gamma(G_{\mu\nu\rho}{}^I, H_{\mu\nu}{}^I) = (0, 0) \ . \tag{2.5}$$

Accordingly, we also have the consistency problem (1.4) solved, by considering the lagrangian

$$\mathcal{L}_1 \equiv -\tfrac{1}{12}\,(G_{\mu\nu\rho}{}^I)^2 - \tfrac{1}{4}\,(H_{\mu\nu}{}^I)^2 - \tfrac{1}{4}\,(F_{\mu\nu}{}^I)^2 \ . \tag{2.6}$$

The total action is also invariant $\delta_\beta I_1 = \delta_\gamma I_1 = 0$. The field equations for B and C-fields are

$$\frac{\delta\mathcal{L}_1}{\delta B_{\mu\nu}{}^I} = +\tfrac{1}{2}\,D_\rho G^{\mu\nu\rho\,I} - \tfrac{1}{2}\,g H^{\mu\nu\,I} \doteq 0 \ , \tag{2.7a}$$

$$\frac{\delta\mathcal{L}_1}{\delta C_\mu{}^I} = -D_\nu H^{\mu\nu\,I} + \frac{1}{2} f^{IJK} F_{\rho\sigma}{}^J G^{\mu\rho\sigma\,K} \doteq 0 \ , \tag{2.7b}$$

The divergence of the B-field equation vanishes now:

$$0 \stackrel{?}{=} D_\nu\left(\frac{\delta\mathcal{L}_1}{\delta B_{\mu\nu}{}^I}\right) = +\frac{1}{2}g\left(\frac{\delta\mathcal{L}_1}{\delta C_\mu{}^I}\right) \doteq 0 \ , \tag{2.8}$$

where the last equality holds because of the C-field equation. In other words, the unwanted FG-term in (1.4) is now cancelled by the contribution of the C-field equation.

Relevantly, the divergence of (2.7b) also vanishes, as it should without any inconsistency:

$$0 \stackrel{?}{=} D_\mu\left(\frac{\delta\mathcal{L}_1}{\delta C_\mu{}^I}\right) = +f^{IJK}F_{\mu\nu}{}^J\left(\frac{\delta\mathcal{L}_1}{\delta B_{\mu\nu}{}^K}\right) \doteq 0 \ . \tag{2.9}$$

We emphasize repeatedly that these invariances have never been accomplished without the peculiar CS terms both in (2.1) and (2.2) [2, 3, 4, 5].

3. Component Formulation of N=1 TM

The supersymmetrization of the purely bosonic system (2.6) has been accomplished in our recent paper [1]. We need three multiplets: (i) A tensor multiplet (TM) $(B_{\mu\nu}{}^I, \chi^I, \varphi^I)$, (ii) A Yang-Mills vector multiplet (YMVM) $(A_\mu{}^I, \lambda^I)$, and (iii) A compensating vector multiplet (CVM) $(C_\mu{}^I, \rho^I)$. Our total action $I \equiv \int d^4x\, g^2 \mathcal{L}$ has the lagrangian [1]

$$\begin{aligned}
\mathcal{L} = & -\frac{1}{12}(G_{\mu\nu\rho}{}^I)^2 + \frac{1}{2}(\overline{\chi}^I \not{D}\chi^I) - \frac{1}{2}(D_\mu\varphi^I)^2 - \frac{1}{2}g^2(\varphi^I)^2 - g(\overline{\chi}^I\rho^I) \\
& - \frac{1}{4}(H_{\mu\nu}{}^I)^2 + \frac{1}{2}(\overline{\rho}^I \not{D}\rho^I) - \frac{1}{4}(F_{\mu\nu}{}^I)^2 + \frac{1}{2}(\overline{\lambda}^I \not{D}\lambda^I) \\
& - \frac{1}{2}g f^{IJK}(\overline{\lambda}^I\chi^J)\varphi^K + \frac{1}{2}f^{IJK}(\overline{\lambda}^I\gamma^\mu\rho^J)D_\mu\varphi^K + \frac{1}{12}f^{IJK}(\overline{\lambda}^I\gamma^{\mu\nu\rho}\rho^J)G_{\mu\nu\rho}{}^K \\
& + \frac{1}{4}f^{IJK}(\overline{\rho}^I\gamma^{\mu\nu}\chi^J)F_{\mu\nu}{}^K - \frac{1}{4}f^{IJK}(\overline{\lambda}^I\gamma^{\mu\nu}\chi^J)H_{\mu\nu}{}^K - \frac{1}{2}f^{IJK}F_{\mu\nu}{}^I H^{\mu\nu\,J}\varphi^K \ ,
\end{aligned} \tag{3.1}$$

up to quartic-order terms $\mathcal{O}(\phi^4)$, and the coupling constant g has the dimension of mass.

The scalar φ^I has its mass g, while there is a mixture between χ^I and ρ^I with the same mass g. As has been mentioned after (2.4), $C_\mu{}^I$ is a compensator field [7], absorbed into the longitudinal component of $B_{\mu\nu}{}^I$. The kinetic term of the C-field becomes the mass term of $B_{\mu\nu}{}^I$. Accordingly, the degrees of freedom (DOF) for the massive TM fields are $B_{\mu\nu}{}^I$ (3), χ with ρ^I (4) and $\varphi^I(1)$.

Our action I is invariant under global $N = 1$ supersymmetry [1]

$$\delta_Q B_{\mu\nu}{}^I = + (\overline{\epsilon}\gamma_{\mu\nu}\chi^I) - 2f^{IJK} C_{[\mu|}{}^J (\delta_Q A_{|\nu]}{}^K) \ , \tag{3.2a}$$

$$\delta_Q \chi^I = + \frac{1}{6}(\gamma^{\mu\nu\rho}\epsilon) G_{\mu\nu\rho}{}^I - (\gamma^\mu\epsilon) D_\mu \varphi^I$$
$$+ \frac{1}{2} f^{IJK}\Big[+ \epsilon(\overline{\lambda}^J \rho^K) - (\gamma_5\gamma^\mu\epsilon)(\overline{\lambda}^J\gamma_5\gamma_\mu\rho^K) - (\gamma_5\epsilon)(\overline{\lambda}^J\gamma_5\rho^K)\Big] \ , \tag{3.2b}$$

$$\delta_Q \varphi^I = + (\overline{\epsilon}\chi^I) \ , \tag{3.2c}$$

$$\delta_Q C_\mu{}^I = + (\overline{\epsilon}\gamma_\mu\rho^I) + f^{IJK}(\overline{\epsilon}\gamma_\mu\lambda^J)\varphi^K \ , \tag{3.2d}$$

$$\delta_Q \rho^I = + \frac{1}{2}(\gamma^{\mu\nu}\epsilon) H_{\mu\nu}{}^I - g\epsilon\varphi^I - \frac{1}{2} f^{IJK}(\gamma^{\mu\nu}\epsilon) F_{\mu\nu}{}^J \varphi^K$$
$$+ \frac{1}{4} f^{IJK}\Big[+ \epsilon(\overline{\lambda}^J\chi^K) - (\gamma^\mu\epsilon)(\overline{\lambda}^J\gamma_\mu\chi^K) + \frac{1}{2}(\gamma^{\mu\nu}\epsilon)(\overline{\lambda}^J\gamma_{\mu\nu}\chi^K)$$
$$- (\gamma_5\gamma^\mu\epsilon)(\overline{\lambda}^J\gamma_5\gamma_\mu\chi^K) - (\gamma_5\epsilon)(\overline{\lambda}^J\gamma_5\chi^K)\Big] \ , \tag{3.2e}$$

$$\delta_Q A_\mu{}^I = + (\overline{\epsilon}\gamma_\mu\lambda^I) \ , \tag{3.2f}$$

$$\delta_Q \lambda^I = + \frac{1}{2}(\gamma^{\mu\nu}\epsilon) F_{\mu\nu}{}^I + \frac{1}{2} f^{IJK}(\gamma_5\epsilon)(\overline{\rho}^J\gamma_5\chi^K) \ , \tag{3.2g}$$

up to $\mathcal{O}(\phi^3)$-terms.

Our tensorial gauge transformation δ_β, and δ_γ-transformation are exactly the same as (2.3) and (2.4), while all the fermionic fields transform only under the usual non-Abelian gauge transformation δ_α, as the B and C-fields do, so that there is no problem with the $\delta_\beta I = 0$ and $\delta_\gamma I = 0$ of the field strengths as in (2.1) and (2.2), *via* (2.5).

The δ_Q-transformations of the field strengths reflect their CS terms:

$$\delta_Q G_{\mu\nu\rho}{}^I = + 3(\overline{\epsilon}\gamma_{[\mu\nu}D_{\rho]}\chi^I) + 3f^{IJK}(\delta_Q A_{[\mu}{}^J) H_{\nu\rho]}{}^K - 3f^{IJK}(\delta_Q C_{[\mu}{}^J) F_{\nu\rho]}{}^K \ , \tag{3.3a}$$

$$\delta_Q H_{\mu\nu}{}^I = - 2(\overline{\epsilon}\gamma_{[\mu}D_{\nu]}\rho^I) + g(\overline{\epsilon}\gamma_{\mu\nu}\chi^I) + 2f^{IJK} D_{[\mu|}\Big[(\delta_Q A_{|\nu]}{}^J)\varphi^K\Big] \ , \tag{3.3b}$$

$$\delta_Q F_{\mu\nu}{}^I = - 2(\overline{\epsilon}\gamma_{[\mu}D_{\nu]}\lambda^I) \ . \tag{3.3c}$$

Since we have *not* added the D-auxiliary field, our YMVM and CVM have *on-shell* DOF 2+2, while *off-shell* DOF 3+4. However, our TM is in the *off-shell* formulation with the total off-shell DOF $4 + 4$, because the off-shell DOF of each field are $[(4-1)\cdot(4-2)]/2 = 3$ for $B_{\mu\nu}$, 4 for χ and 1 for φ.

The field equations for all of our fields are[6] [1]

$$+ \not{D}\lambda^I - \frac{1}{2} g f^{IJK}\chi^J\varphi^K + \frac{1}{2} f^{IJK}(\gamma^\mu\rho^J) D_\mu\varphi^K$$
$$- \frac{1}{4} f^{IJK}(\gamma^{\mu\nu}\chi^J) H_{\mu\nu}{}^K + \frac{1}{12} f^{IJK}(\gamma^{\mu\nu\rho}\rho^J) G_{\mu\nu\rho}{}^K \doteq 0 \ , \tag{3.4a}$$

$$+ \not{D}\chi^I - g\rho^I + \frac{1}{2} g f^{IJK}\lambda^H\varphi^K - \frac{1}{4} f^{IJK}(\gamma^{\mu\nu}\lambda^J) H_{\mu\nu}{}^K + \frac{1}{4} f^{IJK}(\gamma^{\mu\nu}\rho^J) F_{\mu\nu}{}^K \doteq 0 \ , \tag{3.4b}$$

$$+ \not{D}\rho^I - g\chi^I + \frac{1}{2} f^{IJK}(\gamma^\mu\lambda^J) D_\mu\varphi^K$$

[6] These equations are fixed up to $\mathcal{O}(\phi^3)$-terms, due to the quartic fermion terms in the lagrangian.

$$- \frac{1}{12} f^{IJK} (\gamma^{\mu\nu\rho}\lambda^J) G_{\mu\nu\rho}{}^K + \frac{1}{4} f^{IJK} (\gamma^{\mu\nu}\chi^J) F_{\mu\nu}{}^K \doteq 0 \ , \tag{3.4c}$$

$$+ D_\nu F_\mu{}^{\nu\, I} + g f^{IJK} \varphi^J D_\mu \varphi^K + \frac{1}{2} g f^{IJK} (\overline\lambda{}^J \gamma_\mu \lambda^K) + f^{IJK} H_{\mu\nu}{}^J D^\nu \varphi^K$$
$$- \frac{1}{2} f^{IJK} G_{\mu\rho\sigma}{}^J H^{\rho\sigma\, K} + \frac{1}{2} f^{IJK} (\overline\chi{}^J D_\mu \rho^K) + \frac{1}{2} f^{IJK} (\overline\rho{}^J D_\mu \chi^K) \doteq 0 \ , \tag{3.4d}$$

$$+ D_\rho G^{\mu\nu\rho\, I} - g H^{\mu\nu\, I} - \frac{1}{2} f^{IJK} D_\rho (\overline\lambda{}^J \gamma^{\mu\nu\rho} \rho^K)$$
$$+ g f^{IJK} F^{\mu\nu\, J} \varphi^K - \frac{1}{2} g f^{IJK} (\overline\lambda{}^J \gamma^{\mu\nu} \chi^K) \doteq 0 \ , \tag{3.4e}$$

$$+ D_\mu^2 \varphi^I - g f^{IJK} (\overline\lambda{}^J \chi^K) - g^2 \varphi^I - \frac{1}{2} f^{IJK} F_{\mu\nu}{}^J H^{\mu\nu\, K} \doteq 0 \ , \tag{3.4f}$$

$$+ D_\nu H^{\mu\nu\, I} - \frac{1}{2} f^{IJK} F_{\rho\sigma}{}^J G^{\mu\rho\sigma\, K} - \frac{1}{2} f^{IJK} (\overline\chi{}^J D^\mu \lambda^K) - \frac{1}{2} f^{IJK} (\overline\lambda{}^J D^\mu \chi^K)$$
$$+ \frac{1}{2} g f^{IJK} (\overline\lambda{}^J \gamma^\mu \rho^K) - f^{IJK} F^{\mu\nu\, J} D_\nu \varphi^K \doteq 0 \ . \tag{3.4g}$$

In deriving each of these equations, we have also used other field equations.

4. Superspace Reformulation of N=1 TM

We can re-formulate our theory in superspace, as an independent consistency-reconfirmation.

Our superspace BIds for the superfield strengths $F_{AB}{}^I$, $G_{ABC}{}^I$ and $H_{AB}{}^I$ are[7]

$$+ \tfrac{1}{6} \nabla_{[A} G_{BCD)}{}^I - \tfrac{1}{4} T_{[AB|}{}^E G_{E|CD)} - \tfrac{1}{4} f^{IJK} F_{[AB}{}^J H_{CD)}{}^K \equiv 0 \ , \tag{4.1a}$$

$$+ \tfrac{1}{2} \nabla_{[A} H_{BC)}{}^I - \tfrac{1}{2} T_{[AB|}{}^D H_{D|C)}{}^I - g\, G_{ABC}{}^I \equiv 0 \ , \tag{4.1b}$$

$$+ \tfrac{1}{2} \nabla_{[A} F_{BC)}{}^I - \tfrac{1}{2} T_{[AB|}{}^D F_{D|C)}{}^I \equiv 0 \ . \tag{4.1b}$$

Our relevant superspace constraints at the mass dimensions $0 \le d \le 1$ are

$$T_{\alpha\beta}{}^c = + 2(\gamma^c)_{\alpha\beta} \ , \quad G_{\alpha\beta c}{}^I = +2(\gamma_c)_{\alpha\beta}\, \varphi^I \ , \tag{4.2a}$$

$$G_{\alpha bc}{}^I = - (\gamma_{bc}\chi^I)_\alpha \ , \quad H_{\alpha b}{}^I = -(\gamma_b \rho^I)_\alpha - f^{IJK} (\gamma_b \lambda^J)_\alpha\, \varphi^K \ , \tag{4.2b}$$

$$F_{\alpha b}{}^I = - (\gamma_b \lambda^I)_\alpha \ , \quad \nabla_\alpha \varphi^I = - \chi_\alpha{}^I \ , \tag{4.2c}$$

$$\nabla_\alpha \chi_\beta{}^I = - \tfrac{1}{6} (\gamma^{cde})_{\alpha\beta} G_{cde}{}^I - (\gamma^c)_{\alpha\beta} \nabla_c \varphi^I$$
$$- \frac{1}{2} f^{IJK} \Big[+ C_{\alpha\beta} (\overline\lambda{}^J \rho^K) - (\gamma_5\gamma^c)_{\alpha\beta} (\overline\lambda{}^J \gamma_5 \gamma_c \rho^K) - (\gamma_5)_{\alpha\beta} (\overline\lambda{}^J \gamma_5 \rho^K) \Big] \ , \tag{4.2d}$$

$$\nabla_\alpha \rho_\beta{}^I = + \frac{1}{2} (\gamma^{cd})_{\alpha\beta} H_{cd}{}^I + g\, C_{\alpha\beta}\, \varphi^I - \frac{1}{2} f^{IJK} (\gamma^{cd})_{\alpha\beta} F_{cd}{}^J \varphi^K$$
$$- \frac{1}{4} f^{IJK} \Big[+ C_{\alpha\beta} (\overline\lambda{}^J \chi^K) + (\gamma^c)_{\alpha\beta} (\overline\lambda{}^J \gamma_c \chi^K) - \frac{1}{2} (\gamma^{cd})_{\alpha\beta} (\overline\lambda{}^J \gamma_{cd} \chi^K)$$
$$- (\gamma_5\gamma^c)_{\alpha\beta} (\overline\lambda{}^J \gamma_5 \gamma_c \chi^K) - (\gamma_5)_{\alpha\beta} (\overline\lambda{}^J \gamma_5 \chi^K) \ , \tag{4.2e}$$

$$\nabla_\alpha \lambda_\beta{}^I = + \frac{1}{2} (\gamma^{cd})_{\alpha\beta} F_{cd}{}^I - \frac{1}{2} (\gamma_5)_{\alpha\beta}\, f^{IJK} (\overline\rho{}^J \gamma_5 \chi^K) \ . \tag{4.2f}$$

[7] In this superspace section, we use the indices $A = (a,\alpha)$, $B = (b,\beta)$, $\cdots$ for superspace coordinates, where $a, b, \cdots = 0, 1, 2, 3$ (or $\alpha, \beta, \cdots = 1, 2, 3, 4$) are for bosonic (or fermionic) coordinates. In superspace, we use the (anti)symmetrization convention, *e.g.*, $X_{[AB)} \equiv X_{AB} - (-1)^{AB} X_{BA}$, different from our component formulation.

All other components, such as $G_{\alpha\beta\gamma}{}^I$ or $T_{\alpha\beta}{}^\gamma$ *etc.* at $d \leq 1$ are zero.

Although most of technical details associated with superspace formulation are skipped here, we presented a rather independent confirmation for the total consistency of our $N = 1$ non-Abelian tensor multiplet in superspace.

5. Concluding Remarks

In this talk, we have explained how to formulate the $N = 1$ supersymmetrization in 4D of a physical non-Abelian tensor with consistent couplings [1]. This is the supersymmetrization of the special case [4] of the minimal tensor hierarchy [5], which, in turn, is a special case of more general hierarchy in [2][3]. Both the component and superspace formulations of our system are given, as the cross-verification of our system. Our CVM $(C_\mu{}^I, \rho^I)$ plays the role of a compensator multiplet, absorbed into the TM $(B_{\mu\nu}{}^I, \chi^I, \varphi^I)$, making the latter massive.

There exists certain problem for the quantization of Stueckelberg theory [7] for non-Abelian gauge groups [9]. This is because the longitudinal components of the gauge field do not decouple from the physical Hilbert space, so that the renormalizability and unitarity of the system are spoiled [9]. We take rather an optimistic standpoint about this potential problem for the following reasons. First, we mention that our theory is *not* renormalizable due to Pauli couplings. This feature is *not* necessarily a fatal drawback for our theory, because certain theories exist in 4D, such as non-linear sigma models that are *not* renormalizable, but are *not* rejected from the outset. Second, $N = 1$ supersymmetry may well improve quantum behavior of our theory, compared with non-supersymmetric systems. There is good chance that supersymmetries solve the quantum problem of non-Abelian Stueckelberg theories.

The importance of our result [1] is double-fold: (i) A new *supersymmetric* physical system with Stueckelberg mechanism that solves the problem with non-Abelian tensor is presented. (ii) The problem with extra vector fields in the non-singlet representation of a non-Abelian gauge group is now solved. We should also consider the possibility that $N = 1$ supersymmetry may well provide better quantum behavior compared with non-supersymmetric cases.

Acknowledgments

This work is supported in part by DOE grant grant # DE-FG02-10ER41693.

References

[1] Nishino H and Rajpoot S 2012 *Phys. Rev.* **D85** 105017 (*Preprint* 1204.1379)
[2] de Wit B and Samtleben H 2005 *Fortsch. Phys.* **53** 442 (*Preprint* hep-th/0501243)
[3] de Wit B, Nicolai H and Samtleben H 2008 *JHEP* **0802** 044 (*Preprint* 0801.1294)
[4] Chu C-S 2011 *A Theory of Non-Abelian Tensor Gauge Field with Non-Abelian Gauge Symmetry G × G* (*Preprint* 1108.5131)
[5] Samtleben H, Sezgin E and Wimmer R 2011 *Jour. High. Energy. Phys.* **12** 062.
[6] Freedman D Z, Townsend P K 1981 *Nucl. Phys.* B177 282; Ogievetsky V I and Polubarinov I V 1967 *Sov. J. Nucl. Phys.* **4** 156
[7] Stueckelberg E C G 1938 *Helv. Phys. Acta* **11** 225; *For reviews, see, e.g.*, Ruegg H and Ruiz-Altaba M 2004 *The Stuckelberg field. Int. Jour. Mod. Phys.* **19** 3265.
[8] Ferrara S, Freedman D Z and van Nieuwenhuizen P 1976 *Phys. Rev.* 13D 3214; Deser S and Zumino B 1976 *Phys. Lett.* **62B** 335; van Nieuwenhuizen P1981 *Supergravity Phys. Rept.* **68C** 189; Wess J and Bagger J 1992 *'Superspace and Supergravity'*, Princeton University Press
[9] Kunimasa J M and Goto T 1967 *Prog. Theor. Phys.* **37** 452; Slavnov A A 1972 *Theor. Math. Phys.* **10** 99; Veltman M J G1968 *Nucl. Phys.* **B7** 637; Slavnov A A and Faddeev L D 1970 *Theor. Math. Phys.* **3** 312; Vainshtein A I and Khriplovich I B 1971 *Yad. Fiz.* **13** 198; Shizuya K I 1977 *Nucl. Phys.* **B121** 125; Kafiev Y N 1982 *Nucl. Pnys.* **201** 341

Journal of Physics: Conference Series **485** (2014) 012050 doi:10.1088/1742-6596/485/1/012050

A modification of the relativistic energy-momentum relation

Virendra Gupta

Departamento de Física Aplicada, CINVESTAV-Mérida, A.P. 73, 97310 Mérida, Yucatán, México

E-mail: virendra@mda.cinvestav.mx

Abstract. A modification of the accepted relativistic energy momentum relation is suggested. The new relation allows massive particles to have a maximum velocity $c(m)$ greater than the velocity of light c. The effect of the modification suggested here would be most apparent for objects with masses nearly equal to the Planck mass.

Recently the OPERA collaboration reported the detection of superluminal neutrinos [1]. Motivated by this we tried to modify the standard energy, mass and momentum relationship, namely

$$E^2 = m^2c^4 + p^2c^2, \tag{1}$$

to

$$E^2 = m^2c(m)^4 + p^2c(m)^2. \tag{2}$$

In terms of the velocity v,

$$E = \frac{mc(m)^2}{\sqrt{1 - v^2/c(m)^2}}. \tag{3}$$

Here $c(m)$ is a function of a dimensionless variable ζ which depends on the mass m and is defined below. We have simply replaced c by $c(m)$ in the usual relation.

Any modification, in our view, must satisfy the constraint that for $m = 0$, we should have $E = pc$. This means that we require $c(0) = c$, so that for $m = 0$ we have the usual energy-momentum relation for electromagnetic radiation.

For non-relativistic velocities, our modification gives the Newtonian expression for kinetic energy. However, it modifies the rest energy for a massive object.

The crucial question is what is the variable ζ and the function $c(m)$. Massive particles have gravitational interaction and it is interesting to note that one can define a mass, the Planck mass P_M, in terms of the gravitational constant G_N, namely

$$P_M = \sqrt{\hbar c/G_N} \approx 1,22 \times 10^{19} GeV/c^2 \approx 2.18 \times 10^{-5} gr. \tag{4}$$

Here h=Planck's constant and c is the velocity of light.The dimensionless variable ζ is defined to be

$$\zeta = m/P_M \ . \tag{5}$$

Content from this work may be used under the terms of the Creative Commons Attribution 3.0 licence. Any further distribution of this work must maintain attribution to the author(s) and the title of the work, journal citation and DOI.
Published under licence by IOP Publishing Ltd

We define $c(m)$ to be of the form

$$c(m) = c[1 + F(\zeta)] \ , \tag{6}$$

where F is a positive analytic function of ζ with $F(0) = 0$. Positivity of F is necessary for matter to have superluminal velocities. From the definition it is clear that ζ is extremely small for sub-atomic particles (electron, proton, etc.) and is very large for ordinary masses. For example, for $m = 20gr$, $\zeta \sim 10^6$ and enormous for ordinary and stellar objects. Since the value of $F(\zeta)$ will affect the rest mass of the object, one has to choose it carefully. From the practical point of view, we consider

$$F(\zeta) = \zeta^n exp(-\zeta^n), \tag{7}$$

where n is a positive real number. This function has a maximum value $e^{-1} = 0.367879...$ for $\zeta = 1$ independent of the value of n. It is centered around $\zeta = 1$. As n increases the height remains the same, but it becomes narrower and narrower. For extremely large n (tending to infinity), this sequence of functions tends to a vertical line of height e^{-1} located at $\zeta = 1$. In this limit, it is like a "finite Dirac δ function" only non-zero for $\zeta = 1$. The function is plotted in figure 1 for $n = 50, 100$ and 1000. One can see clearly how the function shrinks rapidly as n increases and will be practically narrow as a straight line located at $\zeta = 1$ for much larger n.

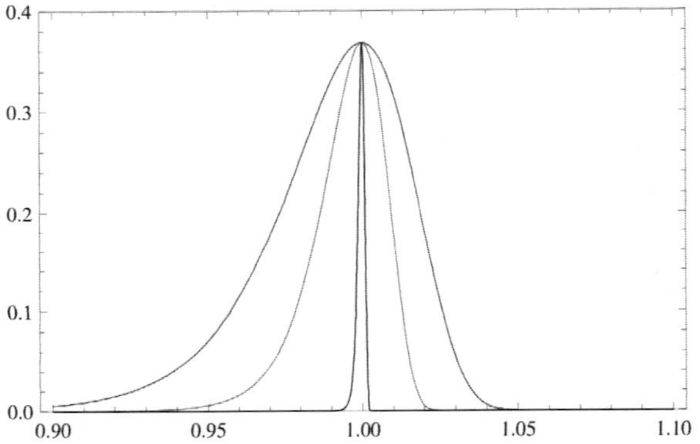

Figure 1. for $n = 50, 100, 1000$ in color blue, red and black respectively. The function narrows rapidly with increasing n. The maximum height at $x = 1$ is independent of n.

Thus, for extremely large n, the value of the usual rest mass mc^2 will not be affected for bodies which are very light (subatomic particles) or very heavy (e.g. planets) compared to the Planck mass, but will affect those with masses of the order of the Planck mass. Such a choice for n is necessary, since our particle accelerators work and we understand planetary motion. These provide extremely stringent tests for the usual energy-momentum relation. However, we do not have stringent tests for masses of the order of the Planck mass. It would be interesting to have experimental tests of the energy-momentum relation for this mass range.

Acknowledgments

It is a pleasure to thank Francisco Larios and Guillermo Cordourier for his unstinting kind help in the preparation of this manuscript.I would like to thank him and Antonio Bouzas for discussions. I would also like to thank Conacyt and SNI for support.

References

[1] OPERA Collaboration 2012 *Measurement of the neutrino velocity with the OPERA detector in the CNGS beam* (*Preprint* `arXiv.1109.4897`). This note was inspired by this report. However, the modification suggested here does not explain their result. In any case, after more stringent analysis they have withdrawn the original claim recently.

[2] Nakamura K. et al. (Particle Data Group) 2010 *Review of Particle Physics, J. Phys.* **G37** 075021

Kaons at CERN: status and prospects

Giuseppina Anzivino [1]

Department of Physics, University of Perugia and INFN, via A. Pascoli, 06123 Perugia (I)

E-mail: giuseppina.anzivino@pg.infn.it

Abstract.
The NA48/2 and NA62 experiments at CERN exploit the decay in flight technique to study rare kaon decays and the underlying physics. Two selected topics will be presented in this paper. The first is a stringent test of the Chiral Perturbation Theory, obtained by new measurements of the $K^{\pm} \to \pi^{\pm}\gamma\gamma$ decay performed by the two experiments. The second one is the measurement of the Branching Ratio of the very rare decay $K^{+} \to \pi^{+}\nu\overline{\nu}$ at 10% level, that will provide a test of the Standard Model and possible extensions beyond it.

1. Introduction

The study of kaons has always been an important source for the investigation of the flavour sector of fundamental interactions and for the understanding of the symmetries of nature, addressing questions related to the observed matter-antimatter asymmetry in the Universe.

Since the discovery of CP violation in 1964 in $K^0 \to \pi^0\pi^0$ [1] kaons remain a privileged observatory for flavour physics phenomena where all the features of flavour physics are present: kaons are a rather simple system of the lightest mesons containing a quark flavour of a generation beyond the first, they have rather long lifetimes and are produced in large quantity. In the era of the high energy frontier, the flavour sector can probe the high energy scale exploiting the precision frontier, complementary to the high energy one. In addition, processes that are suppressed in the Standard Model (SM) are very sensitive to New Physics (NP), therefore the study of rare kaon decays can provide a deeper insight in the fundamental physics.

2. The $K^{\pm} \to \pi^{\pm}\gamma\gamma$ decay

The Chiral Perturbation Theory (ChPT) provides a systematic approach to the intrinsically non-perturbative nature of the strong interactions in kaon physics. Radiative non-leptonic kaon decays provide a crucial test of this theory and can give information on the structure of hadronic interactions at low energy. The contributions of the chiral lagrangian to this decay [2] appear at $\mathcal{O}(p^4)$, where only the $\Delta I = 1/2$ invariant amplitudes $A(z)$ and $C(z)$,with $z = m^2_{\gamma\gamma}/m^2_{K^{\pm}}$, contribute. The decay rate and the spectrum strongly depend on a single parameter $\hat{c}$ predicted to be positive and $\mathcal{O}(1)$. The invariant $m_{\gamma\gamma}$ distribution is favored above $2m_{\pi^+}$ and exibits a cusp at $2m_{\pi^+}$ threshold. Calculations at $\mathcal{O}(p^6)$ [3] show that unitarity correction modify the decay spectrum significantly, leading, in particular, to non-zero differential decay rate at zero di-photon invariant mass .

[1] on behalf of the NA62 and NA48/2 Collaborations

Content from this work may be used under the terms of the Creative Commons Attribution 3.0 licence. Any further distribution of this work must maintain attribution to the author(s) and the title of the work, journal citation and DOI.
Published under licence by IOP Publishing Ltd

The ChPT predictions for the decay spectra for values of $\hat{c} = -2; 0; 2$ are shown in Fig. 1 for $\mathcal{O}(p^4)$ and $\mathcal{O}(p^6)$ parametrizarion: the diphoton mass spectra exhibit a characteristic cusp at twice the pion mass due to the dominant pion loop amplitude.

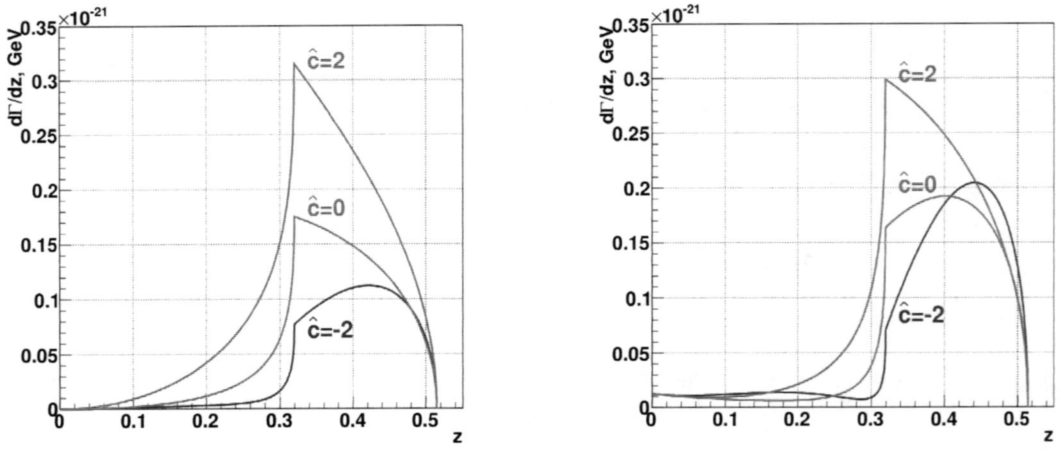

Figure 1. ChPT predictions for $K^\pm \to \pi^\pm\gamma\gamma$ differential rate in terms of $z = m^2_{\gamma\gamma}/m^2_{K^\pm}$ for $\mathcal{O}(p^4)$ (left) and $\mathcal{O}(p^6)$ (right) parametrizarion [3].

On the experimental side, there is only one published result by the BNL E787 experiment [4], based on 31 $K^\pm \to \pi^\pm\gamma\gamma$ candidates.

The preliminary results presented in this paper are based on data collected in two different runs: a 3-day NA48/2 special run in 2004 and a 3-month NA62 run in 2007. The two data sets are comparable in size, since a downscaling factor of $\sim$ 20 on the trigger was active in 2007. Events have been selected in the region $z = m^2_{\gamma\gamma}/m^2_{K^\pm}$>0.2 in order to minimize the $K^\pm \to \pi^\pm\pi^0$ background. We observed 147 (175) decay candidates in the 2004 (2007) data set with a 12% (7%) background contamination.

Fig. 2 shows the $K^\pm \to \pi^\pm\gamma\gamma$ spectra for the invariant mass and the z variable (data and MC expectations) for signal and background for 2004 data set; the cusp at $2m_{\pi^+}$ threshold is evident, as expected by the ChPT predictions. The 2007 data set shows similar results.

The preliminary 2004 and 2007 results and the combined result, that takes into account the correlation of the systematic uncertainties between the two data sets, are reported in Table 1.

Table 1. The preliminary result of the fits to the ChPT parametrization. The model dependent BR values are computed on the full kinematic range.

	2004 data	2007 data	combined
$\hat{c}$, $\mathcal{O}(p^4)$ fit	$1.36 \pm 0.33_{\text{stat.}} \pm 0.07_{\text{syst.}}$	$1.71 \pm 0.29_{\text{stat.}} \pm 0.06_{\text{syst.}}$	$1.56 \pm 0.22_{\text{stat.}} \pm 0.07_{\text{syst.}}$
$\hat{c}$, $\mathcal{O}(p^6)$ fit	$1.67 \pm 0.39_{\text{stat.}} \pm 0.09_{\text{syst.}}$	$2.21 \pm 0.31_{\text{stat.}} \pm 0.08_{\text{syst.}}$	$2.00 \pm 0.24_{\text{stat.}} \pm 0.09_{\text{syst.}}$
BR, $\mathcal{O}(p^6)$ fit	$(0.94 \pm 0.08) \times 10^{-6}$	$(1.06 \pm 0.07) \times 10^{-6}$	$(1.01 \pm 0.06) \times 10^{-6}$

The results are in agreement with the BNL E787ones; the $\mathcal{O}(p^4)$ and $\mathcal{O}(p^6)$ models cannot be discriminated within the current experimental sensitivity.

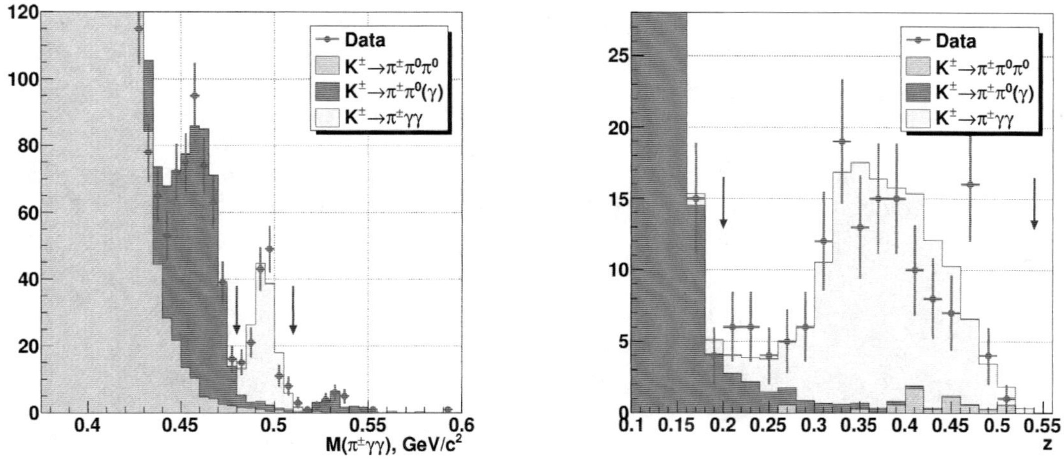

Figure 2. Results for the 2004 data set. Left: the $K^{\pm} \to \pi^{\pm}\gamma\gamma$ invariant mass spectrum with MC expectations for signal and background. Right: the $z = m^2_{\gamma\gamma}/m^2_{K^{\pm}}$ spectrum with MC expectations for signal and background. Arrows indicate the signal region.

3. The $K^+ \to \pi^+\nu\overline{\nu}$ decay

The rare decays $K^+ \to \pi^+\nu\overline{\nu}$ and $K_L \to \pi^0\nu\overline{\nu}$ are extremely attractive processes to study the physics of flavour mixing since both are exceptionally clean modes. These decays are strongly suppressed in the SM, dominated by short-distance dynamics and top quark loop contributions. Moreover, the short-distance amplitude is governed by one single semileptonic operator whose hadronic matrix element can be determined experimentally using the semileptonic kaon decay; therefore the main hadronic uncertainties can be eliminated by experimental data. The two "golden" rare kaon modes offer unique opportunities for testing the Standard Model and deepening our knowledge of the CKM matrix, complementary to those in B decays. Furthermore, they are extremely sensitive to possible new degrees of freedom beyond the Standard Model. At the quark level the two processes arise from the $s \to d\nu\overline{\nu}$ process, which in the SM originates from a combination of the Z^0 penguin and a double $W^{\pm}$ exchange.

The SM prediction[5] for the $K^+ \to \pi^+\nu\overline{\nu}$ channel is $BR = (7.81^{+0.80}_{-0.71} \pm 0.29) \times 10^{-11}$. The error comes mainly from the uncertainty in the CKM matrix elements, while the irreducible theoretical uncertainty amounts to less than 2%. The extreme theoretical cleanness of these decays remains also in new physics scenarios like Minimal Flavour Violation (MFV)[6] or non-MFV models[7] and even not large deviations from the SM value (for example around 20%) can be considered signals of new physics.

From the experimental side, the decay $K^+ \to \pi^+\nu\overline{\nu}$ has been observed by the experiments E787 and E949 at the BNL, using a stopping kaon technique. The combined result from the two experiments for branching ratio is $BR = 1.73^{1.15}_{1.05} \times 10^{-10}$ [8]. The measured central value is about twice as large as, but still compatible with the SM expectations within the errors. It is clear that only a measurement of the branching ratio with at least 10% accuracy can be a significant test of the validity of the SM and of the possible existence of new physics. This is the main goal of the NA62 experiment at CERN-SPS[9]. It aims to collect order of 100 $K^+ \to \pi^+\nu\overline{\nu}$ events in about two years of data taking, keeping a background contamination around 10%.

3.1. Experimental strategy

The measurement of the $K^+ \to \pi^+\nu\overline{\nu}$ Branching Ratio is very challenging due to the weak experimental signature (only one charged track in the final state) and huge background.

The two undetectable neutrinos in the final state require the design of an experiment with redundant measurement of the event kinematics and hermetic vetoes to achieve a background rejection $S/B \simeq 10$. Particular care has to be taken to suppress the two-body decays $K^+ \rightarrow \pi^+\pi^0$ and $K^+ \rightarrow \mu^+\nu$ which have branching ratios up to 10^{10} times larger than the expected signal and can mimic the signal signature if a muon is misidentified as a pion or in case of loss of the photons from the π^0 decay. Backgrounds from K^+ three- and four-body decays are also potentially dangerous. The most frequent K^+ decay modes are listed in Table 2.

Table 2. The most frequent K^+ decay channels.

Decay mode	Branching Ratio
$K^+ \rightarrow \mu^+\nu$	63.5 %
$K^+ \rightarrow \pi^+\pi^0$	20.7 %
$K^\pm \rightarrow \pi^\pm\pi^+\pi^-$	5.6 %
$K^+ \rightarrow \pi^0 e\nu$	5.1 %
$K^+ \rightarrow \pi^0\mu\nu$	3.3 %
$K^\pm \rightarrow \pi^+\pi^-\pi^0$	1.8 %

In order to achieve the design goal, high statistic is needed, keeping at the same time the systematic uncertainties and background at low level. Assuming a 10% signal acceptance and the SM decay rate, at least 10^{13} decays in the fiducial volume are required. This translates in the choice of a high intensity kaon beam: the same extraction line of the CERN SPS used for the NA48 experiment will deliver a 75 GeV/c charged kaon beam, fully adequate for NA62.

The NA62 guiding principles can be summarized as follows. The high rate environment demands high resolution timing. In addition, high resolution and fast tracking is needed to measure the momenta of the incoming kaon and of the daughter particles in order to exploit kinematic rejection by cutting on the missing mass at the decay. The rejection of decays with electrons and muons in the final state requires an efficient particle identification (PID) of all charged particle, namely the positive non-destructive identification of the parent kaon and PID of daughter particles (electrons, muons and pions) and also of muons not coming from the decay. To reject decay modes with π^0 and radiative photons in the final state, hermetic, high efficient γ vetoing is required from 0 to 50 mrad. Furthermore redundancy of information will help to reach the goal.

3.2. Kinematic rejection

The only measurable quantities for the decay under study are the momentum of the incoming kaon, the momentum of the outgoing pion and the angle between the mother and the daughter particles.

It is convenient to use the squared missing mass variable, m^2_{miss}, defined under the hypothesis that the detected charged particle in the final state is a pion:

$$m^2_{miss} \simeq m^2_K\left(1 - \frac{|P_\pi|}{|P_K|}\right) + m^2_\pi\left(1 - \frac{|P_K|}{|P_\pi|}\right) - |P_K||P_\pi|\theta^2_{\pi K} \quad (1)$$

In Fig. 3 the m^2_{miss} for the signal and the kaon decays with the largest branching ratios are shown for $P_K = 75$ GeV/c. If resolution effects are ignored, the $K^+ \rightarrow \pi^+\pi^0$ decay is constrained to a line at $m^2_{miss} = 0$; the m^2_{miss} of the three-pion decays shows a lower bound. The m^2_{miss} of $K^+ \rightarrow \mu^+\nu$ does not appear as a line at $m^2_{miss} = 0$ because it is wrongly evaluated, under the assumption that the track is a pion. In conclusion, about 92% of the background events can be separated from signal by kinematic cuts (see Fig. 3, left). For the remaining (see Fig. 3, right), the background events extend in the signal region and kinematic cuts cannot be used. The suppression has to rely on the photon veto system and PID.

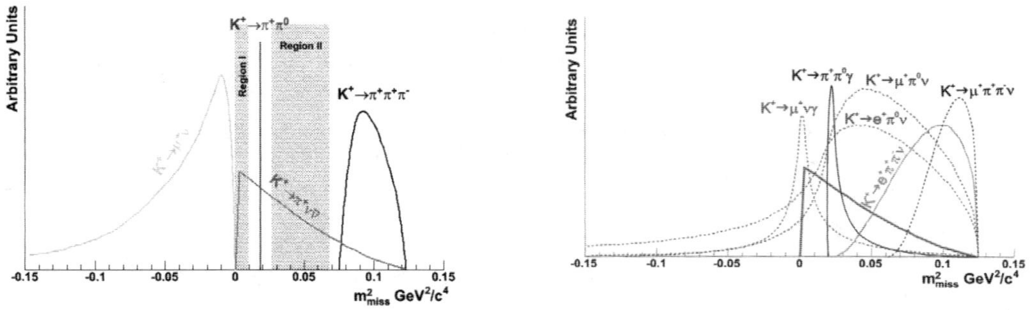

Figure 3. Squared missing mass for background decays that can be separated from the signal using kinematic cuts (left) and for decays where kinematics cannot be used (right).

3.3. Experiment and detectors

The experimental setup is schematically depicted in Fig. 4. A 400 GeV/c primary proton beam extracted from the SPS impinges on a beryllium target to produce a secondary hadron beam at 75 GeV/c with a momentum bite of $\sim$ 1%. Only 6% of the hadrons are charged kaons (unseparated beam). It is crucial to make a positive identification of the minority particles of interest, kaons, in the high rate beam environment before they decay. This is achieved by placing a Cherenkov differential counter (CEDAR) in the incoming beam; the CEDAR is insensitive to pions and protons with minimal accidental mis-tagging. In addition the CEDAR detector is required to achieve a kaon identification efficiency of at least 95% and a time resolution of 100 ps, in conjunction with timing information from other detectors.

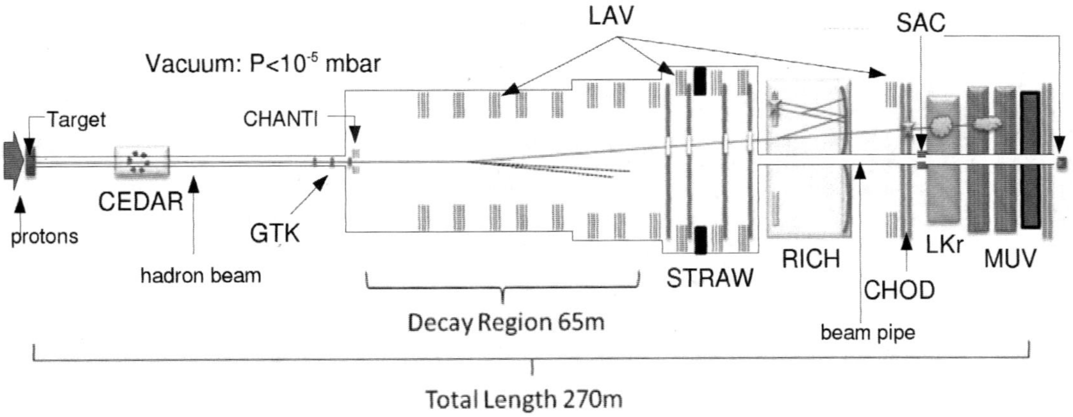

Figure 4. Detector layout

To fulfill the requirements needed for the kinematic rejection, keeping, at the same time, multiple scattering as low as possible, a beam tracker along the beam line, named Gigatracker, and a low mass spectrometer placed in vacuum downstream the decay region have been chosen. The Gigatracker (GKT) will be exposed to an instantaneous rate of $\sim$ 800 MHz. It consists of three Silicon pixel stations, matching the beam size, placed before the decay volume, with 18000 pixels per station. Results from beam tests on prototypes [10] show a time resolution of 150 ps on a single track and a momentum resolution of 0.2%, in agreement with the design requirements.

The direction and the momentum of secondary charged particles originating form the decay region will be measured by the straw tracker. The spectrometer consists of four chambers and a high aperture dipole magnet providing a vertical B-field of 0.36 T. Each chamber is equipped with 1792 straw tubes, which are positioned in four views.

The Photon Veto system is composed by several detectors, covering different angular regions: the Large Angle Veto, made of lead glass blocks arranged in 12 annular rings around the decay volume, the NA48 LKr calorimeter and two small angle calorimeters.

A gas based RICH detector will contribute to the $\pi\mu$ suppression with a rejection factor better than 5×10^{-3} in the momentum range 15-35 GeV/c and will also provide and event time measurement with a resolution better than 100 ps in order to minimize wrong matching with the parent particle measured by an upstream detector. The RICH is a 18 m long segmented tube, 4 m in diameter, filled with Ne at atmospheric pressure and room temperature, equipped with a segmented mirror (17 m focal length) at the downstream end and about 2000 Photomultipliers at the upstream end. Results from the beam tests are well within the expectations, showing a muon suppression factor of 0.7 %, very close to the requested 0.5 % and a time resolution below 100 ps [11]. The detector is complemented with a Charged Hodoscope (CHOD), a Muon Detector (MUV) and a veto detector for charged particles at small angles (CHANTi).

4. Conclusions

Kaons offer unique opportunities to exploit precision physics, complementary to high-energy approach, to study the Standard Model and to search for New Physics. The new measurement of the $K^{\pm} \to \pi^{\pm}\gamma\gamma$ decay significantly improves the precision of the ChPT test with this channel. The NA62 experiment aims to measure the BR of the ultra-rare $K^{+} \to \pi^{+}\nu\overline{\nu}$ decay, collecting $\mathcal{O}(100)$ events in two years of data taking. The construction of the NA62 detector is well advanced; a technical run is in preparation for fall 2012 and the experiment is expected to take data in 2014, according to the SPS schedule.

References

[1] Christenson J H et al 1964 *Phys. Rev. Lett.* **13** 138
[2] Ecker G, Pich A and de Rafael E 1988 *Nucl. Phys. B* **303** 665
[3] D'Ambrosio G and Portoles J 1996 *Nucl. Phys. B* **386** 403
[4] Kitching P et al 1997 *Phys. Rev. Lett.* **79** 4079
[5] Brod J, Gorbahn M and Stamou E 2011 *Phys.Rev. D* **83** 034030
[6] Isidori G, Mescia F, Paradisi P, Smith C, Trine S 2006 *JHEP* **0608** 064 (*Preprint* HEP-PH 0604074)
[7] Blanke M et al 2007 *JHEP* **0701** 066 (*Preprint* HEP-PH 0610298)
[8] Artamonov A V et al 2008 *Phys. Rev. Lett.* **101** 191802
[9] Anelli G et al 2005 (*Preprint* CERN-SPSC-2005-013, SPSC-P-326)
[10] Fiorini M et al 2011 *Nucl. Instrum. Meth.* **A628** 292
[11] Angelucci M et al 2010 *Nucl. Instrum. Meth.* **A621** 205

DisCERNing new physics in $t\bar{t}$-production from top-spin observables

Blaženka Melić

Rudjer Bošković Institute, Theoretical Physics Division, Bijenička 54, HR-10000 Zagreb, Croatia

E-mail: melic@thphys.irb.hr

Abstract. We examine the models of new physics which could potentially explain the significant deviation of the measured forward-backward asymmetry in $t\bar{t}$ production at the Tevatron from the SM prediction, by fitting the models to the all available $t\bar{t}$ production data. To be able further to discriminate among the NP models we calculate predictions for various top spin polarization and top-antitop spin correlation observables at Tevatron and the LHC.

1. Introduction

The top quark physics is interesting due to the several reasons. Top quark is the heaviest of the quarks with a mass comparable with the electroweak symmetry breaking (EWSB) scale. Due to its heaviness it decays very fast and it does not hadronize. Moreover, its coupling to the Higgs is of $O(1)$, and therefore the understanding of the top quark properties provides the insight into the nature of the EWSB mechanism.

Hadron collider experiments at Tevatron and LHC have greatly enriched our knowledge about the top quark physics measuring precisely the $t\bar{t}$ production cross section and its invariant mass ($m_{t\bar{t}}$) production, forward-backward (A_{FB}) or charge asymmetries (A_C) in the production, as well as the single top production with already impressive precision. While other top properties exhibit an agreement with the Standard model predictions, the D0 and CDF experiments are constantly observing anomalously large A_{FB} at $\geq 2\sigma$ level (c.f. [1] for a recent review).

In order to explain these puzzling phenomena at Tevatron, there were several New Physics (NP) models proposed, interfering with the SM contribution by exchange of the new s, t or u-channel resonances. To distinguish among the models, it is important to investigate all possible observables, such as the shape of $A_{FB}(m_{t\bar{t}})$, the resonant peaks in $\sigma_{t\bar{t}}(m_{t\bar{t}})$ or the top quark spin polarization and spin correlations, being the subject of this work [2].

2. Asymmetries and the NP models

Asymmetries in the $t\bar{t}$ production are defined differently for the Tevatron and for the LHC machine. At the Tevatron the A_{FB} is taken as the asymmetry between the production rates in the forward and the backward region, while at the LHC, since it is a symmetric pp machine, there is is no preferred global direction and $A_{FB} = 0$. However, there it is possible to define the charge asymmetry, A_C, since the tops are more forward than antitops (because inside the proton the valence quarks are more boosted than the sea antiquarks). Therefore there is an excess of

Content from this work may be used under the terms of the Creative Commons Attribution 3.0 licence. Any further distribution of this work must maintain attribution to the author(s) and the title of the work, journal citation and DOI.
Published under licence by IOP Publishing Ltd

boosted top quarks along the beam axis and $A_C \neq 0$. For theories with the CP-conserving couplings $A_{FB} = A_C$.

In the SM the asymmetries arise at NLO, and are robust under high-energy QCD corrections [3], while the electroweak corrections contribute to 20% [4]. The SM predicts a small, but non-zero asymmetry in the $t\bar{t}$ production, being $A_{FB}^{SM} \sim 7-9\%$ and $A_C^{SM} \sim 1\%$. While the LHC measurements give $A_C = 0.001 \pm 0.0014$, consistent with the SM prediction, the D0 and CDF are consistently observing anomalously large forward-backward asymmetries:

$$\begin{aligned} A_{FB} &= 0.187 \pm 0.037\,, \\ A_{FB}^{low} &\equiv A_{FB}(m_{t\bar{t}} < 450 \text{ GeV}) = 0.078 \pm 0.054\,, \\ A_{FB}^{high} &\equiv A_{FB}(m_{t\bar{t}} > 450 \text{ GeV}) = 0.296 \pm 0.067\,, \end{aligned}$$

which have to be compared with $A_{FB}^{\rm SM} = 0.066(20)$ and $A_{FB}^{low,\rm SM} \simeq 0.047, A_{FB}^{high,\rm SM} \simeq 0.100$ [3].

To explain these deviations we consider several NP models, which generate asymmetry from the interference with the SM by exchanging new particles in s-channel (G'), t-channel (W', Z', ϕ (the neutral component of the isodublet Φ)), or u-channel (Δ, Σ). In the table 1 we summarize the NP models, together with their lagrangians and couplings considered in the fits. For all details consult [2].

Table 1. New physics models considered.

model	Lagrangian of interaction	fitted coupl.
s-channel: axigluon	$\mathcal{L}_{G'}^{\rm int.} = -\bar{q}(g_V^q - g_A^q\gamma_5)\not{G}'q - \bar{t}(g_V^t - g_A^t\gamma_5)\not{G}'t$	g_A^q, g_A^t
t-channel: Z'	$\mathcal{L}_{Z'}^{\rm int.} = -\bar{u}\gamma_\mu(f_L^{Z'}P_L + f_R^{Z'}P_R)tZ'$	$f_R^{Z'}$
W'	$\mathcal{L}_{W'}^{\rm int.} = -\bar{d}\gamma_\mu(f_L^{W'}P_L + f_R^{W'}P_R)tZ'$	$f_R^{W'}$
scalar doublet Φ	$\mathcal{L}_{\Phi}^{\rm int.} = -y_{ij}^u\bar{q}_{Li}u_{Rj}\Phi - y_{ij}^d\bar{q}_{Li}d_{Rj}\tilde{\Phi}$	y_{ij}^u
u-channel: scalar color triplet Δ	$\mathcal{L}_{\Delta}^{\rm int.} = -g_{(\Delta)ij}\,\epsilon_{abc}\bar{u}_{R,i}^a(u_{R,j}^b)^C\Delta^C$	$g_{(\Delta)ij}$
scalar color sextet Σ	$\mathcal{L}_{\Sigma}^{\rm int.} = -g_{(\Sigma)ij}\ (\bar{u}_{R,i}^a(u_{R,j}^b)^C + \bar{u}_{R,i}^b(u_{R,j}^a)^C)\Sigma^{ab\dagger}$	$g_{(\Sigma)ij}$

3. New physics model selections

New models must explain large asymmetries without significantly changing the cross-section predictions, which are already measured at 7% precision and are in agreement with the SM calculations [5]. To select the models we perform two fits to the available data. One, scenario A, is a global χ^2 fit to the following observables $A = \{A_{FB}, A_C, \sigma_{Tev}, \sigma_{Tev}^{high}, \sigma_{LHC}\}$, while in the scenario B we use the data on $\{A_{FB}^{low}, A_{FB}^{high}, d\sigma_{Tev}/dm_{t\bar{t}}, \sigma_{LHC}\}$ in the fit. We consider NP model as acceptable if it improves upon SM χ^2 ($\chi^2_{A,\rm SM} = 2.3$/d.o.f. in approach A and $\chi^2_{B,\rm SM} = 1.8$/d.o.f. in approach B). The results of the fits are shown in figures 1-4.

For the axigluon model, in the approach B, our model fit selects two parameter regions shown in figure 1: low axigluon mass region with $m_G' < 450$ GeV and $g_A^u g_A^t \sim 0.2$ and the high mass region with $m_G' > 700$ GeV and $g_A^u g_A^t \sim -0.5$.

For the Z' and W' models, we see from the figures 2 that the A_{FB} and A_C measurements are highly correlated and that the tension between them cannot be reduced below 2σ level, and therefore we exclude these NP models from the further consideration.

The neutral field ϕ of the scalar isodoublet model exhibits two allowed parameter regions: the more preferred low m_ϕ mass region, $m_\phi < m_t$ in the approach A, and $m_\phi > 200$ GeV in the scenario B, figure 3.

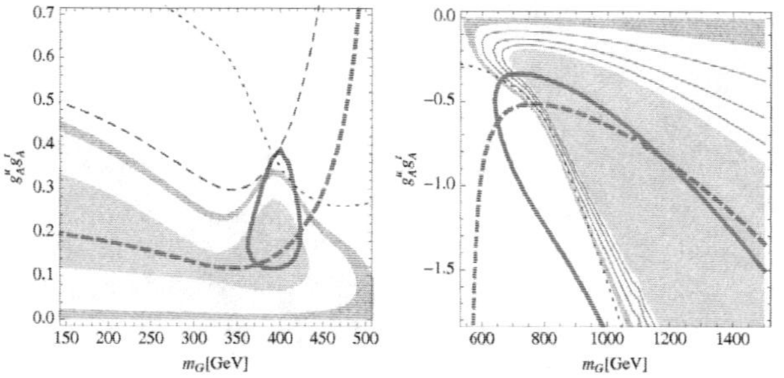

Figure 1. $t\bar{t}$ production constraints on *the axigluon model* in the low (left plot) and high (right plot) mass region in the approach B: binned A_{FB} at 1σ in thick full green line, inclusive A_C at 1σ (2σ) in thick dashed green line (thin dashed red line), $m_{t\bar{t}}$ spectrum at 2σ in thin red dotted line. Parameter regions where the model can improve the SM χ^2 by $-\Delta\chi^2 > 0, 1, 4$ are shaded in red, yellow and green respectively.

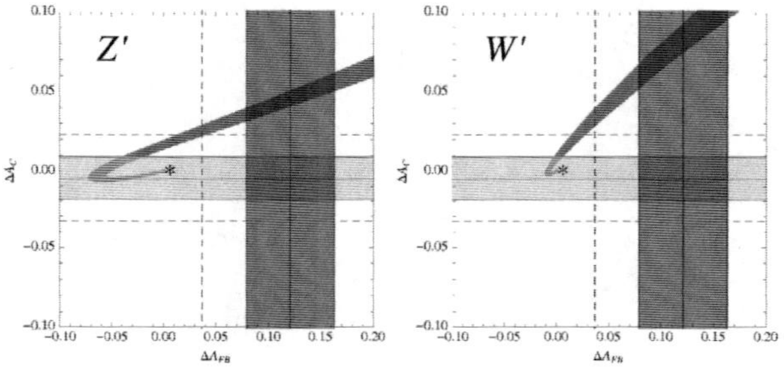

Figure 2. Correlation between the $\Delta A_{FB} = A_{FB} - A_{FB}^{SM}$ and $\Delta A_C = A_C - A_C^{SM}$ contributions of Z' (left plot) and W' (right plot) *models* in shaded narrow gray bands using the approach A. The thickness of the bands is given by the Z' (W') mass variation in the range $[100, 500]$ GeV. The Tevatron average of ΔA_{FB} at 1σ (2σ) is represented by the vertical blue band (dashed vertical lines). The ATLAS measurement of ΔA_C (the SM reference point is marked with "*") at 1σ (2σ) is represented by the horizontal green band (dashed horizontal lines). The tension between the two observables increases with increasing mediator mass (inner edge of the band corresponds to lowest mediator mass).

As for the NP models in the u-channel exchange of new particle, there are some tensions between A_{FB} and the $t\bar{t}$ spectrum measurements, in particular at high $m_{t\bar{t}}$ region, figure 4. However, taking into account the possible caveats of the $d\sigma/dm_{t\bar{t}}$ constraints, we retain the Δ and Σ model parameter regions preferred in the approach A in our analysis of top spin observables.

4. Top spin polarization and spin correlations

Due to the different production mechanisms of top-antitop pairs at Tevatron and LHC, there is a difference in the produced asymmetries. The Tevatron produces about 90% of $t\bar{t}$ pairs in the $q\bar{q}$-annihilation, and just 10% pairs come from the gg-fusion. For the LHC, which is a symmetric proton-proton machine, the situation is just opposite. Furthermore, depending on the production mechanism top quarks are produced in a definite spin configurations. At the threshold, in $q\bar{q}$-annihilation, the produced tops are in $t_L\bar{t}_R$ or $t_R\bar{t}_L$ configuration, while the produced top-antitop pairs from gluons come out in LL and RR spin configuration. Since the top decays before it

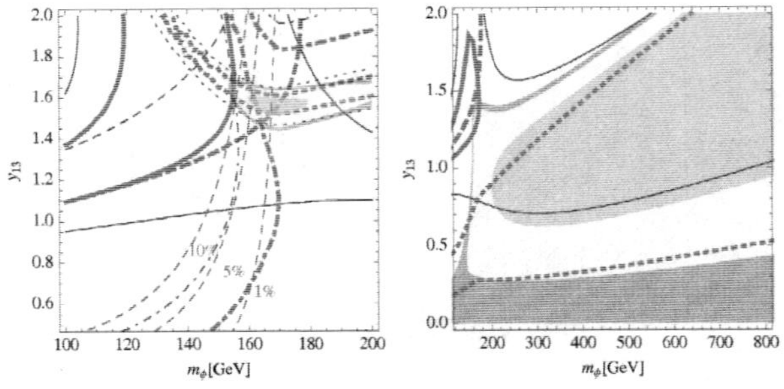

Figure 3. $t\bar{t}$ production constraints on *the scalar isodoublet model* in approach A (left plot) and B (right plot): FBA (inclusive in left plot and binned in right plot) at 1σ (2σ) in thick full green line (thin full red line), inclusive CA at 1σ (2σ) in thick dashed green line (thin dashed red line). In left plot $\sigma_{\rm TEV}$ at 1σ (2σ) in thick green dotted line (thin red dotted line), $\sigma_{\rm LHC}$ at 1σ (2σ) in thick green dash-dotted line (thin red dash-dotted line). In right plot $m_{t\bar{t}}$ spectrum at 1σ (2σ) in thick green dotted line (thin red dotted line). Parameter regions where the model can improve the SM χ^2 by $-\Delta\chi^2 > 0, 1, 4$ are shaded in red, yellow and green respectively. Finally in right plot for $m_\phi < m_t$, the contours of constant branching fraction ${\rm Br}(t \to u\phi^0)$ are displayed in thin gray dashed lines.

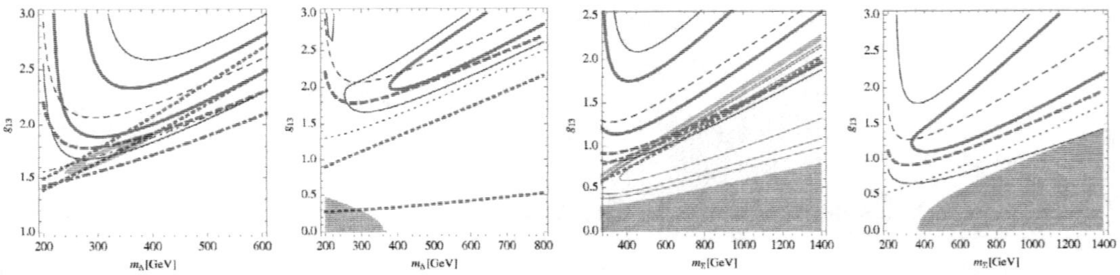

Figure 4. $t\bar{t}$ production constraints on *the scalar color triplet model* and *the scalar color sextet model* in approach A (left plot) and B (right plot): FBA (inclusive in left plot and binned in right plot) at 1σ (2σ) in thick full green line (thin full red line), inclusive CA at 1σ (2σ) in thick dashed green line (thin dashed red line). In left plot $\sigma_{\rm TEV}$ at 1σ (2σ) in thick green dotted line (thin red dotted line), $\sigma^h_{\rm TEV}$ at 2σ in thin red dash-dotted line. In right plot $m_{t\bar{t}}$ spectrum at 1σ (2σ) in thick green dotted line (thin red dotted line). Parameter regions where the model can improve the SM χ^2 by $-\Delta\chi^2 > 0, 1$ are shaded in red and yellow respectively.

hadronizes, almost exclusively in W^+b, the decay products contain information about the top spin. The top events are classified according to the W-decay products as dileptonic (with the branching ratio (BR) of $O(5\%)$ and a low background), leptonic + jets (with BR $\sim O(30\%)$ and a moderate background) and a purely hadronic events, with the largest BR $\sim O(46\%)$, but also a huge (QCD) background. By measuring the angular distributions of the decay products, we can determine the top polarization and top-antitop spin correlations [6] using the double differential angular distribution:

$$\frac{1}{\sigma}\frac{d^2\sigma}{d\cos\theta_f d\cos\theta_{\bar{f}}} = \frac{1}{4}\left(1 + \kappa_f B_t \cos\theta_f + \kappa_{\bar{f}} B_{\bar{t}} \cos\theta_{\bar{f}} - \kappa_f \kappa_{\bar{f}} C \cos\theta_f \cos\theta_{\bar{f}}\right) , \qquad (1)$$

where $\theta_f(\theta_{\bar{f}})$ is the angle between the direction of the top (antitop) spin analyzer f, $(\bar{f})$ (which can be either a direct t $(\bar{t})$ daughter W^+, b $(W^-, \bar{b})$ or a $W^+(W^-)$ decay product $\ell^+(\ell^-), \nu(\bar{\nu})$ or jets) in the t ($\bar{t}$) rest frame. The top (antitop) spin analayzing power factors $\kappa_{f(\bar{f})}$, are different for the different top decaying products, being the largest for the leptons, $\kappa_{l^+(l^-)} = 1$, considered in this work. The coefficients $B_{t(\bar{t})}$ and C are connected with the spin observables $\langle \mathcal{O} \rangle_i$ as

$$\begin{aligned}
\mathcal{O}_1 &= \mathbf{S}_t \cdot \mathbf{S}_{\bar{t}} & &\rightarrow \langle \bar{\mathcal{O}}_1 \rangle = D\,, \\
\mathcal{O}_2 &= \mathbf{S}_t \cdot \hat{\mathbf{a}}\,, \quad \bar{\mathcal{O}}_2 = \mathbf{S}_{\bar{t}} \cdot \hat{\mathbf{b}} & &\rightarrow \langle \mathcal{O}_2 \rangle = B_t \quad \text{and} \quad \langle \bar{\mathcal{O}}_2 \rangle = B_{\bar{t}}\,, \\
\mathcal{O}_3 &= 4(\mathbf{S}_t \cdot \hat{\mathbf{a}})(\mathbf{S}_{\bar{t}} \cdot \hat{\mathbf{b}}) & &\rightarrow \langle \mathcal{O}_3 \rangle = C\,.
\end{aligned} \tag{2}$$

and give the net spin polarization of the top-antitop system, polarization of the (anti) top quark, and the top-antitop spin correlation, both with respect to spin quantization axes $\hat{\mathbf{a}}$ and $\hat{\mathbf{b}}$, respectively, and D is measured in the opening angle distribution

$$\frac{1}{\sigma}\frac{d\sigma}{d\cos\phi} = \frac{1}{2}\left(1 - D\cos\phi_{f\bar{f}}\right)\,, \tag{3}$$

where ϕ is the angle between the direction of flight of the two (top and antitop) spin analyzers, defined in the t and $\bar{t}$ frames, respectively. The spin quantization axes $\hat{\mathbf{a}}$ and $\hat{\mathbf{b}}$ can be chosen to maximize (or minimize) the desired polarization and correlation effects. We have been working with the following choices: $\hat{\mathbf{a}} = -\hat{\mathbf{b}} = \hat{\mathbf{k}}_1$ (helicity basis), $\hat{\mathbf{a}} = \hat{\mathbf{b}} = \hat{\mathbf{p}}$ (beamline basis), and $\hat{\mathbf{a}} = \hat{\mathbf{b}} = \hat{\mathbf{d}}_X$ (off-diagonal basis, specific for model X, which gives almost 100% correlations at Tevatron [7]), where $\hat{\mathbf{p}}$ is the direction of the incoming beam and $\hat{\mathbf{k}}_1$ is the direction of the outgoing top quark, both in the $t\bar{t}$ center of mass frame. The measurements of the spin correlation coefficients C are performed at D0 for the beam and the helicity axis, at the CDF for the off-diagonal axis and at the LHC for the helicity axis, and the good agreement with the SM calculation [6] is confirmed.

5. Results and conclusions

The predictions for the top spin observables at Tevatron and the LHC are shown in figure 5. For the Tevatron, the prediction for the off-diagonal axis are similar to the prediction for the beamline axis. We can see that the beamline axis has a potential to discriminate between the sextet on one side and the the scalar doublet and color triplet model, on the other side. By measuring the top spin along the helicity axis, B_{hel}, on can discriminate between the doublet Φ and the color triplet Δ models. As for the axigluon, to get some information, the precision of $O(2\%)$ is needed. At the LHC, although the helicity axis has a potential to discriminate between the scalar doublet and the axigluon models, generally, the axigluon model would be difficult to probe with the top spin observables. However, at the LHC, the measurement of the spin observables D, C_{beam} and B_{hel} at $O(5\%)$ could help to discern among the scalar models.

To conclude, among the considered models only an axigluon can reproduce all Tevatron observables, without being in severe tension with the recent LHC results on $t\bar{t}$ production cross section, charge asymmetry and top-spin correlations. The models with W' and Z' bosons are excluded as a possible explanation for the Tevatron A_{FB} anomaly by the LHC charge asymmetry measurements. We have demonstrated that future precise measurements of various top spin correlations and especially the top spin polarization could provide a significant constraints on possible explanations of the Tevatron A_{FB} anomaly, in particular in the scalar model channels.

Acknowledgments

The fruitful collaboration with J. Kamenik and S. Fajfer is acknowledged. I would also like to thank the organizers of the PASCOS2012 conference. The work is supported by the Ministry of Education, Science and Sport of the Republic of Croatia under contract No. 098-0982930-2864.

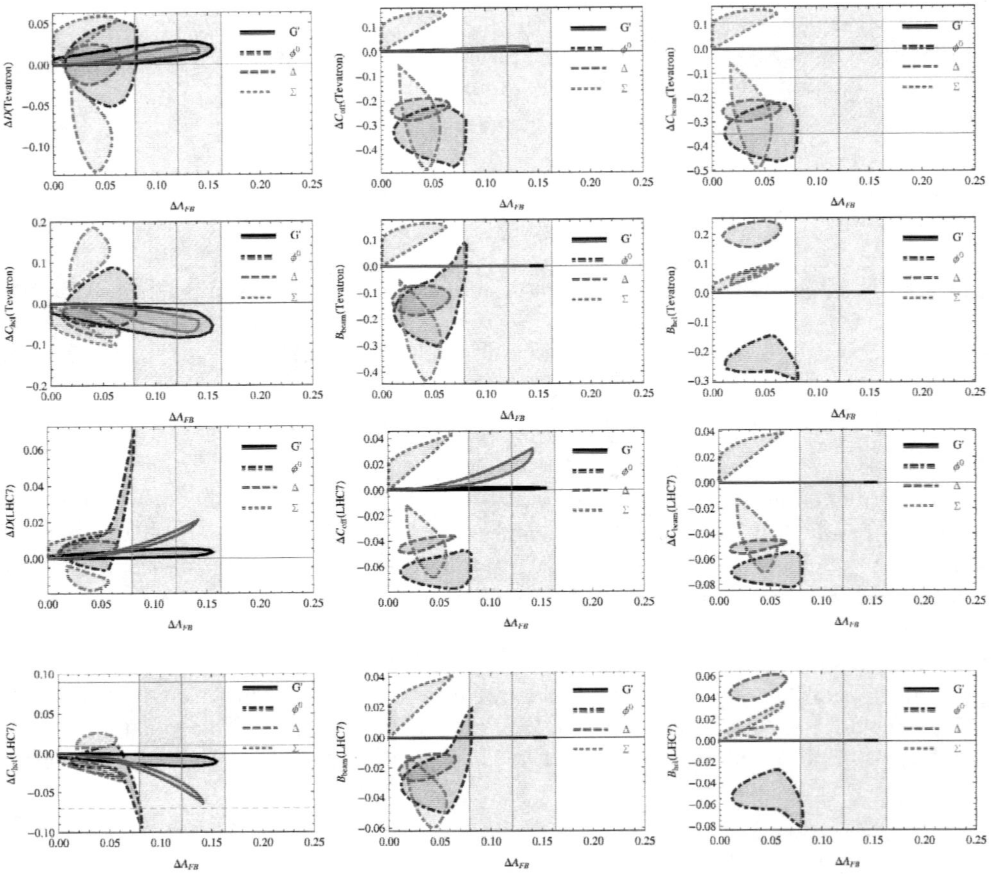

Figure 5. Correlations between the NP contributions to the inclusive A_{FB} and various spin observables at the Tevatron and at the 7 TeV LHC. The present experimental results (68% C.L. regions) are shaded in horizontal and vertical bands. The NP model predictions are determined from the global fit as specified in the text and are bounded by full (axigluon G' in the low ($m_G \lesssim 450$ GeV in black) and high ($m_G \gtrsim 700$ GeV in gray) mass regions), dashed (scalar color triplet Δ), dotted (scalar color sextet Σ) and dot-dashed (neutral component of the scalar isodoublet ϕ^0 in the low ($m_\phi \lesssim m_t$ in darker shade) and high ($m_\phi > 200$ GeV in lighter shade) mass region) contours. For $\Delta C_{\rm hel}$(LHC7) we also show the 95% C.L. contour in thin dashed line.

References

[1] Kamenik J F, Shu J and Zupan J Review of new physics effects in t-tbar production *Preprint* arXiv:1107.5257
[2] Fajfer S, Kamenik J F and Melić B 2012 *JHEP* **8** 114
[3] Ahrens V, Ferroglia A, Neubert M, Pecjak B D and Yang L L 2011 *Phys. Rev.* D **84** 074004
[4] Hollik W and Pagani D 2011 *Phys. Rev.* D **84** 093003
[5] Aguilar-Saavedra J A and Perez-Victoria M 2011 *JHEP* **1105** 034
Aguilar-Saavedra J A and Perez-Victoria M 2011 *Phys. Rev.* D **84** 115013
[6] Bernreuther W, Brandenburg A, Si Z G and Uwer P 2004 *Nucl. Phys.* B **690** 81
[7] Parke S J and Shadmi Y 1996 *Phys. Lett.* B **387** 199
Uwer P 2005 *Phys. Lett.* B **609** 271

PASCOS 2012 – 18th International Symposium on Particles Strings and Cosmology IOP Publishing
Journal of Physics: Conference Series **485** (2014) 012053 doi:10.1088/1742-6596/485/1/012053

Searches for the pair production of scalar leptoquarks at CMS

Darin Baumgartel for the CMS Collaboration

Northeastern University Department of Physics, 360 Huntington Avenue, 111 Dana Research Center, Boston, MA, 02115, USA

E-mail: darin.carl.baumgartel@cern.ch

Abstract. Results are presented for searches for leptoquark (LQ) pair production using 1.8 - 5.0 fb^{-1} of proton-proton collision data at $\sqrt{s} = 7$ TeV collected by the CMS detector at the LHC. First- and second-generation scalar leptoquarks are searched for in final states with either two leptons and two jets ($\ell\ell jj$) or one lepton, missing transverse energy ($\not\!\!E_T$), and two jets ($\ell\nu jj$). Third-generation leptoquarks are searched for in the final state with two b-tagged jets and large $\not\!\!E_T$ ($\nu\nu bb$). No significant excess beyond the standard model predictions is found and 95% confidence level (CL) upper limits are set on the scalar leptoquark pair production cross section in each channel. Limits are calculated for a range of leptoquark mass and for a variable branching fraction (β) of the leptoquark to a charge lepton and a quark. These limits are the most stringent to date.

1. Introduction

Leptoquarks (LQ) are hypothetical particles that carry both baryon number and lepton number and couple to both quarks and leptons. They are predicted by grand unified theories [1, 2], composite models [3], extended technicolor models [4, 5, 6], and superstring-inspired models [7]. LQs are color triplets with fractional electric charge and can be either scalar or vector. They couple to a lepton and a quark with a coupling strength λ, and decay to a charged lepton and a quark with an unknown branching fraction β. To satisfy constraints from bounds on flavor-changing neutral currents and from rare pion and kaon decays [3, 8], it is assumed that LQs couple to quarks and leptons of a single generation. LQs are classified as first-, second-, or third-generation, corresponding to the generation of leptons to which they couple. The dominant processes for LQ pair production at the LHC occur via gluon-gluon fusion and quark-antiquark annihilation, and have been calculated at next-to-leading order (NLO) [9].

Scalar leptoquark pair production is searched for with proton-proton collisions with a center-of-mass energy $\sqrt{s} = 7$ TeV, taken at the Large Hadron Collider (LHC) and recorded with the Compact Muon Solenoid (CMS) Detector, described in detail elsewhere [10]. The data samples used for first- and second-generation LQ searches correspond to an integrated luminosity of 5.0 fb^{-1}, and the third-generation search uses 1.8 fb^{-1}. First- and second-generation searches are performed in two final states. The $\ell\ell jj$ final state requires two high-p_T jets and two high-p_T muons or electrons. The $\ell\nu jj$ final state requires two high-p_T jets, large $\not\!\!E_T$, and one high-p_T muon or electron. Third-generation LQs are searched for in the $\nu\nu bb$ final state requiring large $\not\!\!E_T$ and two high-p_T b-tagged jets.

Content from this work may be used under the terms of the Creative Commons Attribution 3.0 licence. Any further distribution of this work must maintain attribution to the author(s) and the title of the work, journal citation and DOI.
Published under licence by IOP Publishing Ltd

2. Event Reconstruction and Object Identification

In the $eejj$ channel, events are required to pass double-electron or double-photon triggers with a p_T threshold of 33 GeV. Triggers for the $e\nu jj$ channel require an electron, $\not{E}_T$, and two jets, with transverse energy thresholds of 17 - 30 GeV, 15 - 20 GeV, and 25 - 30 GeV respectively. The $\mu\mu jj$ and $\mu\nu jj$ channels require a single-muon trigger with the muon having $|\eta| < 2.1$ and $p_T > 40$ GeV. The search in the $\nu\nu bb$ channel uses a set of triggers based on the razor variables [11, 12] described further in Section 3.

For $\ell\ell jj$ and $\ell\nu jj$ searches, electrons are selected [13] with $p_T > 40$ GeV and $|\eta| < 2.5(2.2)$ in the $eejj(e\nu jj)$ channel, omitting the barrel-endcap transition region $1.44 < |\eta| < 1.57$. Electrons must have a cluster in the electromagnetic calorimeter (ECAL) that is spatially matched to a reconstructed track in the central tracking system, and a shower shape indicative of an electromagnetic shower. Electrons must also be isolated from additional energy deposits in the ECAL and from tracks not belonging to the electron itself. The electron track must have hits in all inner tracker layers. Muons are required to have $p_T > 40$ GeV and $|\eta| < 2.1$, and to be isolated such that the sum of the p_T of all tracker tracks within a cone of $\Delta R = \sqrt{(\Delta\phi)^2 + (\Delta\eta)^2} < 0.3$ around the muon track, excluding the muon track, is less than 0.1. The muon candidates must contain at least 10 hits in the silicon tracker and one hit in the pixel detector, and have a transverse impact parameter with respect to the primary vertex of less than 2 mm.

For the $\nu\nu bb$ analysis, electrons must have $p_T > 20$ GeV and $|\eta| < 2.5$, with pile-up corrected combined isolation less than 15% of the electron p_T. Additional requirements ensure the quality of the shower shape and track-cluster matching in the η and ϕ directions. Muons are selected as for the $\ell\ell jj$ and $\ell\nu jj$ channels, but requiring $|\eta| < 2.4$ and using an isolation considering calorimetric energy deposits and tracker tracks, with a threshold of 15% of the muon p_T.

In the $\ell\ell jj$ and $\ell\nu jj$ searches, jets are reconstructed using a particle-flow algorithm [14], which identifies and measures stable particles by combining information from all CMS sub-detectors. Jets reconstruction uses the anti-k_T [15] algorithm with $R = 0.5$. In the $\nu\nu bb$ search, jets are reconstructed from calorimeter energy deposits using the infrared-safe anti-k_T algorithm with $R = 0.5$. In the $\ell\ell jj$ ($\ell\nu jj$) channel, jets are required to have $p_T > 30(40)$ GeV and be separated from leptons by $\Delta R > 0.3$. Jets energies are calibrated using the methods in [16]. All channels use $\not{E}_T$ reconstructed with particle-flow algorithm. The $\ell\nu jj$ analysis requires $\not{E}_T > 55$ GeV.

The $\nu\nu bb$ analysis also requires that jets are tagged as originating from b-quarks (b-tagged) with the *Track-Counting High-Efficiency* algorithm [17], which identifies jets with 2 good tracks and a three-dimensional impact-parameter significance greater than an optimized threshold. The efficiency of b-jet identification is estimated to be 0.76 ± 0.01

3. Event Selection

3.1. Event selection in the $\ell\ell jj$ and $\ell\nu jj$ channels

For $\ell\ell jj$ and $\ell\nu jj$ searches, events are initially selected requiring the presence of the four leptoquark decay products in each channel. The $\ell\ell jj$ channel requires two electrons or muons and two jets. The $eejj$ ($\mu\mu jj$) analysis requires a dilepton invariant mass ($M_{\ell\ell}$) of at least 60 (50) GeV. Additionally, the scalar sum of the transverse momenta of the decay products, $S_T^{\ell\ell} \equiv p_T(\ell_1) + p_T(\ell_2) + p_T(j_1) + p_T(j_2)$, must be at least 250 GeV.

The $\ell\nu jj$ channel requires two jets, exactly one lepton, and $\not{E}_T$. The leading jet and $\not{E}_T$ must be separated by $\Delta R > 0.5$, and the lepton and $\not{E}_T$ must be separated by $\Delta R > 0.8$. The transverse mass between the lepton and missing transverse energy ($M_T^{\ell\nu}$) must be greater than 50 GeV, and $S_T^{\ell\nu} \equiv p_T(\ell) + \not{E}_T + p_T(j_1) + p_T(j_2)$ must be greater than 250 GeV.

A set final selection optimized thresholds is determined for each leptoquark mass by maximizing the expected signal significance $S/\sqrt{S+B}$. In the $\ell\ell jj$ channel, masses range from 250–900 GeV, and variables thresholds consist of $S_T^{\ell\ell} > 330$–920 GeV, $M_{\ell\ell} > 100$–150 GeV, and the smallest of the two lepton-jet invariant masses $M_{\ell j}^{\min} > 60$–520 GeV. In the $\ell\nu jj$ channel, masses range from 250–850 GeV, and variable thresholds consist of $S_T^{\ell\nu} > 450$–1000 GeV, $\not{E}_T > 100$–240 GeV, and the lepton-jet invariant mass $M_{\ell j} > 150$–540 GeV. For the purpose of calculating lepton-jet invariant masses, leptons are matched to jets in the $\ell\ell jj$ ($\ell\nu jj$) channel by choosing the lepton-jet pairing which minimizes the difference in lepton-jet invariant mass (lepton-jet transverse mass) between the pairs. The sets of optimized thresholds in each channel are detailed further in [18].

3.2. Event selection in the $\nu\nu bb$ channel

In the $\nu\nu bb$ search, events are selected using the razor variable method [11, 12]. Each event is split into two geometric regions, and all of the jets in each region are considered as a singular mega-jet. This defines the dijet topology used to calculate the razor variables, with megajet four-momenta $\vec{p^1}$ and $\vec{p^2}$. The razor mass (M_R) is defined as:

$$M_R \equiv \sqrt{(|\vec{p^1}| + |\vec{p^2}|)^2 - (p_z^1 + p_z^2)^2}, \tag{1}$$

and the razor transverse mass (M_T^R) is defined as:

$$M_T^R \equiv \sqrt{\frac{\not{E}_T(p_T^1 + p_T^2) - \vec{\not{E}_T}\cdot(\vec{p}_T^1 + \vec{p}_T^2)}{2}}, \tag{2}$$

Together, M_R and M_T^R define the razor dimensionless ratio, $R \equiv M_T^R/M_R$.

Initial event selection requires a pair of b-tagged jets, large $\not{E}_T$, no isolated leptons, and at least two jets with $p_T > 60$ which may include the b-tagged jets. The razor Mass M_R is required to be greater than 400 GeV, and R^2 thresholds are places at 0.25, 0.30, and 0.35 for LQ mass hypotheses greater than 200 GeV, 330 GeV, and 340 GeV respectively.

4. Background Modeling and Systematic Uncertainties

4.1. Backgrounds in the $\ell\ell jj$ and $\ell\nu jj$ channels

The $\ell\ell jj$ and $\ell\nu jj$ channel searches rely on MC simulation and data control samples for background estimation. The MC simulated samples used to estimate the contribution from SM background processes are $\mathrm{t\bar{t}}$+jets events, generated with MADGRAPH [19, 20]; single-top events (s, t, and tW channels), generated with POWHEG [21]; Z/γ^*+jets events and W+jets events, generated with SHERPA [22]; and diboson events, generated with PYTHIA [23].

In the $\ell\ell jj$ channel, the Z/γ^*+jets contribution is estimated with MC. A rescaling factor is estimated in the Z-dominated region $80 < M_{\ell\ell} < 100$ GeV. The $\mathrm{t\bar{t}}$ contribution is estimated entirely from data using a control sample of events with one electron, one muon, and at least two jets, weighted to account for the differences in degeneracy, reconstruction and identification efficiency, and trigger acceptance.

In the $\ell\nu jj$ channel, the W+jets and $\mathrm{t\bar{t}}$ contributions are estimated with MC. Rescaling factors are estimated in the region $50 < M_T^{\ell\nu} < 110$ GeV. A W-dominated region with less than four jets is used for the W rescaling factor, whereas a $\mathrm{t\bar{t}}$-dominated region of at least four jets is used for the $\mathrm{t\bar{t}}$ rescaling.

Systematic uncertainties are considered for limit setting. They include integrated luminosity (2.2%); background normalization, based on the statistical uncertainties in the MC and data control samples; background shape, using MC samples with the factorization and renormalization scales and matrix element-parton shower matching threshold varied by a factor of two; number of pileup interactions (8%); momentum and energy scales, including jet energy (4%), muon momentum (1%), and electron energy (1-3%); momentum and energy resolutions, including jet energy (5-14%), muon momentum (4%), and electron energy (1-3%); parton distribution functions; lepton reconstruction, identification, and isolation efficiencies; and trigger acceptance.

4.2. Backgrounds in the $\nu\nu bb$ channel

For the $\nu\nu bb$ channel search, the shapes and scales of the final distributions of M_R and R^2 are determined for background processes to estimate final background contributions. The shape of M_R is well-described by two exponentials with different slope parameters(S_i) [11, 12], which relates to an R^2 threshold R_{cut} as $S_i = A_i + B_i \times R_{cut}$. The final form of the M_R fit is given as:

$$F(M_R) = e^{-(A_1+B_1\times R^2_{min})M_R} + f \times e^{-(A_2+B_2\times R^2_{min})M_R} \qquad (3)$$

Events are considered in 3 disjoint samples to estimate [24] the R^2 and M_R shape and scale of the major backgrounds: heavy flavor multijets, $t\bar{t}$, and W/Z+b-jets. The hadronic sample contains no leptons. The muon and electron samples contain events with one muon or electron with $p_T > 20$ GeV, with leptons treated as neutrinos for the calculation of the razor variables. The muon and electron samples are depleted of LQ signal, but have the same kinematic properties of the background in the hadronic sample.

The lepton samples with at least two b-jets in heavy-flavor-enriched MADGRAPH MC determine the shape of the W+b-jets and Z+b-jets contributions. The shape of the $t\bar{t}$ contribution is determined from the muon sample with at least two b-jets in real data. The contribution from heavy-flavor multijets is estimated with a muon sample in data, using a loosened muon isolation condition to enrich the sample with multijet events. Normalization of the background distributions is determined in a side-band of the signal-like hadronic sample requiring $0.2 < R^2 < 0.25$ and only one b-jet.

Systematic uncertainties considered for limit setting include integrated luminosity (4.5%); trigger efficiencies (2-3%); parton distribution functions; b-tagging efficiency (10%); background shape, from variation within uncertainty on the fit parameters of equation 3; and jet energy scale and resolution, for W and Z backgrounds estimated from MC.

5. Results

In all channels, the observed number of events is in good agreement with the standard-model background expectation. Results are expressed as a 95% CL upper limit on the LQ cross-section using the CL_S modified frequentist approach [25, 26]. The results of the $\ell\ell jj$ and $\ell\nu jj$ searches are in Figures 1 and 2. The results of the $\nu\nu bb$ search is in 3. The $\ell\ell jj$ channel searches exclude first-generation (second-generation) LQs with masses below 830 (840) GeV, assuming $\beta = 1$. Combining the $\ell\ell jj$ and $\ell\nu jj$ results, first-generation (second-generation) LQs with masses below 640 (650) GeV are excluded, assuming $\beta = 1/2$. The third-generation search excludes LQs with masses below 350 GeV, assuming $\beta = 0$. These are the most stringent limits to date.

Acknowledgments

We congratulate our colleagues in the CERN accelerator departments for the excellent performance of the LHC machine. We thank the technical and administrative staff at CERN and other CMS institutes, and acknowledge support from BMWF and FWF (Austria); FNRS and FWO (Belgium); CNPq, CAPES, FAPERJ, and FAPESP (Brazil); MES (Bulgaria); CERN; CAS, MoST, and NSFC (China); COLCIENCIAS (Colombia); MSES (Croatia); RPF (Cyprus); MoER, SF0690030s09 and ERDF (Estonia); Academy of Finland, MEC, and HIP (Finland); CEA and CNRS/IN2P3 (France); BMBF, DFG, and HGF (Germany); GSRT (Greece); OTKA and NKTH (Hungary); DAE and DST (India); IPM (Iran); SFI (Ireland); INFN (Italy); NRF and WCU (Korea); LAS (Lithuania); CINVESTAV, CONACYT, SEP, and UASLP-FAI (Mexico); MSI (New Zealand); PAEC (Pakistan); MSHE and NSC (Poland); FCT (Portugal); JINR (Armenia, Belarus, Georgia, Ukraine, Uzbekistan); MON, RosAtom, RAS and RFBR (Russia); MSTD (Serbia); SEIDI and CPAN (Spain); Swiss Funding Agencies (Switzerland); NSC (Taipei); ThEP, IPST and NECTEC (Thailand); TUBITAK and TAEK (Turkey); NASU (Ukraine); STFC (United Kingdom); DOE and NSF (USA).

Figure 1. (Top) The $\ell\ell jj$ limits on $\sigma \times \beta^2$ versus leptoquark mass. (Bottom) The $\ell\nu jj$ limits on $\sigma \times 2\beta(1-\beta)$ versus leptoquark mass.

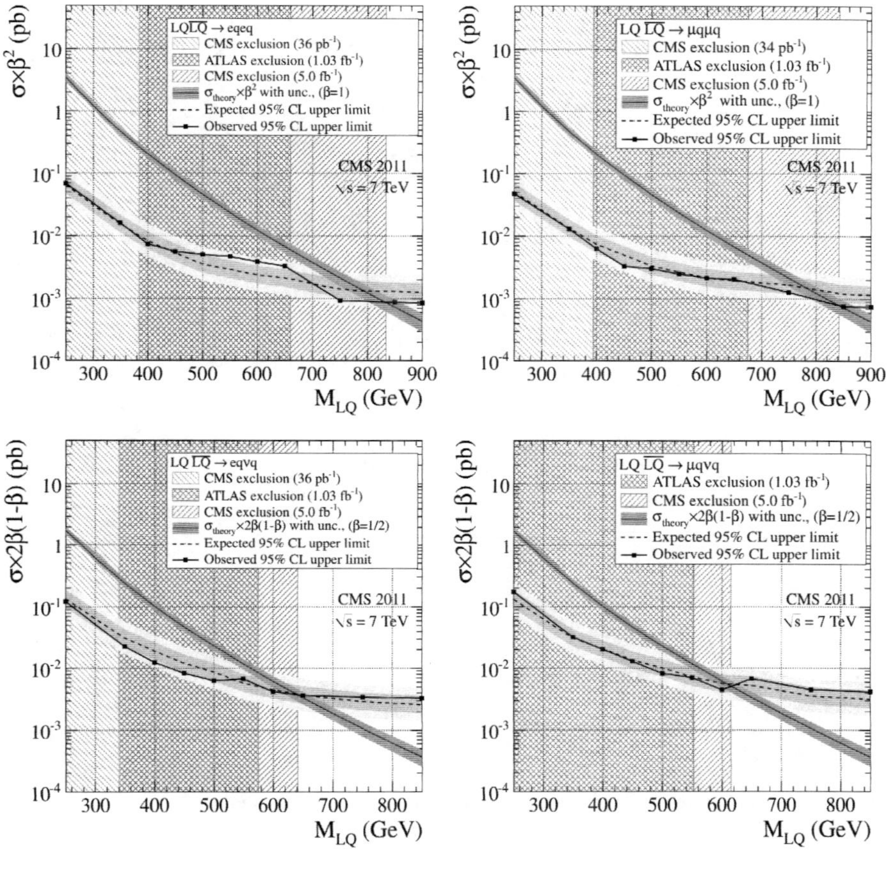

Figure 2. The limits on LQ Mass versus branching fraction β, using the combination of the $\ell\ell jj$ and $\ell\nu jj$. (Left) The $eejj$+$e\nu jj$ channel limits. (Right) The $\mu\mu jj$+$\mu\nu jj$ channel limits.

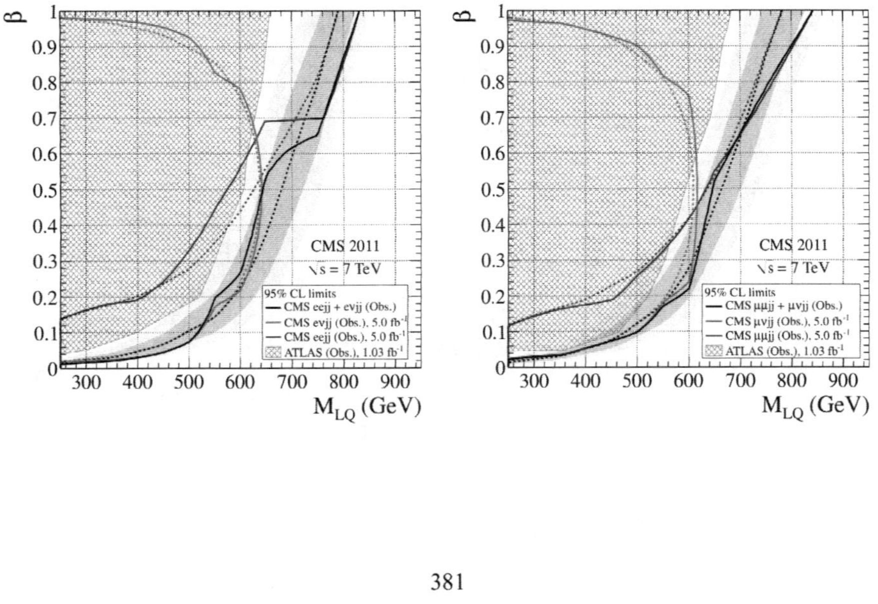

Figure 3. The limits in the $\nu\nu bb$ channel. (Left) The limit on LQ cross-section versus mass, assuming Br(LQ $\rightarrow b\nu_\tau$) = 100%. (Right) The limit on Br(LQ $\rightarrow b\nu_\tau$) versus LQ mass.

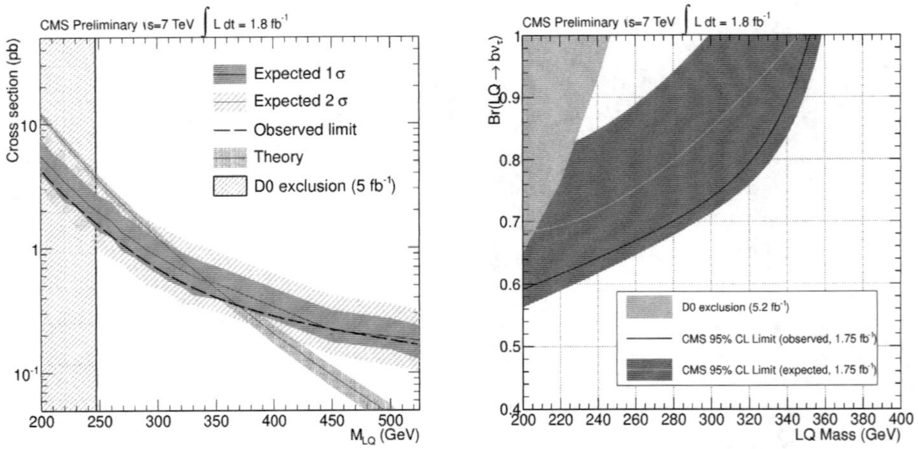

References

[1] Pati J C and Salam A 1974 *Phys. Rev. D* **10** 275
[2] Georgi H and Glashow S 1974 *Phys. Rev. Lett.* **32** 438
[3] Buchmüller W and Wyler D 1986 *Phys. Lett. B* **177** 377
[4] Dimopoulos S and Susskind L 1979 *Nucl. Phys. B* **155** 237
[5] Dimopoulos S 1980 *Nucl. Phys. B* **168** 69
[6] Eichten E and Lane K 1980 *Phys. Lett. B* **90** 85
[7] Angelopoulos V D *et al.* 1987 *Nucl. Phys. B* **292** 59
[8] Shanker O 1982 *Nucl. Phys. B* **204** 375
[9] Krämer M *et al.* 2005 *Phys. Rev. D* **71** 057503 (*Preprint* `hep-ph/0411038`)
[10] Chatrchyan S *et al.* (CMS Collaboration) 2008 *JINST* **3** S08004
[11] Rogan C 2010 (*Preprint* `1006.2727`)
[12] 2012 *Phys. Rev. D* **85**(1) 012004 URL `http://link.aps.org/doi/10.1103/PhysRevD.85.012004`
[13] CMS Collaboration 2010 Electron reconstruction and identification at $\sqrt{s}$ = 7 TeV CMS Physics Analysis Summary CMS-PAS-EGM-10-004 URL `http://cdsweb.cern.ch/record/1299116`
[14] CMS Collaboration 2010 Commissioning of the particle-flow reconstruction in minimum-bias and jet events from pp collisions at 7 TeV CMS Physics Analysis Summary CMS-PAS-PFT-10-002 URL `http://cdsweb.cern.ch/record/1279341`
[15] Cacciari M, Salam G P and Soyez G 2008 *JHEP* **04** 063 (*Preprint* `0802.1189`)
[16] Chatrchyan S *et al.* (CMS) 2011 *JINST* **6** 11002 (*Preprint* `1107.4277`)
[17] 2011
[18] Chatrchyan S *et al.* (CMS Collaboration) 2012 (*Preprint* `1207.5406`)
[19] Maltoni F and Stelzer T 2003 *JHEP* **02** 027 (*Preprint* `hep-ph/0208156`)
[20] Alwall J *et al.* 2007 *JHEP* **09** 028 (*Preprint* `0706.2334`)
[21] Alioli S *et al.* 2010 *JHEP* **06** 043 (*Preprint* `1002.2581`)
[22] Gleisberg T *et al.* 2009 *JHEP* **02** 007 (*Preprint* `0811.4622`)
[23] Sjöstrand T *et al.* 2001 *Comput. Phys. Commun.* **135** 238 (*Preprint* `hep-ph/0010017`)
[24] CMS Collaboration (CMS) 2011 Search for pair production of third-generation scalar leptoquarks using events produced in pp collisions at $\sqrt{s}$= 7 TeV containing b-jets and missing transverse energy CMS Physics Analysis Summary CMS-PAS-EXO-11-030 URL `http://cdsweb.cern.ch/record/1416079`
[25] Junk T 1999 *Nucl. Instrum. Meth. A* **434** 435 (*Preprint* `hep-ex/9902006`)
[26] Read A L 2000 Modified frequentist analysis of search results (the cl_s method) CERN-OPEN 2000-205 1st Workshop on Confidence Limits, CERN, Jan. 2000 URL `http://cdsweb.cern.ch/record/451614`

Journal of Physics: Conference Series **485** (2014) 012054 doi:10.1088/1742-6596/485/1/012054

Dynamical mass generation – Majorana versus Dirac

Otto C W Kong

Department of Physics and Center for Mathematics and Theoretical Physics, National Central University, Chung-Li, Tawian 32054.

E-mail: otto@phy.ncu.edu.tw

Abstract. We review our new simple dynamical symmetry breaking model for (softly broken) supersymmetric theory, with focus on the rich theoretical structure. In particular, the Holomorphic Supersymmetric Nambu–Jona-Lasinio offers competing symmetry breaking options related to the dynamical generation of Majorana or Dirac type superfield masses. It is relevant to the application of the model to electroweak symmetry breaking related to LHC physics.

1. Introduction to NJL Models with Supersymmetry

Let us introduce a sequence of three model Lagrangians : Firstly, there is the Nambu–Jona-Lasinio (NJL) model [1] of 1961 with the Lagrangian

$$\mathcal{L}_\psi = i\bar{\psi}_+\sigma^\mu\partial_\mu\psi_+ + i\bar{\psi}_-\sigma^\mu\partial_\mu\psi_- + g^2\bar{\psi}_+\bar{\psi}_-\psi_+\psi_- \,. \tag{1}$$

Then, there is the first supersymmetric version, or the (old) Supersymmetric Nambu–Jona-Lasinio (SNJL) model introduced in 1982 [2, 3];

$$\mathcal{L}^S_\psi = \int d^4\theta \left(\Phi^\dagger_+\Phi_+ + \Phi^\dagger_-\Phi_-\right)(1-\tilde{m}^2\theta^2\bar{\theta}^2) + \int d^4\theta\; g^2\Phi^\dagger_+\Phi^\dagger_-\Phi_+\Phi_-(1-\tilde{m}^2_c\theta^2\bar{\theta}^2)\,, \tag{2}$$

with a dimension six four-superfield interaction. And finally, our focus here of the Holomorphic Supersymmetric Nambu–Jona-Lasinio (HSNJL) model published in 2010 [4];

$$\mathcal{L}^H_\psi = \int d^4\theta \left(\Phi^\dagger_+\Phi_+ + \Phi^\dagger_-\Phi_-\right)(1-\tilde{m}^2\theta^2\bar{\theta}^2) - \int d^2\theta\; \frac{G}{2}\Phi_+\Phi_-\Phi_+\Phi_-(1+B\theta^2)\,, \tag{3}$$

which has a dimension five four-superfield interaction instead.

The theoretical models are applicable to the dynamical breaking of electroweak symmetry. The NJL model for the top quark gives rises to the so-called top mode Standard Model [5] in connection with the idea of infared (quasi-)fixed point. The symmetry breaking vacuum expectation value (VEV) is a top condensate, giving the prediction then of a heavy top mass above 200 GeV. The SNJL model was similarly applied to give the Minimal Supersymmetry Standard Model (MSSM) also around 1990 [6]. Here with possibly two VEVs v and v', there is no direct mass prediction. The NJL kind of mechanism dictates only a large Yukawa coupling, here also for the top quark. The top mass as $m_t = y_t \cdot v$ can admit a lighter top with a small v lower than the full electroweak symmetry breaking value from $\sqrt{v^2+v'^2}$. The much lighter

Content from this work may be used under the terms of the Creative Commons Attribution 3.0 licence. Any further distribution of this work must maintain attribution to the author(s) and the title of the work, journal citation and DOI.
Published under licence by IOP Publishing Ltd

bottom mass is from a smaller bottom Yukawa coupling; $m_b = y_b \cdot v'$. However, to the 172.1 GeV top, a small tanβ value, < 1.5, as the ratio of the two VEVs is required. The latter has essentially been ruled out. [1] The HSNJL model was constructed with getting around the problem in mind. The practical version has a four-superfield interaction that involves both the top and bottom quarks (superfields). Both scalar and fermion condensates, as part of the superfield condensates, play important roles. It gives an experimentally viable MSSM with $y_t < y_b$ (large tanβ) with other theoretically nice features [4, 7].

1.1. The Holomorphic Model as an Alternative Supersymmetrization

Supersymmetry is an important theme in modern physics. One of the specially attractive feature, in our opinion, is that scalar fields are now part of the chiral superfields with the chiral fermions. The chirality forbids any gauge invariant mass before breaking any symmetry. Moreover, the full matter (super)field spectrum is now strongly constrained by the gauge symmetry and their anomaly cancelation conditions. Introduction of the vectorlike pair of Higgs superfields with their un-natural gauge invariant mass in the usual formulation of the supersymmetric Standard Model looks particularly unattractive from the theoretical perspective. An NJL mechanism, with the Higgs superfield(s) generated as composite and the electroweak scale generated by strong dynamics is hence very appealing. The HSNJL model construction [4, 8] gives exactly such a scenario that looks compatible with all known experimental constraints. It looks like the only surviving model for electroweak symmetry breaking with a NJL type mechanism.

The NJL is a classic model on dynamical symmetry breaking, characterized by simplicity and beauty. The easy way to understand its physics is to assume that large enough coupling g^2, a bi-fermion scalar composite ϕ will form as the auxiliary field in the Lagrangian

$$\begin{aligned} & \mathcal{L}_\psi - (\mu\phi^\dagger + g\psi_+\psi_-)(\mu\phi + g\bar{\psi}_+\bar{\psi}_-) \\ = \quad & i\bar{\psi}_+\sigma^\mu\partial_\mu\psi_+ + i\bar{\psi}_-\sigma^\mu\partial_\mu\psi_- - \mu^2\phi^\dagger\phi - \mu g(\phi^\dagger\bar{\psi}_+\bar{\psi}_- + \phi\psi_+\psi_-) \,. \end{aligned} \tag{4}$$

Equation of motion for ϕ gives $\phi = -g/\mu\,\bar{\psi}_+\bar{\psi}_-$, the composite, showing the Lagrangian to be essentially the same as $\mathcal{L}_\psi$. A nontrivial VEV of ϕ, and hence the bi-fermion condensate, gives the symmetry breaking Dirac mass.

To supersymmetrize the NJL model, one may naively put together all the Lagrangian terms replaced by their superfield versions :

$$\begin{aligned} & i\bar{\psi}_+\sigma^\mu\partial_\mu\psi_+ \longrightarrow \int d^4\theta\ \Phi_+^\dagger\Phi_+\,, \qquad & g^2\bar{\psi}_+\bar{\psi}_-\psi_+\psi_- \longrightarrow \int d^4\theta\ g^2\Phi_+^\dagger\Phi_-^\dagger\Phi_+\Phi_-\,, \\ & -\mu g\,\phi\psi_+\psi_- \longrightarrow \int d^2\theta\ \mu g\,\Phi\Phi_+\Phi_-\,, \qquad & -\mu^2\phi^*\phi \longrightarrow \int d^2\theta\ \frac{\mu}{2}\Phi\Phi\,. \end{aligned}$$

The above does not work. The problem is that the supersymmetrization of the dimension six four-femion interaction is a Kähler potential term while that of the sort of equivalent Higgs mass and Yukawa terms are superpotential terms. To reconcile the conflict, Ref.[2] suggested the introduction of an extra superfield different from the Higgs to play the composite, namely the modifications

$$g^2\bar{\psi}_+\bar{\psi}_-\psi_+\psi_- \longrightarrow \int d^4\theta\ g^2\Phi'^\dagger\Phi' \qquad\qquad -\mu^2\phi^*\phi \longrightarrow \int d^2\theta\ \mu\Phi\Phi'\,. \tag{5}$$

One can see that the equation of motion for Φ gives $\Phi' = -g\,\Phi_+\Phi_-$ as the composite but Φ is the effective Higgs superfield. Soft supersymmetry breaking mass(es) is also needed for a $\langle\Phi\rangle \neq 0$ solution with Dirac mass for $\Phi_+\Phi_-$, and without further modifications, $\langle\Phi'\rangle$ stays zero.

[1] The model has actually other not very nice features from the theoretical point of view [4, 7].

The HSNJL model is an alternative supersymmetrization of the NJL framework without enforcing the exact NJL Lagrangian terms to be contained in its component field expression. A chiral fermion is supersymmetrized to a chiral superfield. A supersymmetric analog for the four-fermion interaction may be a four-superfield interaction. The latter, besides the dimension six option, has a dimension five option which is holomorphic – a superpotential term like the Higgs mass and Yukawa couplings. Another way to think about it is to look at the supersymmetric form of the last line in the expression of the effective NJL Lagrangian of Eq.(4). It can be written, in the presence of soft supersymmetry breaking mass(es), as

$$\mathcal{L}^H_\psi = \int d^4\theta \left[(\Phi_+^\dagger \Phi_+ + \Phi_-^\dagger \Phi_-)(1 - \tilde{m}^2\theta^2\bar{\theta}^2) \right] + \int d^2\theta \left[\frac{\mu}{2}\Phi^2 + \sqrt{\mu G}\Phi\Phi_+\Phi_- \right] + h.c. \, .$$

The structure is exactly what is given by taking $\mathcal{L}^H_\psi$ as given, before the introduction of the auxiliary Higgs superfield, in Eq.(3) with the superpotential $\frac{G}{2}\Phi_+\Phi_-\Phi_+\Phi_-(1+B\theta^2)$ ($B = 0$ for the simplest version). Equation of motion for the Higgs superfield Φ gives $\Phi = -\sqrt{G/\mu}\Phi_+\Phi_-$, *i.e.* it is also the composite. Here one can see that by giving up the four-fermion interaction, we have a supersymmetric model that maintains essentially all other features of the NJL model nicely.

1.2. Towards the MSSM

In applying the NJL mechanism to electroweak symmetry breaking and the related low energy phenomenology, all earlier attempt focused only on the top quark. Only one Dirac pair of fermion (super)field is required to implement the mechanism anyway. The heavy mass of the top makes the natural candidate, only that it eventually turned up to be not heavy enough. To apply our holomorphic SNJL model to the MSSM, it is crucial to note that the Yukawa coupling for the bottom quark may be large, provided that we have a large tanβ. We take both of the third family quarks on the same footing and introduce the holomorphic superpotential term $W = G\,\varepsilon_{\alpha\beta} Q_3^{\alpha a} U_3^{c\,a} Q_3^{\beta b} D_3^{c\,b}(1+B\theta^2)$. The two Higgs superfields of the MSSM may then be introduced as auxiliary superfields to rewrite W, with the $SU(2)$ and color indices suppressed, as

$$\begin{aligned} W &= -\mu\,(H_d - \lambda_t Q_3 U_3^c)(H_u - \lambda_b Q_3 D_3^c)(1+B\theta^2) + G\,Q_3 U_3^c Q_3 D_3^c(1+B\theta^2) \\ &= (-\mu H_d H_u + \mu\lambda_t Q_3 H_u U_3^c + \mu\lambda_b H_d Q_3 D_3^c)(1+B\theta^2)\,, \qquad (\mu\lambda_t\lambda_b = G)\,. \end{aligned} \tag{6}$$

Equation of motion for H_u gives $H_d = \lambda_t Q_3 U_3^c$ while that of H_d yields $H_u = \lambda_b Q_3 D_3^c$. Note that the supersymmetry breaking parameter B gives both the standard A and B parameters. Contrary to the old SNJL based model [6], we have a symmetric role for H_u and H_d as both are quark superfield composites. The symmetric structure does not apply to the soft SUSY breaking sector. Different soft masses for the different superfields would be the original of the nontrivial tanβ value, hence hierarchy between top and bottom masses.

2. Analysis of Dynamical Symmetry Breaking

Dynamical mass generation and symmetry breaking is a very interesting theoretical topic with important phenomenological applications. One of the simplest model of the kind is the NJL model [1]. It is also the first explicit model of spontaneous symmetry breaking. The basic strategy to the non-perturbative analysis is to obtain the gap equation for the self-consistent mass parameter value and look for possible nontrivial solution, typically available under some strong coupling condition (see for example Ref.[9]). For the NJL case with a strong enough (dimension six) four-fermion interaction involving two Weyl fermions, a symmetry breaking Dirac fermion mass would be resulted. The HSNJL model mimics well most basic features of the NJL model in the setting of softly broken supersymmetry, with however a dimension five four-superfield interaction. The symmetry breaking and Dirac mass generation was established recently in Ref.[8].

Such an analysis for a soft broken supersymmetric theory is somewhat tricky, and published only recently in our paper [8] even for the case of the old SNJL model. The first important step there is to consider the (Dirac) mass parameter as one on the superspace [10]. It is like a constant (chiral) superfield, with also an admissible supersymmetry breaking auxiliary part:

$$\mathcal{M} = m - \theta^2 \eta \,, \tag{7}$$

where m is the supersymmetric(/fermion) mass, while η contributes a supersymmetry breaking scalar left-right mixing mass. This is easy to appreciate, for example, by thinking about the mass as coming from a Yukawa term with the Higgs (superfield) taking a VEV. With the parameter, one can obtain the superfield propagators incorporating the soft breaking [11]. The crucial step then is to consider the generating functional and related two-point correlation functions for the theory on the same level – as superspace entities. We have

$$\Gamma = \int \frac{d^4p}{2\pi^4} \int d^2\theta \, \Phi_+(-p,\theta) \, \Gamma^{(2)}_{+-}(p,\theta^2) \, \Phi_-(p,\theta) + h.c. + \cdots \tag{8}$$

giving the gap equation formally as

$$-\mathcal{M} = \Sigma^{(loop)}_{+-}(p,\theta^2)\Big|_{\text{on-shell}} \,. \tag{9}$$

With the approach, we went through the supergraph calculation to obtained, for the first time [8], the gap equation for Dirac mass generation as given by

$$\begin{aligned} m &= 2mg^2 I_1(|m|^2, \tilde{m}^2, |\eta|, \Lambda^2) \,, \\ \eta &= -\eta\, g^2 \tilde{m}_C^2 \, I_2(|m|^2, \tilde{m}^2, |\eta|, \Lambda^2) \,, \end{aligned} \tag{10}$$

for the SNJL model (I_1 and I_2 are a couple of loop integrals the details of which we skip here). Considering the m equation along, assuming $\eta = 0$ or equivalently $\tilde{m}_C^2 = 0$, one obtains the known nontrivial solution obtained in Ref.[3] with a effective potential analysis based on the assumption on the formation of the Higgs superfields as composites and otherwise. The gap equation analysis in that sense proves the composite Higgs formation. And the full set of superspace gap equation offers more interesting general solutions.

For the more interesting case of the HSNJL model, we obtained

$$\begin{aligned} m &= \frac{\bar{\eta} G}{2} \, I_2(|m|^2, \tilde{m}^2, |\eta|, \Lambda^2) \,, \\ \eta &= \bar{m} G \, I_1(|m|^2, \tilde{m}^2, |\eta|, \Lambda^2) + \frac{\bar{\eta} G B}{2} \, I_2(|m|^2, \tilde{m}^2, |\eta|, \Lambda^2) \,, \end{aligned} \tag{11}$$

as a tightly coupled set of equations. One cannot see possible nontrivial solution when neglecting the η part. And at least for some simple limiting cases, we have established the existence of nontrivial solution.

3. Majorana versus Dirac Mass Generations

Unlike the SNJL model, the HSNJL model offers more mass generation options. The possibility of superfield condensate $\langle \Phi_+ \Phi_- \rangle$ is what gives rise to the Dirac mass term $\mathcal{M}\Phi_+\Phi_-$ from the holomorphic four-superfield interaction, explicitly, as $\frac{G}{2} \langle \Phi_+ \Phi_- \rangle \, \Phi_+ \Phi_-$. However, there may be the option of say a condensate of $\langle\Phi_+\Phi_+\rangle$ or $\langle\Phi_-\Phi_-\rangle$ giving Majorana mass terms $\mathcal{M}_-\Phi_-\Phi_- \sim \frac{G}{2}\langle\Phi_+\Phi_+\rangle\,\Phi_-\Phi_-$ and $\mathcal{M}_+\Phi_+\Phi_+ \sim \frac{G}{2}\langle\Phi_-\Phi_-\rangle\,\Phi_+\Phi_+$, respectively.

We have extended the analysis to consider the option [12]. The gap equations are given by

$$\begin{aligned} m_\pm &= \frac{\bar{\eta}_\mp G}{2}\, I_2(|m_\mp|^2, \tilde{m}_\mp^2, |\eta_\mp|, \Lambda^2)\ , \\ \eta_\pm &= \bar{m}_\mp G\, I_1(|m_\mp|^2, \tilde{m}_\mp^2, |\eta_\mp|, \Lambda^2) - \frac{\bar{\eta}_\mp GB}{2}\, I_2(|m_\mp|^2, \tilde{m}_\mp^2, |\eta_\mp|, \Lambda^2)\ . \end{aligned} \tag{12}$$

Here we have allowed for unequal soft supersymmetry breaking masses of the two superfields. The first thing to note is that when the latter are equal, $\tilde{m}_+^2 = \tilde{m}_-^2$, the natural outcome will be $m_+ = m_-$ and $\eta_+ = \eta_-$, in which case the gap equations collapsed to one for the common parameters, which is exactly the same as the one for the Dirac case. Naively, it says the Majorana mass generation is feasible. A more careful thinking leads one to worry what will really be the outcome for the HSNJL model taken without assumption. The completing mass generation scenarios will imply plausibly different symmetry breaking directions.

The above has important implication for the extended model of applying to electroweak symmetry breaking Eq.(6), bringing into question if it does give us what we want. To address the question takes a highly nontrivial effort [13]. However, some careful thinking about the results we have so far gives interesting insight into what should be the basic features. Take the above gap equations for the limiting case that one of the soft mass vanishes, say $\tilde{m}_-^2 = 0$ (and $B = 0$), one can easily see that the Majorana mass generation option is killed. Nontrivial solution is no longer possible. If we consider the Dirac mass gap equation under the same condition, nontrivial solution survives well. In fact, the Dirac mass case is sensitive mostly only the the average of the two soft mass parameters [12]. Explicitly, the gap equations with unequal soft masses are given by

$$\begin{aligned} m &= \frac{\bar{\eta} G}{2}\, I_2(|m|^2, \tilde{m}_{av}^2, |\eta'|, \Lambda^2)\ , \\ \eta &= \bar{m} G\, I_1(|m|^2, \tilde{m}_{av}^2, |\eta'|, \Lambda^2) + \frac{\bar{\eta} GB}{2}\, I_2(|m|^2, \tilde{m}_{av}^2, |\eta'|, \Lambda^2)\ , \end{aligned} \tag{13}$$

with

$$\begin{aligned} \tilde{m}_{av}^2 &= \frac{\tilde{m}_+^2 + \tilde{m}_-^2}{2}\ , \\ |\eta'| &= \sqrt{|\eta|^2 + \left(\frac{\tilde{m}_+^2 - \tilde{m}_-^2}{2}\right)^2}\ . \end{aligned} \tag{14}$$

The line of thinking indicates strongly that a split in the soft supersymmetry breaking masses favors Dirac over Majorana masses. Under that situation, we expect for a certain range of the value of coupling, only Dirac type mass term will arise. This is a positive picture for the application to electroweak symmetry breaking as the phenomenologically required large tanβ scenario fits well into it. [2] The implications for the phenomenological squark masses will be very interesting.

4. Concluding Remarks

The first message is the Holomorphic Supersymmetric Nambu–Jona-Lasinio model works well as a model of dynamical symmetry breaking and mass generation. The structure can be used

[2] The presence of the strong QCD coupling also favors the formation of the colorless condensate for Dirac type masses instead of the color and charge breaking Majorana masses among the quark superfields.

to formulate a version of the MSSM as its low energy effective field theory, in which the Higgs superfields arise as composites of the top and bottom superfields. The version is more interesting compared to the one from the old SNJL model, both theoretically and phenomenologically.

From a superfield theoretical perspective, it is interesting to note the formulation of the analysis based on the consideration of the theoretical notion such as generating functions, correlation functions, and mass parameters as superspace entities with supersymmetric as well as supersymmetry breaking parts. The approach may be interesting, for example, for the application to spontaneous supersymmetry breaking.

The HSNJL model in general has more interesting mass generation, and hence symmetry breaking, options. It may generates both Dirac type and Majorana type mass terms. While we have not finish the most general analysis without prior assumptions, our studies so far has given a clear indication that the splitting in value of the input soft supersymmetry breaking masses favor the Dirac type mass term over the Majorana type ones. The latter goes in line with the application to MSSM where the splitting would be responsible for the hierarchy in the Dirac top and bottom masses generated.

Acknowledgments
The author wants to thank his collaborator, Yan-Min Dai, Gaber Faisel, Dong-Won Jung, and Jae Sik Lee, on the subject matter. The work is partially supported by research grant NSC 99-2112-M-008-003-MY3 of the National Science Council of Taiwan.

References

[1] Nambu Y and Jona-Lasinio G 1961 *Phys. Rev.* **122** 345; *ibid.* **124** 246
[2] Buchmúller W and Love S T 1982 *Nucl. Phys.* B **204** 213
[3] Buchmúller W and Ellwanger U 1984 *Nucl. Phys.* B **245** 237
[4] Jung D-W, Kong O C W, and Lee J S 2010 Phys. Rev. D **81** 031701
[5] Nambu Y 1989 Enrico Fermi Institute Report No. 89-08; Miransky V, Tanabashi M, and Yamawaki K 1989 *Mod. Phys. Lett.* A **4** 1043; Phys. Lett. B **221** 177; Marciano W J 1989 *Phys. Rev. Lett.* **62** 2793; 1990 *Phys. Rev. D* **41** 219; Bardeen W A, Hill C T and M. Lindner M 1990 *ibid.* D **41** 1647
[6] Clark T E, Love S T, and W.A. Bardeen W A 1990 *Phys. Lett.* B **237** 235; Carena M S, Clark T E, Wagner C E M, Bardeen W A and Sasaki K 1992 *Nucl. Phys.* B **369** 33
[7] Kong O C W 2012 *AIP Conf.Proc.* **1200** 1101
[8] Faisel G, Jung D-W, and Kong O C W 2012 *JHEP* **1201** 164
[9] Cvetic G 1999 *Rev. Mod. Phys.* **71** 513
[10] Miller R D C 1983 *Nucl. Phys.* B **228** 316
[11] Helayel-Neto J A 1984 *Phys. Lett.* B **135** 78; Scholl M 1985 *Z. Phys.* C **28** 545
[12] Dai Y-M, Faisel G, Jung D-W, and Kong O C W 2012 *Preprint* arXiv:1208,1017 [hep-th], NCU-HEP-k050
[13] Dai Y-M, Faisel G, Jung D-W, and Kong O C W *work in progress*

Mass matrices in $SU(5) \times Q_6$ SUSY-FUT's

E Jiménez and M Mondragón

Instituto de Física, Universidad Nacional Autónoma de México, Apdo. Postal 20-364, 01000, México D.F., México.

E-mail: ejimenez@fisica.unam.mx

Abstract. In the context of two-loop Finite Supersymmetric Theories with gauge group $SU(5)$, we present parametric solutions to the finiteness conditions using Q_6 as a family symmetry group. Assuming an MSSM scenario with just one pair of light Higgs doublets, we obtain mass matrices at the GUT scale for this class of theories.

1. Introduction

Since the recent discovery of a Higgs-like state in the LHC with mass around 125 GeV [1,2], it is worth looking more closely at beyond the Standard Model theories wich predict a mass for the Higgs particle around this value. N=1 Supersymmetric Finite Unified Theories (FUT's) are theories that are characterized by the fact that, at least at two-loops, UV divergences are absent and can even be made finite to all-loops [3–7]. With $SU(5)$ as a gauge group, a FUT model was proposed in which, restricted to one generation, a top mass at ~ 172 GeV and a Higgs mass at $\sim 122 - 126$ GeV were predicted, in good agreement with experimental data [8–10] (for a recent review of finite theories and their phenomenological implications see ref. [11]). In this paper, we analyse $SU(5)$ FUTs with the discrete group Q_6 as family symmetry group, and present the resulting mass matrices. A study with different family symmetry group can be found in [12].
Section 2 is devoted to general aspects of finite theories and the form that the finiteness conditions take in $SU(5)$. In section 3 we very briefly review the Q_6 properties to proceed with the assignment of the fields with this symmetry group. Then, in Section 4 we present the boundary conditions that the finiteness conditions impose on the MSSM, ending with some concluding remarks in section 5.

2. SUSY-FUT's in $SU(5)$

For a SUSY theory with a gauge simple group G the absence of UV divergences at two-loop level is guaranteed if it satisfies two requirements [13]. First the Yukawa couplings and the gauge couplings are related as

$$\sum_{ij} Y_{ijk} Y^{ijl} = g^2 \delta^l_k, \tag{1}$$

where $Y^{ijl} = Y^*_{ijl}$. The second condition is that the assignment of matter content satisfies the following relation

$$\sum_i T(R_i) - 3C_2(G) = 0\,, \tag{2}$$

Content from this work may be used under the terms of the Creative Commons Attribution 3.0 licence. Any further distribution of this work must maintain attribution to the author(s) and the title of the work, journal citation and DOI.
Published under licence by IOP Publishing Ltd

where $T(R_i)$ is the Dynkin-index associated with the i-field that it is in the R_i representation of the gauge group G. The remarkable feature of these Finiteness Conditions (FC), is that if one finds a non-degenerate and isolated solution to eq.(1), then finiteness to all order in perturbation theory can be achieved [3, 4].
For $G = SU(5)$, a solution to eq. (2) is obtained with the following matter content [14, 15]:

$$\mathbf{5} : X_a\,, \quad \bar{\mathbf{5}} : \bar{X}_a\,, \quad \mathbf{24} : \Sigma\,, \quad \mathbf{10} : 10_i\,, \quad \bar{\mathbf{5}} : \bar{5}_i\,. \qquad (a = 1,2,3,4;\quad i = 1,2,3) \tag{3}$$

The 10_i and $\bar{5}_i$ fields are interpreted as the fields where the three families of quarks and leptons are. The most general $SU(5)$ Yukawa superpotential with this matter content is [16]

$$\begin{aligned} f_y \;=\; & \frac{1}{2} Y_{1010X}\, 1010X \,+\, Y_{10\bar{5}\bar{X}} 10\bar{5}\bar{X} \,+\, Y_{X\bar{X}\Sigma} X\Sigma\bar{X} \,+\, \frac{1}{6} Y_{\Sigma^3}\, \Sigma^3 \\ & + \frac{1}{2} Y_{\bar{5}\bar{5}10} \bar{5}\bar{5}10 \,+\, \frac{1}{2} Y_{\bar{X}\bar{X}10}\, \bar{X}\bar{X}10 \,+\, Y_{X\Sigma\bar{X}}\, X\Sigma\bar{5}\,, \end{aligned} \tag{4}$$

where generation and gauge indices are omitted. The last three terms would generate renormalizable terms that violate leptonic and baryonic number, so we make $Y_{\bar{5}\bar{5}10} = Y_{\bar{X}\bar{X}10} = Y_{X\Sigma\bar{5}} = 0$, we can achieve this either as a result of the family group assigment we will discuss below or because of an extra Abelian simmetry that has been imposed. We denote the family part of the Yukawas as f_{ija}, g_{ija}, h_{ab} and p for the first four terms in equation (4), respectively. The general finiteness conditions (FC) reduce to [17]

$$X_a^b = \quad 3\, f_{ija}\, f^{ijb} + \frac{24}{5}\, h_{ac}\, h^{bc} \quad = \quad \frac{24}{5}\, g^2 \delta_a^b \tag{5}$$

$$\bar{5}_i^j = \quad 4\, g_{kia}\, g^{kja} \quad = \quad \frac{24}{5}\, g^2\, \delta_i^j \tag{6}$$

$$\bar{X}_a^b = \quad 4\, g_{ija}\, g^{ijb} + \frac{24}{5}\, h_{ca}\, h^{cb} \quad = \quad \frac{24}{5}\, g^2\, \delta_a^b \tag{7}$$

$$10_i^j = \quad 3\, f_{ika}\, f^{jka} + 2\, g_{ika}\, g^{jka} \quad = \quad \frac{36}{5}\, g^2\, \delta_i^j \tag{8}$$

$$\Sigma = \quad h_{ab}\, h^{ab} + \frac{21}{5} p^* p \quad = \quad \frac{36}{5}\, g^2\,, \tag{9}$$

In principle there are 55 quadratic equations with 89 complex unknows taking into account the three families. Thus family symmetries become mandatory, but before, we can obtain some useful conditions on the solutions. From eqs. (7) and (8) we have:

$$\sum_a^4 \bar{X}_a^a - \sum_i^3 \bar{5}_i^i = \sum_{a,b}^4 |h_{ab}|^2 = g^2, \tag{10}$$

so from this last equation and eq. (9) one can see that

$$|p|^2 \quad = \quad \frac{15}{7}\, g^2. \tag{11}$$

If only m ($m \leq 4$) fields of $\bar{X}_a$ are coupled in the $10\bar{5}\bar{X}$ term, (i.e., $g_{kib} = 0$ unless $b = b^1, \cdots, b^m$), then, taking the sum $\bar{X}_b^b$ over b ($b = b^1, \cdots, b^m$) minus the sum $\bar{5}_i^i$ over I ($i = 1, 2, 3$), implies the following relation

$$\sum_b^{b^1,\cdots,b^m} \bar{X}_b^b - \sum_i^3 \bar{5}_i^i = \sum_b^{b^1,\cdots,b^m} \sum_c^4 |h_{cb}|^2 = (m-3) g^2. \tag{12}$$

Similarly for the $1010X$ sector; $f_{ija} = 0$ unless $a = a^1, \ldots, a^n$,$(n \leq 4)$, then

$$\sum_{a}^{a^1,\cdots,a^n} X_a^a - (\sum_i^3 10_i^i - \frac{1}{2}\sum_i^3 \bar{5}_i^i) = \sum_{a}^{a^1,\cdots,a^n}\sum_{c}^{4} |h_{ac}|^2 = (n-3)g^2, \tag{13}$$

so we need at least three fields $\bar{X}'s$ $(X's)$ to be coupled to the $10\bar{5}$ (1010) terms. If $m = 3\,(n = 3)$ the only non-vanishing elements of h_{ab} are those of the column $b \neq \{b^1, b^2, b^3\}$ $(a \neq \{a^1, a^2, a^3\})$. The converse is also true, if the only non-vanishing elements of h_{ab} are just of one column(row), then $m = 3(n = 3)$. When $(m = n = 3)$, h_{ab} is reduced to having only one non-vanishing element.

3. Q_6 symmetry group assignment

Since all (Bi)Dihedrical groups have only one and two dimensional irreducible representations (irreps), their study becomes important to explain the observed hierarchy of masses, where we have one heavy plus two light generations. The Q_6 group has four one-dimensional irreps (denoted as $\mathbf{1}, \mathbf{1}', \mathbf{1}', \mathbf{1}'''$) and two dimension two irresp (denoted as **2**, **2**'). In tables 1 and 2 the tensor products of the two-dimensional vectors are presented showing explicitely their tensor matrices, also it is shown the assignment necessary for other fields to couple to these tensors in order to form a Q_6 invariant term.

$\tau_0 = \begin{pmatrix} 1 & 0 \\ 0 & 1 \end{pmatrix}$		$\tau_1 = \begin{pmatrix} 0 & 1 \\ 1 & 0 \end{pmatrix}$	$\tau_2 = \begin{pmatrix} 0 & -1 \\ 1 & 0 \end{pmatrix}$		$\tau_3 = \begin{pmatrix} 1 & 0 \\ 0 & -1 \end{pmatrix}$
$(\mathbf{2}\otimes\mathbf{2})_{\mathbf{1}'}$	$(\mathbf{2}'\otimes\mathbf{2}')_{\mathbf{1}}$	$(\mathbf{2}\otimes\mathbf{2}')_{\mathbf{1}''}$	$(\mathbf{2}\otimes\mathbf{2})_{\mathbf{1}}$	$(\mathbf{2}'\otimes\mathbf{2}')_{\mathbf{1}'}$	$(\mathbf{2}\otimes\mathbf{2}')_{\mathbf{1}'''}$
$\mathbf{1}'$	$\mathbf{1}$	$\mathbf{1}'''$	$\mathbf{1}$	**1'**	$\mathbf{1}''$

Table 1. The first row represents the tensor matrices that appear in the tensor products of two doublets that form one-dimensional irreps, which are shown in the second row. The third row tells us which one irrep forms a Q_6 invariant when multiplied with a tensor of the second row.

$(\mathbf{2}\times\mathbf{2})_{\mathbf{2}'}$ $\mathbf{2}'$	$\begin{pmatrix} -\tau_1 \\ \tau_3 \end{pmatrix}$	$(\mathbf{2}'\times\mathbf{2}')_{\mathbf{2}'}$ $\mathbf{2}'$	$\begin{pmatrix} -\tau_3 \\ \tau_1 \end{pmatrix}$	$(\mathbf{2}\times\mathbf{2}')_{\mathbf{2}}$ $\mathbf{2}$	$\begin{pmatrix} \tau_0 \\ \tau_2 \end{pmatrix}$

Table 2. Matrices of the two-dimensional part of the tensor product of two doublets. The doublet assignment needed to form an invariant term is also shown.

If the 10_i $(i = 1, 2, 3)$ fields are all in one dimensional irreps, we can form invariant tensors with X_a, $(a = 1, 2, 3, 4)$ (in the sector $1010X$), only if some of these are also in one-dimensional irreps of Q_6. Since the main motivation of introducing non-Abelian groups as Q_6 is to relate matrix elements for the Yukawa elements, we will take 10_1 and 10_2 to be a doublet and also we will assign at least one doublet structure to each of the remaining fields, 5_i, X_a and $\bar{X}_a$. Thus,

$$10 \equiv \begin{pmatrix} 10_1 \\ 10_2 \end{pmatrix}, \quad 5 \equiv \begin{pmatrix} \bar{5}_1 \\ \bar{5}_2 \end{pmatrix}, \quad X \equiv \begin{pmatrix} X_1 \\ X_2 \end{pmatrix}, \quad \bar{X} \equiv \begin{pmatrix} \bar{X}_1 \\ \bar{X}_2 \end{pmatrix}, \tag{14}$$

are doublets of Q_6, whereas 10_3 and $\bar{5}_3$ have to be in one of the four one-dimensional irreps. In order to work both sectors at the same time, we define

$$\begin{aligned} \Psi^A &= 10(10)\,, \quad \Psi^B = 10(\bar{5})\,, \quad \Psi_3^A = 10_3(10_3) \quad \Psi^C = X(\bar{X}), \\ \Psi_2^C &= X_2(\bar{X}_2), \quad \Psi_3^B = 10_3(\bar{5}_3), \quad \Psi_3^C = X_3(\bar{X}_3), \quad \Psi_4^C = X_4(\bar{X}_4). \end{aligned} \tag{15}$$

the field before (inside of) the parenthesis applies to the $1010X$ ($10\bar{5}\bar{X}$) sector. We will further assume the same assignment in both sectors, with doublet structure ($\mathbf{2'}\times\mathbf{2'}\times\mathbf{2'}$) and ($\mathbf{2}\times\mathbf{2}\times\mathbf{2'}$) for ($\Psi^A\times\Psi^B\times\Psi^C$). If Ψ_3^C and Ψ_4^C are in a one-dimensional irrep then the most general Yukawa part of the superpotential is of the form

$$\Psi^A\Psi^B(a\Psi_3^C + a'\Psi_4^C) + b\,\Psi^A\Psi_3^B\Psi^C + c\,\Psi_3^A\Psi^B\Psi^C + d\,\Psi^A\Psi^B\Psi^C + \Psi_3^A\Psi_3^B(e\Psi_3^C + e'\Psi_4^C), \tag{16}$$

where the tensor τ matrices are not show explicitely. We examine for simplicity the case where $a' = e' = 0$, then one has the situation where $n = m = 3$, so the solution for h_{ab} needs to be

$$h_{ab} = \mathsf{diag}\{0, 0, 0, g\}. \tag{17}$$

In this case the finiteness conditions read

$$|b_{f,g}|^2 + |c_{f,g}|^2 + 2|d_{f,g}|^2 = (l_{f,g})^2, \quad 2|a_{f,g}|^2 + |e_{f,g}|^2 = (l_{f,g})^2, \tag{18}$$

$$|a_g|^2 + |c_g|^2 + 2\,|d_g|^2 = (l_g)^2, \quad 2\,|b_g|^2 + |e_g|^2 = (l_g)^2, \tag{19}$$

$$\begin{aligned} 3(\,|a_f|^2 + |b_f|^2 + 2\,|d_f|^2\,) + 2(\,|a_g|^2 + |b_g|^2 + 2\,|d_g|^2\,) &= 3(l_f)^2 + 2(l_g)^2, \\ 3\,(\,2|c_f|^2 + |e_f|^2) + 2\,(\,2|c_g|^2 + |e_g|^2) &= 3(l_f)^2 + 2(l_g)^2. \end{aligned} \tag{20}$$

with

$$(l_f)^2 = \frac{8}{5}\,g^2, \quad (l_g)^2 = \frac{6}{5}\,g^2, \tag{21}$$

where the f and g couplings correspond to the $1010X$ and $10\bar{1}\bar{0}\bar{X}$ sectors, respectively. All the non-diagonal conditions are identically zero. The most general solution to this set of quadratic equations is

$$|a_{f,g}| = |b_{f,g}| = |c_{f,g}| = \sqrt{\frac{1}{2}}\,(l_{f,g})\sin(\theta_{f,g}), \tag{22}$$

$$|e_{f,g}| = \sqrt{2}|\,d_{f,g}| = (l_{f,g})\cos(\theta_{f,g}), \tag{23}$$

where θ_g and θ_f are arbitrary free parameters. By looking at tables 1 and 2, we can find assigments that forbid (a, b, c) or (e, d), in these cases we can achieve all-loop finiteness.

4. Boundary conditions in the MSSM

Taking an $SU(5)$ spontaneous symmetry breaking of the form $\langle\Sigma\rangle = w\,\mathsf{diag}\{-2,-2,-2,+3,+3\}$, we assume that we are left with the MSSM matter content, with only one pair of light Higgses. Thus the FC become boundary conditions at M_{GUT}. We transform the X_a and $\bar{X}_a$ sectors by applying an U_X and $U_{\bar{X}}$ unitary transformation, respectively [17,18]. Denoting the doublet part of $SU(2)$ of X_a ($\bar{X}_a$) by H_a ($\bar{H}_a$), and taking the fourth component in this new basis as the lightest, one has that the Yukawa couplings take the form

$$\left(\sum_i^3 f_{ijk}v_k\right), \quad \left(\sum_i^3 g_{ijk}\bar{v}_k\right) \tag{24}$$

with

$$v_i = \langle H_i\rangle = (U_X^*)_{i4}v_u, \quad \bar{v}_i = \langle\bar{H}_i\rangle = (U_{\bar{X}})_{i4}v_d, \tag{25}$$

where v_u and v_d are the vacuum expectation values of the up and down Higgs fields, in an obvious notation. We now show explicitely the resulting mass matrices:

$$\begin{pmatrix} -dw_1 + aw_3 & dw_2 & bw_1 \\ dw_2 & dw_1 + aw_3 & bw_2 \\ b\,w_1 & bw_2 & ew_3 \end{pmatrix}, \quad (\mathbf{2'}\times\mathbf{2'}\times\mathbf{2'}), \quad \Psi_3^A = \Psi_3^B = \Psi_3^C = \mathbf{1}, \tag{26}$$

and

$$\begin{pmatrix} dw_2 + aw_3 & -2dw_1 & bw_1 \\ -2dw_1 & -dw_2 + aw_3 & bw_2 \\ bw_1 & bw_2 & ew_3 \end{pmatrix}, \quad (\mathbf{2} \times \mathbf{2} \times \mathbf{2}'), \quad \Psi_3^A = \Psi_3^B = \mathbf{1}'', \ \Psi_3^C = \mathbf{1}', \qquad (27)$$

where w_i represents v_i or $\bar{v}_i$, since we are assuming the same structure for both sectors. This form of the mass matrices appears also in the MSSM-Q_6 model of ref. [19], as well as in an S_3 invariant extension of the SM with three Higgs doublets [20], but with the difference that the mass matrices obtained here are at the GUT scale. The case where a' and e' are different from zero in eq. (16) has been worked out, with the result that the FC are always overdetermined. In principle it is still possible to find a different class of solutions if one considers the three fields $\bar{5}_i$ in a one-dimensional irrep.

5. Conclusions

In this article we have obtained the boundary conditions at the M_{GUT} scale for the MSSM mass matrices that come from the finiteness conditions and Q_6 as a family symmetry group. The assignments that we have made of the fields in the Q_6 irreps, which consist of taking at least one doublet of Q_6 for each of the matter fields, allow only two-loop finite solutions (although there are all-loop finite solutions for this class of field assignments, corresponding to $\theta_f = 0$ and $\theta_g = \pi/2$ values (or viceversa), which generate mass matrices that are too simple to be considered in a realistic MSSM scenario). To study the consequences of these results in low energy physics one has to run the mass matrices using the renormalization group equations to the electroweak scale.

Acknowledgments

E. J. thanks the Instituto de Física-UNAM for supporting his attendance to this interesting and stimulating conference. We acknowledge support from the Mexican grants: PAPIIT IN113712 and CONACyT-132059.

References

[1] ATLAS-Collaboration *Physics Letters B* 1–29 ISSN 0370-2693
[2] CMS-Collaboration *arXiv:1207.7235 [hep-ex]*
[3] Lucchesi C, Piguet O and Sibold K 1988 *Helv. Phys. Acta* **61** 321
[4] Lucchesi C, Piguet O and Sibold K 1988 *Phys. Lett.* **B201** 241
[5] Kazakov D I 1987 *Mod. Phys. Lett.* **A2** 663–674
[6] Ermushev A V, Kazakov D I and Tarasov O V 1987 *Nucl. Phys.* **B281** 72–84
[7] Lucchesi C and Zoupanos G 1997 *Fortschr. Phys.* **45** 129–143 (*Preprint* hep-ph/9604216)
[8] Kapetanakis D, Mondragon M and Zoupanos G 1993 *Z. Phys.* **C60** 181–186 (*Preprint* hep-ph/9210218)
[9] Heinemeyer S, Mondragon M and Zoupanos G 2008 *JHEP* **07** 135 (*Preprint* 0712.3630)
[10] Heinemeyer S, Mondragon M and Zoupanos G 2012 *Int.J.Mod.Phys.Conf.Ser.* **13** 118–126
[11] Mondragon M and Zoupanos G 2011 *Phys.Part.Nucl.Lett.* **8** 173–186
[12] Babu K S, Enkhbat T and Gogoladze I 2003 *Phys. Lett.* **B555** 238–247 (*Preprint* hep-ph/0204246)
[13] Jones D R T and Mezincescu L 1984 *Phys. Lett.* **B138** 293
[14] Hamidi S, Patera J and Schwarz J H 1984 *Phys. Lett.* **B141** 349
[15] Jiang X d and Zhou X j 1987 *Phys. Lett.* **B197** 156
[16] Björkman J, Jones D and Raby S *Nuclear Physics B* 503–532 ISSN 0550-3213
[17] Hamidi S and Schwarz J H 1984 *Phys. Lett.* **B147** 301
[18] Leon J, Perez-Mercader J, Quiros M and Ramirez-Mittelbrunn J 1985 *Phys. Lett.* **B156** 66
[19] Babu K S and Kubo J 2005 *Phys.Rev.* **D71** 056006 (*Preprint* hep-ph/0411226)
[20] Kubo J, Mondragon A, Mondragon M and Rodriguez-Jauregui E 2003 *Prog.Theor.Phys.* **109** 795–807 (*Preprint* hep-ph/0302196)

Extracting the fragmentation functions with global analyses

D de Florian and R J Hernández-Pinto

Universidad de Buenos Aires, FCEyN (1428), Pabellón 1, Ciudad Universitaria, Capital Federal, Argentina

E-mail: `deflo@df.uba.ar`, `roger@df.uba.ar`

Abstract. We discuss preliminary results on the fragmentation functions obtained in a combined NLO fit to data of single-inclusive hadron production in electron-positron annihilation, proton-proton collisions, and deep-inelastic lepton-proton scattering with either pions or kaons identified in the final state.

1. Introduction

Fragmentation functions are fundamental objects which describe the collinear transition of a quark i into a hadron H with a fraction z of its momentum, and it is usually named as $D_i^H(z)$. These fragmentation functions can only be obtained by performing global fits. In here, we describe the extraction of the fragmentation functions by using different set of data.

A field theoretical definition of the fragmentation in terms of bi-local operators was given, up to kinematical pre-factors, as

$$D_i^H(z) \approx \int dx^- e^{-iP_H^+ x^-/z} \, \mathrm{Tr}\left[\gamma^+ \langle 0|\Psi(0)\mathcal{P}|H(P_H^+)X\rangle\langle H(P_H^+)X|\mathcal{P}'\bar{\Psi}(x)|0\rangle\right], \quad (1)$$

where $\mathcal{P}$ and $\mathcal{P}'$ denotes the gauge-links to render it gauge-invariant. Since the fragmentation functions are non perturbative objects, they cannot be computed from first principles and they need to be extracted by fitting the experimental data of different kind of processes. However, the scale dependence of the fragmentation functions can be obtained in pertubative QCD (pQCD) and can be determined by renormalization group equations, similar to those for parton densities (PDF).

In order to have a consistent description of the fragmentation functions, we assume momentum conservation of the fragmenting parton in the hadronization process, summarized by a sum rule stating that

$$\sum_H \int_0^1 dz z D_i^H(z, Q^2) = 1. \quad (2)$$

The main goal of the global fit is to extract the universal fragmentation functions from all available processes such that their predictions are well consistent with future experiments. Moreover, different kind of processes might help with the extraction of the fragmentation of

Content from this work may be used under the terms of the Creative Commons Attribution 3.0 licence. Any further distribution of this work must maintain attribution to the author(s) and the title of the work, journal citation and DOI.
Published under licence by IOP Publishing Ltd

different flavor of quarks and they will be better constrained with more data sets. So, the processes that we are going to use for the extraction of the fragmentation are going to be described in the following.

1.1. Single-inclusive e^+e^- annihilation

The cross section for the single-inclusive e^+e^- annihilation (SIA) into a specific hadron H, $e^+e^- \to (\gamma, Z) \to H$, can be written as,

$$\frac{1}{\sigma_{tot}}\frac{d\sigma^H}{dz} = \frac{\sigma_0}{\sum_q \hat{e}_q^2}\left[2F_1^H(z,Q^2) + F_L^H(z,Q^2).\right] \tag{3}$$

where $Q/2 = \sqrt{s}/2$ and $z_H \equiv 2p_H \cdot q/Q^2 = 2E_H/\sqrt{s}$.

$$\sigma_{tot} = \sum_q \hat{e}_q^2 \sigma_0 \left[1 + \frac{\alpha_s(Q^2)}{\pi}\right] \tag{4}$$

is the total cross section including its NLO $\mathcal{O}(\alpha_s(Q^2))$ correction and $\sigma_0 = 4\pi\alpha_s^2(Q^2)/s$. The unpolarized "time-like" structure functions F_1^H and F_L^H are given at NLO by,

$$2F_1^H(z,Q^2) = \sum_q \hat{e}_q^2 \left\{\left[D_q^H + D_{\bar{q}}^H\right](z,Q^2) + \frac{\alpha_s(Q^2)}{2\pi}\left[C_q^1 \otimes (D_q^H + D_{\bar{q}}^H) + C_g^1 \otimes D_g^H\right](z,Q^2)\right\} \tag{5}$$

$$F_L^H(z,Q^2) = \frac{\alpha_s(Q^2)}{2\pi}\sum_q \hat{e}_q^2 \left[C_q^L \otimes (D_q^H + D_{\bar{q}}^H) + C_g^L \otimes D_g^H\right](z,Q^2), \tag{6}$$

with $\otimes$ denoting a standard convolution and the $C_{q,g}^{1,L}$ coefficients in the $\overline{\text{MS}}$ scheme can be found in App. A of Ref. [1]

1.2. Semi-Inclusive Deep-Inelastic Scattering

The cross section for the semi-inclusive deep-inelastic production of a hadron, $eN \to e'HX$, can be written in a factorized form in a similar way to the fully inclusive DIS case,

$$\frac{d\sigma^H}{dxdydz_H} = \frac{2\pi\alpha^2}{Q^2}\left[\frac{(1+(1-y)^2)}{y}2F_1^H(x,z_H,Q^2) + \frac{2(1-y)}{y}F_L^H(x,z_H,Q^2)\right] \tag{7}$$

with x and y denoting the usual DIS scaling variables ($Q = sxy$), and where $z_H \equiv p_H \cdot p_N / p_N \cdot q$. The structure functions are given at NLO by,

$$\begin{aligned} 2F_1^H(x,z_H,Q^2) &= \sum_{q,\bar{q}} \hat{e}_q \left\{ q(x,Q^2)D_q^H(z_H,Q^2) + \frac{\alpha_s(Q^2)}{2\pi} \times \right. \\ &\times \left. \left[q \otimes C_{qq}^1 \otimes D_q^H + q \otimes C_{gq}^1 \otimes D_g^H + g \otimes C_{qg}^1 \otimes D_q^H\right](x,z_H,Q^2)\right\} \end{aligned} \tag{8}$$

$$F_L^H(x,z_H,Q^2) = \frac{\alpha_s(Q^2)}{2\pi}\sum_{q,\bar{q}} \hat{e}_q^2 \left[q \otimes C_{qq}^L \otimes D_q^H + q \otimes C_{gq}^L \otimes D_g^H + g \otimes C_{qg}^L \otimes D_q^H\right](x,z_H,Q^2) \tag{9}$$

with the NLO ($\overline{\text{MS}}$) coefficient functions $C_{ij}^{1,L}$ [1, 2].

1.3. Hadron-Hadron Collisions

The single-inclusive production of a hadron H at high transverse momentum p_T in hadron-hadron collisions is also amenable to QCD perturbative theory. The differential cross section can be written in factorized form as,

$$E_H \frac{d^3\sigma}{dp_H^3} = \sum_{a,b,c} f_a \otimes f_b \otimes d\hat{\sigma}_{ab}^c \otimes D_c^H, \tag{10}$$

where the sum runs over all partonic channels $a + b \to c + X$, with $d\hat{\sigma}_{ab}^c$ the associated partonic cross section. $d\hat{\sigma}_{ab}^c$ can be expanded as a power series in the strong coupling α_s. The $\mathcal{O}(\alpha_s^3)$ NLO corrections are available in the literature [3]

2. Parametrization and extraction of the fragmentation functions

In the global analysis we will determine individual fragmentation functions for all partons (quarks and gluons) from data. We adopt for the fragmentations the following input distributions [4],

$$D_i^H(z,\mu_0) = \frac{N_i z^{\alpha_i}(1-z)^{\beta_i}[1+\gamma_i(1-z)^{\delta_i}]}{B[2+\alpha_i,\beta_i+1]+\gamma_i B[2+\alpha_i,\beta_i+\delta_i+1]}, \tag{11}$$

where $B[a,b]$ represents the Euler Beta-function and N_i is a normalization factor such that the sum rule in Eq. (5) is satisfied. We have imposed certain relations upon the individual fragmentation functions for pions and kaons. For instance, we impose isospin symmetry for the sea fragmentation functions to charged pions, i.e.

$$D_{\bar{u}}^{\pi^+} = D_{\bar{d}}^{\pi^+} \tag{12}$$

but we allow a different normalization in the $q + \bar{q}$ sum,

$$D_{d+\bar{d}}^{\pi^+} = N D_{u+\bar{u}}^{\pi^+}. \tag{13}$$

For strange quarks it is assumed that

$$D_s^{\pi^+} = D_{\bar{s}}^{\pi^+} = N' D_{\bar{u}}^{\pi^+} \tag{14}$$

with N' independent of z.

For charged kaons we fit $D_{u+\bar{u}}^{K+}$ and $D_{s+\bar{s}}^{K+}$ independently. For the unfavored fragmentation the data are unable to discriminate between flavors and, consequently, we assume that all distributions have the same functional form:

$$D_{\bar{u}}^{K^+} = D_s^{K^+} = D_d^{K^+} = D_{\bar{d}}^{K^+}. \tag{15}$$

We adopt the functional form of Eq. (14) also for the fragmentation of heavy charm and bottom quarks into charged pions and kaons but setting $\gamma_i = 0$. We assume that $D_c^H = D_{\bar{c}}^H$ and $D_b^H = D_{\bar{b}}^H$ for $H = \pi^+, K^+$.

The parameters describing the fragmentation functions for pions and kaons at scale μ_0 are determined by a standard χ^2 minimization for N data points,

$$\chi^2 = \sum_{i=1}^{N} \frac{(T_i - E_i)^2}{\delta E_i^2} \tag{16}$$

where E_i is the measured value for a given observable, δE_i the error associated with this measurement, and T_i is the corresponding theoretical estimate for a given set of parameters. For the associated error we take, as usual, the statistical and systematical errors in quadrature in δE_i. In order to compute the observable in a fast way, we are performing all the computation using the Mellin technique.

2.1. Mellin Technique

The integro-differential evolution equations can be straightforwardly solved analytically in Mellin n-moment space along the lines described in Ref. [5]. The Mellin moments of, for instance, the fragmentation functions $D_i^H(z,Q^2)$, are defined as,

$$D_i^H(n,Q^2) \equiv \int_0^1 dz z^{n-1} D_i^H(z,Q^2), \tag{17}$$

and can be expressed in terms of the Euler Beta functions for our ansatz at scale μ_0. The evolved fragmentation functions in z-space are re-obtained by an inverse Mellin transformation given by

$$D_i^H(z,Q^2) = \frac{1}{2\pi i}\int_{\mathcal{C}_n} dn z^{-n} D_i^H(n,Q^2), \tag{18}$$

where $\mathcal{C}_n$ denotes an appropriately chosen contour in the complex n-plane. This method is implemented for the extraction of pion and kaon fragmentation functions.

3. Data Sets

We use charged pion and kaon production data in SIA from TPC [6] at $\sqrt{s}$ =29 GeV, SLD [7], ALEPH [8], DELPHI [9], and OPAL [10], all at $\sqrt{s} = M_Z$. We also consider the data from TASSO [11] at intermediate c.m.s. energies of $\sqrt{s}$ =33 and 44 GeV.

Due to the conceptual problem with the fragmentation functions at small z, the cut $z_{min} = 0.05(0.1)$ is imposed for all pion (kaon) data sets.

ALEPH, DELPHI and TPC provide tagged results distinguishing between the sum of light u, d, s quarks, charm, and bottom events.

To further constraint the fragmentation functions of different flavors, as well as to separate favored (valence) and unfavored (sea) fragmentations, we include experimental information of SIDIS. More specifically, we make use of charged pion and kaon multiplicities from the HERMES [12] and COMPASS (preliminary) [13] experiments. These data also provide an important consistency check of the flavor separation obtained from flavor tagged SIA "data".

A wealth of data on single-inclusive hadron production from RHIC experiments [14, 15, 16, 17] have also been included in our global analysis. These encompass the p_T spectrum of neutral pions at central rapidities $|\eta| \leq 0.35$ by PHENIX [14] and at two different forward rapidities $\langle\eta\rangle = 3.3$ and 3.8 by STAR [15].

BRAHMS have also published p_T spectra for identified pions and kaons at two values of rapidities $\eta = 2.95$ and 3.3 [16] of which we use only the former in the fit as it has large theoretical uncertainties due to small p_T values probed. In addition, there are data on K_s^0 production for central rapidities $|\eta| \leq 0.5$ from STAR [17]. To accommodate the K_S^0 data in the fit, we assume that $K_S^0 = (K^+ + K^-)/2$ with $u \to K^+$ and $d \to K^+$ fragmentation functions interchanged.

4. Results

Now, we present the preliminary results for pions and kaons fragmentation functions, and this analysis is an update and cross check of the previous DSS fit of fragmentation functions [4] where we have included a preliminary SIDIS data coming from the COMPASS experiment. We also modified the PDFs with the extraction presented by the MSTW collaboration [18].

4.1. Pion Results

The COMPASS data set have helped us to test the prediction of the fragmentation functions at $Q^2 \approx 40$ GeV with good agreement. In Figure 1 we present the SIDIS results of the global pion fits. The SIA and hadron-hadron plots have not presented a notorious difference to those shown in Ref. [4]

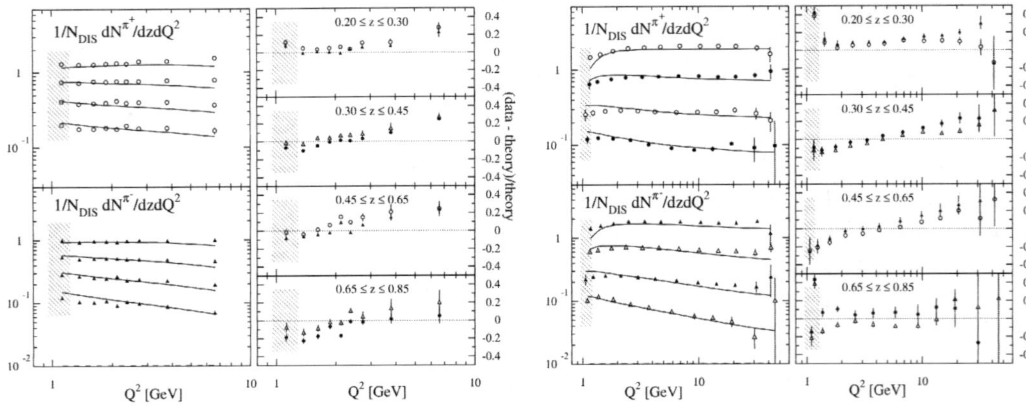

Figure 1. Comparison of the NLO results for charged pion multiplicities in SIDIS, with HERMES (left panel) and prelimiary COMPASS data (right panel).

4.2. Kaon Results

For charged kaons, the SIDIS results have been more complicated to fit. The multiplicities are very sensitive to the PDF strange quark distribution and the net result is shown in Figure 2. As it can be seen from the plot, we have also not considered multiplicities for values of Q^2 larger than 25 GeV because of their large uncertainties and also because of the unusual behavior that they present. Similarly to the pion fit, the SIA and hadron-hadron plots have not changed dramatically to those presented by DSS.

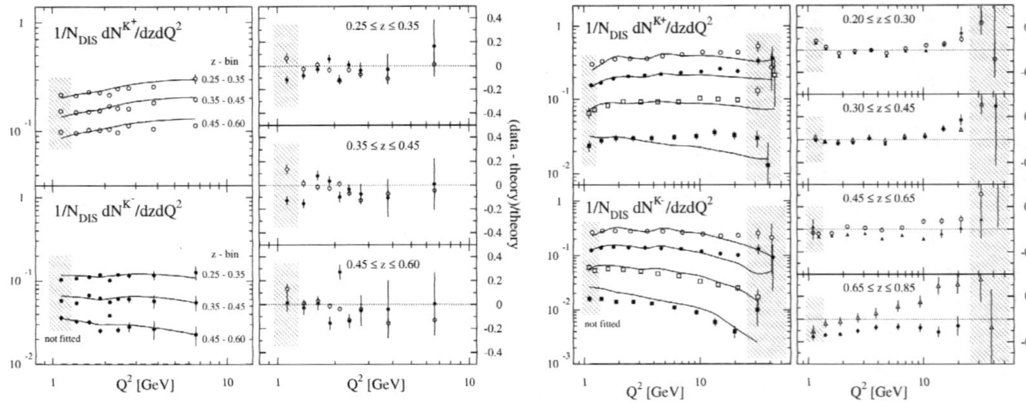

Figure 2. Same as in Fig. 2 but for charged kaons.

There are new set of data recently released or presented in conferences, for instance Ref. [19], which will be analyzed and we expect these issues to be clarified with more experimental data.

Acknowledgments
This work was supported by the the Research Executive Agency (REA) of the European Union under the Grant Agreement number PITN-GA-2010-264564 (LHCPhenoNet). RJHP work is also supported by CONACyT, México, grant 132067.

References
[1] de Florian D, Stratmann M and Vogelsang W 1998 *Phys. Rev* **D57** 5811
[2] Altarelli G, Ellis R K, Martinelli G and Pi S Y 1979 *Nucl. Phys.* **B160** 301; Nason P and Webber B 1996 *Nucl. Phys.* 1994 **B421** 473; *ibid.* 1996 **B480** 755(E); Furmanski W and Petronzio R 1982 *Z. Phys.* **C11** 293; Graudenz D 1994 *Nucl. Phys.* **B432** 351
[3] Jäger B, Schäffer A, Stratmann M and Vogelsang W 2003 *Phys. Rev.* **D67** 054005; Aversa F, Chiappetta P, Greco M and Guillet J 1989 *Nucl. Phys.* **B327** 105; de Florian D 2003 *Phys. Rev.* **D67** 054004
[4] de Florian D, Sassot R and Stratmann M 2007 *Phys. Rev.* **D75** 114010
[5] Glück M, Reya E and Vogt A 1990 *Z. Phys.* **C48** 471; *ibid.* 1992 *Phys. Rev.* **D45** 3986
[6] Aihara H *et al.* (TPC Collaboration) 1987 *Phys. Rev. Lett.* **61** 1263; *ibid.* 1987 *Phys. Lett.* **B184** 299; Lu X Q 1986 *Ph.D. thesis, John Hopkins University* (*Preprint* UMI-87-07273)
[7] Abe K *et al* (SLD Collaboration) 1999 *Phys. Rev.* **D59** 052001
[8] Buskulic D *et al.* (ALEPH Collaboration) 1995 *Z. Phys.* **C66** 355
[9] Abreu P *et al.* (DELPHI Collaboration) 1998 *Eur. Phys. J.* **C5** 585
[10] Akers R *et al.* (OPAL Collaboration) 1994 *Zeit. Phys.* **C63** 181; Abbiendi G *et al.* (OPAL Collaboration) 2000 *Eur. Phys. J.* **C16** 407
[11] Braunschweig W *et al.* (TASSO Collaboration) 1989 *Zeit. Phys.* **C42** 189
[12] Hillenbrand A (HERMES Collaboration) 2005 "Measurement and Simulation of the Fragmentation Process at HERMES" *Ph.D. thesis, Erlangen Univ., Germany*; private communications
[13] Makke N (COMPASS collaboration) 2011 "Measurement of the polarization of strange quark in the nucleon and determination of quark fragmentation functions into hadrons" *Ph. D. thesis, Université Paris Sud XI, Paris, France*
[14] Adler S S *et al.* (PHENIX Collaboration) 2003 *Phys. Rev. Lett.* **91** 241803; Barish K (PHENIX Collaboration) 2006 "The PHENIX Spin Program: Recent Results and Future Prospects" talk presented at SPIN 2006, Kyoto, Japan.
[15] Adams J *et al.* (STAR Collaboration) *Phys. Rev. Lett.* **97** 152302
[16] Arsene I *et al.* (BRAHMS Collaboration) 2007 *Phys. Rev. Lett.* **98** 252001
[17] Abelev B I *et al.* (STAR Collaboration) 2007 *Phys. Rev.* **C75** 064901
[18] Martin A D, Stirling W J, Thorne R S and Watt G 2009 *Eur. Phys. J.* **C63** 189
[19] Leitgab M (BELLE Collaboration), 2012 "Measurement of the Fragmentation Functions (FFs) at Belle", talk presented at DIS2012, Bonn, Germany; Abelev B *et al.* (ALICE Collaboratio) 2012 (*Preprint* arXiv:1205.5724)

PASCOS 2012 – 18th International Symposium on Particles Strings and Cosmology IOP Publishing
Journal of Physics: Conference Series **485** (2014) 012057 doi:10.1088/1742-6596/485/1/012057

A Grand Unified model with Q_6 as the flavour symmetry

J C Gómez-Izquierdo[1], F González Canales[1,2] and M Mondragón[1]

[1] Instituto de Física, Universidad Nacional Autónoma de México
México 01000, D.F., México
[2] Facultad de Ciencias de la Electrónica, Benemérita Universidad Autónoma de Puebla, Apdo. Postal 157, 72570, Puebla, Pue., México

E-mail: jcarlos@fisica.unam.mx, ffelix@fisica.unam.mx, myriam@fisica.unam.mx

Abstract. We present a non-minimal renormalizable SUSY $SU(5)$ model where the flavour sector exhibits a Q_6 flavour symmetry, and R-parity is conserved. We find that the CKM matrix has an NNI and a *Fritzsch* texture for the down and up sectors respectively, both of which are known to be compatible with the experimental CKM values. At the same time, in the leptonic side, the model predicts a strong inverted hierarchy spectrum and a sum rule among the neutrino masses. As main results, the atmospheric ($\theta_{23}^{\ell^{th}} = 46.18^{+0.66}_{-0.65}$) and solar ($\theta_{12}^{\ell^{th}} = 36.62 \pm 4.06$.) mixing angles are found to be consistent with the experimental data. However, the reactor mixing angle value, $\theta_{13}^{\ell^{th}} = 3.38^{+0.03}_{-0.02}$, is small and not in good agreement with the global experimental fits, but it is consistent with the MINOS experiment and fairly large in comparison to the tribimaximal scenario.

1. Introduction

The Q_6 flavour symmetry group has been proposed as responsible for the textures in the quark as well as in the leptonic sectors [1–5]. This appealing flavour symmetry allows the appearance of the NNI textures in the quark mass matrices [6, 7], so that the mixing can be accommodated in good agreement with the experimental data [1, 5]. The rich phenomenology Q_6 provides in supersymmetry (SUSY) scenarios is remarkable, such as prohibiting the dangerous terms that mediate fast proton decay [2, 3]. Thus, an immediate question that arises is how to combine the Q_6 flavour symmetry with a GUT framework, in particular, with the SUSY $SU(5)$ model. Here we will address this problem and we will see that it is possible to accommodate the masses and mixings of fermions in such a framework.

2. SUSY $SU(5) \otimes U(1) \otimes Q_6$ MODEL

The assigned matter content under SUSY $SU(5) \otimes U(1) \otimes Q_6$ is displayed on Table 1. Let us comment on our notation and the matter content: ϕ and $\bar{\phi}$ are singlet scalars under the gauge group, the former gives mass to the right-handed neutrinos and the latter is introduced in order to cancel anomalies in the $U(1)$ abelian group. In addition, ϕ breaks the SUSY $SU(5)\otimes U(1)\otimes Q_6$ gauge group into SUSY $SU(5) \otimes Q_6$; Φ^a_b stands for the **24** adjoint scalar representation which breaks the SUSY $SU(5)$ gauge group to the MSSM; there are three families of Higgs type H^u_i and H^d_j. On the other hand, we do need to include H_{45} and $H_{\bar{45}}$ scalar representations so as to

Content from this work may be used under the terms of the Creative Commons Attribution 3.0 licence. Any further distribution of this work must maintain attribution to the author(s) and the title of the work, journal citation and DOI.
Published under licence by IOP Publishing Ltd

	$SU(5)$	Q_6	$U(1)$
(H_1^d, H_2^d)	$\bar{\mathbf{5}}$	$\mathbf{2}_1$	$-x$
H_3^d	$\bar{\mathbf{5}}$	$\mathbf{1}_{+,2}$	$-x$
(H_1^u, H_2^u)	$\mathbf{5}$	$\mathbf{2}_1$	x
H_3^u	$\mathbf{5}$	$\mathbf{1}_{+,2}$	x
(F_1, F_2)	$\bar{\mathbf{5}}$	$\mathbf{2}_2$	$\frac{3x}{2}$
F_3	$\bar{\mathbf{5}}$	$\mathbf{1}_{-,3}$	$\frac{3x}{2}$
(T_1, T_2)	$\mathbf{10}$	$\mathbf{2}_2$	$-\frac{x}{2}$
T_3	$\mathbf{10}$	$\mathbf{1}_{-,3}$	$-\frac{x}{2}$
(N_1^c, N_2^c)	$\mathbf{1}$	$\mathbf{2}_2$	$-\frac{5x}{2}$
N_3^c	$\mathbf{1}$	$\mathbf{1}_{-,1}$	$-\frac{5x}{2}$
ϕ	$\mathbf{1}$	$\mathbf{1}_{+,2}$	$5x$
$\bar{\phi}$	$\mathbf{1}$	$\mathbf{1}_{+,2}$	$-5x$
$H_{\overline{45}}$	$\overline{\mathbf{45}}$	$\mathbf{1}_{+,2}$	$-x$
H_{45}	$\mathbf{45}$	$\mathbf{1}_{+,2}$	x
Φ	$\mathbf{24}$	$\mathbf{1}_{+,0}$	0

Table 1. Matter content in the SUSY $SU(5) \otimes U(1) \otimes Q_6$ model.

fix the incorrect relation $\mathbf{M}_d = \mathbf{M}_e^T$, this can also be achieved by means of higher dimensional operators [8–10]. Regarding the fermion sector, N_i^c denotes the right-handed neutrino, which is a singlet under the $SU(5)$ gauge group; F_i and T_j stand for the **5**-plets and the **10** antisymmetric-plet, respectively. Here, a, b, c are $SU(5)$ indices, and i, j are family indices. More explicitly,

$$\begin{aligned} F_{ia} &= (d^c, L); \quad T_j^{ab} = \frac{1}{\sqrt{2}}(u^c, q, \ell^c); \quad L = \begin{pmatrix} \nu_\ell \\ \ell \end{pmatrix}; \quad q = \begin{pmatrix} u \\ d \end{pmatrix}, \\ H^{ua} &= \begin{pmatrix} \mathtt{H}^u \\ \mathbf{H}^u \end{pmatrix}; \quad H_b^d = \begin{pmatrix} \mathtt{H}^d \\ \mathbf{H}^d \end{pmatrix}; \quad \mathbf{H}^u = \begin{pmatrix} h^{+u} \\ h^{0u} \end{pmatrix}; \quad \mathbf{H}^d = \begin{pmatrix} h^{0d} \\ h^{-d} \end{pmatrix}. \end{aligned} \tag{1}$$

Here, $\mathtt{H}^u$ and $\mathtt{H}^d$ are the coloured triplet scalars; $\mathbf{H}^d$ and $\mathbf{H}^u$ are identified as the weak doublets of the MSSM group. Since we are only interested in studying the masses and mixings, we do not dwell on the details of the full gauge symmetry breaking from SUSY $SU(5) \otimes U(1) \otimes Q_6$ to the MSSM group. We assume that the coloured Higgs triplets are very heavy, in order to avoid fast proton decay. Such subtle issues are part of a wider study of this particular model, that is still in progress. For the scalars fields, we use the following vacuum expectation values (vev's)

$$\begin{aligned} \langle H^u \rangle &= \begin{pmatrix} \mathbf{0} \\ \langle \mathbf{H}^u \rangle \end{pmatrix}; \quad \left\langle H^d \right\rangle = \begin{pmatrix} \mathbf{0} \\ \langle \mathbf{H}^d \rangle \end{pmatrix}; \quad \langle H_{45} \rangle_\alpha^{\alpha 5} = v_{45}; \quad \langle H_{45} \rangle_4^{45} = -3v_{45}, \\ \langle H_{\overline{45}} \rangle_{\alpha 5}^\alpha &= v_{\overline{45}}; \quad \langle H_{\overline{45}} \rangle_{45}^4 = -3v_{\overline{45}}; \quad \langle \phi \rangle = v_s; \quad \left\langle \bar{\phi} \right\rangle = \bar{v}_s; \quad \alpha, \beta = 1, 2, 3. \end{aligned} \tag{2}$$

The superpotential, invariant under the Q_6 discrete group, is

$$\begin{aligned} W &= \sqrt{2}y_1^d(F_1T_2 - F_2T_1)H_3^d + \sqrt{2}y_2^d(F_1T_3H_2^d - F_2T_3H_1^d) + \sqrt{2}y_3^dF_3(T_1H_2^d - T_2H_1^d) \\ &+\sqrt{2}y_4^dF_3T_3H_3^d + \frac{y_1^u}{4}(T_1T_2 - T_2T_1)H_3^u + \frac{y_2^u}{4}(T_1T_3H_2^u - T_2T_3H_1^u) + \frac{y_3^u}{4}T_3(T_1H_2^u - T_2H_1^u) \\ &+\frac{y_4^u}{4}T_3T_3H_3^u + \sqrt{2}Y_1(F_1T_2 - F_2T_1)H_{\overline{45}} + \sqrt{2}Y_2F_3T_3H_{\overline{45}} + \sqrt{2}\tilde{Y}_1(T_1T_2 - T_2T_1)H_{45} \\ &+\sqrt{2}\tilde{Y}_2T_3T_3H_{45} + y_1^n(N_1^cF_2 - N_2^cF_1)H_3^u + y_2^n(N_1^cF_3H_2^u - N_2^cF_3H_1^u) \\ &+y_3^nN_3^c(F_1H_2^u + F_2H_1^u) + y_1^m(N_1^c\phi N_1^c + N_2^c\phi N_2^c) + y_2^mN_3^c\phi N_3^c. \end{aligned} \tag{3}$$

There is a missing term which is invariant under SUSY $SU(5) \otimes U(1) \otimes Q_6$, namely $y_{ij} N_i^c \bar{\phi} N_j^c$, however, it is not allowed by the $U(1)$ symmetry. From the above superpotential, using the scalar vev's given in Eq.(2), one must obtain the mass terms; working in the $SU(5)$ basis the Lagrangian for the mass terms is

$$\begin{aligned} \mathcal{L}^q &= -\bar{d}_{iR} (\mathbf{M}_d)_{ij} d_{jL} - \bar{u}_{iR} (\mathbf{M}_u)_{ij} u_{jL} + h.c. , \\ \mathcal{L}^l &= -\bar{\ell}_{iR} (\mathbf{M}_\ell)_{ij} \ell_{jL} - \bar{N}_{iR} (\mathbf{M}_D)_{ij} \nu_{jL} - \tfrac{1}{2} \bar{N}_{iR} (\mathbf{M}_R)_{ij} N_{jR}^c + h.c. , \end{aligned} \tag{4}$$

where the fermion mass matrices have the following structures:

$$\begin{aligned} \mathbf{M}_d &= \begin{pmatrix} 0 & y_1^d h_3^{0d} + 2Y_1 v_{\bar{45}} & y_2^d h_2^{0d} \\ -y_1^d h_3^{0d} - 2Y_1 v_{\bar{45}} & 0 & -y_2^d h_1^{0d} \\ y_3^d h_2^{0d} & -y_3^d h_1^{0d} & y_4^d h_3^{0d} + 2Y_2 v_{\bar{45}} \end{pmatrix}; \; \mathbf{M}_u = \begin{pmatrix} 0 & -2\tilde{Y}_1 v_{45} & \bar{y}^u h_2^{0u} \\ 2\tilde{Y}_1 v_{45} & 0 & -\bar{y}^u h_1^{0u} \\ \bar{y}^u h_2^{0u} & -\bar{y}^u h_1^{0u} & y_4^u h_3^{0u} \end{pmatrix} \\ \mathbf{M}_\ell &= \begin{pmatrix} 0 & -\left(y_1^d h_3^{0d} - 6Y_1 v_{\bar{45}}\right) & y_3^d h_2^{0d} \\ y_1^d h_3^{0d} - 6Y_1 v_{\bar{45}} & 0 & -y_3^d h_1^{0d} \\ y_2^d h_2^{0d} & -y_2^d h_1^{0d} & y_4^d h_3^{0d} - 6Y_2 v_{\bar{45}} \end{pmatrix}; \; \mathbf{M}_D = \begin{pmatrix} 0 & y_1^n h_3^{0u} & y_2^n h_2^{0u} \\ -y_1^n h_3^{0u} & 0 & -y_2^n h_1^{0u} \\ y_3^n h_2^{0u} & y_3^n h_1^{0u} & 0 \end{pmatrix} \end{aligned} \tag{5}$$

and $\bar{y}^u \equiv (y_2^u + y_3^u)/2$. At the same time, we get $\mathbf{M}_R = diag(M_{R_1}, M_{R_1}, M_{R_3})$. Therefore, after the type I see-saw mechanism one obtains the $\mathbf{M}_\nu = \mathbf{M}_D^T \mathbf{M}_R^{-1} \mathbf{M}_D$ effective neutrino mass matrix.

3. Masses and mixings

3.1. Quark and lepton masses

If we suppose that $h_2^{0u} = h_1^{0u} \equiv h^{0u}$ and $h_2^{0d} = h_1^{0d} \equiv h^{0d}$ in Eq. (5), then one realizes that the $\mathbf{M}_{(d,\ell)}$ and $\mathbf{M}_u$ mass matrices contain implicitly the NNI and Fritzsch textures respectively, which appear explicitly as follows: the above mass matrices are diagonalized by $\mathbf{U}_{f(R,L)}$ unitary matrices according to Eq.(4), thus one obtains, $\tilde{\mathbf{M}}_f = \mathbf{U}_{fR}^\dagger \hat{\mathbf{M}}_f \mathbf{U}_{fL}$, in general. Here, $\tilde{\mathbf{M}}_f = diag(\tilde{m}_{f_1}, \tilde{m}_{f_2}, 1)$ and $\hat{\mathbf{M}}_f = \mathbf{M}_f / m_{f_3}$ are dimensionless mass matrices.

Then, taking $\mathbf{U}_{f(R,L)} = \mathbf{U}_{\pi/4} \mathbf{u}_{f(R,L)}$, one can easily get $\tilde{\mathbf{M}}_f = \mathbf{u}_{fR}^\dagger \mathbf{m}_f \mathbf{u}_{fL}$ where

$$\mathbf{m}_f = \mathbf{U}_{\pi/4}^T \hat{\mathbf{M}}_f \mathbf{U}_{\pi/4} = \begin{pmatrix} 0 & \pm\tilde{A}_f & 0 \\ \mp\tilde{A}_f & 0 & -\sqrt{2}\tilde{B}_f \\ 0 & -\sqrt{2}\tilde{C}_f & \tilde{D}_f \end{pmatrix}, \text{ and } \mathbf{U}_{\pi/4} = \begin{pmatrix} \frac{1}{\sqrt{2}} & -\frac{1}{\sqrt{2}} & 0 \\ \frac{1}{\sqrt{2}} & \frac{1}{\sqrt{2}} & 0 \\ 0 & 0 & 1 \end{pmatrix} \tag{6}$$

Here, $\pm$ the upper (lower) sign corresponds to the $\mathbf{m}_d$ ($\mathbf{m}_\ell$ and $\mathbf{m}_u$) mass matrices. In addition, the $\mathbf{m}_f$ dimensionless coefficients can be read directly from Eq. (5), and we have to keep in mind that $\tilde{B}_u = \tilde{C}_u$ for the up quarks. The degeneracy in the vev's for two scalar fields modifies the effective neutrino mass matrix which is given explicitly as

$$\mathbf{M}_\nu = \begin{pmatrix} A_\nu^2 + B_\nu^2 & B_\nu^2 & A_\nu C_\nu \\ B_\nu^2 & A_\nu^2 + B_\nu^2 & A_\nu C_\nu \\ A_\nu C_\nu & A_\nu C_\nu & 2C_\nu^2 \end{pmatrix} \quad \text{with} \quad A_\nu = \sqrt{x} y_1^n h_3^{0u}, B_\nu = \sqrt{y} y_3^n h^u \text{ and } C_\nu = \sqrt{x} y_2^n h^u \tag{7}$$

3.2. Quark and lepton mixings

We will describe briefly how to diagonalize the mass matrices, $\mathbf{m}_{(d,\ell)}$ and $\mathbf{m}_u$, respectively. Let us focus on the former one which has the NNI texture [6, 7]. We will follow the prescription applied to $\mathbf{m}_u$ in the case of Fritzsch mass textures [11, 12]. For a pedagogical review on how to diagonalize this type of matrices see ref. [13].

Going back to the expression $\tilde{\mathbf{M}}_f = \mathbf{u}_{fR}^\dagger \mathbf{m}_f \mathbf{u}_{fL}$, we are interested in obtaining the $\mathbf{u}_{fL}$ left-handed matrices necessary to construct the CKM matrix, for this we must build the bilineal form: $\tilde{\mathbf{M}}_f^\dagger \tilde{\mathbf{M}}_f = \mathbf{u}_{fL}^\dagger \mathbf{m}_f^\dagger \mathbf{m}_f \mathbf{u}_{fL}$. From this relation, we can factorize the CP phases that come from $\mathbf{m}_f^\dagger \mathbf{m}_f = \mathbf{Q}_f (\mathbf{m}_f^\dagger \mathbf{m}_f) \mathbf{Q}_f^\dagger$ see [13], such that, $\mathbf{Q}_f = \mathrm{diag}\,(1, \exp(-i\eta_{f_2}), \exp(-i\eta_{f_3}))$. Thus, we appropriately choose $\mathbf{u}_{fL} = \mathbf{Q}_f \mathbf{O}_{fL}$ where $\mathbf{O}_{fL}$ is the real orthogonal matrix that diagonalizes the $(\mathbf{m}_f^\dagger \mathbf{m}_f)$ matrix which we have not written here. Notice that $\mathbf{O}_{fL} = (\mid f_1\rangle, \mid f_2\rangle, \mid f_3\rangle)$, where the three eigenvectors have the following form

$$\mid f_i\rangle = N_{f_i} \begin{pmatrix} (\tilde{m}_{f_i}^2 - |\tilde{A}_f|^2 - |\tilde{C}_f|^2)|\tilde{A}_f||\tilde{B}_f| \\ (\tilde{m}_{f_i}^2 - |\tilde{A}_f|^2)|\tilde{C}_f||\tilde{D}_f| \\ (\tilde{m}_{f_i}^2 - |\tilde{A}_f|^2)(\tilde{m}_{f_i}^2 - |\tilde{A}_f|^2 - |\tilde{C}_f|^2) \end{pmatrix}, \mid f_3\rangle = N_{f_3} \begin{pmatrix} (1 - |\tilde{A}_f|^2 - |\tilde{C}_f|^2)|\tilde{A}_f||\tilde{B}_f| \\ (1 - |\tilde{A}_f|^2)|\tilde{C}_f||\tilde{D}_f| \\ (1 - |\tilde{A}_f|^2)(1 - |\tilde{A}_f|^2 - |\tilde{C}_f|^2) \end{pmatrix}. \quad (8)$$

Here, N_{f_i} $(i = 1, 2)$ and N_{f_3} stand for the normalization factors whose definition must be read directly from the above expression. Three dimensionless free parameters can be fixed in terms of the physical masses and $|\tilde{D}_f| \equiv y_f^2$, which is the only dimensionless free parameter [6, 7]. Explicitly, these are given by

$$|\tilde{A}_f| = \frac{q_f}{y_f};\ |\tilde{B}_f| = \sqrt{\frac{1 + P_f - y_f^4 - R_f}{2} - \left(\frac{q_f}{y_f}\right)^2};\ |\tilde{C}_f| = \sqrt{\frac{1 + P_f - y_f^4 + R_f}{2} - \left(\frac{q_f}{y_f}\right)^2}, \quad (9)$$

where

$$P_f = \tilde{m}_{f_1}^2 + \tilde{m}_{f_2}^2,\ q_f = \sqrt[4]{\tilde{m}_{f_1}^2 \tilde{m}_{f_2}^2},\ R_f = \sqrt{\left(1 + P_f - y_f^4\right)^2 - 4\left(P_f + q_f^4\right) + 8 q_f^2 y_f^2}. \quad (10)$$

Hence, there are two free parameters, y_f $(f = d, \ell)$, in the leptonic sector. So far, we have found the $\mathbf{U}_{(d,\ell)L}$ left-handed matrices.

Let us now focus on $\mathbf{m}_u$, where we must remember that $|\tilde{B}_u| = |\tilde{C}_u|$. In this case, one can fix the three free parameters in terms of the physical masses. Explicitly, we obtain

$$|\tilde{A}_u| = \sqrt{\frac{\tilde{m}_u \tilde{m}_c}{1 - \tilde{m}_c + \tilde{m}_u}},\ |\tilde{B}_u| = \sqrt{\frac{(1 - \tilde{m}_c)(1 + \tilde{m}_u)(\tilde{m}_c - \tilde{m}_u)}{1 - \tilde{m}_c + \tilde{m}_u}},\ |\tilde{D}_u| = 1 - \tilde{m}_c + \tilde{m}_u. \quad (11)$$

Following the analysis previously described, the $\mathbf{O}_{uL}$ orthogonal matrix that diagonalizes $(\mathbf{m}_u^\dagger \mathbf{m}_u)$ is fixed in terms of the above mentioned parameters. Thus, using the expression given in Eq. (8), we get

$$\mathbf{O}_{uL} = \begin{pmatrix} -\sqrt{\frac{\tilde{m}_c(1-\tilde{m}_c)}{(1-\tilde{m}_u)(\tilde{m}_c+\tilde{m}_u)(1-\tilde{m}_c+\tilde{m}_u)}} & -\sqrt{\frac{\tilde{m}_u(1+\tilde{m}_u)}{(1+\tilde{m}_c)(\tilde{m}_c+\tilde{m}_u)(1-\tilde{m}_c+\tilde{m}_u)}} & \sqrt{\frac{\tilde{m}_u\tilde{m}_c(\tilde{m}_c-\tilde{m}_u)}{(1-\tilde{m}_u)(1+\tilde{m}_c)(1-\tilde{m}_c+\tilde{m}_u)}} \\ -\sqrt{\frac{\tilde{m}_u(1-\tilde{m}_c)}{(1-\tilde{m}_u)(\tilde{m}_c+\tilde{m}_u)}} & \sqrt{\frac{\tilde{m}_c(1+\tilde{m}_u)}{(1+\tilde{m}_c)(\tilde{m}_c+\tilde{m}_u)}} & \sqrt{\frac{(\tilde{m}_c-\tilde{m}_u)}{(1-\tilde{m}_u)(1+\tilde{m}_c)}} \\ \sqrt{\frac{\tilde{m}_u(1+\tilde{m}_u)(\tilde{m}_c-\tilde{m}_u)}{(1-\tilde{m}_u)(\tilde{m}_c+\tilde{m}_u)(1-\tilde{m}_c+\tilde{m}_u)}} & -\sqrt{\frac{\tilde{m}_c(1-\tilde{m}_c)(\tilde{m}_c-\tilde{m}_u)}{(1+\tilde{m}_c)(\tilde{m}_c+\tilde{m}_u)(1-\tilde{m}_c+\tilde{m}_u)}} & \sqrt{\frac{(1+\tilde{m}_u)(1-\tilde{m}_c)}{(1-\tilde{m}_u)(1+\tilde{m}_c)(1-\tilde{m}_c+\tilde{m}_u)}} \end{pmatrix} \quad (12)$$

Therefore, the full left-handed unitary matrices are given, in general, by $\mathbf{U}_{fL} = \mathbf{U}_{\pi/4}\mathbf{u}_{fL}$, where $\mathbf{u}_{fL} = \mathbf{Q}_f \mathbf{O}_{fL}$. Then, the CKM matrix may be completely determined by $\mathbf{V}_{CKM} = \mathbf{U}_{uL}^\dagger \mathbf{U}_{dL} = \mathbf{O}_{uL}^T \mathbf{Q}_q \mathbf{O}_{dL}$, where we have defined $\mathbf{Q}_q = \mathbf{Q}_u^\dagger \mathbf{Q}_d$. In this way, the CKM matrix can be obtained either analytically or numerically. However, in here we are just interested in getting a numerical expression for it.

In the leptonic sector, we have that $\mathbf{U}_{\ell L} = \mathbf{U}_{\pi/4}\mathbf{Q}_\ell \mathbf{O}_{\ell L}$, but we need to obtain the neutrino mixing contribution in order to have a complete $PMNS$ matrix which is defined as $\mathbf{V}_{PMNS} = \mathbf{U}_{\ell L}^\dagger \mathbf{U}_{\nu L}\mathbf{K}$, we shall neglect the $\mathbf{K}$ Majorana phases. The $\mathbf{M}_\nu$ mass matrix given in Eq. (7) is diagonalized by $\mathbf{U}_\nu = u_{\pi/4}\mathbf{P}_\nu^\dagger \mathbf{O}_\nu$, that is, $\tilde{\mathbf{M}}_\nu = \mathrm{diag}\,(m_{\nu_1}, m_{\nu_2}, m_{\nu_3}) = \mathbf{O}_\nu^T \hat{\mathbf{M}}_\nu \mathbf{O}_\nu$ where $\mathbf{P}_\nu = diag(\exp i\eta_{\nu_1}, \exp i\eta_{\nu_2}, \exp i\eta_{\nu_1})$, and

$$u_{\pi/4} = \begin{pmatrix} \frac{1}{\sqrt{2}} & 0 & -\frac{1}{\sqrt{2}} \\ \frac{1}{\sqrt{2}} & 0 & \frac{1}{\sqrt{2}} \\ 0 & 1 & 0 \end{pmatrix};\quad \hat{\mathbf{M}}_\nu = \begin{pmatrix} |A_\nu|^2 + 2|B_\nu|^2 & \sqrt{2}|A_\nu||C_\nu| & 0 \\ \sqrt{2}|A_\nu||C_\nu| & 2|C_\nu|^2 & 0 \\ 0 & 0 & |A_\nu|^2 \end{pmatrix}. \quad (13)$$

A necessary condition to factorize the phases in a similar way to the quark sector, is that the A_ν^2 and B_ν^2 phases must be aligned [13]. On the other hand, $|A_\nu|$, $|B_\nu|^2$ and $|C_\nu|^2$ are determined by the physical masses

$$|A_\nu|^2 = m_{\nu 3};\ |B_\nu|^2_{\mp} = \frac{1}{4}\left(m_{\nu_2} + m_{\nu_1} - m_{\nu_3} \mp R_\nu\right);\ |C_\nu|^2_{\pm} = \frac{1}{4}\left(m_{\nu_2} + m_{\nu_1} - m_{\nu_3} \pm R_\nu\right), \quad (14)$$

where $R_\nu = \sqrt{(m_{\nu_2} + m_{\nu_1} - m_{\nu_3})^2 - 4m_{\nu_2}m_{\nu_1}}$. A straightforward analysis on the inverted and normal hierarchies among the neutrino masses rules out the latter one. For the inverted case ($m_{\nu_2} > m_{\nu_1} > m_{\nu_3}$), $|B_\nu|^2_-$ and $|C_\nu|^2_+$ turn out to be real and positive, if and only if, m_{ν_3} is tiny. Actually, from R_ν we obtain the following sum rule, $m_{\nu_3} \leq (\sqrt{m_{\nu_2}} - \sqrt{m_{\nu_1}})^2$, where the equality in the above expression means an upper bound for the m_{ν_3} mass. Having fixed $|B_\nu|^2_-$ and $|C_\nu|^2_+$ in terms of the physical neutrino masses, the $\mathbf{O}_\nu$ matrix is well determined by them. Explicitly, we obtain

$$\mathbf{O}_\nu = \begin{pmatrix} \sqrt{\dfrac{m_{\nu_3}\left(m_{\nu_2} + m_{\nu_1} - m_{\nu_3} + R_\nu\right)}{\left(m_{\nu_2} - m_{\nu_1}\right)\left(m_{\nu_2} - m_{\nu_1} + m_{\nu_3} - R_\nu\right)}} & \sqrt{\dfrac{m_{\nu_3}\left(m_{\nu_2} + m_{\nu_1} - m_{\nu_3} + R_\nu\right)}{\left(m_{\nu_2} - m_{\nu_1}\right)\left(m_{\nu_2} - m_{\nu_1} - m_{\nu_3} + R_\nu\right)}} & 0 \\ -\sqrt{\dfrac{m_{\nu_2} - m_{\nu_1} + m_{\nu_3} - R_\nu}{2\left(m_{\nu_2} - m_{\nu_1}\right)}} & \sqrt{\dfrac{m_{\nu_2} - m_{\nu_1} - m_{3\nu} + R_\nu}{2\left(m_{\nu_2} - m_{\nu_1}\right)}} & 0 \\ 0 & 0 & 1 \end{pmatrix}. \quad (15)$$

Therefore, $\mathbf{V}_{PMNS} = \mathbf{U}_\ell^\dagger \mathbf{U}_\nu = \mathbf{O}_{\ell L}^T \mathbf{Q}_\ell^\dagger \mathbf{S}_{23} \mathbf{P}_\nu^\dagger \mathbf{O}_{\nu L}$, where $\mathbf{S}_{23} = \mathbf{U}_{\pi/4}^T u_{\pi/4}$. Comparing this matrix with the standard parametrization given in [14], we find that the reactor, atmospheric and solar mixing angles are completely determined as follows

$$|\sin\theta_{13}| = |O_{21\ell}|;\ |\sin\theta_{23}| = \frac{|O_{22\ell}|}{\sqrt{1 - |O_{21\ell}|^2}};\ |\tan\theta_{12}|^2 = \frac{|O_{11\ell}O_{12\nu} + O_{31\ell}O_{22\nu}\exp\left(i\bar{\eta}_{3e}\right)|^2}{|O_{11\ell}O_{11\nu} + O_{31\ell}O_{21\nu}\exp\left(i\bar{\eta}_{3e}\right)|^2}. \quad (16)$$

There is a remarkable coincidence between the above formulas and those presented in refs. [15,16].

4. Results

4.1. CKM mixing matrix

As we showed before, the CKM matrix is defined as $\mathbf{V}_{CKM} = \mathbf{U}_{uL}^\dagger \mathbf{U}_{dL} = \mathbf{O}_{uL}^T \mathbf{Q}_q \mathbf{O}_{dL}$, where the orthogonal matrices depend on the physical quark masses, to be more explicit, they depend on their ratios, that is on m_u/m_t, m_c/m_t and m_d/m_b, m_s/m_b for the up and down quarks, respectively. Then, knowing those ratios, we just need to tune the free parameters y_d and the two CP-violating phases to get a reliable CKM matrix. This numerical analysis is still in progress.

4.2. PMNS mixing matrix

In order to fix the y_e free parameter, we performed a χ^2 analysis using the theoretical expressions for the reactor and atmospheric angles given in Eq. (16) and the following experimental values: $\sin^2 2\theta_{13}^\ell = 0.076 \pm 0.068$ [17] and $\sin^2\theta_{23}^\ell = 0.52 \pm 0.06$ [18]. Also, we consider the following values for the charged lepton masses values: $m_e = 0.51099$ MeV, $m_\mu = 105.6583$ MeV and $m_\tau = 1776.82$ MeV [14]. As a result, we obtain that $y_e = 0.8478^{+0,0045}_{-0.0046}$ is the best fit at 1σ, moreover, the best theoretical values (at 1σ) for the reactor and atmospheric angles that come out from this analysis are

$$\theta_{23}^{\ell th} = 46.18^{+0.66}_{-0.65} \quad \text{and} \quad \theta_{13}^{\ell th} = 3.38^{+0.03}_{-0.02} \quad \text{where} \quad \chi^2_{bf} = 0.85 \quad (17)$$

The solar angle, which now depends on the neutrino masses and one Dirac phase, can be determined using the neutrino mass sum rule. This is written in terms of the observables $\Delta m^2_{\odot}$ and Δm^2_{ATM} as

$$m_{\nu_3} \leq \left(\sqrt[4]{m^2_{\nu_3} + \Delta m^2_{\odot} + \Delta m^2_{ATM}} - \sqrt[4]{m^2_{\nu_3} + \Delta m^2_{ATM}} \right)^2 . \tag{18}$$

Using the experimental results on $\Delta m^2_{\odot}$ and Δm^2_{ATM} [18], we obtain, $0 \leq m_{\nu_3} \leq 4 \times 10^{-6}$ eV. As a result, the m_{ν_2} and m_{ν_1} neutrino masses are easily calculated taking $m_{\nu_3} = 3.9 \times 10^{-6}$ eV. Therefore, $m_{\nu_2} = 0.05080$ eV and $m_{\nu_1} = 0.04987$ eV.

Having done that, the neutrino masses are not any more free parameters in the solar mixing angle expression given in Eq. (16). As consequence, it is easy to calculate the solar mixing angle with a particular value for the $\bar{\eta}_{3e} = \pi$ Dirac phase and using the y_e value at 90% at C.L, we obtain

$$\theta^{\ell th}_{12} = 36.62 \pm 4.06. \tag{19}$$

Although, the atmospheric and solar angle are in good agreement with the experimental data, the reactor angle is not consistent with the global fits, although it is not completely negligible and it is large in comparison to the tribimaximal scenario, $\theta^{\ell th}_{13} = 3.38^{+0.03}_{-0.02}$. This value may be enhanced to the current experimental one by considering that the first and second right-handed neutrinos are not degenerate in mass, that is, $M_{R_1} \neq M_{R_2}$ in the right handed neutrino mass matrix. These results will be presented very soon.

Acknowledgments

This work was partially supported by a PAPIIT grant IN113412. JCG participation was supported by ICyTDF and also JCGI thanks Red de Altas Energías-CONACYT for the financial support.

References

[1] Babu K S and Kubo J 2005 *Phys. Rev.* **D71** 056006 (*Preprint* hep-ph/0411226)
[2] Kajiyama Y, Itou E and Kubo J 2006 *Nucl. Phys.* **B743** 74–103 (*Preprint* hep-ph/0511268)
[3] Kajiyama Y 2007 *JHEP* **04** 007 (*Preprint* hep-ph/0702056)
[4] Babu K and Meng Y 2009 *Phys.Rev.* **D80** 075003 (*Preprint* 0907.4231)
[5] Babu K, Kawashima K and Kubo J 2011 *Phys.Rev.* **D83** 095008 (*Preprint* 1103.1664)
[6] Harayama K and Okamura N 1996 *Phys.Lett.* **B387** 614–622 (*Preprint* hep-ph/9605215)
[7] Harayama K, Okamura N, Sanda A and Xing Z Z 1997 *Prog.Theor.Phys.* **97** 781–790 (*Preprint* hep-ph/9607461)
[8] Berezhiani Z, Tavartkiladze Z and Vysotsky M 1998 (*Preprint* hep-ph/9809301)
[9] Bajc B, Fileviez Perez P and Senjanovic G 2002 131–139 (*Preprint* hep-ph/0210374)
[10] Bajc B, Fileviez Perez P and Senjanovic G 2002 *Phys.Rev.* **D66** 075005 (*Preprint* hep-ph/0204311)
[11] Fritzsch H 1978 *Phys.Lett.* **B73** 317–322
[12] Fritzsch H, Xing Z z and Zhou Y L 2011 *Phys. Lett.* **B697** 357–363 (*Preprint* 1101.4272)
[13] Gonzalez Canales F and Mondragon A 2011 *J. Phys. Conf. Ser.* **287** 012015 (*Preprint* 1101.3807)
[14] Nakamura K *et al.* (Particle Data Group) 2010 *J. Phys.* **G37** 075021
[15] Felix O, Mondragon A, Mondragon M and Peinado E 2007 *AIP Conf. Proc.* **917** 383–389 (*Preprint* hep-ph/0610061)
[16] Mondragon A, Mondragon M and Peinado E 2007 *Phys. Rev.* **D76** 076003 (*Preprint* 0706.0354)
[17] Adamson P *et al.* (MINOS Collaboration) 2011 *Phys.Rev.Lett.* **107** 181802 5 pages, 3 figures (*Preprint* 1108.0015)
[18] Schwetz T, Tortola M and Valle J 2011 *New J.Phys.* **13** 063004 * Temporary entry * (*Preprint* 1103.0734)

Cabibbo-Kobayashi-Maskawa matrix: rephasing invariants and parameterizations

H Pérez R,[1] P Kielanowski[2] and S R Juárez W[3]

[1,2] Departamento de Física, Centro de Investigación y de Estudios Avanzados del IPN, Apartado Postal 14-740, 07000 México D.F., Mexico

[3] Departamento de Física, Escuela Superior de Física y Matemáticas, Instituto Politécnico Nacional, U.P. "Adolfo López Mateos". C.P. 07738, México D.F.

E-mail: [1]`hperezr@fis.cinvestav.mx`, [2]`kiel@fis.cinvestav.mx`, [3]`rebeca@esfm.ipn.mx`

Abstract. In this work we study two topics: the first one considers the general phase invariant monomials built out of the CKM matrix elements and their conjugates. We show, that there exist 30 fundamental phase invariant monomials and 18 of them are a product of 4 CKM matrix elements and 12 are a product of 6 CKM matrix elements. In our main result we show that all rephasing invariant monomials can be expressed as a product of at most 5 factors with positive powers. Next, we propose a general method of a recursive construction of the CKM matrix for any number of generations. This allows to construct a parameterization with desired properties. As an application we generalize the Wolfenstein parameterization to the case of 4 generations and obtain restrictions on the CKM suppression of the fourth generation.

1. Introduction

It is well known that the quark fields are defined up to a phase, therefore it is possible to introduce the states $u_\alpha \to e^{i\varphi_\alpha}u_\alpha$ and $d_i \to e^{i\phi_i}d_i$ so that the elements of the CKM matrix become

$$V_{\alpha i} \to e^{i(\phi_i-\varphi_\alpha)}V_{\alpha i}.$$

With this redefinition one can easily check that the two simplest rephasing invariant (RI) functions of the CKM matrix are

$$W_{\alpha i} \equiv |V_{\alpha i}|^2, \tag{1}$$

$$Q_{\alpha i\beta j} \equiv V_{\alpha i}V_{\beta j}V^*_{\alpha j}V^*_{\beta i}, \quad \alpha \neq \beta \quad i \neq j. \tag{2}$$

In order to avoid $Q_{\alpha i\beta j} = W_{\alpha i}W_{\beta j}$ the conditions $\alpha \neq \beta$ and $i \neq j$ must be satisfied. The CKM RIs have been studied in detail by many authors [1, 2, 3].

Higher order invariants can be generated, for example,

$$V_{11}V_{22}V_{33}V^*_{13}V^*_{21}V^*_{32}, \tag{3}$$

$$V_{11}V^2_{13}V^2_{21}V_{22}V^2_{32}V_{33}\left(V^3_{12}V^3_{23}V^3_{31}\right)^*, \tag{4}$$

In [4] is demonstrated the possibility of expressing all higher rephasing invariants in terms of the RIs in Eqs. (1) and (2), but not all the powers of these RIs are positive, for example

$$V_{11}V_{22}V_{33}V^*_{13}V^*_{21}V^*_{32} = W^{-1}_{12}V_{12}V^*_{12}V_{11}V_{22}V_{33}V^*_{13}V^*_{21}V^*_{32} = W^{-1}_{12}Q_{1122}Q_{3312}. \tag{5}$$

Content from this work may be used under the terms of the Creative Commons Attribution 3.0 licence. Any further distribution of this work must maintain attribution to the author(s) and the title of the work, journal citation and DOI.
Published under licence by IOP Publishing Ltd

We will proof in section one that all higher order rephasing invariants of the CKM matrix can be expressed as product of fundamental rephasing invariants taken to positive powers and the product of the squares of the absolute values of the CKM matrix elements also taken to positive powers. This result is independent of the unitary of the CKM matrix. The discussion of the RIs depends on the number of generations and we will discuss these invariants for 3 generations.

In section two we outline how to construct and parameterize the $(n \times n)$ CKM matrix $V^{(n)}$, if we know the CKM matrix $V^{(n-1)}$. Our algorithm allows to set up a parameterization with desired properties. As an application we generalize the Wolfenstein parameterization to the case of 4 generations and obtain restrictions on the CKM suppression of the fourth generation.

2. A new form to construct rephasing invariants

Let us denote by $\mathrm{P}(m,n)$ the most general monomial constructed from the CKM matrix elements and its conjugates

$$\mathrm{P}(m,n) = A \prod_{\alpha i} (V_{\alpha i})^{m_{\alpha i}} \prod_{\beta j} (V^*_{\beta j})^{n_{\beta j}}, \tag{6}$$

here m and n are 3×3 matrices with integer non negative matrix elements $[m]_{\alpha i} = m_{\alpha i}$ and $[n]_{\beta j} = n_{\beta j}$. We can assume, without lost of generality $A = 1$. The condition that the elements of the matrices m and n are integers may be relaxed, but the CKM observables are monomials that contain only integer powers.

The monomials $\mathrm{P}(m,n)$ fulfills the following properties

a. $\mathrm{P}(m_1,n_1) \cdot \mathrm{P}(m_2,n_2) = \mathrm{P}(m_1+m_2, n_1+n_2)$.

b. $\mathrm{P}(m,n)^* = \mathrm{P}(n,m)$.

c. $\mathrm{P}(m,m) = \prod_{\alpha i} \left(|V_{\alpha i}|^2\right)^{m_{\alpha i}}$.

The monomial $\mathrm{P}(m,n)$ in general is not rephasing invariant. Suppose that we make the following phase transformation of the CKM matrix

$$V_{\mathrm{CKM}} \to \mathrm{diag}(e^{i\phi_1}, 0, 0) V_{\mathrm{CKM}}, \tag{7}$$

then the monomial $\mathrm{P}(m,n)$ is transformed in the following way

$$\mathrm{P}(m,n) \to e^{i\phi_1(m_{11}+m_{12}+m_{13}-n_{11}-n_{12}-n_{13})}\,\mathrm{P}(m,n), \tag{8}$$

so we see that $\mathrm{P}(m,n)$ is invariant under the transformation in Eq. (7) only if the sum of the elements of the first row of the matrices m and n are equal. From this one obtains theorem 1.

Theorem 1 *The monomial* $\mathrm{P}(m,n)$ *is rephasing invariant if the sums of the elements of the corresponding rows and columns of the matrices* m *and* n *are equal. It means that for the rephasing invariant monomial* $\mathrm{P}(m,n)$ *the matrices* m *and* n *fulfill the following conditions*

$$\sum_{i=1}^{3} m_{\alpha i} = \sum_{i=1}^{3} n_{\alpha i}, \quad \sum_{i=1}^{3} m_{i\alpha} = \sum_{i=1}^{3} n_{i\alpha}, \quad \alpha = 1,2,3. \tag{9}$$

Definition 1 *The rephasing invariant monomial of the CKM matrix which cannot be factored out into the product of the absolute values of the elements of the CKM matrix and other invariant is called the* pure rephasing invariant *(PRI).*

Example 1.- The rephasing invariant $V_{11}V_{22}V_{33}V^*_{12}V^*_{21}V^*_{33} = |V_{33}|^2 V_{11}V_{22}V^*_{12}V^*_{21}$ is not a PRI because it has the factor $|V_{33}|^2$. The remaining part $V_{11}V_{22}V^*_{12}V^*_{21}$ is a PRI.

Table 1. Fundamental rephasing invariants.

4-th order FRIs ($J_1, J_2, \ldots, I_{18}$)	6-th order FRIs ($I_1, I_2, \ldots, I_{12}$)
$J_1 = V_{11}V_{22}V_{12}^*V_{21}^*$	$I_1 = V_{11}V_{22}V_{33}V_{13}^*V_{21}^*V_{32}^*$
$J_5 = V_{11}V_{33}V_{13}^*V_{31}^*$	$I_4 = V_{11}V_{23}V_{32}V_{13}^*V_{22}^*V_{31}^*$
$J_2 = V_{11}V_{23}V_{13}^*V_{21}^*$	$I_2 = V_{11}V_{22}V_{33}V_{12}^*V_{23}^*V_{31}^*$
$J_6 = V_{12}V_{33}V_{13}^*V_{32}^*$	$I_5 = V_{12}V_{23}V_{31}V_{13}^*V_{21}^*V_{32}^*$
$J_3 = V_{12}V_{23}V_{13}^*V_{22}^*$	$I_3 = V_{11}V_{23}V_{32}V_{12}^*V_{21}^*V_{33}^*$
$J_7 = V_{21}V_{32}V_{22}^*V_{31}^*$	$I_6 = V_{12}V_{21}V_{33}V_{13}^*V_{22}^*V_{31}^*$
$J_4 = V_{11}V_{32}V_{12}^*V_{31}^*$	$I_{6+\ell} = (I_\ell)^*, \quad \ell = 1, \ldots, 6$
$J_8 = V_{21}V_{33}V_{23}^*V_{31}^*$	
$J_9 = V_{22}V_{33}V_{23}^*V_{32}^*$	
$J_{9+k} = (J_k)^*, \quad k = 1, \ldots, 9$	

The PRIs can be represented by two matrices m and n but it can also be represented by

$$\mathrm{B}(p) = \prod_{p_{\alpha i} \geq 0} (V_{\alpha i})^{p_{\alpha i}} \prod_{p_{\beta j} \leq 0} (V_{\beta j}^*)^{-p_{\beta j}}, \tag{10}$$

where p is a 3×3 matrix with the following properties:

1. The matrix elements of p are integers (positive, negative or 0).
2. The sum of the elements of p in each row and column is equal to 0.
3. A permutation of the rows and columns of the p matrix is reversible and the resulting matrix is also the p matrix of pure rephasing invariant.

Example 2.- The matrix p for the PRI, defined in Eq. (4) is

$$p = \begin{pmatrix} 1 & -3 & 2 \\ 2 & 1 & -3 \\ -3 & 2 & 1 \end{pmatrix}.$$

The matrix p in the example 2 can be decomposed into the sum of matrices, all of them with the same properties that p, for example

$$p = \begin{pmatrix} 1 & -1 & 0 \\ 0 & 0 & 0 \\ -1 & 1 & 0 \end{pmatrix} + \begin{pmatrix} 0 & -1 & 1 \\ 0 & 1 & -1 \\ 0 & 0 & 0 \end{pmatrix} + \begin{pmatrix} 0 & 0 & 0 \\ 1 & 0 & -1 \\ -1 & 0 & 1 \end{pmatrix} + \begin{pmatrix} 0 & -1 & 1 \\ 1 & 0 & -1 \\ -1 & 1 & 0 \end{pmatrix}. \tag{11}$$

If one carefully analyses the decomposition in the previous example Eq. (11), we can see that exist a subset within the PRIs from which all rephasing invariants can be constructed. It leads us to the following definition.

Definition 2 *The* fundamental rephasing invariant (FRI) *is such a pure rephasing invariant monomial that is the product of 4 or 6 CKM matrix elements and its complex conjugates.*

There are 30 fundamental rephasing invariants Table 1, 18 of them are the products of 4 CKM matrix elements and the others 12 are the products of 6 CKM matrix elements. For each FRI in there corresponds a p matrix, e.g.,

$$J_1 \to p_{J_1} = \begin{pmatrix} 1 & -1 & 0 \\ -1 & 1 & 0 \\ 0 & 0 & 0 \end{pmatrix}. \quad I_1 \to p_{I_1} = \begin{pmatrix} 1 & 0 & -1 \\ -1 & 1 & 0 \\ 0 & -1 & 1 \end{pmatrix}, \text{ etc.} \tag{12}$$

Journal of Physics: Conference Series **485** (2014) 012058 doi:10.1088/1742-6596/485/1/012058

All the matrices p_{J_i} and p_{I_j} corresponding to the invariants in Table 1 can be obtained by the permutations of the rows and columns of the matrices p_{J_1} and p_{I_1} that are given in Eq. (12).

Theorem 2 *Any pure rephasing invariant can be expressed in a unique way as the product of positive powers of at most 4 fundamental rephasing invariants. Not more than one of these invariants can be from the 6-th order FRIs and the remaining are from the 4-th order FRIs.*

To prove the theorem we analyze the matrices p_{J_i} and p_{I_j} and we show that the matrix corresponding to a PRI can be decomposed in a unique way as a linear combination with positive coefficients of at most 4 p matrices corresponding to the fundamental rephasing invariants defined in table 1. The inverse theorem is not true, the product of two or more FRIs is rephasing invariant, but it does not have to be the PRI.

From the unitarity of the CKM matrix it follows that the 6-th order FRIs in table 1 can be expressed by the 4-th order FRIs and the squares of the CKM matrix elements, e.g.,

$$I_1 = V_{11}V_{22}V_{33}V_{13}^*V_{21}^*V_{32}^* = |V_{22}|^2 V_{12}V_{33}V_{13}^*V_{32}^* - |V_{13}|^2 V_{22}V_{33}V_{23}^*V_{32}^* = |V_{22}|^2 J_6 - |V_{13}|^2 J_9, \quad (13)$$

and there are analogous formulas for the remaining I_j's. It should be emphasized that without the unitarity of the CKM matrix there are no simple relations between the invariants of the 4-th and 6-th order. Thus relation in Eq. (13) is also a test of the unitarity of the CKM matrix.

3. Recursive construction of the CKM matrix

The presented algorithm does not impose any conditions on the parameterization of the matrix $V^{(n-1)}$ and assume that the explicit form of the matrix $V^{(n-1)}$ is known. We write the CKM matrix $V^{(n)}$ in terms of column vectors

$$V^{(n)} = \left(\mathbf{v}_1^{(n)}, \mathbf{v}_2^{(n)}, \ldots, \mathbf{v}_n^{(n)}\right) \qquad \text{with} \qquad \mathbf{v}_k^{(n)} = \begin{pmatrix} V_{1k}^{(n)} \\ V_{2k}^{(n)} \\ \vdots \\ V_{nk}^{(n)} \end{pmatrix}, \quad k = 1, 2, ..., n., \quad (14)$$

The CKM matrix $V^{(n)}$ is built from $V^{(n-1)}$ in two steps:

a. We construct n *real* column vectors $\mathbf{e}_1, \ldots, \mathbf{e}_n$, normalized to 1 and orthogonal $\mathbf{e}_i \cdot \mathbf{e}_j = \delta_{ij}$, with n rows that depend on $(n-1)$ independent parameters.

b. The columns of the CKM matrix $V^{(n)}$ are obtained from the vectors $\mathbf{e}_k$ and the elements of the matrix $V_{ij}^{(n-1)}$ in the following way

$$\begin{aligned} \mathbf{v}_k^{(n)} &= V_{1k}^{(n-1)}\mathbf{e}_1 + \textstyle\sum_{\ell=2}^{n-1} V_{\ell k}^{(n-1)} e^{-i\delta_{\ell-1,n}}\mathbf{e}_\ell, \quad k = 1, \ldots, n-1, \\ \mathbf{v}_n^{(n)} &= \mathbf{e}_n, \end{aligned} \qquad n \geq 2, \quad (15)$$

The matrix $V^{(n)}$ constructed in such a way is unitary and it depends on parameters of the matrix $V^{(n-1)}$, $(n-1)$ parameters of the vectors $\mathbf{e}_1, \ldots, \mathbf{e}_n$ and on $(n-2)$ phase factors $e^{-i\delta_{\ell,n}}$ from Eq. (15).

4. Wolfenstein parameterization for 4 quark generations

We shall generalized the Wolfenstein parameterization [5] to the case of 4 quark generations using the method outlined earlier.

$$V^{(3)} = \begin{pmatrix} 1 - \frac{\lambda^2}{2} & \lambda & A\lambda^3(\rho - i\eta) \\ -\lambda & 1 - \frac{\lambda^2}{2} & A\lambda^2 \\ A\lambda^3(1 - \rho - i\eta) & -A\lambda^2 & 1 \end{pmatrix}. \quad (16)$$

The Wolfenstein parameterization given in Eq. (16) is approximate. It can be given an exact meaning by assuming a one to one correspondence between the Wolfenstein parameters A, λ, ρ and η and the parameters of the standard parameterization $s_{12} \equiv \lambda$, $s_{23} \equiv A\lambda^2$ and $s_{13}e^{i\delta_{13}} \equiv A\lambda^3(\rho + i\eta)$ [6, 7].

First we have to construct the vectors $\mathbf{e}_i$, which are expressed in the spirit of the Wolfenstein parameterization in terms of the powers of λ. The vectors $\mathbf{e}_i$ are real and are chosen in the following way

$$\mathbf{e}_1 = N_1 \begin{pmatrix} 1 + z_2^2 + z_3^2 \\ -z_1 z_2 \\ -z_1 z_3 \\ -z_1 \end{pmatrix}, \quad \mathbf{e}_2 = N_2 \begin{pmatrix} 0 \\ 1 + z_3^2 \\ -z_2 z_3 \\ -z_2 \end{pmatrix}, \quad \mathbf{e}_3 = N_3 \begin{pmatrix} 0 \\ 0 \\ 1 \\ -z_3 \end{pmatrix}, \quad \mathbf{e}_4 = N_4 \begin{pmatrix} z_1 \\ z_2 \\ z_3 \\ 1 \end{pmatrix}. \tag{17}$$

Here $z_i = A_i^{(4)} \lambda^{k_i}$, $A_i^{(4)} \sim 1$, $k_i \geq 1$ are integers and N_i are suitable normalization factors. The powers k_i are considered to be constants (suppression factors) and $A_i^{(4)}$, $i = 1, 2, 3$ are free parameters. It is easy to verify that the vectors $\mathbf{e}_i$ are normalized to 1 and orthogonal.

The vectors $\mathbf{e}_i$ were constructed to fulfill the following two additional properties

$$(\mathbf{e}_i)_j|_{A_i=0} = \begin{cases} 1 & \text{if } i = j \\ 0 & \text{otherwise} \end{cases}, \qquad V^{(4)}\Big|_{\substack{A_i=0 \\ \delta_{1,4}=0 \\ \delta_{2,4}=0}} = \begin{pmatrix} V^{(3)} & 0 \\ 0 & 1 \end{pmatrix}, \quad i = 1, 2, 3. \tag{18}$$

The columns of the CKM matrix for 4 generations are then equal

$$\begin{aligned} \mathbf{v}_1^{(4)} &= V_{11}^{(3)} \mathbf{e}_1 + e^{-i\delta_{1,4}} V_{21}^{(3)} \mathbf{e}_2 + e^{-i\delta_{2,4}} V_{31}^{(3)} \mathbf{e}_3, \quad \mathbf{v}_2^{(4)} = V_{12}^{(3)} \mathbf{e}_1 + e^{-i\delta_{1,4}} V_{22}^{(3)} \mathbf{e}_2 + e^{-i\delta_{2,4}} V_{32}^{(3)} \mathbf{e}_3, \\ \mathbf{v}_3^{(4)} &= V_{13}^{(3)} \mathbf{e}_1 + e^{-i\delta_{1,4}} V_{23}^{(3)} \mathbf{e}_2 + e^{-i\delta_{2,4}} V_{33}^{(3)} \mathbf{e}_3, \quad \mathbf{v}_4^{(4)} = \mathbf{e}_4, \end{aligned} \tag{19}$$

and the CKM matrix is

$$V^{(4)} = \left(\mathbf{v}_1^{(4)}, \mathbf{v}_2^{(4)}, \mathbf{v}_3^{(4)}, \mathbf{v}_4^{(4)} \right). \tag{20}$$

The matrix $V^{(4)}$ in Eq. (20) is described by 9 parameters: λ, A, ρ, η of the matrix Eq. (16), introduced by Wolfenstein, $A_1^{(4)}$, $A_2^{(4)}$, $A_3^{(4)}$ of the vectors in Eqs. (17) and $\delta_{1,4}$, $\delta_{2,4}$ from Eq. (15). Not all these parameters can be determined from the experimental data. On the other hand we can derive some restrictions on the powers k_i of the suppression factors for the 4-th generation from the experimental information for the CKM matrix for 3 generations. We have the following information

$$\begin{aligned} &|V_{12}| \sim \lambda, \quad |V_{21}| \sim \lambda, \quad |V_{23}| \sim \lambda^2, \quad |V_{32}| \sim \lambda^2, \quad |V_{13}| \sim \lambda^3, \\ &|V_{31}| \sim \lambda^3, \quad (|V_{12}| - |V_{21}|) \sim \lambda^3, \quad (|V_{23}| - |V_{32}|) \sim \lambda^4. \end{aligned} \tag{21}$$

Now, using the information in Eq. (21) with the explicit representation of the 4×4 CKM matrix in Eqs. (19) and (20) we obtain the following restrictions on the powers k_i

$$k_i \geq 1, \quad k_1 + k_2 \geq 3, \quad k_2 + k_3 \geq 4, \quad k_1 + k_3 \geq 3, \tag{22}$$

which can be resolved and give $k_1 \geq 1$, $k_2 \geq 2$, $k_3 \geq 2$ and the vector $\mathbf{v}_4$ for the minimal values of k_i has the following form

$$\mathbf{v}_4 = N_4 \begin{pmatrix} A_1^{(4)} \lambda \\ A_2^{(4)} \lambda^2 \\ A_3^{(4)} \lambda^2 \\ 1 \end{pmatrix}. \tag{23}$$

In the case of 3 generations the suppression has totally different structure. However, we would like to note that the real suppression may be different, because we have only obtained the *lower limits* of the suppression powers.

5. Conclusions

From theorem 2 follows that any rephasing invariant monomial of the CKM matrix for 3 generations is the product of no more than 5 factors: 4 fundamental rephasing invariants taken to positive powers and the product of the squares of the absolute values of the CKM matrix elements also taken to positive powers. Another important conclusion that can be obtained from theorem 2 and the unitarity of the CKM matrix is that the imaginary part of any rephasing invariant monomial is proportional to the Jarlskog invariant or equal to 0. Let us note that general rephasing invariant monomials constructed from the CKM matrix elements appear at higher orders of the renormalization group equations for the CKM matrix elements. The systematic analysis of such equations based on the results obtained in this paper will be published elsewhere.

From the theoretical point of view all exact parameterizations of the CKM matrix are equivalent. From the practical point of view the situation is less obvious, because certain experimental facts can be presented in a more transparent way in one parameterization than in the other. The scheme of the construction of the parameterization of the CKM matrix presented in this work allows to adjust properties of the parameterization according to the needs. Such an approach has not been discussed before and it can facilitate the discussion of the properties of the CKM in the Standard Model or its extensions.

Acknowledgments

The authors wish to thank the organizers, specially Liliana Velasco-Sevilla, for inviting us to participate in the 18th International Symposium on Particles Strings and Cosmology. H. Pérez R. and P. Kielanowski thank Professor Maria Krawczyk for hospitality in the Institute of Theoretical Physics at Warsaw University, Poland. This work was partly supported by Polish Ministry of Science and Higher Education Grant N N202 230337, CONACyT and Proyecto SIP: 20120639, Secretaría de Investigación y Posgrado, Beca EDI y Comisión de Operación y Fomento de Actividades Académicas (COFAA) del IPN.

References

[1] Jarlskog C 1985 *Phys. Rev. Lett.* **55** 1039
[2] Greenberg O W 1985 *Phys. Rev.* D **32** 1841
[3] Dunietz I, Greenberg O W and Wu D d 1985 *Phys. Rev. Lett.* **55** 2935
[4] Branco G C, Lavoura L and Silva Joao P 1999 *CP Violation* International Series of Monographs on Physics vol 103 (USA:Oxford Science Publications) p 103
[5] Wolfenstein L, 1983 *Phys. Rev. Lett.* **51** 1945
[6] Buras A J, Lautenbacher M E and Ostermaier G 1994 *Phys. Rev.* D **50** 3433
[7] Nakamura N and Group P D 2010 *Journal of Physics G: Nuclear and Particle Physics* **37** 075021

Renormalization group running of couplings and masses in the basic extension of the standard model

D Morales C[1,a], S R Juárez[1,b] and P Kielanowski[2,c]
[1]Departamento de Física, ESFM-IPN, México D.F., México
[2]Departamento de Física, CINVESTAV-IPN, México D.F., México
E-mail: [a]damoralesc@hotmail.com,[b]rebecajw@gmail.com,[c]kiel@fis.cinvestav.mx

Abstract. We discuss the range of validity of the basic two Higgs doublet model by numerically solving the renormalization group equations for the quartic couplings and masses. We consider parameter-values that satisfy the vacuum stability and triviality principles, some of them used in the literature.

1. Introduction

The theory that is associated with the sectors of the gauge bosons and fermions of the standard model (SM), is in excellent agreement with a large amount of experimental data on low energy phenomena [1, 2, 3]. Currently, the most interesting part of the SM is the Higgs boson, which generates the masses of fermions and bosons $W^{\pm}$ and Z^0 [4, 5]. A minimal extension of the Higgs sector of the SM consists in the addition of one scalar doublet SU(2) with exactly the same quantum numbers as the original. Here we focus our attention on the simplest possible extension of the SM, the two Higgs doublet model (2HDM).

2. The 2HDM model

The Higgs sector of the 2HDM consists of two identical (hypercharge-one) scalar doublets Φ_1 and Φ_2. The 2HDM is determined by the choice of the Higgs potential and the Yukawa couplings of the two scalar doublets with the elements of the three generations of quarks and leptons. We denote the two complex Higgs doublet fields by

$$\Phi_i = \begin{pmatrix} \phi_i^+ \\ \phi_i^0 \end{pmatrix}, \tag{1}$$

with i=1,2. The most general Higgs potential, which can be constructed with two scalar doublets, which is gauge invariant and renormalizable is given by

$$\begin{aligned} V \; = \; & -\mu_1^2\Phi_1^\dagger\Phi_1 - \mu_2^2\Phi_2^\dagger\Phi_2 - \left\{\mu_{12}^2\left(\Phi_1^\dagger\Phi_2\right) + \text{h.c.}\right\} \\ & +\lambda_1\left(\Phi_1^\dagger\Phi_1\right)^2 + \lambda_2\left(\Phi_2^\dagger\Phi_2\right)^2 + \lambda_3\left(\Phi_1^\dagger\Phi_1\right)\left(\Phi_2^\dagger\Phi_2\right) \\ & +\lambda_4\left(\Phi_1^\dagger\Phi_2\right)\left(\Phi_2^\dagger\Phi_1\right) + \frac{1}{2}\left[\lambda_5\left(\Phi_1^\dagger\Phi_2\right)^2 + \lambda_5^*\left(\Phi_2^\dagger\Phi_1\right)^2\right] \\ & +\left\{\left[\lambda_6\left(\Phi_1^\dagger\Phi_1\right) + \lambda_7\left(\Phi_2^\dagger\Phi_2\right)\right]\left(\Phi_1^\dagger\Phi_2\right) + \text{h.c.}\right\}. \end{aligned} \tag{2}$$

Content from this work may be used under the terms of the Creative Commons Attribution 3.0 licence. Any further distribution of this work must maintain attribution to the author(s) and the title of the work, journal citation and DOI.
Published under licence by IOP Publishing Ltd

The number of parameters in eq. (2) is reduced to five λ_i s by imposing the Z_2 discrete symmetry [6, 7, 8, 9, 10], preserving CP symmetry.

2.1. Constraints for the λ_i parameters

The vacuum stability constraints arise from the requirement of a positive potential for large values of the fields in an arbitrary direction in the plane (Φ_1, Φ_2) imposed on the quartic potential. Furthermore, the masses are defined by the quartic couplings and the vacuum expectation values.

$$M^2_{H^0,h^0} = \lambda_1 v_1^2 + \lambda_2 v_2^2 \pm [(\lambda_1 v_1^2 - \lambda_2 v_2^2)^2 + 4v_1^2 v_2^2 \Lambda^2]^{\frac{1}{2}}, \tag{3}$$

$$M^2_{A^0} = -v^2\lambda_5, \quad M^2_{H^\pm} = -(1/2)v^2(\lambda_4 + \lambda_5), \tag{4}$$

where $\Lambda = (1/2)(\lambda_3 + \lambda_4 + \lambda_5)$, v_1 and v_2 are the vacuum expectation values of the scalar fields, which are related in the following way $v = v_1^2 + v_2^2 \approx (246 \text{ GeV})^2$.

The restrictions on the quartic couplings may also be obtained directly from the mass spectrum of 2HDM. A rigorous way to obtain additional restrictions for the quartic couplings is through the use of the Lagrange multipliers method and by the implementation of the vacuum stability condition. The problem we consider involves constraints that are formulated by equality and inequality relations [11, 12, 13]. The Lagrange function we propose has the following form

$$L(x, \Lambda, \Gamma) = F(x) + \sum_{i=1}^{n} \Lambda_i g_i(x) + \sum_{j=i}^{m} \Gamma_j h_j(x), \tag{5}$$

where $F(x)$ is the quartic potential, Λ_i and Γ_i are the Lagrange multipliers and the explicit form of the constraints $g_i(x)$ and $h_i(x)$ is

$$g(x) = x_1 + x_2 - v^2 = 0, \qquad h(x) = x_1 x_2 - x_3^2 - x_4^2 \geq 0, \tag{6}$$

where

$$x_1 = \Phi_1^\dagger \Phi_1, \qquad x_2 = \Phi_2^\dagger \Phi_2, \qquad x_3 = \mathrm{Re}(\Phi_1^\dagger \Phi_2), \qquad x_4 = \mathrm{Im}(\Phi_1^\dagger \Phi_2). \tag{7}$$

Applying the stability condition

$$L(x_{\min}) > 0, \tag{8}$$

and the Karush-Kuhn-Tucker restraints [12, 14], we obtain the following requirements to be satisfied by the parameters of the potential

$$\lambda_1 > 0, \qquad \lambda_1 > 0, \tag{9}$$

$$\lambda_5 < 0, \qquad \lambda_4 + \lambda_5 < 0, \tag{10}$$

$$\lambda_3 + \lambda_4 \pm |\lambda_5| < \lambda_1 + \lambda_2, \tag{11}$$

$$-2\sqrt{\lambda_1 \lambda_2} < \lambda_3 + \lambda_4 \pm |\lambda_5|. \tag{12}$$

3. Results

The analysis based on the renormalization group equations [RGE] is of great importance in the study of the physics of the standard model and its extensions [15, 16]. The knowledge of the energy evolution of the mass and coupling parameters is essential to analyze electroweak experiments. For the numerical evaluation we use data that has been published recently in relation with the charged Higgs mass [17, 18, 19]. Several scenarios are considered in the analysis.

In a first stage, considering [17], we have fixed the values of the Higgs masses as $M_{H^\pm}$= 609 GeV, M_{A^0}= 622 GeV. In this case, in which the masses are large, to obtain reasonable results in the evolution of the couplings, the appropriate initial values for the scalars are λ_1= 12, λ_2= 3, λ_3= 11.5, λ_4= -5.5, λ_5= -6 and $\tan\beta = v_2/v_1$=1.4 (β is the mixing angle between the charged states). The behavior of these parameters may be observed in Figures 1 and 2. Here we can see that the range of validity of the model is limited, since the energy extent before reaching singularities is very short (M_t < E < 292 GeV, in terms of the variable t, 0 < t < 0.52, where t is defined as t = $\ln(\frac{E}{m_t})$, E is the renormalization point energy and m_t is mass of the top quark).

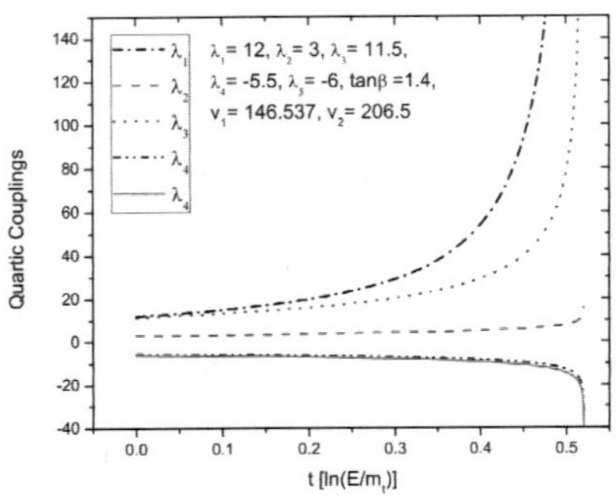

Figure 1. Evolution of quartic couplings.

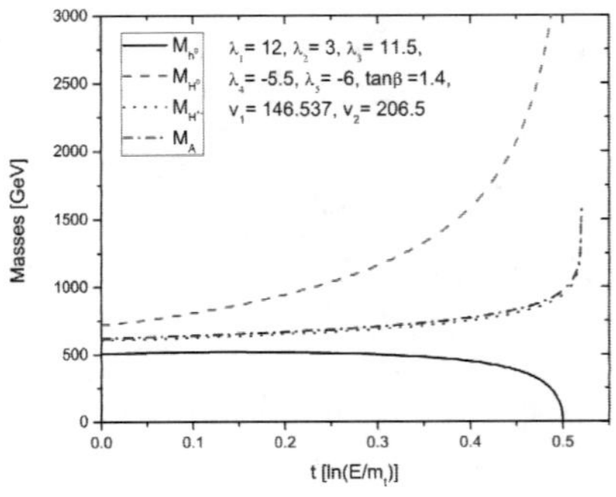

Figure 2. Evolution of the masses.

In a second stage rested on [18], we fixed the values of the masses as $M_{H^{\pm}}$= 129.4 GeV, M_{A^0}= 131.9 GeV. Here we consider the following values for the parameters λ_1=λ_2= 0.18, λ_3= 3, λ_4= -0.25, λ_5= -0.27 and $\tan\beta$= 41.2. The behavior of quartic couplings and masses is exhibited in Figures 3 and 4. In this case we can see that the model is valid in the whole energy range, i.e. up to the energy scale of electroweak unification. For a more numerical detail see Table 1.

Table 1. Data for Figures 3 and 4.

	E[GeV] = 173.1	E[GeV] = 7000	E[GeV] = 1.2×10^{13}
$\tan\beta$	41.21	41.2	41.2
ν_1	6.16	5.97	5.71
ν_2	253.73	245.87	234.47
ν	253.81	245.94	234.53
g_t	0.965	0.817	0.508
g_b	0.965	0.820	0.524
λ_1	0.18	0.135	0.123
λ_2	0.18	0.135	0.115
λ_3	0.3	0.19	0.095
λ_4	− 0.25	− 0.087	0.56
λ_5	− 0.27	− 0.32	− 0.675
m_{h^0}	2.92	1.93	2.8
m_{H^0}	152.26	127.65	112.27
m_{A^0}	129.42	110.20	55.84
$m_{H^{\pm}}$	131.88	138.01	192.64

In Table 1, we consider a column that is refereed to a 7 TeV energy in which the M_{H^0}= 127.7.

This value is very close to the one reported in [20, 21, 22, 23] as an evidence of the existence of the SM Higgs particle.

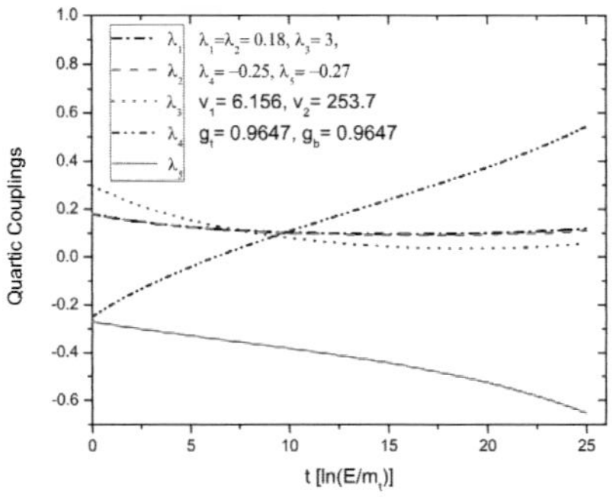

Figure 3. Evolution of quartic couplings.

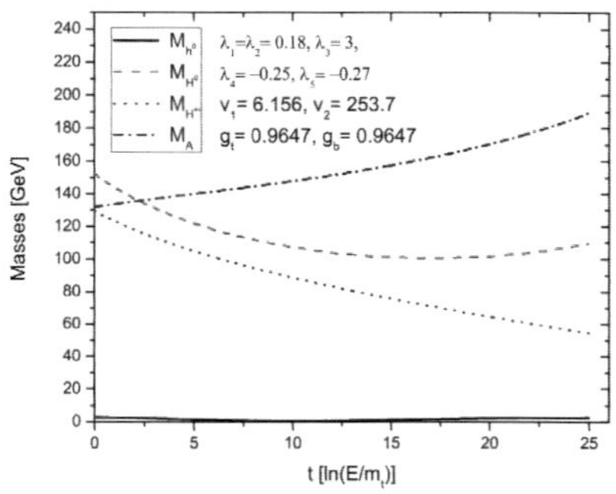

Figure 4. Evolution of the masses.

In a third scenario, a unification of the Yukawa couplings associated with the quarks t and b ($g_t = g_b$) is considered at low energies. Figures 5, 6 and 7 show the behavior of the Yukawa couplings, quartic couplings and Higgs masses. The model is valid in the whole energy range. Table 2 shows the numerical data for this figures.

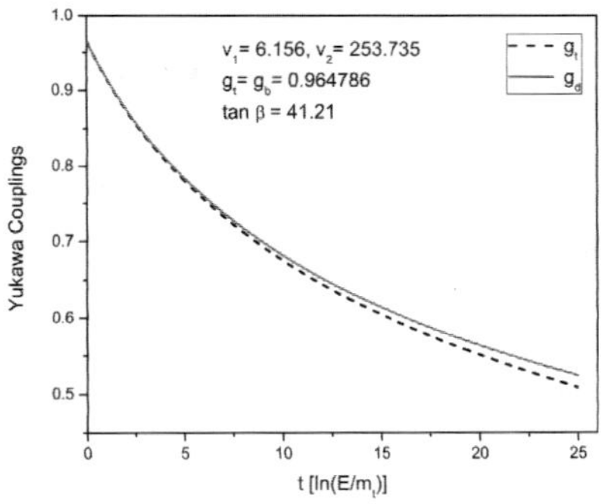

Figure 5. Evolution of the Yukawa couplings, with unification at low energy.

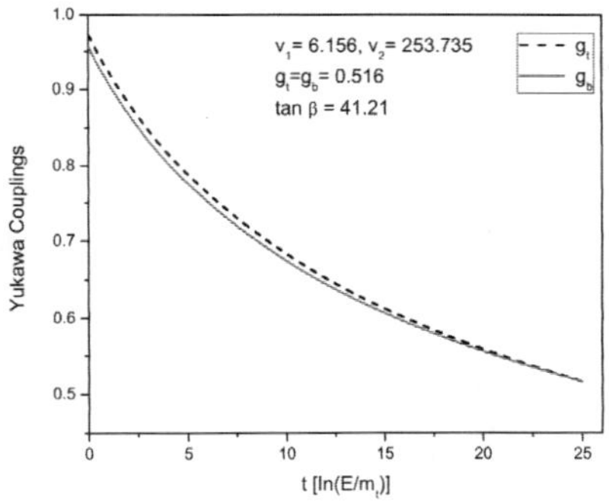

Figure 6. Evolution of the Yukawa couplings, with unification at high energy.

In the last case we consider a unification of the Yukawa couplings at high energies. Figures 8, 9 and 10 show the evolution of the parameters. As can be seen, the model is also valid in the whole energy range. The numerical values of the parameters for this case are presented in Table 3.

4. Conclusions

We have considered the scalar potential in the 2HDM model, which is the simplest extension of the standard model, in which the CP symmetry is preserved. Through the vacuum stability principle and the Lagrange multipliers method we have obtained conditions to be satisfied by the quartic couplings in the potential. We have solved numerically the renormalization group equations for the quartic couplings and masses in the 2HDM model. We have considered various scenarios for which, we have obtained the limits of validity of the model.

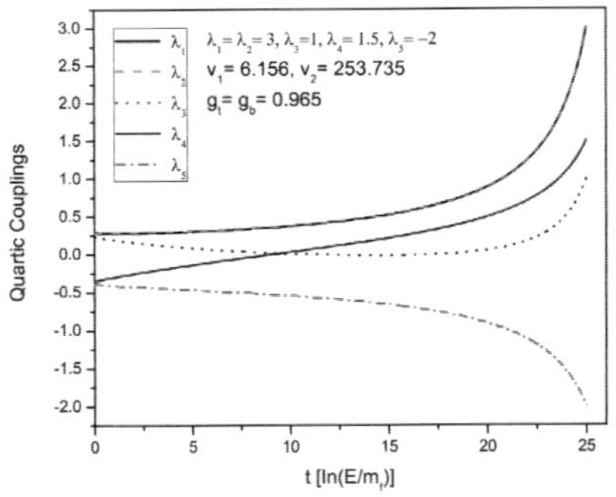

Figure 7. Evolution of quartic couplings.

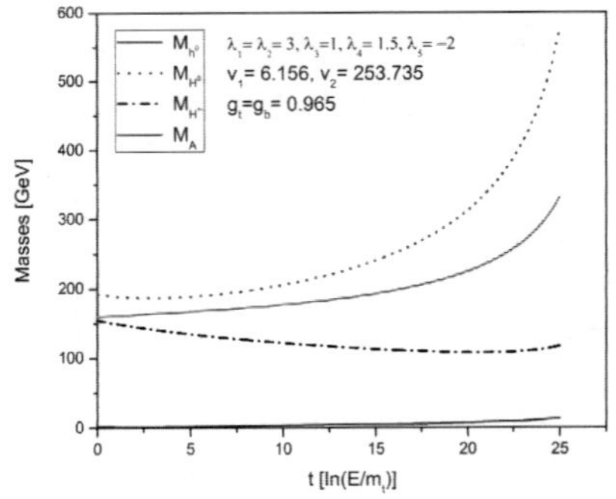

Figure 8. Evolution of the masses.

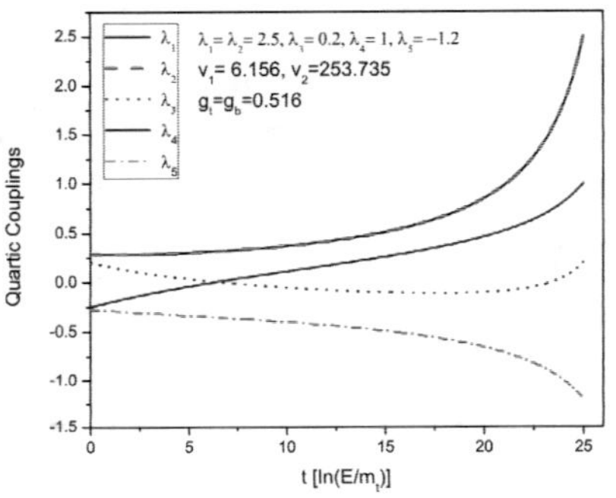

Figure 9. Evolution of quartic couplings.

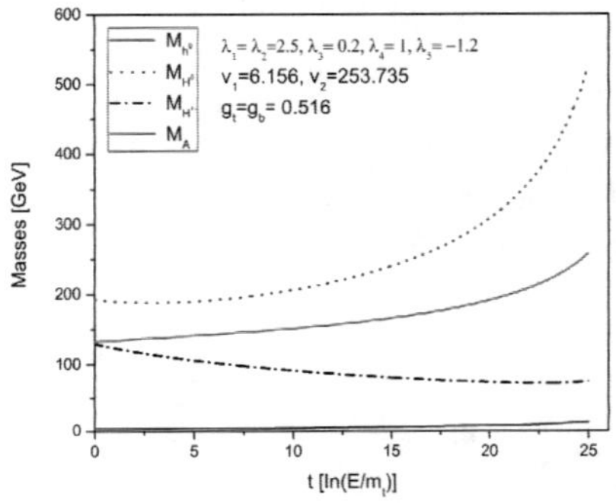

Figure 10. Evolution of the masses.

Table 2. Data for Figures 5, 7 and 8.

	E[GeV] = 173.1	E[GeV] = 1.2×10^{13}
$\tan\beta$	41.21	41.02
ν_1	6.156	5.715
ν_2	253.74	234.465
ν	253.81	234.534
g_t	0.9647	0.5082
g_b	0.9647	0.5238
λ_1	0.2849	3.0
λ_2	0.2852	3.0
λ_3	0.2347	1.0
λ_4	− 0.3487	− 1.5
λ_5	− 0.393	− 2.0
m_{h^0}	2.1252	13.9518
m_{H^0}	191.684	574.32
m_{A^0}	159.115	331.682
$m_{H^\pm}$	154.569	117.267

Table 3. Data for figures 6, 9 and 10.

	E[GeV] = 173.1	E[GeV] = 1.2×10^{13}
$\tan\beta$	41.21	41.43
ν_1	6.156	5.6869
ν_2	253.735	235.641
ν	253.81	235.709
g_t	0.9729	0.516
g_b	0.9567	0.516
λ_1	0.2915	2.5
λ_2	0.2876	2.5
λ_3	0.2043	0.2
λ_4	− 0.2498	1.0
λ_5	− 0.2735	− 1.2
m_{h^0}	3.9232	12.7164
m_{H^0}	192.458	526.909
m_{A^0}	132.749	258.207
$m_{H^\pm}$	129.838	74.5378

Acknowledgments
S.R.J.W. thanks to the "Comisión de Operación y Fomento de Actividades Académicas" (COFAA) and EDI from Instituto Politécnico Nacional. D.M.C. thanks to CONACyT-México, and to Dr. Myriam Mondragón for her hospitality and for the fellowship to participate in the PASCOS 2012.

References
[1] Branco G C, *et al* 2012 *Physics Reports: Theory and phenomenology of two-Higgs-doublet models* doi:10.1016/j.physrep.2012.02.002.
[2] ALEPH, DELPHI, L3, OPAL Collaborations and LEP *Electroweak Working Group "A combination of preliminary electroweak measurements and constraints on the standard model"* arXiv:hep-ex/0511027.
[3] Kraml S,*et al* 2006. CERN-CPNSH *Workshop May 2004 - Dec 2005* GENEVA Ed. By Sabine Kraml.
[4] Higgs P W 1967 *Phys. Lett.* **12** 132.
[5] Guralnik G, Hagen C, Kibble T 1964 *Phys. Rev. Lett.* **13** 585.
[6] Glashow S L, Iliopoulos J, Maiani L 1977 *Phys. Rev.* D **2** 1285.
[7] Glashow S L, Weinberg S 1977 *Phys. Rev.* D **15** 1958.
[8] Deshpande N G, Ma E 1978 *Phys. Rev.* D **18** 2574.
[9] Babu K S, Ma E 1985 *Phys. Rev.* D **31**, 2861.
[10] Ferreira P M, Jones D R 2009 JHEP08 069.
[11] Ioffe A D and Tihomirov V M 1979 *Theory of Extremal Problems*, (North-Holland Publishing Company).
[12] Lange K 2004 *Optimization* (Springer-Verlag NY LLC).
[13] Snyman J A 2005 Practical Mathematical Optimization (Springer Science + Business media Inc. NY).
[14] Wu Hsien-Chung 2007 *Eur. J. Operational Research* **176** 46-59.
[15] Inoue K, Kakuto A, Nakano Y 1980 Prog. Theor. Phys. **63** 234.
[16] Das C R, Parida M K 2001 *Eur. Phys. J.* C **20** 121.
[17] Mahmoudi F, Stal O 2010 P*hys. Rev.* D **81** 035015.
[18] Hashemi M, 2012 *Eur. Phys. J.* C **72** 1994.
[19] Rozanska M, Brein O 2010 *International Workshop on Prospects for Charged Higgs Discovery at Collider* (CHARGES 2010)/20.
[20] http://www.atlas.ch/news/2012/latest-results-from-higgs-search.html.
[21] CMS-PAS-HIG-11-023, ATLAS-CONF-2011-157.
[22] The CMS Collaboration, arXiv:1207.7235v1[hep-ex].
[23] The ATLAS Collaboration, arXiv:1207.7214v1[hep-ex].

PASCOS 2012 – 18th International Symposium on Particles Strings and Cosmology IOP Publishing
Journal of Physics: Conference Series **485** (2014) 012060 doi:10.1088/1742-6596/485/1/012060

Higgs mass in the MSSM with flavor mixing soft terms

J E Barradas-Guevara[1], O Félix-Beltrán[2], F Flores-Baez[3] and M Gómez-Bock[4]

[1] FCFM, BUAP, Apdo. Postal 1364, C.P. 72000 Puebla, Pue., México.
[2] FCE, BUAP, Apdo. Postal 542, C. P. 72570 Puebla, Pue., México.
[3] FCFM, UANL. Av. Universidad S/N, Ciudad Universitaria, San Nicolás de los Garza, Nuevo León, C.P. 66451.
[4] DAFM, UDLAP, Ex Hda. Sta. Catarina Mártir, C.P. 72810, Cholula, Pue., México.

E-mail: barradas@fcfm.buap.mx, olga_flix@ece.buap.mx, fflores@fcfm.uanl.mx, melina.gomez@udlap.mx

Abstract. The Minimal Supersymmetric Standard Model assumes scalar masses universality as a simplification to avoid undesired flavor changing neutral currents (FCNC). In this work, we consider that the trilinear SUSY soft-terms have a hierarchical structure of flavor mixing within generations, leading to a non-degeneration on sfermion masses preserving FCNC under control. We consider this kind of mixing for the u-type squark mass matrix and analyze the consequences on the radiative corrections to the Higgs mass. We work under the consideration of m_h^{max} benchmark scenario, in which the parameter space is set to maximize the radiative contributions to the Higgs mass. Based on recent experimental results on possible Higgs signals, we calculate the mass of the neutral light CP-even Higgs taking into account flavor mixing contributions.

1. Introduction

At present, the Higgs mechanism is an ingredient generally accepted in the Standard Model (SM), or any other theoretical model beyond, as a generator of gauge boson and fermion masses. Full identification of the Higgs particle is close to be proven experimentally. The mass generation problem is intrinsically related to flavor physics, current experimental results on the Higgs boson can set bounds to Flavor Violation (FV) parameters. Namely, it is relevant in order to study non-vanishing flavor violation processes within the experimental current results. Going beyond the SM, through a supersymmetric theory in its minimal structure, the MSSM has a relevant success in stabilizing the EW scale, it unifies three of the interaction couplings, generates candidates for dark matter, and could account for FV couplings through SUSY at one-loop. This last part is what we try to exploit here, as in other works Flavor Violation may phenomenologically bound some parameters of the models. There is much work done in this direction within SUSY models and we mention some diverse examples on flavor violation on leptons and quarks [1, 2, 3], and directly on Higgs sector as [4, 5].

2. Minimal Supersymmetric Standard Model

The Minimal Supersymmetric Standard Model (MSSM) contains the least number of Higgs doublets needed to define a supersymmetric structure of the model in order to spontaneously

Content from this work may be used under the terms of the Creative Commons Attribution 3.0 licence. Any further distribution of this work must maintain attribution to the author(s) and the title of the work, journal citation and DOI.
Published under licence by IOP Publishing Ltd

break EW symmetry and give masses to both u-and d-type fermions [6], *i.e.* two Higgs complex doublets, widening the Higgs particle spectrum as three neutral h^0, H^0, A^0 and two charged $H^{\pm}$ Higgs bosons. Spontaneous Symmetry Breaking (SSB) is carried out when the scalar fields develop nonzero vacuum expectation values breaking the $SU(2)_L \times U(1)$, and each Higgs doublet couples with one type of fermions. All the fields have canonical kinetic terms with the usual D_μ and field strengths $F_{\mu\nu}$. The only freedom that one has is the choice of the superpotential W, from which the form of the scalar potential and the Yukawa interactions between fermion and scalar fields can be obtained [6, 7]. The expression of W is given by:

$$W_{MSSM} = \frac{1}{2}M^{ij}\phi_i\phi_j + \frac{1}{6}y^{ijk}\phi_i\phi_j\phi_k. \tag{1}$$

By renormalization, only bilinear and trilinear terms are permitted [8]. For the Next-to-MSSM and considering Z_3 invariance, the bilinear terms are present only after the SSB. For this model, also the mixing on the trilinear soft terms is possible and could be implemented as in the MSSM.

2.1. Phenomenological restrictions in pMSSM

One MSSM version which uses the phenomenological restrictions for to reduce the parameters number of the model, is known as pMSSM [9]. This model basically assumes three issues concerning parameters of the MSSM: (*i*) CP-conserving (no extra source), (*ii*) no FCNC, and (*iii*) $m_{\tilde{f}1} \approx m_{\tilde{f}2}$ to accomplish $K^0 - \bar{K}^0$ mixing. Then ending up with 22 input parameters: $\tan\beta$; m_1^2, m_2^2; M_1, M_2, M_3; $\tilde{m}_q$, $\tilde{m}_{uR}$, $\tilde{m}_{dR}$, $\tilde{m}_l$, $\tilde{m}_{eR}$; $\tilde{m}_{Qt}$, $\tilde{m}_{tR}$, $\tilde{m}_{bR}$, $\tilde{m}_{L\tau}$, $\tilde{m}_{\tau R}$; $A_{u,c}$, $A_{d,s}$, $A_{e,\mu}$; A_t, A_b, A_τ. We propose a modification on the trilinear soft terms: instead of decoupling of the first two families from the third, we assume that the second family contribution could be relevant and decouple only the first family to calculate the possible consequences of this assumption on mixing second and third squarks families.

3. Flavor structure ansatz on MSSM soft-SUSY breaking terms

We assume CP-conservation, that is non-complex parameter in squark mass matrix. Then, from the soft mass terms, and F and D-terms coming from the superpotential, the u-type squark mass matrix in the MSSM is given by

$$M_{\tilde{u}}^2 = \begin{pmatrix} m_{\tilde{u}L}^2 + m_u^2 + M_Z^2\cos 2\beta(I_3^q - Q_q s_w^2) & X_u \\ X_u & m_{\tilde{u}R}^2 + m_u^2 + M_Z^2\cos 2\beta Q_q s_w^2 \end{pmatrix}, \tag{2}$$

where

$$X_u = A_u - \mu\cot\beta \quad \text{and} \quad m_{\tilde{u}R}^2 \simeq m_{\tilde{u}L}^2 \simeq \tilde{m}_0^2\mathbf{I}_{3x3}. \tag{3}$$

We consider a flavor ansatz, as it also has been considered in other models [10] as well as in Supersymmetric models at GUT scale in [11]. Here we propose a flavor mixing ansatz for the trilinear *Soft-SUSY Breaking* terms at low energy scale [12], considering a complete mixing within the two SM-heavy families, different from the one presented in [13]. In a general MSSM, the *Soft-SUSY Breaking* terms are expressed in the Lagrangian

$$-\mathcal{L}_{trilinear} = \sum_{i,j=gen} A_{ij}^u \tilde{Q}_i H_2 \tilde{u}_{Rj}^* + A_{ij}^d \tilde{Q}_i H_1 \tilde{d}_{Rj}^* + A_{ij}^l \tilde{L}_i H_1 \tilde{l}_{Rj}^*, \tag{4}$$

where A_{ij}^f is a 6×6 matrix, if we have the three families coupled together in the trilinear terms. Then, we consider that one family is decoupled at first order, that is

$$A_u^{ij} = A_{uLO}^{ij} + \delta A_u^{ij}, \tag{5}$$

$$A_{uLO} = \begin{pmatrix} 0 & 0 & 0 \\ 0 & w & z \\ 0 & y & 1 \end{pmatrix} A_0. \tag{6}$$

Furthermore, the squark u-type mass matrix has the explicit form

$$\tilde{M}_{\tilde{u}}^2 = \left(\begin{array}{cc|cccc} \tilde{m}_L^2 & 0 & 0 & 0 & 0 & 0 \\ 0 & \tilde{m}_R^2 & 0 & 0 & 0 & 0 \\ \hline 0 & 0 & \tilde{m}_L^2 & X_2 & 0 & A_z \\ 0 & 0 & X_2 & \tilde{m}_R^2 & A_y & 0 \\ 0 & 0 & 0 & A_y & \tilde{m}_L^2 & X_t \\ 0 & 0 & A_z & 0 & X_t & \tilde{m}_R^2 \end{array}\right), \tag{7}$$

with $X_2 = A_0 w - \mu m_c \cot\beta$ and $X_t = A_0 - \mu m_t \cot\beta$. For simplicity we consider $z = y$, which implies that after rotation the physical masses for u-type squarks are given by

$$\begin{aligned} m_{\tilde{c}_1}^2 &= \frac{1}{2}(2\tilde{m}_0^2 + X_2 + X_t - R), \quad m_{\tilde{c}_2}^2 = \tfrac{1}{2}(2\tilde{m}_0^2 - X_2 - X_t + R), \\ m_{\tilde{t}_1}^2 &= \frac{1}{2}(2\tilde{m}_0^2 - X_2 - X_t - R), \quad m_{\tilde{t}_2}^2 = \tfrac{1}{2}(2\tilde{m}_0^2 + X_2 + X_t + R), \end{aligned} \tag{8}$$

where $R = \sqrt{4A_y^2 + (X_2 - X_t)^2}$. So, the rotation matrix for the two mixed heavy SM squarks flavors which diagonalize the 4×4 squark mass matrix is given by

$$\mathcal{O}_{\tilde{u}} = \frac{1}{\sqrt{2}} \begin{pmatrix} \Theta & \sigma^1\Theta \\ -\Theta & \sigma^1\Theta \end{pmatrix}, \tag{9}$$

where σ^1 is a Pauli matrix and

$$\Theta = \begin{pmatrix} -\sin\frac{\psi}{2} & \cos\frac{\psi}{2} \\ -\cos\frac{\psi}{2} & -\sin\frac{\psi}{2} \end{pmatrix}, \tag{10}$$

$$\tan\psi = \frac{2A_y}{(X_2 - X_t)}. \tag{11}$$

We get that the six physical states are obtained by rotating the EW states $\tilde{M}_{diag}^2 = O_{\tilde{u}}^\dagger M_{\tilde{u}}^2 O_{\tilde{u}}$, and we obtain

$$\begin{pmatrix} \tilde{u}_L \\ \tilde{c}_L \\ \tilde{t}_L \\ \tilde{u}_R \\ \tilde{c}_R \\ \tilde{t}_R \end{pmatrix} = \frac{1}{\sqrt{2}} \begin{pmatrix} 1 & 0 & 0 & 0 & 0 & 0 \\ 0 & -\sin\frac{\psi}{2} & -\cos\frac{\psi}{2} & 0 & \sin\frac{\psi}{2} & \cos\frac{\psi}{2} \\ 0 & \cos\frac{\psi}{2} & -\sin\frac{\psi}{2} & 0 & -\cos\frac{\psi}{2} & \sin\frac{\psi}{2} \\ 0 & 0 & 0 & 1 & 0 & 0 \\ 0 & -\sin\frac{\psi}{2} & \cos\frac{\psi}{2} & 0 & -\sin\frac{\psi}{2} & \cos\frac{\psi}{2} \\ 0 & \cos\frac{\psi}{2} & \sin\frac{\psi}{2} & 0 & \cos\frac{\psi}{2} & \sin\frac{\psi}{2} \end{pmatrix} \begin{pmatrix} \tilde{u}_1 \\ \tilde{c}_1 \\ \tilde{t}_1 \\ \tilde{u}_2 \\ \tilde{c}_2 \\ \tilde{t}_2 \end{pmatrix}. \tag{12}$$

On the other hand, in order to obtain the non-mixing limit, *i.e.* decoupling between squark families, we only need to consider that $y \to 0$, which implies $A_y \to 0$, $\sin\psi \to 0$, $\cos\psi \to -1$ and $\psi = \pi$.

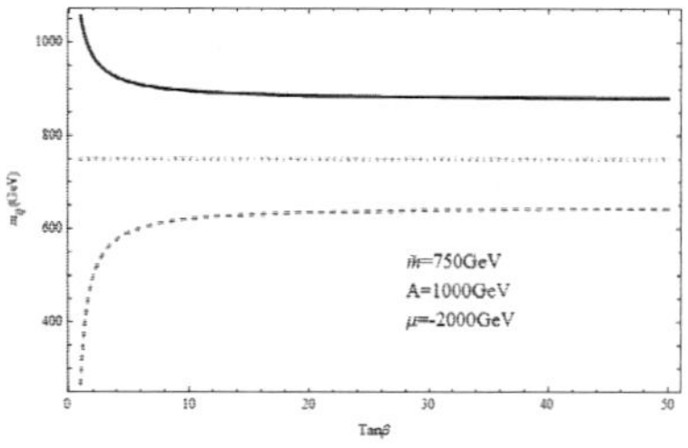

Figure 1. Second and third family u-type squarks masses.

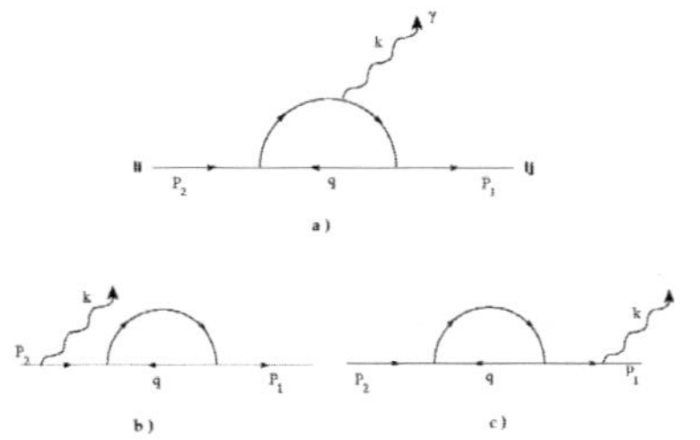

Figure 2. Flavor violating processes in quark sector through squarks-neutralino loops.

3.1. Phenomenological consequences

In addition to the above, now we mention the possible phenomenological consequences of the mixing in the trilinear soft terms, although we only concentrate on the last one.

(i) **Sfermion sector, non degenerate squark masses**, Figure 1.

(ii) **FV in quark sector.** Flavor violation in the u-type quark sector implies SUSY loops with squarks running in it, *i.e.* $BR(B \to X_s\gamma)$ and $BR(t \to c\gamma)$, as in Figure 2.

(iii) **Higgs sector, radiative corrections to Higgs mass**. We focus on this possibility, and we calculate 1-loop self-energy contribution with scharm-stop inside the loop, Figure 3. There has been work which considers mixing in the leptonic sector to achieve neutrino mixing, bounding the mixing parameters using the renormalized Higgs mass as 125 GeV [14].

At leading order, the CP-even neutral Higgs masses are related to the CP-odd mass m_{A^0} which is taken as free parameter [15]:

$$\begin{aligned} m_{h,H}^2 &= \frac{1}{2}(m_A^2 + m_Z^2) \mp \frac{1}{2}\sqrt{(m_A^2 + m_Z^2)^2 - 4m_A^2 m_Z^2 \cos^2 2\beta}, \\ m_H^{\pm} &= m_A^2 + \cos^2\theta_w m_Z^2. \end{aligned} \qquad (13)$$

The relations among MSSM parameters impose, at tree level, a strong hierarchical mass spectrum: $m_h < m_Z$, $m_A < m_H$ and $m_W < m_{H^\pm}$, which is broken by radiative corrections. The elements of the mass matrix M_h^2 are constructed explicitly from self-energies contributions diagrams: Σ_h.

Using the FeynHiggs code [16] we obtained radiative corrections to m_{h^0} up to 2-loops, Figure 4, for $X_t = 2$ TeV, $\mu = 200$ GeV and a SUSY breaking scale $M_{\rm SUSY} = 1$ TeV. In this figure, the Higgs mass is calculated in the m_h^{max} benchmark scenario (where parameters of the higher-order corrections are chosen to give the maximum value for m_h). In Figure 4 we set the MSSM parameters and one can see that for this combination of parameters, the 2-loop correction to the Higgs mass yields $M_h \approx 125$ GeV for a value of $\tan\beta \approx 7$ (all uncertainties neglected). In what follows, we will find how much does the FV 1-loop may contribute to the radiative correction. We calculate the self-energy with flavor violation within second and third families of u-type squarks, Figure 3.

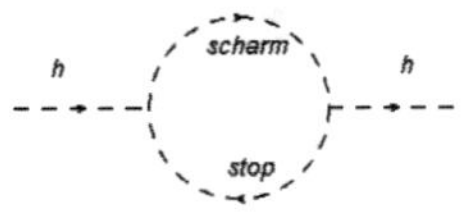

Figure 3. Flavor violating Higgs self-energy diagram in the squark u-type sector.

$$\begin{aligned}\Sigma_{h0}^{\tilde{c}\tilde{t}} &= \frac{\cos^2\frac{\psi}{2}\sin^2\frac{\psi}{2}e^2}{8\pi^2 M_W^2 s_W^2 \sin^2\beta} \times \\ & \Big\{ B_0\left[0, \tilde{m}_{c1}^2, \tilde{m}_{t2}^2\right] (\cos\alpha(m_c+m_t)(A_0+2m_c-2m_t)+\mu\sin(\alpha)(m_c-m_t))^2 \\ & + B_0\left[0, \tilde{m}_{c2}^2, \tilde{m}_{t1}^2\right] (\cos\alpha(m_c+m_t)(A_0-2m_c+2m_t)+\mu\sin(\alpha)(m_c-m_t))^2 \Big\}. \quad (14)\end{aligned}$$

We numerically evaluate this expression, avoiding the divergent part [1], to obtain a quantification of the possible contribution to the mass which is plotted in Figure 5. We can see that the dominant FV contribution could lower the value of the Higgs mass at one-loop level, compared to the MSSM case with no flavor violation. Then the value for $\tan\beta$ could increase for a Higgs mass of ≈ 125 GeV.

4. Conclusions

In this work, we obtained analytical expressions for the mixing angles for squarks considering a possible mixing within the 2nd and 3rd families, this gives a comprehensive and easy manipulation of the parameters. We obtained the flavor violation couplings performing a complete rotation to physical masses instead of the commonly used Mass Insertion Approximation (MIA) [17], in order to achieve flavor violating processes. The MIA method comes from intuitive applications of a Taylor expansion on masses in a non-physical basis, in practice, this results on splitting propagators, and taking the dominat terms in the mass related functions. Once this mass functions are set, it is nevertheless easy to apply and useful to broadly analyze supersymmetric flavor violation processes. For this work, we only focused in the u-type squark sector to obtain the dominant viable contributions to the Higgs mass radiative corrections. We found that the dominant m_t^4 is not finite and would need to be renormalized. Having experimentally set the Higgs mass, this additional contribution to the Higgs mass should be taken into account to properly bound the MSSM parameters.

Acknowledgments

O F-B, E B-G, and M G-B gratefully acknowledge the provision of funds provided by VIEP-BUAP projects. The work of M G-B was partially supported by DGAPA-UNAM under contract PAPIIT IN113712.

[1] A renormalization procedure should be done to cancel the UV divergent part.

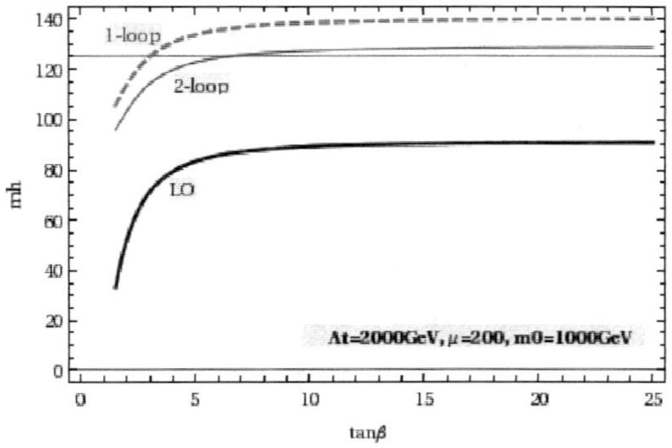

Figure 4. Light neutral MSSM Higgs mass with $\tan\beta$ dependence, complete 1-loop and up to main 2-loops using the FeynHiggs code [16].

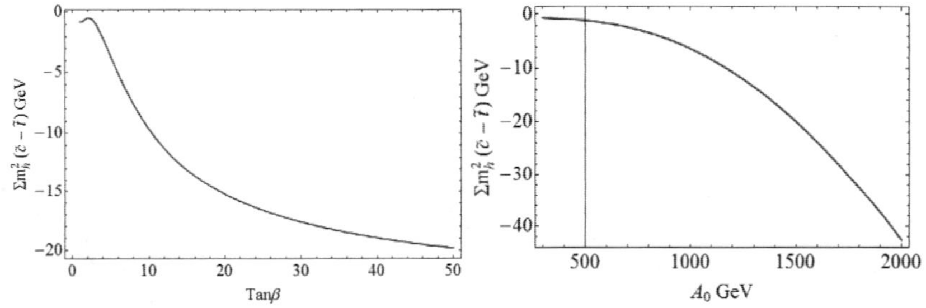

Figure 5. Dominant FV squarks radiative corrections to h^0 mass assuming mixing in second and third families of squarks. Dependence on $\tan\beta$ and in the trilinear term A_0 are shown in the first and second graph, respectively.

References

[1] Paradisi P 2005 *J. High Energy Phys.* JHEP **0510** (2005) 006 (*Preprint* arXiv:hep-ph/0505046)
[2] Okumura K and Leszek Roszkowski L 2003 *J. High Energy Phys* JHEP **0310** (2003) 024
[3] Olive K A and Velasco-Sevilla L 2008 *J. High Energy Phys* JHEP **0805** (2008) 052 (*Preprint* arXiv:0801.0428)
[4] Diaz-Cruz J L 2003 *J. High Energy Phys* JHEP **0305** (2003) 036 (*Preprint* arXiv:hep-ph/0207030).
[5] Diaz-Cruz J L, Gomez-Bock M, Noriega-Papaqui R and Rosado A 2009 *RMF* **55**(4) (2009) 270
[6] Nilles H P 1984 *Phys. Rep.* **110** (1984) 1-162; Haber H E and Kane G L 1985 *Phys. Rep.* **117** (1985) 75; Barbieri R 1988 *Riv. Nuovo Cim.* **11** (1988) 1
[7] Djouadi A 2008 Phys. Rept. **459**, 1
[8] For a review see: Martin S P 1997 *In *Kane, G.L. (ed.): Perspectives on supersymmetry II* 1-153*
[9] Strubig A, Caron S and Rammensee M 2012 *J. High Energy Phys* JHEP **1205**, 150 (*Preprint* arXiv:1202.6244)
[10] Cheng T P and Sher M 1987 *Phys. Rev.* D **35**, 3484 Fritzsch H and Xing Z Z 2000 *Prog. Part. Nucl. Phys.* **45** 1 Fritzsch H and Xing Z Z 2003 *Phys. Lett.* B **555**, 63 (*Preprint* arXiv:hep-ph/0212195)
[11] King S F, Peddie I N R, Ross G G, Velasco-Sevilla L and Vives O 2005 *J. High Energy Phys* JHEP **0507** (2005) 049 (*Preprint* hep-ph/0407012)
[12] Gomez-Bock M 2008 *RMF* **54** (2008) 30 (*Preprint* arXiv:0810.4309)
[13] Diaz-Cruz J L, He H J and Yuan C P 2002 *Phys. Lett.* B **530**, 179 (*Preprint* arXiv:hep-ph/0103178)
[14] Arana-Catania M, Heinemeyer S, Herrero M J and Penaranda S 2012 (*Preprint* arXiv:1201.6345)
[15] Haber H E 2001 *Nucl. Phys. Proc. Suppl.* **101** (2001) 217 (*Preprint* arXiv:hep-ph/0103095)
[16] Frank M, Heinemeyer S, Hollik W and Weiglein G 2010 *Nucl. Phys. Proc. Suppl.* **205-206**, 152 (2010)
[17] See for example: Hagelin J S, Kelley S and Tanaka T 1994 *Nucl. Phys.* B **415** (1994) 293. Moroi T 1996 *Phys. Rev.* D **53** 6565 [Erratum-ibid. D **56** (1997) 4424] (*Preprint* hep-ph/9512396).

Implications of Yukawa textures in the neutral Higgs decays within the 2HDM-III

J E Barradas-Guevara[1], H Bello-Martínez[1], O Félix-Beltrán[2], and J Hernández-Sánchez[2,3]

[1] Facultad de Ciencias Físico-Matemáticas, Benemérita Universidad Autónoma de Puebla, Apdo. Postal 1364, C.P. 72000 Puebla, Pue., México.
[2] Facultad de Ciencias de la Electrónica, Benemérita Universidad Autónoma de Puebla, Apdo. Postal 542, C. P. 72570 Puebla, Pue., México.
[3] Dual C-P Institute of High Energy Physics, Puebla, Pue., México.

E-mail: clever_884@hotmail.com,barradas@fcfm.buap.mx, olga_flix@ece.buap.mx, jaimeh@ece.buap.mx

Abstract. We discuss the implications of assuming a four-zero Yukawa ansatz for the neutral Higgs decays, within the context of the general 2-Higgs Doublet Model of type III. We begin by presenting a detailed analysis of the neutral Higgs boson couplings with fermions and gauge bosons and the resulting effects on its decays. In particular, we are interested on the possibility of the neutral Higgs boson production in current colliders.

1. Introduction

Two open problems in particle physics are the Higgs mechanism [1] which generates fermion masses, and the flavour problem [2]. Thus, the search for the Higgs boson is one of the major goals for the experiments at the Large Hadron Collider (LHC). Moreover, in the Standard Model (SM), the structure of Yukawa couplings is not well understood and neither are their origin nor the underlying principles [2, 3, 4]. Phenomenologically, the SM parameterizes the values of the Yukawa couplings, however, theoretically, we are in the darkness. This is one of the reasons why many physicists agree that the SM should be considered as an effective theory that remains valid up to an energy scale of $O(\text{TeV})$. One way to extend the SM is including an additional isodoublet SU(2) Higgs field. This model is known as the Two Higgs Doublet Model (2HDM) [5, 6, 7, 8], with an scalars spectrum extended (three neutral Higgs bosons and two charged Higgs bosons) and including the unwanted FCNC at tree level.

In the so called version III of the model [9, 10, 11], FCNC are kept under control by imposing a certain shape for the Yukawa matrices that reproduce the observed fermion masses and mixing angles [12]. The use of *texture forms* [13] permits to establish a direct relation from the elements of the matrix with the mixing parameter used in calculating the branching ratios, without dropping in advance terms proportional to the lighter fermion masses [14].

What we know so far about the Yukawa terms is that they should reproduce the fermion masses, the mixing angles of the CKM matrix and keep the FCNC under control, as they are restricted by experimental data. From the Lagrangian form, it is natural to consider that the flavor violation or the mixtures between families, could arise directly from the Yukawa terms,

Content from this work may be used under the terms of the Creative Commons Attribution 3.0 licence. Any further distribution of this work must maintain attribution to the author(s) and the title of the work, journal citation and DOI.
Published under licence by IOP Publishing Ltd

which, in general, are not diagonal. In this work we are interested in the study of the Higgs boson decays $H \to f_i f_j$, as a possible signal of fermion flavor violation for the complete spectrum of neutral Higgs bosons h^0, H^0, A^0 of the 2HDM-III.

A four texture zero ansatz for the Yukawa matrices in the context of 2HDM-III to evaluate the branching ratios of the neutral Higgs bosons decaying into fermion pairs is considered. Moreover, the model predictions and its test at future Large Hadron Collider (LHC) data are explored. Hence, our starting point is an Hermitian four texture ansatz in order to construct the mass matrices, which have been found to be in agreement with the observed data [15, 16]. The relation from the matrix elements to the mixing parameter used to calculate FCNC processes in the Higgs sector is developed in [17]. Some Higgs phenomenology has been addressed in the literature, mainly aimed to describe the light Higgs boson [18, 19].

2. Structure of the Yukawa sector in 2HM-III

In general, the Higgs boson couplings to fermions are given as

$$\mathcal{L}_Y = Y_a^i \overline{\phi}_L^i \Phi_a \phi_R^i + \text{h. c.}, \tag{1}$$

where ϕ_L denotes the left fermion doublet, ϕ_R is right-handed fermion field, Φ_a are the two Higgs doublets ($a = 1, 2$), and the coefficient Y_a^i ($i = l, u, d$) can be expressed as a 3×3 matrix (three generations), Yukawa matrix. Here we consider massless neutrinos. Then, in the Yukawa sector of the 2HDM-III, both Higgs doublets may couple with the two types of fermions, *i.e.*, *up* and *down*, so that we have two different Yukawa terms for each doublet [9, 10, 11].

After spontaneous symmetry breaking, one can derive the fermion mass matrices, namely

$$M_f = \frac{1}{\sqrt{2}}(v_1 Y_1^f + v_2 Y_2^f), \qquad f = u, d, l. \tag{2}$$

Here, we are taking into account the fact that working with a *hierarchical ansatz* for the mass matrix and by means of equation (2), the simplest case is to consider that both Yukawa couplings $Y_{1,2}^f$ possess the same structure (without anomalous cancellation of any of the elements of the matrices). Particularly, we use a Hermitian four texture zero ansatz, and because of eq. (2) the complete mass matrix inherits this structure. The mass matrix is diagonalized through the bi-unitary matrices $V_{L,R}$, though each Yukawa matrices are not diagonalized by this transformation. The diagonalization is performed in the following way $\bar{M}_f = V_{fL}^\dagger M_f V_{fR}$.

The fact that M_f is Hermitian, under the considerations given above (Hermitian Yukawa matrices), directly implies that $V_{fL} = V_{fR}$, and the mass eigenstates for the fermions are given by

$$u = V_u^\dagger u' \qquad d = V_d^\dagger d' \qquad l = V_l^\dagger l'. \tag{3}$$

Then eq. (2) in this basis is given as

$$\bar{M}_f = \frac{1}{\sqrt{2}}(v_1 \tilde{Y}_1^f + v_2 \tilde{Y}_2^f), \tag{4}$$

where $\tilde{Y}_i^f = V_{fL}^\dagger Y_i^f V_{fR}$ and for the quark case we may write

$$\tilde{Y}_1^d = \frac{\sqrt{2}}{v \cos\beta} \bar{M}_d - \tan\beta \tilde{Y}_2^d, \qquad \tilde{Y}_2^u = \frac{\sqrt{2}}{v \sin\beta} \bar{M}_u - \cot\beta \tilde{Y}_1^u. \tag{5}$$

In the lepton case we perform the usual substitution $d \to l$.

By using the redefined fields as the physical states, and considering the Yukawa couplings in this basis, the Lagrangian is obtained to carry out the phenomenology analysis. As we said earlier, it is assumed that each of the Yukawa matrices has the same shape, Hermitian four texture zero ansatz [12], and we see that the three matrices have the same hierarchy and can be parameterized in the same manner.

Accordingly, the $V_{L,R}$ matrices are constructed as the product of two matrices, one of which contains the complex phases[1]. Furthermore, as is given in [17], we impose the condition $m_{f_1} \ll m_{f_2}, m_{f_3}, \mid A^f \mid (f = u, d, l)$, with $A^f = m_{f_3} - \beta^f m_{f_2}$ and β^f a number within the interval $[0,1]$. Therefore, the couplings $\tilde{Y}_2^{d,l}$ and $\tilde{Y}_1^u$, that appear in eq. (5) acquire a simple structure given by $(\tilde{Y}_2^{d,l,u})_{ij} = \frac{\sqrt{m_i^{d,l} m_j^{d,l}}}{v} \tilde{\chi}_{ij}^{d,l,u}$. Then we keep most of the FCNC processes under control provided that $\mid \tilde{\chi}_{ij}^f \mid \leq O(10^{-1})$. In order to carry out the phenomenological study, we rewrite the Lagrangian in terms of the parameter of the model $(\tilde{\chi}_{ij}^f)$. We display here only the leptonic part of the Lagrangian of Yukawa interactions

$$\begin{aligned}
\mathcal{L}_Y^l &= \frac{g}{2}\bar{l}_i \left[\left(\frac{m_{l_i}}{m_W}\right)\frac{\cos\alpha}{\cos\beta}\delta_{ij} + \frac{\sin(\alpha-\beta)}{\sqrt{2}\cos\beta}\left(\frac{\sqrt{m_{l_i}m_{l_j}}}{m_W}\right)\tilde{\chi}_{ij}^l\right] l_j H^0 \\
&+\frac{g}{2}\bar{l}_i \left[-\left(\frac{m_{l_i}}{m_W}\right)\frac{\sin\alpha}{\cos\beta}\delta_{ij} + \frac{\cos(\alpha-\beta)}{\sqrt{2}\cos\beta}\left(\frac{\sqrt{m_{l_i}m_{l_j}}}{m_W}\right)\tilde{\chi}_{ij}^l\right] l_j h^0 \\
&+\frac{ig}{2}\bar{l}_i \left[-\left(\frac{m_{l_i}}{m_W}\right)\tan\beta\delta_{ij} + \frac{1}{\sqrt{2}\cos\beta}\left(\frac{\sqrt{m_{l_i}m_{l_j}}}{m_W}\right)\tilde{\chi}_{ij}^l\right] \gamma^5 l_j A^0.
\end{aligned}$$

We consider that the model parameter is complex in general, $\tilde{\chi}_{ij}^f = \chi_{ij}^f \exp(\imath\vartheta_{ij})$; with the real part $\chi_{ij}^f = |\tilde{\chi}_{ij}^f|$ and the effect of the phase ϑ_{ij} would be included as a variation of -1 to 1 on χ_{ij}^f. Having obtained the couplings in terms of the model parameter χ_{ij}^f, we are ready to calculate the Higgs branching ratios.

3. Branching ratios of the neutral Higgs bosons

Within the SM, we do not have flavor violation decays at tree level, the SM decay width of the Higgs boson to fermions at tree level is given by [8, 20, 21]. In order to evaluate the Higgs branching ratios, we need to include the dominant decay modes, in addition to the fermionic ones. Although we are working within an extension of the SM, we may get some hints for its possible application or connection to a more fundamental theory, by performing phenomenological analysis of Higgs decays. More specifically, we use this model to evaluate the branching ratios for the three neutral Higgs bosons in the 2HDM-III.

In this model, the angles α (the mixing angle in the $CP-even$ Higgs sector), and β (which is the ratio of the vacuum expectation value of the two doublets), are free parameters. Unlike the case of the MSSM, where one can fix α in terms of $\tan\beta$ and m_{A^0}. However, we consider three different scenarios that depend on how these angles are related to each other: (A) $\beta - \alpha = \pi/2$, which it obtain for h^0 the SM-like decays; (B) $\beta - \alpha = 0$, by obtaining for h^0 no flavor violation; and (C) $\beta - \alpha = \pi/3$, which take an angle in between these extreme cases.

Because for h^0 the flavor violation couplings vanish for $\alpha - \beta = 0$, we only show results for scenarios A and C, $\tan\beta = 5, 15, 30$ with $\chi_{ij} = -1.0,\ 0,\ 1.0$. In the latter case: large $\tan\beta$, the value is chosen where the behavior of the branching ratios becomes, in general, quite independent of $\tan\beta$ for different channels of the light Higgs boson.

In Figs. 1 and 2 the dependence of the BR on the light Higgs mass h^0 is displayed, which correspond to scenario A and C, respectively. In these cases, the vector boson decay channel is open for almost the entire mass range, in fact it is the main decay mode, leaving bb fermionic

[1] The complete process can be found in Ref. [13] and the explicit expression of these matrices is given in Refs. [15] and [17].

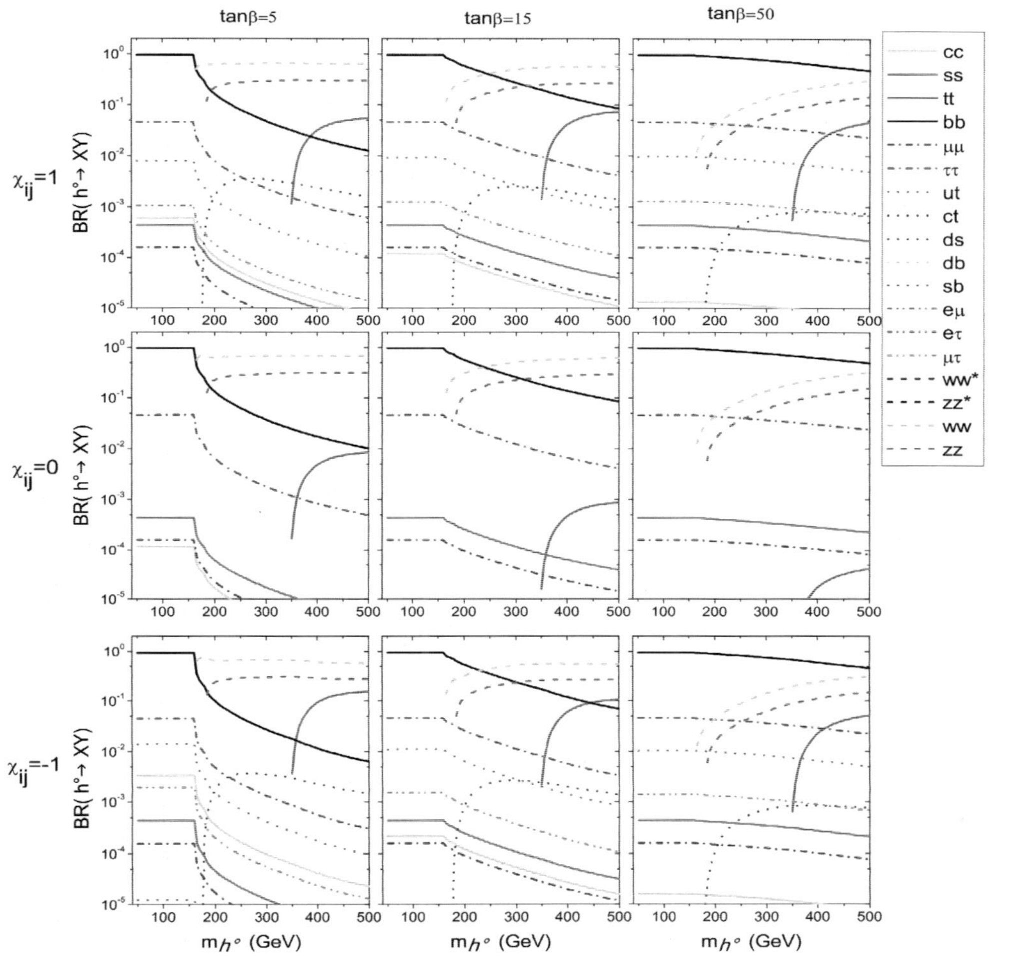

Figure 1. Branching ratios of the fermionic and bosonic processes $h^0 \to XY$ in scenario A, within the Higgs mass range $50\,\text{GeV} < m_{h^0} < 500\,\text{GeV}$, for $\tan\beta = 5, 15, 50$, and $\chi_{ij} = -1.0, 0, 1.0$.

mode as dominant contribution of the decay rate when $50\,\text{GeV} \leq m_{h^0} \leq 100\,\text{GeV}$ with $\tan\beta = 5$, however, this channel is given dominant within $50\,\text{GeV} \leq m_{h^0} \leq 500\,\text{GeV}$ for $\tan\beta = 50$. However, there is a region where the fermion decay becomes important, when the top-threshold is reached. In this case, there is a strong dependence, mostly on large values of $\tan\beta$ and a model parameter close to 1: $\tan\beta > 15$ and $\chi_{ij} \to 1$. On the other hand, the flavor violating decays are reduced in the region of parameter space where $\tan\beta$ is small and χ_{ij} is close to zero as can be seen in the Fig. 1. Here we have fixed the heavy Higgs mass at 300GeV and taken the more favored scenario for the enhancement of these decays, $\beta - \alpha = \pi/2$.

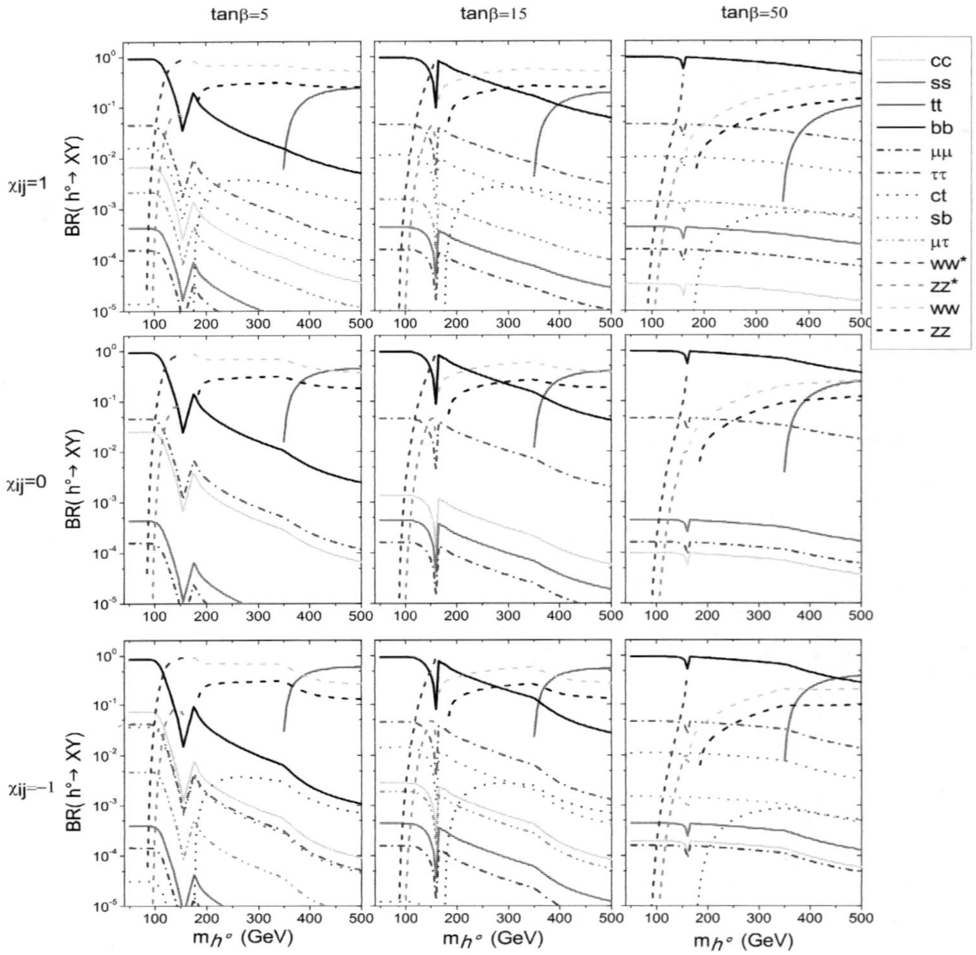

Figure 2. Branching ratios of the fermionic and bosonic processes $h^0 \to XY$ in scenario C, within the Higgs mass range $50\,\text{GeV} < m_{h^0} < 500\,\text{GeV}$, for $\tan\beta = 5, 15, 50$, and $\chi_{ij} = -1.0, 0, 1.0$.

4. Conclusions

We studied the 2HDM type III and calculated the corresponding Lagrangian of Yukawa interactions with four texture zero matrices like a form of the Yukawa matrices, and using the Cheng-Sher *anzats*. Also, we obtained the Feynman rules to tree level for the neutral Higgs bosons and calculated the width of decay as well as the corresponding branching ratios for the main decay modes. The parameter space was scanned by defining three scenarios, in which we were considered that $\tan\beta = 5, 15, 50$, $\chi_{ij} = -1.0, 0, 1.0$, and $50\,\text{GeV} < m_{h^0} < 500\,\text{GeV}$. Thus, we have explored the parameter space of this model in order to determinate the areas where h^0 reachs maximal branching ratios. Studying these modes at current and future colliders could be important to find new Higgs signals.

Acknowledgments
H B-M, O F-B, E B-G, and J H-S gratefully acknowledge the provision of funds provided by VIEP-BUAP projects.

References
[1] Peter W Higgs P W 1964 *Phys. Rev. Lett.* vol13 p 508
[2] Frogatt C D and Nielsen H B 1979 *Nucl. Phys. B* vol 147 p 277
[3] For example see: Cheng T P and Li L F 1984 *Gauge theory of elementary particle physics* (Oxford University Press)
[4] Fritzsch H *The physics of flavor is the flavor of physics* arXiv:hep-ph/0111051
[5] Glashow S and Weinberg S 1977 *Phys. Rev. D* vol 15 p 1958
[6] Maniatis M von Manteuffel A Nachtmann O and Nagel F 2006 *Eur. Phys. J. C* vol 48 pp 805-823
[7] Krawczyk M 2006 *PoS HEP2005* p 335
[8] For a review see: Gunion J F, Haber H E, Kane G L and Dawson S 1990 *The Higgs Hunter's Guide* (Addison-Wesley, Reading, MA)
[9] Gómez-Bock M and Noriega-Papaqui R 2006 *J. Phys. G* vol 32 p 761
[10] Barradas Guevara J E Cazarez Bush F Cordero Cid A Felix Beltran O Hernandez Sanchez J and Noriega Papaqui R 2010 *J. Phys. G* vol 37 p 115008
[11] Diaz Cruz J L Noriega Papaqui R and Rosado A 2004 *Phys. Rev. D* vol 71 p 015014
[12] Cheng T P and Sher M 1987 *Phys. Rev. D* vol 35 p 3484
[13] Fritzsch H and Xing Z 2000 *Prog. Part. Nucl. Phys.* vol 45 p 1
[14] See for example: Atwood D Reina L and Soni A 1997 *Phys. Rev. D* vol 55 p 3156; Hou S W 1992 *Phys. Lett. B* vol 296 p 179; Luke M E and Savage M J 1993 *Phys. Lett. B* vol 307 p 387
[15] Fritzsch H and Xing Z Z 2003 *Phys. Lett. B* vol 555 p 63
[16] Xing Z Z and Zhang H 2004 *J. Phys. G* vol 30 p 129
[17] Diaz-Cruz J L Noriega-Papaqui R and Rosado A 2004 *Phys. Rev. D* vol 69 p 095002
[18] Diaz-Cruz J L Noriega-Papaqui R and Rosado A 2005 *Phys. Rev. D* vol 71 p 015014
[19] Martinez R Rodriquez J A and Milanes D A 2005 *Phys. Rev. D* vol 72 p 035017
[20] Keung W Y and Marciano W J 1984 *Phys. Rev. D* vol 30 p 248
[21] Barradas E Diaz-Cruz J L Gutierrez A and Rosado A 1996 *Phys. Rev. D* vol 53 p 1678; Diaz-Cruz J L and Perez M A 1986 *Phys. Rev. D* vol 33 p 273

PASCOS 2012 – 18th International Symposium on Particles Strings and Cosmology IOP Publishing
Journal of Physics: Conference Series **485** (2014) 012062 doi:10.1088/1742-6596/485/1/012062

General 4-zero texture mass matrix parametrizations

J Barranco, D Delepine and L Lopez-Lozano

División de Ciencias e Ingenierías, Universidad de Guanajuato, Campus León, C.P. 37150, León, Guanajuato, México.

E-mail: lao-tse@fisica.ugto.mx

Abstract. It is performed the diagonalization of a non-Hermitian four-zero texture Yukawa matrix with a general formalism. This procedure leads to 3 possibilities to parametrize the relation between the fermion masses and the elements of the corresponding Yukawa matrix. Then, the matrices that diagonalize each Yukawa mass matrix are combined in order to obtain 9 different theoretical CKM or PMNS mixing matrices [1]. Through a χ^2 analysis, we have constrained the values of the remaining free parameters such as the theoretical mixing matrix matches the latest experimental measurements of the mixing matrices. This analysis was done without assuming any approximations. In the case of the quark sector, it is found that only four different theoretical mixing matrices are compatible with the actual high precision experimental measurement of the CKM matrix elements. For the lepton sector, where the masses of neutrinos are not known, we found that independently of the parametrization that have been chosen, the updated experimental measurements of the mixing angles in the PMNS matrix, imply a mass for the heaviest left-handed neutrino to be ~ 0.05eV.

1. Introduction

None of the Standard Model (SM) symmetries explain the mass spectrum of fermions or the number o families. This give to fermion masses and the mixing matrix elements the status of free parameters of the model that have to be fixed trough experimental measurements. Nevertheless, experimental measurements of this free parameters can be done. Once measured, such values can be used as inputs to make new predictions for different non-measured processes. Surprisingly, most of those predictions have been now confirmed experimentally, turning the SM in the most successful description of the elementary particles at the electroweak scale. In last years the mixing in both, the quark and the leptonic sectors, that are given in terms of the Cabbibo-Kobayashi-Maskawa (CKM) matrix elemenst or the Pontecorvo-Maki-Nakagawa-Sakata (PMNS) matrix elements , have been measured with increasingly precision. It is expected that future B factories, like SuperB [2], will improve the measurements of CKM parameters with a precision less than 1%. In a similar way, current and future neutrino facilities will improve the precision of the PMNS elements. In other words, any extension of the SM that would like to explain the question about the mass spectrum and mixing, should have to reproduce the mass and mixing data at least at the same precision. A more general theory, with an underlying symmetry between flavors will predict a Yukawa matrix in the flavor basis that could reduce the number of free parameters because it exists a close relation between the masses of fermion and the mixing matrix elements [3, 4, 5, 6, 7, 8]. There are many models Beyond the Standard Model (BSM) that lead to Yukawa matrices with structures that could reveal possible underlying symmetries between flavor of fermions. From a phenomenological point of view, it is possible to

Content from this work may be used under the terms of the Creative Commons Attribution 3.0 licence. Any further distribution of this work must maintain attribution to the author(s) and the title of the work, journal citation and DOI.
Published under licence by IOP Publishing Ltd

analyze if there is an underlying symmetry with a different approach, which consist in proposing a reasonable phenomenological form of the Yukawa matrix (*a Texture*) and restrict its entrances with experimental information. One of the early proposes was made by H. Fritzsch [9, 10] with a texture mass matrix with the smallest number of free parameters that can deal with experimental bounds. Following the same philosophy, the 4-zero texture mass matrix has been one of the most used to parametrize the Yukawa couplings in the flavor basis [7, 11, 12, 13, 9, 4, 5, 14]. This matrix can arise from extended models with discrete symmetries or GUT at low energies, thus it is important to understand the properties and scopes of this kind of matrices in order to interpret the possible new physics effects that are predicted by extended models leading to this type of textures. It is important to stress that the number of the introduced free parameters at the Yukawa matrices can not be greater than the relation found in the SM if we want to have new information about flavor. In this sense, the search of Yukawa matrices that can describe very general relation between flavor mixing and masses with the minimum number of free parameters is an essential task to parametrize new physics effect coming from several models. In this work we point out that, although in many works has been demonstrated that the hermitian 4-zero Yukawa matrix is a good framework to find phenomenologically consistent relation between mass and mixing angles, the hermitcity condition is not needed. A particular non-hermitian matrices can be considered with the same number of free parameters and conditions phases of the matrix elements. This open the possibility to consider models BSM that naturally predict non-hermitian Yukawa matrices.

2. Parametrizations of the non-hermitian four-zero Yukawa matrix

The general 4-zero matrix textures has the form

$$M^f = \begin{pmatrix} 0 & C_f & 0 \\ C'_f & D_f & B_f \\ 0 & B'_f & A_f \end{pmatrix}. \tag{1}$$

In hermitian case, one has to assume that $C'_f = C^*_f$ and $B'_f = B^*_f$. Here we only assume that $|C_f| = |C'_f|$ and $|B_f| = |B'_f|$ but the phases can be different [15, 16], is in this sense that we are using the term general. Because this matrix is in general non-hermitian, the diagonalitazion process must be done with a bi-linear transformation of the form

$$U_L^{f\dagger} M^f U_R^f = \text{diag}(m_1^f, m_2^f, m_3^f), \tag{2}$$

where $f = u, d, \ell$ is a flavor index, $\{U_L^f, U_R^f\}$ are $SU(3)$ matrices and the eigenvalues m_1^f, m_2^f, m_3^f are already known because they are the fermion masses.

Only the left-handed matrix is observable once the CKM or the PMNS matrix elements are measured, thus we only need to find the matrix U_L^f solving the equation

$$U_L^{f\dagger} H^f U_L^f = \text{diag}(m_1^{f2}, m_2^{f2}, m_3^{f2}), \tag{3}$$

where is defined the matrix $H^f \equiv M^f M^{f\dagger}$. From now on, we omit the index f for short.

The key question in the reduction of parameters is to observe that H^f is hermitian by construction, thus it is possible to separate the phases of its non-diagonal element with a unitary transformation as usual, writing $B(B') = be^{i\phi_B(\phi_{B'})}$, $C(C') = ce^{i\phi_C(\phi_{C'})}$, $D = de^{i\phi_D}$ and $A = ae^{i\phi_A}$, *i.e.*

$$H = P^\dagger \tilde{H} P, \tag{4}$$

where $P = e^{-\frac{i}{2}\Xi}\text{diag}\left(e^{\frac{i}{2}\Xi}, e^{i(\phi_C - \phi_D)}, e^{i(\phi_C + \phi_{B'} + \Xi)}\right)$ and

$$\Xi = \arctan\left[\frac{a\sin(\phi_B + \phi_{B'} - \phi_A - \phi_D)}{d + a\cos(\phi_B + \phi_{B'} - \phi_A - \phi_D)}\right]. \tag{5}$$

Then the real and symmetric matrix to diagonalize is given by

$$\tilde{H} = \begin{pmatrix} c^2 & cd & bc \\ cd & c^2+d^2+b^2 & b|d+a\delta| \\ bc & b|d+a\delta^*| & a^2+b^2 \end{pmatrix}, \tag{6}$$

where $\delta = e^{i(\phi_D-\phi_{B'}-\phi_B+\phi_A)}$. Although this matrix is more complicated than the matrix obtained in the hermitian case, it depends on the same 4 parameters and only one additional parameter δ that takes values in terms of the phases. It is straightforward to find the eigenvectors $\mathbf{v}(m_i^2)$ of $\tilde{H}$ solving the equation $(\tilde{H}-m_i^2 1_{3\times3})\mathbf{v}(m_i^2)=\mathbf{0}$, nevertheless if one restrict the solution to obey the invariants relations under unitary transformations for the matrix $\tilde{H}$, we get that δ must be real and positive to obtain consistent solutions with the hierarchy of fermion masses, that is $\Xi_f = 0$ for $f = u, d, \ell, \nu$. Thus the matrix of phases can be written as $P^f = \mathrm{diag}(1, e^{i\phi_1^f}, e^{i\phi_2^f})$, where $\phi_1^f \equiv \phi_{C_f} - \phi_{D_f}$ and $\phi_2^f \equiv \phi_{C_f} + \phi_{B'_f}$ for $f = u, d, \ell, \nu$. As expected, the hermitian case can be reached if the phases are constrained to $\phi_{D_f} = 0$ and $\phi_{B_f} = -\phi_{B'_f}$. It is important to stress that the relations between the elements of the mixing matrix and the masses are not the same than in the hermitian case and the specific mass spectrum of fermions imposes relations on the phases of the matrix (1), this is important because we have obtain that the same number of phases needed in the hermitian case is the same number of phases needed in the non-hermitian case imposing a specific spectrum of masses.

In order to reduce free parameters in $\tilde{H}$ using the fact that squared fermions masses are the eigenvector of $\tilde{H}$, the equations

$$\begin{aligned} \mathrm{Tr}(\tilde{H}) &= m_1^2+m_2^2+m_3^2, \\ \mathrm{Tr}^2(\tilde{H})-\mathrm{Tr}(\tilde{H}^2) &= 2m_1^2m_2^2+2m_1^2m_3^2+2m_2^2m_3^2, \\ \mathrm{Det}(\tilde{H}) &= m_1^2m_2^2m_3^2, \end{aligned} \tag{7}$$

are solved for b, c and d in terms of the eigenvalues m_i^2 and the parameter a. The system (7) has formally 32 solutions that are closely related by chiral transformations that basically change the sign of the masses m_i, thus without loss of generality we can assumed that $m_i > 0$, where $i = 1, 2, 3$, and adjust the interval of possible values that the free parameters can take. Once the restrictions on parameters and the normal order of masses $m_1 < m_2 < m_3$ are introduced, the number of independent solution is reduce to three. This means that exist three independent parametrizations for U_L^f that diagonalize H. Such parametrizations are given by

$$\text{Parametrization 1} \qquad \tilde{m}_1 \le a' \le \tilde{m}_2 \qquad \begin{cases} b' &= \sqrt{\frac{(a'-\tilde{m}_1)(-a'+\tilde{m}_2)(\tilde{m}_3+a')}{a'}} \\ d' &= -a'+\tilde{m}_1+\tilde{m}_2-\tilde{m}_3 \end{cases} \tag{8}$$

$$\text{Parametrization 2} \qquad \tilde{m}_1 \le a' \le \tilde{m}_3 \qquad \begin{cases} b' &= \sqrt{\frac{(a'-\tilde{m}_1)(a'+\tilde{m}_2)(\tilde{m}_3-a')}{a'}} \\ d' &= -a'+\tilde{m}_1-\tilde{m}_2+\tilde{m}_3 \end{cases} \tag{9}$$

$$\text{Parametrization 3} \qquad \tilde{m}_2 \le a' \le \tilde{m}_3 \qquad \begin{cases} b' &= \sqrt{\frac{(a'+\tilde{m}_1)(a'-\tilde{m}_2)(\tilde{m}_3-a')}{a'}} \\ d' &= -a'-\tilde{m}_1+\tilde{m}_2+\tilde{m}_3 \end{cases} \tag{10}$$

where for all cases $c' = \sqrt{\frac{\tilde{m}_1\tilde{m}_2\tilde{m}_3}{a'}}$. Here the parameters $a' = \frac{a}{m_3}$, $b' = \frac{b}{m_3}$, $c' = \frac{c}{m_3}$ and $d' = \frac{d}{m_3}$ have been defined. Likewise the scaled masses are $\tilde{m}_i = \frac{m_i}{m_3}$ for $i = 1, 2, 3$; here one can see that $\tilde{m}_3 = 1$.

Parametrization	a'_u 1 σ C.L.	a'_d 1 σ C.L.	ϕ_{bfp}	χ^2_{min}
$(2,2)$	$0.993 < a'_u < 0.997$	$0.9982 < a'_d < 1.0$	$\phi_{\text{bfp}} = 1.5723$	1.27
$(2,3)$	$0.9950 < a'_u < 0.9977$	$0.99915 < a'_d < 1.0$	$\phi_{\text{bfp}} = 1.568$	1.48
$(3,2)$	$0.9945 < a'_u < 0.9978$	$0.9989 < a'_d < 1.0$	$\phi_{\text{bfp}} = 1.568$	2.13
$(3,3)$	$0.9958 < a'_u < 0.9978$	$0.99949 < a'_d < 1.0$	$\phi_{\text{bfp}} = 1.5723$	2.163

Table 1. Allowed values of the parameters a'_u and a'_d at 68% C.L.

3. χ^2 analysis

We define the quark and lepton flavor mixing matrices, U_{PMNS} and V_{CKM} with the help of the matrices O_f, $f = l, \nu, u, d$ that diagonalize the Yukawa mass matrices of charged leptons, left-handed neutrinos, u-type quarks and d-type quark respectively. By incorporating the phases again we have that the theoretical mixing matrix obtained from the four-zero texture Yukawa mass matrices are:

$$V^{\text{th}}_{\text{CKM}} = O_u^T P^{u-d} O_d \,, \qquad U^{\text{th}}_{\text{PMNS}} = O_l^T P^{l-\nu} O_\nu \,, \tag{11}$$

where $P^{u-d} = \text{diag}[1, e^{i(\phi_1^u - \phi_1^d)}, e^{i(\phi_2^u - \phi_2^d)}]$ and $P^{l-\nu} = \text{diag}[1, e^{i(\phi_1^\ell - \phi_1^\nu)}, e^{i(\phi_2^\ell - \phi_2^\nu)}]$.

The matrices V_{CKM} and U_{PMNS} can be expressed as explicit functions of the masses of quarks and leptons and the few free parameters $a'_u, a'_d, \phi, a'_l, a'_\nu, m_{\nu 3}$. We fix $\phi \equiv \phi_1^u - \phi_1^d = \phi_2^u - \phi_2^d$ and $\Phi_1 \equiv \phi_1^\ell - \phi_1^\nu = \phi_2^\ell - \phi_2^\nu$.

By using the current values of the unitary angles and the CKM entries [17] together with the current values of the PMNS matrix , including a large lepton mixing angle θ_{13} provided by T2K, MINOS and Double Chooz experiments [18], we have performed a simple χ^2 analysis. For quarks the χ^2 function is given by:

$$\chi^2_{\text{quarks}}(a'_u, a'_d, \phi) = \sum_{i=1}^{9} \left(\frac{V^{th}_{\text{CKM}}(a'_u, a'_d, \phi) - |V_{\text{CKM}}|}{\delta V_{\text{CKM}}} \right)^2 + \chi^2_{\text{angles}} \,. \tag{12}$$

with

$$\chi^2_{\text{angles}} = \left(\frac{\alpha^{th} - \alpha^{Exp}}{\delta \alpha^{Exp}} \right)^2 + \left(\frac{\beta^{th} - \beta^{Exp}}{\delta \beta^{Exp}} \right)^2 + \left(\frac{\gamma^{th} - \gamma^{Exp}}{\delta \gamma^{Exp}} \right)^2 \,. \tag{13}$$

The allowed values for a'_u and a'_d 68% C.L. are shown in Table 1. It is interesting to observe that the best fit point (bfp) for the phase angle ϕ is very close to $\sim \pi/2$ in all cases. Furthermore, the strong hierarchy in the CKM matrix structure demands an strong hierarchy in the Yukawa mass matrix, so strong that practically it is impossible to see a difference between the different matrices obtained for the each parametrization.

For the leptonic case, the χ^2 is also a function of the mass of the heaviest neutrino $m_{\nu 3}$. We perform the χ^2 analysis only with the moduli of the U_{PMNS} matrix

$$\chi^2_{\text{leptons}}(a'_l, a'_\nu, \Phi_1, m_{\nu 3}) = \sum_{i=1}^{9} \left(\frac{U^{th}_{\text{PMNS}}(a'_l, a'_\nu, \Phi_1, m_{\nu 3}) - |U_{\text{PMNS}}|}{\delta U_{\text{PMNS}}} \right)^2 \,. \tag{14}$$

In this case we have an extra parameter, that is, the mass of the heaviest neutrino. The best fit points for the different parametrizations after minimizing the χ^2 function are shown in Table 2. From this table, it can be noticed that each parametrization can lead to different structure in the Yukawa mass matrices. Actually, the CP phase angle Φ_1 has now different values, even as small as zero. But it is also important to stress that all 9 possibilities leads to almost the same mass for the heaviest neutrino $m_{\nu_3} \sim 0.05$ eV.

Parametrization	$m_{\nu_3}^{\text{bfp}}$	Φ_1^{bfp}	a_ν^{bfp}	a_l^{bfp}	χ^2_{min}
(1, 1)	4.99×10^{-2}	$\Phi_1 = 3.14159$	$a_\nu = 0.2125$	$a_l = 0.0000$	$\chi^2_{\text{min}} = 4.6932$
(2, 2)	4.96×10^{-2}	$\Phi_1 = 2.80233$	$a_\nu = 0.6750$	$a_l = 0.0375$	$\chi^2_{\text{min}} = 0.0033$
(3, 3)	4.99×10^{-2}	$\Phi_1 = 1.679911$	$a_\nu = 0.4250$	$a_l = 0.9875$	$\chi^2_{\text{min}} = 0.7727$
(1, 2)	4.96×10^{-2}	$\Phi_1 = 2.366666$	$a_\nu = 0.2250$	$a_l = 0.8375$	$\chi^2_{\text{min}} = 0.0008$
(2, 1)	4.99×10^{-2}	$\Phi_1 = 0.000000$	$a_\nu = 0.2125$	$a_l = 0.0000$	$\chi^2_{\text{min}} = 4.6931$
(3, 2)	4.96×10^{-2}	$\Phi_1 = 3.141592$	$a_\nu = 0.3250$	$a_l = 0.0000$	$\chi^2_{\text{min}} = 0.0063$
(1, 3)	4.96×10^{-2}	$\Phi_1 = 0.000000$	$a_\nu = 0.2125$	$a_l = .02500$	$\chi^2_{\text{min}} = 2.0005$
(2, 3)	4.99×10^{-2}	$\Phi_1 = 1.466076$	$a_\nu = 0.4250$	$a_l = 0.9875$	$\chi^2_{\text{min}} = 0.6872$
(3, 1)	4.95×10^{-2}	$\Phi_1 = 3.141592$	$a_\nu = 0.0375$	$a_l = 0.1625$	$\chi^2_{\text{min}} = 2.8631$

Table 2. Best fit points for the leptonic sector.

3.0.1. The relevance of large $\sin\theta_{13}$ The increasing precision in the determination of the mixing angles is a powerful tool in order to analyze if certain texture is able to reproduce the mixing pattern. In our analysis this can be seen in Fig. 1. In black solid line it is shown the projection of the χ^2 function in the m_{ν_3} plane, obtained when the small, old value of $\sin\theta_{13}$. It can be seen that at $\sim 1\sigma$ C.L. there is only a lower bound on the neutrino mass m_{ν_3}. On the other hand, if we update the value on $\sin\theta_{13}$ to the latest experimental value [18], we can perform the same χ^2 analysis and we obtain the red line shown in the same Fig. 1. It can be seen now that the χ^2 function as a function of the neutrino mass m_{ν_3} have been narrowed pointing out to a favored mass of $m_{\nu_3} = 0.05$.

4. Conclusions

Here we have shown that when a non-hermitian four-zero mass matrix is considered, the explicit inclusion of the known eigenvalues as the masses of fermions leads to relations on the phases. This relations reduce the number of phenomenologically relevant free parameters to be the same as in the hermitian case. Although this analysis can not determine the specific values of phases, it demonstrates that the same χ^2 analysis that is done for the hermitian case, can be use for more general matrices as the four-zero mass matrix with that leads to different relations and bounds. The general four-zero mass matrix can arise from extensions of the SM and here we have shown without approximations the implications on the relations between fermion masses and mixing matrix elements. The χ^2 analysis shows that a small variation on the experimental measurements can have important consequences on the prediction of free parameters. In the lepton sector an updated measurement of the neutrino mixing angle θ_{13} leads to a prediction for the mass of the heaviest neutrino to be ~ 0.05eV.

Aknowledgments

This works has been partially supported by Conacyt Proyecto 156618, Conacyt-SNI and PIFI project of the SEP (Secretaria de Educacion Publica, Mexico) and Guanajuato University through DAIP project.

References

[1] Barranco J, Delepine D and Lopez-Lozano L 2012 (*Preprint* 1205.0859)
[2] Aushev T, Bartel W, Bondar A, Brodzicka J, Browder T *et al.* 2010 (*Preprint* 1002.5012)
[3] Branco G, Emmanuel-Costa D and Gonzalez Felipe R 2000 *Phys.Lett.* **B477** 147–155 (*Preprint* hep-ph/9911418)
[4] Fritzsch H 1979 *Nucl.Phys.* **B155** 189
[5] Fritzsch H and Xing Z z 2000 *Prog.Part.Nucl.Phys.* **45** 1–81 (*Preprint* hep-ph/9912358)

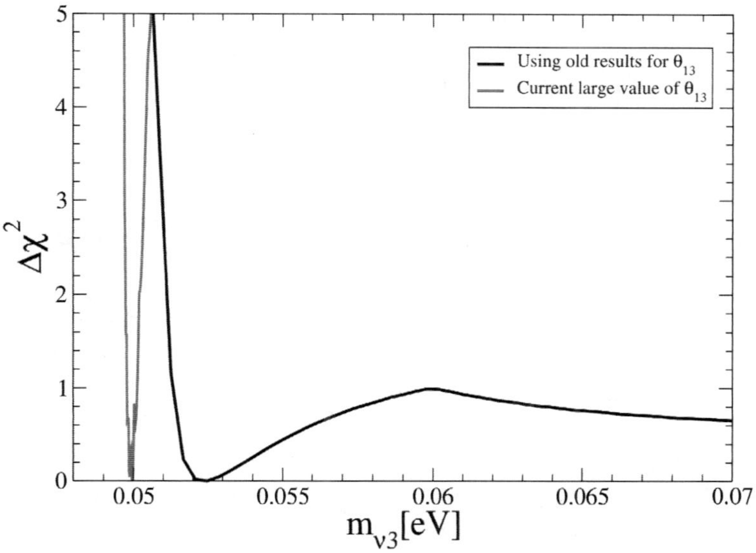

Figure 1. Effect of the new measurement on the neutrino mixing angles in the determination of the neutrino mass for a 4-zero mass matrix texture.

[6] Harari H, Haut H and Weyers J 1978 *Phys.Lett.* **B78** 459
[7] Hu L J, Dulat S and Ablat A 2011 *Eur.Phys.J.* **C71** 1772
[8] Ibanez L E and Ross G G 1994 *Phys.Lett.* **B332** 100–110 (*Preprint* `hep-ph/9403338`)
[9] Fritzsch H 1978 *Phys.Lett.* **B73** 317–322
[10] Fritzsch H 1977 *Phys. Lett.* **B70** 436
[11] Xing Z z 2002 *Phys.Lett.* **B530** 159–166 laTex 11 pages. Slight changes. Phys. Lett. B (in printing) Report-no: BIHEP-TH-2002-5 (*Preprint* `hep-ph/0201151`)
[12] Barranco J, Gonzalez Canales F and Mondragon A 2010 *Phys.Rev.* **D82** 073010 (*Preprint* `1004.3781`)
[13] Gonzalez Canales F and Mondragon A 2011 *J.Phys.Conf.Ser.* **287** 012015 presented at XIV Mexican School on Particles and Fields, 4-13 November 2010, Morelia México (*Preprint* `1101.3807`)
[14] Fritzsch H and Xing Z z 2003 *Phys.Lett.* **B555** 63–70 (*Preprint* `hep-ph/0212195`)
[15] Branco G and Silva-Marcos J 1994 *Phys.Lett.* **B331** 390–394
[16] Mondragon A, Mondragon M and Peinado E 2011 *Phys.Atom.Nucl.* **74** 1046–1054
[17] Nakamura K *et al.* (Particle Data Group) 2010 *J.Phys.G* **G37** 075021
[18] Fogli G, Lisi E, Marrone A, Palazzo A and Rotunno A 2011 *Phys.Rev.* **D84** 053007 slightly revised text/ results unchanged. To appear in Phys. Rev. D (*Preprint* `1106.6028`)

S_3 as a unified family theory for quarks and leptons

F González Canales[1,2], A Mondragón[1], U J Saldaña Salazar[1] and L Velasco-Sevilla[3]

(1) Instituto de Física, Universidad Nacional Autónoma de México,
Apdo. Postal 20-364, 01000, México D.F., México.
(2) Facultad de Ciencias de la Electrónica, Benemérita Universidad Autónoma de Puebla,
Apdo. Postal 157, 72570, Puebla, Pue., México.
(3) CINVESTAV-IPN, Apdo. Postal 14-740, 07000, México D.F., México.

E-mail: ulisesjesus@fisica.unam.mx

Abstract. We present an S_3-invariant extension of the SM which is able to account for the mixing both in the quark and the lepton sector. We focus here on the quark sector and present different realizations of the model, according to how many Higgs fields are involved. We then confront the models with the up-to-date values of masses and mixing in the quark sector.

1. Introduction

The long awaited remaining piece of the Standard Model (SM), *i.e.* the Higgs particle, has been probably recently detected [1, 2]. Despite this great discovery, the SM, as was originally formulated, needs to be extended in order to accommodate the small masses of neutrinos. Moreover, clearly the number of free parameters needed to account for the values of fermion masses and mixing parameters (12 masses, 6 mixing angles, and 2 (4) Dirac phases (Dirac plus Majorana phases)), reflects our lack of understanding of the flavour puzzle. Nevertheless, at present, mathematical relations among masses and mixing angles [3–6] seem to indicate a common origin for the pattern of mixing and the hierarchical spectrum of masses. This origin could be justified by introducing additional symmetries into the theory, which are commonly called *horizontal or family symmetries* because they act in the generation or family space of fermions.

The main goal of this work is to present an extension of the SM where successful and exact mathematical relations among fermion masses and mixing angles can be obtained. In section 2, we give some brief arguments on how easy the S_3 symmetry can be interpreted as a family symmetry. In section 3, we discuss and introduce the construction of the S_3-invariant extension of the Standard Model (S_3-SM) we are considering. In section 4, we contrast the different realizations of this model with the most up to date experimental information. In section 5, we summarize and conclude with some remarks.

2. S_3 as a family symmetry

In our universe, atoms and atomic processes, as beta decay, require a single *family* of fundamental fermions (e, ν_e, d, u). However, in subatomic processes nature seems to need two more families of fundamental particles, replicas of the first family but with different masses. The couplings of the three families to the force carriers is universal, making no distinction between particles

Content from this work may be used under the terms of the Creative Commons Attribution 3.0 licence. Any further distribution of this work must maintain attribution to the author(s) and the title of the work, journal citation and DOI.
Published under licence by IOP Publishing Ltd

with the same quantum charges but which belong to a different family. The SM Lagrangian, except for the terms describing the Yukawa interactions, is invariant under the permutation of the family index, that is, any permutation of the family index should leave the system invariant. The latter statement can be treated as having applied the following symmetry transformations,

$$\begin{pmatrix}1&0&0\\0&1&0\\0&0&1\end{pmatrix}\begin{pmatrix}0&1&0\\1&0&0\\0&0&1\end{pmatrix}\begin{pmatrix}0&0&1\\0&1&0\\1&0&0\end{pmatrix}\begin{pmatrix}1&0&0\\0&0&1\\0&1&0\end{pmatrix}\begin{pmatrix}0&0&1\\1&0&0\\0&1&0\end{pmatrix}\begin{pmatrix}0&1&0\\0&0&1\\1&0&0\end{pmatrix}, \tag{1}$$

which leave invariant an equilateral triangle whose vertices are the unitary basis vectors. This last statement expresses what we call an isomorphism between the permutation group of three objects, S_3, and the equilateral triangle symmetry group, D_3. We will later refer to this basis, where each family is undistinguishable, as the *universal-basis.*

2.1. S_3 theoretical aspects

S_3 is the smallest non-abelian discrete symmetry group. It has $3! = 6$ elements and has two classes of non-trivial transformations: even and odd permutations, A and B, respectively. The two kind of generators, A and B, fulfill the relations

$$A^2 = B^3 = 1, \qquad ABA = B^{-1}. \tag{2}$$

When taking into account the trivial transformation, it is found that S_3 contains three classes of transformations; thus, it has three irreducible representations (irreps): a doublet **2**, and two singlets, $\mathbf{1}_S$ and $\mathbf{1}_A$, symmetric and antisymmetric, respectively.

The Kronecker products between irreps may be decomposed as a direct sum of irreps:

$$\begin{gathered}\mathbf{1}_S\otimes\mathbf{1}_S=\mathbf{1}_S,\quad \mathbf{1}_A\otimes\mathbf{1}_A=\mathbf{1}_S,\quad \mathbf{1}_A\otimes\mathbf{1}_S=\mathbf{1}_A,\\ \mathbf{1}_S\otimes\mathbf{2}=\mathbf{2},\quad \mathbf{1}_A\otimes\mathbf{2}=\mathbf{2},\quad \text{and}\quad \mathbf{2}\otimes\mathbf{2}=\mathbf{1}_A\oplus\mathbf{1}_S\oplus\mathbf{2}.\end{gathered} \tag{3}$$

2.2. The flavour-basis

Fermions have different masses, in fact, so different that are commonly referred as hierarchical masses,

$$\begin{pmatrix}m_e\\m_{\nu_1}\\m_d\\m_u\end{pmatrix} << \begin{pmatrix}m_\mu\\m_{\nu_2}\\m_s\\m_c\end{pmatrix} << \begin{pmatrix}m_\tau\\m_{\nu_3}\\m_b\\m_t\end{pmatrix}, \tag{4}$$

where we are assuming, just for illustration, a normal hierarchy and non degenerate neutrino masses. After the introduction of the Yukawa interactions, each fermion should couple differently to the Higgs field, such that, when the electroweak symmetry gets spontaneuosly broken, each fermion gains a different mass. A unitarity transformation takes us from the *universal-basis* to the S_3 *symmetry adapted basis*, where all the elements in Equation 1 appear now as a direct sum of irreducible representations, $\mathbf{2}\oplus\mathbf{1_S}$. It is in this new basis, where the interaction eigenstates are defined, the new eigenstates may also be called flavour eigenstates (in regard to flavour oscillations). Hereafter, we will refer to this basis as the *flavour basis.*

The interaction eigenstates in terms of the universal eigenstates are defined as

$$F_{\mathbf{D}} = \begin{pmatrix}F_1\\F_2\end{pmatrix} = \begin{pmatrix}\frac{1}{\sqrt{2}}(f_1-f_2)\\ \frac{1}{\sqrt{6}}(f_1+f_2-2f_3)\end{pmatrix} \quad\text{and}\quad F_{\mathbf{S}} = \frac{1}{\sqrt{3}}(f_1+f_2+f_3). \tag{5}$$

Note that we have not yet defined which flavour eigenstate corresponds to a certain family. We will make this choice later.

3. An S_3-invariant extension of the Standard Model

Models with S_3 as their family symmetry have been widely studied, see references therein [6]. Our approach is somewhat similar to the one used in [7], where different models generated by a different number of Higgs $SU(2)_L$ doublets were explored. Nevertheless, we consider a different assignment of fermions to the irreducible representations of S_3. Here we will show that a weak-basis transformation may help us to reduce the number of free parameters, such that, mixing angles can be expressed in terms of mass ratios. We will focus on the general case of adding three more Higgs $SU(2)_L$ doublets aside from the SM one, briefly mentioning how it compares to the case of the addition of two Higgs doublets [1]. The four Higgs fields are accommodated such that they occupy all the irreps of S_3: $H_{DW} = (H_{1W}, H_{2W})^T$, H_{SW}, and H_{AW}, where the subindexes D, S, and A refer respectively to the doublet, symmetric and antisymmetric S_3 representations, and the subindex W means that the field behaves as a doublet of $SU(2)_L$.

3.1. Assignments between S_3 irreps and families

We need to decide how the assignment between the three families with the S_3 irreps should be done. For it, we recall the following fact: for each kind of fermion, there exist three identical particles, except for their mass. In fact, if we divide the mass of each quark over the mass of the heaviest quark of each kind, we can detect a clear hierarchy among their masses: the first two families are in fact really close to each other when compared to the third one. This suggests that the fermion masses of each type could be reflecting an S_3 symmetric pattern: $\mathbf{2} \oplus \mathbf{1}$.

Hence, we generically assign the first two families, $f_{I(L,R)}$ and $f_{II(L,R)}$, to the doublet representation, $\mathbf{2}$, and the third family, $f_{III(L,R)} \equiv g_{L,R}$, to the symmetric singlet representation, $\mathbf{1}_S$,

$$\begin{pmatrix} f_{I(L,R)} \\ f_{II(L,R)} \end{pmatrix} \sim \mathbf{2}; \quad f_{III(L,R)} \equiv g_{L,R} \sim \mathbf{1_S},$$

respectively. Here I, II or III represent the family index of a left or right-handed fermion field $f_{(L,R)}$, and $f_{(L,R)}$ may represent any SM fermion field.

3.2. A weak-basis transformation: the hierarchichal-basis

Fermions of the SM become massive after the Higgs mechanism. The implied generic mass matrix form for both quarks and leptons is

$$\mathcal{M}^f_{S_3} = \begin{pmatrix} \sqrt{2}Y_2^f v_S + Y_3^f w_2 & Y_3^f w_1 + \sqrt{2}Y_4^f v_A & \sqrt{2}Y_5^f w_1 \\ Y_3^f w_1 - \sqrt{2}Y_4^f v_A & \sqrt{2}Y_2^f v_S - Y_3^f w_2 & \sqrt{2}Y_5^f w_2 \\ \sqrt{2}Y_6^f w_1 & \sqrt{2}Y_6^f w_2 & 2Y_1^f v_S \end{pmatrix}, \tag{6}$$

where the Y_I^f are complex Yukawa couplings and w_1, w_2, v_S, and v_A are the vacuum expectation values of the S_3 doublet components, the S_3 symmetric singlet, and the S_3 antisymmetric singlet, respectively. We remind the reader the main goal of this work, that is, to confront the up-to-date values of quark masses and mixing to the expressions that relate mixing angles to mass ratios. These relations are obtained by using the three invariants of a mass matrix. However, the abundance of free parameters makes the calculation cumbersome. It is therefore convenient to find first an equivalent form which could simplify the problem.

The weak charged current is the only manifestation in the SM of the interaction of two particles of the same sector (quark or leptonic) but with different electric charge. We call weak-basis transformations (WBT) the similarity transformations of mass matrices that leave invariant the weak charged currents. It has been already noticed that a suitable WBT may help to reduce the number of free parameters of a mass matrix by introducing a set of zeroes in its entries [8, 9]. In our case, we find that a WBT in the S_3 doublet space could exhibit one

[1] On a separate work we will present a study of all the possible cases with different number of Higgs $SU(2)_L$ doublets.

of the two desired null matrix elements that reproduce the 4-zero Fritzsch-like texture [10–12]. This kind of mass matrices are considered to be, along with the Nearest-Neighbour Interaction mass matrix form, one of the two kind of mass matrices that allow a similar treatment of quarks and leptons such that they have a viable phenomenology ([13] and [14, 15]), which we call here *unified*. A further reduction on the number of free parameters may be achieved by demanding Hermiticity to the mass matrix, which is always possible within any extension of the SM that does not introduce flavour changing right-handed currents [3]. For these reasons, Hermicity is often invoked. Here, we then consider the following generic Hermitian mass matrix for Dirac fermions:

$$\mathcal{M}^f_{Hier.} = \begin{pmatrix} |\mu^f_1| + |\mu^f_2|\cos^2\theta(1-3\tan^2\theta) & |\mu^f_2|\sin\theta\cos\theta(3-\tan^2\theta) + i|\mu^f_5| & 0 \\ |\mu^f_2|\sin\theta\cos\theta(3-\tan^2\theta) - i|\mu^f_5| & |\mu^f_1| - |\mu^f_2|\cos^2\theta(1-3\tan^2\theta) & \mu^f_8\sec\theta \\ 0 & \mu^{f*}_8\sec\theta & |\mu^f_3| \end{pmatrix}, \tag{7}$$

where we have denoted $\mu^f_1 \equiv \sqrt{2}Y^f_2 v_S$, $\mu^f_2 \equiv Y^f_3 w_2$, $\mu^f_3 \equiv 2Y^f_1 v_S$, $\mu^f_5 \equiv \sqrt{2}Y^f_4 v_A$, and $\mu^f_8 \equiv \sqrt{2}Y^f_6 w_1$. Hermiticity implies a particular phase value for the complex Yukawa couplings, while the nullity of the matrix elements, $(1,3)$ and $(3,1)$, is controlled by the relation $\tan\theta = w_1/w_2$.

In order to reproduce the 4-zero Fritzsch-like texture we need a zero $(1,1)$ entry. When a unified treatment is pursued, by using the same mass matrix for all fermions, there is not a possible combination of WBTs that could allow having at the same time zero $(1,3)$, $(3,1)$ and $(1,1)$ entries [9]. Nevertheless, here we show that the origin of the null entry $(1,1)$ could be in fact much simpler. For simplicity, consider a Hermitian n-dimensional matrix $\mathcal{A}_{n\times n}$ which is diagonalized by the unitary matrix $\mathcal{U}$,

$$\mathcal{U}\mathcal{A}_{n\times n}\mathcal{U}^{-1} = diag(a_1, a_2, ..., a_n), \tag{8}$$

we may see that the unitary matrix diagonalizing $\mathcal{A}_{n\times n}$ still diagonalizes the resulting matrix $\overline{\mathcal{A}}_{n\times n}$ after the extraction of any of its diagonal elements,

$$\mathcal{A}_{ii}\mathbb{1}_{n\times n} + \mathcal{U}\overline{\mathcal{A}}_{n\times n}\mathcal{U}^{-1} = diag(a_1, a_2, ..., a_n), \tag{9}$$

$$\mathcal{U}\overline{\mathcal{A}}_{n\times n}\mathcal{U}^{-1} = diag(\overline{a}_1, \overline{a}_2, ..., \overline{a}_n), \tag{10}$$

where we have denoted $\overline{a}_j \equiv a_j - \mathcal{A}_{ii}$. This shows how the unitary matrix $\mathcal{U}$ is independent of the removed entry $\mathcal{A}_{ii}$. Therefore, by diagonalizing $\mathcal{A}_{n\times n}$ or $\overline{\mathcal{A}}_{n\times n}$ we arrive to the same unitary matrix, but now with an important difference: the new set of eigenvalues is shifted. Hence, we will shift the set of eigenvalues to produce the $(1,1)$ entry equal to zero. This basis, where the mass matrix with the 4-zero Fritzsch-like texture appears is usually called the *hierarchical-basis*.

For these proceedings, we focus on the quark sector of the models. When reparameterizing the mass matrix, the produced shift Δ_f, in the set of eigenmasses, is in fact much smaller than one. This allows us to consider the "shifted" approach as a generalization to the standard approach, where no shift is made but where the zero entries are imposed by new conditions. For the purpose of illustrating the relations among mixing parameters and mass ratios, we will consider that the set of shifted eigenmasses is equal to the no-shifted ones, *i.e.* $\Delta_f = 0$. The complete analysis of the model, $\Delta_f \neq 0$, will be presented in a separate work. The main success of this kind of models is that they allow us to express the elements of the quark mixing matrix V^{th}_{CKM} as a explicit function of the quark mass ratios

$$\begin{aligned}
V^{th}_{ud} &= \sqrt{\frac{\widetilde{m}_c\widetilde{m}_s\xi^u_1\xi^d_1}{\mathcal{D}_{1u}\mathcal{D}_{1d}}} + \sqrt{\frac{\widetilde{m}_u\widetilde{m}_d}{\mathcal{D}_{1u}\mathcal{D}_{1d}}}\left(\sqrt{(1-\delta_u)(1-\delta_d)\,\xi^u_1\xi^d_1} + \sqrt{\delta_u\delta_d\xi^u_2\xi^d_2}e^{i\phi_2}\right)e^{i\phi_1}, \\
V^{th}_{us} &= -\sqrt{\frac{\widetilde{m}_c\widetilde{m}_d\xi^u_1\xi^d_2}{\mathcal{D}_{1u}\mathcal{D}_{2d}}} + \sqrt{\frac{\widetilde{m}_u\widetilde{m}_s}{\mathcal{D}_{1u}\mathcal{D}_{2d}}}\left(\sqrt{(1-\delta_u)(1-\delta_d)\,\xi^u_1\xi^d_2} + \sqrt{\delta_u\delta_d\xi^u_2\xi^d_1}e^{i\phi_2}\right)e^{i\phi_1}, \\
V^{th}_{ub} &= \sqrt{\frac{\widetilde{m}_c\widetilde{m}_d\widetilde{m}_s\delta_d\xi^u_1}{\mathcal{D}_{1u}\mathcal{D}_{3d}}} + \sqrt{\frac{\widetilde{m}_u}{\mathcal{D}_{1u}\mathcal{D}_{3d}}}\left(\sqrt{(1-\delta_u)(1-\delta_d)\,\delta_d\xi^u_1} - \sqrt{\delta_u\xi^u_2\xi^d_1\xi^d_2}e^{i\phi_2}\right)e^{i\phi_1}, \\
V^{th}_{cd} &= -\sqrt{\frac{\widetilde{m}_u\widetilde{m}_s\xi^u_2\xi^d_1}{\mathcal{D}_{2u}\mathcal{D}_{1d}}} + \sqrt{\frac{\widetilde{m}_c\widetilde{m}_d}{\mathcal{D}_{2u}\mathcal{D}_{1d}}}\left(\sqrt{(1-\delta_u)(1-\delta_d)\,\xi^u_2\xi^d_1} + \sqrt{\delta_u\delta_d\xi^u_1\xi^d_2}e^{i\phi_2}\right)e^{i\phi_1},
\end{aligned} \tag{11}$$

$$
\begin{aligned}
V_{cs}^{th} &= \sqrt{\frac{\widetilde{m}_u\widetilde{m}_d\xi_2^u\xi_2^d}{\mathcal{D}_{2u}\mathcal{D}_{2d}}} + \sqrt{\frac{\widetilde{m}_c\widetilde{m}_s}{\mathcal{D}_{2u}\mathcal{D}_{2d}}}\left(\sqrt{(1-\delta_u)(1-\delta_d)\xi_2^u\xi_2^d} + \sqrt{\delta_u\delta_d\xi_1^u\xi_1^d}e^{i\phi_2}\right)e^{i\phi_1}, \\
V_{cb}^{th} &= -\sqrt{\frac{\widetilde{m}_u\widetilde{m}_d\widetilde{m}_s\delta_d\xi_2^u}{\mathcal{D}_{2u}\mathcal{D}_{3d}}} + \sqrt{\frac{\widetilde{m}_c}{\mathcal{D}_{2u}\mathcal{D}_{3d}}}\left(\sqrt{(1-\delta_u)(1-\delta_d)\delta_d\xi_2^u} - \sqrt{\delta_u\xi_1^u\xi_1^d\xi_2^d}e^{i\phi_2}\right)e^{i\phi_1}, \\
V_{td}^{th} &= \sqrt{\frac{\widetilde{m}_u\widetilde{m}_c\widetilde{m}_s\delta_u\xi_1^d}{\mathcal{D}_{3u}\mathcal{D}_{1d}}} + \sqrt{\frac{\widetilde{m}_d}{\mathcal{D}_{3u}\mathcal{D}_{1d}}}\left(\sqrt{\delta_u(1-\delta_u)(1-\delta_d)\xi_1^d} - \sqrt{\delta_d\xi_1^u\xi_2^u\xi_2^d}e^{i\phi_2}\right)e^{i\phi_1}, \\
V_{ts}^{th} &= -\sqrt{\frac{\widetilde{m}_u\widetilde{m}_c\widetilde{m}_d\delta_u\xi_2^d}{\mathcal{D}_{3u}\mathcal{D}_{2d}}} + \sqrt{\frac{\widetilde{m}_s}{\mathcal{D}_{3u}\mathcal{D}_{2d}}}\left(\sqrt{\delta_u(1-\delta_u)(1-\delta_d)\xi_2^d} - \sqrt{\delta_d\xi_1^u\xi_2^u\xi_1^d}e^{i\phi_2}\right)e^{i\phi_1}, \\
V_{tb}^{th} &= \sqrt{\frac{\widetilde{m}_u\widetilde{m}_c\widetilde{m}_d\widetilde{m}_s\delta_u\delta_d}{\mathcal{D}_{3u}\mathcal{D}_{3d}}} + \left(\sqrt{\frac{\xi_1^u\xi_2^u\xi_1^d\xi_2^d}{\mathcal{D}_{3u}\mathcal{D}_{3d}}} + \sqrt{\frac{\delta_u\delta_d(1-\delta_u)(1-\delta_d)}{\mathcal{D}_{3u}D_{3d}}}e^{i\phi_2}\right)e^{i\phi_1}.
\end{aligned}
$$

with

$$\xi_1^{u,d} = 1 - \widetilde{m}_{u,d} - \delta_{u,d}, \quad \xi_2^{u,d} = 1 + \widetilde{m}_{c,s} - \delta_{u,d}, \tag{12}$$

$$
\begin{aligned}
\mathcal{D}_{1(u,d)} = (1-\delta_{u,d})(\widetilde{m}_{u,d}+\widetilde{m}_{c,s})(1-\widetilde{m}_{u,d}), \quad \mathcal{D}_{2(u,d)} &= (1-\delta_{u,d})(\widetilde{m}_{u,d}+\widetilde{m}_{c,s})(1+\widetilde{m}_{c,s}), \\
\mathcal{D}_{3(u,d)} &= (1-\delta_{u,d})(1-\widetilde{m}_{u,d})(1+\widetilde{m}_{c,s}).
\end{aligned} \tag{13}
$$

Here $\widetilde{m}_u = m_u/m_t$, $\widetilde{m}_c = m_u/m_t$, $\widetilde{m}_d = m_d/m_b$ and $\widetilde{m}_s = m_s/m_b$. The parameters δ_u and δ_d reflect the mixing between doublet and singlet fermion representations. We note that when the phase ϕ_2 is equal to zero, we recover the model of three $SU(2)_L$ Higgs doublets, two in the doublet representation and one in a singlet symmetric representation of S_3. For the general case, when we have the extra Higgs doublet in the anti-symmetric singlet representation of S_3, the preferred solution is when $\phi_1 = 0$, and such that the parameters δ_u and δ_d can be treated essentially as free parameters. In this case then, CP violation is achieved by a non zero value of ϕ_2.

4. χ^2 analysis for the different models (quark sector)

For the χ^2 fit we proceed as follows. We construct the χ^2 function as

$$\chi^2 = \frac{\left(V_{ud}^{\rm th} - V_{ud}\right)^2}{\sigma_{V_{ud}}^2} + \frac{\left(V_{us}^{\rm th} - V_{us}\right)^2}{\sigma_{V_{us}}^2} + \frac{\left(V_{ub}^{\rm th} - V_{ub}\right)^2}{\sigma_{V_{ub}}^2} + \frac{\left(\mathcal{J}_q^{\rm th} - \mathcal{J}_q\right)^2}{\sigma_{\mathcal{J}_q}^2}, \tag{14}$$

where the quantities with super-index "th" are the complete expressions for the CKM elements, as given by the S_3 models, and those without, the experimental quantities along their uncertainty, are

$$
\begin{aligned}
2011: \quad & V_{ud} = 0.97428 \pm 0.00015, & 2012: \quad & V_{ud} = 0.97427 \pm 0.00015, \\
& V_{us} = 0.2253 \pm 0.007, & & V_{us} = 0.2253 \pm 0.007, \\
& V_{ub} = 0.00347 \pm 0.00014, & & V_{ub} = 0.00351 \pm 0.00015, \\
& J = (2.91 \pm 0.155) \times 10^{-5}, & & J = (2.96 \pm 0.18) \times 10^{-5}.
\end{aligned} \tag{15}
$$

We present a comparison of fits using the available data known before July 2012 (labeled as 2011) and the up-to-date data presented by the PDG, given the important change in the uncertainty of the mass of the strange quark. Since we assume unitarity of the CKM mixing matrix, we need to fit just to four observables. The theoretical expressions of the parameters of the CKM elements are given in terms of the mass ratios, $\widetilde{m}_i$, hence the minimization of the defined χ^2 is a function of the *parameters* $\widetilde{m}_i$, δ_u, δ_d, $\cos\phi_1$ and $\cos\phi_2$. That means, that as a result of the minimization, there is a *best fit value*, for each of those quantities, where χ^2 achieves its minimum. Of course, the mass ratios $\widetilde{m}_i$ are not free parameters. The limits that we use correspond to the 3σ experimental/lattice regions. The used values for the parameters $\widetilde{m}_i$ are given in Table 1. We used MINUIT and varied all the parameters $\widetilde{m}_i$, within the 3σ range given in Table 1, and δ_u, δ_d, as really free parameters. The condition that $\phi_1 = 0$ is imposed

	2011	2012
$\widetilde{m_u}\,(M_Z)$	0.0000082 ± 0.0000027	0.0000083 ± 0.0000030
$\widetilde{m_c}\,(M_Z)$	0.0036 ± 0.0004	0.0037 ± 0.00008
$\widetilde{m_d}\,(M_Z)$	0.00098 ± 0.00018	0.00098 ± 0.00017
$\widetilde{m_s}\,(M_Z)$	0.0205 ± 0.0056	0.0190 ± 0.0014

Table 1. Comparison of the values of the mass ratios, at M_Z, reported up to 2011 and the newest values reported by the PDG.

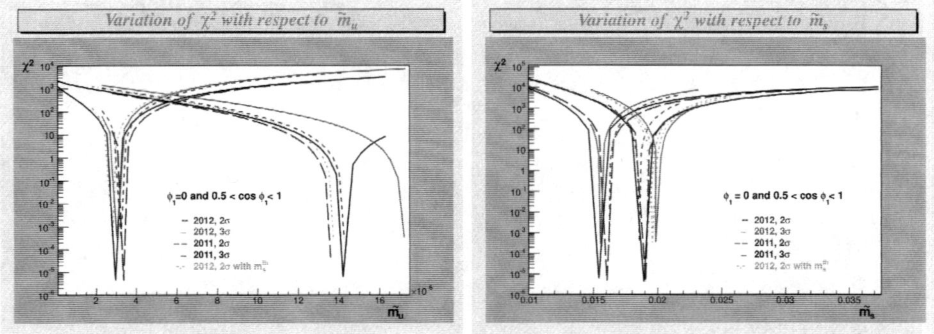

Figure 1. Results of the χ^2 fit as a function of $\widetilde{m_u}$ and $\widetilde{m_s}$. The values of the mass ratios $\widetilde{m_u}$ and $\widetilde{m_s}$ reported by the PDG before July 2012 had an uncertainty of about 30% of their central value. Therefore, they were difficult to fit. In July 2012, the up-to-date values of the quark mass ratios were reported by the PDG. The main change was the reduction in the uncertainty of the mass of the strange quark, driven by lattice determinations, which is now about 5% of its central value. We have also made a fit using only an average of the theoretical determination of m_s, $m_s(2\text{GeV}) = 0.101 \pm 0.011$, in order to asses the impact of the reduction in the uncertainty of m_s. We also show the χ^2 as a function of $\widetilde{m_u}$ because of the size of its uncertainty. In the figure corresponding to $\widetilde{m_u}$, the plots centered around 3×10^{-6} are the result of the fit where ϕ_1 is fixed to 0, which corresponds to a version of the model with four Higgs $SU(2)_L$ doublets, two in the doublet representation of S_3 and the other two in singlet representations. In this case, CP violation is a result of the non zero value of ϕ_2. In the figure of χ^2 as a function of $\widetilde{m_s}$, the plots centered around 1.6×10^{-2} correspond to the fit where $\phi_1 = 0$. The other plots, correspond to the fits of the model with three Higgs fields.

by consistency in finding appropriate solutions for the case of the four Higgs fields that we are considering here. In this case, CP violation arises as a result of having $\phi_2 \neq 0$. Minimization with MINUIT relies on starting the process with a seed, for all parameters, close enough to the minimum. Therefore, if the range of variation is large, it is difficult to find a best fit point. On the other hand, to check for global minima, one should remove the limits of the "free" parameters. Unfortunately, if we perform the fit leaving completely free the parameters $\widetilde{m_i}$, the quality of the fit decreases, and most importantly, it would turn out that $\widetilde{m_u}$ would be of order 10^{-3}. In Figure 1 we present the result of the fit for the case that we considered here and we compare it to the case of the model with three Higgs fields. For the case of four Higgs doublets, the values found for δ_u, δ_d and ϕ_2 are respectively 7.9×10^{-2}, 3.9×10^{-2} and 0.469 radians. For the case of three Higgs doublets, the values found for δ_u, δ_d and ϕ_1 are respectively 2.5×10^{-2}, 4.09×10^{-2} and 1.047 radians.

Conclusions

In order to explore the Higgs sector of S_3 family models, we studied a model containing four Higgs $SU(2)_L$ doublets belonging to the S_3 representation $4 = \mathbf{2} \oplus \mathbf{1}_S \oplus \mathbf{1}_A$. We were able to identify the conditions under which the 4-zero Fritzsch-like texture mass matrices are obtained.

This allowed us to reduce the number of parameters needed to describe the generic model with four Higgs $SU(2)_L$ doublets and to find an exact parameterization of the mass matrices in terms of their mass eigenvalues. We provided also exact formulas for mixing angles of the CKM matrix in terms of quark mass ratios. This line of work was developed in [16], without a particular model in mind, but it was clear the usefulness in classifying mass matrix patterns according to their transformation to a 4-zero Fritzsch-like texture. The reduction to this form, then allowed us to make a direct comparison among different models and to perform a χ^2 analysis to the CKM elements. A particular case of the general model presented, is a case for which only three Higgs $SU(2)_L$ doublets are allowed and for which we effectively reduced one free parameter. Finally, we have then proceeded to perform a χ^2 analysis to the CKM elements, both to the general model, S_3-SM, and to the restricted one. In both cases, we have obtained impressive values of the minimum of the χ^2 function, 3.3×10^{-4} and 3.4×10^{-4}, respectively.

Acknowledgments
U. J. S. S. thanks the Instituto de Física-UNAM for supporting his attendance to this interesting and stimulating workshop. We acknowledge support from the Mexican grants: PAPIIT IN113712 and CONACyT-132059.

References
[1] Chatrchyan S *et al.* (The CMS Collaboration) 2012 (*Preprint* 1207.7235)
[2] Aad G *et al.* (The ATLAS Collaboration) 2012 (*Preprint* 1207.7214)
[3] Fritzsch H and Xing Z z 2000 *Prog.Part.Nucl.Phys.* **45** 1–81 (*Preprint* hep-ph/9912358)
[4] Ishimori H, Kobayashi T, Ohki H, Shimizu Y, Okada H *et al.* 2010 *Prog.Theor.Phys.Suppl.* **183** 1–163 (*Preprint* 1003.3552)
[5] Altarelli G and Feruglio F 2010 *Rev.Mod.Phys.* **82** 2701–2729 (*Preprint* 1002.0211)
[6] Hirsch M, Meloni D, Morisi S, Pastor S, Peinado E *et al.* 2012 Long author list - awaiting processing (*Preprint* 1201.5525)
[7] Yahalom R 1984 *Phys.Rev.* **D29** 536
[8] Branco G, Emmanuel-Costa D and González Felipe R 2000 *Phys.Lett.* **B477** 147–155 (*Preprint* hep-ph/9911418)
[9] Branco G, Emmanuel-Costa D, González Felipe R and Serodio H 2009 *Phys.Lett.* **B670** 340–349 (*Preprint* 0711.1613)
[10] Weinberg S 1977 *Trans.New York Acad.Sci.* **38** 185–201
[11] Fritzsch H 1977 *Phys.Lett.* **B70** 436
[12] Fritzsch H 1978 *Phys.Lett.* **B73** 317–322
[13] Gupta M and Ahuja G 2011 *Int.J.Mod.Phys.* **A26** 2973–2995
[14] Babu K S and Kubo J 2005 *Phys.Rev.* **D71** 056006 (*Preprint* hep-ph/0411226)
[15] Simoes C 2011 *J.Phys.Conf.Ser.* **335** 012031
[16] Barranco J, González Canales F and Mondragón A 2010 *Phys.Rev.* **D82** 073010 (*Preprint* 1004.3781)

Journal of Physics: Conference Series **485** (2014) 012064 doi:10.1088/1742-6596/485/1/012064

Global D-brane models with stabilised moduli and light axions

Michele Cicoli

ICTP, Strada Costiera 11, Trieste 34014, Italy and INFN Sezione di Trieste, Italy

E-mail: mcicoli@ictp.it

Abstract. We review recent attempts to try to combine global issues of string compactifications, like moduli stabilisation, with local issues, like semi-realistic D-brane constructions. We list the main problems encountered, and outline a possible solution which allows globally consistent embeddings of chiral models. We also argue that this stabilisation mechanism leads to an axiverse. We finally illustrate our general claims in a concrete example where the Calabi-Yau manifold is explicitly described by toric geometry.

1. Introduction

Two longstanding problems of string compactifications are moduli stabilisation and the derivation of realistic D-brane models. Type II theories are promising to address these issues because of local sources. On one side, D-branes provide non-Abelian gauge bosons whereas, on the other side, local sources allow to evade no-go theorems which prevent the turning on of background fluxes. In turn, these fluxes generate a potential for most of the moduli. Moreover, in this *brane-world scenario*, finding a solution to these two crucial problems seems to be somewhat easier since moduli fixing is a *global* issue whereas model building is a *local* issue. Thus one can hope that, at least at leading order, a separate study of the two issues makes sense.

So far, people have found viable mechanisms to fix the moduli especially in type IIB where the backreaction of the fluxes is very mild and semi-realistic D-brane models can also be constructed. It is therefore now time to combine the two solutions. The first attempts to do so have shown that the two issues are not completely decoupled since one generically faces the following problems:

- Tension between moduli fixing by non-perturbative effects and chirality [1];
- Tension between moduli fixing by non-perturbative effects and the cancellation of Freed-Witten anomalies [2];
- Possible shrinking of various divisors induced by D-terms (especially the four-cycles supporting the visible sector) [2, 3].

The main goal is then to overcome these difficulties keeping control over the 4D supergravity theory and stabilising the moduli inside the Kähler cone. Another non-trivial requirement is getting, at the same time, also an interesting particle phenomenology and cosmology.

In this review, we focus on type IIB/F-theory compactifications with D3/D7-branes and O3/O7-planes where the compact Calabi-Yau three-fold can be explicitly described in terms of toric geometry [4]. We perform a choice of brane set-up and world-volume fluxes that yields

Content from this work may be used under the terms of the Creative Commons Attribution 3.0 licence. Any further distribution of this work must maintain attribution to the author(s) and the title of the work, journal citation and DOI.
Published under licence by IOP Publishing Ltd

a chiral MSSM-like model. We also check the global consistency of our model by ensuring the cancellation of D7-tadpoles, torsion charges and Freed-Witten anomalies (leaving enough space in the D3-tadpole to turn on three-form fluxes). The Kähler moduli are fixed within the Kähler cone and the regime of validity of the effective low-energy theory in a way compatible with chirality. Moreover, the D-terms do not induce the shrinking of any divisor.[1] We also show how this moduli stabilisation mechanism (the LARGE Volume Scenario) leads naturally to an axiverse whose main feature is the presence of many axions with exponentially small masses [6, 7]. Our model is also able to reproduce a visible sector gauge coupling of the correct size and interesting phenomenological scales like TeV-scale supersymmetry and an intermediate scale decay constant for the QCD axion (realised as a local closed string mode, see [8]) for natural values of the underlying parameters.

2. Kähler moduli stabilisation

The closed string moduli of type IIB compactifications are the axio-dilaton S, the complex structure moduli U_α, $\alpha = 1, ..., h^{2,1}$ and the Kähler moduli $T_i = \tau_i + \mathrm{i}\, c_i$, where $\tau_i = \mathrm{Vol}(D_i)$ and $c_i = \int_{D_i} C_4$, $i = 1, ..., h^{1,1}$. The background fluxes $G_3 = F_3 + \mathrm{i} S H_3$ generate a tree-level superpotential $W_{\rm tree}(S, U)$ which fixes S and U at $D_{S,U} W = 0$. On the other hand, due to the *no-scale structure*, the T-moduli are flat at tree-level, and so when one studies Kähler moduli stabilisation, the S and U-moduli can be fixed at their flux-stabilised values. Thus we can consider $W_0 = \langle W_{\rm tree}\rangle$ and $K_{\rm tree} = -2 \ln \mathcal{V}$, where $\mathcal{V}$ is a function of the T-moduli which gives the volume of the internal manifold. There are several sources for Kähler moduli stabilisation:

$$V = V_D + V_F^{\rm tree} + V_F^{\rm pert} + V_F^{\rm np} \,. \tag{1}$$

In the previous expression $V_D \sim \mathcal{O}(1/\mathcal{V}^2)$ is D-term potential (generated by fluxes on D7-branes), $V_F^{\rm tree} \sim \mathcal{O}(1/\mathcal{V}^2) = 0$ is the tree-level F-term potential which vanishes due to the no-scale structure, $V_F^{\rm pert} \lesssim \mathcal{O}(1/\mathcal{V}^3)$ contains perturbative effects which come from α' and g_s corrections to K, and finally $V_F^{\rm np} \lesssim \mathcal{O}(1/\mathcal{V}^3)$ is derived from non-perturbative corrections to W (E3-instantons or gaugino condensation on a D7-stack). Our strategy will be to focus on the large volume limit, $\mathcal{V} \gg 1$, and study the behaviour of V at each order in an inverse volume expansion. At leading order in $1/\mathcal{V}$, we look for supersymmetric solutions by imposing $V_D = 0$. At subleading order, we instead minimise V_F leading to stable vacua which break supersymmetry.

Let us now briefly discuss the main problems faced when trying to freeze the T-moduli in a way consistent with the presence of chirality and world-volume fluxes.

2.1. Non-perturbative effects and chirality

There is a known tension between Kähler moduli fixing by non-perturbative effects and chirality. This is because chirality is induced by non-zero fluxes on intersections of branes, implying that the visible sector has $\mathcal{F} \neq 0$. Moreover, if there are chiral modes on the intersection between the cycle supporting non-perturbative effects $T_{\rm np}$ and the visible sector, the prefactor A of the non-perturbative superpotential $W_{\rm np} = A\, e^{-a T_{\rm np}}$ depends on visible modes ϕ. In order to preserve the visible gauge group, $\langle\phi\rangle$ has to vanish, implying $A = 0$ and no contribution to $W_{\rm np}$.

This tension constraints the flux choice: there has to be no chirality at the intersection between the cycles supporting non-perturbative effects and the visible sector. As pointed out in [3], this implies that the best place for non-perturbative effects is a 'diagonal' del Pezzo divisor since it is rigid and it has only self intersections.

[1] For the global embedding of models with fractional branes at del Pezzo singularities see [5].

2.2. Non-perturbative effects and Freed-Witten anomaly

As shown in [9], one has to turn on a half-integer flux on any *non-spin* four-cycle D ($c_1(D)$ is odd) to cancel Freed-Witten (FW) anomalies: $F = f^i\eta_i + \frac{1}{2}c_1(D)$, with $f^i \in \mathbb{Z}$ and $\eta_i \in H^2(D,\mathbb{Z})$. Moreover, an $O(1)$ E3-instanton contributes to $W_{\rm np}$ if the E3 wraps a rigid cycle which is transversally invariant under the orientifold and has total flux $\mathcal{F} = F - B = 0$ (similar considerations apply also for a gaugino condensation stack). However, if the cycle is non-spin, the cancellation of FW anomalies requires $F \neq 0$. One can then cancel F by a proper choice of B but once B is fixed to cancel a half-integral flux on stack a, generically $\mathcal{F} \neq 0$ on a second non-spin stack b (unless they do not intersect). Hence the cancellation of FW anomalies generically prevents to have more than one non-perturbative effect to fix the T-moduli.

Our strategy to overcome this problem is the simplest one: fix the T-moduli by only one non-perturbative effect. This leads to the LARGE Volume Scenario (LVS) where the presence of just a single diagonal del Pezzo with non-perturbative effects is enough to ensure the stabilisation of the Calabi-Yau volume by setting $W_0 \sim \mathcal{O}(1)$ and exploiting α'-corrections to K [10].

2.3. D-term problem

As we have seen, world-volume fluxes are necessary for chirality and to cancel worldsheet anomalies. In turn, these fluxes generate Fayet-Iliopoulos (FI) terms of the form $\xi_a \propto \int_{D_a} J \wedge \mathcal{F}_a = k_{ajk}\mathcal{F}_a^k t^j$. If the VEV of all charged fields vanishes, the D-term conditions imply $\xi_a = 0$, generically forcing the shrinking of some four-cycles. Let us explain this problem more in detail. In general, $\xi_a = 0$ is a system of homogeneous linear equations in all the $h^{1,1}$ Kähler moduli. However, if the non-perturbative effects are supported on $n_{\rm np}$ diagonal del Pezzo divisors, these would not enter in $\xi_a = 0$ since they have no chiral intersections. Thus the generic number of unknowns is $n = h^{1,1} - n_{\rm np}$. If the matrix of the system $\xi_a = 0$ has $d = n$, the only solution is the one where all the t's collapse to zero size. In order to avoid this shrinking, one has to choose the brane configuration and the fluxes such that $d < n$. In this case, the D-terms fix d directions and leave $(n-d)$ flat directions. If $n - d = 1$, all the t's have to be of the same size which must be small to obtain the right value of the visible gauge coupling: $g^{-2} \sim t^2$. Thus in this case one cannot realise the LVS since this is characterised by an exponentially large volume. The only way to get LVS is to ensure that $n - d = 2$. If $d = 1$, the minimal n to allow for LVS is then $n = 3$, leading to models with $h^{1,1} = 4$ for $n_{\rm np} = 1$.

3. The axiverse and moduli stabilisation

The previous analysis has taught us that the Kähler moduli have to be fixed by a combination of different effects (for $W_0 \sim \mathcal{O}(1)$). Let us now study what is the implication of this stabilisation scheme for the axions of the effective field theory:

- d combinations of T-moduli are fixed by the leading D-term potential, and so d axions get eaten up by anomalous $U(1)$s;
- $n_{\rm np}$ 'diagonal' del Pezzo divisor are fixed by non-perturbative effects, and so the corresponding axions become heavy since they obtain the same mass of order $m_{3/2}$;
- The remaining $n_{\rm ax} = h^{1,1} - n_{\rm np} - d \geq 2$ moduli have to be fixed perturbatively, and so the corresponding axions remain massless. In particular, the volume is frozen by α' corrections to K whereas all the remaining moduli are stabilised by subleading g_s corrections to K.

This analysis implies that chiral models with the visible sector in the geometric regime have always $n_{\rm ax} \geq 2$ light axions with possibly a very large $n_{\rm ax}$ for $h^{1,1} \sim \mathcal{O}(100)$. One axion has to be the QCD axion and develops a mass via QCD instanton effects. On the other hand, all the other axions get a tiny mass via higher order non-perturbative effects. This is the typical picture of the *axiverse* where there is a plethora of ultra-light axions whose mass spectrum is logarithmically hierarchical [6, 7].

4. Explicit example

Let us now illustrate our general claims in an explicit example where the internal manifold is built by means of toric geometry.

4.1. A K3-fibred Calabi-Yau

The Calabi-Yau three-fold is a hypersurface in a 4D toric ambient variety. Its weight matrix is:

z_1	z_2	z_3	z_4	z_5	z_6	z_7	z_8	$D_{\rm CY}$
1	1	1	0	0	0	1	4	8
1	1	0	0	0	1	0	3	6
0	1	1	1	0	0	0	3	6
0	1	0	0	1	0	0	2	4

. (2)

The toric variety is the moduli space of the corresponding gauged linear sigma model where the size of the FI terms ξ_i, $i = 1, ..., 4$, is given by the Kähler moduli. Thus this manifold has $h^{1,1} = 4$ (and $h^{1,2} = 106$). The Calabi-Yau is a divisor of the ambient variety described by the last column in (2). The topological data are obtained from PALP [11]:

- Basis of $H_4(\mathrm{CY}, \mathbb{Z})$: $\Gamma_1 = D_7\,, \quad \Gamma_2 = D_2 + D_7\,, \quad \Gamma_3 = D_1\,, \quad \Gamma_4 = D_5\,;$
- Intersection form: $I_3 = 2\Gamma_1^3 + 4\Gamma_2^3 + 4\Gamma_4^3 + 2\Gamma_2^2\Gamma_3 - 2\Gamma_4^2\Gamma_3\,.$

The internal manifold is a K3 fibration with a diagonal del Pezzo [3]. The K3 fibre is D_1 whereas the 'diagonal' dP_7 corresponds to $\Gamma_1 = D_7$. There are also three other rigid (but not del Pezzo) divisors: D_4, D_5 and D_6. Expanding the Kähler form as $J = \sum_{i=1}^4 t_i \Gamma_i$, the Calabi-Yau volume becomes $\mathcal{V} = \frac{1}{3}\left[2t_2^3 + 3t_2^2 t_3 + t_4^2(2t_4 - 3t_3) + t_1^3\right]$ while the volume of D_7 is simply $\tau_7 = t_1^2$. The Kähler cone is instead defined as:

$$r_1 \equiv -t_1 > 0\,, \qquad r_2 \equiv t_1 + t_2 + t_4 > 0\,, \qquad r_3 \equiv t_3 - t_4 > 0\,, \qquad r_4 \equiv -t_4 > 0\,. \quad (3)$$

The reason why we focused on K3-fibred manifolds is because they are promising for particle physics and cosmology. In fact, they can lead to anisotropic compactifications that realise scenarios with supersymmetric large extra dimensions [12], light hidden $U(1)$s with kinetic mixing with the photon [13] and natural string models of quintessence [14]. Moreover, compactifications where the size of the overall volume is controlled by more than one divisor are the starting point to build both single-field [15] and multi-field [16] inflationary models.

4.2. Model building with D7-branes

Let us briefly review how semi-realistic models can be built via D7-branes wrapped around smooth four-cycles. In order to get an $\mathcal{N} = 1$ low-energy theory, one has to consider orientifold projections $O = (-1)^{F_L}\Omega_p\sigma$ where σ is a holomorphic involution. By construction, we shall always consider divisors D which are transversally invariant, and so N_a D7-branes (plus their images) wrapped around D support an $Sp(2N_a)$ gauge group. If we switch on a diagonal flux $\mathcal{F}$ on the branes wrapping D, the gauge group breaks down to $Sp(2N_a) \to SU(N_a) \times U(1)$, so obtaining phenomenologically interesting unitary groups. The diagonal $U(1)$ instead becomes massive via the Stückelberg mechanism. Moreover, the D7-brane flux generates chiral modes. The number of chiral zero-modes in symmetric and antisymmetric $U(N_a)$ representations is:

$$I_a^{(S,A)} = \mp\frac{1}{2}\int_{\rm CY}[D7_a] \wedge [O7] \wedge \mathcal{F}_a - \int_{\rm CY}[D7_a] \wedge [D7_a] \wedge \mathcal{F}_a\,. \quad (4)$$

On the other hand, chiral zero-modes in bi-fundamental representations $(N_a, \overline{N}_b)$ and (N_a, N_b) live at the intersection between two different stacks (a and b). Their number is given by:

$$I_{a\bar{b}} = \int_{\rm CY}[D7_a] \wedge [D7_b] \wedge (\mathcal{F}_a - \mathcal{F}_b)\,, \qquad I_{ab} = \int_{\rm CY}[D7_a] \wedge [D7_b] \wedge (\mathcal{F}_a + \mathcal{F}_b)\,.$$

4.3. Charge cancellation

In order to have a globally consistent compactification, the homological charges have to be cancelled. These are encoded into the following quantities:

$$\Gamma_{D7} = [D7] + [D7]\wedge\mathcal{F} + [D7]\wedge\left(\frac{1}{2}\mathcal{F}\wedge\mathcal{F} + \frac{\chi(D7)}{24}\right), \quad \Gamma_{O7} = -8[O7] + [O7]\wedge\frac{\chi(O7)}{6}.$$

The D7-charge cancellation requires $\Sigma_{D7}[D7] = 8[O7]$, while the total D5-charge is zero by construction since all branes and image-branes wrap the same divisor and $\mathcal{F}' = -\mathcal{F}$. The total D3-charge gets contributions from the geometry and the fluxes. We shall not explicitly turn on H_3 and F_3 but we shall check that the D3-tadpole leaves enough space to include them. Also K-theoretic torsion charges must sum to zero. This is equivalent to the absence of $SU(2)$ gauge anomalies on any probe Sp-brane [17].

4.4. Orientifold projection and D7-brane stacks

We perform the following choice for the holomorphic involution σ: $z_8 \mapsto -z_8$. This gives an O7-plane at $z_8 = 0$ but no O3-planes. All the divisors are invariant, implying that $h^{1,1}_- = 0$ and $h^{1,1}_+ = h^{1,1}$. In order to cancel the D7-charge of the O7, we need to have a D7-configuration on the divisor class $8[D_8]$ which is described by the polynomial $\eta^2 - z_8^2\chi = 0$. This corresponds to a *Whitney brane* with double intersection with the O7. If we want to have different stacks, this polynomial has to factorise. Choosing η and χ of the form $\eta = z_i^m\tilde{\eta}$ and $\chi = z_i^{2m}\tilde{\chi}$, the initial polynomial reduces to $z_i^{2m}\left(\tilde{\eta}^2 - z_8^2\tilde{\chi}\right)$. This corresponds now to one $Sp(2m)$ stack along $z_i = 0$ plus a Whitney brane. More generally, requiring N_a branes on D_4, N_b on D_5, N_{k3} on D_1 and N_{gc} on D_7 (plus their images), we get $z_1^{2N_{k3}} z_4^{2N_a} z_5^{2N_b} z_7^{2N_{gc}}\left(\tilde{\eta}^2 - z_8^2\tilde{\chi}\right)$. One can easily check that there is no further factorisation if $N_{gc} \leq 4$, $N_{gc} + N_{k3} \leq 4 + N_a$ and $N_a - N_b \leq N_{gc}$. In order to have $\mathcal{F}_{gc} = 0$ on the gaugino condensation stack N_{gc}, we need to choose $B = F_{gc}$, where F_{gc} is the FW flux. Finally the K-theory constraints are satisfied if N_b is an even number.

4.5. Example with one D-term

Let us perform an explicit choice of brane set-up which satisfies all the previous constraints: $N_a = 3$, $N_{k3} = 1$, $N_{gc} = 3$ and $N_b = 0$ with world-volume fluxes $\mathcal{F}_a^\sigma = \mathcal{F}_a$, $\sigma = 1, 2, 3$, (diagonal flux) on D_4 and $\mathcal{F}_{k3} = 0$ on D_1. The gauge group is broken to:

$$Sp(6)\times SU(2)\times Sp(6) \to SU(3)\times U(1)\times SU(2)\times Sp(6) \to SU(3)\times SU(2)\times Sp(6).$$

Given that $N_b = 0$ and $\mathcal{F}_{k3} = 0$, only one D-term is generated.[2] The chiral modes of the model are $I_a^{(A)} = 2\beta_a - \nu$, $I_a^{(S)} = -2\beta_a + 3\nu$, $I_{ak3} = 2\alpha_a$, $I_{aW} = 4(4\beta_a - \alpha_a) + 8\,\nu$, $I_{k3W} = 0$ and $I_{agc} = \nu$ where α_a, β_a and ν are integral combinations of flux numbers. A possible choice of flux numbers consistent with all our requirements is $\alpha_a = 1$, $\beta_a = -1$ and $\nu = 0$. Then, the total D3-charge turns out to be $Q^{\rm tot}_{(D3)} = -606$ and the only non-zero chiral intersections are $I_a^{(A)} = -2$, $I_a^{(S)} = 2$, $I_{ak3} = 2$ and $I_{aW} = -20$.

4.6. D-term potential

The model involves just one non-trivial FI-term: $\xi_a \propto \int_{\rm CY}[D7_a]\wedge J\wedge\mathcal{F}_a = [(\beta_a - \alpha_a)(r_1 + r_2) + 2\alpha_a r_3]$. Substituting our flux choice, the solution to $\xi_a = 0$ is $\tau_4 = 3\,(\tau_1 - \tau_5) - \tau_7$. Plugging this relation into the Calabi-Yau volume, this reduces to $\mathcal{V} = \frac{1}{6}\sqrt{\tau_1 - \tau_5}\,(10\tau_1 - \tau_5) - \frac{1}{3}\tau_7^{3/2}$. Since $\alpha_{\rm vis}^{-1} = \tau_4$ and the del Pezzo divisor τ_7 will be fixed at small size, also the combination $\tau_s \equiv (\tau_1 - \tau_5)$ has to be fixed small. Defining the 'big' cycle as $\tau_b \equiv (10\tau_1 - \tau_5)/2$, the volume form further simplifies to $\mathcal{V} = \frac{1}{3}\left(\sqrt{\tau_s}\,\tau_b - \tau_7^{3/2}\right)$.

[2] For a GUT-like example with gauge group $SU(5)\times U(1)\times Sp(8)$ and two D-terms see [4].

4.7. F-term potential

The F-term potential depends on τ_s, τ_b and τ_7. The leading order contributions to the potential come from α' and non-perturbative corrections which give rise to (at leading order in $1/\mathcal{V}$):

$$V \simeq 2\pi^2 A^2 \frac{\sqrt{\tau_7}}{\mathcal{V}}\, e^{-\pi\tau_7} - 2\,\pi A W_0\, \frac{\tau_7}{\mathcal{V}^2}\, e^{-\frac{\pi\tau_7}{2}} + \frac{3W_0^2\hat{\xi}}{4\mathcal{V}^3}\,. \qquad (5)$$

This potential depends only on $\mathcal{V}$ and τ_7, and so there is one flat direction left over. Equation (5) is the typical LVS potential which has a SUSY-breaking minimum at $\mathcal{V} \simeq \left[W_0\sqrt{\tau_7}/(2\pi A)\right] e^{\frac{\pi\tau_7}{2}}$ and $\tau_7 \simeq (3\xi/2)^{2/3}\, g_s^{-1}$. For natural values of the parameters, $W_0 = 1$, $A = 0.1$ and $g_s = 0.05$, one finds $\langle\tau_7\rangle \simeq 16$ and $\langle\mathcal{V}\rangle \simeq 10^{12}$, justifying the validity of our approximations. This choice gives TeV-scale supersymmetry since the gravitino mass is $m_{3/2} = \sqrt{g_s/(8\pi)} W_0 M_P/\mathcal{V} \simeq$ 100 TeV while the soft terms generated by gravity mediation are of the order $M_{\rm soft} \simeq m_{3/2}/\ln\left(M_P/m_{3/2}\right) \simeq 3$ TeV. The string scale is intermediate, $M_s \simeq M_P/\sqrt{4\pi\mathcal{V}} \simeq 10^{11}$ GeV, giving a perfect decay constant for local axions: $f_a \sim M_s$ [6, 8]. String loop corrections can lift the remaining flat direction at subleading order, stabilising τ_s small ($\tau_s \simeq 30$) and τ_b large ($\tau_b \simeq 10^{11}$) well inside the Kähler cone [4]. Thus the Calabi-Yau is very anisotropic. Note that a slightly different value of g_s ($g_s = 0.02$) gives a huge value of the volume, $\mathcal{V} \simeq 10^{29}$, which leads to stringy scenarios with $M_s \simeq 2$ TeV and two micron-sized extra dimensions [12].

5. Conclusions

We outlined a general strategy to combine Kähler moduli stabilisation with chiral D-brane models in type IIB compactifications. We presented a concrete chiral model by means of toric geometry. This allowed to make a specific choice of brane set-up and fluxes that gave rise to an MSSM-like model where several consistency constraints have been checked. We also computed the scalar potential and minimised it, obtaining interesting phenomenological scales. This is the first explicit realisation of an LVS model. Finally we argued that this moduli fixing mechanism leads to an axiverse. Crucial issues that we leave for future research are: (*i*) an analysis of background fluxes to fix the S and U-moduli; (*ii*) the derivation of the right spectrum and Yukawa couplings; (*iii*) a systematic search through the PALP list of Calabi-Yau three-folds.

Acknowledgments

I would like to thank Mark Goodsell, Christoph Mayrhofer, Andreas Ringwald and Roberto Valandro for the fruitful collaboration on the topics covered in this review.

References

[1] Blumenhagen R, Moster S and Plauschinn E 2008 *JHEP* **0801** 058
[2] Blumenhagen R *et al* 2009 *Nucl. Phys.* B **815** 1; Collinucci A *et al* 2009 *JHEP* **0907** 074
[3] Cicoli M, Kreuzer M and Mayrhofer C 2012 *JHEP* **1202** 002
[4] Cicoli M, Mayrhofer C and Valandro R 2012 *JHEP* **1202** 062
[5] Cicoli M, Krippendorf S, Mayrhofer C, Quevedo F and Valandro R 2012 *Preprint* arXiv:1206.5237 [hep-th]
[6] Cicoli M, Goodsell M and Ringwald A 2012 *Preprint* arXiv:1206.0819 [hep-th]
[7] Arvanitaki A *et al* 2010 *Phys. Rev.* D **81** 123530; Acharya B S *at al* 2010 *JHEP* **1011** 105
[8] Ringwald A 2012 *Preprint* arXiv:1209.2299 [hep-ph]; Conlon J P 2006 *JHEP* **0605** 078
[9] Minasian R, Moore G W 1997 *JHEP* **9711** 002; Freed D S, Witten E 1999 *Asian J. Math* **3** 819
[10] Balasubramanian V *et al* 2005 *JHEP* **0503** 007; Cicoli M *et al* 2008 *JHEP* **0810** 105
[11] Kreuzer M, Skarke H 2004 *Comput. Phys. Commun.* **157** 87-106
[12] Cicoli M, Burgess C P and Quevedo F 2011 *JHEP* **1110** 119
[13] Cicoli M, Goodsell M, Jaeckel J and Ringwald A 2011 *JHEP* **1107** 114
[14] Cicoli M, Pedro F G and Tasinato G 2012 *JCAP* **1207** 044
[15] Cicoli M *et al* 2009 *JCAP* **0903** 013; Cicoli M, Pedro F G and Tasinato G 2011 *JCAP* **1112** 022
[16] Burgess C P *et al* 2010 *JHEP* **1008** 045; Cicoli M *et al* 2012 *JCAP* **1205** 039
[17] Uranga A M 2001 *Nucl. Phys.* **B598** 225-246

PASCOS 2012 – 18th International Symposium on Particles Strings and Cosmology IOP Publishing
Journal of Physics: Conference Series **485** (2014) 012065 doi:10.1088/1742-6596/485/1/012065

Automated Systematic Generation of Flat Directions in Free Fermionic Heterotic Strings

Jared Greenwald

Baylor University, Department of Physics, One Bear Place #97316, Waco, TX 76798-7316, USA

E-mail: Jared_Greenwald@Baylor.edu

Abstract. It has been shown that string moduli are not completely stabilized in traditional weakly coupled free fermionic heterotic string flat direction phenomenology. This necessitates systematic investigation of the VEV moduli space in order to compare with reality. We summarize current efforts at Baylor University to enhance the efficiency and depth of the current systematic free fermionic heterotic string investigations to include the calculation of flat direction derived phenomenological properties through supercomputing. We focus on the necessary assumptions, complexity and open questions that arise while making a fully-automated flat direction analysis program.

1. Introduction

As a leading candidate for the unification of gravity and gauge theories, string theory has revealed the landscape, a moduli space which yields some 10^{100} to 10^{1000} models. Sifting through the various models to search for the Standard Model (SM) or its Grand Unified embedding necessitates choosing a particular low-energy string construction method. Heterotic string theory received a lot of early attention due to its ease of generating GUT models [1]. Particular success was found among the free fermionic constructions since its naturally generated $SO(10)$ GUT group can be broken to a variety of phenomenologically realistic models [2] - [9].

In the weakly-coupled free fermionic (WCFF) approach, the NAHE basis vector set is a natural start to model building [7]. These vectors, and subsequent additions, represent the boundary conditions of fermions propagating around the string worldsheet. Hence, the vectors represent the compactification process. One shortcoming to the WCFF construction is the inability to map a given model, as generated by the addition of more vectors to the NAHE set, to its low energy effective field theory (LEEFT). Thus, some groups prefer to utilize string theory constructions where it is easier to, *a priori*, generate a desirable model. This set-back is compensated by the ease in which the WCFF landscape subregions can be computationally analyzed. It is in this vein that we proceed to discuss some of the advantages and difficulties of automating WCFF heterotic sting (WCFFHS) searches.

The model construction consists of two parts: model generation and flat direction analysis. Early work done in these two areas was quite successful using a combination of computational (FORTRAN) and analytic techniques [10, 11]. However, a fully automated search was not possible until recent advances in supercomputing technology. Most recently, the development of the C++ based FF and Gauge Frameworks at Baylor University have allowed for the generation

Content from this work may be used under the terms of the Creative Commons Attribution 3.0 licence. Any further distribution of this work must maintain attribution to the author(s) and the title of the work, journal citation and DOI.
Published under licence by IOP Publishing Ltd

of large sets of WCFFHS models. The second aspect of model building, flat direction analysis, has yet to be fully automated. This is an area of active research at Baylor University and the remainder of this article will focus on flat direction calculations.

The importance of flat directions should not be overlooked as all of the interaction and mass information is revealed only though these calculations. Therefore, a 'true' string model can exist only after the flat directions have been investigated. Often the term 'model' is used loosely among string phenomenologists, so context must guide the reader as to the intended definition. Here, we will use the term to mean a set of models specified by a given collection of gauge groups which have the same charges. In other words, a model is a charge matrix. This matrix can yield many actual models due to the freedom allowed by post-compactification moduli stabilization.

2. Charge Matrix

To study flat directions, analysis of a charge matrix, Q, is required. This encodes the actual gauge symmetry information of the model where the columns are the various quantum fields which are charged under the gauge groups of the theory (the rows). The value for the charge is the dimension of the specific representation that field is charged under. Therefore, $Q_{ij} \in \mathbb{Z}$.

2.1. General Structure

An $M \times N$ charge matrix will consist of N fields charged under M gauge groups. Since $M < N$, Q will usually be singular. Typically, a model will consist of 50 to 100 fields charged under 5 to 22 gauge groups. This yields: $M \in [50, 100]$, $N \in [5, 22]$, and $M - N \in [28, 95]$.

Some fields have charges that can be paired. These paired fields are called "vector-like". For Abelian groups or complex NA representations, fields are pairable if they have opposite (non-trivial, non-singlet) charges. For real reps of NA groups, the charges must be the same for both fields. Let the number of paired fields in a given model be: N_p, with N_s and N_{NA} representing the unpaired NA singlet and non-singlet fields, respectively. It is convention to rearrange and relabel the fields as:

$$\overbrace{\Phi_1, \ldots, \Phi_{N_p}}^{\text{vector-like}}, \overbrace{\Phi_{N_p+1}, \ldots, \Phi_{N_p+N_s}}^{\text{NA singlet}}, \overbrace{\Phi_{N_p+N_s+1}, \ldots, \Phi_M}^{\text{NA}}$$

3. Flat Directions

In particle physics, the Vacuum Expectation Value (VEV) of a field, denoted as $\langle\Phi\rangle$, is a parameter of the model. As VEV's are specified, a surface or point in the parameter space governed by the model's scalar potential is chosen. This then dictates the physics (particle masses, interaction types, strengths of the forces *etc* ...) of the model.

3.1. SUSY and Anomalous $U(1)$'s

Supersymmetry (SUSY) is a pivotal aspect of the SM. Since SUSY particles are as yet undetected, SUSY must be broken at or above 1 TeV. The choice of SUSY-breaking scale determines how the coupling constants of the forces run and meet. SUSY breaking can occur due to the anomalous $U(1)$ charges, $U(1)_A$, which are unbalanced charges that give rise to the VEV of the dilaton field. The D-flat directions must cancel the anomalous charge.

3.2. D-flatness

For the case of $U(1)$-charged fields, D-flatness requires: $\langle D_A\rangle = \langle D_\alpha\rangle = 0$ where:

$$D_A = \sum Q_{Ak} |\langle \Phi_k \rangle|^2 + \xi \; ; \tag{1}$$

$$D_\alpha = \sum Q_{\alpha k} |\langle \Phi_k \rangle|^2 \; , \; \alpha \neq A \; ; \tag{2}$$

$$\xi = \frac{g^2 \mathrm{Tr}(Q_A)}{192\pi^2} M_{\mathrm{Pl}}^2 . \tag{3}$$

The Φ_k are the fields which acquire VEVs of order $\sqrt{\xi}$. The Q_{Ak} and $Q_{\alpha k}$ denote the anomalous and non-anomalous charges and $M_{\mathrm{Pl}} \approx 2 \times 10^{15}$ TeV denotes the reduced Planck mass. With the $\{Q_{ij}\}$ constituting the charge matrix, Q, we can re-express the above as a system of linear equations where the solution space is constrained according to: $|\langle \Phi_i \rangle|^2 \in \mathbb{Z}, \; 1 \leq i \leq N_p$ and $|\langle \Phi_j \rangle|^2 \in \mathbb{N}^0, \; N_p < j \leq M$. According to our convention, the first N_p fields are vector-like and therefore a negative norm-squared of the VEV represents the pair-partner field gaining a VEV rather than the original field. If we define $x_i \equiv |\langle \phi_i \rangle|^2$ then we can write the D-flatness equation as:

$$Q \cdot \vec{x} = \begin{pmatrix} -\xi \\ \vec{0} \end{pmatrix} \tag{4}$$
$$x_i \in \mathbb{Z}, \; 1 \leq i \leq N_p$$
$$x_j \in \{\mathbb{Z} \geq 0\} \, , \; N_p < j \leq M$$

3.2.1. D-flat Solutions Due to the singular nature of Q, it is necessary to construct a pseudoinverse in order to solve the system of equations. Additionally, the singularness will allow for multiple (usually, infinite) solutions. A useful method in dealing with matrix equations of this form is Singular Value Decomposition (SVD). This process decomposes a given matrix into the product of three matrices:

$$Q = U \Sigma V^T$$
$$\begin{pmatrix} q_{11} & \cdots & q_{1n} \\ \vdots & \ddots & \vdots \\ q_{m1} & \cdots & q_{mn} \end{pmatrix} = (\vec{u}_1, \cdots, \vec{u}_m) \, diag\,(\sigma_1, \cdots, \sigma_l, 0, \cdots, 0) \begin{pmatrix} \vec{v}_1^T \\ \vdots \\ \vec{v}_n^T \end{pmatrix} . \tag{5}$$

Given that the diagonal matrix, Σ, has l nonzero elements, the first l column vectors of U: $\{\vec{u}_1, \cdots, \vec{u}_l\}$ then form a basis over the range of Q. Similarly, the last $n - l$ columns of V: $\{\vec{v}_{l+1}, \cdots, \vec{v}_n\}$ are a basis for the nullspace of Q. Using SVD to solve a system of equation will yield one solution and additional solutions can be generated by adding to it linear combinations of the nullspace of Q.

Letting $\vec{x}_i^{null} \equiv \vec{v}_{l+i}, \forall i \in [1, n-l]$, then any solution to Eq. 4 can be represented by:

$$\vec{x} = \vec{x}_{SVD} + \sum_{i=1}^{n-l} \alpha_i \vec{x}_i^{null} . \tag{6}$$

The coefficients,$\{\alpha_i\}$, are not without constraint. According to Eq.4, any given vector-like field can take on both negative and positive while non-vector-like fields must be strictly positive. This means that the allowed range of the $\{\alpha_i\}$ will be determined by

$$\sum_{j=1}^{l} x_{ij}^{null} \alpha_j \geq -x_i^{SVD} , \tag{7}$$
$$N_p + 1 \leq i \leq M,$$
$$\alpha_j \in \mathbb{Z}$$

It is readily apparent that the coefficient space will likely be unbounded. Therefore, fully automated searches will be done in a piecewise manner. Additional constraints may come from specifying allowable VEV ranges for the fields before D-flat directions are even calculated. These constraints would be nicely applied to visible sectors but could be relaxed, in varying degrees, to the hidden sector. In fact, by controlling these ranges we may be able to tune the order at which SUSY is broken. Investigations into methods to *a priori* constraining the coefficient space (for example, using Linear/Integer Programming (LP)) are on-going.

3.2.2. NA D-flatness It should be noted that our lack of attention to NA D-flatness constraints is intentional. In fact, it is not necessary to perform the full D-flatness calculation before moving onto the F-terms. In [12] and elsewhere, it was claimed that the space of gauge invariant monomials is in one-to-one correspondence to the space of D-flat stabilizing moduli. This was shown to be to too strong of a condition in [13]. There it was shown that this mapping is, in fact, surjective and that there exist D-flat solutions which do not correspond to gauge invariant monomials. In our current build approach, we take the middle ground by requiring D-flatness, through the $U(1)$ charges, and will ensure that superpotential terms are gauge invariant. This means that our searches will not be exhaustive for the moment. Extension of the framework is certainly possible and searching for gauge variant terms could then be performed in future runs.

3.3. F-flatness

The F-flatness constraints are imposed on the superpotential, W. The n-th order term of W is a sum of all possible terms whose powers add up to n:

$$W_n = \sum \Phi_1^{r_1} \ldots \Phi_M^{r_M},$$

$$n = \sum_{i=1}^{M} r_i.$$

The requirements for F-flatness are:

$$\langle F_i \equiv \frac{\partial W}{\partial \Phi_i} \rangle = 0; \tag{8}$$

$$\langle W \rangle = 0. \tag{9}$$

General superpotential terms begin at order one and range through infinity. In stringy potentials, however, first and second orders do not exist. Additionally, the order at which the superpotential fails to be F-flat (i.e. does not satisfy Eq. 8) corresponds to the scale at which SUSY is broken. At 3rd through 4th order SUSY is broken at about the Planck scale: 10^{15} TeV and for each superpotential order increase, the energy scale decreases by a factor of 10. SUSY breaking at 1 TeV then corresponds to $n = 17$. This is a desirable condition although having the superpotential flat to all orders is also acceptable since there are other mechanisms that allow for SUSY breaking.

3.3.1. Stringent Flat Directions The difficulty in calculating F-flatness is the sheer number of terms that need to be calculated if one were to perform an order-by-order analysis and unfortunately, flatness at one order does not ensure it will remain at higher orders. Another difficulty arises in the allowance for cross-cancellations. Eq. 8 produces a sum of many terms which, when VEV'ed, can only be equal to zero if: 1) the individual terms are zero, 2) there is cancellation between them or 3) some combination of the two. One way to circumvent a truncated order-by-order analysis is to search for stringent flat directions. These are sets of flat solutions where no cross-cancellation is allowed. While not exhaustive, it is believed that all other solutions branch off or lie near these 'root'-like stringent flat directions. This has been observed in previous calculations. If possible, though, non-stringent solutions should be calculated despite their inherit 'fine-tuning' problem. Additionally, all-order results can be calculated by investigating potentially dangerous superpotential terms. These are terms which contain less than two factors of non-VEV'ed fields. This can be realized by having, at minimum, two unique non-VEV'ed fields or one non-VEV'ed field raised to a power greater than one.

Since stringent F-flatness is solely concerned with the presence of the VEV'ed and non-VEV'ed fields the solutions space can be represented in terms of powers of the individual F-terms:

$$\Phi_1^{r_1} \dots \Phi_M^{r_M} \longrightarrow \vec{R} \equiv \begin{pmatrix} r_1 \\ r_2 \\ \vdots \\ r_M \end{pmatrix}. \tag{10}$$

The mapping we need to investigate is:

$$Q \cdot \vec{R} = \vec{0}. \tag{11}$$

In other words we need to consider the nullspace of Q. The values of the components of $\vec{R}$ are of no consequence except for those r_i that correspond to solutions of the D-flat equation (Eq. 4) where $x_i = 0$. Let the columns of $Ns(Q)$ be the basis vectors that span the nullspace(Q). The investigation of potentially dangerous F-terms can then be recast into two possibilities. The first can be seen as satisfying Eq. 8 and is represented as:

$$Ns(Q)_{red} \cdot P = \mathbb{I} \tag{12}$$

where $Ns(Q)_{red}$ is the reduced nullspace basis for the charge matrix constructed by keeping the rows of $Ns(Q)$ which correspond to $x_i = 0$ and removing all others. The columns of P, $col(P)$, will be the coefficients to be used with the $Ns(Q)$ basis vectors to see if that combination stays in the nullspace:

$$Q \cdot \sum_i P_{ij} \; col(Ns(Q))_i \stackrel{?}{=} \vec{0} \tag{13}$$

where j is equal to the number of x_k that are equal to zero and $i \in [0, n-l]$ as dictated by Eq. 5.

The second F-term constraint that must be tested is the null case, which is related to Eq. 9 (the supergravity constraint):

$$Ns(Q)_{red} \cdot \vec{p} = \vec{0}. \tag{14}$$

This case is more difficult since it may be severely underdetermined. If there were a finite number of solutions, they could be assembled into the matrix P as before and tested as in Eq. 13. In general, there will be zero, one or infinite solutions to Eq. 14. However, after integerizing the solutions, only a finite number should exist within the physically realizable order range of the superpotential: 3 - 17. In the case of all order infinite solutions existing beyond order seventeen, we find ourselves severely handicapped at how to approach SUSY breaking without relying on some other mechanism such as, gravity-mediation, kinetic mixing or shadow charges [14, 15].

For either case (Eqs. 12,14), if a particular set of nullspace basis vector coefficients, $col(P)$, is shown to be mapped outside of the nullspace then we have found a potentially dangerous term which will then be tested against worldsheet selection rules [16]. If it survives, the vector with the smallest value of $\sum_i |r_i|$ is the energy scale of SUSY breaking closest to 1 TeV.

4. Conclusion

Fully automated flatness analysis is computationally feasible and within reach. Great strides have been made at Baylor University in creating a C++ framework which will perform stringent searches of the WCFFHS models. As way of generating flat directions, we have seen that SVD is a fast and reliable method to solving the $U(1)$ D-flatness constraints. Additionally, SVD plays a larger role in the C++framework in directly producing D-flat solutions than in the previous FORTRAN code. Due to the unboundedness of the D-flat solutions, a form of 'rough tuning' for allowable VEV ranges will need to be initially applied until a more effective method can be constructed and coded. One viable option is Linear/Integer Programming which investigates the population densities for the constrained coefficient space.

The primary goal for a these flat direction searches will be to investigate phenomenologically viable models. Masses (and mass hierarchies), interactions strengths, proton stability and SUSY-breaking scales will be some of the first calculations performed on each model generated. The modularity and popularity afforded by C++ will allow for easy extension of the Flat Direction Framework. Future additions will likely include non-stringy model capabilities, a LP coefficient generator, Kähler potential calculations, NA D-flat constraints and Non-stringent F-flatness. The latter three will probably require a more robust numerical solver likely supplied through numerical polynomial homotopy continuation algorithms. This is an avenue also under current investigation.

References

[1] Candelas P, Horowitz G, Strominger A and Witten E 1985 *Nucl. Phys.* B **258** 46.
[2] Antoniadis I, Leontaris G and Rizos J 1990 *Phys. Lett.* B **245** 161;
Leontaris G and Rizos J 1999 *Nucl. Phys.* B **554** 3.
[3] Antoniadis I, Bachas C, and Kounnas C 1987 *Nucl. Phys.* B **289** 87);
Kawai H, Lewellen D, and Tye S 1987 *Nucl. Phys.* B **288** 1.
[4] Faraggi AE and Nanopoulos DV 1993 *Phys. Rev.* D **48** 3288;
Faraggi AE 1992 *Nucl. Phys.* B **387** 239;
Faraggi AE 1999 *Intl.J Mod. Phys.* **14** 1663.
[5] Faraggi AE, Nanopoulos D and Yuan K 1990 *Nucl. Phys.* B **335** 347;
Faraggi AE 1992 *Phys. Rev.* D **46** 3204.
[6] Faraggi AE 1992 *Phys. Lett.* B **274** 47;
Faraggi AE 1992 *Phys. Lett.* B **278** 131.
[7] Antoniadis I, Ellis J, Hagelin J and Nanopoulos D 1989 *Phys. Lett.* B **231** 65;
Lopez L, Nanopoulos D and Yuan K 1993 *Nucl. Phys.* B **399** 3.
[8] Antoniadis I, Leontaris G and Rizos J 1990 *Phys. Lett.* B **245** 161;
Leontaris G and Rizos J 1999 *Nucl. Phys.* B **554** 3.
[9] Cleaver GB, Faraggi AE and Savage C 2001 *Phys. Rev.* D **63** 066001.
[10] Cleaver GB, Faraggi AE and Nanopoulos D 1999 *Phys. Lett.* B **455** 135;
Cleaver GB, Faraggi AE and Nanopoulos D 2001 *Intl. J. Mod. Phys.* **16** 425.
[11] Cleaver G, Faraggi AE, Greenwald J, Moore D, Pechan K, Remkus E and Renner T 2011 *Euro. Phys. J.* C **71** 12.
[12] Luty M and Taylor IV M 1996 *Phys. Rev.* D **53** 3399.
[13] Cleaver GB , Nanopoulos DV, Perkins JT and Walker J 2008 *Int.J. Mod. Phys.* A **23** 3461-92.
[14] Goodsell M,Ramos-Sanchez S and Ringwald A 2011 Kinetic Mixing of U(1)s in Heterotic Orbifolds *Preprint* hep-th/1110.6901
[15] Andreas S, Goodsell MD and Ringwald A 2011 Dark matter and Dark Forces from a supersymmetric hidden sector *Preprint* hep-ph/1109.2869
[16] Kalara S, Lopez J and Nanopoulos D 1991 *Nucl. Phys.* B **353** 650.

The systematic construction of free fermionic heterotic string gauge models

D Moore

Department of Physics, Baylor University, Waco, TX 76798, USA

E-mail: douglas_moore1@baylor.edu

Abstract. It has been shown that the string landscape consists of roughly 10^{500} string vacua. While the construction of all of these models is currently made infeasible by technological limits, systematic construction schemes can be employed to explore regions of the landscape. These surveys can provide insight into the theory that purely analytical analysis cannot. We discuss one such systematic survey scheme being developed and deployed at Baylor University as well as results of a recently launched survey of layer 2 gauge models.

1. Introduction

It is well known that the number of possible string derived models is on the order of 10^{500} [1, 2]. Consequently, any efforts to explore this landscape of string vacua require the use of high-performance computing and a choice of construction method. Each construction method has access to different, overlapping regimes of the landscape; here we will focus on the weakly coupled free fermionic heterotic string (WCFFHS) construction formalism [3–7]. The WCFFHS formalism has produced some of the most phenomenologically viable models to date [8–42] and is ideal for computer construction. Random examinations of the landscape, using this formalism, have been performed in the past [43, 44]; however, due to the many-to-one nature of this construction a random survey of the input parameters has many endemic problems that are non-trivial to address [45]. One way to deal with these problems is to *systematically* survey the valid input parameters.

Two software frameworks, currently under development at Baylor University, are being designed and used specifically for the purpose of performing such systematic surveys of the WCFFHS landscape. One such framework, the Gauge Framework, focuses on systematically building gauge models in four to ten large spacetime dimensions. These surveys serve multiple purposes including aiding in attempts at understanding and reducing the redundancies inherent to the construction method.

A detailed account of the WCFFHS formalism from the perspective of gauge model building is provided in [46]. The term "gauge model" and the definition of "uniqueness" used in this study were defined therein, and are thus omitted here for brevity. Presently we are extending the layer 1 survey previously considered to 6, 8, and 10 large spacetime dimensions. To that end we review how the input space was modified to produce such models, section 2. From there a few basic results from that survey are presented, section 3, focusing on the distribution of unique models over the allowed orders and dimensions as well as the occurrence of GUT groups. We conclude with a some preliminary results from a recently launched layer 2 survey, section 4.

Content from this work may be used under the terms of the Creative Commons Attribution 3.0 licence. Any further distribution of this work must maintain attribution to the author(s) and the title of the work, journal citation and DOI.
Published under licence by IOP Publishing Ltd

2. Gauge Model Building

The Gauge Framework focuses on the construction of gauge models, as described in [46]. These models are in many ways the most simplistic models that one can build, and can be thought of as a basis from which more complex models can be built. This makes them interesting as a starting point for systematic surveys because they can be used to guide further searches. Here the WCFFHS construction method is reviewed with gauge models in mind.

Within the free fermionic framework two inputs are required, the set of basis vectors, $\boldsymbol{A}$, and the GSO projection coefficient matrix, $\boldsymbol{k}$. In order to systematically build these models, it is necessary to systematically build the input pairs, $(\boldsymbol{A}, \boldsymbol{k})$, ensuring that all of the modular invariance constraints are met. Our primary deviation from [46] is in the subset of possible basis vector sets considered which is explained below. Beyond this the process follows as before; no changes have been made to the construction of states nor the possible GSO projection matrices, $\boldsymbol{k}$.

2.1. Basis Vectors

In this study, all basis vector sets, $\boldsymbol{A}$, are considered in $D = 4, 6, .., 10$ large spacetime dimensions. For clarity, a basis vector set is defined as:

$$\boldsymbol{A} = \left\{ \vec{\alpha}_i \mid \vec{\alpha}_i \in \mathbb{Q}^{40-2D} \cap (-1, 1]^{40-2D} \right\}, \tag{1}$$

where $i = 1, 2, \ldots, l$[1]. For our purposes we will always take $l \geq 3$ and will refer to $L = l - 2$ as the layer. Each of these basis vectors represents the boundary conditions of complex worldsheet fermion degrees of freedom. By convention, α_i^j with $j = 1, \ldots, (14-D)$ are taken to represent the boundary conditions of the left-moving supersymmetric string and α_i^j with $j = 11, \ldots, (26-D)$ to represent the right-moving bosonic string boundary conditions. The order of each basis vector, N_i, is defined to be the smallest integer such that

$$N_i \; \alpha_i^j = 0 \pmod 2). \tag{2}$$

Of course, the possible basis vectors are constrained by modular invariance in such a way that

$$N_i \; \vec{\alpha}_i^2 = \begin{cases} 0 \pmod 8) & \text{if } N_i \text{ even} \\ 0 \pmod 4) & \text{if } N_i \text{ odd} \end{cases} \tag{3}$$

and

$$N_{ij} \; \vec{\alpha}_i \cdot \vec{\alpha}_j = 0 \pmod 4), \tag{4}$$

where $N_{ij} \equiv \mathrm{LCM}(N_i, N_j)$.

Since the basis vector sets are $L = 1$, there are three basis vectors, two of which will always be the same for every basis vector set generated:

- The first basis vector, denoted $\mathbb{1}$, is the all periodic basis vector: $(1^{14-D} \mid\mid 1^{26-D})$.
- The second basis vector is the SUSY generator, $\boldsymbol{S}$, which is $(1^4 \; 0^{10-D} \mid\mid 0^{26-D})$.

Generating these basis vectors can introduce excessive redundancy thereby hindering systematic search algorithms. However, an algorithm for efficiently generating only the modular invariant basis vectors was developed and implemented in [47].

[1] We do not allow for chiral Ising models here. So all left- or right-moving real fermions can be paired to form left- or right-moving complex fermions.

3. Layer 1 Results

Traditionally, the collection of string derived, low energy effective field theories (LEEFTs) is referred to as the landscape. However, because we are interested less in the field theories and more in the mapping from the WCFFHS input space to this landscape, only those LEEFTs that are mapped to by the particular input subspace defined above are considered. For these purposes it is sufficient to refer to the LEEFTs to which these inputs map as the "FFHS gauge landscape."

Using the WCFFHS formalism, all unique, layer 1 gauge models from order 2 through 22 in 4, 6, 8 and 10 large spacetime dimensions have been constructed. The number of unique models generated as well as how they are distributed amongst the possible number of spacetime supersymmetries is given in Table 1.

Table 1: Number of Unique Models by Dimension - The values represent the number of unique models. Models of "both flavors" are models whose gauge group comes in both the $\mathcal{N}=0$ and $\mathcal{N}=\mathcal{N}_{max}$ varieties. The total number of models constructed, unique or otherwise, is also provided for each value of D.

	$D=10$	$D=8$	$D=6$	$D=4$
$\mathcal{N}=\mathcal{N}_{max}$	2	13	18	68
$\mathcal{N}=0$	6	50	73	509
Groups in Both Flavors	1	6	18	50
Total Number of Models	4,953,930	12,493,632	29,079,534	85,859,757

There is significant growth in both the total number of models and the number of unique models as the space is compactified. This comes as no surprise and results from an increase in the number of fermionic degrees of freedom, that is, computationally the input space has grown because the basis vectors are longer. However, something of particular interest is that the growth in the number of unique models is faster than that in the total number of models. To clarify, the overall efficiency, while still abysmal, improves as the space is compactified. Concretely, roughly 1.6×10^{-4} % of the models generated at $D=10$ are unique while at $D=4$ the number is a bit higher, 6.7×10^{-4} %. This is not necessarily something one might expect and could either be a general property of the formalism or may result from the mechanisms used to reduce redundancy. As of now, this is unclear and further work is to be done.

In addition to the way the unique models distribute across the number of large spacetime dimensions, we can look at the distribution across the order, Figure 1. In all of the dimensions considered, the production of new, unique, supersymmetric models falls off by order 22 while the production of $\mathcal{N}=0$ models falls off by order 24. This results from the fewer constraints on the models and the fact that there are no odd order $\mathcal{N}=0$ models, [46]. One might suspect that beyond order 24 no unique models will be found. To address this, a higher order survey is already underway, the results of which will be released in a future work. All together these results suggest that, while the input space is effectively infinite, the redundancy in the many-to-one mapping from input space to model space is such that a general systematic survey is reasonable. Further, these results conform to the conclusion that the landscape is in fact finite.

We go on to consider the GUT groups occurring in each of the dimensions in question, Table 2. Specific focus is given to the Flipped-SU_5, E_6, SO_{10}, $SU_4\times SU_2\times SU_2$ (PS), $SU_3\times SU_2\times SU_2$ (LRS), and $SU_3\times SU_2\times U_1$ gauge groups. As this study focuses primarily on the gauge content alone, we have chosen to consider models that have promising groups for unification but have

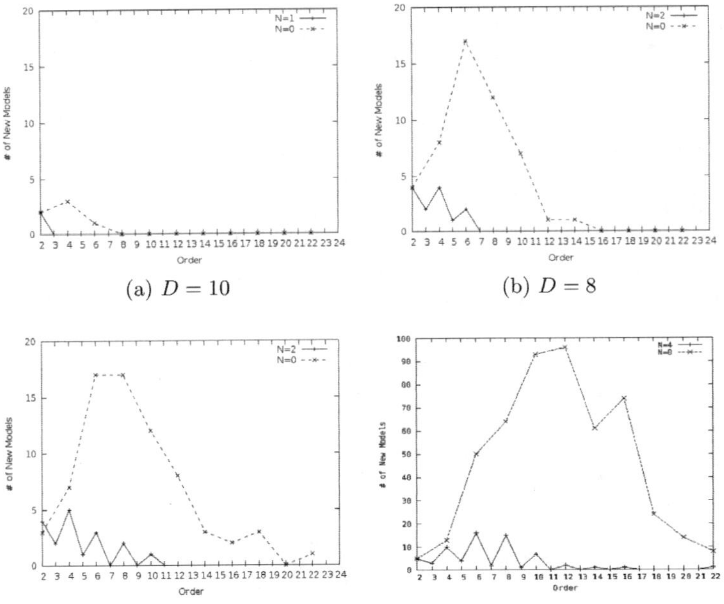

(a) $D = 10$ (b) $D = 8$

(c) $D = 6$ (d) $D = 4$

Figure 1: Number of Unique Models by Order

neglected the matter content. Strictly speaking, this means that the models constructed and presented here may not, and in all likelihood are not, true GUT models. However, it may be possible to construct models with favorable matter content without changing the gauge group by adding nontrivial left-movers to these gauge models.

Table 2: Occurrence of GUT Groups by Dimension - The values represent the percentage of unique models possessing at least one such group factor.

	$\mathcal{N} = \mathcal{N}_{max}$				$\mathcal{N} = 0$			
	$D = 10$	$D = 8$	$D = 6$	$D = 4$	$D = 10$	$D = 8$	$D = 6$	$D = 4$
$\mathcal{F}SU_5$	0%	0%	0%	5.9%	0%	0%	5.5%	20.9%
$E6$	0%	7.6%	11.1%	8.8%	0%	8%	11.0%	9.6%
$SO10$	0%	0%	11.1%	13.2%	0%	8%	13.7%	13.9%
PS	0%	0%	0%	5.9%	0%	2%	8.2%	16.3%
LRS	0%	0%	0%	0%	0%	0%	0%	8.7%
MSSM	0%	0%	0%	0%	0%	0%	0%	14.7%

One thing to note here is the occurrence of the LRS and MSSM in $D = 4$ alone. This may be specific to gauge models or it may suggest that those gauge groups are more easily produced in $D = 4$. This is a possibility due to the increased number of degrees of freedom. Surveys looking at generic models to determine whether this is, in fact, the case are being prepared and will likely launch after more of the gauge landscape has been explored.

4. Layer 2 Results

Of particular interest in this study are the preliminary results from the ongoing layer 2 survey. To date, the survey has built all models from order $(2,2)$ to order $(2,7)$ in $D=4$. This amounts to approximately 149 million models, significantly more than was found in the $D=4$, layer 1, order 2 to 22 survey. This is expected as adding additional layers increases the number of degrees of freedom exponentially. However, while one might expect this to increase the number of unique models generated, current results suggest that this is not the case. In fact, not a single unique model has been generated beyond those found in the previous layer 1, $D=4$ survey. To some degree this redundancy is expected. For example, we know that the number of unique models generated at order $(2,3)$ is exactly the same as at order 6. However, this only accounts for layer 2 models whose orders share no non-unit divisors. Because very little work has been done to build models of high order, this redundancy has never been observed and could prove quite fundamental to understanding the WCFFHS mini-landscape.

5. Conclusion

While many gauge surveys are still underway, results are already being produced that are significant from a model building standpoint. We are finding that the free fermionic region of the landscape is smaller than was originally expected. Additionally we are seeing indicators that $D=4$ is somewhat favorable to model production as compared to $D=6$, 8 and 10; the increased efficiency of unique model generation as well as the increased occurrence of left-right symmetric and standard model gauge groups make $D=4$ more productive. Further still, results were presented that suggest that increasing layer may not yield unique models beyond those which layer 1 can generated. Future work will primarily focus on computationally exploring the layer 2 region of the landscape as well as analytic considerations of the layer 2 results presented herein.

References

[1] Bousso R and Polchinski J 2000 *JHEP* **06** 006 (*Preprint* hep-th/0004134)
[2] Ashok S and Douglas M R 2004 *JHEP* **01** 060 (*Preprint* hep-th/0307049)
[3] Antoniadis I, Bachas C P and Kounnas C 1987 *Nucl. Phys.* **B289** 87
[4] Antoniadis I and Bachas C 1988 *Nucl. Phys.* **B298** 586
[5] Kawai H, Lewellen D C and Tye S H H 1986 *Phys. Rev.* **D34** 3794
[6] Kawai H, Lewellen D C and Tye S H H 1987 *Nucl. Phys.* **B288** 1
[7] Kawai H, Lewellen D C, Schwartz J A and Tye S H H 1988 *Nucl. Phys.* **B299** 431
[8] Cleaver G B, Faraggi A E, Nanopoulos D V and Walker J W 2001 *Nucl. Phys.* **B593** 471–504 (*Preprint* hep-ph/9910230)
[9] Lopez J L, Nanopoulos D V and Yuan K j 1993 *Nucl. Phys.* **B399** 654–690 (*Preprint* hep-th/9203025)
[10] Faraggi A E, Nanopoulos D V and Yuan K j 1990 *Nucl. Phys.* **B335** 347
[11] Faraggi A E 1992 *Nucl. Phys.* **B387** 239–262 (*Preprint* hep-th/9208024)
[12] Antoniadis I, Leontaris G K and Rizos J 1990 *Phys. Lett.* **B245** 161–168
[13] Leontaris G K and Rizos J 1999 *Nucl. Phys.* **B554** 3–49 (*Preprint* hep-th/9901098)
[14] Faraggi A E 1992 *Phys. Lett.* **B278** 131–139
[15] Faraggi A E 1993 *Nucl. Phys.* **B403** 101–121 (*Preprint* hep-th/9208023)
[16] Faraggi A E 1993 *Nucl. Phys.* **B407** 57–72 (*Preprint* hep-ph/9210256)
[17] Faraggi A E 1992 *Phys. Lett.* **B274** 47–52
[18] Faraggi A E 1993 *Phys. Rev.* **D47** 5021–5028
[19] Faraggi A E 1996 *Phys. Lett.* **B377** 43–47 (*Preprint* hep-ph/9506388)
[20] Faraggi A E 1997 *Nucl. Phys.* **B487** 55–92 (*Preprint* hep-ph/9601332)
[21] Cleaver G B 1998 *Nucl. Phys. Proc. Suppl.* **62** 161–170 (*Preprint* hep-th/9708023)
[22] Cleaver G B and Faraggi A E 1999 *Int. J. Mod. Phys.* **A14** 2335–2356 (*Preprint* hep-ph/9711339)
[23] Cleaver G, Cvetic M, Espinosa J R, Everett L L and Langacker P 1998 *Nucl. Phys.* **B525** 3–26 (*Preprint* hep-th/9711178)
[24] Cleaver G, Cvetic M, Espinosa J R, Everett L L and Langacker P 1999 *Nucl. Phys.* **B545** 47–97 (*Preprint* hep-th/9805133)

[25] Cleaver G *et al.* 1999 *Phys. Rev.* **D59** 055005 (*Preprint* hep-ph/9807479)
[26] Cleaver G *et al.* 1999 *Phys. Rev.* **D59** 115003 (*Preprint* hep-ph/9811355)
[27] Cleaver G B 1998 332–340 (*Preprint* hep-ph/9812262)
[28] Cleaver G B, Faraggi A E and Nanopoulos D V 1999 *Phys. Lett.* **B455** 135–146 (*Preprint* hep-ph/9811427)
[29] Cleaver G B, Faraggi A E and Nanopoulos D V 2001 *Int. J. Mod. Phys.* **A16** 425–482 (*Preprint* hep-ph/9904301)
[30] Cleaver G B, Faraggi A E, Nanopoulos D V and Walker J W 2001 *Nucl. Phys.* **B593** 471–504 (*Preprint* hep-ph/9910230)
[31] Cleaver G B 1999 (*Preprint* hep-ph/9901203)
[32] Cleaver G B, Faraggi A E, Nanopoulos D V and Walker J W 2000 *Mod. Phys. Lett.* **A15** 1191–1202 (*Preprint* hep-ph/0002060)
[33] Cleaver G B, Faraggi A E and Savage C 2001 *Phys. Rev.* **D63** 066001 (*Preprint* hep-ph/0006331)
[34] Cleaver G B, Faraggi A E, Nanopoulos D V and Walker J W 2002 *Nucl. Phys.* **B620** 259–289 (*Preprint* hep-ph/0104091)
[35] Cleaver G B, Clements D J and Faraggi A E 2002 *Phys. Rev.* **D65** 106003 (*Preprint* hep-ph/0106060)
[36] Cleaver G B, Faraggi A E and Nooij S 2003 *Nucl. Phys.* **B672** 64–86 (*Preprint* hep-ph/0301037)
[37] Cleaver G B 2002 (*Preprint* hep-ph/0210093)
[38] Cleaver G *et al.* 2003 *Phys. Rev.* **D67** 026009 (*Preprint* hep-ph/0209050)
[39] Perkins J *et al.* 2007 *Phys. Rev.* **D75** 026007 (*Preprint* hep-ph/0510141)
[40] Cleaver G B, Faraggi A E, Manno E and Timirgaziu C 2008 *Phys. Rev.* **D78** 046009 (*Preprint* 0802.0470)
[41] Greenwald J, Pechan K, Renner T, Ali T and Cleaver G 2011 *Nucl.Phys.* **B850** 445–462 (*Preprint* 0912.5207)
[42] Cleaver G, Faraggi A, Greenwald J, Moore D, Pechan K *et al.* 2011 *Eur.Phys.J.* **C71** 1842 (*Preprint* 1105.0447)
[43] Dienes K R 2006 *Phys. Rev.* **D73** 106010 (*Preprint* hep-th/0602286)
[44] Dienes K R, Lennek M, Senechal D and Wasnik V 2007 *Phys. Rev.* **D75** 126005 (*Preprint* 0704.1320)
[45] Dienes K R and Lennek M 2007 *Phys. Rev.* **D75** 026008 (*Preprint* hep-th/0610319)
[46] Moore D, Greenwald J, Renner T, Robinson M, Buescher C *et al.* 2011 *Mod.Phys.Lett.* **A26** 2411–2426 (*Preprint* 1107.5758)
[47] Robinson M, Cleaver G and Hunziker M B 2009 *Mod. Phys. Lett.* **A24** 2703–2715 (*Preprint* 0809.5094)

AUTHOR INDEX

AUTHOR INDEX

Institute of Physics
Dirac House, Temple Back
Bristol BS1 6BE UK

CURRAN ASSOCIATES INC
proceedings.com